Lecture Notes in Electrical Engineering

Volume 984

The book series *Lecture Notes in Electrical Engineering* (LNEE) publishes the latest developments in Electrical Engineering—quickly, informally and in high quality. While original research reported in proceedings and monographs has traditionally formed the core of LNEE, we also encourage authors to submit books devoted to supporting student education and professional training in the various fields and applications areas of electrical engineering. The series cover classical and emerging topics concerning:

- Communication Engineering, Information Theory and Networks
- Electronics Engineering and Microelectronics
- Signal, Image and Speech Processing
- Wireless and Mobile Communication
- Circuits and Systems
- Energy Systems, Power Electronics and Electrical Machines
- Electro-optical Engineering
- Instrumentation Engineering
- Avionics Engineering
- Control Systems
- Internet-of-Things and Cybersecurity
- Biomedical Devices, MEMS and NEMS

For general information about this book series, comments or suggestions, please contact leontina.dicecco@springer.com.

To submit a proposal or request further information, please contact the Publishing Editor in your country:

China

Jasmine Dou, Editor (jasmine.dou@springer.com)

India, Japan, Rest of Asia

Swati Meherishi, Editorial Director (Swati.Meherishi@springer.com)

Southeast Asia, Australia, New Zealand

Ramesh Nath Premnath, Editor (ramesh.premnath@springernature.com)

USA, Canada

Michael Luby, Senior Editor (michael.luby@springer.com)

All other Countries

Leontina Di Cecco, Senior Editor (leontina.dicecco@springer.com)

**** This series is indexed by EI Compendex and Scopus databases. ****

Prasenjit Chatterjee · Dragan Pamucar ·
Morteza Yazdani · Dilbagh Panchal
Editors

Computational Intelligence for Engineering and Management Applications

Select Proceedings of CIEMA 2022

Springer

Editors
Prasenjit Chatterjee
Department of Mechanical Engineering
MCKV Institute of Engineering
Howrah, West Bengal, India

Morteza Yazdani
Universidad Internacional de Valencia
Valencia, Spain

Dragan Pamucar
Department of Operations Research and Statistics
Faculty of Organizational Sciences
University of Belgrade
Belgrade, Serbia

Dilbagh Panchal
Department of Mechanical Engineering
National Institute of Technology Kurukshetra
Kurukshetra, Haryana, India

ISSN 1876-1100 ISSN 1876-1119 (electronic)
Lecture Notes in Electrical Engineering
ISBN 978-981-19-8492-1 ISBN 978-981-19-8493-8 (eBook)
https://doi.org/10.1007/978-981-19-8493-8

This Springer imprint is published by the registered company Springer Nature Singapore Pte Ltd.
The registered company address is: 152 Beach Road, #21-01/04 Gateway East, Singapore 189721, Singapore

The editors would like to dedicate this book to their parents, life partners, children, students, scholars, friends and colleagues.

Organization Committees

Steering Committees

Patrons

Prof. Dr. Abhijit Lahiri, Principal, MCKV Institute of Engineering, West Bengal, India
Prof. Dr. Lazar Z. Velimirović, Mathematical Institute of the Serbian Academy of Sciences and Arts, Belgrade, Serbia

Conference Chairs

Prof. Dr. Anjali Awasthi, Concordia Institute for Information Systems Engineering, Montreal, Canada
Prof. Dr. Shankar Chakraborty, Department of Production Engineering, Jadavpur University, West Bengal, India

Organizing Chair

Prof. Dr. Arghya Sarkar, Dean, Academic and Administration, MCKV Institute of Engineering, West Bengal, India

Organizing Conveners

Dr. Dragan Pamucar, Associate Professor, Military Academy, Department of Logistics, University of Defence in Belgrade, Belgrade, Serbia
Dr. Prasenjit Chatterjee, Dean (Research and Consultancy), MCKV Institute of Engineering (An Autonomous Institute), West Bengal, India

Organizing Co-conveners

Dr. Dilbagh Panchal, Department of Mechanical Engineering, National Institute of Technology Kurukshetra, Kurukshetra, Haryana, India
Dr. Jaideep Dutta, Head of the Department, Department of Mechanical Engineering, MCKV Institute of Engineering, West Bengal, India
Prof. Suprakash Mondal, Mallabhum Institute of Technology, Bishnupur, West Bengal, India

Publication Chairs

Dr. Dragan Marinković, Faculty of Mechanical and Transport Systems, Technische Universität Berlin, Germany
Dr. Željko Stević; University of East Sarajevo, Republic of Srpska, B&H

Publicity Chairs

Dr. Morteza Yazdani, ESIC Business and Marketing School, Madrid, Spain
Dr. Harjit Pal Singh, Assistant Director (Research and Innovation), CT Institute of Engineering, Management and Technology, Jalandhar, India
Prof. Dr. S. B. Goyal, Faculty of Information Technology, City University, Malaysia

Organizing Committee

Prof. Dr. Goran Ćirović, Faculty of Technical Sciences, University of Novi Sad, Novi Sad, Serbia
Prof. Dr. Hanaa Hachimi, Sultan Moulay Slimane University, Beni Mellal, Morocco
Dr. Anupam Deep Sharma, CT Institute of Engineering, Management and Technology, Jalandhar, India

Dr. Sarfaraz Hashemkhani Zolfani, School of Engineering, Department of Industrial Engineering, Catholic University of the North, Chile
Dr. Darko Božanić, Military Academy, University of Defence in Belgrade, Serbia;
Prof. Dr. Golam Kabir, Industrial Systems Engineering, Faculty of Engineering and Applied Science, University of Regina, Canada
Prof. Dr. Hacer Guner Goren, Pamukkale University, Department of Industrial Engineering, Turkey
Prof. Dr. Irik Mukhametzyanov, Ufa State Petroleum Technological University, Russian Federation
Prof. Dr. Samarjit Kar, National Institute of Technology, Durgapur, Department of Mathematics, India
Prof. Kristina Marković, Faculty of Engineering, University of Rijeka, Croatia

Advisory Committee

Prof. Simonov Kusi-Sarpong, Department of Decision Analytics and Risk, Southampton Business School, University of Southampton, UK
Prof. Dr. Valentina Emilia Balas, Automation and Applied Informatics, Aurel Vlaicu University of Arad, Romania
Prof. Dr. Florentin Smarandache, University of New Mexico, USA
Prof. Dr. Ahmed Mohammed, Faculty of Transport and Logistics, Muscat University, Oman
Dr. Irfan Ali, Department of Statistics and Operations Research; Aligarh Muslim University, Aligarh, India
Prof. Dr. Hùng Bùi Thanh, Institute of Engineering and Technology, Thu Dau Mot University, Vietnam
Dr. Fatih Ecer, Department of Business Administrative, Afyon Kocatepe University, Afyonkarahisar, Turkey

Technical Program Committee

Prof. Valentin Popov, Department of System Dynamics and Friction Physics, Technische Universität Berlin, Germany
Prof. Radu-Emil Precup, Department of Automation and Applied Informatics, Politehnica University of Timisoara, Romania
Prof. Dr. Alfred Chee Ah Chill, Faculty of Business and Management, City University, Malaysia
Prof. Jurgita Antuchevičienė, Faculty of Civil Engineering, Department of Construction Management and Real Estate, Vilnius Gediminas Technical University, Lithuania

Dr. Kali Charan Rath, Department of Mechanical Engineering, GIET University, Gunupur, India
Dr. K. Muthumanickam, Associate Professor, Kongunadu College of Engineering and Technology (Autonomous), India
Dr. Kukatlapalli Pradeep Kumar, Department of Computer Science and Engineering, Christ University, Bengaluru, India
Dr. L. Sujihelen, Sathyabama Institute of Science and Technology, India
Dr. M. Gurusamy, Professor, Brindavan College, Bengaluru, India
Dr. Wojciech Sałabun, Faculty of Computer Science and Information Technology, West Pomeranian University of Technology in Szczecin, Poland
Prof. Navnish Goel, S. D. College of Engineering and Technology, India
Prof. Om Prakash Jena, Ravenshaw University, Odisha, India
Prof. Partha Protim Das, Sikkim Manipal Institute of Technology, Majhitar, Sikkim, India
Dr. Pramoda Patro, Amrutvahini College of Engineering, Sangamner, India
Dr. Raghunathan Krishankumar, Department of Computer Science and Engineering, Amrita School of Engineering, Coimbatore, Amrita Vishwa Vidyapeetham, Coimbatore, Tamil Nadu, India
Dr. Kattur Soundarapandian Ravichandran, Rajiv Gandhi National Institute of Youth Development, Sriperumbudur, Tamil Nadu, India
Dr. Piyush Kumar Shukla, Associate Professor, UIT RGPV Bhopal, India
Dr. P. Ponmurugan, Head, R&D, Sengunthar Engineering College, Tiruchengode, India
Dr. Pushpdant Jain, VIT Bhopal University, India
Prof. Rahul Bhanubhai Chauhan, Parul Institute of Business Administration, Parul University, Gujarat, India
Dr. R. M. Suresh, Pro Vice Chancellor, Bharath University, Chennai
Dr. Ramesh Cheripelli, G. Narayanamma Institute of Technology and Science, Hyderabad, India
Prof. Sarvesh Kumar, Integral University, Lucknow, India
Dr. Sonali Vyas, University of Petroleum and Energy Studies, Dehradun, India
Prof. Shashikant Patil, Department of Computer Science and Engineering (AI and ML), Vishwaniketan's Institute of Management Entrepreneurship and Engineering Technology, India
Dr. Sri Ram Chandra Polisetty, Godavari Institute of Engineering and Technology (Autonomous), India
Dr. Sheilza Jain, J. C. Bose University of Science and Technology, YMCA, Faridabad
Dr. P. Asha, Sathyabama Institute of Science and Technology, Chennai, India
Dr. Sourabh Shastri, University of Jammu, India
Dr. Sunil L. Bangare, Sinhgad Academy of Engineering, Pune, India
Dr. Suchismita Satapathy, KIIT Deemed to be University, India
Dr. Sunil Kumar Chawla, Chandigarh University Mohali, India
Dr. Tanupriya Choudhury, University of Petroleum and Energy Studies, Dehradun, India

Dr. Yogesh Chabra, CT Institute of Engineering, Management and Technology, Jalandhar, India
Dr. R. Vijaya Kumar Reddy, Prasad V. Polturi Siddhartha Institute of Technology, India
Dr. Avinash Sharma, Maharishi Markandeshwar (Deemed to be University), India
Dr. Souvik Pal, Global Institute of Management and Technology, India
Prof. X. Agnello J. Naveen, Thanthai Roever Institute of Agriculture and Rural Development (TRIARD)
Naveen Kumar, Research Engineer, Railenium, Villeneuve d'Ascq, France
Dr. Mercy Paul Selvan, Sathyabama Institute of Science and Technology, Chennai, India
Dr. Rohit Tanwar, Department of Computer Science, University of Petroleum and Energy Studies, Dehradun, India
Dr. Reshmi Priyadarshini, Sharda University, India
Dr. Praveen Shukla, National Institute of Technology, Raipur, India
Prof. Dr. Sandra Klinge, Chair of Structural Analysis, Technische Universität Berlin, Germany
Dr. Praveen Kumar, VIT University, Bhopal, India
Dr. S. K. Althaf Hussain Basha, Krishna Chaitanya Institute of Technology and Sciences, Markapur, India
Dr. Shaik Nagul, Lendi Institute of Engineering and Technology, Andhra Pradesh, India
Dr. Navnish Goel, S. D. College of Engineering and Technology, Uttar Pradesh, India
Dr. Georgios Eleftherios Stavroulakis, Department of Production Engineering and Management, Technical University of Crete, Greece
Dr. Ibrahim Badi, Faculty of Engineering, Misurata University, Libya
Dr. Tamara Nestorović, Department of Mechanics of Adaptive Systems, Ruhr-Universität Bochum, Germany
Dr. Sonja Jozić, Department of Manufacturing Engineering, University of Split, Croatia
Dr. Iakov Lyashenko, Department of Modeling of Complex Systems, Summy State University, Ukraine
Dr. Kanak Kalita, Vel Tech University, Chennai, India
Dr. Manoj Sharma, SIRT, RGPV Bhopal, Madhya Pradesh, India
Shubham Joshi, SVKM'S NMIMS University, Mumbai, MPSTME, Shirpur Campus, India
Dr. Gujar Anantkumar Jotiram, D. Y. Patil College of Engineering and Technology, Kolhapur, Maharashtra, India
Dr. Sujay Chakraborty, National Institute of Technology Raipur, India
Dr. Subrata Sahana, Sharda University, India
Dr. Krishnendu Rarhi, Chandigarh University, India
Dr. Pankaj Bhambri, Guru Nanak Dev Engineering College, Ludhiana, India
Prof. Sanjin Troha, Faculty of Engineering, University of Rijeka, Croatia
Dr. Joshila Grace L. K., School of Computing, Sathyabama Institute of Science and Technology, Tamil Nadu, India

Dr. Mohammad Irfan Alam, School of Mechanical Engineering, VIT Bhopal University, India
Dr. Papiya Debnath, Department of Mathematics, Techno International New Town, Kolkata, India
Dr Akash Tayal, Department of Electronics and Communication, Indira Gandhi Delhi Technical University for Women, Delhi, India
Dr. Markus Hess, Department of Structural Dynamics and Physics of Friction, TU Berlin, Germany
Dr. Rajdeep Chakraborty, Department of Computer Science and Engineering, Netaji Subhash Engineering College, Kolkata, India
Dr. Szabolcs Fischer, Department of Transport Infrastructure and Water Resources Engineering, Széchenyi István University, Hungary
Dr. Golam Kibria, Department of Mechanical Engineering, Aliah University, Kolkata, India
Dr. Mandeep Kaur, Sharda University, India
Dr. Arti Jain, Department of Computer Science and Engineering, Jaypee Institute of Information Technology, Uttar Pradesh, India

Preface

Regional Association for Security and Crisis Management, Serbia; Mathematical Institute of the Serbian Academy of Sciences and Arts; MCKV Institute of Engineering, West Bengal, India; and CT Institute of Engineering, Management and Technology, Punjab, India; presented CIEMA-2022, the 1st International Conference on Computational Intelligence for Engineering and Management Applications.

The fundamental goal of CIEMA-2022 was to create an international venue for academics, scholars, academicians, and industry professionals to present, discuss, and exchange knowledge regarding theoretical and applied computational intelligence research in all areas of engineering and management applications. Catholic University of the North in Chile, the European Centre for Operational Research (ECOR) in Serbia, Sultan Moulay Slimane University in Morocco, City University in Malaysia, The Institution of Green Engineers (IGEN), and the Society for Data Science (S4DS) in India supported in organizing CIEMA-2022.

Many excellent keynote addresses by notable speakers were delivered at the conference including Dr. Lazar Velimirovic, Mathematical Institute of the Serbian Academy of Sciences and Arts, Belgrade, Serbia; Prof. Valentina E. Balas, Department of Automatics and Applied Software, Faculty of Engineering, University Aurel Vlaicu, Arad, Romania; Prof. Dr. Bui Thanh Hung, Director of Artificial Intelligence Laboratory, Faculty of Information Technology, Ton Duc Thang University, Vietnam; Prof. Dr. Samarjit Kar, Department of Mathematics, National Institute of Technology Durgapur, India; and Dr. K. S. Ravichandran, Rajiv Gandhi Institute of Youth Development, Sriperumbudur, India.

Different application areas including computational intelligence for fundamental engineering applications; computational intelligence for advanced engineering applications and computational intelligence for management applications were the main tracks in CIEMA-2022. Researchers and professors from highly reputed international institutes and universities graced CIEMA by giving their consents to serve in different committees. All submitted abstracts of CIEMA-2022 were initially screen by internal committee members, and then, recommended abstracts were reviewed by conference committee members along with external reviewers. Full papers of CIEMA-2022 had undergone rigorous peer review process backed by experts in respective fields, chosen from committee members and external reviewers. Each paper of CIEMA has been reviewed by at least two experts, and papers were selected purely on the basis of originality, significance of contribution, diversity of theme, approaches and reviewers' recommendations. CIEMA had received a total of 391 submissions, and finally, 155 papers were accepted by the reviewers. The overall acceptance rate was around only 39%. CIEMA-2022 received papers from 19 different countries across the globe including Albania, Bosnia and Herzegovina, Brazil, China, Chile, Ecuador, Ethiopia, Germany, India, Iran, Japan, Malaysia, Morocco, Nigeria, Serbia, South Korea, Turkey, Russia, and Vietnam.

The select proceedings of CIEMA-2022 includes some of the high-quality papers presented during CIEMA-2022. The volume is divided into the following parts based on themes and application domains considered in the papers.

Computational Intelligence in Energy/Logistics/Manufacturing/Power Applications
Computational Intelligence in Healthcare Applications
Computational Intelligence in Image/Gesture Processing
Reviews in Computational Intelligence
Various Aspects of IOT, Machine Learning and Cyber-Network Security
Computational Intelligence in Special Applications
Computational Intelligence in Management Applications
Computational Optimization for Decision Making Applications

Howrah, India — Dr. Prasenjit Chatterjee
Belgrade, Serbia — Dr. Dragan Pamucar
Valencia, Spain — Dr. Morteza Yazdani
Haryana, India — Dr. Dilbagh Panchal

Contents

About the Editors

Dr. Prasenjit Chatterjee is currently a Professor of Mechanical Engineering and Dean (Research and Consultancy) at MCKV Institute of Engineering, West Bengal, India. He has over 5400 citations with an h-index of 38 and 120 research papers in various international journals and peer reviewed conferences. He has authored and edited more than 30 books on intelligent decision-making, fuzzy computing, supply chain management, optimization techniques, risk management and sustainability modelling. He is the Editor-in-Chief of *Journal of Decision Analytics and Intelligent Computing*. He has also been the Guest Editor of several special issues in different SCIE/Scopus/ESCI (Clarivate Analytics) indexed journals. Dr. Chatterjee is one of the developers of two multiple-criteria decision-making methods called Measurement of Alternatives and Ranking according to COmpromise Solution (MARCOS) and Ranking of Alternatives through Functional mapping of criterion sub-intervals into a Single Interval (RAFSI).

Dragan Pamucar is an Associate Professor at the University of Belgrade, Faculty of Organizational Sciences, Department of Operations Research and Statistics, Serbia. He received a Ph.D. in Applied Mathematics with a specialization in multi-criteria modeling and soft computing techniques, from the University of Defence in Belgrade, Serbia, in 2013 and a M.Sc. degree from the Faculty of Transport and Traffic Engineering in Belgrade in 2009. His research interest areas are computational intelligence, multi-criteria decision-making problems, neuro-fuzzy systems, fuzzy, rough, and intuitionistic fuzzy set theory, neutrosophic theory. He has authored/co-authored over 70 papers published in SCI-indexed international journals including experts' systems with applications, computational intelligence, computers, and industrial engineering technical gazette, sustainability, symmetry, water, etc.

Morteza Yazdani currently works at the Universidad Internacional de Valencia, Madrid. Before, he worked as an Assistant Professor at Universidad Autonoma de Madrid, ESIC University and Universidad Loyola Andalucia, Spain. He has also worked at the University of Toulouse and the European University of Madrid as a post-doctoral researcher. He participated in the editorial board of the *International*

Journal of Production Management and Engineering and is Reviewer of several international journals such as *Expert Systems with Applications, Journal of Cleaner Production, Soft Computing* and *Applied Soft Computing*, etc. His main research areas are decision-making modeling and fuzzy decision system for humanitarian supply chain and energy systems. He has published over 50 research papers in high-impact journals and peer-reviewed conferences.

Dilbagh Panchal is an Assistant Professor (Grade-I) in the Department of Mechanical Engineering, National Institute of Technology Kurukshetra, Haryana, India. He works in reliability and maintenance engineering, fuzzy decision-making, supply chain management, and operation management. He obtained his Bachelor (Hons.) in Mechanical Engineering from Kurukshetra University, Kurukshetra, India, in 2007 and Master's in Manufacturing Technology in 2011 from Dr. B. R. Ambedkar National Institute of Technology Jalandhar, India. He has done his Ph.D. from the Indian Institute of Technology Roorkee, India, in 2016. He is presently supervising two Ph.D. scholars; five M.Tech. dissertations have been guided by him, and two are in progress. He has published 25 research papers in SCI/Scopus-indexed journals. Five book chapters and one book have been also published by him under a reputed publisher. With this, six international conferences have been also attended by him. He also edited a book on *Advanced Multi-criteria Decision-Making for Addressing Complex Sustainability Issues*. He is currently a part of Associate Editors team of the *International Journal of System Assurance and Engineering Management*. He is currently serving as Active Reviewer of many reputed international journals.

Computational Intelligence in Energy/Logistics/Manufacturing/Power Applications

An Integrated Approach for Robot Selection Under Utopia Environment

Bipradas Bairagi

Abstract In the constantly changing global scenario, modern industrial organizations are searching suitable automated manufacturing systems having industrial robot as an integrated part. Proper selection of robots for industrial purpose considering multiple conflicting criteria is very critical task for decision makers. An integrated approach for robot selection under utopia environment has been introduced in the current research work. This paper reflects the coefficient of decision-making attitude and employs Hurwicz Criterion to determine robot selection index (RSI) using cost index, utility index and relative closeness. The present work investigates the consistent in the consequence of robot selection problem obtained by the proposed method with that of existing method. To analyze the proposed algorithm, an illustrative example on selection of industrial robots having multiple conflicting criteria has been cited and solved using the proposed method. Consequently, sensitivity analysis of robot selection indices has been carried out for assisting the decision makers to make decision under varying decision-making attitude. Finally, the ranking of the robots under consideration is tabulated using the proposed methods.

Keywords Coefficient of decision-making attitude · Robot selection index · Hurwitz criterion · MCDM · TOPSIS · Cost index relative closeness

1 Introduction

Decision making is a selection process of the best possible alternative or action of course from a number of available alternatives. At every moment in our real life, decision is made for the selection of best alternatives on the basis of some selection criteria. Decision making starts from the selection of hospital before the birth of a child and is associated with selection of name, cloths, foods, school, college, university, tutors, profession, location selection of house and so on.

B. Bairagi (✉)
Department of Mechanical Engineering, Haldia Institute of Technology, Haldia, India
e-mail: bipradasbairagi79@gmail.com

P. Chatterjee et al. (eds.), *Computational Intelligence for Engineering and Management Applications*, Lecture Notes in Electrical Engineering 984,
https://doi.org/10.1007/978-981-19-8493-8_1

The selection criteria on the basis of which all these decisions are made are objective, subjective and critical in nature. Objective criteria are those which are both measurable and quantitative such as cost, weight, speed, distance and lifecycles measured in suitable unit, reliability. Subjective criteria are those which are qualitative but neither measurable nor quantifiable. Subjective criteria are associated with imprecision and vagueness and realized by human perception such as looks of an alternative. Critical criteria are those which decide the requirement of further evaluation of data of an alternative. If critical criteria of an alternative are not satisfied, then the associated alternative is not considered for further evaluation or processing. Political stability, social or communal situation are examples of critical criteria.

In last few decades, extensive researches on industrial robot selection have been executed. Goswami and Behera applied two well-known MCDM approaches, ARAS and COPRAS, for the evaluation and selection of conveyors, AGV and robots as material handling equipment [1]. Soufi et al. introduced an AHP-based MCDM methodology for the evaluation and selection of material handling equipment to be utilized in manufacturing systems [2]. Satyam et al. applied a multi attribute decision making approach for evaluation and selection of conveyors as material handling equipment [3]. Nguyen et al. advocated a combined multi-criteria decision-making model for the evaluation and selection of conveyor as the material handling equipment based on fuzzy analytical hierarchy process and fuzzy ARAS with vague and imprecision information [4]. Mathewa and Sahua made a comparison among the novel multi-criteria decision-making approaches by solving a problem on material handling equipment selection [5].

Shih [6] has evaluated performances of robots on their incremental benefit–cost ratio, and in this work, cost is shown in two representations. They have not demonstrated the effectiveness of their model. MCDM technique is widely used in ranking one or more alternatives from a set of available alternatives with respect to several criteria [7–12]. The study examines the modification in the most favorable result of a perturbation in a diversity of constraint or substitution rates, the criteria weights or the vagueness on performance procedures [13, 14].

Decision makers always like to know which option is best among a set of several feasible alternatives [15]. In order to determine the tactical decision-making about twofold market high machinery goods, a model on mixed integer goal programming is introduced to smooth the progress of the procedure of marketing media selection [16]. A framework is constructed using MCDM method for the sustainable technologies for generation of electricity in Greece to elaborate more realistic and transparent outcomes [17].

As per the global literature, the decision support framework formulation must be adequately flexible to consider variation in institutional settings, original capabilities and motivation as well as style of decision making [18, 19]. An integrated approach is introduced for considering the vendor selection procedure. In the beginning, the authors formulated the vendor selection problem using multiple criteria decision-making methodology [20]. Then, the modified TOPSIS is used to select challenging products in terms of their general performances. TOPSIS is a constructive tool for working out multiple criteria decision making problems. In TOPSIS method, the optimal alternative is one which has the shortest distance from the positive ideal solution and the farthest distance from the negative ideal solution.

The gap analysis of the above literature survey clearly shows that there is still absolute necessity of further investigation for proper selection of the robots considering variable decision-making attitude with the objective of aiding managerial decision makers.

The rest of the paper is planed as follows. Section 2 is dedicated for representing integrated and extended algorithm. Section 3 gives the detailed description of a case study along with illustration, calculation and discussion. Section 4 is decorated with conclusion.

2 Integrated Algorithm with Extension

The following algorithm has been proposed to select the best alternative by solving the above problem.

Step 1: Establish a decision matrix (M_d^e), where e = order of expert. M_d^e is represented as the following matrix.

$$M_d^e = \begin{array}{c} \\ A_1 \\ A_2 \\ \dots \\ A_i \\ \dots \\ A_m \end{array} \begin{array}{c} \begin{array}{cccccc} C_1 & C_2 & \dots C_j & \dots C_n \end{array} \\ \begin{bmatrix} x_{11}^e & x_{12}^e & \dots & x_{1j}^e & \dots & x_{1n}^e \\ x_{21}^e & x_{22}^e & \dots & x_{2j}^e & \dots & x_{2n}^e \\ \dots & \dots & \dots & \dots & \dots & \dots \\ x_{i1}^e & x_{i2}^e & \dots & x_{ij}^e & \dots & x_{in}^e \\ \dots & \dots & \dots & \dots & \dots & \dots \\ x_{m1}^e & x_{m2}^e & \dots & x_{mj}^e & \dots & x_{mn}^e \end{bmatrix} \end{array} \quad (1)$$

where A_i = ith alternative, i = 1, 2, …, m; m is number of alternatives; C_j = jth criterion, j = 1, 2, …, n; n is the number of criteria; x_{ij}^e is the performance rating of alternative A_i with respect to criterion C_j by eth decision maker (expert), e = 1, 2, …, E; E is the total number of decision makers or experts. The criteria are divided into two parts; one is benefit criteria, and the other is cost criteria; let the number of benefit criteria be 'a' and the number of cost criteria be 'b', so that $a + b = n$.

Step 2: Build normalized decision matrix M_{nd}^e from performance rating assigned by each expert, $e = 1, 2 \dots E$.

$$M_{nd}^{e} = \begin{matrix} A_1 \\ A_2 \\ \cdots \\ A_i \\ \cdots \\ A_m \end{matrix} \begin{bmatrix} y_{11}^e & y_{12}^e & \cdots & y_{1j}^e & \cdots & y_{1a}^e \\ y_{21}^e & y_{22}^e & \cdots & y_{2j}^e & \cdots & y_{2a}^e \\ \cdots & \cdots & \cdots & \cdots & \cdots & \cdots \\ y_{i1}^e & y_{i2}^e & \cdots & y_{ij}^e & \cdots & y_{ia}^e \\ \cdots & \cdots & \cdots & \cdots & \cdots & \cdots \\ y_{m1}^e & y_{m2}^e & \cdots & y_{mj}^e & \cdots & y_{ma}^e \end{bmatrix} \tag{2}$$

where y_{ij}^e = normalized value of x_{ij}^e and $0 \le y_{ij}^e \le 1$. y_{ij}^e is determined using the equation

$$y_{ij}^e = \frac{x_{ij}^e}{\sqrt{\sum_{i=1}^{n} x_{ij}^e}}, \quad \text{where } i = 1, \ldots, m; \quad \text{and} \quad j = 1, \ldots, a. \tag{3}$$

Step 3: A weight vector W^e is assigned by each expert for the benefit

$$W^e = \left[w_1^e \ldots w_j^e \ldots w_5^e \right] \tag{4}$$

where $\sum_{j=1}^{a} w_j^e = 1$ and $j = 1, 2, \ldots, a$.

Step 4: Construct weighted normalized decision matrix U_b^h, $h = 1, 2 \ldots H$ by each decision maker

$$M_{mnd}^{e} = \begin{matrix} A_1 \\ A_2 \\ \cdots \\ A_i \\ \cdots \\ A_m \end{matrix} \begin{bmatrix} y_{11}^e w_1^e & y_{12}^e w_2^e & \cdots & y_{1j}^e w_j^e & \cdots & y_{1n}^e w_n^e \\ y_{21}^e w_1^e & y_{22}^e w_2^e & \cdots & y_{2j}^e w_j^e & \cdots & y_{2n}^e w_{bn}^e \\ \cdots & \cdots & \cdots & \cdots & \cdots & \cdots \\ y_{i1}^e w_1^e & y_{i2}^e w_2^e & \cdots & y_{ij}^e w_j^e & \cdots & y_{in}^e w_n^e \\ \cdots & \cdots & \cdots & \cdots & \cdots & \cdots \\ y_{m1}^e w_1^e & y_{m2}^e w_2^e & \cdots & y_{mj}^e w_j^e & \cdots & y_{mn}^e w_n^e \end{bmatrix}$$

$$= \begin{matrix} A_1 \\ A_2 \\ \cdots \\ A_i \\ \cdots \\ A_m \end{matrix} \begin{bmatrix} z_{11}^e & z_{12}^e & \cdots & z_{1j}^e & \cdots & z_{1n}^e \\ z_{21}^e & z_{22}^e & \cdots & z_{2j}^e & \cdots & z_{2n}^e \\ \cdots & \cdots & \cdots & \cdots & \cdots & \cdots \\ z_{i1}^e & z_{i2}^e & \cdots & z_{ij}^e & \cdots & z_{in}^e \\ \cdots & \cdots & \cdots & \cdots & \cdots & \cdots \\ z_{m1}^e & z_{m2}^e & \cdots & z_{mj}^e & \cdots & z_{mn}^e \end{bmatrix} \tag{5}$$

where $z_{ij}^e = y_{ij}^e w_j^e$; that is product of y_{ij}^e and w_j^e; and the value of $z_{ij\ j}^e$ lies between 0 and 1; that is, $0 \le z_{ij}^e \le 1$, $i = 1, 2, 3 \ldots m$, $j = 1, 2, 3 \ldots a$.

Step 5: Determination of positive ideal solution V^{e+} (PIS) and negative ideal solution V^{e-} (NIS)

$$V^{e+} = [v_1^{e+} \ldots v_a^{e+}]\{\max v_{ij}^{e+};\ i = 1, \ldots, 2, \ldots, m,\ j = 1, 2, \ldots, a;\} \tag{6}$$

$$V^{e-} = [v_1^{e-} \ldots v_a^{e-}]\,\{\min v_{ij}^{e-};\ i = 1, 2, \ldots, m,\ j = 1, 2, \ldots, a;\} \tag{7}$$

Step 6: Calculation of separation measures for each alternative S_i^{e+} and S_i^{e-} of the benefit criteria from PIS and NIS. Individual separation measures are calculated using the following formula.

$$S_i^{e+} = \left\{ \sum_{j=1}^{n} \left| v_{ij}^{e} - v_{ij}^{e+} \right|^2 \right\}^{1/2} \quad \text{where } i = 1, 2, \ldots, m; \tag{8}$$

$$S_i^{e-} = \left\{ \sum_{j=1}^{n} \left| v_{ij}^{e} - v_{ij}^{e-} \right|^2 \right\}^{1/2} \quad \text{where } i = 1, 2, \ldots, m; \tag{9}$$

Step 7: Determination of group separation measure $\overline{S}_{bi}^{h+}$ and $\overline{S}_{bi}^{h-}$ from the above-calculated value of individual separation measure of alternatives. As number of decision makers is *e*, so each of S_i^{e+} and S_i^{e-} is a set of *e* number of alternatives. For group separation measures, the following formulas are used.

$$\overline{S}_i^{e+} = S_i^{1+} \otimes S_i^{2+} \otimes \cdots \otimes S_i^{e+} \quad \text{where } i = 1, 2, \ldots, m; \tag{10}$$

$$\overline{S}_i^{e-} = S_i^{1-} \otimes S_i^{2-} \otimes \cdots \otimes S_i^{e-} \quad \text{where } i = 1, 2, \ldots, m; \tag{11}$$

Geometrical mean (GM) is preferred to arithmetic mean (AM) for performing the above operation. If geometrical mean (GM) is followed, then the abovementioned operations take the forms.

$$\overline{S}_i^{+} = \left\{ \prod_{k=1}^{k=e} S_i^{k+} \right\}^{1/e} \quad ; \quad \text{where } i = 1, \ldots, m; \tag{12}$$

$$\overline{S}_i^{-} = \left\{ \prod_{k=1}^{k=e} S_i^{k-} \right\}^{1/e} \quad ; \text{where } i = 1, \ldots .m. \tag{13}$$

If arithmetic mean (AM) is followed, then expression takes the forms

$$\overline{S}_i^{+} = \frac{1}{e} \sum_{k=1}^{k=e} S_i^{k^+}; \quad \text{where } i = 1, \ldots, m; \tag{14}$$

$$\overline{S}_i^{+} = \frac{1}{e} \sum_{k=1}^{k=e} S_i^{k^+}; \quad \text{where } i = 1, \ldots, m. \tag{15}$$

Step 8: Finding of group relative closeness using both types of group separation measures (GM and AM). The following formulas are used for group relative closeness.

$$\overline{C_i^g} = \frac{\overline{S_i^-}}{\overline{S_i^-} + \overline{S_i^+}} \quad \text{where } i = 1, 2, \ldots, m. \tag{16}$$

$$\overline{C_i^a} = \frac{\overline{\overline{S_i^-}}}{\overline{\overline{S_i^- + S_i^+}}} \quad \text{where } i = 1, 2, \ldots, m. \tag{17}$$

$\overline{C_i^g}$ denotes closeness coefficient measured from separation measures associated from geometric mean. $\overline{C_i^a}$ denotes closeness coefficient measured from separation measures associated from arithmetic mean. It is clear that the value of the relative closeness $\overline{C_i^g}$ and $\overline{C_i^a}$ lies between 0 and 1; in other words, $0 \leq \overline{C_i^g} \leq 1$ and $0 \leq \overline{C_i^a} \leq 1$. It should be noted the greater the value of the relative closeness, the performance of the concerned alternative will be better.

Step 9: Each decision maker evaluates the utility for cost criteria. Numerical value of money is straightforwardly used for assessment. For multiple costs, combined cost is used. Determination of utility for criteria of cost category is made by the decision makers to consider the attitude for risk. The following exponential utility function utilized is advocated in this regard.

$$U_i^e = U(T_i) = T\left\{1 - e^{-\frac{T_i}{T}}\right\} \tag{18}$$

where T_i is the total cost of the alternative i and T is the risk tolerance of DMs.

Step 10: Utility index: In case incremental analysis, the utility information necessarily is less than unity. This is ensured by dividing the utilities by greatest value of their column-wise cost. The equation is

$$U_i^{e'} = \frac{U_i^e}{(U_i^e)_{\max}}; \quad \text{where } i = 1, 2, \ldots, m; \quad \text{and} \quad 0 \leq U_i^{e'} \leq 1. \tag{19}$$

Step 11: Calculation of group utility indices $U_i^{e'}$ by using utility indices and the following expression of geometrical mean (GM) for each alternative

$$\overline{U_i^{e'}} = \left\{\prod_{k=1}^{e} U_i^{e'}\right\}^{1/e}, \quad \text{where } i = 1, 2, \ldots, m; \ k = 1, 2 \ldots e. \tag{20}$$

Step 12: Calculation of group utility indices $U_i^{e'}$ by using utility indices and the following expression of arithmetical mean (GM) for each alternative

$$\overline{U_i^{e'}} = \frac{1}{e}\sum_{k=1}^{k=e} U_i^k; \quad i = 1, 2 \ldots m; \; k = 1, 2 \ldots e. \tag{21}$$

Step 13: Determination of cost index using the following mathematical formula

$$\text{CI}_i = \left[T_i\left(\sum_{i=1}^{i=m}\frac{1}{T_i}\right)\right]^{-1} \tag{22}$$

CI_i = Cost index for alternative *i*, T_i = total cost for alternative *i*, $i = 1, \ldots, m$ (Fig. 1).

Step 14: Determination of robot selection index (RSI)

$$\text{RSI}_i = \alpha \times \text{Relative Closeness (AM)} + (1-\alpha) \times \text{Cost index (AM)} \tag{23}$$

$$\text{RSI}_i = \alpha \times \text{Relative Closeness (GM)} + (1-\alpha) \times \text{Cost index (GM)} \tag{24}$$

$$\text{RSI}_i = \alpha \times \text{Relative Closeness (AM)} + (1-\alpha) \times \text{Group Utility Index (AM)} \tag{25}$$

$$\text{RSI}_i = \alpha \times \text{Relative Closeness (GM)} + (1-\alpha) \times \text{Grout Utility Index (GM)} \tag{26}$$

RSI_i stands for robot selection index for *i*th alternative, α = Coefficient of decision-making attitude, and subscripts AM and GM denote arithmetic mean and geometric mean.

Step 15: Arrange the alternatives according to the descending order of their robot selection index.

3 Case Study

A robot selection problem has been taken from Ref. [6]. A set of four robots are preliminary screened. Five experts in a group are employed for the selection of best robot. The expert group considers four objective criteria, namely velocity, load carrying capacity, cost and repeatability, as well as two subjective criteria, namely vendor's service quality and programming flexibility for further assessment and decision making. Step-by-step calculation procedure and concerned stepwise results have been shown in the following phase.

Step 1: Formation of decision matrix by E1, E2 E3, E4 and E5 is shown in Tables 1, 2, 3, 4 and 5, respectively.

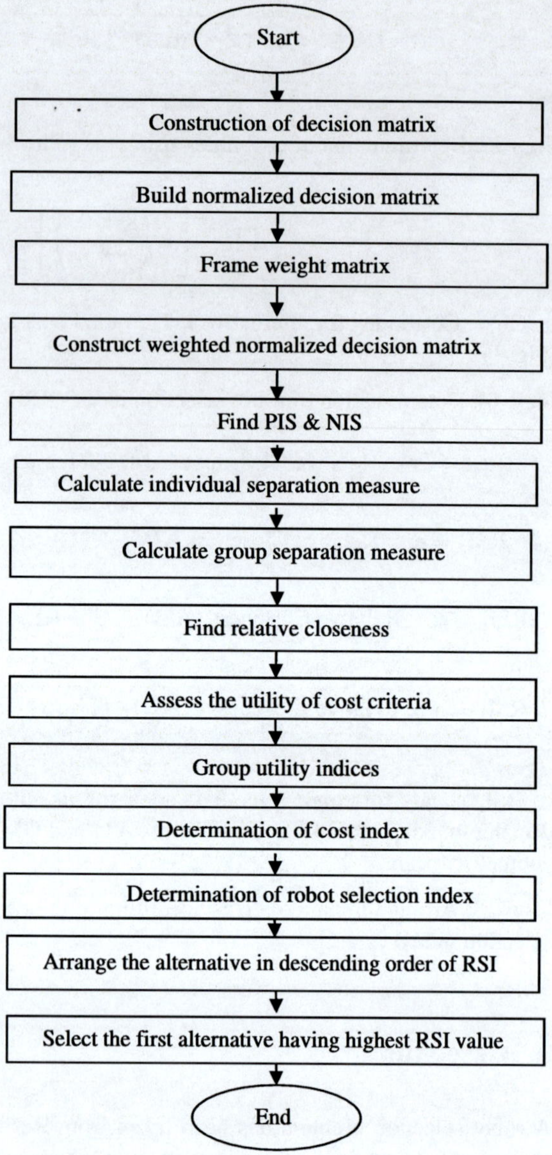

Fig. 1 Flowchart of robot selection framework

Robot selection index is calculated from relative closeness and utility index (utility index is accepted from Hsu-Shih Shih article) (Tables 6, 7 and 8; Figs. 2, 3, 4, 5, 6, 7, 8 and 9).

Table 1 Decision matrix by E1

Robots	Velocity (m/s)	Load capacity (kg)	Costs ($)	Repeatability (mm)	VSQ	PF
Robot 1	1.8	90	9500	0.45	2	9
Robot 2	1.4	80	5500	0.35	3	8
Robot 3	0.8	70	4500	0.20	4	7
Robot 4	0.8	60	4000	0.15	5	2

Table 2 Decision matrix by E2

Robots	Velocity (m/s)	Load capacity (kg)	Costs ($)	Repeatability (mm)	VSQ	PF
Robot 1	1.8	90	9500	0.45	9	3
Robot 2	1.4	80	5500	0.35	8	4
Robot 3	0.8	70	4500	0.20	7	5
Robot 4	0.8	60	4000	0.15	3	8

Table 3 Decision matrix by E3

Robots	Velocity (m/s)	Load capacity (kg)	Costs ($)	Repeatability (mm)	VSQ	PF
Robot 1	1.8	90	9500	0.45	8	4
Robot 2	1.4	80	5500	0.35	7	5
Robot 3	0.8	70	4500	0.20	6	6
Robot 4	0.8	60	4000	0.15	4	7

Table 4 Decision matrix by E4

Robots	Velocity (m/s)	Load capacity (kg)	Costs ($)	Repeatability (mm)	VSQ	PF
Robot 1	1.8	90	9500	0.45	7	5
Robot 2	1.4	80	5500	0.35	6	6
Robot 3	0.8	70	4500	0.20	5	7
Robot 4	0.8	60	4000	0.15	5	6

Table 5 Decision matrix by E5

Robots	Velocity (m/s)	Load capacity (kg)	Costs ($)	Repeatability (mm)	VSQ	PF
Robot 1	1.8	90	9500	0.45	9	3
Robot 2	1.4	80	5500	0.35	8	4
Robot 3	0.8	70	4500	0.20	7	5
Robot 4	0.8	60	4000	0.15	3	8

Table 6 Relative closeness, cost index, utility index and cost index

Robot	Relative closeness (AM)	Relative closeness (GM)	Cost index (GM, AM)	Utility index (GM and AM)	Cost index by formula (22) (AM and GM)
Robot 1	0.4563	0.7455	0.7588	0.9978	0.1386
Robot 2	0.7196	0.5835	0.4393	0.5782	0.2395
Robot 3	0.6067	0.3524	0.3594	0.4732	0.2927
Robot 4	0.3972	0.2288	0.3195	0.4207	0.3295

Table 7 Summary of robot selection index

Robot selection index						
Robot	Arithmetic mean of relative closeness			Geometric mean of relative closeness		
	Cost index	Utility index	Cost index calculated by formula (22)	Cost index	Utility index	Cost index calculated by formula (22)
Robot 1	0.5561	0.6350	0.3515	0.7499	0.8287	0.5452
Robot 2	0.6271	0.6729	0.5612	0.5359	0.5817	0.4670
Robot 3	0.5251	0.5626	0.5031	0.3546	0.3923	0.3327
Robot 4	0.3716	0.4050	0.3749	0.2587	0.2921	0.2620

Table 8 Ranking and summary for robot selection

Robot selection						
Rank	Arithmetic mean of relative closeness			Geometric mean of relative closeness		
	Cost index	Utility index	Cost index by Eq. (22)	Cost index	Utility index	Cost index by Eq. (22)
1	Robot 2	Robot 1	Robot 2	Robot 1	Robot 1	Robot 1
2	Robot 1	Robot 2	Robot 3	Robot 2	Robot 2	Robot 2
3	Robot 3	Robot 3	Robot 4	Robot 3	Robot 3	Robot 3
4	Robot 4	Robot 4	Robot 1	Robot 4	Robot 4	Robot 4

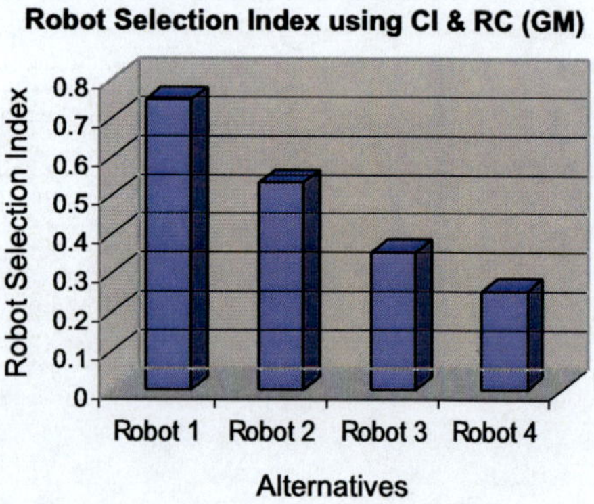

Fig. 2 Robot selection index using robot CI and RC (GM)

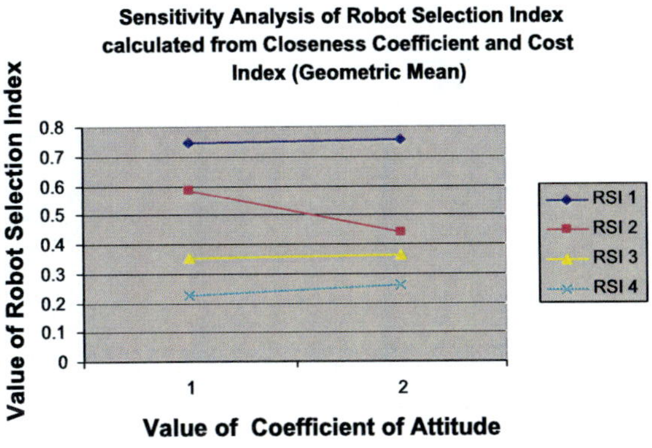

Fig. 3 Sensitivity analysis of robot selection index (GM)

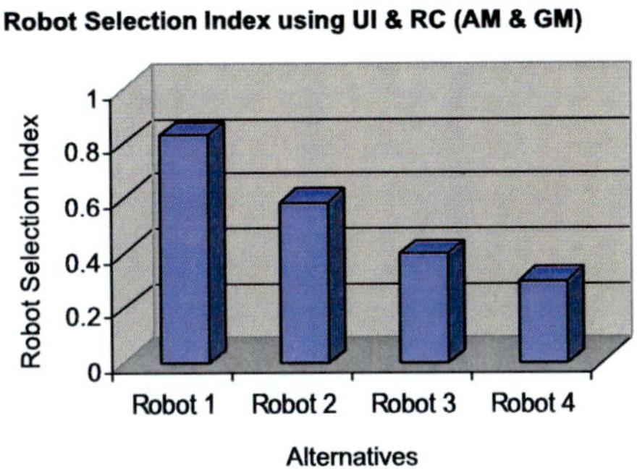

Fig. 4 Robot selection index using UC and RC (AM and GM)

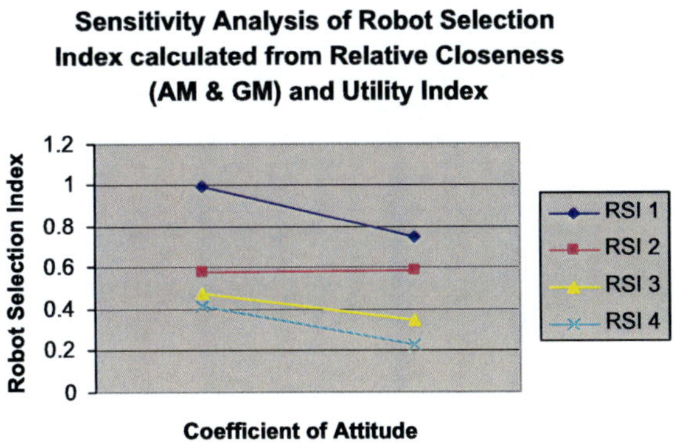

Fig. 5 Sensitivity analysis of robot selection index (AM and GM)

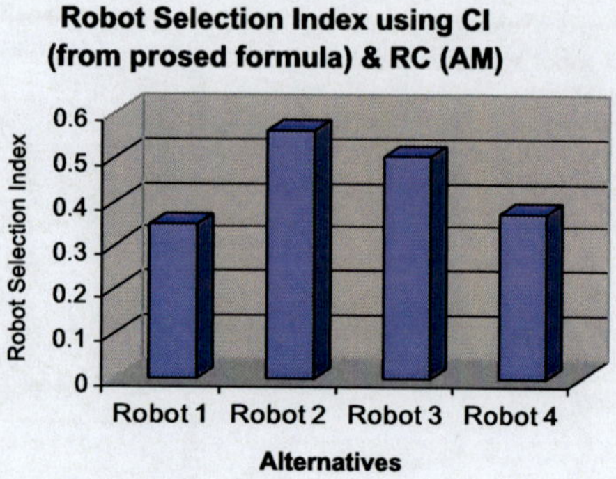

Fig. 6 Robot selection index from CI (using proposed formula) and RC (AM)

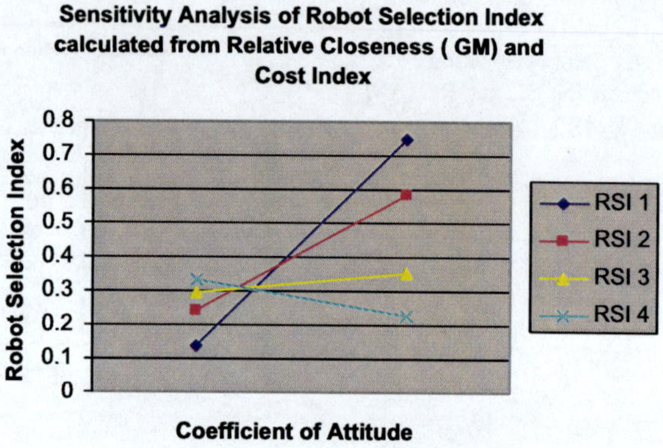

Fig. 7 Sensitivity analysis of robot selection index (GM and GM)

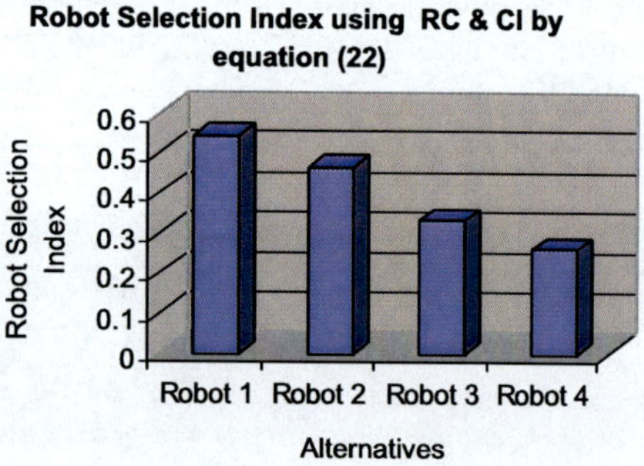

Fig. 8 Robot selection index from RC and CI using Eq. (22)

4 Conclusions

This study examines the consistency in the consequence of the robot selection problem unraveled by proposed algorithm. In the proposed methods, robot selection index is the key factors or key measures of ranking the robots. The proposed

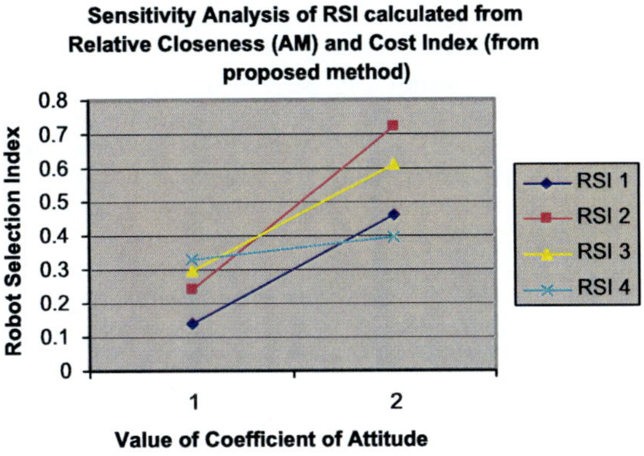

Fig. 9 Sensitivity analysis of robot selection index from RC (AM) and CI (proposed method)

method first divides all the criteria into benefit and cost category. Benefit criteria are processed to determine relative closeness, and cost criteria are processed to determine cost index and utility index. By proper combination of cost index or utility index with corresponding relative closeness (AM and GM), six sets of robot selection index have been measured. It is evident that the ranking orders of robots are consistent for all cases in which geometric mean of relative closeness is used. This proves the proposed procedure is robust, reliable and effective. In addition to this, the result associated to arithmetic mean of utility index gives same ranking order. The remaining two sets of result have the consistency to select the best robot. Sensitivity analysis of robot selection index shows the effectiveness of the proposed algorithm.

References

1. Goswami SS, Behera DK (2021) Solving material handling equipment selection problems in an industry with the help of entropy integrated COPRAS and ARAS MCDM techniques. Process Integr Optim Sustain 5:947–973
2. Soufi Z, David P, Yahouni Z (2021) A methodology for the selection of material handling equipment in manufacturing systems. IFAC Pap Online 54:122–127
3. Satyam F, Satywan K, Avinash K (2021) Application of multi-attribute decision-making methods for the selection of conveyor. Res Square. https://doi.org/10.21203/rs.3.rs-1033410/v1
4. Nguyen H-TN, Siti D, Nukman Y, Hideki A (2016) An integrated MCDM model for conveyor equipment evaluation and selection in an FMC based on a fuzzy AHP and fuzzy ARAS in the presence of vagueness. PLOS ONE 11. https://doi.org/10.1371/journal.pone.0153222
5. Mathewa M, Sahua S (2018) Comparison of new multi-criteria decision making methods for material handling equipment selection. Manage Sci Lett 8:139–150
6. Shih H-S (2008) Incremental analysis for MCDM with an application to group TOPSIS. Eur J Oper Res 186:720–734
7. Wang T-C, Lee H-D (2009) Developing a fuzzy TOPSIS approach based on subjective weights and objective weights. Expert Syst Appl 36:8980–8985
8. Ding J-F, Liang G-S (2005) Using fuzzy MCDM to select partners of strategic alliances for liner shipping. Inf Sci 173:197–225

9. Belton V, Stewart TJ (2002) Multiple criteria decision analysis: an integrated approach. Kluwer Academic Publishing, Boston
10. Dyer JS, Fishburn PC, Steuer RE, Wallenius J (1992) Multiple criteria decision making, multiattribute utility theory: the next ten years. Manag Sci 38:645–654
11. Gal T, Stewart TJ, Hanne T (eds) (1999) Multicrteria decision making: advances in MCDM models, algorithms, theory, and applications. Kluwer Academic Publishing, Norwell
12. Liu D, Stewart TJ (2004) Integrated object-oriented framework for MCDM and DSS modeling. Decis Support Syst 38:421–434
13. Steuer RE (1986) Multiple criteria optimization: theory computation and application. Wiley
14. Taha HA (2003) Operations research: an introduction. Pearson. Upper Saddle River
15. Hwang CL, Yoon K (1981) Multiple attribute decision making. Springer, Berlin
16. Kwak NK, Lee CW, Kim JH (2005) An MCDM model for media selection in the dual consumer/industrial market. Eur J Oper Res 166:255–265
17. Doukas H, Patlitzianas KD, Psarras J (2006) Supporting sustainable electricity technologies in Greece using MCDM. Resour Policy 31:129–136
18. Romero C (1999) Determination of the optimal externality: efficiency versus equity. Eur J Oper Res 113:183–192
19. Ehtamo H, Kettunen E, Hamalainen RP (2001) Searching for joint gains in multi-party negotiations. Eur J Oper Res 130:54–69
20. Shyura H-J, Shih HS (2006) A hybrid MCDM model for strategic vendor selection. Taiwan Math Comput Model 44:749–761

A Novel Soft-Computing Technique in Hydroxyapatite Coating Selection for Orthopedic Prosthesis

Bipradas Bairagi and Kunal Banerjee

Abstract In the recent era of advanced technology, the use of bio-coatings in prosthetic application has extensively been increasing. The selection of bio-coatings depends upon a number of qualitative and quantitative characteristics. Therefore, the right selection of bio-coating in particular application requires suitable mathematical and scientific basis. This study explores a new fuzzy-based mathematical algorithm for analysis and evaluation of performance characteristics of bio-coatings for proper decision makings. The proposed method is illustrated with a suitable example on evaluation and selection of bio-coating for orthopedic prosthesis under fuzzy multi-criteria decision-making environment. In this technique, diverse important criteria are considered and their importance weights are estimated applying experience and opinion of the experts involved in the decision-making process. Based on the criteria, alternative bio-coatings are screened and respective performance ratings are evaluated for further processing. Subjective importance is assigned to each expert based on respective experience and capability. Thereafter, performance ratings of alternative bio-coating, weights of criteria, and importance weights of experts are integrated with soft-computing tool in logical manner. The result and analysis of the problem under consider reveal that the proposed technique is completely capable of evaluating and selecting the best bio-coating in prosthetic application. The comparison of the result obtained by the proposed method with those found by conventional decision-making techniques for the solution of the problem validates the proposed problem as a useful technique in the field.

Keywords Soft-computing technique · FMCDM · Hydroxyapatite coating selection · Bio-coating

---

B. Bairagi (✉) · K. Banerjee
Department of Mechanical Engineering, Haldia Institute of Technology, Haldia, India
e-mail: bipradasbairagi79@gmail.com

P. Chatterjee et al. (eds.), *Computational Intelligence for Engineering and Management Applications*, Lecture Notes in Electrical Engineering 984,
https://doi.org/10.1007/978-981-19-8493-8_2

1 Introduction

In the era of globalization, medical science has attained a paradigm shift introducing new techniques and methodologies in every layer of its wide spread field. Application of biomaterials, natural origin or synthesized in a laboratory, implanted to replace missing tissue, is an important issue. It would directly come in touch of body fluid and environment. Biomaterials for orthopedic prosthesis are being extensively used, especially in joint replacement like knee joints and hip joints. The physiological impact on the users' body may be an important factor as well as its functions. Nowadays, artificial heart valve, prosthesis, heart stents, artificial teeth, etc., are placed within the body. But the materials of that artificial part should be tested properly and should minimize the risk due to introduction of foreign material. Researchers have tried fruitful approach of producing body material from outside. Many material are frequently used as natural material of bone such as hydroxyapatite $[Ca_{10}(PO_4)_6(OH)_2]$ which is produced outside the body by chemical process. These natural body materials act as an important source for this functional material. Since, the environment of body parts differs and the chemical as well as physical structure in most of the biomaterial is complex, so it is very challenging task to design a material for particular bio-application. The complex criteria-based application can be overcome with the help of soft-computing technique. Biomaterials have external use like spectacles, die-color of hair and internal use like orthopedic prosthesis. Biomaterials may be degradable and non-degradable. The medicine used for maintaining therapeutic level is one kind of biomaterials which is soluble and degradable in body fluid. But a prosthetic material should be non-degradable in nature.

There are multiple conflicting criteria to be considered for selection of prosthetic materials in particular application of joint and bone replacement. However, while selecting a proper orthopedic prosthesis material for application in a particular joint, the consideration of all essential and associated performance criteria is a rigorous computational work which is beyond the capacity of conventional mathematical model. With the advancement of technology, decision makers show interest of judging biomaterial from diverse point of view with new attitude. As a result of this, new qualitative characteristics creep into the lists of criteria under consideration. Naturally, the traditional methodology unavoidably fails to capture the required potential impact of those new criteria, especially when they are vague, imprecise, uncertain, partially true or approximate in fuzzy environment.

Soft-computing technique can be fruitfully applied in evaluation of multiple criteria in fuzzy environment. Soft-computing includes genetic algorithm, fuzzy logic and artificial neural network methods. However, fuzzy logic may be most suitable as it can compute the qualitative measurements. A proper decision in selection of biomaterial reduces probability of malfunctioning in in vivo and in vitro tests considering the qualitative criteria also. In case of clinical trials, consideration of all related factors is essential because each factor has positive and negative physiological impact on human body. In selection of proper biomaterials in bone joints, multiple attributes must be considered with proper importance. In this purpose, multi-criteria decision-making (MCDM) techniques in fuzzy environment may be significantly employed.

Rahim et al. developed a fuzzy TOPSIS-based MCDM approach for selection of material considering health, safety and environmental risk [1]. Mehmood et al. performed a case study on evaluation and selection of material and described two challenges for most favorable MEMS sensors using Ashby's method [2]. Dalmarinho et al. implemented 2-tuple linguistic-based MCDM methodology for the evaluation, ranking and selection of materials for specific purpose [3]. Yadav et al. introduced a new TOPSIS- and PSI-based integrated MCDM method for selection of materials to be used in marine engineering purposes [4].

Reduce space searching algorithm is applied to obtain conflicting property [5]. This paper considers little number of criteria. These criteria are associated with bio-metal and not associated with bio-coating. Still hydroxyapatite bio-coating is essential for enhanced mechanical strength A suitable negotiation between mechanical properties and porosity be generally developed [6–8]. In coatings on implant materials, too much porosity is really undesirable [9–11]. Huang et al. [12] made a comparison between MIPS-HAP coatings and X-ray diffraction patterns of as-sprayed MAPS-HAP on SS316L. In the literature, a lot of different methods are reported such macro-plasma spraying [13, 14], micro-plasma spraying [15, 16] laser-related processes [17, 18], sputtering [19], electrophoretic deposition [20], thermal spraying techniques [21], biomimetics [22] and EHDA spraying.

From the gap analysis of the above literature survey, it is clear that previous researchers have not addressed performance analyses of bio-coating using soft-computing technique. Proper selection of bio-coatings depends upon a number of qualitative and quantitative characteristics. Therefore, the right selection of bio-coating in particular application requires suitable mathematical and scientific basis. This study explores a new fuzzy-based mathematical algorithm for analysis and evaluation of performance characteristics of bio-coatings for proper decision makings.

The objective of the paper is to explore a novel decision-making soft-computing technique which is useful and effective for performance evaluation of bio-coating under MCDM environment with a view to assist.

We organize remaining part of the paper into some sections. Section 2 describes the proposed algorithm, Sect. 3 defines the problem to be solved for making decision, Sect. 4 presents the calculation and discussion on the results, and Sect. 5 provides essential concluding remarks.

2 Proposed Algorithm

In this section, the step-by-step procedure of the proposed algorithm is presented.

Step 1: Form a decision-making committee consisting of a set of experts of different sections of organization like technical, finance, marketing and management. The set of experts is denoted by $E = \{E_1, \ldots E_i, \ldots E_p\}$, where p is the total number of experts involved in evaluation process.

Step 2: Determine a set of the selection criteria to be considered for the evaluation process. The set of criteria is denoted by $C = \{C_1 \ldots C_j \ldots C_n\}$, where n is the total number of evaluating criteria recognized by experts committee.

Step 3: Search for alternative and make a short list consisting of the same by conducting a screening test on the basis of recognized criteria for further analysis. The list of alternatives may be denoted by $A = [A_1, \ldots, A_i, \ldots, A_m]^T$. T denotes transpose matrix.

Step 4: For assessing the rating of alternatives based on subjective criteria, importance weights of both objective and subjective criteria and the judicious capability of the experts qualitatively, use the following unique set of linguistic variables.

Step 5: Make each expert construct a decision matrix by assessing each alternative with respective to each criterion.

By expert E_i	Subjective criteria					
	Benefit criteria (BC)		Non-benefit criteria (NC)		Criteria with fixed optimum value (CO)	
Alterntive	C_1	…	C_j	…	…	C_n
A_1	LV_{11}	…	LV_{1j}	…	…	LV_{1n}
…	…	…	…	…	…	…
A_i	LV_{i1}	…	LV_{ij}	…	…	LV_{in}
…	…	…	…	…	…	…
A_m	LV_{m1}	…	LV_{mj}	…	…	LV_{mn}

A_i denotes the ith alternative 'm' denotes the total number of alternatives. This rating may be available from various tests designed for the purpose, past data, design handbooks, magazines, producers/manufactures, users, market surveys or any others reliable sources. Since the type of criteria is the sole decision of decision makers or experts based on their requirement, therefore, it may vary with decision makers and kind of problems. There will be p numbers of such matrices evaluated by each expert $E_1, E_2, \ldots, E_p$, respectively. Subjective criteria are qualitative and assessed with linguistic variables.

Step 6: Calculate the average decision matrix encompassing all individual decision matrix using the following formula

$$(LV_{ij})_{av} = \left(\frac{1}{p}\sum_{i=1}^{p} l_{ij}, \frac{1}{p}\sum_{i=1}^{p} m_{ij}, \frac{1}{p}\sum_{i=1}^{p} u_{ij}\right) \quad (1)$$

where p is the total number of experts involved in evaluation process, $i = 1, 2, \ldots, m$; and $j = 1, 2, \ldots, n$.

Step 7: Determine the criteria weight matrix assessed by all members of the expert committee in linguistic variables based on their past experience, knowledge, opinion and feelings, keeping the actual requirements in view.

Experts	Subjective criteria					
	Benefit criteria (BC)		Non-benefit criteria (NC)		Criteria with fixed optimum value (CO)	
	C_1	...	C_j	...	...	C_n
E_1	w_{11}	...	w_{1j}	...	...	w_{1n}
...	...		...	...	...	...
E_i	w_{i1}	...	w_{ij}	...	...	w_{in}
...	...	...	...	...	...	...
E_m	w_{m1}	...	w_{mj}	...	...	W_{mn}

BC denotes benefit criteria, NC denotes non-benefit criteria, and CO denotes criteria with optimum value. Here, w_{ij} linguistic variable expressing weight of criterion j assessed by expert E_j.

Step 8: Construct the experts' judicious capability matrix.

Experts	Criteria					
	Benefit criteria (BC)		Non-benefit criteria (NC)		Criteria with fixed optimum value (CO)	
	C_1	...	C_j	...	...	C_n
E_1	a_{11}	...	a_{1j}	...	...	a_{1n}
...	...	...	...	...	...	...
E_i	a_{i1}	...	a_{ij}	...	...	a_{in}
...	...	...	...	...	...	...
E_m	a_{m1}	...	a_{mj}	...	...	a_{mn}

where a_{ij} denotes linguistic variable expressing acceptance weight of Expert E_i assessed by expert E_j.

Step 9: Convert the performance ratings, criteria weight and experts' judicious capability from linguistic variable into fuzzy triangular number using the conversion table in Step 4. Homogenize the performance ratings using the following formula.

$$\tilde{h}_{ij} = \left(\overline{l}_{ij}^r, \overline{m}_{ij}^r, \overline{u}_{ij}^r\right) = \left(\frac{l_{ij}}{\min l_{ij}^r + \max u_{ij}^r}, \frac{m_{ij}^r}{\min l_{ij}^r + \max u_{ij}^r}, \frac{u_{ij}^r}{\min l_{ij}^r + \max u_{ij}^r}\right), \quad C_j \in \mathrm{BC} \tag{2}$$

$$\tilde{h}_{ij} = \left(\bar{l}^r_{ij}, \overline{m}^r_{ij}, \bar{u}^r_{ij}\right) = \left(1 - \frac{l_{ij}}{\min l^r_{ij} + \max u^r_{ij}}, 1 - \frac{m^r_{ij}}{\min l^r_{ij} + \max u^r_{ij}}, 1 - \frac{u^r_{ij}}{\min l^r_{ij} + \max u^r_{ij}}\right), \quad C_j \in \mathrm{NC} \tag{3}$$

$$\tilde{h}_{ij} = \left(\bar{l}^r_{ij}, \overline{m}^r_{ij}, \bar{u}^r_{ij}\right) = \left(\mathrm{e}^{-\left(\frac{l^r_{ij} - l^*_j}{l^*_j}\right)}, \mathrm{e}^{-\left(\frac{m^r_{ij} - m^*_j}{m^*_j}\right)}, \mathrm{e}^{-\left(\frac{u^r_{ij} - u^*_{ij}}{u^*_j}\right)}\right), \quad C_j \in \mathrm{CO} \tag{4}$$

BC denotes benefit criteria, NC denotes non-benefit criteria, CO denotes criteria with optimum value. Equations (2), (3) and (4) are recommended for normalization of performance rating of benefit criteria, non-benefit criteria and criteria with optimum value, respectively. $i = 1, 2, \ldots, m$; and $j = 1, 2, \ldots, n$.

Step 10: Normalize the experts' judicious capability matrix consisting of triangular fuzzy numbers using the following formula

$$A = \left(\tilde{a}^N_{ij}\right)_{p \times p}$$

where

$$\tilde{a}^N_{ij} = \left(\frac{l^a_{ij}}{\min l^a_{ij} + \max u^a_{ij}}, \frac{m^a_{ij}}{\min l^a_{ij} + \max u^a_{ij}}, \frac{u^a_{ij}}{\min l^a_{ij} + \max u^a_{ij}}\right) = \left(\bar{l}^a_{ij}, \overline{m}^a_{ij}, \bar{u}^a_{ij}\right) \tag{5}$$

Step 11: Normalize the criteria weight matrix consisting of triangular fuzzy numbers using the following formula

$$W = \left(\tilde{w}^N_{ij}\right)_{p \times n}$$

where

$$\tilde{w}^N_{ij} = \left(\frac{l^w_{ij}}{\min l^w_{ij} + \max u^w_{ij}}, \frac{m^w_{ij}}{\min l^w_{ij} + \max u^w_{ij}}, \frac{u^w_{ij}}{\min l^w_{ij} + \max u^w_{ij}}\right) = \left(\bar{l}^w_{ij}, \overline{m}^w_{ij}, \bar{u}^w_{ij}\right) \quad i = 1, 2, \ldots, m; \quad \text{and} \quad j = 1, 2, \ldots, n. \tag{6}$$

Step 12: Normalize the expert judicious capability in fuzzy triangular number using the following formula.

$$\tilde{a}_i = \left(\left(\prod_{j=1}^{p} \bar{l}^a_{ij}\right)^{\frac{1}{p}}, \left(\prod_{j=1}^{p} \overline{m}^a_{ij}\right)^{\frac{1}{p}}, \left(\prod_{j=1}^{p} \bar{u}^a_{ij}\right)^{\frac{1}{p}}\right)$$

$$= \left(\overline{l}_i^a, \overline{m}_i^a, \overline{u}_i^a\right) \quad i = 1, 2, \ldots, m; \quad \text{and} \quad j = 1, 2, \ldots, n. \tag{7}$$

Step 13: Determine the weight of each criterion in fuzzy triangular numbers using the following formula.

$$\tilde{\omega}_j = \left(-\log\frac{1}{p}\sum_{i=1}^{p}\overline{l}_i^a\overline{l}_{ij}^w, -\log\frac{1}{p}\sum_{i=1}^{p}\overline{m}_i^a\overline{m}_{ij}^w, -\log\frac{1}{p}\sum_{i=1}^{p}\overline{u}_i^a\overline{u}_{ij}^w\right)$$
$$= \left(\overline{l}_j^w, \overline{m}_j^w, \overline{u}_j^w\right) \tag{8}$$

where $j = 1, 2, \ldots, n$.

Step 14: Performance index (PI) of alternatives is calculated using the following formula.

$$\text{PI}_i = -\frac{1}{n}\sum_{j=1}^{n}\log\left(\overline{l}_j^w l_{ij}^{ro}\overline{m}_j^w m_{ij}^{ro}\overline{u}_j^w u_{ij}^{ro}\right)$$
$$i = 1, 2, \ldots, m; \quad \text{and} \quad j = 1, 2, \ldots, n. \tag{9}$$

PI_i is performance index of alternative i where g is the number of objective criteria; $i = 1, 2, \ldots, m$. m denotes the number of alternatives.

Step 15: Determine the modified performance index using the formula

$$\text{MPI}_i = \text{PI}_i - [\min(\text{PI}_i)] \tag{10}$$

[.] denotes the greatest integer functions where $i = 1, 2, \ldots, m$. Arrange the alternative in the increasing order of their performance index. Select the alternative having the highest performance index values as the best alternative.

3 Numerical Example

An Eastern Indian Organization plans for evaluation and selection of bio-coating for specific prosthetic application. For smooth conduction of evaluation process, a committee of five experts from finance (E_1), marketing (E_2), technical (E_3), managerial (E_4) and purchasing (E_5) departments are employed by the organization. The group of five experts of the evaluation committee is responsible for decision-making process. The committee unanimously selects a set of eleven conflicting criteria important for selection of bio-coatings in prosthetic application. The members of the committee also evaluate mutual judgment capability based on experience and knowledge. The set of eleven criteria selected by the committee is listed as follows: Coefficient of friction (C_1), Penetration depth (C_2), Residual depth (C_3), Nano-hardness

(C_4), Crystallinity (C_6), Microstructure (C_7), Coating Stability (C_8), Toxicity effect (C_9), Osteointegration (C_{10}) and Porosity (C_{11}).

The experts primarily screen five boi-coatings. The set of bio-coatings are listed as follows: Coating 1: HAP/SS316L (LSA 400W), Coating 2: HAP + CNT/Ti6Al4V (LSA 400W, Coating 3: HAP/SS316L (MIPS1.5 W), Coating 4: HAP/SS316L (MAPS 40 kW) and Coating 5: HAP/Ti6Al4V (MAPS 40 kW).

An image of scanning electron micrograph of hydroxyapatite coating is shown in Fig. 2. The next section is dedicated for calculation and discussion of the above numerical example on prosthetic bio-coatings evaluation and selection.

4 Calculation and Discussion

All the abovementioned eleven criteria have been estimated qualitatively by the experts due to inadequacy of required information, existence of fuzziness, lack of completeness and ease of criteria assessment. Nine degree linguistic variables are employed for assessing the fuzzy rating of alternatives with respect to fuzzy multiple criteria. The linguistic variables are described as Extremely High (EH), Very High (VH), High (H), Slightly High (SH), Medium (M), Slightly Low (SL), Low (L),Very Low (VL) and Extremely Low (EL). Corresponding triangular fuzzy numbers are (80, 90, 100), (70, 80, 90), (60, 70, 80), (50, 60, 70), (40, 50, 60) (30, 40, 50), (20, 30, 40), (10, 20, 30) and (0, 10, 20), respectively. The abovementioned performance ratings in terms of linguistic variables assessing qualitative value of alternative with respect to criteria assigned by experts involved in the decision-making process are furnished in Table 1.

The subjective performance ratings of alternative coatings in terms of linguistic variables assessed by expert E_1 are shown in Table 2. It is seen that the coefficient of friction (C_1) of alternative A_1 (Coating 1: HAP/SS316L (LSA 400W) is '*M*' which stands for medium. The other abbreviations bear similar meanings. The performance rating matrix consisting of abbreviations of linguistic variables differs from experts to

Table 1 Linguistic variable, abbreviation and triangular fuzzy number

Description	Abbreviation	Triangular fuzzy number
Extremely high	EH	(80, 90, 100)
Very high	VH	(70, 80, 90)
High	H	(60, 70, 80)
Slightly high	SH	(50, 60, 70)
Medium	M	(40, 50, 60)
Slightly low	SL	(30, 40, 50)
Low	L	(20, 30, 40)
Very low	VL	(10, 20, 30)
Extremely low	EL	(0, 10, 20)

experts because of possessing unequal knowledge, experience and diverse decision-making attitudes. If we analyze the Table 2 where expert 1 has estimated the coatings with linguistic variables, many conflicting properties arises for consideration. Here for coating 1, the microstructure and porosity are near about desired one but the criteria like crystallinity, toxicity effect, etc., are not good in respect of physiological application. The properties of Coating 3 in respect of crystalline, osteointegration are satisfactory but porosity level, toxicity, is not up to the mark. Table 3 presents subjective performance rating of alternative coatings in terms of linguistic variables assessed by E_2. Experts E_3 estimates subjective performance rating of alternative coatings in terms of linguistic variables which are shown in Table 4 which is the expression of individual judgments of the concerned expert.

Table 5 Matrix of subjective performance rating of all alternative coatings. The ratings are expressed in terms of linguistic variables assessed by E_4. Table 6 gives the subjective performance rating of alternative coatings in terms of linguistic variables assessed by experts denoted by E_5. In the current study, triangular fuzzy numbers (TFN) are used for the conversion of linguistic variables

As per this conversion system, the subjective performance rating of alternative coatings assessed by expert E_1 is converted into corresponding TFN. Similarly, the corresponding TFN of subjective performance rating in linguistic variables for alternative coatings is estimated by expert E_2. The conversion of subjective performance rating of alternative coating in linguistic variables (LV) into TFN is accomplished by E_3. The triangular fuzzy numbers corresponding to subjective performance rating of coatings is measured by expert denoted by E_4. The matrix of triangular fuzzy number

Table 2 Subjective performance rating of coatings in terms of LVs assessed by E_1

Coating	C_1 (T)	C_2 (−)	C_3 (−)	C_4 (+)	C_5 (+)	C_6 (+)	C_7 (+)	C_8 (+)	C_9 (-)	C_{10} (+)	C_{11} (T)
A_1	M	L	M	M	L	L	SH	EL	H	L	SL
A_2	H	M	SL	L	M	SL	EL	EL	VH	SL	EH
A_3	M	H	VL	H	VH	EH	EH	VH	H	SH	VL
A_4	VH	VL	H	SL	VL	VH	VH	VL	SL	VH	SL
A_5	VL	M	L	L	SL	EL	SL	SL	L	L	L

Table 3 Subjective performance rating of coatings in terms of LVs assessed by E_2

A_i	C_1 (T)	C_2 (−)	C_3 (−)	C_4 (+)	C_5 (+)	C_6 (+)	C_7 (+)	C_8 (+)	C_9 (−)	C_{10} (+)	C_{11} (T)
A_1	L	H	M	SL	VH	L	L	SL	H	VH	H
A_2	M	SH	H	EH	EH	SL	SL	EH	VH	EL	VH
A_3	VH	M	H	H	M	L	EL	L	L	SL	L
A_4	VL	H	VH	SL	L	VH	SL	VL	SL	VH	VH
A_5	SL	SL	M	M	VL	SH	EH	VL	H	EH	H

Table 4 Subjective performance rating of coatings in terms of LVs assessed by E_3

Coating	C_1 (T)	C_2 (−)	C_3 (−)	C_4 (+)	C_5 (+)	C_6 (+)	C_7 (+)	C_8 (+)	C_9 (−)	C_{10} (+)	C_{11} (T)
A_1	VH	SL	L	H	M	H	SH	EL	H	L	SL
A_2	EH	EH	M	M	H	VH	VH	SL	SL	VH	SL
A_3	M	H	VH	H	H	L	SL	EH	VH	EL	VH
A_4	L	SL	VL	VH	VH	SL	EH	VL	H	EH	H
A_5	VL	M	SL	M	M	H	SH	EL	H	L	SL

Table 5 Subjective performance rating of coatings in terms of LVs assessed by E_4

Coating	C_1 (T)	C_2 (−)	C_3 (−)	C_4 (+)	C_5 (+)	C_6 (+)	C_7 (+)	C_8 (+)	C_9 (−)	C_{10} (+)	C_{11} (T)
A_1	L	M	VH	SL	H	L	H	EL	EH	VL	L
A_2	M	SL	EH	EH	SH	SL	EL	EL	L	L	SL
A_3	VH	VL	M	H	M	EH	EH	VH	H	M	EH
A_4	VL	H	L	SL	SH	VH	VH	VL	VL	VH	VH
A_5	SL	L	VL	M	SL	EL	SL	SL	L	L	EL

Table 6 Subjective performance rating of coatings in terms of LVs assessed by E_5

Coating	C_1 (T)	C_2 (−)	C_3 (−)	C_4 (+)	C_5 (+)	C_6 (+)	C_7 (+)	C_8 (+)	C_9 (−)	C_{10} (+)	C_{11} (T)
A_1	VH	SL	SL	SL	L	SH	SH	EL	H	L	SL
A_2	EH	EH	EH	EH	M	EL	EL	EL	VH	SL	EH
A_3	M	H	H	H	VH	EH	EH	SH	H	SH	VL
A_4	L	SL	SL	SL	VL	VH	VH	L	SL	VH	SL
A_5	VL	M	M	M	SL	SL	L	EH	L	L	L

corresponding to subjective performance rating in linguistic variables of alternative coatings is constructed by expert committee member E_5.

Average decision matrix consisting subjective performance rating of alternatives is determined. Calculation procedure for average decision matrix is shown below.

$$(l_{11}, m_{11}, u_{11})_{av} = \left(\frac{40+20+70+20+70}{5}, \frac{50+30+80+30+80}{5}, \frac{60+40+90+40+90}{5}\right) = (44, 54, 64)$$

Normalization decision matrix is calculated by using the equation of Step 8. Calculation procedure for normalization of criteria with target (optimum) value is shown below.

$$\tilde{h}_{11} = \left(e^{-\left(\frac{44-40}{40}\right)^{0.5}}, e^{-\left(\frac{54-50}{50}\right)^{0.5}}, e^{-\left(\frac{64-60}{50}\right)^{0.5}}\right) = (0.73, 0.75, 0.77)$$

Here (40, 50, 60) is the target (optimum) value of the criteria, and (44, 54, 64) is the fuzzy rating of the alternative with respect to criterion.

The normalized value of a decision matrix for subjective benefit criteria with fuzzy number (38, 48, 58) is illustrated below.

$\tilde{h}_{14} = \left(\frac{38}{18+80}, \frac{48}{18+80}, \frac{58}{18+80}\right) = (0.39, 0.49, 0.59)$ Weight matrix consisting of criteria weights in terms of triangular fuzzy number is assessed by each expert of the committee based on their experience, knowledge. Weight matrix consisting subjective weight assessed by each member of expert committee is homogenized (normalized). The calculation procedure for normalization of fuzzy number (60, 70, 80) assigned for C_1 by $E1$ is shown below.

$$\tilde{w}_{11}^{N} = \left(\frac{60}{20+100}, \frac{70}{20+100}, \frac{80}{20+100}\right) = (0.5, 0.58, 0.67)$$

where $\min l_{ij}^{w} = 20$ and $\max u_{ij}^{w} = 100$.

The expert acceptance matrix or the judging capability matrix in triangular fuzzy number is presented in Table 7 as per the conversion matrix produced in Step 4.

The normalization of expert acceptance matrix is carried out by using Eq. (7). The calculation procedure for normalization of fuzzy number (60, 70, 80) assigned for C_1 by E_1 is shown below. Homogenization (normalization) the triangular fuzzy number is presented in the Table 8. Weight of each criterion in fuzzy triangular numbers combining the effects is calculated.

Table 9 shows individual criteria basis performance of alternative in crisp values. The aggregate performance indices as well as modified performance indices are calculated using Eqs. (9) and (10), respectively, which are depicted in Table 10. The ranking order as per the proposed method is given with decreasing order of performance indices and is shown in Table 10. Figure 1 depicts the modified performance indices if the alternative bio-coatings.

A comparative analysis of the results in terms of ranking order obtained by the proposed methodology with those obtained by some well-established conventional methodologies is carried out, and the results are summarized in Table 11 as well as depicted in Fig. 2. It clearly shows that the result obtained by the proposed method

Table 7 Expert acceptance matrix

Experts	E_1	E_2	E_3	E_4	E_5
E_1	(60, 70, 80)	(30, 40, 50)	(40, 50, 60)	(70, 80, 90)	(60, 70, 80)
E_2	(50, 60, 70)	(80, 90, 100)	(60, 70, 80)	(80, 90, 100)	(70, 80, 90)
E_3	(40, 50, 60)	(60, 70, 80)	(60, 70, 80)	(40, 50, 60)	(40, 50, 60)
E_4	(50, 60, 70)	(30, 40, 50)	(70, 80, 90)	(20, 30, 40)	(50, 60, 70)
E_5	(30, 40, 50)	(40, 50, 60)	(40, 50, 60)	(10, 20, 30)	(60, 70, 80)

Table 8 Homogenize the expert acceptance matrix in triangular fuzzy number

Experts	E_1	E_2	E_3	E_4	E_5
E_1	(0.55, 0.64, 0.73)	(0.27, 0.36, 0.45)	(0.36, 0.45, 0.55)	(0.64, 0.73, 0.82)	(0.55, 0.64, 0.73)
E_2	(0.45, 0.55, 0.64)	(0.73, 0.82, 0.91)	(0.55, 0.64, 0.73)	(0.73, 0.82, 0.91)	(0.64, 0.73, 0.82)
E_3	(0.36, 0.45, 0.55)	(0.55, 0.64, 0.73)	(0.55, 0.64, 0.73)	(0.36, 0.45, 0.55)	(0.36, 0.45, 0.55)
E_4	(0.45, 0.55, 0.64)	(0.27, 0.36, 0.45)	(0.64, 0.73, 0.82)	(0.18, 0.27, 0.36)	(0.45, 0.55, 0.64)
E_5	(0.27, 0.36, 0.45)	(0.36, 0.45, 0.55)	(0.36, 0.45, 0.55)	(0.09, 0.18, 0.27)	(0.55, 0.64, 0.73)

Table 9 Performance index with ranking order

A_i	C_1	C_2	C_3	C_4	C_5	C_6	C_7	C_8	C_9	C_{10}	C_{11}	PI	MPI	Rank
A_1	0.59	0.57	0.56	0.65	0.54	0.44	0.41	0.50	0.50	0.40	0.40	5.56	0.56	2
A_2	0.68	0.71	0.69	0.56	0.48	0.45	0.50	0.38	0.48	0.39	0.39	5.71	0.71	1
A_3	0.64	0.63	0.61	0.56	0.48	0.38	0.38	0.26	0.44	0.50	0.38	5.26	0.26	5
A_4	0.65	0.57	0.55	0.66	0.59	0.37	0.36	0.44	0.37	0.36	0.50	5.42	0.42	4
A_5	0.69	0.56	0.50	0.67	0.62	0.50	0.45	0.33	0.37	0.45	0.37	5.51	0.51	3

Table 10 Performance rating, modified performance rating and rank

Alternative	PI	MPI	Rank
A_1	2.08	0.27	2
A_2	2.12	0.31	1
A_3	1.81	0.00	5
A_4	1.90	0.09	4
A_5	1.96	0.15	3

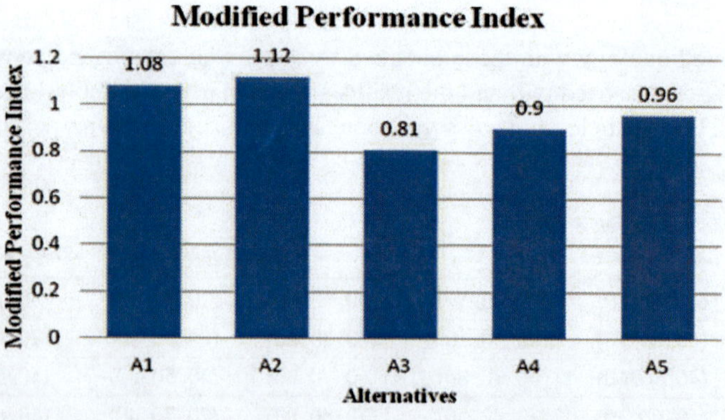

Fig. 1 Modified performance indices of the alternative bio-coatings

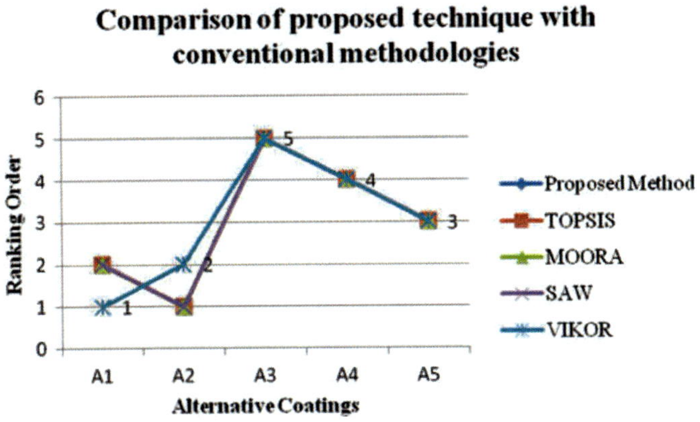

Fig. 2 Comparison of proposed technique with conventional methodologies

is $A_2 > A_1 > A_5 > A_4 > A_3$ which is fully identical with those obtained by TOPSIS, MOORA and SAW with spearman rank correlation coefficients 1. Therefore, Coating 2 that is A_2: HAP + CNT/Ti6Al4V (LSA 400 W) is unanimously the best bio-coating for the specified purpose of application. Though the ranking order of the alternative bio-coatings slightly differs with those obtained by VIKOR method, yet the ordinal ranking orders providing by proposed method and VIKOR show a significant degree of linear association with spearman rank correlation coefficient $r_s = 0.9, p < 0.001$.

Table 11 Comparison of ranks with proposed and existing methods

Alternatives	Rank by proposed method	Rank by TOPSIS	Rank by MOORA	Rank by SAW	Rank by VIKOR
A_1	2	2	2	2	1
A_2	1	1	1	1	2
A_3	5	5	5	5	5
A_4	4	4	4	4	4
A_5	3	3	3	3	3
Spearman rank correlation coefficient of the proposed method with	1	1	1	1	0.9

5 Conclusions

This investigation meets the requirement by developing an appropriate technique in the domain of decision making. The finding of the investigation is that the proposed technique is capable of measuring performance index of alternatives by integrating ratings of alternatives, weights of criteria and subjective importance of experts involving in the decision-making procedure. Additionally, the technique is applicable in evaluation, ranking and selection of hydroxyapatite coatings for orthopedic prosthesis under fuzzy environment, enabling the technique helpful in proper decision making. The proposed method can concurrently judge any number of alternatives, experts as well as qualitative bio-coating selection criteria to calculate performance indices in evaluating and ranking alternative bio-coating for a given decision-making problem under fuzzy environment. The methodology recommends a general technique that can be appropriate to various selection problems encountered in practice.

References

1. Rahim AAA, Musa SN, Ramesh S, Lim MK (2021) Development of a fuzzy-TOPSIS multi-criteria decision-making model for material selection with the integration of safety, health and environment risk assessment. Int J Mater Des Appl 235(7):1532–1550
2. Mehmood Z, Haneef I, Udrea F (2020) Material selection for optimum design of MEMS pressure sensors. Microsyst Technol 26:2751–2766
3. Setti D, Verona MN, Medeiros BB, Restelli A (2019) Materials selection using a 2-tuple linguistic multi-criteria method. Mater Res 22. https://doi.org/10.1590/1980-5373-MR-2018-0846
4. Yadav S, Pathak VK, Gangwar S (2019) A novel hybrid TOPSIS-PSI approach for material selection in marine application. Sådhanå 44(58):1–12
5. Datta S, Mahfouf M, Chattopadhyay PP, Sultana N (2016) Imprecise knowledge based design and development of titanium alloys for prosthetic applications. J Mech Behav Biomed Mater 53:350–365
6. Kweh SWK, Khor KA, Cheang P (2000) Plasma-sprayed hydroxyapatite (HA) coatings with flame-spheroidized feedstock: microstructure and mechanical properties. Biomaterials 21:1223–1234
7. Dey A, Mukhopadhyay K (2010) Anisotropy in nano-hardness of microplasma sprayed hydroxyapatite coating. Adv Appl Ceram 109:346–354
8. Gross C, Berndt C (2002) Biomedical application of apatites. Rev Mineral Geochem 48:631–672
9. Mancini CE, Berndt CC, Sun L, Kucuk A (2001) Porosity determinations in thermally sprayed hydroxyapatite coatings. J Mater Sci 36:3891–3896
10. Katti S (2004) Biomaterials in total joint replacement. Colloids Surf B 39:133–142
11. Pawlowski L (1995) The science and engineering of thermal spray coatings. Wiley, Chichester
12. Huang J, Jayasinghe SN, Best SM, Edirisinghe MJ, Brooks RA, Bonfield W (2004) Electrospraying of a nano-hydroxyapatite suspension. J Mater Sci 39:1029–1032
13. Cheng GJ, Pirzada D, Cai M, Mohanty P, Bandyopadhyay A (2005) Bioceramic coating of hydroxyapatite on titanium substrate with Nd-YAG laser. Mater Sci Eng C 25:541–547
14. Mohammadi A, Moayyed AZ, Mesgar ASM (2007) Adhesive and cohesive properties by indentation method of plasma-sprayed hydroxyapatite coatings. Appl Surf Sci 253:4960–4965

15. Dey A, Mukhopadhyay AK, Gangadharan S, Sinha MK, Basu D (2009) Characterization of microplasma sprayed hydroxyapatite coating. J Therm Spray Technol 18:578–592
16. Dey A, Mukhopadhyay AK (2011) Fracture toughness of microplasma sprayed hydroxyapatite coating by nanoindentation. Int J Appl Ceram Technol 8:572–590
17. Cheng K, Zhang S, Weng W, Khor KA, Miao S, Wang Y (2008) The adhesion strength and residual stress of colloidal-sol gel derived [beta]-tricalcium-phosphate/fluoridated-hydroxyapatite biphasic coatings. Thin Solid Films 516:3251–3255
18. Arias MB, Mayor J, Pou Y, Leng B, Leon M (2003) Perez-Amora, micro-and nano-testing of calcium phosphate coatings produced by pulsed laser deposition. Biomaterials 24:3403–3408
19. Nieh G, Jankowsk AF, Koike J (2001) Processing and characterization of hydroxyapatite coatings on titanium produced by magnetron sputtering. J Mater Res 16:3238–3245
20. Guo X, Gough J, Xiao P (2007) Electrophoretic deposition of hydroxyapatite coating on Fecralloy and analysis of human osteoblastic cellular response. J Biomed Mater Res Part A 80:24–33
21. Gross A, Samandari SS (2007) Nano-mechanical properties of hydroxyapatite coatings with a focus on the single solidified droplet. J Aust Ceram Soc 43:98–101
22. Chakraborty J, Sinha MK, Basu D (2007) Biomolecular template induced biomimetic coating of hydroxyapatite on an SS316L substrate. J Am Ceram Soc 90:1258–1312

Remote Production Monitoring System

Anila Baby, Akshada Shinde, and Komal Dandge

Abstract Nowadays, the manufacturing industry plays an important role in India. In manufacturing industry, it is important to keep track of manufactured quantity, which is a time-consuming and laborious procedure. This quantity is tracked manually for maintaining a record of finished goods tested products. Currently, this process of data insertion is done manually which may lead to errors and is also a time-consuming process. So, the proposed system which is based on the modern technologies like Internet of Things (IoT) and cloud computing helps in overcoming this problem of manual maintaining of data and has also taken recently arising pandemic situation under consideration. It uses the microcontroller ESP8266, to create connection and exchange data between the cloud and physical system. Due to use of cloud storage, the chances of data loss are minimized, while the IoT plays an important role by helping in contactless information sharing. This system uses ThingSpeak Platform for cloud computing where stored data, i.e., count of tested product, can be downloaded via any device anytime, anywhere in an Excel Sheet along with a date and time stamp for each tested product. Thus, user can remotely monitor the production quantity and can also retrieve the data anytime. This is proposed remote production monitoring system.

Keywords ESP8266 · IoT · ThingSpeak · Cloud computing · Excel · Remotely

A. Baby · A. Shinde (✉) · K. Dandge
Deogiri Institute of Engineering and Management Studies, Aurangabad, India
e-mail: s.akshada1999@gmail.com

A. Baby
e-mail: anilababy799@gmail.com

K. Dandge
e-mail: komaldandge@dietms.org

P. Chatterjee et al. (eds.), *Computational Intelligence for Engineering and Management Applications*, Lecture Notes in Electrical Engineering 984,
https://doi.org/10.1007/978-981-19-8493-8_3

1 Introduction

The manufacturing industry depends on the number of manufactured products per day. This number varies on the basis of human resource, availability of raw materials and according to customers need. Thus, making an importance in the economy of the industry, the need to keep track of data is crucial. The Remote Production Monitoring System introduces IoT in production field for easy accessibility by keeping track of data on a time–time basis, thus helping to overcome the problems arising due to manual data recording. It is also important that the manufactured goods are tested to find defects and overcome the same to provide reliability to the customers, which in turn increases the industry's goodwill. The proposed system makes use of modern technologies like Internet of Things and cloud computing to keep track of these finished goods tested products. The use of IoT in this system has led to contactless information sharing. The Cloud platform is used to store this data and can be viewed anytime by the user, and it also displays real-time data. The ability of network devices to sense and collect data from different places around the world, and then to share these data across the Internet, to process and utilize for various purposes represents the Internet of Things or IoT. We know that Internet connects people from various places anytime, anywhere, and we call the same communication. In the same way, these sensors and electronic devices communicate with each other and provide us with data like climate of a certain place or the even turning off a light switch while we are outdoors. By representing future of computing and communications, IoT has become a technical revolution. The development of IoT depends on in a number of important dynamic technical innovation fields like wireless sensors and nanotechnology. They are going to tag each object for controlling, automating, monitoring and identifying. Various types of connections such as Bluetooth, Wi-Fi, RFID and ZigBee, in addition to allowing wide area connectivity with the help of technologies like GSM, GPRS, 3G and LTE, are used by IoT sensors. The information shared by the IoT-enabled things are about condition of things and the surrounding environment with people, software systems and other machines [1]. Cloud computing has become the most well-known technology in this developing era. Cloud infrastructure consists of cloud management software, servers, storage, network, deployment software and platform virtualization. The features of cloud not only include data anytime, anywhere with varying levels of security according to the users' need but also help to host and run applications, software for low cost without actually the need of buying the infrastructure and resources for the same. For the same cloud has various deployment models: Software as a Service (SaaS), Platform as a Service (PaaS) and Infrastructure as a Service (IaaS). Depending upon the requirement of the user data security and accessibility cloud has four types: public cloud, private cloud, community cloud and hybrid cloud [2]. The proposed system uses ThingSpeak Cloud which has easy accessibility and is a cost-effective cloud platform.

2 Literature Survey

Production department in some manufacturing industries uses manual methods to keep count of tested. This is a laborious process and may lead to manual errors in counting. In some cases, the testing boards in wire harness manufacturing industries contain an inbuilt counter which gives the number of tested products. But these inbuilt counters reset to '0' after certain number of units due to limited EEPROM storage available in microcontrollers which are used in these testing boards. So, using modern technologies like IoT and cloud computing, correct count can be monitored, stored and retrieved anytime by the user. These modern technologies offer the potential to transform way of production and increasing availability of the information using networked sensors. Number of companies and research organizations are expecting a wide range of impact of the Internet of Things on Internet and the economy in coming decades. Companies like Huawei and research organizations like Manyika et al., etc. expect 100 billion IoT connections which results in economic growth from 3.9 to 11 trillion dollars annually by 2025. Internet of Things leads to a world of technology and to a new era where different kinds of sensors can communicate, calculate and transform information quickly. This technology has attracted many researchers to work on new efficient ideas daily and bring up new own research [3].

The IoT basically implements and uses the concept of cloud computing which make it easily accessible to all kinds of users, e.g., individuals, originations and industries, which will make user to access and store resources and data in the various platform without actually having the physical storage and hardware or server. Multinational companies like Google, Microsoft, Amazon, Apple, etc. provide different storage options Google Drive, OneDrive, etc. Additionally, in business operations, cloud computing survey (as cited in Knorr, E., 2020) which was held in year 2020 revealed that 92 percent of the organizations are shifted to the cloud to carry out quick business operation and utilize storage facilities. In company spendings, the budget for cloud computing has taken third place. Subsequently, MNC like Accenture in the year 2021 stated that the cloud service industry has been rising annually worldwide and in 2020 was valued at 370 billion USD. According to the (Columbus, L., 2018) by year 2020, 83 percent of organizations, industries and enterprises' workloads will be transformed and potentially migrated to cloud technologies. As workload is the brain of the organization that consists of various support information systems to distribute, collect, analyze and keeping customers' data as a priority (IST, n.d.). With cloud-enabled workload solutions, the systems used in the business' operations are accessed and integrated through the web technologies. Notably, Sid Nag of Gartner stated that the adapting cloud computing technologies is now mainstream (Gartner, 2019) [4].

3 System Development

The main aim of the proposed system is to remotely access the count of tested products on a production line in manufacturing industry and share the data via Wi-Fi to a cloud platform for maintaining a record of the same.

3.1 Block Diagram

The block diagram for remote production monitoring system is shown in Fig. 1 which consists of NodeMCU ESP8266 microcontroller, OLED display, and output signal from testing device, push button, and DC power supply.

a. Microcontroller: The controller used for the system is ESP8266, which is also called NodeMCU. It is low-cost IoT platform which is used to connect the physical system to connect to a network and send to the channel on a cloud. Its built-in Wi-Fi/Bluetooth, and it has 128 KB RAM and 4 MB of flash memory to store data and programs. It operates at 5 V DC and an adjustable clock frequency of 80–160 MHz.
b. OLED display: The OLED is organic light-emitting diode. It displays deep black levels by working with no backlight. It works in response to an electric current to emit light. It is light in weight and smaller in size than the liquid crystal display (LCD). It is also easy to interface with other devices due to less number of pins. OLED displays driven by SSD1306 driver IC requires less number of external components and has reduced power consumption. It operates at 3.3 V DC. The proposed system uses 1.3 in. OLED display.
c. Input Signal: The production line has testing devices for testing the products; after testing an OK product, a signal is transmitted by the board to the microcontroller, for counter. This count increases by one for each new signal from the board.
d. Push button: Push buttons are used for the function of model selection and resetting the data stored in the microcontroller.

3.2 Software

a. Arduino IDE: The Arduino Integrated Development Environment or Arduino Software (IDE) is an official software introduced by Arduino.cc, that is mainly used for editing, compiling, and uploading the code in the Arduino device. Almost all Arduino modules are compatible with this software that is an open source and is readily available to install and start compiling the code on the go [5].
b. ThingSpeak Cloud: The ThingSpeak is a cloud platform which helps in aggregating, visualizing, and analyzing live data streams in the cloud. It provides a

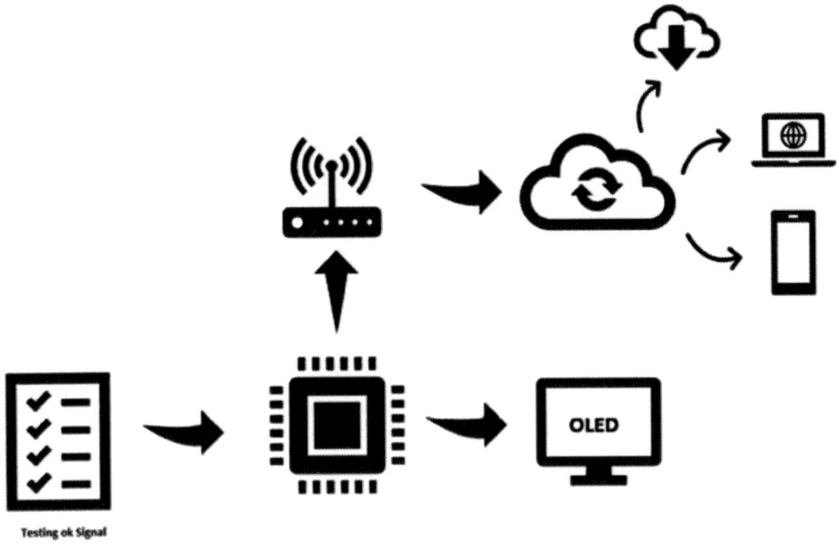

Fig. 1 Block diagram of remote production monitoring system

communication between your devices and ThingSpeak in the form of visuals of data posted by the device. It also executes MATLAB code so that you can perform online analysis and processing of data.

c. ThingView: The ThingView is an app version of ThingSpeak cloud which can be downloaded on the user's android phone and can be used to visualize the data on the cloud.

3.3 *Circuit Diagram*

The circuit diagram for the system as shown in Fig. 2 is simple and robust. It consists of ESP8266 microcontroller (Wi-Fi module) which is operated on 5 V DC. The Wi-Fi module or NodeMCU is used for establishing a connection with a network and to exchange data between the physical system and cloud platform, i.e., the ThingSpeak cloud platform. The NodeMCU is connected to testing device to receive the pulse for an OK tested product, at each new signal due to new tested product the counter for tested products increments its value by one, as well as this data are automatically updated on the cloud. Thus, providing with the number of tested finished goods. The system uses push buttons UP, DOWN, and ENTER for the purpose of model selection, for a certain production line, as many models get tested on a single testing board on the line. Thus, enabling the system to work on many models. Another push button RESET is used to reset the counter-value at the end of the day, so next day the value again starts from zero and helps the user in retrieving day to day information. These push buttons are also connected to the pins of the controller. Four pins of the

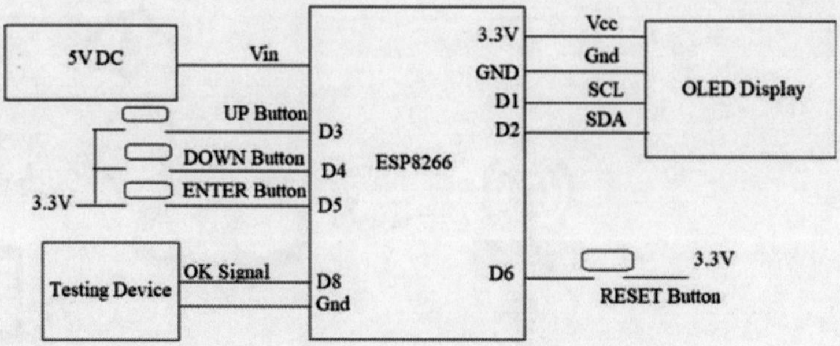

Fig. 2 Circuit diagram of remote production monitoring system

microcontroller are connected to the pins of the OLED display which is operated at 3.3 V DC. The OLED display (1.3 in.) is used to display the selected model during testing and number of tested products for the same, which increases with each new pulse from the testing board. The ThingSpeak cloud platform is used to store and display the data online, which can be accessed by creating an account on the platform. The data can be viewed anytime, anywhere by signing in to the users account. The same can be viewed even on android phones on ThingsView application just by signing in. Thus, the user gets the count for finished goods for the day only by a touch of a finger.

4 Algorithm

Step 1: Power ON the system and turn on the Internet.
Step 2: Wait till display shows connected status, i.e., until the NodeMCU is connected to hotspot network.
Step 3: Choose the model with the help of UP and DOWN buttons.
Step 4: Press Enter.
Step 5: Make sure the correct model is chosen.
Step 6: When the product is tested, the signal from testing board for OK tested product increases the count value.
Step 7: The counter-value is displayed on the OLED display and also uploaded to the cloud.
Step 8: At the end of the day, push RESET button provided at the back to reset the counter.

5 Performance Analysis

The proposed remote production monitoring system is a cost-effective way to keep track of finished goods per day. At each new pulse from testing device for OK tested product, the value of the counter increases by one and simultaneously gets updated on the channel created by the user. The count can be monitored either from a PC (shown in Fig. 3) or from mobile phone with ThingView application (shown in Fig. 4). The data can be downloaded in the form of an Excel Sheet anywhere, anytime along with a date and time stamp for each OK tested product (shown in Fig. 5).

Figure 3 shows the view of data on the ThingSpeak cloud platform. The count can be displayed either in the form of numbers or as graph. Thus, providing ease for data monitoring according to the users requirement. This can be done by selecting a widget to display the data. Once after the user creates a ThingSpeak account, they need to create number of fields on the channel depending on the number of models to be tested.

Figure 4 depicts the data as viewed by the user on the mobile application ThingView. The graph is plotted as date v/s count. Thus, showing the change in count on the basis of days.

The data can be downloaded by the user as shown in Fig. 5 in an Excel Sheet which has a date and time stamp along with the count. The column named as field can be changed to the name of the model being tested. On one Excel Sheet, the number of field, i.e., the number of models that can be viewed depends on the number of fields assigned in a certain channel.

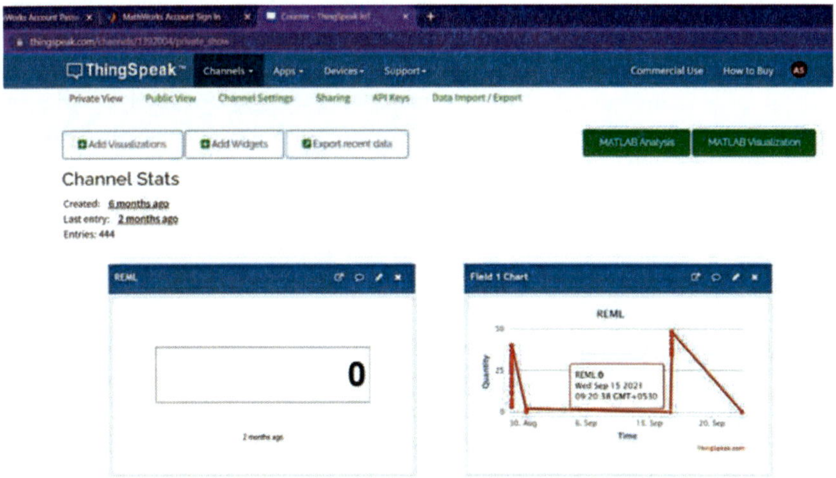

Fig. 3 View on ThingSpeak cloud

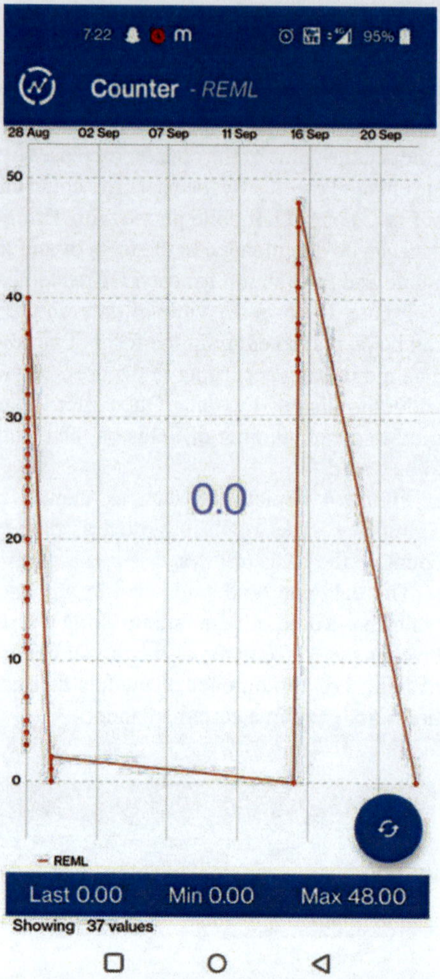

Fig. 4 View on ThingView application

6 Conclusion

The proposed system is designed in such a way that it can be useful to maintain the data of finished goods in manufacturing industries and access that data remotely. This system is not only useful in maintaining the data but can also play a role in monitoring. As this single system can be used on single testing devices with multiple models, it becomes more convenient.

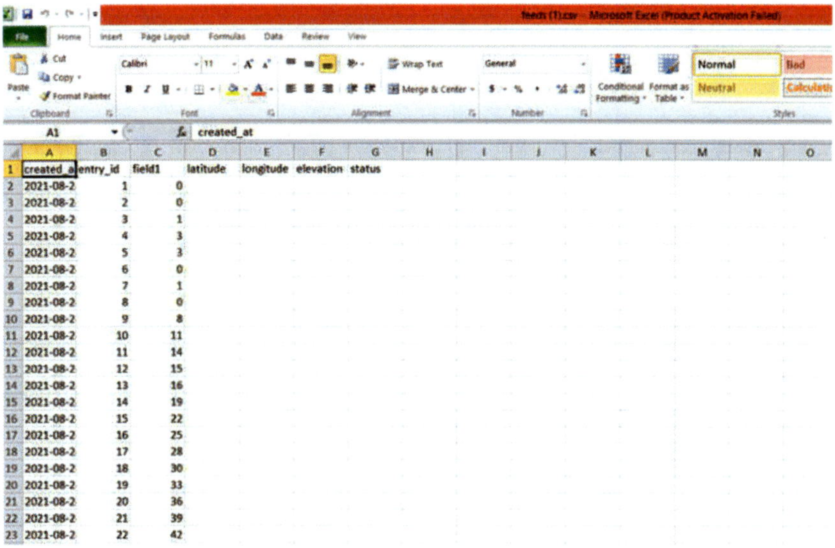

created_at	entry_id	field1	latitude	longitude	elevation	status
2021-08-2	1	0				
2021-08-2	2	0				
2021-08-2	3	1				
2021-08-2	4	3				
2021-08-2	5	3				
2021-08-2	6	0				
2021-08-2	7	1				
2021-08-2	8	0				
2021-08-2	9	8				
2021-08-2	10	11				
2021-08-2	11	14				
2021-08-2	12	15				
2021-08-2	13	16				
2021-08-2	14	19				
2021-08-2	15	22				
2021-08-2	16	25				
2021-08-2	17	28				
2021-08-2	18	30				
2021-08-2	19	33				
2021-08-2	20	36				
2021-08-2	21	39				
2021-08-2	22	42				

Fig. 5 Data downloaded in the form of excel sheet

References

1. Bharat (2017) A study on Internet of things its applications, challenges and related future technologies. Int J Innov Res Sci Eng Technol (IJIRSET) 6(12)
2. Tyagi A (2017) A review paper on cloud computing. Int J Eng Res Technol (IJERT) 5(23):1–2
3. Mouha RA (2021) Internet of things (IoT). J Data Anal Inform Process 9(2)
4. Garcia GJV (2021) Past, present, and future of cloud computing: an innovative case study. ResearchGate Publication. https://doi.org/10.13140/RG.2.2.16826.93128
5. Fezari M, Al Dahoud A (2018) Integrated development environment "IDE" for Arduino. ResearchGate Publication

Application of Wavelet Neural Network for Electric Field Estimation

Suryendu Dasgupta, Arijit Baral, and Abhijit Lahiri

Abstract Estimation of electric stress along the electrode and the insulator surfaces using wavelet neural networks (WNNs) has been carried out in this work. The application example considered here is an electrode-spacer arrangement which is used in gas-insulated substations (GISs). Four different WNNs using Gaussian, Morlet, Mexican hat and Shannon wavelet have been used to estimate electric stress over the electrode-insulator arrangement under study. Electric field computations have been carried out by applying boundary element method (BEM). The WNN is trained using a training set comprising of 71 data, and consequently, the trained network is tested with a testing set consisting of 15 data. Root mean squared error is a metric used for ascertaining the accuracy of the trained network while the testing accuracy is determined with the help of mean absolute error (MAE). For a given wavelet function, three parameters of the network, viz., the number of wavelons (N_w), number of iterations (N_{it}) and the learning factor (γ_k) are exhaustively varied, and the combination of these three parameters that yields least value of RMSE is determined, and the corresponding network is considered as the optimum WNN architecture. Hence, for the given application example corresponding to four wavelet functions, there will be four optimum WNN architectures. Among these four optimum architectures, the one that produces the least MAE will be considered as the best estimator for the application example under study.

Keywords Wavelet neural network · Electric field · Gaussian wavelet · Morlet wavelet · Mexican hat wavelet · Shannon wavelet

S. Dasgupta · A. Baral
Department of Electrical Engineering, IIT(ISM) Dhanbad, Dhanbad, Jharkhand, India
e-mail: arijit@iitism.ac.in

A. Lahiri (✉)
MCKV Institute of Engineering, Howrah, West Bengal, India
e-mail: jit_lahiri@yahoo.com

P. Chatterjee et al. (eds.), *Computational Intelligence for Engineering and Management Applications*, Lecture Notes in Electrical Engineering 984,
https://doi.org/10.1007/978-981-19-8493-8_4

1 Introduction

Power quality and reliability are the two major issues of power transmission. To ensure reliability, the power apparatus and the components of a transmission system should be sustainable with high endurance. This requires periodic condition monitoring of the apparatus and the components of the system. One of the important tasks, in this context, is to have an understanding about the distribution of electric field over the electrodes and the insulator surface that are used in HV applications.

Appreciable efforts have been put by various researchers in the past few years to optimize electrode-spacer contour with the objective of minimization of electric field intensity over HV equipments. Identification of those contour dimensions having maximum effect on the stress profile of the system and estimation of the same with the variation of one or more of these dimensions are required to optimize the electric stress distribution over the spacer and/or electrode surfaces. The geometrical parameters that effect the electric stress over the spacer and/or electrode surfaces are referred to as critical dimensions in this work.

Various numerical methods are available to calculate electric field for complex models with complex geometries [1–5]. Kumara et al. [6] applied models based on the empirical relations for the electric field and temperature-dependent electrical conductivities as well as models based on bipolar charge transport for the calculation of electric field in the case of HVDC cables and their comparison. Guo et al. [7] applied finite element method for calculation and analysis of distribution of electric field of a power line carrier reactor for a $\pm$ 1100 kV indoor DC yard model. But, when the problem is to optimize the electric stress by optimizing the geometry of the arrangement, then it requires estimation of electric stress as a function of the critical dimensions of the arrangement with higher degree of accuracy.

Function estimation is about developing a regression model [8] of an unknown function from an input/output pairs of data which represent the function itself. The difficulty that arises in developing a statistical regression model is in identifying the nature of the variation of the output pattern with any variation in the input pattern. This is because of the fact that, in real life, not all data are linearly correlated, and it becomes difficult to assess the nature of nonlinearity that exists between the response and the predictor variables.

Another pitfall of regression analysis is presence of either structural or data multicollinearity or both [8]. In either of the cases, the variance in the coefficient estimates increases sharply, and a minor change in the model results in a large swing of the estimates thereby reducing the precision of the coefficient estimates and weakening the regression model.

Over the last decade, a substantial development has been made in machine learning. Nonlinear networks reported by Poggio et al. [9], Hornik et al. [10] and Park et al. [11] proved to be efficient estimators of continuous functions. Lahiri et al. applied artificial neural network (ANN) to estimate three-dimensional electric stresses for optimizing them applying genetic algorithm [12] and simulated annealing algorithm [13], respectively. Support vector machine was applied by Banerjee et al. [14] to estimate electric stresses that were symmetric in nature, for optimizing support insulator configurations that are used in HV systems.

In the recent past, transition from wavelet theory to wavelet transform has offered an alternative approach toward function estimation problems [15]. Supported by wavelet transform, Zhang et al. [16] proposed wavelet network, a new learning network, as an alternative to feed forward neural network for approximating general nonlinear static functions from input–output observations. The rudimentary idea was replacement of the neurons of neural network with wavelons thereby building the computing unit by coupling in succession a multidimensional wavelet and an affine transform [16].

Zhang et al. [17] designed a WNN for function approximation that follows a radial basis function (RBF) network having the RBF replaced by orthonormal scaling functions that may or may not be radially symmetric. Zhang et al. [17] tested the efficiency of the proposed network both from theoretical analysis and experimental results and observed that a better performance is offered by the proposed WNN over RBF and multilayer perceptron (MLP) networks.

Oonsivilai et al. [18] successfully applied WNN for estimating short-term commercial load requirement by using Morlet and Mexican hat wavelets to generate transfer functions of the hidden layer neurons of the network architecture.

The present work is about estimation of maximum resultant electric stress (E^{Rm}) over the live electrode as a function of critical dimensions of the example under study by WNN. In case the electrode-spacer arrangement has a regular geometry, electric stress can be calculated by using mathematical relations that involves the geometry of the arrangement. But, it is not possible in the cases in where the electrode-spacer arrangement has got an irregular geometry. In those cases, estimation of electric field as a function of critical dimensions is carried out by machine learning algorithms. The critical dimensions can be adjusted to achieve minimum electric stress over the live electrode to ensure power system reliability and cost effectiveness of the electrode-spacer arrangement. The major advantages of WNN over other neural networks (NNs) lie in the fact that assigning the number of nodes, number of hidden layers and initialization of weight matrix are more convenient. If the dimension of the training data increases, other NN gets trained at the cost of the rate of convergence and with the risk of getting stuck to a local minimum. To achieve the same degree of accuracy, WNN requires fewer numbers of training data and lesser number of neurons compared to multilayer perceptron (MLP) networks. The objective of using WNN is to exploit its better convergence capability in three-dimensional space as compared to conventional neural networks [18].

Another classifier that has the ability to perform regression analysis is support vector machine (SVM) [19, 20]. Thus SVM can be one of the tools for function estimation. The major drawback that SVM suffers is tuning of its penalty parameter and gamma parameter if radial basis function (RBF) kernel is used, and the task depends only on experience and has no standard solution.

The drawback that WNN suffers is in selecting the mother wavelet. Hence, in this work, four different mother wavelets, namely Gaussian wavelet, Morlet wavelet, Mexican hat wavelet and Shannon wavelet, have been considered. The performance of the wavelet network using each of these wavelet functions will be studied to choose the best network for estimating E^{Rm} of the example under study.

2 Problem Formulation

In this paper, the application example under study is an electrode-spacer arrangement which is symmetrical about the axis of rotation employed in GIS. The arrangement is shown in Fig. 1 [21]. It is normally preferred for ratings of 12 kV ≤ V ≤ 420 kV and above.

Fig. 1 Application example

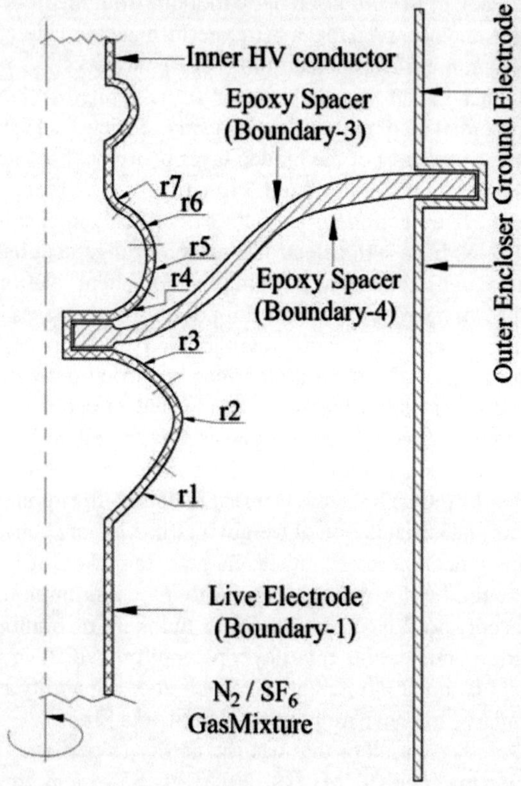

A cylindrical metal enclosure being maintained at the ground potential is used to enclose the spacer and the live electrode that is being kept at a normalized reference voltage of 1 V. The metallic parts of the arrangement are made of aluminum, and the material used for spacer is epoxy cast resin of relative permittivity 5.3. An insulating gas mixture comprising of 20% N_2 and 80% SF_6 by proportion of relative permittivity 1.005 fills up the space between the live and the ground electrodes [22]. Figure 2 shows the three-dimensional views of (a) the live electrode (b) the epoxy spacer and (c) the assembled arrangement.

The electrode-spacer arrangement under consideration comprises of four boundaries:

(a) Boundary-1: between the insulating gas mixture and the live electrode.
(b) Boundary-2: between the insulating gas mixture and the ground electrode.

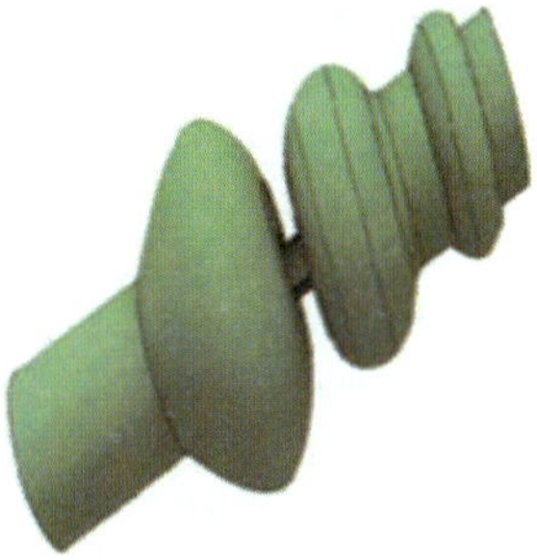

a) Electrode Configuration.

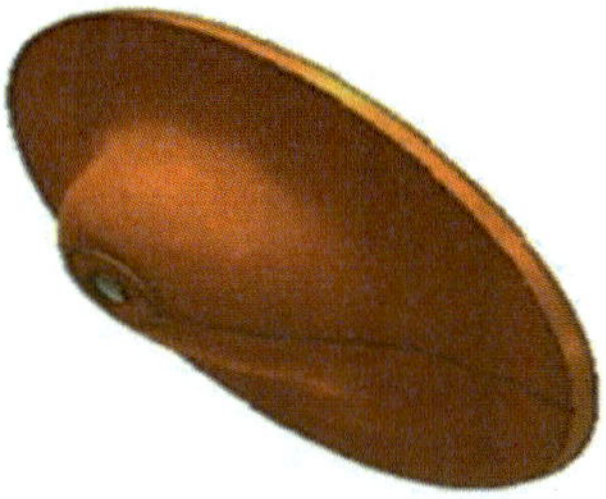

b) Spacer Configuration.

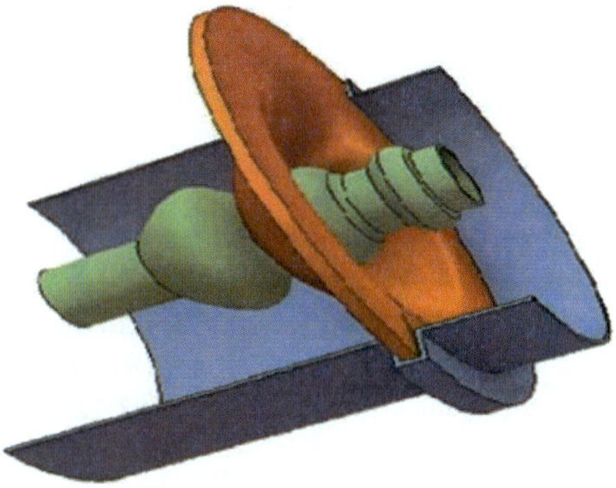

c) Cross-sectional view of assembled model.

Fig. 2 Electrode-spacer arrangement

Table 1 Maximum and minimum values of critical dimensions

Critical dimension	Maximum value (m)	Minimum value (m)
r_4	3.0	0.5
r_5	6.0	0.5
r_6	17.5	2.5
r_7	4.5	0.1

(c) Boundary-3: between the gas mixture and the concave surface of the epoxy spacer and
(d) Boundary-4: between the convex surface of the epoxy spacer and the gas mixtures.

At boundary-1, on the convex side of the spacer, the curved portion of the live electrode comprises of three segments which are elliptical in shape of radius of curvature $r_1 = 0.1$ m, $r_2 = 0.2$ m, $r_3 = 0.05$ m. On the other hand, on the concave side of the spacer r_4, r_5, r_6 and r_7 are the radii of the four elliptical segments forming the curved portion of the live electrode.

Exploratory field computation revealed that E^{Rm} over the live electrode boundary changes significantly with changes in r_4 to r_7. Thus r_4 to r_7 are altered within an extensive range with the objective of studying the effect of such variations on the electric stress profile over the live electrode boundary while r_1, r_2, r_3 are kept fixed at the abovementioned values. The limits within which the radii r_4 to r_7 are varied is shown in Table 1.

Electric field is computed using BEM [23] with the objective to prepare the training dataset intended for training the WNN. The critical dimensions, viz., r_4, r_5, r_6 and r_7 constitute the input vector while E^{Rm} is the output vector.

3 Wavelet Neural Network

Based on wavelet theory, wavelet network can be considered as a modified feed forward neural network that emerged as an efficient estimator of any arbitrary nonlinear function. Owing to this fact, WNN is not only capable of analyzing non-stationary signals, but it can also be used as a classifier. Unlike the neurons of an ANN, WNN has a collinear combination of translation, rotation, dilation and wavelet that are termed as wavelons.

The basic objectives of a WNN are mainly to conserve the universal approximation property by providing a class of networks that will exhibit the same density property to establish a definite link between the network coefficients and some approximation transformations and finally to attain the same approximation with a network that is more compact in size.

Since for the example under study, the number of input parameters is 4; hence, a multidimensional wavelet network (MDWN) has to be considered and the 4 radii, mentioned earlier, along with E^{Rm} forms the input vector and the output vector of the MDWN, respectively.

Equation (1) represents the structure of a multidimensional wavelet network (MDWN), having N_w number of wavelons:

$$g(x) = \sum_{i=1}^{i} w_i \times \varphi(\Delta_i \times R_i \times (x - \tau_i)) + \overline{\theta} \tag{1}$$

where w_i denotes the weight matrix, φ represents the wavelet function, Δ_i is the dilation matrix that are diagonal matrices derived from dilation vectors, R_i is the rotation matrix, τ_i is the translation vectors, $\overline{\theta}$ is an additional parameter introduced for approximating functions that have non-zero mean because φ being distributed about zero mean cannot take care of these functions. w_i, Δ_i, R_i, τ_i and $\overline{\theta}$ are the parameters of a WNN, and the convergence of any WNN depends on the initialization of these parameters. The structure of a WNN is represented in Fig. 3a. Figure 3b represents the detail of the jth wavelon where the rotation matrix, denoted as R_j which is a function of the rotational angle ϑ, is used to restore the orthogonality of the wavelons that are distorted while translational parameter τ is vectorially added during feed forward calculation.

R matrix is given by:

$$R = \begin{pmatrix} \cos\vartheta_1 & -\sin\vartheta_1 & 0 & \cdots\cdots & 0 \\ \sin\vartheta_1 & \cos\vartheta_1 & 0 & \cdots\cdots & 0 \\ 0 & 0 & 1 & \cdots\cdots & 0 \\ \vdots & 0 & 0 & 1 \ \cdots & 0 \\ \vdots & \vdots & \vdots & \vdots \ \cdots & \vdots \\ 0 & 0 & 0 & \cdots\cdots & 1 \end{pmatrix} \times \begin{pmatrix} \cos\vartheta_2 & 0 & -\sin\vartheta_2 & 0 & \cdots & 0 \\ 0 & 1 & 0 & 0 & \cdots & 0 \\ \sin\vartheta_2 & 0 & \cos\vartheta_2 & & \cdots\cdots & 0 \\ \vdots & \vdots & \vdots & & \cdots\cdots & \vdots \\ \vdots & \vdots & \vdots & & \cdots\cdots & \vdots \\ 0 & 0 & 0 & & \cdots\cdots & 1 \end{pmatrix} \cdots \begin{pmatrix} 1 & 0 & 0 & \cdots & \cdots & 0 \\ 0 & 1 & 0 & \cdots & \cdots & 0 \\ \vdots & \vdots & \cdots\cdots & & \vdots & \vdots \\ \vdots & \vdots & \cdots\cdots & & \vdots & \vdots \\ 0 & 0 & 0 & \cdots & \cos\vartheta_k & -\sin\vartheta_k \\ 0 & 0 & 0 & \cdots & \sin\vartheta_k & \cos\vartheta_k \end{pmatrix} \tag{2}$$

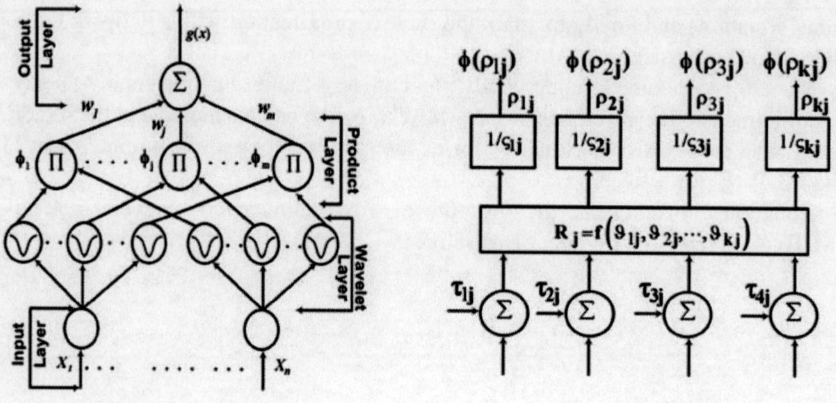

a) WNN Architechture b) j[th] Wavelon

Fig. 3 Schematic diagram of a WNN

If ν is the number of input dimensions, then $k = 0.5\upsilon \times (\upsilon - 1)$. In the present case, since there are 4 inputs, hence, the rotational matrix will be the product of 6 matrices each of order 4×4.

$$R = (\alpha, \beta)\text{ plane} \times (\alpha, \gamma)\text{ planerotation} \times (\alpha, \delta)\text{ planerotation} \\ \times (\beta, \gamma)\text{ planerotation} \times (\beta, \delta)\text{ planerotation} \times (\gamma, \delta)\text{ planerotation} \quad (3)$$

and the respective notations are obtained from Eq. (2). Thus for a 4 input-1 output architecture, the rotation matrix will be as follows:

$$R = \begin{pmatrix} R_1 & 0 & 0 & 0 \\ 0 & R_2 & 0 & 0 \\ 0 & 0 & R_3 & 0 \\ 0 & 0 & 0 & R_4 \end{pmatrix} \quad (4)$$

where

$$R_i = \begin{pmatrix} R_{i11} & R_{i12} & R_{i13} & R_{i14} \\ R_{i21} & R_{i22} & R_{i23} & R_{i24} \\ R_{i31} & R_{i32} & R_{i33} & R_{i34} \\ R_{i41} & R_{i42} & R_{i43} & R_{i44} \end{pmatrix} \quad (5)$$

The dilation matrix is given by:

$$\Delta = \begin{pmatrix} \Delta_1 & 0 & 0 & \cdots\cdots & 0 \\ 0 & \Delta_2 & 0 & \cdots\cdots & 0 \\ 0 & 0 & \Delta_3 & \cdots\cdots & 0 \\ 0 & 0 & 0 & \Delta_4 \cdots & 0 \\ \vdots & \vdots & \vdots & \vdots \; \cdots & 0 \\ 0 & 0 & 0 & \cdots\cdots & \Delta_n \end{pmatrix} \tag{6}$$

$$\Delta_i = \begin{pmatrix} \frac{1}{\zeta_{1i}} & 0 & 0 & \cdots\cdots & 0 \\ 0 & \frac{1}{\zeta_{2i}} & 0 & \cdots\cdots & 0 \\ 0 & 0 & \frac{1}{\zeta_{3i}} & \cdots\cdots & 0 \\ 0 & 0 & 0 & \frac{1}{\zeta_{4i}} \cdots & 0 \\ \vdots & \vdots & \vdots & \vdots \; \cdots & \vdots \\ 0 & 0 & 0 & \cdots\cdots & \frac{1}{\zeta_{ki}} \end{pmatrix} \tag{7}$$

where ζ denotes the scaling function of the wavelet. Thus, the product term $\Delta_i \times R_i \times (x - \tau_i)$ of Eq. (1) will be obtained for the ith wavelon by multiplying the matrices R_i, Δ_i with the vector $\overline{\rho}$ given by:

$$\overline{\rho} = \begin{pmatrix} x_1 - \tau_{1i} \\ x_2 - \tau_{2i} \\ \vdots \\ \vdots \\ x_k - \tau_{ki} \end{pmatrix} \tag{8}$$

and the product will result in a vector $\overline{\zeta}$ given by:

$$\overline{\zeta} = \begin{pmatrix} \zeta_{1i} \\ \zeta_{1i} \\ \vdots \\ \vdots \\ \zeta_{1i} \end{pmatrix} \tag{9}$$

Once $\overline{\zeta}$ is known the mother wavelet family can be defined in the following manner.

The Gaussian wavelet family is defined as:

$$\begin{pmatrix} \varphi(\zeta_{1i}) \\ \varphi(\zeta_{2i}) \\ \varphi(\zeta_{3i}) \\ \vdots \\ \vdots \\ \varphi(\zeta_{ki}) \end{pmatrix} = \begin{pmatrix} \zeta_{1i}\mathrm{e}^{\frac{\zeta_{1i}^2}{2}} \\ \zeta_{2i}\mathrm{e}^{\frac{\zeta_{2i}^2}{2}} \\ \zeta_{3i}\mathrm{e}^{\frac{\zeta_{3i}^2}{2}} \\ \vdots \\ \vdots \\ \zeta_{ki}\mathrm{e}^{\frac{\zeta_{ki}^2}{2}} \end{pmatrix} \tag{10}$$

The Morlet wavelet family is defined as:

$$\begin{pmatrix} \varphi(\zeta_{1i}) \\ \varphi(\zeta_{2i}) \\ \varphi(\zeta_{3i}) \\ \vdots \\ \vdots \\ \varphi(\zeta_{ki}) \end{pmatrix} = \begin{pmatrix} \cos(5\zeta_{1i})\mathrm{e}^{\frac{\zeta_{1i}^2}{2}} \\ \cos(5\zeta_{2i})\mathrm{e}^{\frac{\zeta_{2i}^2}{2}} \\ \cos(5\zeta_{3i})\mathrm{e}^{\frac{\zeta_{3i}^2}{2}} \\ \vdots \\ \vdots \\ \cos(5\zeta_{ki})\mathrm{e}^{\frac{\zeta_{ki}^2}{2}} \end{pmatrix} \tag{11}$$

Mexican hat wavelet family is given by:

$$\begin{pmatrix} \varphi(\zeta_{1i}) \\ \varphi(\zeta_{2i}) \\ \varphi(\zeta_{3i}) \\ \vdots \\ \vdots \\ \varphi(\zeta_{ki}) \end{pmatrix} = \begin{pmatrix} 0.867325(1-\zeta_{1i}^2)\mathrm{e}^{\frac{\zeta_{1i}^2}{2}} \\ 0.867325(1-\zeta_{2i}^2)\mathrm{e}^{\frac{\zeta_{2i}^2}{2}} \\ 0.867325(1-\zeta_{3i}^2)\mathrm{e}^{\frac{\zeta_{3i}^2}{2}} \\ \vdots \\ \vdots \\ 0.867325(1-\zeta_{ki}^2)\mathrm{e}^{\frac{\zeta_{ki}^2}{2}} \end{pmatrix} \tag{12}$$

The Shannon wavelet family is defined as:

$$\begin{pmatrix} \varphi(\zeta_{1i}) \\ \varphi(\zeta_{2i}) \\ \varphi(\zeta_{3i}) \\ \vdots \\ \vdots \\ \varphi(\zeta_{ki}) \end{pmatrix} = \begin{pmatrix} \frac{2(\cos\pi\zeta_{1i}-\sin\pi\zeta_{1i})}{\pi(2\zeta_{1i}-1)} \\ \frac{2(\cos\pi\zeta_{2i}-\sin\pi\zeta_{2i})}{\pi(2\zeta_{2i}-1)} \\ \frac{2(\cos\pi\zeta_{3i}-\sin\pi\zeta_{3i})}{\pi(2\zeta_{3i}-1)} \\ \vdots \\ \vdots \\ \frac{2(\cos\pi\zeta_{ki}-\sin\pi\zeta_{ki})}{\pi(2\zeta_{ki}-1)} \end{pmatrix} \tag{13}$$

The four wavelet functions discussed above are used to study the performance of the network using each of these wavelet functions in order to choose the best network for estimating E^{Rm}. The product $\zeta_{1i}\zeta_{2i}\zeta_{3i} \ldots \zeta_{ki}$ when multiplied with w_i and summed up for all the wavelons will produce the output when it is added with $\overline{\theta}$.

In this work, the guidelines set by Zhang et al. [16] have been followed for initializing the network parameters. The rotation angles are selected in such a way that R_i is initialized as identity matrix [16], w_i is initially set as a null matrix, the mean of the assigned target values are treated as the initial value of $\overline{\theta}$. Δ_i and τ_i are also initialized according to the guidelines provided in [16].

Error convergence is better if the network parameters are properly initialized in comparison to the case when their values are selected randomly. Moreover, proper normalization of the input–output data leads to a better convergence in error. Choice of mother wavelet and also the nature of input–output data determines the accuracy of estimation.

Another factor that affects the convergence of the network is rate of learning (γ_k). Small values of γ_k result in small changes in the synaptic weights of the network during transition between two consecutive iterations. Thus, smaller the value of γ_k, i.e., slower the rate of learning, smoother is the trajectory in the weight space. In contrary to this, if γ_k is large, then the rate of learning accelerates causing large changes in the synaptic weights and may result in an unstable network.

Training accuracy is measured in terms of RMSE which measures the deviation between the analytical and the estimated values of the output and is given by:

$$\text{RMSE} = \sqrt{\frac{1}{N}\sum_{i}\left(E_{\text{trg}}^{Rm}(x_i) - E_{\text{an}}^{Rm}(x_i)\right)^2}, \quad i = 1, 2, \ldots, N \tag{14}$$

where $E_{\text{trg}}^{Rm}(x_i)$ is the ith estimated value of E^{Rm} returned by the WNN during training and $E_{\text{an}}^{Rm}(x_i)$ is the analytical value of E^{Rm} obtained by actually calculating the electric field, i.e., the output data fed to the training algorithm and N is the number of training sets.

To measure the accuracy of the training, another parameter is estimated which is the MAE and is given by:

$$\text{MAE} = \frac{1}{N}\sum_{i}\left|E_{\text{es}}^{Rm}(x_i) - E_{\text{an}}^{Rm}(x_i)\right|; \quad i = 1, 2, \ldots, N \tag{15}$$

where $E_{\text{es}}^{Rm}(x_i)$ is the ith estimated value returned by the trained network during testing and N is the number of testing sets.

4 Results and Discussions

This section deals with the process of obtaining the optimum network with each of the four wavelet functions. This is done by varying the three parameters N_w, N_{it} and γ_k over a wide range, varying one parameter at a time while keeping the other two parameters constant at a particular value to obtain the minimum value of RMSE. Here, the process of obtaining the optimum network using Mexican hat wavelet is discussed in details. Following the same procedure, the optimum network corresponding to each of the other three wavelet functions is obtained.

In order to find an optimum network using Mexican hat wavelet, initially, N_{it} and γ_k are assigned constant values of 50 and 0.1, respectively, and N_w is varied from 5 to 30 at an interval of 5. The effect this variation is shown in Table 2. It is apparent from Table 2 that the value of RMSE increases as N_w increases and finally becomes constant from N_w equal to 20. Since the value of RMSE is least for N_w equal to 5, hence, the optimum value of N_w is chosen to be 5.

Now, keeping N_w constant at 5 and γ_k equal to 0.1, N_{it} is varied from 50 to 300 in steps of 50. During this variation, it is seen that the value of RMSE is constant till N_{it} equals to 150, but with further increment in N_{it}, RMSE has started increasing gradually. Hence, the value of N_{it} is chosen to be 50 so that the computational time is minimized. The results are given in Table 3.

Finally, keeping N_w and N_{it} constant, at 5 and 50, respectively, the value of γ_k is varied from 0.1 to 0.0003. The impact of this variation on RMSE is shown in Table 4 from where it is observed that RMSE is oscillatory for γ_k between 0.1 and 0.002. On further decreasing γ_k beyond 0.002, RMSE started increasing and becomes constant

Table 2 Effect of variation of N_w on RMSE using Mexican hat wavelet

N_w	N_{it}	γ_k	RMSE
5	50	0.1	0.102
10	50	0.1	0.103
15	50	0.1	0.105
20	50	0.1	0.115
25	50	0.1	0.115
30	50	0.1	0.115

Table 3 Effect of variation of N_{it} on RMSE using Mexican hat wavelet

N_w	N_{it}	γ_k	RMSE
5	50	0.1	0.102
5	100	0.1	0.102
5	150	0.1	0.102
5	200	0.1	0.103
5	250	0.1	0.104
5	300	0.1	0.105

when γ_k is further decreased beyond 0.001. Table 4 reveals that minimum RMSE, i.e., 0.072 is obtained for γ_k equal to 0.004 as well as 0.003. In order to ensure stability of the network architecture, the smaller value between the two is chosen, i.e., 0.003.

Following the same method as used for Mexican hat wavelet, the optimum network parameters for the other three wavelet functions, viz., Gaussian, Morlet and Shannon are obtained. The values of the parameters of the optimum network obtained with the four wavelets along with the corresponding values of RMSEs are provided in Table 5.

As mentioned earlier, 15 sets of data are used for the purpose of testing the trained networks regarding their performance. The MAE produced by each of the four networks is summarized in Table 6. It is apparent from Table 6 that in terms of

Table 4 Effect of variation of γ_k on RMSE using Mexican hat wavelet

N_w	N_{it}	γ_k	RMSE
5	50	0.1	0.102
5	50	0.09	0.101
5	50	0.08	0.100
5	50	0.07	0.098
5	50	0.06	0.097
5	50	0.05	0.095
5	50	0.04	0.094
5	50	0.03	0.093
5	50	0.02	0.094
5	50	0.01	0.098
5	50	0.009	0.091
5	50	0.008	0.078
5	50	0.007	0.073
5	50	0.006	0.074
5	50	0.005	0.074
5	50	0.004	0.072
5	50	0.003	0.072
5	50	0.002	0.075
5	50	0.001	0.079
5	50	0.0009	0.080
5	50	0.0008	0.080
5	50	0.0007	0.080
5	50	0.0006	0.080
5	50	0.0005	0.080
5	50	0.0004	0.080
5	50	0.0003	0.080

Table 5 Comparison of optimum network parameters for different wavelets

Mother wavelet	N_w	N_{it}	γ_k	RMSE
Gaussian	5	50	0.01	0.086
Morlet	5	50	0.001	0.085
Mexican hat	5	50	0.003	0.072
Shannon	5	50	0.01	0.085

Table 6 MAE obtained from each of the networks during testing

Mother wavelet	MAE
Gaussian	0.0212
Morlet	0.0162
Mexican hat	0.0076
Shannon	0.0085

MAE the best fitted WNN to estimate E^{Rm} for the example under study is the one that uses Mexican hat wavelet.

A comparative study between the analytical value and the estimated value of E^{Rm} during testing is represented in Fig. 4. From Fig. 4, it is apparent that all the four networks are efficient in estimating E^{Rm} for the application example considered in the present study, but still, Mexican hat wavelet will be used for the purpose of estimating E^{Rm} for being the best fitted WNN in terms of MAE.

5 Conclusions

In the present study, different WNN architectures have been used with the objective to estimate the value of E^{Rm} over the live electrode surface of a GIS arrangement. The critical dimensions of the live electrode have been identified and are altered over a wide range maintaining the overall constraint of dimensionality to get the dataset for training. Similarly, the network parameters that have been found to affect the training error the most are identified.

Using four different wavelet functions, the optimum network corresponding to each of them are obtained, and finally, it is observed that the maximum error in predicting the value of E^{Rm} is 2.12% as obtained from the network using Gaussian wavelet function while the minimum error in estimating the value of E^{Rm} is 0.76% as obtained from the network using Mexican hat wavelet function. This trained WNN can now be used to estimate E^{Rm} in optimization problems where the critical dimensions can be optimized to optimize E^{Rm}. Thus the effectiveness of WNN in estimating electric stress as a function of the dimensional parameters of electrode-spacer arrangements used in practical purposes is established in this work.

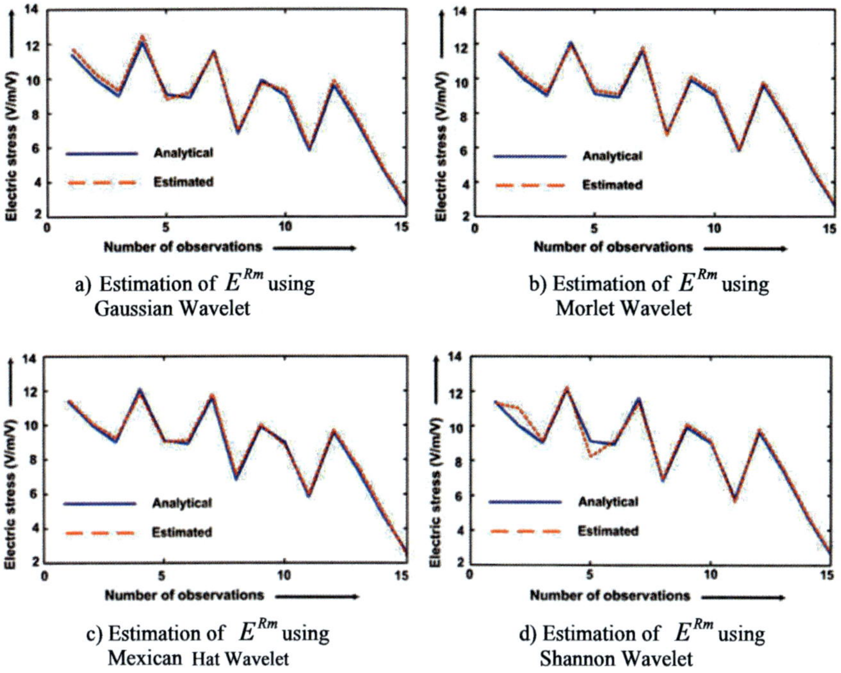

a) Estimation of E^{Rm} using Gaussian Wavelet

b) Estimation of E^{Rm} using Morlet Wavelet

c) Estimation of E^{Rm} using Mexican Hat Wavelet

d) Estimation of E^{Rm} using Shannon Wavelet

Fig. 4 Performance testing using four different wavelets

References

1. Sharma MS (1970) Potential functions in electromagnetic field problems. IEEE Trans Magn 6:513–518
2. Andersen OW (1973) Laplacian electrostatic field calculations by finite elements with automatic grid generation. IEEE Trans Power Appar Syst 96:1156–1160
3. Singer H, Steinbigler H, Weiss PA (1974) Charge simulation method for the calculation of high voltage fields. IEEE Trans Power Appar Syst 93:1660–1668
4. Blaszczyk A, Steinbigler H (1994) Region-oriented charge simulation. IEEE Trans Magn 30:2924–2927
5. Chakravorti S, Steinbigler H (1998) Capacitive-resistive field calculation on HV bushings using the boundary-element method. IEEE Trans Dielectr Electr Insul 5:237–244
6. Kumara S, Serdyuk YV, Jeroense M (2021) Calculation of electric fields in HVDC cables: comparison of different models. IEEE Trans Dielectr Electr Insul 28:1070–1078
7. Guo Y, Zhao Z, Zhang W, Yan G, Li Y, Peng Z (2021) Optimization design on shielding electrodes for PLC reactor in UHV indoor DC yard. In: IEEE international conference on the properties and applications of dielectric materials (ICPADM). https://doi.org/10.1109/ICPADM49635.2021.9493891
8. Jim F, Regression analysis: an intuitive guide for using and interpreting linear models 1st edn. www.statisticsbyjim.com
9. Poggio T, Girosi F (1990) Networks for approximation and learning. Proc IEEE 78:1481–1497
10. Hornik K, Stinchcombe M, White H (1989) Multilayer feed forward networks are universal approximators. Neural Netw 2:359–366

11. Park J, Sandberg IW (1991) Universal approximation using radial-basis-function networks. Neural Comput 3:246–257
12. Lahiri A, Chakravorti S (2004) Electrode-spacer contour optimization by ANN aided genetic algorithm. IEEE Trans Dielectr Electr Insul 11:964–975
13. Lahiri A, Chakravorti S (2005) A novel approach based on simulated annealing coupled to artificial neural network for 3D electric field optimization. IEEE Trans Power Delivery 20:2144–2152
14. Banerjee S, Lahiri A, Bhattacharya K (2007) Optimization of support insulators used in HV systems using support vector machine. IEEE Trans Dielectr Electr Insul 14:360–367
15. Delyon B, Juditsky A, Benveniste A (1995) Accuracy analysis for wavelet approximations. IEEE Trans Neural Netw 6:332–348
16. Zhang Q, Benveniste A (1992) Wavelet networks. IEEE Trans Neural Netw 3:889–898
17. Zhang J, Walter GG, Miao Y, Lee WNW (1995) Wavelet neural network for function learning. IEEE Trans Signal Process 43:1485–1497
18. Oonsivilai A, EI-Hawary ME (1999) Wavelet neural network based short term load forecasting of electric power system commercial load. In: engineering solutions for the next millennium, IEEE Canadian conference on electrical and computer engineering (Cat. No. 99TH8411), pp 1223–1228. https://doi.org/10.1109/CCECE.1999.804865
19. Schölkopf B, Burges CJC, Smola AJ (1999) Advances in kernel methods—support vector learning. MIT Press, Cambridge
20. Smola AJ (1996) Regression estimation with support vector learning machines. Technical report. TechnischeUniversität, München, München, Germany
21. Dasgupta S, Lahiri A, Baral A (2016) Optimization of electrode-spacer geometry of a gas insulated system for minimization of electric stress using SVM. In: Frontiers in computer, communication and electrical engineering. Taylor & Francis Group, London, pp 501–506. https://doi.org/10.1201/b20012-98
22. Christophorou LG, Burnt RJ (1995) SF_6/N_2 mixtures basic and HV insulation properties. IEEE Trans Dielectr Electr Insul 2:925–1002
23. Gutfleisch F, Singer H, Foerger K, Gomollon J (1994) A calculation of HV fields by means of the boundary element method. IEEE Trans Power Delivery 9:743–749

Development of an Industrial Control Virtual Reality Module for the Application of Electrical Switchgear in Practical Applications

Kevin R. Atiaja, Jhon P. Toapanta, and Byron P. Corrales

Abstract This work proposes the implementation of a multilevel virtual reality (VR) tool applied to electromechanical engineering students through the use of Oculus Rift glasses, to complement the learning process in the manipulation of different elements of electrical equipment in the laboratory, in addition to acquiring practical experiences in industrial control. The VR module developed is made up of 3 levels: easy, normal, and difficult, each with its activities; to advance to the next scenario, it is necessary to complete all the activities proposed by each level; to evaluate the application 30 students belonging to the Technical University of Cotopaxi participated divided into two groups, the second group was provided with this tool as a compliment, while the first group did not have it. A questionnaire was carried out to determine the degree of learning and usability by the user at the time of completing their interaction with the virtual environment. The results show an improvement on the part of the students belonging to group B in the recognition of control and maneuver elements, reduction of errors in the connection as well as the reduction of the assembly time of the circuits.

Keywords Virtual reality · Learning · Engineering · Industrial control · Electricity

1 Introduction

In countries such as China, the prayer hall for the good harvest has an extremely high historical value and the sudden increase in tourists puts at risk the relics of the hall; this leads the authorities to develop the Digital Hall of Prayer for Good Harvest

K. R. Atiaja (✉) · J. P. Toapanta · B. P. Corrales
Universidad Técnica de Cotopaxi, Latacunga, Ecuador
e-mail: kevin.atiaja8513@utc.edu.ec

J. P. Toapanta
e-mail: jhon.toapanta1882@utc.edu.ec

B. P. Corrales
e-mail: byron.corrales@utc.edu.ec

P. Chatterjee et al. (eds.), *Computational Intelligence for Engineering and Management Applications*, Lecture Notes in Electrical Engineering 984,
https://doi.org/10.1007/978-981-19-8493-8_5

software through the application of VR and OpenGL technology, presenting images in real time to people; in this way, they better understand the historical culture of that country [1]. Distance education is used for teaching in science, medicine, and even engineering; in these branches, virtual environments play an essential role since they allow the student to connect with the knowledge that is intended to be acquired using scenarios similar to the real ones [2, 3]. In classrooms or other teaching scenarios, VR helps to strengthen the learning outcomes of students, incorporating electronic and electrical experiments which have improved experimental education resulting in being an effective model [4]. The implementation of a laboratory where a workstation is simulated with virtual, remote, and simulated experiments in real time helps not only students but also engineers to develop their skills in the field of automation, and in this way, technology is offered for industrial development [5, 6]. The implementation of a virtual tool applied in the field of automotive engineering, to enhance the education process so that students can achieve educational results [7]. In Paleontology, the analysis and visualization of fossils in 3 dimensions has revolutionized the knowledge about these organisms; the new protocols allow to obtain a superficial reconstruction including soft tissues from incomplete remains [8]. The development of VR simulators for the training and evaluation of orthopedic surgeons is directly related to a significant improvement when carrying out orthopedic surgical procedures [9, 10]. The development of virtual environments together with 3D modeling software creates spaces where valves can be calibrated, which provides a high degree of realism through the use of HTC Vive glasses [11]. At present, interactive virtual environments are developed to implement industrial processes; for the inspection and verification of bottles, an artificial camera is implanted that allows detection failures in the product [12]. Virtual reality is having a great reception in the process of simulation, concern in the operation, or behavior of the industrial sector in the field of engineering; 3D graphic engines are quite similar to the real ones [13]. The 3D virtual environment for dual-distillation towers, thanks to the Hardware in Loop technology, allows trained users to interact with the processes. The virtual environment includes elements, equipment, graphics, and sounds of good quality [14, 15]. The implementation of a 3D virtual environment for the intelligent calibration of pressure transmitters allows indicating the level, as an executable computer application for the Windows system. To visualize the physical variables present in the virtual module, a differential and absolute pressure transmitter is integrated [16]. A virtual automation system for fluid filling developed in Unity, complemented by the "Hardware In The Loop" simulation to form signals from sensors and actuators provides a realistic tridimensional environment [17]. A 3D virtual system based on bicycle tracks is presented; after examining the usability of the system, the results show that the children presented an improvement in muscle activation, in addition to feeling immersed and enjoying the game [18].

The automotive sector presents an alternative focused on virtual environments to analyze the user's impressions with the design elements and then apply it to car interiors; in this way, users can explain themselves well through design elements [19]. In modern industry and manufacturing engineering, VR is used to design, test, and control the parts of a machine; in addition to providing a quick response to changes in machining requirements, in this way, the virtual tool can be used for training, learning, and operation of different machines [20]. Virtual reality consists of environments for the supervision, monitoring and control of industrial processes in real time, as well as applications for the control of variables such as level, flow, temperature and pressure [21].

2 Structure of the Virtual Module

2.1 Problematic

The pandemic caused by COVID 19 for more than a year has forced teachers to look for new teaching tools, among them is the use of virtual reality applications as a substitute for the laboratories located within the universities. In the process of education where teachers teach classes to students, the interaction between both parties is fundamental for teaching, being necessary that the student must strengthen the knowledge acquired when it comes to purely practical and theoretical issues, but in the field of engineering, it is important to bring both together. For this, it is of vital importance to have fully equipped laboratories which lead not only to considerable investments but also to a great effort on the part of academic institutions to maintain the spaces and work equipment in optimal conditions, to this is added the deterioration of the components over the years thus reducing their useful life and causing a malfunction in practices, which entails a periodic disbursement for the acquisition of new elements. One solution is the implementation of a VR application for the industrial control category, the same one that has elements similar to the real ones to progress in their skills at a low cost. The time for laboratory practices is subject to pre-established schedules within each curricular mesh, thus limiting the number of hours used by each student, where they are not enough to fully cover all the content that integrates the proposed study plan at the beginning of the academic period. With the implementation of a VR module, students will develop skills in their practices leaving aside the fear of receiving an electric shock or making a mistake in the connections that because a short circuit, this translated into real laboratory practices and without having the appropriate security measures can affect the physical integrity of the student.

2.2 Description of the Proposal

Figure 1 shows the degree of interaction between teacher and student, the meeting point being the VR environment, the user accesses the stage through the use of Oculus Rift glasses, and the teacher observes the student's activities through a monitor.

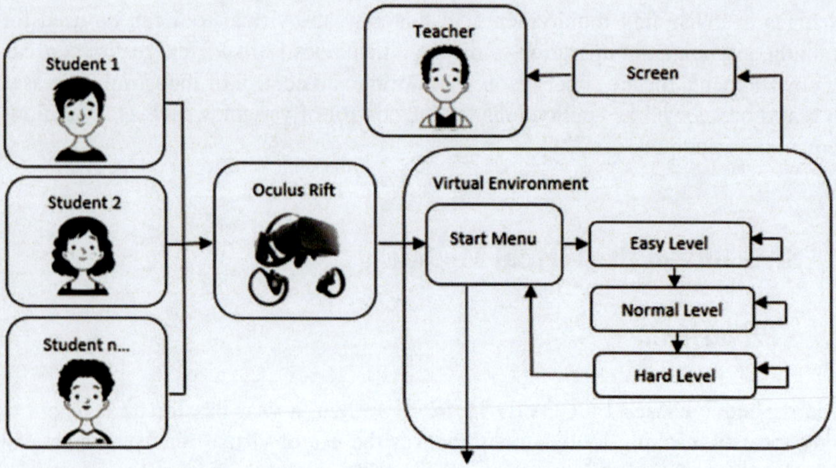

Fig. 1 Teacher–student diagram

The virtual reality module is made up of two scenarios; the first corresponds to a start menu, and the second belongs to the virtual laboratory, consisting of a room of (2 × 2) m, a control board, three-phase electric motor, voltmeter, elements of the electric switchgear (din rail, pilot lights, N/A-N/C pushbuttons, emergency stoppages, thermomagnetic switches, contactors, thermal overload relays, timers, and signals) (Fig. 2).

Within the VR environment, the elements that make it up to have the possibility of being manipulated, thus offering an immersive and interactive experience for the user in addition to having 3 levels which are detailed below:

The "easy" level consists of a welcome menu and an activity message as shown in (Fig. 3).

This section details the activities to be carried out, such as the measurement of voltage using the voltmeter as shown in (Fig. 4) the user must obtain two measurements 110 and 220 V.

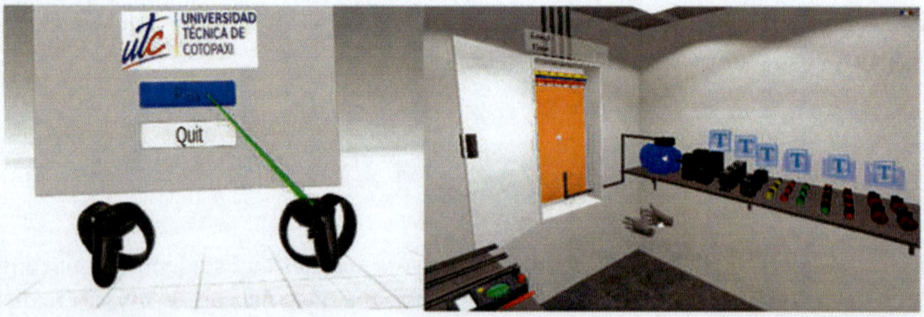

Fig. 2 Scenarios

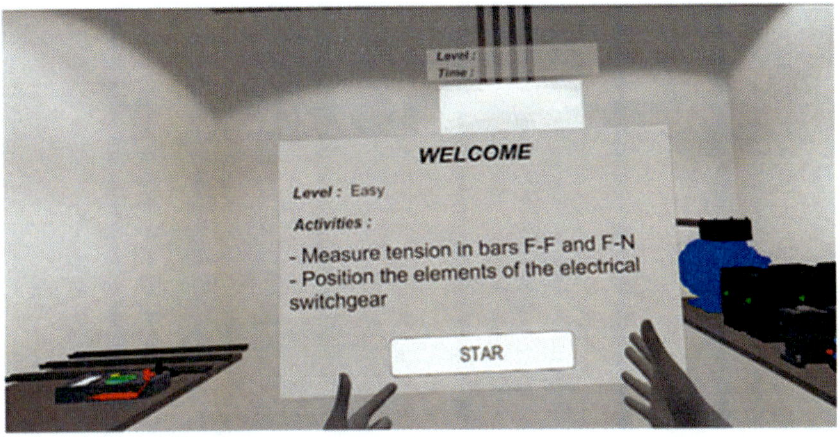

Fig. 3 Easy level start menu

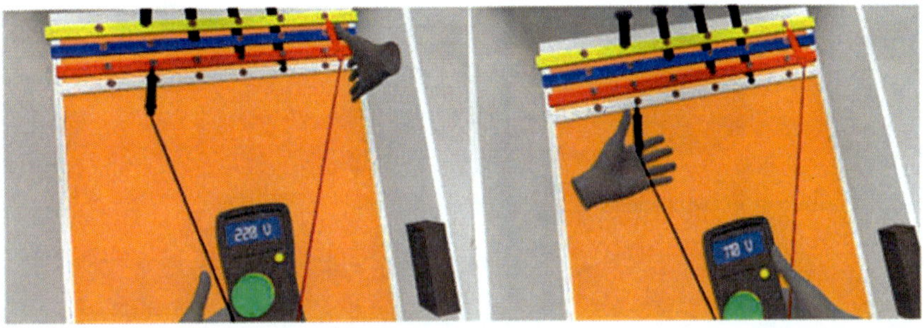

Fig. 4 Voltage measurement

Subsequently, the positioning of the different elements within the control board is visualized; there is a breaker, pushbuttons among others, as can be seen in (Fig. 5).

The "normal" level consists of a welcome menu (Fig. 6); the tasks to be performed involve what has been learned in the easy level; the proposed circuit will be carried out, corresponding to the on and off of a pilot light (Fig. 7).

Within the "difficult" level is a welcome menu (Fig. 8), covering what was learned in the previous levels, now the student must make the connection of the three-phase motor inside the terminal box in the star and triangle configuration (Fig. 9) in addition to a proposed circuit belonging to a direct start (Fig. 10).

All levels feature a sign at the top of the board indicating the degree of difficulty present in addition to a real-time counter indicating the number of minutes within the scene; scenarios 2 and 3 contain an algorithm based on wiring logic to identify misconnected elements; in this way, a short circuit noise will be reproduced, and later, a sign with a message of "Game Over" is displayed if it were to act, see (Fig. 11). Once the activities of each level have been completed, a notice will be displayed in

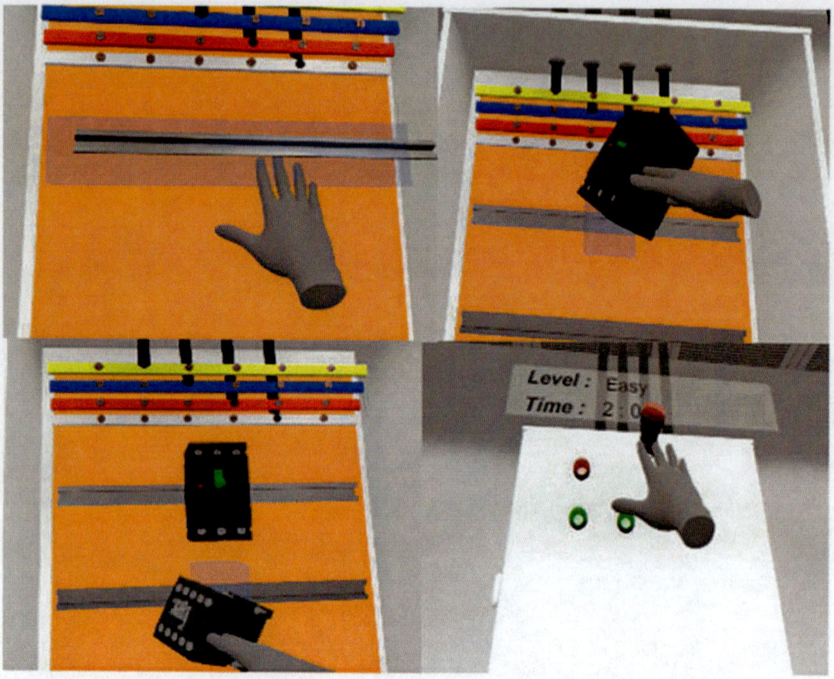

Fig. 5 Positioning of the elements in the electrical panel

Fig. 6 Normal level start menu

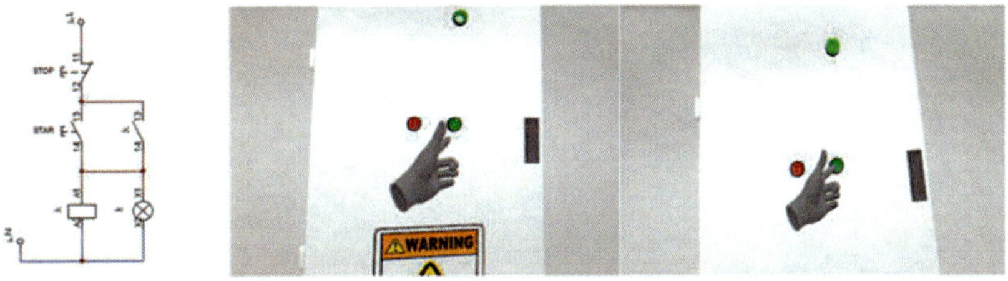

Fig. 7 Circuit proposed for the "Normal" level

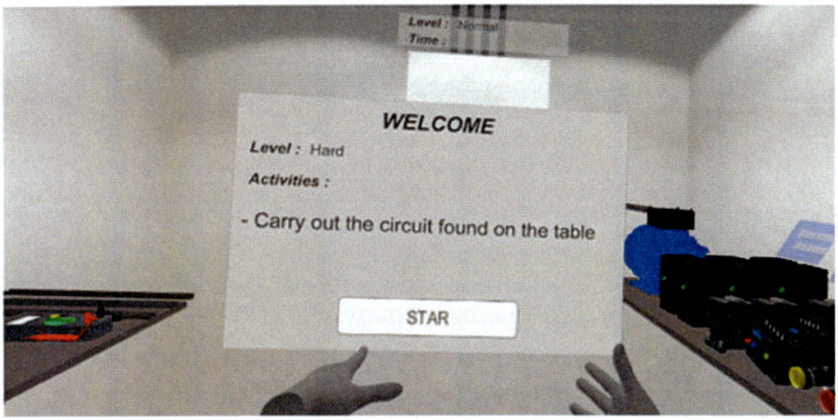

Fig. 8 Hard level start menu

Fig. 9 Star-triangle connection

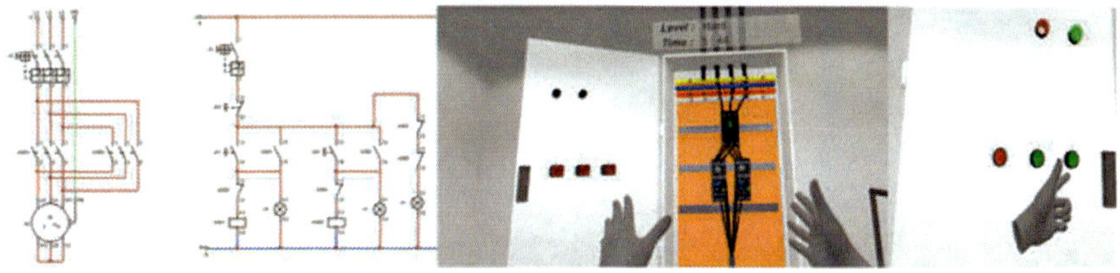

Fig. 10 Circuit proposed for the "Hard" level

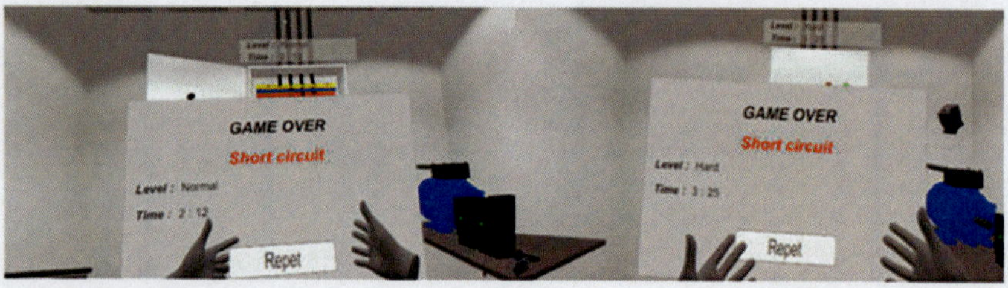

Fig. 11 Short circuit dialog box

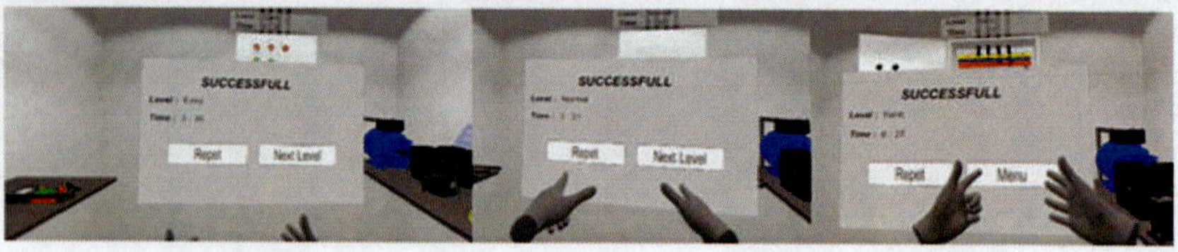

Fig. 12 Message completed

which the student is informed of the time it took to overcome them, in addition to having the possibility of repeating the activity as many times as required or failing that, advance to the next scenario or return to the start menu see (Fig. 12).

The levels are presented progressively and developed in an orderly manner; for this, the scripts play a crucial role within the VR environment to respond to the operations of different processes associated with the control board as shown in (Fig. 13).

The input and output devices that allow the recognition of the virtual environment are as follows: Oculus Rift VR glasses, HTC Vive, and Gear VR as haptic inputs, this allows immersion and control within the application. The handling of different input and output devices in the application requires that the configuration of the code is universal so that it is compatible with several computers without the need to reassemble the project and automatically detect the aforementioned devices. The scripting phase manages the link of each of the devices to be used and that interact with the virtual environment; then, the interface is implemented where the student has different ways of interacting with the VR module.

2.3 Analysis of Results

In this section are the educational specifications on the ability and skill that students learn within the virtual laboratory, to carry out their practices so that students are trained and prepared to assemble different control circuits, you can also visualize the simulation in real time of the operation of the electrical switchgear through the

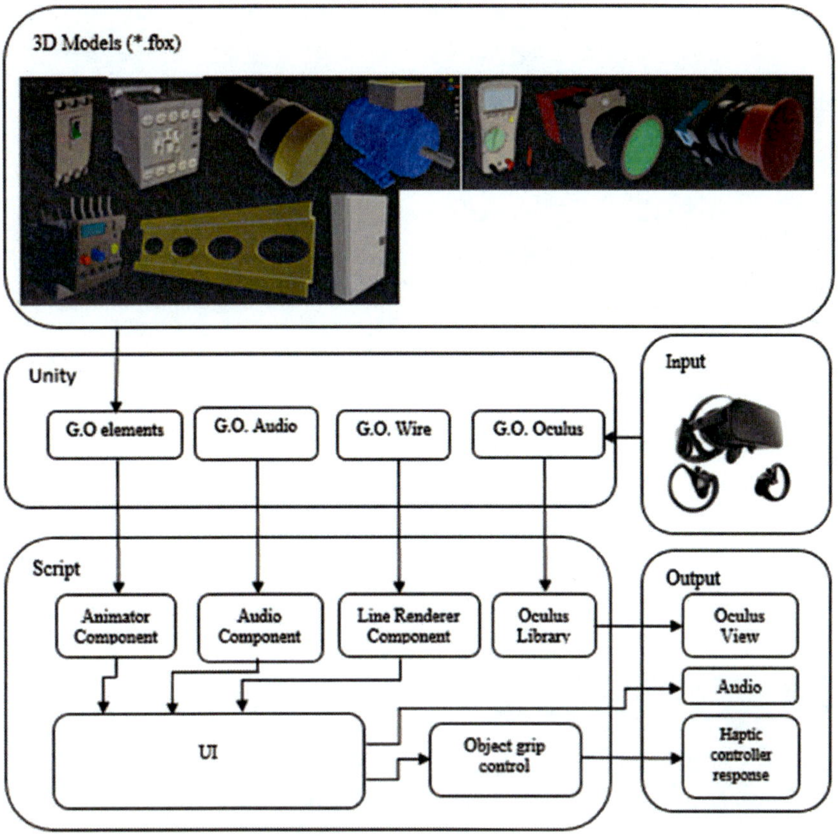

Fig. 13 Component interaction diagram

use of components and VR devices such as Oculus Rift, HTC Vive, and Gear VR essential to interact with the virtual environment; this application can be installed on computers that have the minimum requirements accepted by VR glasses. Figure 14 shows the student interacting with the virtual industrial control laboratory.

To corroborate the functionality of the VR laboratory, a practice focused on the direct start of a three-phase motor was carried out; two groups of students A and B were formed, and who in advance received the theoretical foundation pertinent to the subject. Group A receives the laboratory guide and goes to the industrial control workshop. On the other hand, group B previously interacted with the virtual module before heading to the workshop. The assessment applies to group B.

I. Evaluation of the VR module is based on the percentage of contingencies presented by Group A concerning Group B.
II. Usability assessment is established through a questionnaire according to the level of satisfaction perceived by teachers and students.

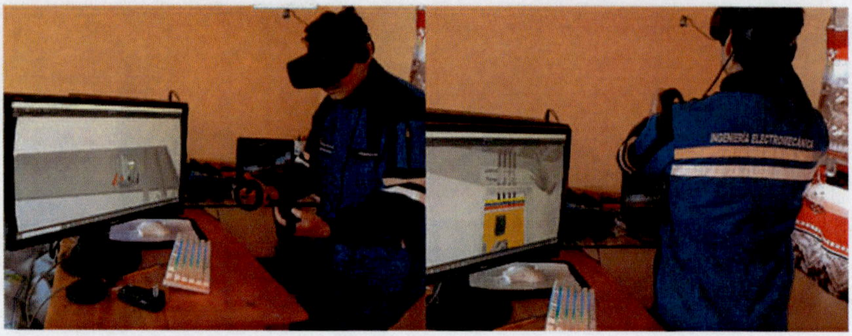

Fig. 14 User interaction-environment

A. Evaluation of the module

Group B had to face several levels with a certain limited number of activities to be fulfilled, the last being the assembly of a control circuit for the direct start of a three-phase motor in a star configuration; later, they went to make the same circuit in the industrial control workshop (Fig. 15).

Fig. 15 Student in the industrial control workshop

B. Usability evaluation

This is focused on the ease with which users can interact with the VR laboratory; the evaluation involves both teachers and students using the application and corroborating the similarity that exists between the elements that are in the environment, compared to the real ones thus evaluating the different levels of learning provided by the VR module to perform a three-phase motor start in a star configuration; the questionnaire was applied to 30 students of the subject of industrial control and 3 teachers of the Electromechanical Engineering career; in (Fig. 16), the data of the questionnaires are visualized; in (Table 1), questions are shown for teachers (Ds) and students (Es); once they have completed their virtual experience, a weighting will be requested where 10 is the highest score and 0 the lowest score.

The two tables of data obtained from the people consulted show that there is a high degree of acceptance of the virtual industrial control module, where students present a 90% acceptance considering that they exceed a weighting of 9 claiming

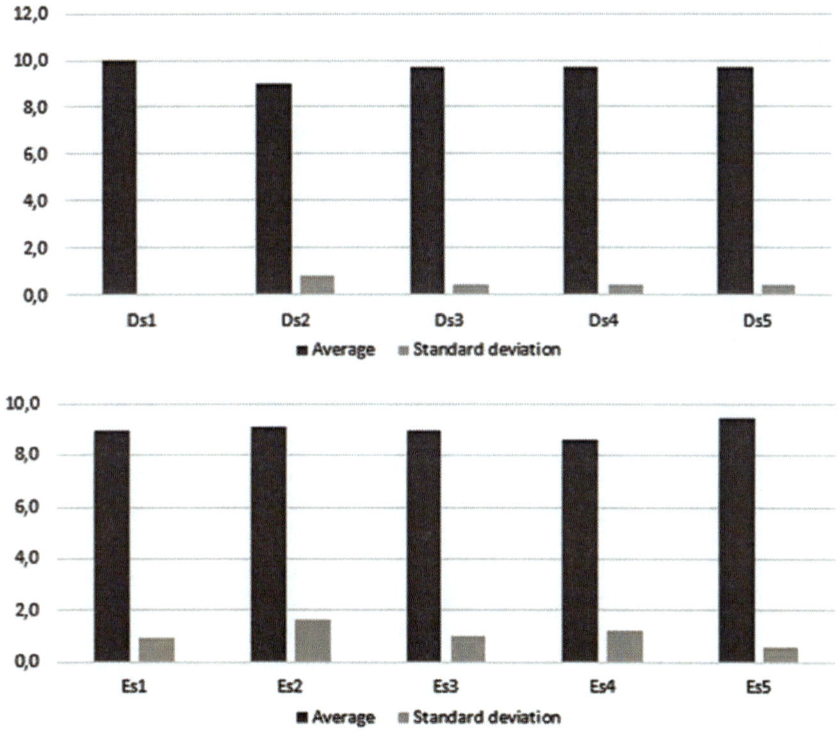

Fig. 16 Survey data

Table 1 Questions to evaluate the usability of virtual environments

	Questionnaire
Es1	How much are you informed about virtual environments?
Es2	How easy do you find it to manage virtual reality environments?
Es3	How difficult did you perceive when interacting with the VR environment?
Es4	How realistic did you find the VR environment?
Es5	How likely are you to recommend this environment for continued use within industrial control practices?
Ds1	How knowledgeable are you about VR environments in education?
Ds2	What is the degree of motivation that students present when using this type of tool?
Ds3	How likely are you as a teacher to implement this tool in your classes?
Ds4	How important do you think it is to incorporate virtual reality modules into other areas of the faculty?
Ds5	How likely are you to use this tool as a student assessment method?

that the system is very practical as useful and that they would not have problems interacting with the VR module; the data of the teachers show an 96% in which they agree that the application will help to improve the practical part and agree that the virtual environment can be used in the laboratory of the other faculties as a means of teaching and learning with these types of projects the students show enthusiasm for their practical classes; this helps to complement the education process within the Electromechanical Engineering career.

3 Conclusions

The use of this type of virtual reality module encourages students to manipulate the elements that make up the electric switchgear, developing skills in the assembly of control circuits; in this way, it helps students to improve their practices and reduce the fear of receiving an electric shock thus affecting their physical integrity, due to a bad connection caused by the student leading to a short circuit or damage to some control; the results indicate that the virtual module is easy and practical; the teachers state that it is a very interactive application being of great help for those people who start in the area of industrial control; this type of virtual modules takes on great importance in the face of unexpected events such as the appearance of COVID 19.

References

1. Ni Z, Gao Z (2015) Developing digital hall of prayer for good harvest software to promote historical culture by applying virtual reality technology. Cult Comput 217–218
2. Potkonjak V, Gardner M, Callaghan V, Mattila P, Guetl, C, Petrovic V, Jovanovic K (2016) Virtual laboratories for education in science, technology, and engineering: a review. Comput Educ 95:309–327
3. Neira Tovar L, Castañeda E, Ríos Leyva V, Leal D (2020) Work-in-progress—a proposal to design of virtual reality tool for learning mechatronics as a smart industry trainer education. In: 6th international conference of the immersive learning research network (iLRN), pp 381–384
4. Hao C, Zheng A, Wang Y, Jiang B (2021) Experiment information system based on an online virtual laboratory. Smart Syst Infrastruct Appl 13(2):13–27
5. Popescu D, Ionete C, Aguridan R, Popescu L, Meng Q, Ionete A (2009) Remote vs. simulated, virtual or real-time automation laboratory. In: IEEE international conference on automation and logistics, pp 1410–1415
6. Torres L, Galan D, Cabrerizo F, Herrera E, Dormido S (2016) Virtual and remote labs in education: a bibliometric analysis. Comput Educ 98:14–38
7. Ortiz J, Sánchez J, Velasco P, Sánchez C (2017) Teaching-learning process through VR applied to automotive engineering. In: Proceedings of the 2017 9th international conference on education technology and computers, pp 36–40
8. Cunningham J, Rahman I, Lautenschlager S, Rayfield EJ, Donoghue PC (2014) A virtual world of paleontology. Trends Ecol Evol 29:347–357
9. Vaughan N, Dubey V, Wainwright T, Middleton RG (2015) Does virtual-reality training on orthopaedic simulators improve performance in the operating room. In: 2015 science and information conference, pp 51–54
10. Vaughan N, Dubey VN, Wainwright TW, Middleton RG (2015) Can virtual-reality simulators assess experience and skill level of orthopaedic surgeons. In: Science and information conference, vol 2, pp 105–108
11. Ibáñez P, Pruna E, Escobar I, Ávila G (2021) 3D virtual system for control valve calibration. In: Augmented reality, virtual reality, and computer graphics, vol 12980, pp 260–572
12. Pila B, Alcoser E, Pruna E, Escobar I (2021) Inspection and verification training system of production lines in automated processes, through virtual environments. In: Augmented reality, virtual reality, and computer graphics, vol 2980, pp 603–620
13. Garcés R, Lomas J, Pilatasig J, Andaluz V, Tutasig A, Zambrano A, Varela J (2021) Virtual control of a double effect evaporator for teaching-learning processes. In: Augmented reality, virtual reality, and computer graphics, vol 12980, pp 690–700
14. Pruna E, Ballares G, Teneda H (2021) 3D virtual system of a distillation tower, and process control using the hardware in the loop technique. In: Augmented reality, virtual reality, and computer graphics, vol 12980, pp 621–638
15. Alpúsig S, Pruna E, Escobar I (2021) Virtual environment for control strategies testing: a hardware-in-the-loop approach. In: Augmented reality, virtual reality, and computer graphics, vol 12980, pp 588–602
16. Rochar V, Rochar K, Pruna E (2021) 3D virtual environment for calibration and adjustment of smart pressure transmitters. In: Augmented reality, virtual reality, and computer graphics, vol 12980, pp 639–654
17. Aguilar I, Correa J, Pruna E (2021) 3D virtual system of a liquid filling and packaging process, using the hardware in the loop technique. In: Augmented reality, virtual reality, and computer graphics, vol 12980, pp 573–587
18. Pruna E, Escobar I, Quevedo W, Acurio A, Pilatásig M, Pilatásig L, Bucheli M (2018) 3D virtual system based on cycling tracks for increasing leg strength in children. Adv Intell Syst Comput 746(2):1009–1019

19. Kim C, Lee C, Lehto MR, Hwan Yun M (2010) Affective evaluation of user impressions using virtual product prototyping. Hum Factors Ergonomics Manuf Serv Ind 21:1–13
20. Lin W, Fu J (2006) Modeling and application of virtual machine tool. In: 16th international conference on artificial reality and telexistence—workshops, vol 1, pp 16–19
21. Andaluz V, Castillo-Carrion D, Miranda R, Alulema J (2017) Virtual reality applied to industrial processes. In: Augmented reality, virtual reality, and computer graphics, pp 59–74

Coordination of Wind Turbines and Battery Energy Storage Systems in Microgrid

B. Sravan Kumar and L. Ramesh

Abstract The potential of energy storage systems in power system and small wind farms has been investigated in this work. Wind turbines along with battery energy storage systems (BESSs) can be used to reduce frequency oscillations by maintaining a balance between active power and load consumed. Compared to doubly fed induction generators, permanent magnet synchronous generator (PMSG) is used to integrate power from the wind power systems to the grid as nacelle is of less weight, and it operates easily at low speeds, and gear box is not required. The impact of BESS in microgrid is studied. Two individual wind turbines along with their respective BESS are connected to the microgrid. BESS behaviour when more wind is present at one of the wind turbines and BESS behaviour when one of the wind turbines turned off is studied. Load was successfully shared between two wind turbines and BESS in the case of failure or fault in the microgrid.

Keywords BESS · PMSG · Wind farm · Coordination

1 Introduction

Wind flow is due to the asymmetrical heating of Earth's surface as air above the land heats fast and rises upward whereas the air above the water heats slowly. The air above the water replaces the air above the land resulting in wind flow [1]. This wind flow can be converted into electricity using wind turbines. Wind turbine rotor blades rotate due to the difference in the air pressure on both sides of the blades. Mechanical energy provided by the spinning rotor blades is converted into electrical energy using electrical generators. Amongst non-conventional energy sources, usage of wind turbines in the microgrid has tremendously increased due to its competitiveness over fossil fuels in terms of terms of expenditure incurred for producing electric power [2]. Also, the advancement in wind turbine technologies, sustainability nature of wind and superior power electronic converters has resulted in increased grid interfacing

B. S. Kumar (✉) · L. Ramesh
Dr. M.G.R. Educational and Research Institute, Chennai, India
e-mail: bsravan2193@gmail.com

P. Chatterjee et al. (eds.), *Computational Intelligence for Engineering and Management Applications*, Lecture Notes in Electrical Engineering 984,
https://doi.org/10.1007/978-981-19-8493-8_6

of wind turbines. In spite of these advancements, interfacing of wind turbines poses a major challenge for grid operators [3–5].

After thoroughly studying the various research papers relating to the grid connected wind turbines, we have found that most of the wind turbines use BESS to store the excess amount of energy generated during excess wind flow and to maintain the balance between active power and load consumed [6–8]. BESS also does a variety of functions such as load shifting, load levelling, peak shaving, frequency control, power quality control, transient stability control and unbalanced load compensation [5]. Reference [9] concentrated on utilizing the BESS to compensate the voltage drop in the microgrid caused due to the integration of microgrid. Research papers [10–12] showed that microgrid instability can be stabilized using BESS by adjusting active and reactive power.

In this paper, we have presented modelling of wind turbines and case study analysis considering shutting down of one of the wind turbines and considering more wind at one of the wind turbines. Load sharing and behaviour of BESS is shown.

2 Simulation Model

Proposed Simulink model consists of two wind turbines and battery energy storage system connected to the microgrid. Initially, wind source model was created by considering average wind speed, noise and periodic disturbances. Wind turbine was modelled by considering the following equations.

Power in the wind P_{in} is a function of air density (ρ), the blade swept area (A) and the cube of the wind velocity (V)

$$P_{in} = 1/2 \times \rho \times A \times V^3$$

Hence, maximum power captured by wind turbine P_{out} is a function of air density (ρ), the blade swept area (A) and the cube of the wind velocity (V) and performance coefficient of wind turbine (C_P).

$$P_{out} = 1/2 \times \rho \times A \times V^3 \times C_p(\alpha, \beta)$$

where α is tip-speed ratio and β is blade pitch angle

$$\alpha = (\text{blade tip speed})/(\text{wind speed})$$
$$\text{blade tip speed} = (\text{rotational speed} \times \pi \times D)/60$$

where D is the diameter of the turbine.

Power coefficient (C_p) is a measure of wind turbine efficiency and is given as

$$C_p = P_{in}/P_{out}$$

C_p was modelled by using the following equation

$$C_p(\alpha, \beta) = C_1\left(\left(\frac{C_2}{\delta}\right) - (C_3 x\beta) - (C_4 x\beta^x) - C_5\right)e^{-C_6\frac{1}{\delta}}$$

where δ is calculated from the below equation

$$\frac{1}{\delta} = \frac{1}{\alpha + 0.0.8\beta} - \frac{0.035}{1 + \beta^3}$$

$C_1, C_2, C_3, C_4, C_5, C_6$ are constants given by the manufacturer. x is a constant and depends on the type of the turbine.

Using the above equations, wind turbine was modelled as shown in Fig. 1. Constant generator speed, constant pitch angle and constant or variable wind speeds are provided as input to the wind turbine. Using a manual switch, wind speed is changed from variable to constant and vice versa. Base wind speed is considered as 12 m/s, and the turbine power characteristics is as shown in Fig. 2. PMSG type of generator is used to convert mechanical energy to electrical energy as it can operate at variable rotor speeds. Generated power is converted into DC using universal bridge converter and is fed to the BESS.

Three-phase salient pole permanent magnet synchronous generator (PMSG) receives mechanical torque as an input from the wind turbine. Three-phase output from the PMSG is converted to DC using bridge rectifier circuit. This DC is fed to nickel metal hydride battery having a nominal voltage of 300 V, rated capacity of 6.5 Ah, initial state of charge of 60% and battery response time of 30 s. To control

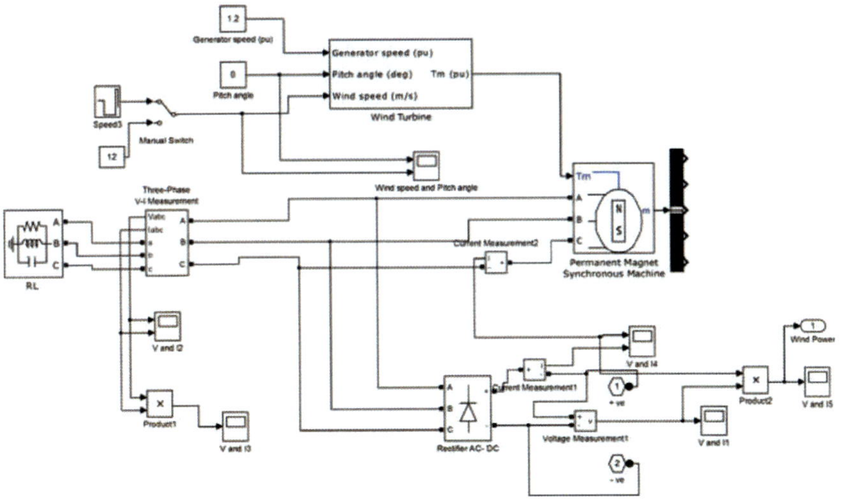

Fig. 1 Modelling of wind turbine

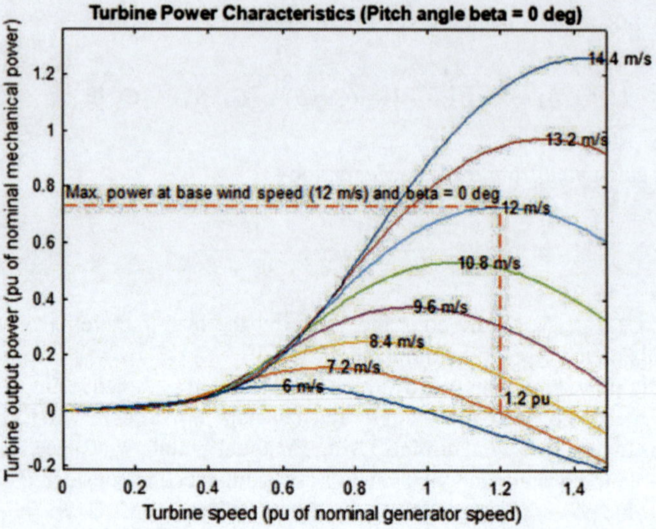

Fig. 2 Turbine power characteristics

charging and discharging of BESS, two IGBT switches in parallel with a RC snubber circuit are connected as shown in Fig. 3. Since IGBT cannot conduct in the reverse direction, freewheeling diode is connected in antiparallel so as to allow the reverse current to pass through it. Gate signal for these IGBT switches is controlled by using the control circuit as shown in Fig. 4.

Switches turn ON when a voltage greater than zero appears across collector-emitter terminals, and a signal greater than zero is applied at the gate input. Switches are turned OFF when a voltage less than zero is applied across collector-emitter terminals. Voltage source control for universal bridge converter is designed. During load changes and faults in microgrid or during fluctuating nature of wind, current controller is designed to control the current as shown in Fig. 5. Phase locked loop is used for the frequency measurements of the grid.

3 Design of BESS, Wind Turbines and Microgrid

To illustrate the effectiveness of the proposed model, we have presented two individual 1.5 MW PMSG wind turbines connected to two individual 200 V, 6.5 Ah nickel metal hydride batteries, respectively. Power output from wind turbines is converted into DC using pulse-width modulation-based universal bridge converter, and then, DC is fed to BESS. This DC is again converted into AC and is fed to the three-phase series RLC branch, which acts as a filter to reduce the harmonic contents in the AC

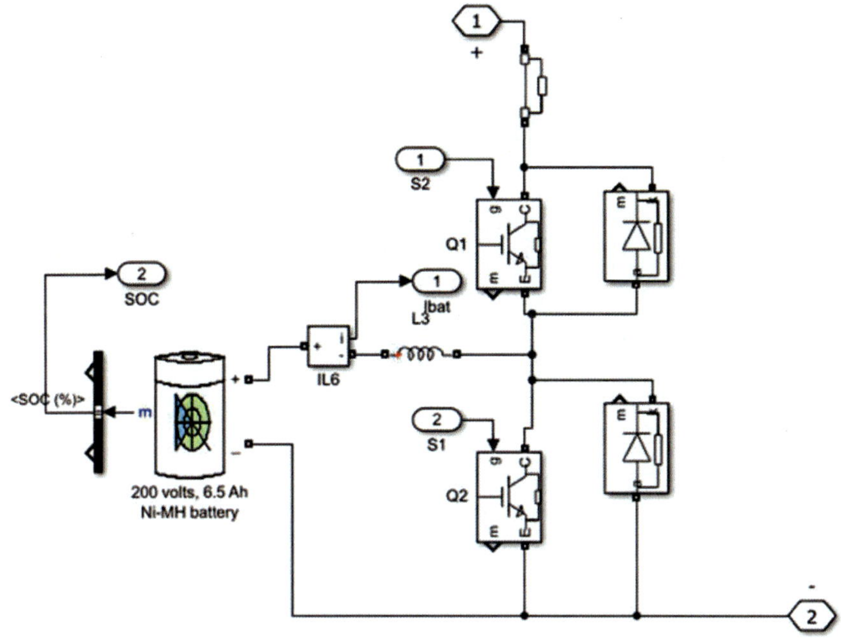

Fig. 3 Modelling of BESS

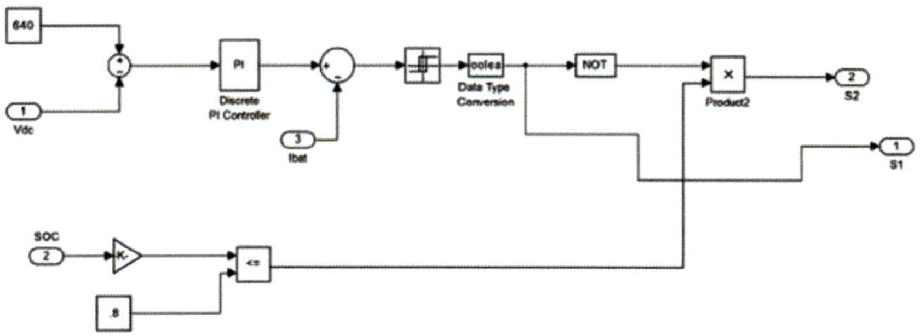

Fig. 4 Control circuit for switches

output. Three-phase voltage source of 25 kV is considered as base voltage of microgrid. Shunt combination of R, L, C is used as a load connected to the microgrid, which exhibits a constant impedance for a particular frequency. Proposed complete Simulink model is as shown in Fig. 6.

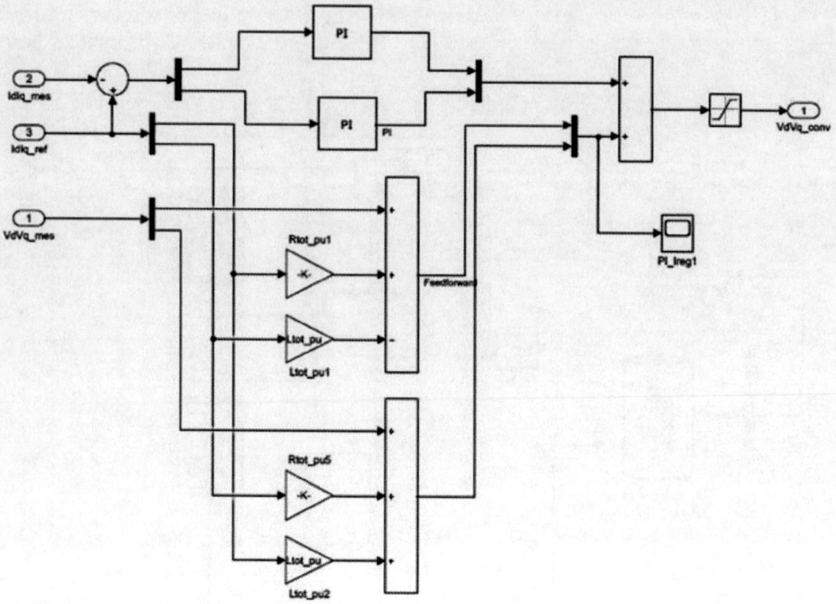

Fig. 5 Current controller design

4 Results

We assumed two different cases to establish the coordination between BESS and wind turbines.

Case 1: BESS Behaviour When More Wind Is Present at Second Wind Turbine Figure 7 shows the simulation result for the proposed system to show the power management strategies implemented with MPPT controller. Here, the load sharing is chosen between two wind turbines, batteries and grid system according to their generations. The wind speed for wind plant-1 is 10 m/s during 0–4 s, 9 m/s applied during 4–5 s, and later, it changed to 10 m/s for 5–6 s and for wind plant-2, between 0–4 s speed of 13 m/s, between 4–5 s speed of 8 m/s and between 5–6 s 13 m/s is applied. Under these conditions, the performance of wind turbines and sharing of power to various load conditions is verified. We can observe that battery power increases in proportional to increase in wind power of second wind turbine.

Case 2: BESS Behaviour When One of Wind Turbine Is Effected by Zero Wind During this case, the wind plant-1 is effected by low wind speed, and wind speed for wind plant-2 is taken as 12 m/s between 0–2 s, 10 m/s between 2–4 s, 6 m/s between 4–5 s and from 5 to 6 s as 10 m/s. During 0–2 s, wind speed at wind turbine 1 is very low, and the complete load requirement is shared by wind plant-2 and BESS as shown in Fig. 8.

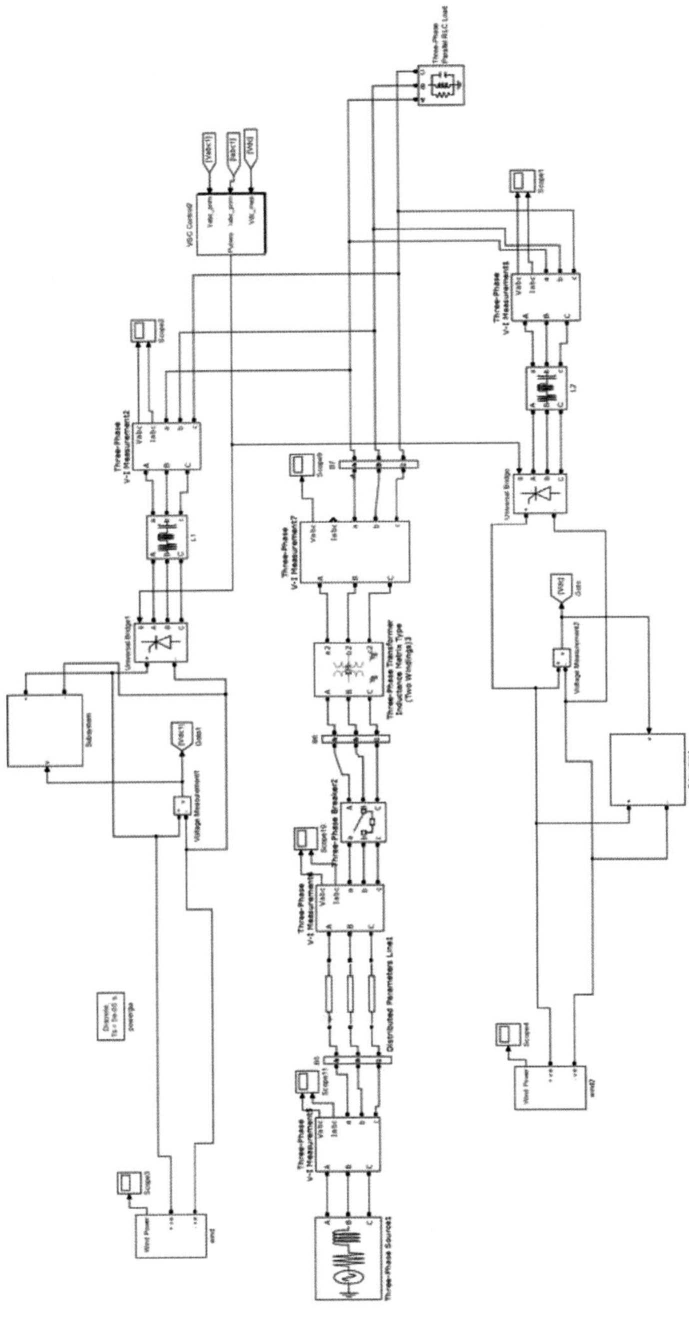

Fig. 6 Entire Simulink model

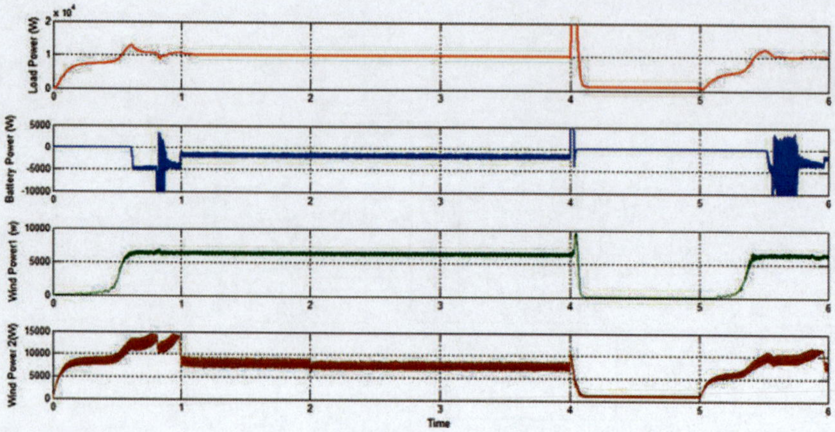

Fig. 7 Simulation result for more wind at the second wind turbine

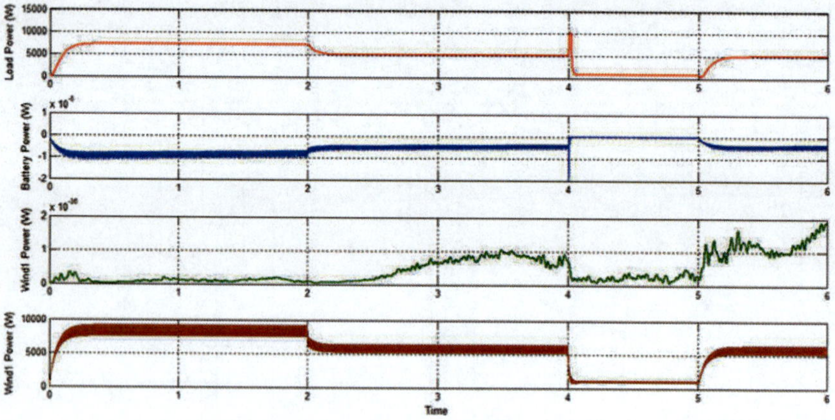

Fig. 8 Simulation result for less wind at first wind turbine

5 Conclusion

In this work, two wind turbines along with respective BESS connected to the micro-grid were successfully modelled in MATLAB, and the targeted goals such as sharing the load between BESS and wind turbines in the case of failure of any one of the wind turbines were successfully met.

References

1. Bokde N, Feijoo A, Villanueva D, Kulat K (2019) A review on hybrid empirical mode decomposition models for wind speed and wind power prediction. Energies 10:3390
2. Herbert GMJ, Iniyan S, Sreevalsan E, Rajapandian S (2007) A review of wind energy technologies. Renew Sustain Energy Rev 11:1117–1145
3. Kusakana K (2015) Optimal scheduled power flow for distributed photovoltaic/wind/diesel generators with battery storage system. IET Renew Power Gener 9(8):916–924
4. Guo L, Yu Z, Wang C, Li F, Schiettekatte J, Deslauriers JC, Bai L (2016) Optimal design of battery energy storage system for a wind, diesel off-grid power system in a remote Canadian community. IET Gen Transm Distrib 10(3):608–616
5. Nguyen CL, Lee HH (2016) A novel dual-battery energy storage system for wind power applications. EEE Trans Ind Electron 63(10):6136–6147
6. Sravan Kumar B, Ramesh L (2019) Review and key challenges in battery to battery power management system. In: 5th international conference on computing, communication, control and automation (ICCUBEA)
7. Faisal M, Hannan MA, Ker PJ, Hussain A, Mansor MB, Blaabjerg F (2018) Review of energy storage system technologies in microgrid applications: issues and challenges special section on advanced ES technologies and their applications
8. Moness M, Moustafa AM (2015) A survey of cyber-physical advances and challenges of wind energy conversion systems: prospects for internet of energy. IEEE IoT J
9. Avendaño N, Celeita D, Hernandez M, Ramos G (2017) Impact analysis of wind turbine and battery energy storage connection in power systems. IEEE
10. Tani A, Camara MB, Dakyo B (2015) Energy management in the decentralized generation systems based on renewable energy using ultracapacitors and battery to compensate the wind/load power fluctuations. IEEE Trans Ind Appl 51(2):1817–1827
11. Weihua L, Songqi F, Weichun G, Zhiming W (2012) Research on the control strategy of large-scale wind power energy storage system. In: IEEE PES innovative smart grid technologies, pp 1–4
12. Shokrzadeh S, Jozani MJ, Bibeau E, Molinski T (2015) A statistical algorithm for predicting the energy storage capacity for baseload wind power generation in the future electric grids. Energy 89:793–802

Weather-Aware Selection of Wireless Technologies for Neighborhood Area Network of Indian Smart Grid

Jignesh Bhatt, Omkar Jani, and V. S. K. V. Harish

Abstract As a contemporary bidirectional intelligent instrumentation telemetry system, wireless technology is pivotal in serving crucial communication requirements of smart grid. As a result, research into the influence of weather variations on communication system availability is critical. The goal of this study is to investigate and examine the influence of weather variations on communication system's run-time availability, as well as to recommend better alternatives for the particular region. The proposed methodology was validated on a real-time dataset from an Indian smart grid installation and found to be simple, adaptable, and viable for field engineers and design engineers planning for modifications and future expansions, respectively.

Keywords Communication design · Evaluation · Instrument telemetry · Smart cities · Smart grids · Weather impacts · Wireless technologies

---

J. Bhatt (✉)
Department of Instrumentation and Control Engineering, Faculty of Technology, Dharmsinh Desai University, Nadiad 387001, India
e-mail: jigneshgbhatt@gmail.com

Department of Electrical Engineering, School of Technology, Pandit Deendayal Energy University, Gandhinagar 382007, India

O. Jani
Department of Research and Culture, Kanoda Energy Systems Pvt. Ltd., Ahmedabad 380015, India

V. S. K. V. Harish
Department of Electrical Engineering, Netaji Subhas University of Technology (NSUT), Dwarka Sector-3, Dwarka, Delhi 110078, India
e-mail: vskvharish@ieee.org

P. Chatterjee et al. (eds.), *Computational Intelligence for Engineering and Management Applications*, Lecture Notes in Electrical Engineering 984,
https://doi.org/10.1007/978-981-19-8493-8_7

1 Introduction

Governments throughout the world are promoting smart city initiatives to upgrade existing cities into smart cities, with the goal of increasing inhabitants' quality of life. Smart cities provide a wide range of unique applications that not only ease routine tasks but also provide much more user-friendly environment for living and working. There are a few interesting studies available that provide a comparative analysis of Indian smart cities based on a variety of parameters [1]. Uninterrupted, high-quality electricity is unquestionably a necessary consumable for residents' daily lives, lifestyle development, and long-term economic growth. As a result, a new smart grid is emerging as a smart, self-diagnostic, and self-healing alternative to the present archaic and non-intelligent electric grid, which is prone to frequent malfunctions. The smart grid, in reality, is the backbone of the smart city's energy system [2].

The integration of distributed energy resources and, as a result, the facilitation of variable costs of electric units utilized during peak and off-peak hours are facilitated by smart grid for smart city users. This enables and encourages users to plan and reschedule electric equipment usage to balance their need, urgency, and budget. In engineering design, a variety of computational and data-driven selection procedures are recommended [3], especially when dealing with multi-criteria decision-making challenges influenced by unforeseen and unanticipated uncertainties [4]. As a result, the patterns of energy consumption vary, and the peak during high load situations can be reduced. The communication system of the smart grid seamlessly provides live updates on unit costs of electric supply and customer decisions based on the same, resulting in the activation of actions from utility automatic control systems. As a result, in order to achieve a sustainable smart city, a dependable smart grid and, therefore, a reliable communication system are required. The design of a highly available, quick, resilient, and bidirectional wireless data transmission system is one of the most significant challenges for smart grid design engineers [5]. Importantly, changes in weather are one of the leading causes of a sudden and unexpected reduction in performance dependency of wireless communication systems, resulting in an insufficient supply of electricity. "How to choose the suitable wireless technology that might be more reliable (available) against weather impacts?" is the research question attempted to address in this study. The assessment of the impact of weather fluctuations on the availability of wireless technology placed in a functioning smart grid in real time is a challenging task. Because of the varying weather patterns in the area, smart grid installations may be one-of-a-kind. Furthermore, the possibility of unanticipated climatic calamities could pose further obstacles to the communication system. However, the architectural and technological resemblances between the smart grid communication system and the instrumentation telemetry system need additional investigation into its associated applications and characteristics [2].

Therefore, in light of the above, various research gaps identified are as follows:

(i) How should the effects of weather variations be studied on functioning wireless technologies installed in existing smart grid installations?
(ii) How can the impacts described in (i) above be measured and quantified?

(iii) How should other wireless technology alternatives be evaluated for replacement if the existing technology is not performing satisfactorily?
(iv) How to examine and determine which wireless communication technology options are capable of meeting the communication needs?
(v) Which wireless communication technology should be selected from the alternatives listed in (iv) above as the most appropriate?

This study aims to provide a useful methodology as well as customizable mathematical tools toward weather-aware optimal choice of wireless technology for deployment to cater the smart grid. This study examines the influence of weather variations on the availability of installed wireless technology and suggests better alternatives. An evaluation of an Indian case study validates the recommended methodology proposed. The rest of the research presented in this study has been structured in different sections. A survey of related published works is presented in Sect. 2. Section 3 proposes a methodology for evaluating the influence of weather variations on installed wireless technology. In Sect. 4, the testing of suggested methodology on the dataset of a functional smart grid pilot of an Indian case study is discussed. Finally, Sects. 5 and 6 present the results and conclusions respectively.

2 Related Works

A review of relevant literature was conducted to develop the work presented in this study, and its findings are presented in this section. The study of smart grid needs a thorough examination of critical elements such as important applications, supporting technologies, and so on, with a focus on smart grid communication [6]. The smart grid architectural model [7] specifies communication that is structurally analogous to bidirectional data connection between the field and the control center in a distributed industrial control system [8]. The smart grid's reliability is reliant on the communication system's robustness [9]. As a result, designing a resilient and effective communication system is a key challenge in the development of smart grids and smart cities [10]. Several research efforts have been discovered that investigate the effects of weather changes on the performance of various smart grid subsystems, as well as providing appropriate improved approaches. Weather-sensitive battery storage predictive control [11], adaptive DC microgrid protection [12], and so on. The scope of these studies is confined to a few subsystems, and the examination of effects of weather variations on the smart grid's communication system still remains a relatively under-explored area. Wireless technologies are gaining traction in the development of smart grid communication networks. As a result, more research into the effects of changing weather conditions on wireless technology performance is necessary. A number of published works are available for study. For instance, a study investigated the impact of rainfall and wind speed attenuations on GSM [13] and 2G cellular networks [14], a study of the effects of humidity changes on RF wireless sensor networks [15], etc. The findings of such experimental studies revealed a considerable drop in wireless communication network performance, accuracy, and reliability. As a result, it is necessary to examine, evaluate, and study the effects of weather fluctuations on the robustness of wireless technologies installed in a functional smart grid. For evaluating communication performance reliability, availability is a critical criterion [16]. Several research studies explored into the impact of availability as a key reliability metric to investigate the impact on smart grid communication performance. For example, a study of the reliability of synchrophasor-based dynamic systems based on thermal ratings and system integrity protection methods while subjected to changes in communication network availability [17], and so on. In various published works, average values of time periods between failures (or mean time between failures—MTBF), average values of time periods to complete maintenance (or mean time to repair—MTTR), failure rate, and other characteristics are used to calculate availability. For example, the determination of run-time availability of wireless technology deployed in Indian smart grid installation was evaluated using these parameters [18].

In fact, a systematic methodology is needed from a design and operational standpoints to examine the impacts of changing weather on wireless technology's performance to cater the applications of the smart grid domain and, if necessary, suggest better efficient alternatives. One such optimal wireless technology selection approach is discussed with a data-driven decision-making tool, toward the design of a sustainable communication system [19], and verified using case studies of Indian smart grid pilots [20]. The performance and dependability of wireless communication networks operating in outdoor environments can be degraded by changes in weather conditions. As a result, it's vital to consider weather conditions and climatic variations throughout the year when choosing a communication technology for smart grid installation in a particular region. In light of the foregoing discussion, this study presents a methodology for evaluating the effects of weather fluctuations on the availability of wireless technology deployed in an operational smart grid and recommending better alternatives if the installed one is found to be inadequate. The proposed methodology is tested using an Indian case study, and its outcomes are reported.

3 Methodology

Figure 1 depicts the proposed methodology graphically. Validation based on MTBF–MTTR-based availability analysis is used to assess the impacts of weather variations on the fitness of the installed communication technology. If the installed communication technology is found to be weak or unfit, viable best-fit communication technology alternatives are determined and recommended for replacement of the existing one using Cost Function (CF) optimization-based assessment for the same communication requirements. The suggested approach is validated using a run-time communication dataset provided by a utility company and discussed in Sect. 4 with numerical findings and graphical illustrations, which is an important attribute of this study. For this work, the performance of deployed wireless technology was evaluated using availability analysis as described in [18]. Following that, a decision-support tool presented in [19] was used to recommend the top three best-fit wireless technology alternatives. The performance of installed communication technologies was evaluated using the communication datasheets of an Indian utility company. The communication data, availed in the form of data packets, was then categorized, suitably filtered, and analyzed. To calculate communication availability in various scenarios, the availability metrics, MTBF and MTTR parameters [18], as well as Eqs. (1) to (3), were used.

Mean Time Between Failures,

$$\text{MTBF} = \text{Operational Availability}\,(P) - \text{Total downtime}\,(H) \tag{1}$$

Mean Time To Repair,

$$\text{MTTR} = \text{Total downtime}\,(H)/\text{Failure frequency}\,(F) \tag{2}$$

Availability,

$$A = \text{Mean Time Between Failures (MTBF)}/\text{Operational availability}\,(P) \tag{3}$$

Here, P = Operational availability, H = Total downtime recorded in a month, F = Failure frequency recorded for a month.

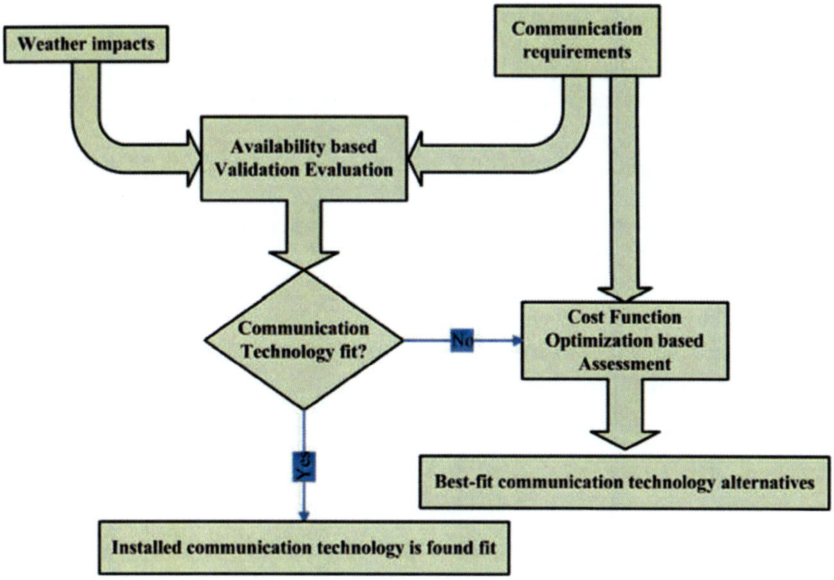

Fig. 1 Graphical illustration of the methodology suggested

For comparison, standard communication availability values for various smart grid critical applications were referred from [21]. If the installed communication technology is deemed to be unsuitable or insufficient, suitable alternatives are suggested.

The communication needs of installed/selected critical applications were evaluated. Following that, the specifications of all wireless technologies were examined to ensure that they could meet the communication requirements for the applications installed, and the top there most suitable wireless technology alternatives were proposed as per [20] and Eq. (4).

$$\text{Cost Function } \mathrm{CF}_{ij} = \left(\frac{1}{\sum_{q=1}^{N_{KPiu}} W_{qi}} * \sum_{q=1}^{N_{KPiu}} \left(W_{qi} N_{qij} \right) \right) \tag{4}$$

Here

i = Index for critical applications
j = Index for wireless technologies
R_{Ni} = required data rate to satisfy the ith application's communication needs
M = Highest value among the data rates for all smart grid applications chosen
$P_{\mathrm{bpsNET}j}$ = Proportional value of data rate achievable for a specific fixed bandwidth, while utilizing jth wireless technology
$NWLAT_i$ = Highest value of delay tolerable to cater the ith application's communication needs
MAX_{NLAT} = Highest delay among all the wireless technology options investigated
$TotLat_{ij} = TP + RTT_i$ = Overall value of delay; where TP is Processing Time (s) and RTT_i is Round Trip Time (s)
$W_{\mathrm{DR}ij} = R_{Ni}/M$ = Data rate's weighted value, if ith application is catered by jth technology
$N_{\mathrm{DR}ij} = R_{Ni}/P_{\mathrm{bpsNET}j}$ = Data rate's normalized value, if ith application is catered by jth technology
$W_{\mathrm{delay}ij} = [1 - (NWLAT_i/MAX_{NLAT})]$ = Delay's weighted value, if ith application is catered by jth technology
$N_{\mathrm{delay}ij} = [1 - (TotLat_{ii}/NWLAT_i)]$ = Delay's normalized value, if ith application is catered by jth technology
CF_{ij} = Cost Function to evaluate the balance ith application's communication requirements by jth wireless technology

4 An Indian Case Study

The under mentioned data regarding the installation are obtained from [22].

(a) Participating customer strength: 21,824. Table 1 presents weather-classified availability of different categories of consumers.

(b) Applications commissioned: Distributed Generation Management (DGM), Outage Management (OM), Supervisory Control and Data Acquisition (SCADA), Peak Load Management (PLM), Advanced Metering Infrastructure (AMI), Demand Response (DR), Meter Data Management System (MDMS)—Total 07 applications, with their communication needs specified in Table 2 [23]
(c) Wireless technologies employed: Cellular 2.5G GPRS and RF ZigBee 868 MHz, with their specifications summarized in Table 3 [24] with reference column data collected from Universal Telecom Council (UTC)
(d) Daily rate of sampling data packets: 04 samples/hour
(e) Size of communication data packet: ~ 1 kb.

It has been learnt from [25] that the location of smart grid pilot generally experiences Summer during February to May months, Monsoon during June to October months, and Winter during November to January months. It is also observed that the warmest month is May, wettest month is July and the coolest month is January.

The complete communication dataset is characterized in the following categories:

(a) *based on economic classification*: Rural and urban—data analyzed and summarized in Table 1, Fig. 2a, b.
(b) *based on customer types*: Residential, commercial, industrial, others (like banks, religious places, parks, staff quarters, etc.)—data analyzed and summarized in Table 1, Fig. 2c–f.
(c) *based on impact of climate changes:* Historical weather and climatic conditions-related data for the location have been obtained from [25]. After analysis of field data with respect to weather data, summary has been recorded in Table 1, Fig. 2g–i.

Based on Eq. (4), Tables 2 and 3, a Cost Function (CF) optimization technique was used to assess the smart grid installation, and the top three apt communication technology alternatives were evaluated and graphically depicted in Fig. 3.

From the assessment results graphically displayed in Fig. 3, it can be seen that 4G cellular Long-Term Evolution (LTE), WLAN, Satellite (LEO), and Worldwide Interoperability for Microwave Access (WiMAX), respectively, have the lowest CF values, making them superior alternatives.

5 Results and Discussions

The lowest CF values were observed for LTE 4G cellular, WLAN, Satellite (LEO), and WiMAX, respectively, according to the results of the assessment graphically depicted in Fig. 3, making them better alternatives. In terms of economic classification, urban areas have better communication availability than rural areas, which have more frequent availability fluctuations. In terms of consumer classification, others and industrial customers have more consistent and high communication availability

Table 1 Impact of climate changes on the availability of communication

S. No.	Weather	Year-month	Location's economic classification-based availability (%)		Availability (%) based on consumer classification			
			Rural consumers	Urban consumers	Residential consumers	Industrial consumers	Commercial consumers	Others
1	Summer	2018–02	88.0208	99.0327	76.6642	98.1275	82.8075	94.3552
		2018–03	68.6156	97.7151	87.6165	97.7263	91.7428	97.4574
		2018–04	84.6528	91.3194	74.2361	98.2778	95.0069	96.7639
		2018–05	74.5296	64.8006	77.8472	98.0242	93.4028	95.3338
		2019–02	5.1538	0.0694	71.9370	17.8968	17.0437	17.1925
		2019–03	0.5914	0.0694	88.2482	92.6658	95.1747	89.4982
		2019–04	7.2269	0.0694	92.3704	95.2755	96.2940	93.8380
		2019–05	50.0672	0.0694	93.7881	98.3020	97.5314	94.1420
2	Monsoon	2018–06	79.7222	66.1806	78.3032	74.7523	96.0370	93.6528
		2018–07	83.8710	99.7984	70.8042	0.0694	96.6420	96.5659
		2018–08	61.1044	82.1237	83.4655	36.2186	90.1210	95.4816
		2018–09	35.4907	96.6690	79.0556	96.9815	90.1343	94.2361
		2018–10	81.4539	32.1931	80.8109	96.4583	92.6815	96.0775
		2018–11	85.2778	93.3380	94.5208	98.3264	97.1736	97.3056
3	Winter	2018–12	70.1635	80.7236	87.9525	93.7343	92.0632	93.5596
		2018–01	89.5161	49.4624	89.6147	99.6461	93.8306	98.1362
		2019–01	5.1165	90.3293	86.9579	97.1931	90.0202	94.5856

Table 2 Communication needs of installed applications of smart grid [23]

S. No.	Smart grid application	Reference (UTC)		Selected	
		Data rates (kbps)	Delay (s)	Data rates (kbps)	Delay (s)
1	AMI	10–100 (500 for backbone)	2–15	500	2
2	OM	56	2	56	2
3	PLM	56	2	56	2
4	DGM	9.6–100	0.1–2.0	100	0.1
5	DR	14–100	0.5–180	100	0.5
6	SCADA	10–30	2–4	30	2
7	MDMS	56	2	56	2

Table 3 Specifications of wireless technologies [24]

S. No.	Wireless communication technology	Reference data (source: UTC)		Selected data for testing		Spectral efficiency (in *b/s/Hz*)
		Data rates (*kbps*)	Delay (*s*)	Data rates (*kbps*)	Delay (*s*)	
1	GPRS	40–50	0.7–1	50	1	1.35
2	RF	20	0.03	20	0.03	0.5
3	LTE	300,000	0.005–0.010	300,000	0.01	3.6

than residential and commercial customers, who have large variations in communication availability, especially during the monsoon. Residential, commercial, and other types of connections outperformed the industrial and commercial categories in terms of communication availability all year long, in all seasons. Very high communication drops have been seen during peak monsoon (July) and peak winter (Feb.). If the applications are increased from the current 07 to all 19, the minimum data rate need rise from 900 to 3200 kbps (minimum 3.56 times). The data rate is projected to increase by 145 times if installation capacity (customer base) is expanded from 21,824 to 3,050,512, which is a huge increase. If the number of applications as well as the number of consumers increases, data rate requirements would likely go up by 520–600 times, which will be impossible to meet with current technologies RF (ZigBee 868 MHz) or GPRS. In the event of an increase in applications and/or user base, LTE with 300 Mbps and 0.01 *s* delay can easily meet all communication requirements. In the event that LTE is not feasible for any reason, Wi-Fi should be prioritized, followed by Satellite (Low Earth Orbit), and finally WiMAX technology, in order of feasibility and viability.

6 Conclusions

The availability of commissioned technologies such as cellular 2.5G GPRS and RF in a functional smart grid installation was examined in this study and determined to be inadequate. To meet current and future communication needs, better wireless technology alternatives—LTE 4G cellular, Wi-Fi (or Wireless LAN), Satellite (Low

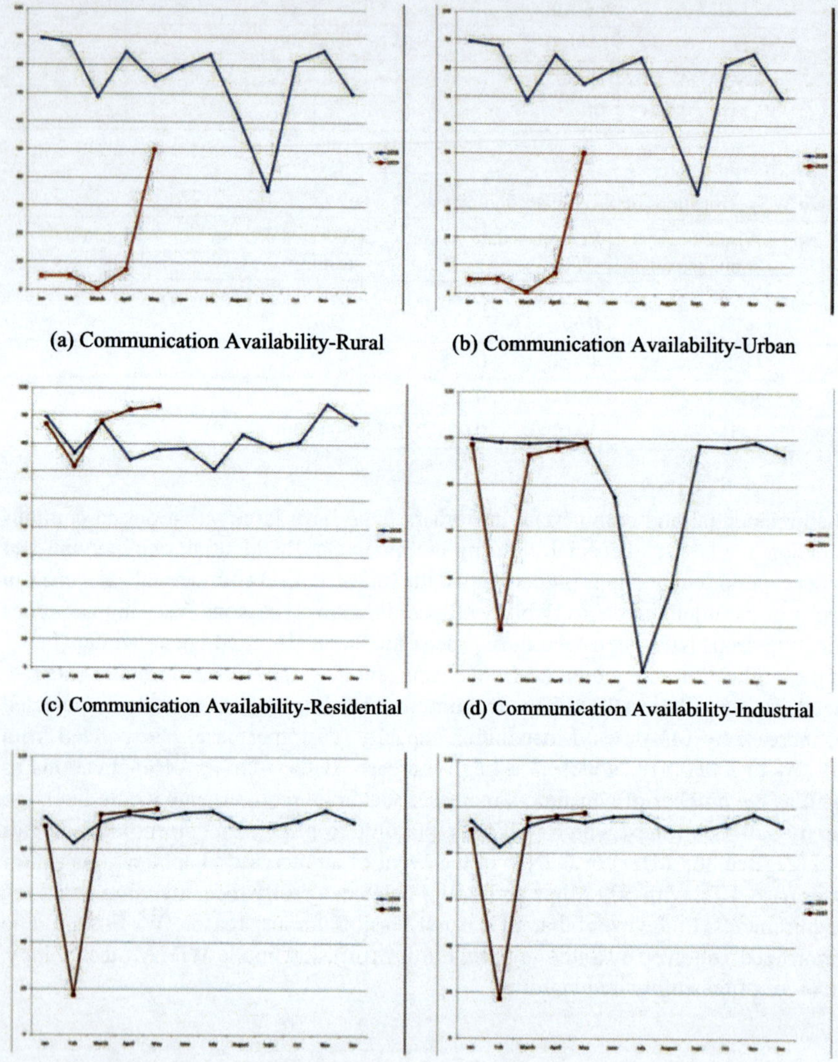

(a) Communication Availability-Rural

(b) Communication Availability-Urban

(c) Communication Availability-Residential

(d) Communication Availability-Industrial

(e) Communication Availability-Commercial

(f) Communication Availability-Others

Fig. 2 Availability analysis for different categories of communication dataset

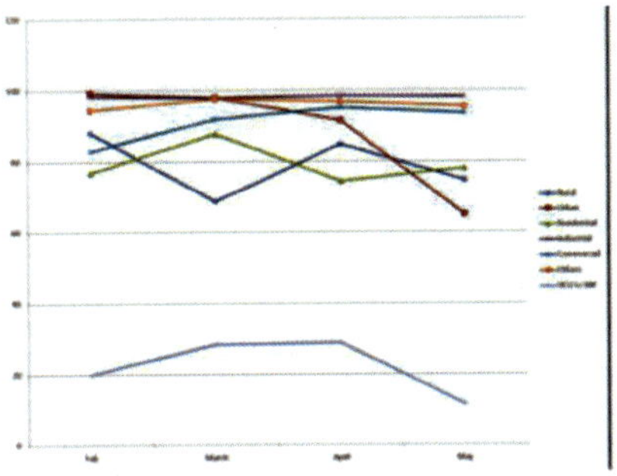

(g) Communication Availability-Summer

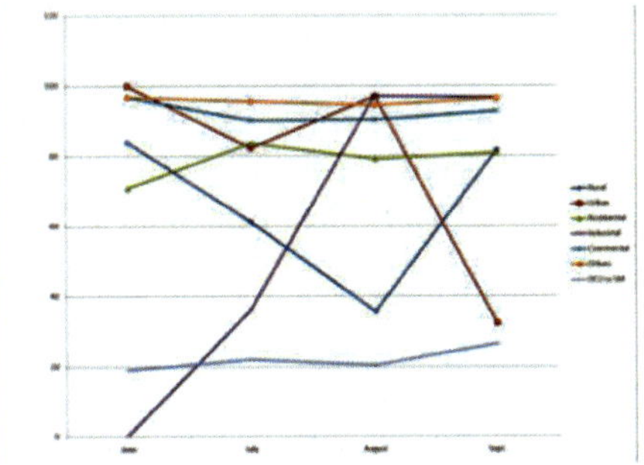

(h) Communication Availability-Monsoon

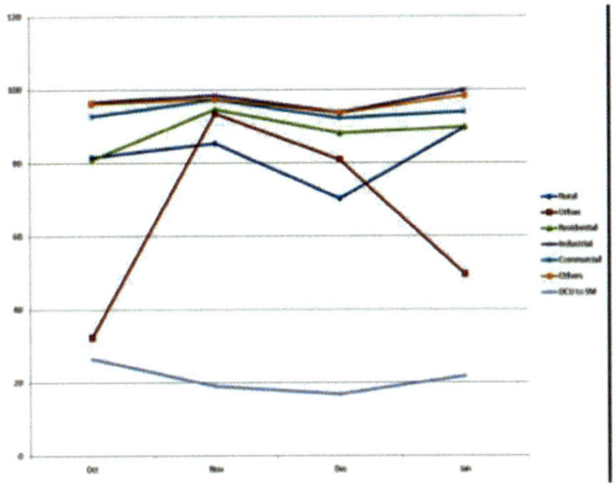

(i) Communication Availability-Winter

Fig. 2 (continued)

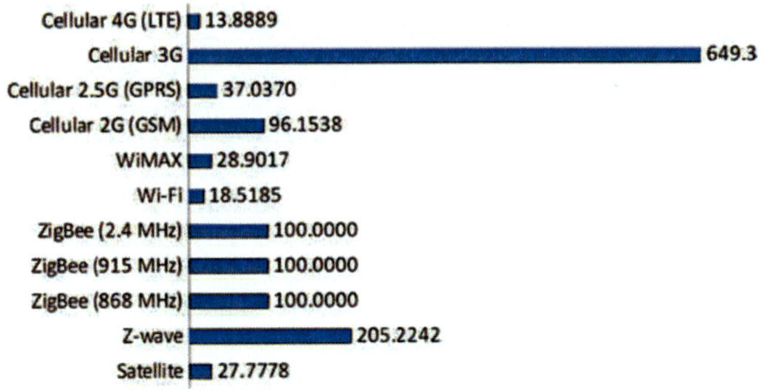

Fig. 3 Assessment for AMI application

Earth Orbit), and WiMAX were recommended. Communication technology performance is likely to be constrained by differences in types of geographic locations as well as year-round climate changes in those locations. Such impacts are difficult to predict using simulations or other similar techniques; nevertheless, the suggested methodology would be more trustworthy due to verified employing real-time field datasets. With currently deployed technologies, the rural, residential, and business subscribers might be dissatisfied with the smart grid as a result of underperformance in communication; further these subscribers make for a significant segment of the consumerbase. A rapid reduction in performance as a result of climate change could also be a cause for concern. A case study with seven applications supports the findings presented. The work can be extended to include all nineteen applications for full capability verification. Future studies could include further exploring more varieties of wireless technology options. Cost can also be added in the future works in addition to data rate and delay as key performance indicators. Analyzing smart meter and data concentrator unit communication could also be similar interesting study.

Acknowledgements The invaluable guidance received from Dr. Chetan Bhatt (Prof.-IC Engineering, GES) is gratefully acknowledged. Well-timed cooperation and support received from Mrs. Kumud Wadhwa (Sr. GM, NSGM) by issuing permission for data access and publication our analysis is acknowledged with thanks. The authors thank the support provided by the Indian utility by providing us the datasheets of communication.

References

1. Ranjan R, Chatterjee P, Panchal D, Pamucar D (2019) Performance evaluation of sustainable smart cities in India. In: Advanced multi-criteria decision making for addressing complex sustainability issues. IGI Global, pp 14–40. https://doi.org/10.4018/978-1-5225-8579-4.ch002
2. Bhatt J, Shah V, Jani O (2014) An instrumentation engineer's review on smart grid: critical applications and parameters. Renew Sustain Energy Rev 40:1217–1239. https://doi.org/10.1016/j.rser.2014.07.187
3. Yazdani M, Chatterjee P, Zavadskas EK, Streimikiene D (2018) A novel integrated decision-making approach for the evaluation and selection of renewable energy technologies. Clean Technol Environ Policy 20:403–420. https://doi.org/10.1007/s10098-018-1488-4
4. Abdel-Basset M, Gamal A, Chakrabortty RK, Ryan M (2021) Development of a hybrid multi-criteria decision-making approach for sustainability evaluation of bioenergy production technologies: a case study. J Clean Prod 290:125805. https://doi.org/10.1016/j.jclepro.2021.125805
5. Salkuti SR (2020) Challenges, issues and opportunities for the development of smart grid. Int J Electr Comput Eng 10:1179–1186. https://doi.org/10.11591/ijece.v10i2.pp1179-1186
6. Dileep G (2020) A survey on smart grid technologies and applications. Renew Energy 146:2589–2625. https://doi.org/10.1016/j.renene.2019.08.092
7. Panda DK, Das S (2021) Smart grid architecture model for control, optimization and data analytics of future power networks with more renewable energy. J Clean Prod 301:126877. https://doi.org/10.1016/j.jclepro.2021.126877
8. Matoušek P, Ryšavý O, Grégr M, Havlena V (2020) Flow based monitoring of ICS communication in the smart grid. J Inf Secur Appl 54. https://doi.org/10.1016/j.jisa.2020.102535

9. Das L, Munikoti S, Natarajan B, Srinivasan B (2020) Measuring smart grid resilience: methods, challenges and opportunities. Renew Sustain Energy Rev 130:109918. https://doi.org/10.1016/j.rser.2020.109918
10. Yohanandhan RV, Elavarasan RM, Pugazhendhi R, Premkumar M, Mihet-Popa L, Terzija V (2022) A holistic review on cyber-physical power system (CPPS) testbeds for secure and sustainable electric power grid—part—I: background on CPPS and necessity of CPPS testbeds. Int J Electr Power Energy Syst 136:107718. https://doi.org/10.1016/j.ijepes.2021.107718
11. Gutierrez-Rojas D, Mashlakov A, Brester C, Niska H, Kolehmainen M, Narayanan A, Honkapuro S, Nardelli PHJ (2021) Weather-driven predictive control of a battery storage for improved microgrid resilience. IEEE Access 9:163108–163121. https://doi.org/10.1109/ACCESS.2021.3133490
12. Tiwari SP, Koley E, Ghosh S (2021) Communication-less ensemble classifier-based protection scheme for DC microgrid with adaptiveness to network reconfiguration and weather intermittency. Sustain Energy Grids Netw 26:1–11. https://doi.org/10.1016/j.segan.2021.100460
13. Fang SH, Yang YHS (2016) The impact of weather condition on radio-based distance estimation: a case study in GSM networks with mobile measurements. IEEE Trans Veh Technol 65:6444–6453. https://doi.org/10.1109/TVT.2015.2479591
14. Luomala J, Hakala I (2015) Effects of temperature and humidity on radio signal strength in outdoor wireless sensor networks. In: Proceedings of the 2015 federated conference on computer science and information systems (FedCSIS 2015), pp 1247–1255. https://doi.org/10.15439/2015F241
15. Sabu S, Renimol S, Abhiram D, Premlet B (2017) Effect of rainfall on cellular signal strength: a study on the variation of RSSI at user end of smartphone during rainfall. TENSYMP 2017—IEEE international symposium technology for smart cities. https://doi.org/10.1109/TENCONSpring.2017.8070024
16. Hossain E, Roy S, Mohammad N, Nawar N, Roy D (2021) Metrics and enhancement strategies for grid resilience and reliability during natural disasters. Appl Energy 290:1–24
17. Jimada-Ojuolape B, Teh J (2021) Impacts of communication network availability on synchrophasor-based DTR and SIPS reliability. IEEE Syst J 1–12. https://doi.org/10.1109/JSYST.2021.3122022
18. Bhatt J, Jani O, Harish VSKV (2022) Development of a mathematical framework to evaluate and validate the performance of smart grid communication technologies: an Indian case study. In: Mathematical modeling, computational intelligence techniques and renewable energy. Springer Science and Business Media LLC, pp 335–349. https://doi.org/10.1007/978-981-16-5952-2_29
19. Bhatt J, Jani O, Harish VSKV (2021) Optimal wireless technology selection approach for sustainable indian smart grid. Strateg Plan Energy Environ 40:255–278
20. Bhatt J, Jani O, Harish VSKV (2021) Development of an assessment tool to review communication technologies for smart grid in India. In: Advances in clean energy technologies. Springer Science and Business Media LLC, pp 563–576. https://doi.org/10.1007/978-981-16-0235-1_43
21. Kounev V, Lévesque M, Tipper D, Gomes T (2016) Reliable communication networks for smart grid transmission systems. J Netw Syst Manag 24:629–652. https://doi.org/10.1007/s10922-016-9375-y
22. National smart grid mission: SG pilot projects. https://www.nsgm.gov.in/en/sg-pilot. Last accessed 18 Feb 2022
23. Ghorbanian M, Dolatabadi SH, Masjedi M (2019) Communication in smart grids: a comprehensive review on the existing and future communication and information infrastructures. IEEE Syst J 1–14. https://doi.org/10.1109/JSYST.2019.2928090

24. Mai TT, Haque ANMM, Vo T, Nguyen PH, Pham MC (2018) Development of ICT infrastructure for physical LV microgrids. In: 2018 IEEE international conference on environment and electrical engineering and 2018 IEEE industrial and commercial power systems Europe (EEEIC/I&CPS Europe). IEEE, Palermo, Italy, pp 1–6. https://doi.org/10.1109/EEEIC.2018.8493788
25. World Weather and Climate Information (2022) Climate and average monthly weather in Mysore (Karnataka), India. https://weather-and-climate.com/average-monthly-Rainfall-Temperature-Sunshine,mysore,India. Last accessed 18 Feb 2022

Computational Intelligence in Healthcare Applications

Intelligent System for Diagnosis of Pulmonary Tuberculosis Using Ensemble Method

Siraj Sebhatu, Pooja, and Parma Nand

Abstract In this research, model development is carried out under supervised learning, as the system tries to correct and update itself by comparing the outcome with the target result. After all, only one model category is used, enhanced model performance through substituting the selected features with high sensitivity and low accuracy in clinical knowledge. The experimental analysis shows that the gradient boosting (GB) XG boosting model achieves the best result using the original dataset to predict PTB-disease. The ensemble model composed of the AdaBoost, bagging, random forest, GB, and multi-layer perceptron models is the best to detect. The ensemble model reaches 97% accuracy, which exceeds each classification's accuracy. The model is used to help doctors analyze and evaluate medical cases to validate the diagnosis and minimize human error. It effectively mitigates clinical diagnosis in such difficult challenges as microscopic scanning and reduces the likelihood of misdiagnosis. The model differentiates the patient using a voting method of different machine learning classifiers to provide accurate solutions from having only one model. The novelty of this approach lies in its adaptability to the ensemble model that is continually optimizing itself based on data.

Keywords Ensemble learning · Optimization · Stacking · Pulmonary tuberculosis disease

S. Sebhatu (✉) · Pooja · P. Nand
Department of Computer Science and Engineering, Sharda University, Plot No. 32–34, KP III, Greater Noida, Uttar Pradesh 201310, India
e-mail: siraj.sebhatu@wsu.edu.et

Pooja
e-mail: Pooja.1@sharda.ac.in

P. Nand
e-mail: parma.nand@sharda.ac.in

P. Chatterjee et al. (eds.), *Computational Intelligence for Engineering and Management Applications*, Lecture Notes in Electrical Engineering 984,
https://doi.org/10.1007/978-981-19-8493-8_8

1 Introduction

Tuberculosis is a disease caused by mycobacterium tuberculosis that can damage respiratory organs and affect other internal parts of the body. It usually extends throughout the air and highest-burden worldwide as one of the human infectious diseases. Especially, the risk of TB attack is higher for patients with the human immune deficiency virus (HIV). In India, it is also a significant health issue on pulmonary tuberculosis diagnosis that was always a problem. Most of the detection methods need high cost and complete power devices. Depending on the diagnosis, delayed or inappropriate treatment can result in unsatisfactory results, including the exacerbation of clinical symptoms, poor quality of life, and increased disease prevalence. Still, these devices require time and hard work [1], leading to low detection rates and incorrect diagnosis, even for experienced pathologists. Some new technologies were developed, including PCR and RNA scopes, to improve effectiveness and sensitivity in diagnosing TB bacilli, but neither has been successful and widely accepted so far [1]. Moreover, the effective diagnosis of the disease is a necessary first step toward eradicating TB. In small-income countries, where the disease is predominantly present, the method of diagnosis should be fast, precise, and easy to use. The rapid request for tuberculosis infection control measures reflects new diagnostic methods [2]. Classification of health data is essential in detecting and screening any disease. It even lets doctors make decisions about their diagnostics and treatments.

We proposed an ensemble voting approach to compare the efficiency of ensemble learning classifiers on pulmonary tuberculosis physical examination, clinical, and key-population factor health data. Moreover, we used bagging, boosting, blending, and stacking. A designed model helps to detect pulmonary tuberculosis by using the trained model. The most important cause of failure to control tuberculosis disease global effort is delaying appropriate treatments and misdiagnosis. Due to this reason, the patient is exposed to long-term lung damage. Studies reported that a therapy delay of more than four months (12.1 weeks) would result in a larger proportion of patients with chronic TB, an increased mortality rate, and a higher failure in treatment. The timeframe between the patient's initial diagnosis and first contact with a care provider was determined as the patient's delay. The delay in diagnosis was stated as the period between first medical assessment and diagnosis [3] cough, hemoptysis, night sweats, fever, and weight loss is suggestive symptoms of TB. Such signs include not only tuberculosis but also other diseases. According to these difficult circumstances, requests to enact potential alternative approaches for PTB diagnosis are important, bringing down the cost and time resources, and improving prediction accuracy. Some studies have been done to address these problems related to the diagnosis of TB that have been implemented using sound, images, blood miRNA profiles, and variables as input parameters. Diagnosis and treatment parameters for microscopy and limited period for traditional cultivation approaches have already been focused on designing accelerated methods for mycobacterium. Bovis detection of mycobacterial isolates in clinical specimens and early identification [4]. The main disadvantages are difficulty, lengthy use, and the lack of suitable techniques for bovine tuberculosis (PTB). It is highly difficult to analyze and interpret the results; it is not unique responses triggered by other mycobacterial organisms.

This research aims to build a classification model for the initial screening of pulmonary tuberculosis (PTB) disease. This model can be applied to determine whether an individual has been infected with PTB or not, based on clinical symptoms and other key factors in the population [5]. The physical examination and clinical active pulmonary tuberculosis symptoms are coughs, fever, hemoptysis, weight loss, night sweat, duration_predominant symptom its duration (PDSD), visual appearance of sputum, HIV, diabetics, etc. [6]. Key population around the individual exposed to the number of factors like as age, gender, contact TB person, tobacco, prison inmates, miner, migrant, refugee, urban slum, healthcare workers associated with medical expertise have been used to train the model [7, 8] but it does have limitations such as low accuracy and long observation time. Consequently, an efficient TB screening method is needed [9]. The content of the paper is structured accordingly. The first part provides a detailed description of tuberculosis, the problem area, background, the aim of the research, and the relevance of the research findings. The second part describes the previous research related to this study. The third part will discuss the methodologies and proposed methods used to carry out this research. Eventually, the fourth part discusses the result and conclusions.

2 Related Work

The latest improvements in ensemble methods and massive datasets have supported algorithms to perform numerous diagnostic tasks for respiratory diseases, such as PTB classification [10]. This model significantly improved the disease's screening accuracy compared with a ruled-based method. The study used deep learning and classic machine learning based on physical indicators of symptoms, biochemical observations to diagnose adult asthma. Their research has lung and bronchial challenge check accuracy, 60% SVM accuracy, and 65% logistic analysis accuracy [11]. Other work considered ensemble technique enhancing the integration of various classifiers, performing high accuracy on a single ensemble model, and a model provides a solution to reduce the isolation of PTB patients [2]. The ensemble classifiers' prediction accuracy was tested using cross-validation (tenfold), and findings were evaluated to achieve the best prediction accuracy, for instance, bagging and AdaBoost. The results show that bagging achieves the highest accuracy with 97%, random forest 93%, and AdaBoost 96% [12].

Several classification methods have been discussed. The C4.5 tree of decision and support vector machine was not statistically significant, compared with the SVM with Naive Bayes and the K-nearest neighbor, statistically significant. Retroviral pulmonary tuberculosis (RPTB) and pulmonary tuberculosis (PTB) together with AIDS with appropriate learning classifiers [10]. Models based on ANN help to support a tuberculosis diagnosis study under limited resources. Analyzing important information from the database MLP detection performance achieved 97% of sensitivity and applying cluster techniques using SOM network comparing three risk group detecting the disease the algorithm perform 89% of sensitivity [11]. According to the analysis, which applies support vector machine (SVM) and decision tree (C5.0) based on a multi-objective gradient evaluation, medical test PTB can easily detect infection and achieve a more reliable diagnostic outcome [12].

On the one hand, classification of PTB based on a random forest model performs early diagnostic optimization 81% of the area under the curve (AUC) [13], and on the other study in 2017 using SVM, C5.0 shows the results of the model perform as better accuracy of 85.54% [14]. At the same year assessment of PTB characteristics using the neural network, the model performs with the highest accuracy using the pruning method [11]. The decision tree also optimizes the time to verify results compared to the K-nearest neighbors [15]. A classification model based on a single MLP with a better accuracy performance obtained a sensitivity of 83.3% and specificity of 94.3% [16]. The most significant factors found by the clustering are the TB evaluation findings indicating that patients with hemoglobin used their age, sex, smoking, and alcohol.

3 Methods

3.1 *Ensemble Methods for Prediction of Pulmonary Tuberculosis Diagnosis*

Ensemble methods (EM) represent people's co-decision process in the treatment of hard decisions. The core idea in machine learning is to create strong predictors with weak but distinct models combined [17]. In most cases, EM's goals are to achieve more effective and robust solutions than alternative individual models to solve complex problems [18]. Generally, the study includes three main phases as follows: processing of specific classifiers, selection of members, and specification of the decision process. Independent errors on the models generated to integrate the group should occur to optimize ensemble performance, which does not result in approaching strategies that explore data, structural variables diversity [18]. This optimization is expected to have small clustered errors in various models, thus improving committee performance and complementing each other. Key approaches to creating diverse models include using various training sets, model hyperparameters, and classification methods. Data interoperability is among the most efficient methods, based on most of its extensions on the well-known random forest, bagging, and boosting algorithms [17]. Bagging is roughly focused on preparing various trainings to design "supposed" models. A specific training set is generated from an original random sample of bootstrap (BS), i.e., a replacement sampling. A baseline classifier is created for each BS, and the committee's output is an average of the outputs of each model. For each BS, a different classifier generates a base algorithm, and the result of the committee consists of an average of all outputs of the models. In later years, several bagging combinations were advocated, most of them considering strategies to reduce the number of baseline classifiers involved in the groups by the sequential backward selection, genetic algorithms, and clustering.

3.2 Data Source and Data Descriptions

Further experiments have shown that an appropriate selection method is more compatible with class members and mitigates the class difference effects. Alternative decision fusion mechanisms are often discussed, such as reducing the ensemble output variance. In this work, standard datasets (from the Wolaita Sodo University Teaching Referral Hospital) are employed to evaluate an intelligent system for pulmonary tuberculosis diagnosis. In the following, the datasets are listed briefly described below in Table 1. The medical data we describe contains 3252 individual TB patient records. The whole file with many documents is stored in one file. That record is consistent with one patient's most relevant data. Initial doctor inquiries as symptoms and necessary patient test details were major attributes. A class has 19 symptoms, such as gender, age, cough, fever, hemoptysis, weight loss, night sweat, predominant period symptom (PDSD), the visual appearance of sputum, HIV, diabetics, and other main population factors contact TB person tobacco, inmates, miners migrants, refugees, urban slums, healthcare workers, and also the class outcome of the attribute, namely early diagnosis. Table 1 shows the names of 19 different attributes listed along with their categorical and numerical together with the data type.

3.3 Feature Selection

We recommend three methods of selection as follows: wrapper selection procedure, filtering of features using decision trees, and removing highly correlated technical attributes. We will compare and contrast the accuracy of the classification with the features chosen using each of the techniques. Finally, we pick the most appropriate set of features [19]. Feature selection with reverse function elimination is a greedy algorithm for searching. It continues with all available features and then drops one feature at a time [20]. We use the reverse isolation of ranking attributes to make this algorithm more effective. The optimal subset function may not be special, as specific function sets may achieve the same accuracy (e.g., two correlated features may replace each other) [20]. Backward isolation can more easily capture interacting features using backward elimination.

Table 1 List and data types of attributes

No.	Name of variables	Data types
1	Age	Numerical
2	Gender	Categorical
3	Contact TB person	Categorical
4	Tobacco	Categorical
5	Prison inmates	Categorical
6	Miner	Categorical
7	Migrant	Categorical
8	Refugee	Categorical
9	Urban slum	Categorical
10	Healthcare worker	Categorical
11	Cough	Categorical
12	Fever	Categorical
13	Hemoptysis	Categorical
14	Weight loss	Categorical
15	Night sweat	Categorical
16	Duration of symptom	Numerical
17	Sputum a. Mucopurulent b. Saliva c. Bloodstained	Categorical
18	Human immune deficiency virus (HIV)	Categorical
19	Diabetics	Categorical

3.4 Missing Value Treatment Method

Treatment of missing values using the imputation technique [21, 22] is the popular missing data treatment technique in which missing values are replaced by a certain approximate value in combination with the available dataset. The purpose of this technique is to use correlations to help estimate missing values, which can be contained in the valid values of the dataset. In [23], investigators investigated the efficiency indicator of zero substitution and trajectories, the mean trajectory, singular decomposition value (SVD), and a weighted *K*NN method to replace missing values in medical datasets and stated that the *k*NN method is equivalent to the inference of missing values. The nearest neighbors, the Euclidean distance measurements, are determined by minimizing the distance function [23].

3.5 Combination of Ensemble Voting Technique

The first word used, bagging, is slang for the bootstrapping and aggregating combination. Bootstrapping is a technique for reducing the classification variance and minimizing overfitting by re-examining training sets with the same cardinality as the entire set. The produced model should be less extreme than a single model. High variance is not good for a model, so its efficiency is sensitive to the data provided. Thus, the model can work poorly, even if more data is given. And the variance of our model may not even be reduced [17]. Suppose a set of simple learners is generated, and not after trying to find the best single pupil. In that case, ensemble methods depend on combining these methods to achieve the best generalization results. The fundamental reasons are the statistical issue and computational issue.

Consequently, the statistical problem is generally too large to explore hypotheses for limited training dataset and many different hypotheses that give training data accuracy. Suppose the learner algorithm chooses one of the hypotheses; moreover, the importance of combining the hypotheses is reducing the wrong choice of the hypothesis [17, 18]. Additionally, computational often conducts a certain different local search which is stuck in optimal local numbers with many learning algorithms. It can still be very challenging to find the best hypothesis even if sufficient training data is available; using the combination could provide a better estimate of a true unknown hypothesis from a local search from several different starting points.

3.6 Designing Different Types of Bootstrap Samples

The majority of design a set of new ensembles approach focuses exclusively on model selection, and our proposal provides for the generation of a set of various bootstrap samples, using a low-cost computational procedure. Some bootstrap samples are the fundamental concept (BSs) (βT, $\beta > 1$) applicants, selecting the T most different from the existing training dataset (ETS), where T is a user-defined parameter, and n attributes denoted by C_1, C_2 ... C_n classifier, BS_{-1}, BS_{-2} ... BS_{-n} bootstrap. We designed a mechanism to determine the degree of similarity of the existing training dataset (ETS) and certain bootstrap samples (BS) [17, 18].

3.7 Soft Voting

The majority voting plurality and weighted votes for individual classifiers generating class labels can be used, while the soft vote is usually the option for individual classifiers producing class likelihood outcomes. With the soft weighted vote, with each classifier, we measure a percentage weight. For each model of record, a predicted class likelihood is obtained and increased by the weight of the classifier and is finally averaged [21, 22]. The final label of the class comes with the highest average likelihood from the class label. Besides, weights are difficult to find if you only offer your best estimates of which model you assume should be more or less weighted [2]. A deterministic optimization equation or neural net can be built to counteract this subjective process to determine the right weighting of each model to optimize the higher performances of ensemble model accuracy. Here, the individual classifier hi outputs, a one (l)-dimensional vector $(h_i^j(X), \ldots, h_i^j(X))^{\mathrm{T}}$ for the instance X, where $h_i^j(X) \epsilon [0, 1]$ can be regarded as a prediction of the posterior likelihood $P(\varsigma j \mid X)$. If all the individual equivalent consideration of classifiers, the simple soft voting method generates the combined output by simply averaging all the individual outputs, and the final output for class C_j is given by $T^1 \sum_{X=1}^{T} h_i^j(X)$. Soft voting for homogenous assemblies is commonly used. The class probabilities created by different types of learners cannot necessarily be compared for heterogeneous classes without careful calibration, the class probability outputs are often converted to class label outputs by setting $h_i^j(X)$ to 1 if, $h_i^j(X) = \max j(X)$ and 0 otherwise, and the voting methods for crisp labels can be applied [17, 21].

3.8 Bagging

Bagging is a whole approach used for various training datasets to achieve different classifications, which used the existing learning algorithm. A bootstrap technique for re-sampling the training dataset improves the difference of the training datasets [23]. This approach eliminates noisy data, outliers, and variance, as Fig. 1 shows. Every classifier then receives training on a re-sample of instances, which assigns to these instances a predicted class. The estimates of the various classifiers (with equal weight) are then combined by majority voting. The significance of this approach to support boosts a model disease detection accuracy and consistency of a machine learning algorithm[18].

The top-level model takes the low-level performance and allows the estimation of the stacking algorithm presented in Fig. 1. The initial data is processed as input to several individual models in stacking. Then, the meta classifier estimates the input and output of the respective model, and the meta classifiers support using the weights of each classifier for prediction [17, 22]. The highest performing model results are selected, and the left remains. Bagging combines multiple base classifiers trained by using different learning algorithms L on the existing dataset S, utilizing a meta classifier.

Let $\mathrm{D} = \{\mathrm{bs}_{-1}, \mathrm{bs}_{-2}, \mathrm{b}_{-3}, \mathrm{bs}_{-4}, \mathrm{bs}_{-5} \ldots \mathrm{bs}_{-\mathrm{n}}\}$ be the given dataset
$\mathrm{E} = \{\}$, the set of ensemble classifiers
$\mathrm{C} = \{\mathrm{c}_1, \mathrm{c}_2, \mathrm{c}_3 \ldots \mathrm{c}_\mathrm{n}\}$, the classifiers
$\mathrm{X} =$ the training set, $\mathrm{X} \epsilon \mathrm{D}$
$\mathrm{Y} =$ the test set, $\mathrm{Y} \epsilon \mathrm{D}$
$\mathrm{L} = \mathrm{n}(\mathrm{D})$

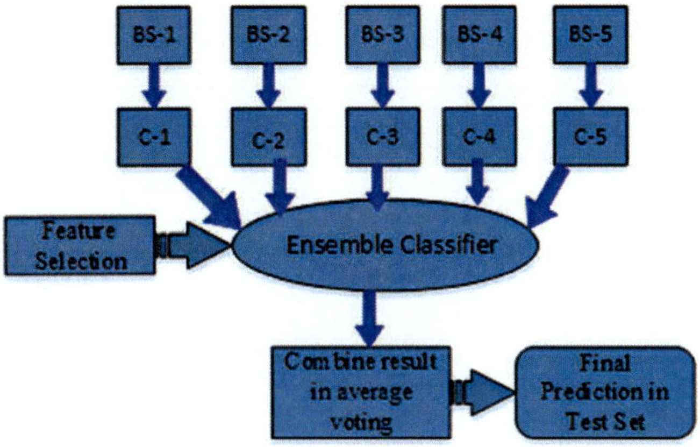

Fig. 1 Ensemble model process

For I = I to L do
S (i) = {Bootstrap sample I with replacement} 1 ς X
E = E ∪ C (i)
Next I
For I = 1 to L
R (i) = Y classified by E (i)
Next i
Result = max (R (i): i = 1, 2 ... n)

4 Experimental Analysis and Result

4.1 The Ensemble Classifier Performance

A comparative study is conducted on the original dataset of various classification algorithms. Some algorithms have a strong precision, while others are weak. Multiple ensemble classifiers are used to boost the effectiveness of the weak classifiers. This study used group algorithms such as bagging and stacking. The bagging algorithm Table 2a, b executes an ensemble with the bagging random forest classifier, bagging extra trees classifier, bagging *K* neighbors classifier, bagging SVC, bagging ridge classifier algorithms (Table 3's base classifier; the decision region analysis of voting is shown in Fig. 2).

We have improved the accuracy (0.978 vs. 0.976) and decreased the variance (std: (±) 0.015 against std: (±) 0.04) so that we function according to our ensemble modeling by integrating all the different models in one.

Table 2 Improvement in boosting accuracy

Accuracy	Standard deviation (std)	Classifier
(a)		
Mean of: 0.968	(±) 0.0012	Bagging random forest classifier
Mean of: 0.896	(±) 0.002	Bagging extra trees classifier
Mean of: 0.896	(±) 0.001	Bagging *K* neighbors classifier
Mean of: 0.945	(±) 0.001	Bagging SVC
Mean of: 0.936	(±) 0.001	Bagging ridge classifier
Mean of: 0.968	(±) 0.015	Bagging ensemble
(b)		
Mean of: 0.952	(±) 0.015	Random forest classifier
Mean of: 0.888	(±) 0.014	Extra trees classifier
Mean of: 0.883	(±) 0.014	*K* neighbors classifier
Mean of: 0.956	(±) 0.001	SVC
Mean of: 0.938	(±) 0.005	Ridge classifier
Mean of: 0.97.6	(±) 0.04	Ensemble

Table 3 Performance analysis of our conceptual model

Accuracy	Standard deviation (std)	Classifier
0.95	(± 0.01)	Random forest
0.88	(± 0.01)	Extra trees
0.88	(± 0.01)	*K* neighbors
0.956	(± 0.00)	SVC
0.938	(± 0.01)	Ridge classifier
0.97	(± 0.01)	Ensemble

4.2 Performance Analysis of Our Conceptual Meta-Ensemble Platform (Design)

From the findings of the last section, it is understandable that applying various outputs of different classifiers improves classification accuracy over the current independent classifier in the combination. Nonetheless, it does not do as well as boost—the values by optimizing processes specifically to minimize the value of errors, thus integrating works indirectly. Since our proposed model works so well to generate

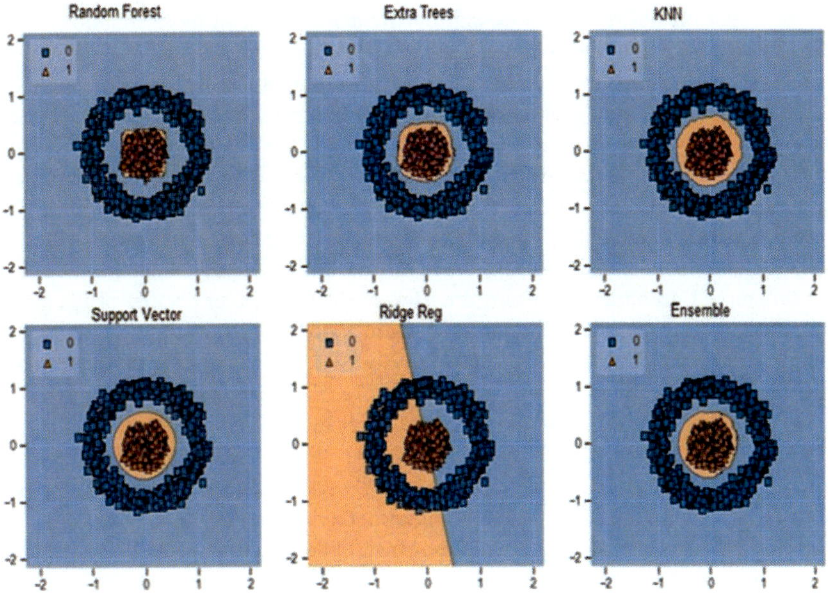

Fig. 2 Classifiers on plot decision regions

the best results from the combination, we have used this approach to combine our ensemble results with boost performance, stacking, and bagging and form a meta-ensemble architecture platform Figs. 1 and 3.

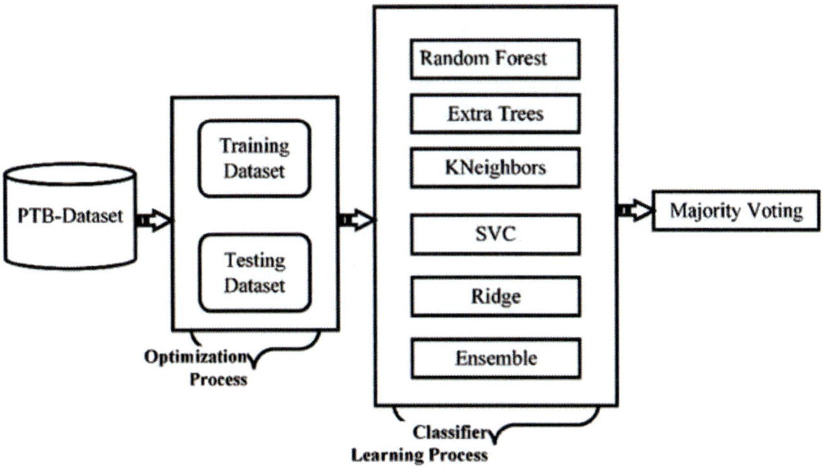

Fig. 3 Ensemble voting architecture

The experimental results show that our model is better in most situations than choosing the best classifier in the combined effect. We have also proposed an effective way of combining the ensemble's classification outcomes, but it depends on each classifiers' class outputs in the combined [22]. The experimental findings demonstrate that our approach performs better in certain cases than choosing the most appropriate single classification. Therefore, our model is tested in this way on a single class dataset.

A comparative analysis has been made on the original dataset of various classification algorithms. Some algorithms are highly accurate, although some have limited performance. Ensemble algorithms are used to increase the efficiency of the weak classifiers. This research used algorithms to enhance voting, boosting, and stacking, including bagging. The bagging algorithm executes an ensemble with the bagging random forest classifier, bagging extra trees classifier, bagging K neighbors classifier, and bagging SVC, bagging ridge classifier algorithms for boosting. For this experiment, sets are generated using the random forest classifier, extra trees classifier, K neighbors classifier bagging, and SVC sacking classifier for boosting use. Machine learning algorithms' most popular method for making classifications and applying voting is to find normal outputs. Much of the votes has often been included as ensemble techniques. Voting through the majority is another ensemble strategy that combines several classifications for improved accuracy [22]. In this proposed model, the K neighbors classifier weak classifiers for the original dataset and less accuracy. Ridge classifier and random forest performed well and had better classification accuracy. It is inferred from Fig. 4 that a party of weak classifiers with high majority vote classifiers significantly increases the accuracy of a weak classifier. Ensemble multilayer perceptron, gradient boosting, and SVM improved the accuracy of strong classifiers.

Bagging and boosting enhancing comparative analysis are seen in Fig. 4. The findings indicate that both bagging and boosting weak classifiers are useful to increase the accuracy of weak classifiers. The calculation period is measured as the sum of 100 loops, and the value is in seconds. There is a measure of the classifier comparable estimation period for bagging and boosting methods which are shown in Table 4.

4.3 Decision Boundaries

The decision method was developed by Kuncheva. In this approach, the expected outcome of the classifiers on instance x is organized in the decision method as the matrix.

$$\mathrm{DP}(x) = \begin{pmatrix} hi^{1}(x) & h^{J}(x) & h^{I}(x) \\ h\dot{i}(x) & h^{J}\dot{i}(x) & h\dot{i}^{I}(x) \\ h\underline{t}(x) & h\underline{t}^{J}(x) & h^{I}\underline{t}(x) \end{pmatrix}$$

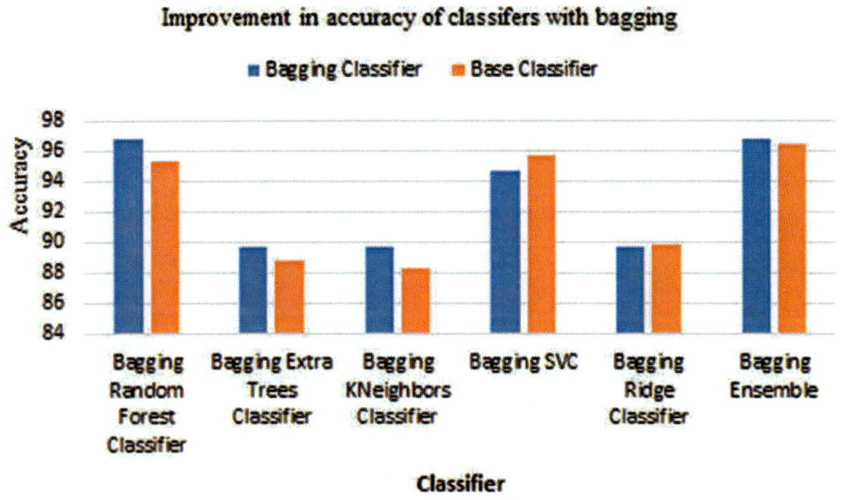

Fig. 4 Bagging classifiers accuracy improvement

Table 4 Bagging and boosting computing time comparison

Classification algorithm	Run time	Stacking run time	Bagging run time
K neighbors	0.221157	0.299157	0.288229
Random forest	1.886573	0.656246	0.699129
GaussianNB	0.046866	0.082776	0.072806
Extra trees	1.870072	0.616354	0.583439
SVM	0.717640	2.108359	2.074450
Ridge classifier	0.062484	0.080791	0.094748
Logistic regression	0.527627	0.274266	0.270278

Based on the training dataset $D = \{(x, y) \ldots (x_m, y_m)\}$, the decision approach is estimated as the expected DP(x). i.e.

$$\mathrm{DT}_k = \sum \mathrm{DP}(x), \quad k = 1, \ldots, I$$

Boosting on the optimization.

Boosting uses a different re-sampling method. The sample selection is based on a current training dataset in this scenario. The initial dataset is the first classification in which each sample has equal weight (Fig. 5). The weight will be reduced when the sample has properly been classified in the previous training dataset; otherwise, it will increase if the sample is misclassified [22]. The committee choice uses a weighted voting technique to ensure a more accurate classification which is given a higher weighting than a less accurate classification in shown Fig. 5.

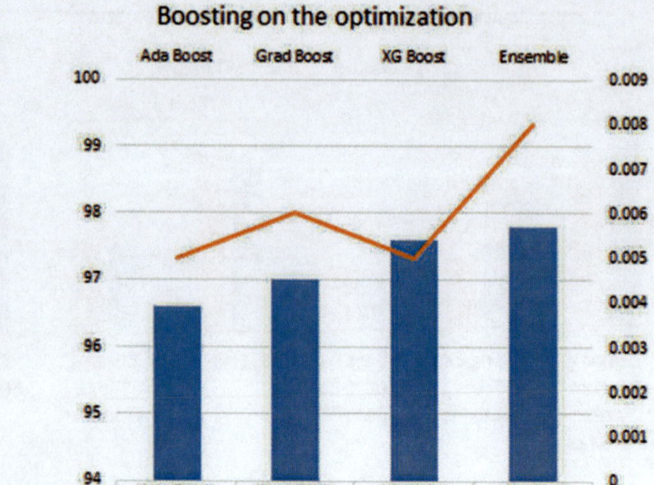

	Ada Boost	Grad Boost	XG Boost	Ensemble
Accuracy	96.6	97	97.6	97.8
Standard	0.005	0.006	0.005	0.008

Fig. 5 Performance of boosting analysis

This optimization can be overcome for gradient descent, and gradient Vanilla is used to reduce the number of parameters. Estimating parameters appear easy if we have a smoother convex parameter, but not all problems make such an easy path. Our problem is that it generates a dynamic gradient with many categorical and binary variables and local minima to hold in during the optimization process. It can use a different type of descent called boosting for these problems.

4.4 AdaBoost

AdaBoost's algorithm is a linear combination of a "simple" and "weak" classifier for the building of a "strong" classifier. Rather than re-sampling, each sample uses a weight to measure the likelihood of selecting a training set. The final classified algorithm is based upon weighted voting by the weak classifiers. This classifier is more sensitive to noisy data and outliers [17, 22]. Nevertheless, it may be less exposed to overfitting in certain challenges than other learning algorithms.

4.5 Machine Learning Classifiers for Stacking

The machine learning (ML) algorithm ensemble uses gradient-enhancing decision-makers, which depend on the combination of weak individual classifier's performance [24]. That model consists of decision trees with logical structures and a leaf describing a weight of probability [25]. With dichotomous attributes, directions differ from "yes" (present) to "no" (absences) replies; and for continual attributes, cutoffs have decision-making restrictions to guide pathways. The final probability estimate is the total weight of all trees in the model.

4.6 Stacking

Stacking is a technique used to combine many classification models through a meta classifier. The multiple layers are arranged one by one. Each model converts its predictions into the above model, and the top layer model takes decisions based on the following models below. The result shown from the original dataset is given to the bottom layer models [17, 22].

The fundamental concept is to train first-level learners using learner training datasets and create a new dataset for second-level learners. First-level learner performance is still called input features. At the same time, the original labels continue to be seen as new training data labels. The first-level learners are often created by applying different study algorithms so that stacked classes are often heterogeneous. However, they can also create homogenous stacked classes [22].

Let $D = \{bs_{-1}, bs_{-2}, b_{-3}, bs_{-4}, bs_{-5} \ldots bs_{-n}\}$ be the given dataset
$E = \{E_1, E_2, E_3, \ldots, E_n\}$, the set of ensemble classifiers
$C = \{c_1, c_2, c_3, \ldots, c_n\}$, the classifiers
X = the training set, X ϵ D
Y = the test set, Y ϵ D
L = n (D)
For I = I to L do
M (I) = Model trained using E (i) on X
Next i
M = M ∪ K
Result = Y classified by M

The pseudocode of the stacking algorithm.

It has been designed to measure the performance of a model for computational analysis. Boosting is one of the ensemble approaches to adjust the errors of the existing model and to improve the new model [1]. Recurrent versions should be introduced before no major changes can be made. Boosting algorithm is designed to create a new model that predicts the residuals of prior models and then is added together to make the final prediction. A descent algorithm significantly reduces failure when new models are implemented. This method encourages classification and regression, and efficiency has changed significantly [24]. This algorithm has been published in the library of Python SciKit-learn and comes with new regularization techniques. The model must be closely configured to obtain optimum efficiency. A tuning boost can be an extremely challenging process due to its hyperparameter level. Such parameters may be divided into general parameters booster, analysis function, and command line. A grid search will do tuning.

This study uses an optimal grid method with a large parameter size. This method can effectively be achieved by integrating parameters with rational parametric values in a smaller combination. In the selection process, K cross-validation is used to test the model's consistency [23, 24]. We used scikit-learn Python modules and resources for our simulation for research and experimental analysis. The model achieved the performance of a true positive rate is high, false discovery rate low, and $F1$-score test with ten-fold cross-validations. Our proposed model automatically performed better than existing models based on machine learning.

Results indicate that certain algorithms have better efficiency of detection than others. Table 5 lists the classification methods used to predict various classifying algorithms for the best accuracy. Such measures would be the most relevant criterion for classifying the health informatics field as the best algorithm. The prediction results in precision between single classifiers, random forest and ensemble, LGBM, and extra trees are the best. Graphically equivalent in Fig. 6 is other measures like F-measure and ROC of the above-noted classifiers. The average F-score and ROC of the two divisions are seen. These classifiers are seen in Fig. 7 for prediction accuracy.

The ROC curve is a descriptive plot determining a binary classifier frame's predictive efficiency, as its threshold edge has fluctuated. The ROC curve is made by plotting the true positive rate (TPR) against the false positive rate (FPR) at different limit settings. The true positive rate is sensitivity. For machine learning, the true positive increase is considered sensitive. Otherwise, the false positivity rate is considered the drop-out or possibility of a negative outcome (1-specificity). The ROC curves and the proposed model for the baseline classifier are related. The ROC curve takes false positive rate indicating the ratio of the wrong classification on positive class, and the percentage of correct classification on positive class, indicating true positive rate. The ROC curve shows the likelihood that a true positive instance will be better

Table 5 Performance of stacking classifiers

Accuracy	Standard	Classifier
0.84	(± 0.01)	*K* neighbors
0.95	(± 0.05)	Random forest
0.52	(± 0.01)	GaussianNB
0.93	(± 0.04)	Extra trees
0.935	(± 0.00)	SVC
0.94	(± 0.05)	Ridge classifier
0.96	(± 0.00)	Stacking

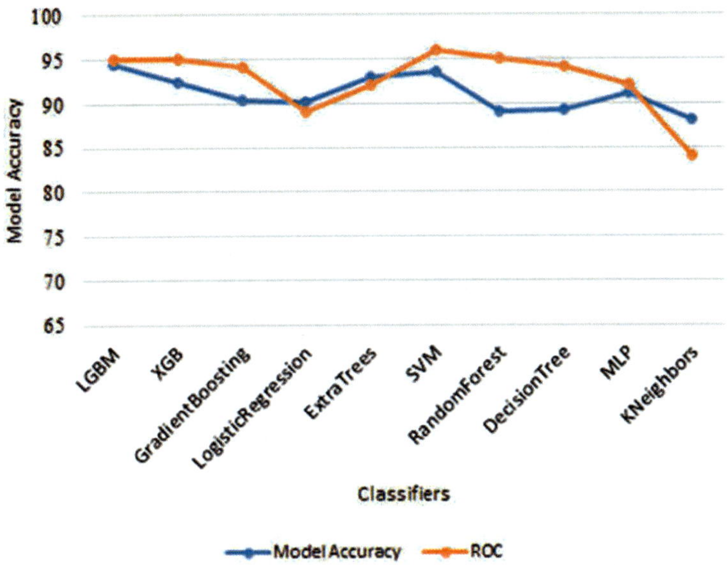

Fig. 6 Comparison of average *F*-score and ROC area

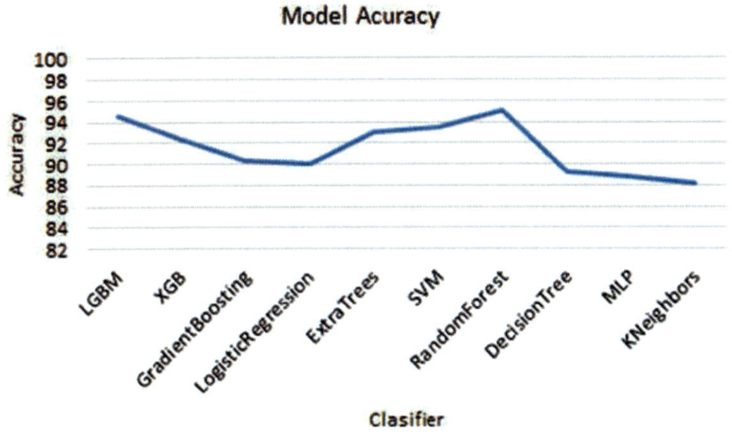

Fig. 7 Comparing the prediction accuracy of all classifiers

estimated than an actual negative instance by the classifier. As the classification efficiency is strong, the complete ROC curves for the baseline classifier seen on the PTB original dataset, the proposed model, and the diagnostic model accuracy measures are in Figs. 6 and 7.

5 Conclusions

In this research study, we propose to improve the consistency of the classification of a data integration ensemble model. We conclude that applying the functional specification for the building and configuration of the model nominated improves its accuracy. Currently, tuberculosis-artificial analytical know-how has not been acquired, and thus, a clinical diagnosis remains needed to conclude that this method has achieved diagnostic results. Our findings indicate that the model integrating all relevant data types effectively output performs models that consider one data type. These approaches may be transformed as better methods of incorporating state-of-the-art technology because several models may be used to locate details about each technology category, such that certain knowledge is not accessible. After all, only one model category is used. Enhanced design performance methods substitute selected features with high sensitivity and low accuracy in clinical knowledge. The ensemble model reaches 97% accuracy, which exceeds each classification's accuracy. The model is used to help doctors analyze and evaluate medical cases to validate the diagnosis and minimize human error. It effectively mitigates clinical diagnosis in such complex challenges as microscopic scanning and reduces the likelihood of misdiagnosis.

Acknowledgements We thank the referees for Sharda University hospital and Wolaita Sodo University Teaching Referral hospital.

References

1. Xiong Y, Ba X, Hou A, Zhang K, Chen L, Li T (2018) Automatic detection of mycobacterium tuberculosis using artificial intelligence. J Thoracic Dis 10(3):1936. https://doi.org/10.21037/jtd.2018.01.91
2. Alves E, Souza Filho JB, Kritski AL (2019) An ensemble approach for supporting the respiratory isolation of presumed tuberculosis inpatients. Neurocomputing 331:289–300. https://doi.org/10.1016/j.neucom.2018.11.074
3. Sreeramareddy CT, Qin ZZ, Satyanarayana S, Subbaraman R, Pai M (2014) Delays in diagnosis and treatment of pulmonary tuberculosis in India: a systematic review. Int J Tubercul Lung Dis 18(3):255–266. https://doi.org/10.5588/ijtld.13.0585
4. Sahli H, Mouelhi A, Diouani MF, Tlig L, Refai A, Landoulsi RB, Essafi M (2018) An advanced intelligent ELISA test for bovine tuberculosis diagnosis. Biomed Signal Process Control 46:59–66. https://doi.org/10.1016/j.bspc.2018.05.031
5. Sarin R, Vohra V, Khalid UK, Sharma PP, Chadha V, Sharada MA (2018) Prevalence of pulmonary tuberculosis among adults in selected slums of Delhi city. Indian J Tuberc 65(2):130–134. https://doi.org/10.1016/j.ijtb.2017.08.007
6. Mithra KS, Emmanuel WS (2018) GFNN: Gaussian-fuzzy-neural network for diagnosis of tuberculosis using sputum smear microscopic images. J King Saud Univ Comput Inf Sci. https://doi.org/10.1016/j.jksuci.2018.08.004
7. Dande P, Samant P (2018) Acquaintance to artificial neural networks and use of artificial intelligence as a diagnostic tool for tuberculosis: a review. Tuberculosis 108:1–9. https://doi.org/10.1016/j.tube.2017.09.006

8. Global Laboratory Initiative (2018) GLI model TB diagnostic algorithms
9. Sohn H (2016) Improving tuberculosis diagnosis in vulnerable populations: impact and cost-effectiveness of a novel, rapid molecular assays. Doctoral dissertation, McGill University Libraries
10. Asha T, Natarajan S, Murthy KB (2012) Estimating the statistical significance of classifiers used in the prediction of tuberculosis. IOSR J Comput Eng (IOSRJCE) 5(5)
11. Orjuela-Cañón AD, Mendoza JEC, García CEA, Vela EPV (2018) Tuberculosis diagnosis support analysis for precarious health information systems. Comput Methods Programs Biomed 157:11–17. https://doi.org/10.1016/j.cmpb.2018.01.009
12. Jahantigh FF, Ameri H (2017) Evaluation of TB patients characteristics based on predictive data mining approaches. J Tuberc Res 5(1):13–22. https://doi.org/10.4236/jtr.2017.51002
13. Zulvia FE, Kuo RJ, Roflin E (2017) An initial screening method for tuberculosis diseases using a multi-objective gradient evolution-based support vector machine and C5.0 decision tree. In: 2017 IEEE 41st annual computer software and applications conference (COMPSAC), vol 2. IEEE, pp 204–209. https://doi.org/10.1109/COMPSAC.2017.57
14. Wu Y, Wang H, Wu F (2017) Automatic classification of pulmonary tuberculosis and sarcoidosis based on random forest. In: 2017 10th International congress on image and signal processing, biomedical engineering and informatics (CISP-BMEI). IEEE, pp 1–5. https://doi.org/10.1109/CISP-BMEI.2017.8302280
15. Benbelkacem S, Atmani B. and Benamina, M (2013) Treatment tuberculosis retrieval using a decision tree. In 2013 international conference on control, decision and information technologies (CoDIT). IEEE, pp 283–288. https://doi.org/10.1109/CoDIT.2013.6689558
16. Alves EDS, Souza Filho JB, Galliez RM, Kritski A (2013) Specialized MLP classifiers to support the isolation of patients suspected of pulmonary tuberculosis. In: 2013 BRICS congress on computational intelligence and the 11th Brazilian congress on computational intelligence. IEEE, pp 40–45. https://doi.org/10.1109/BRICS-CCI-CBIC.2013.18
17. Schwenker F (2013) Ensemble methods: foundations and algorithms [book review]. IEEE Comput Intell Mag 8(1):77–79. https://doi.org/10.1109/MCI.2012.2228600
18. Ren Y, Zhang L, Suganthan PN (2016) Ensemble classification and regression-recent developments, applications and future directions. IEEE Comput Intell Mag 11(1):41–53. https://doi.org/10.1109/MCI.2015.2471235
19. Shah SMS, Batool S, Khan I, Ashraf MU, Abbas SH, Hussain SA (2017) Feature extraction through parallel probabilistic principal component analysis for heart disease diagnosis. Phys A 482:796–807. https://doi.org/10.1016/j.physa.2017.04.113
20. Yu L, Liu H (2003) Feature selection for high-dimensional data: a fast correlation-based filter solution. In: Proceedings of the 20th international conference on machine learning (ICML-03), pp 856–863
21. Bania RK, Halder A (2020) R-ensemble: a greedy rough set based ensemble attribute selection algorithm with *k*NN imputation for classification of medical data. Comput Methods Programs Biomed 184:105122. https://doi.org/10.1016/j.cmpb.2019.105122
22. Zhou ZH (2012) Ensemble methods: foundations and algorithms. CRC Press
23. Syafrullah M (2019) Diagnosis of smear-negative pulmonary tuberculosis using ensemble method: a preliminary research. In: 2019 6th international conference on electrical engineering, computer science and informatics (EECSI). IEEE, pp 112–116. https://doi.org/10.23919/EECSI48112.2019.8976920
24. Kim J, Chang H, Kim D, Jang DH, Park I, Kim K (2020) Machine learning for prediction of septic shock at initial triage in the emergency department. J Crit Care 55:163–170. https://doi.org/10.1016/j.jcrc.2019.09.024
25. Evora LHRA, Seixas JM, Kritski AL (2017) Neural network models for supporting drug and multidrug-resistant tuberculosis screening diagnosis. Neurocomputing 265:116–126. https://doi.org/10.1016/j.neucom.2016.08.151

Alcoholic Addiction Detection Based on EEG Signals Using a Deep Convolutional Neural Network

Chunouti Vartak and Lochan Jolly

Abstract Alcoholism is a significant issue that can lead to a variety of serious illnesses. Depression, stress, anxiety, brain, heart, and liver disorders, unhappiness, loss of excellent health, and financial troubles are just a few of the major challenges that can arise as a result of alcohol abuse. Electroencephalography (EEG) signals can be used to diagnose alcoholism in a straightforward manner. The EEG readings indicate the electrical activities of the brain. It is tough to understand them because they are random in nature, but they hold the most important information regarding the brain's state. Because of the complexity of EEG signals, accurately classifying alcoholism with only a short dataset is difficult. Artificial neural networks, particularly convolutional neural networks (CNNs), give efficient and accurate solutions in a number of pattern-based categorization applications. The EEG dataset from the University of California at Irvine Machine Learning (UCI-ML) was used. We were able to obtain 98% average accuracy using CNN on raw EEG data by improving a baseline CNN model and exceeding it in a variety of performance evaluation parameters. Dropout, batch normalisation, and kernel regularisation processes were used to improve the baseline model, as well as a comparison of the two models.

Keywords Alcoholism · EEG · CNN

1 Introduction

Alcohol is a psychoactive substance that has been widely utilised in our society for ages and has the ability to cause addiction. This problem is most common in people between the ages of 18 and 30, although it also affects people over the age of 65. It is too difficult to characterise and diagnose alcohol use disorder (AUD) at a young age

C. Vartak (✉) · L. Jolly
Department of Electronics and Telecommunication, Thakur College of Engineering and Technology, Mumbai, India
e-mail: vartak.chunouti@gmail.com

L. Jolly
e-mail: lochan.jolly@thakureducation.org

P. Chatterjee et al. (eds.), *Computational Intelligence for Engineering and Management Applications*, Lecture Notes in Electrical Engineering 984,
https://doi.org/10.1007/978-981-19-8493-8_9

[1] because the lifestyle of youngsters involves the daily use of alcohol. Alcohol's impact on the body starts from the moment they take their first sip. Difficulty in walking, blurred vision, slurred speech, slow down reflexes, and impaired memory are some of the signs that can be seen in an alcoholic person. It strongly affects the human brain. Some of the symptoms are detectable just after your first or second drink, but it can resolve quickly when people stop drinking. Besides this, if they continue to drink heavily, they may have cognitive problems that last long after they have achieved sobriety. It is difficult to focus on daily life pressures when those who drink alcohol turn to their passion. Daily drinking can develop a person's habit into an addiction, causing them to reach the stage of alcoholism [2]. The majority of them have no idea when they crossed the line from habit to addiction, or from addiction to alcoholism. The excess use of alcohol is a root of disease, which increases the death ratio, it affects socially and economically to our community. Most alcoholics are unwilling to accept aid from others to get out of their addiction; this puts their lives at risk. Alcohol-related deaths account for 5.9% of all deaths each year, with 3.3 million people dying every year [3].

Under the covering cortex of the human brain, millions of neurons are at work, controlling cognition, memory, action, and behaviour directly or indirectly. When a synaptic connection between neurons is active, it produces an electrical signal as well as a magnetic field. The mind of a child or adolescent is too immature to distinguish between different types of dependency growth, whether it is about food, a habit, or an activity. Technology has evolved in biomedical science over time. For reading such activities, a number of new and improved tools are now available. One of the tools is electroencephalography (EEG), which is commonly utilised in various lab setups. This type of recording is beneficial for analysing and detecting functional or structural changes in the brain that occur as a result of any addiction or condition. Traditional psychological methods of addiction diagnosis were the only ones accessible in the past, but they required a lot of skill to diagnose and assess. Brain cells work together to transfer information in the form of a signal from one brain region to another, and they are in charge of both regular physiology and behaviour inhibition. If any change due to neurodevelopmental, neuromechanical, or addiction cause, the behavioural or physiological change occurs. Traditional psychological approaches employed by psychologists are difficult to detect neurological illnesses, especially behavioural disorders, but it is critical to detect them early in their development to minimise the global burden of disease (GBD) [4]. To a large extent, neuroimaging aids in the diagnosis of activity or substance addiction. Relapse, loss of control, anxiety, depression, and mood modification are some of the frequent signs that distinguish psychological addiction. Addiction to substances and behaviours is more common. The use of alcohol is included in the concept of substance abuse. Moderate and high consumption can be distinguished, with moderate consumption falling under societal or cultural norms, whereas excessive consumption might lead to binge drinking or disorder. Data shows that alcohol consumption varies by country, with lower levels in North Africa and the Middle East but high levels in Europe. In Eastern Europe, there was a lot of consumption. Belarus, Russia, the Czech Republic, and Lithuania are examples of such countries [5]. A human's cognitive state can become apparent

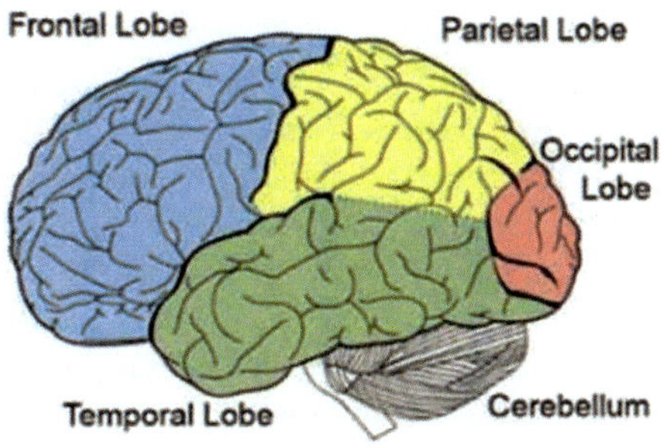

Fig. 1 Human brain structure

by the subjective experiences such as feeling and mood. Hence, measuring cognitive congruency in human would be beneficial to analyse human behaviour for certain activities based on person's mood.

Over the years, neuroscience has made significant progress in understanding the specific brain structure of humans using electrophysiological methods such as EEG, fMRI, and other techniques. Any conscious or unconscious brain activity can be described in terms of generated electrical activities, and it should be underlined. EEG signals can help provide a better understanding of the participant's cognitive state at the time of observations because electrophysiological methods like EEG are highly temporal in nature. The voltage fluctuations induced by ionic current flows within brain neurons are measured by EEG signals. This electrical activity is then utilised to figure out 'how people think'. EEG signals are routinely monitored at several parts of the scalp and range from 10 to 100 μV. As seen in Fig. 1, the human brain is divided into four main parts as follows: frontal, parietal, temporal, and occipital [5].

Each part is related with information processing in a distinct way. The frontal lobe is usually associated with working memory and decision-making during visual information processing. The parietal lobe is responsible for action-related activities, whilst the temporal lobe is responsible for object recognition. The occipital lobe is responsible for processing visual information related to attention. The slow and high-frequency bands of the EEG waveform, i.e., gamma (> 30 Hz), beta (12–30 Hz), alpha (8–12 Hz), theta (4–8 Hz), and delta (4 Hz), can be used to assess a person's cognitive state. In comparison with high-frequency bands, slow waves have a higher amplitude and lower frequency content. Signals observed at scalp sites are dominant in certain frequency bands only in certain cognitive states of mind. As a result, these frequency ranges are frequently utilised to study human brain reactions and cognitive activities. Slow waves, such as theta and delta, are commonly connected with the subconscious mind, but alpha waves, which are more prominent in the occipital and parietal lobes, signify a calm state of mind. At the frontal and other parts of the brain, high-frequency waves such as beta and gamma waves represent an active state of mind. These frequency waves at specific sites on the scalp demonstrate significant cognitive behaviour. To analyse alcohol addicts, we attempted to extract cognitive traits based on certain EEG bands and scalp locations in this work.

The analysis and classification of EEG data for the detection of alcoholism in this study are of particular interest to us. Based on a collection of EEG signals, we should be able to determine if an EEG signal belongs to a healthy person or someone who has an alcoholic inclination.

This research report is divided into six sections. Section 2 described all of the work that was done in conjunction with it. In Sect. 3, the methodology is discussed. We described our recommended method in Sect. 4 of this paper. We wrote experimental results in Sect. 5, and then, we wrapped up our report in Sect. 6, which included a quick summary of our study.

2 Related Work

There are two types of methods for EEG-based classification: deep learning methods and feature extraction using traditional machine learning classifiers. Time domain, frequency domain, time–frequency analyses, wavelet analyses, entropy analyses, and energy distribution studies have all been proposed as traditional feature extraction techniques for EEG data [6], or a mix of two or more of these approaches [7]. The downside of feature extraction is that it is not only computationally expensive but also difficult and time-consuming. Furthermore, because human data processing is very subjective, other researchers are unlikely to reproduce the findings. Despite these shortcomings, manual artefact removal was utilised in more than a quarter of the research analysed by [8]. Machine learning has recently been used to enhance traditional signal classification methods [9–11]. Jiajie et al. [12] employed approximate entropy (AE) and sample entropy (SE) as feature extractors, as well as SVM and *K*NN as classification approaches, to develop a clinical decision support system for alcoholism categorization. Siuly et al. [13] provides an introduction of numerous feature extraction algorithms that use various classifiers, as well as a comparison of their performance on the UCI-ML dataset. Pre-processing techniques such as independent component analysis for artefact removal, principal component analysis, and local Fisher's discriminant analysis are employed before any classification algorithms are applied to the EEG data [14]. Ren and Han [15] employed class separability methods to eliminate redundant features from an EEG signal by combining linear approaches with a nonlinear feature extraction method [16]. In the instance-based learning method *K*NN, ICA beat PCA when used with a deep learning (bidirectional long short-term memory) model. As a result, whilst choosing a feature selection technique, the type of classification method to be employed must be carefully evaluated. Fourier feature maps [17] and 3D grids [18] are examples of feature-based techniques that employ visual elements of the signals. Saminu et al. [6] presents a description of EEG signal categorization methodologies that combine traditional feature extraction techniques with machine learning classifiers. DNNs are a useful tool for classifying complex nonlinear systems. CNNs have been shown to be the best deep learning architectures in situations like EEG signal processing [19]. We show how CNNs have been used to classify EEG signals in a few different ways. Chaabene et al.

used CNN to identify tiredness [20]. They chose 14 channels for EEG signals then pre-processed the signals for noise removal and band annotations in the improved signals. To categorise a person as drowsy or awake, they employed four convolution layers, one max-pooling layer, and two fully linked layers to form a network with about 14 million parameters. Without network optimization, they had the highest test accuracy of 79%. For EEG data classification of alcoholism [21] used a multichannel pyramidal convolutional neural network (MPCNN). They started by assessing the performance of each channel separately, starting with 61 channels from five different brain locations. Out of five created models tested with various topologies and many parameters, the best model uses 19 best-performing channels as input and gives an accuracy of 100% with 14,066 parameters. [22] took a somewhat different strategy, developing two novel activation functions to increase CNN's performance in EEG categorization. On the alcohol EEG dataset, they were able to attain an accuracy of 92.3% using one of the activation functions, which is an improvement above either the softmax or sigmoid default activation functions. Xu et al. [23] used the VGG-16 model for the classification of motor imagery (MI) EEG signals, which was initially created for the general image classification task. Except for the final output layer, which is fine-tuned in the target model using the EEG dataset, the newer model has the same initial layers as VGG-16. Before being used as input into the target model, the EEG signals are transformed into time–frequency spectral pictures using the short-time Fourier transform (STFT). On these photos, 2D CNN is used to perform categorization. For all subjects, the average stated accuracy was 74.2%. This is a 2.8% improvement over the CNN they designed. With the translation of EEG signals into 2D pictures, Srabonee et al. [24] obtained 98.13% classification accuracy. In addition to image processing, the photographs were subjected to Pearson's correlation analysis before being used as input into the CNN model.

From the above discussion, we may deduce that there are a range of ways for classifying EEG data, and the race to identify the best method is still on. The best strategy, in our opinion, should be chosen based on flexibility and efficiency. Unfortunately, many of the systems presented only provide a few evaluation criteria, such as accuracy, whilst ignoring other metrics, such as precision and recall, which may be less important, resulting in a diluted total performance.

3 Methodology

Our model was written in Python and used the Keras deep learning API with TensorFlow as the backend. Begleiter [25] created the EEG signal dataset, which was received from the University of California at Irvine Machine Learning repository [26]. There are 122 participants in all, each with 120 trials with two separate stimuli. The participants were split into two groups: alcoholics and non-drinkers. Each subject was given either a single stimulus or two stimuli in either a matched or nonmatched condition [27], with $S1$ being identical to $S2$ and $S2$ being different. The EEG data was captured using the International 10/20 system [12], which involved the insertion of 64 electrodes on the skull. The 10/20 system elements represent the frontal polar (FP), frontal (F), temporal (T), central (C), parietal (P), ground (G), and occipital (O) areas. Electrode results are quite sensitive to noise. As a result, each electrode result was amplified before being sent through a filter having a pass-band of (0.02, 50 Hz). This band-pass filter limits the signal bandwidth whilst also preventing low-frequency baseline wander noise. The data was then sampled at a frequency of 256 Hz with a 12-bit resolution analogue-to-digital converter (ADC). With the exception of basic normalisation, we did not do any specific pre-processing operations on the raw input data, which is a significant component of the current work.

3.1 Feature Selection Using CNN

CNNs are a type of artificial neural network that is used to analyse data that is typically in a series in one dimension, such as speech or EEG/ECG signals, or in two dimensions, such as images. Convolution is a specific type of linear operation on two functions of a real-valued parameter [28], and it is the main operation in CNN. The input is the initial function of the CNN operation, followed by the kernel and the output, which is known as the feature map. CNN requires fewer parameters owing to sparse connectivity because the kernel is smaller than the input, which not only decreases the model's memory requirements but also increases its statistical efficiency [28]. At distinct input locations, one sort of information can be extracted by a kernel of the convolution function. To extract several types of information, we use numerous convolution functions in a single CNN model. The convolution function is usually followed by a pooling function, which down samples the output of the convolution layer [28]. A single convolution block is frequently used to represent both convolution and pooling processes. Fully connected (dense layers) follow the convolution block. For better differentiation, multiple of these layers may be required. Each layer may have a different activation function depending on the situation. For binary classification, the sigmoid/logistic regression activation function is used. The hyperparameters can be tweaked manually or automatically, and the range of their values can have an impact on the algorithm's time and running costs. Once a model has been trained from the training set, based on the validation set, the hyperparameter values are changed. The final set of hyperparameters is set on the test set after observing the generalisation error.

3.2 Performance Metrics and Evaluation

The optimum metric for binary classification is accuracy. All of the optimizations were mostly carried out with accuracy as the primary metric. We also present the cross-validation results for the final model. Precision, recall, $F1$-score, Cohen's kappa, and area under the curve (AUC) were also determined.

4 CNN Architecture for EEG Classification

The research uses CNN to determine if an EEG signal belongs to an alcoholic or a healthy person. 1D convolutional layers and a fully connected layer are the model's major components. The chosen model included four convolution layers, the first of which contained 16 filters, the second 32, and the final layer 64. In all levels, the kernel size was set to 15. Every layer's convolution stride was also set to two steps. For each convolution layer, the (ReLU) activation function was utilised to provide nonlinearity to the process. Any drop in the value of any of these factors resulted in a decrease in performance, whilst increasing the value had no effect.

Feature selection is handled by the convolutional/pooling layers. A single fully connected layer was used to categorise the EEG signals, with the sigmoid activation function for binary classification and the binary cross-entropy approach for loss minimization. The architecture of the model is shown in Fig. 2. After each epoch, a tiny validation set is utilised to change the weights as the model learns. We set the validation set to be 20% of the total dataset size, and the model chooses validation instances for each iteration automatically. The model takes input examples in the same dimension as the training data during testing. The developed model's loss and accuracy are shown in Fig. 3a, b.

As a result, to get the best results on the validation set, we had to fine-tune the CNN model. Once we have achieved the appropriate validation accuracy, we can run the model on the test data to evaluate its final performance. We used a previously created model as a baseline and improved it in order to obtain. On the validation set, there is a high level of accuracy. A technique known as regularisation was used to prevent from overfitting on training data. Dropout, batch normalisation, and $L1/L2$ regularisation were utilised as regularisation approaches.

5 Result

There were only 100 epochs in all for the training. For a single run, Table 1 compares the results of the two models on the test set: the unregularised model vs the regularised model.

To objectively examine the performance of the final, regularised model, we employed K-fold cross-validation [28] with $K = 3, 5, 10$. When the supplied dataset is not particularly large, the preferred method for model evaluation is K-fold cross-validation. The results of K-fold cross-validation and batch size variation are shown in Table 2.

Table 2 demonstrates that the accuracy improves as the number of training samples increases from $K = 3$ to $K = 10$. This is in line with the general principle of machine learning, which asserts that the amount of samples in a model enhances its accuracy [28].

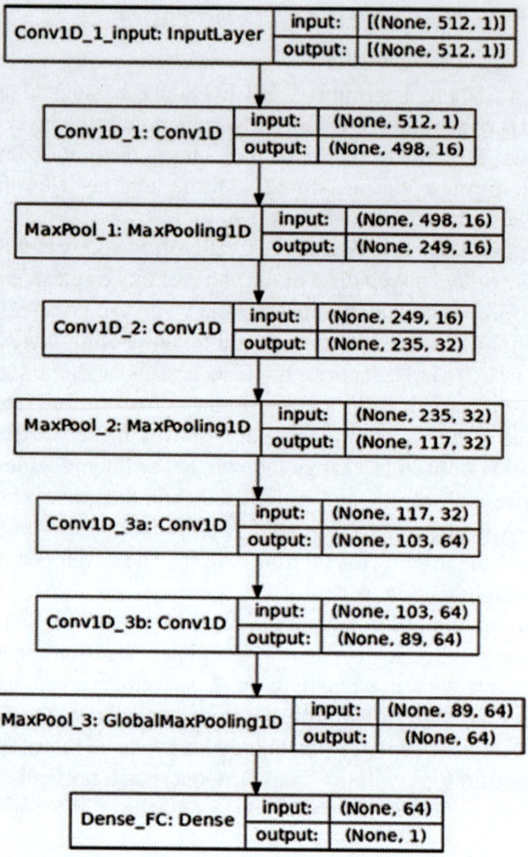

Fig. 2 CNN model architecture

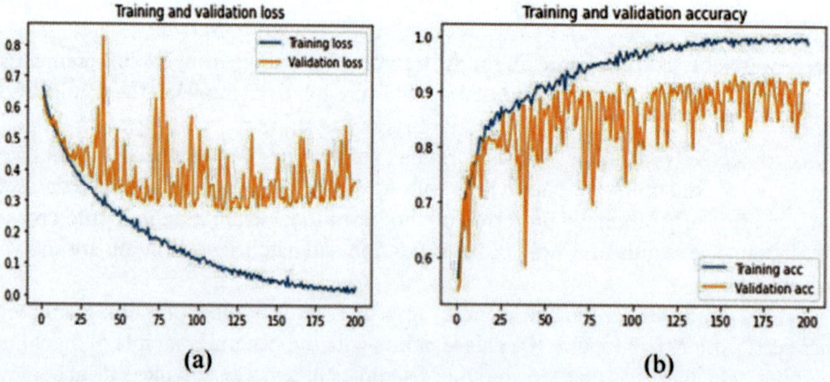

Fig. 3 Training and validation loss and training and validation accuracy of the baseline CNN model

Table 1 Comparing the performance of the baseline and regularized CNN model using various metrics

CNN model	Accuracy (%)	Precision (%)	Recall (%)	$F1$-score (%)	AUC (%)	Kappa (%)
Baseline	91.15	92.22	89.24	90.71	91.08	82.25
Regularized	98.43	100	96.77	98.36	98.38	96.87

Table 2 Result of K-fold cross-validation

	3-fold		5-fold		10-fold		
	Validation	Test	Validation	Test	Validation	Test	Best run
Samples	256	192	153	192	76	192	
Batch size	$\mu(\sigma)$		$\mu(\sigma)$		$\mu(\sigma)$		
4	0.92 (0.01)	0.96	0.92 (0.01)	0.93	0.95 (0.01)	0.97	0.97
8	0.93 (0.01)	0.94	0.94 (0.02)	0.95	0.94 (0.02)	0.96	0.97
16	0.92 (0.01)	0.94	0.94 (0.01)	0.96	0.94 (0.01)	0.98	0.96
32	0.93 (0.02)	0.95	0.93 (0.03)	0.95	0.94 (0.03)	0.96	0.97
64	0.92 (0.01)	0.95	0.94 (0.01)	0.95	0.94 (0.03)	0.95	1.0
128	0.92 (0.01)	0.90	0.92 (0.02)	0.95	0.95 (0.03)	0.95	0.99
256	0.88 (0.02)	0.89	0.90 (0.03)	0.90	0.90 (0.04)	0.89	0.97

6 Conclusion

Alcoholism detection is a significant social issue. The electroencephalogram (EEG) is a useful technique for detecting alcoholism. This research describes how CNN may be used to classify EEG signals in order to diagnose alcoholism. By detailing alternative regularisation procedures, other academics and practitioners can build on current knowledge to construct more efficient and better-performing CNN designs. The techniques discussed in the article are not limited to EEG classification; they can be used in a variety of CNN applications. We got up to 98% accuracy with the present approaches on the UCI-ML dataset, with good precision, recall, AUC, and other metrics. Data segmentation, input perturbation, and weight initialisation decisions may require more research, which has been listed as a priority for the near future.

References

1. Alcohol use disorder I psychology today. [Online]. Available: https://www.psychologytoday.com/conditions/alcohol-use-disorder. Accessed: 13 Sept 2017
2. Alcohol abuse I SASC. [Online]. Available: http://www.sasc-dbq.org/alcohol-abuse. Accessed: 14 Sept 2017
3. Schuckit MA (2014) Recognition & management of withdrawal delirium (delirium tremens). N Engl J Med 371(22):2109–2113
4. Feigin VL Abajobir AA, Abate KH (2017) Global, regional, and national burden of neurological disorders during 1990–2015: a systematic analysis for the global burden of disease study 2015. Lancet Neurol 16:877–897. https://doi.org/10.1016/Sl474-4422(17)30299-5
5. Boden MA (2008) Mind as machine: a history of cognitive science. Oxford University Press
6. Saminu S, Xu G, Shuai Z, Abd El Kader I, Jabire AH, Ahmed YK, Karaye IA, Ahmad IS (2021) A recent investigation on detection and classification of epileptic seizure techniques using EEG signal. Brain Sci 11:668
7. Orosco L, Correa AG, Laciar E (2013) A survey of performance and techniques for automatic epilepsy detection. J Med Biol Eng 33:526–537
8. Craik A, He Y, Contreras-Vidal JL (2019) Deep learning for electroencephalogram (EEG) classification tasks: a review. J Neural Eng 16:031001
9. Anuragi A, Sisodia DS (2019) Alcohol use disorder detection using EEG signal features and flexible analytical wavelet transform. Biomed Signal Process Control 52:384–393
10. Zhu G, Li Y, Wen PP, Wang S (2014) Analysis of alcoholic EEG signals based on horizontal visibility graph entropy. Brain Inform 1:19–25
11. Shri TP, Sriraam N (2017) Pattern recognition of spectral entropy features for detection of alcoholic and control visual ERP's in multichannel EEGs. Brain Inform 4:147–158
12. Acharya JN, Hani AJ, Cheek J, Thirumala P, Tsuchida TN (2016) American clinical neurophysiology society guideline 2: guidelines for standard electrode position nomenclature. Neurodiagn J 56:245–252
13. Siuly S, Bajaj V, Sengur A, Zhang Y (2019) An advanced analysis system for identifying alcoholic brain state through EEG signals. Int J Autom Comput 16:737–747
14. Velu P, de Sa VR (2013) Single-trial classification of gait and point movement preparation from human EEG. Front Neurosci 7:84. Arshad J, Townend P, Xu J (2011) A novel intrusion severity analysis approach for clouds. Future Gen Comput Syst. https://doi.org/10.1016/j.future.2011.08.009
15. Ren W, Han M (2019) Classification of EEG signals using hybrid feature extraction and ensemble extreme learning machine. Neural Process Lett 50:1281–1301
16. Rahman S, Sharma T, Mahmud M (2020) Improving alcoholism diagnosis: comparing instance-based classifiers against neural networks for classifying EEG signal. In: Proceedings of the international conference on brain informatics, Padova, Italy, 18–20 Sept 2020. Springer, Berlin, pp 239–250
17. Abbas W, Khan NA (2018) DeepMI: deep learning for multiclass motor imagery classification. In: Proceedings of the 2018 40th annual international conference of the IEEE engineering in medicine and biology society (EMBC), Honolulu, HI, USA, 17–21 July 2018. IEEE, Piscataway, pp 219–222
18. Wei X, Zhou L, Chen Z, Zhang L, Zhou Y (2018) Automatic seizure detection using three-dimensional CNN based on multi-channel EEG. BMC Med Inform Decis Mak 18:71–80
19. Roy Y, Banville H, Albuquerque I, Gramfort A, Falk TH, Faubert J (2019) Deep learning-based electroencephalography analysis: a systematic review. J Neural Eng 16:051001. Arshad J, Townend P, Xu J (2011) A novel intrusion severity analysis approach for clouds. Future Gen Comput Syst. https://doi.org/10.1016/j.future.2011.08.009
20. Chaabene S, Bouaziz B, Boudaya A, Hökelmann A, Ammar A, Chaari L (2021) Convolutional neural network for drowsiness detection using EEG signals. Sensors 21:1734
21. Qazi E-U-H, Hussain M, AboAlsamh HA (2021) Electroencephalogram (EEG) brain signals to detect alcoholism based on deep learning. CMC Comput Mater Contin 67:3329–3348

22. Bhuvaneshwari M, Kanaga EGM (2021) Convolutional neural network for addiction detection using improved activation function. In: Proceedings of the 2021 5th international conference on computing methodologies and communication (ICCMC), Tamil Nadu, India, 8–10 Apr 2021, pp 996–1000. Loganayagi B, Sujatha S (2012) Enhanced cloud security by combining virtualization and policy monitoring techniques. Proc Eng 30:654–661
23. Xu G, Shen X, Chen S, Zong Y, Zhang C, Yue H, Liu M, Chen F, Che W (2019) A deep transfer convolutional neural network framework for EEG signal classification. IEEE Access 7:112767–112776. Arshad J, Townend P, Xu J (2011) A novel intrusion severity analysis approach for clouds, Future Gen Comput Syst. https://doi.org/10.1016/j.future.2011.08.009
24. Srabonee JF, Peya ZJ, Akhand M, Siddique N (2020) Alcoholism detection from 2D transformed EEG signal. In: Proceedings of the international joint conference on advances in computational intelligence, Dhaka, Bangladesh, 20–21 Nov 2020. Springer, Singapore, pp 297–308
25. Begleiter H (2021) Multiple electrode time series EEG recordings of control and alcoholic subjects. Available online: https://kdd.ics.uci.edu/databases/eeg/. Accessed on 4 Aug 2021
26. Begleiter H (1999) EEG database data set. In: Ingber L (ed) UCI machine learning repository. University of California at Irvine, Irvine. Available online: https://archive.ics.uci.edu/ml/datasets/EEG+Database. Accessed on 4 Aug 2021
27. Zhang XL, Begleiter H, Porjesz B, Wang W, Litke A (1995) Event related potentials during object recognition tasks. Brain Res Bull 38:531–538. Wei X, Zhou L, Chen Z, Zhang L, Zhou Y (2018) Automatic seizure detection using three-dimensional CNN based on multi-channel EEG. BMC Med Inform Decis Mak 18:71–80
28. Goodfellow I, Bengio Y, Courville A (2016) Deep learning, vol 1. MIT Press, Cambridge. Abbas W, Khan NA (2018) DeepMI: deep learning for multiclass motor imagery classification. In: Proceedings of the 2018 40th annual international conference of the IEEE engineering in medicine and biology society (EMBC), Honolulu, HI, USA, 17–21 July 2018. IEEE, Piscataway, pp 219–222

Application of Machine Learning Algorithms for Cataract Prediction

Soumyadeep Senapati, Kanika Prasad, Rishi Dwivedi, Ashok Kumar Jha, and Jogendra Jangre

Abstract The usage of revolutionary technology like artificial intelligence (AI) in the healthcare system has improved service management and medical decision-making. Nowadays, in healthcare, AI is utilized to develop an automated system which is used for early detection and regular diagnostic applications. In medium-income countries like India, there is a backlog of over several millions blind people, and cataract is one of the primary reasons for blindness in majority of them. Various factors, such as higher eye surgery frequency, early intervention, and population aging, have led to an increase in the number of cataract surgeries. Many people suffering from cataract in India embodies an exclusive set of difficulties and contribute toward opportunity with respect to available data. Machine learning (ML) algorithms can develop models to analyze data quickly and show results without utilizing much time. Therefore, in this work, an AI-based model is developed which can predict whether the patient has cataract or not. This model is based on different measures like accuracy, precision, recall, $F1$-measure on significant ML algorithms. In this work, Jupyter Notebook is used for data prepossessing, numerical simulation, statistical modeling. Finally, the results obtained using logistic regression (LR), Naïve Bayes (NB), and K-nearest neighbor (KNN) algorithms are compared based on different measures. The comparative result obtained shows that the developed model can predict cataract from operation records using KNN algorithm effectively.

Keywords Healthcare · Cataract · Artificial intelligence (AI) · Machine learning (ML) · Algorithms

S. Senapati · K. Prasad (✉) · A. K. Jha · J. Jangre
Department of Production and Industrial Engineering, National Institute of Technology, Jamshedpur 831014, India
e-mail: kprasad.prod@nitjsr.ac.in

R. Dwivedi
Department of Finance, Xavier Institute of Social Service, Ranchi 834001, India

P. Chatterjee et al. (eds.), *Computational Intelligence for Engineering and Management Applications*, Lecture Notes in Electrical Engineering 984,
https://doi.org/10.1007/978-981-19-8493-8_10

1 Introduction

Cataract is a long-lasting disease which cannot be cured. It is the key cause of blindness globally which poses a major challenge to the healthcare sector. In cataract, a cloudy area is developed in the lens of the eye which is caused as a result of breakdown of proteins in eye lens. This can lead to decrease in vision, trouble in driving, reading, or recognizing faces, increased risk of falling and depression, and can even cause blindness if not treated well in time. A variety of factors such as diabetes, trauma, skin disease, radiation, genetics, use of corticosteroid medication, smoking tobacco, prolonged exposure to sunlight, and alcohol are responsible for cause of cataract in a person. According to statistics, several million people worldwide turn blind due to cataracts every year, and this figure is expected to rise exponentially in the coming future. Although there is no long-term cure for cataract, it can be treated if diagnosed early enough, and complications associated with the condition can be avoided. There is a confirmation from recent research and doctors that recovery from the surgery will be faster if the disease is diagnosed early. During recent times, machine learning (ML) algorithms have proved to be suitable for initial detection of various diseases and analyze them with technological advancements [1]. ML is an effective technique for analyzing the data which is a subdivision of artificial intelligence (AI). It requires data to build a model. The trends and pattern are actually extracted by ML from the dataset, and anticipated result is accomplished. ML is divided into four groups which are supervised learning, semi-supervised learning, unsupervised learning, and reinforcement learning. In the current work, supervised learning algorithms are used to build the model that can be used for cataract prediction. Algorithms like logistic regression (LR), Naïve Bayes (NB), and *K*-nearest neighbor (*K*NN) have been applied to make the prediction if a patient has cataract. The organization of this research work as follows: Some of the previous related works are presented in section two. Different methodologies and algorithms used to develop the model are presented in section three. The fourth section outlines the various evaluation measures that were utilized to assess the developed model. The experimental approach, results obtained, and analysis of the output are shown in section five. Conclusions of the work are mentioned at the end.

2 Literature Review

The past researchers already applied different ML algorithms in healthcare sector and other service industries in order to improve the services provided by them.

Saha et al. discovered that using a random forest (RF) model with certain hyperparameter values is the most effective technique to categorize high-risk patients who are likely to be readmitted to the hospital [2]. Sunarti et al. suggested utilizing AI in healthcare to improve patient diagnosis, prevention, and treatment while also enhancing cost efficiency and equality in health services [3]. The role of AI in improving customer experience was investigated by Daqar and Smoudy. The study's findings demonstrated that AI and customer experience had a good and significant association [4]. Dogan and Birant applied ML and data mining approach in manufacturing to improve the processes with recent advances in manufacturing by grouping them under four main subjects such as scheduling, monitoring, quality, and failure [5]. Paturi and Cheruku reviewed application and performance of ML techniques in manufacturing sector from the past two decades [6]. Haleem et al. proposed applications of AI for cardiology during the recent pandemic [7]. Rawat et al. proposed application of ML and data visualization techniques for decision support in the insurance sector that can help insurance companies in deducing patterns in various segments [8]. Tognetto et al. presented a review on application of AI for cataract management. The proposed AI-driven diagnosis seemed to be comparable and even exceed the accuracy of experienced clinicians in classifying disease [9]. Cecula et al. proposed applications of AI to improve patient flow on mental health inpatient units [10]. Haleem et al. conducted a study on application of AI in medical field. The study presented five significant technologies and ten primary applications of AI in the medical field [11]. Jin Long et al. focused-on development of an AI management system to improve product quality and production efficiency in furniture manufacturing [12]. Mojjada et al. proposed ML models for COVID-19 future forecasting. The results obtained indicated that the LR model is an effective tool in predicting new cases of corona, death numbers, and recovery among other ML algorithms [13]. Dharani et al. proposed evaluation of performance of an LR and support vector regression (SVR) models to predict COVID-19 pandemic [14]. Gothai et al. predicted growth of COVID-19 and trend using ML approach. The experimental setup with the LR, SVR, and time series algorithms showed that time series Holt's model outperforms LR and SVR algorithms [15]. Carrasco et al. proposed a framework based on ML to detect relationships between banking operation records, taking into account the large volume of data involved in the process with the help of precision, recall, and $F1$-measure [16]. Pratap and Kokil developed a computer-aided automatic cataract detection approach based on fundus pictures gathered from different open access datasets to detect various stages of the cataract, such as normal, mild, moderate, and severe [17]. Lin et al. developed a model that can predict cataracts. RF and adaptive boosting approaches were used to create cataract identification models based on birth circumstances, family medical history, and family environmental factors [18]. Caixinha et al. applied ultrasound technique using ML to detect the presence of cataract. NB, *K*NN, SVM, and fisher linear discriminant classifiers were tested. The experimental results obtained demonstrated that the SVM shows the highest performance (90.62%) for initial versus severe cataract classification [19].

Although a considerable amount of work has been done in past related to the application of AI in healthcare, there is very limited work related to cataract prediction using ML algorithms. Additionally, the comparison of results obtained based on these methods is scarce. Moreover, the standard data required for cataract prediction through ML algorithms is not publicly available for research.

Therefore, the main objective of this work is to predict cataract using different ML algorithms. The dataset with necessary details for conducting the study is obtained from private eye hospital that would be used for determining the probability of need for a cataract patient to undergo surgery or not. Jupyter Notebook would be used to predict the accuracy of output obtained using LR, NB, and *K*NN algorithms and identify the most efficient algorithm to get the most precise output.

3 Research Methodology

The several steps followed in the research study are presented in Fig. 1. In the first step, dataset from a local private eyecare center for last 6 months (June–November 2021) is collected. Then, in the second step, the dataset gathered is preprocessed to create the prediction model. The training dataset is then subjected to several machine learning algorithms. Finally, using a test dataset, the performance of the approaches is tested in order to establish the best classifier for cataract prediction. The details of each stage are mentioned below.

3.1 Dataset

The model's main goal is to determine whether or not the patient has cataracts. Python and different machine learning tools such as Jupyter Notebook, Scikitlearn, and pandas are used for data preprocessing. The 30% of dataset is used for testing

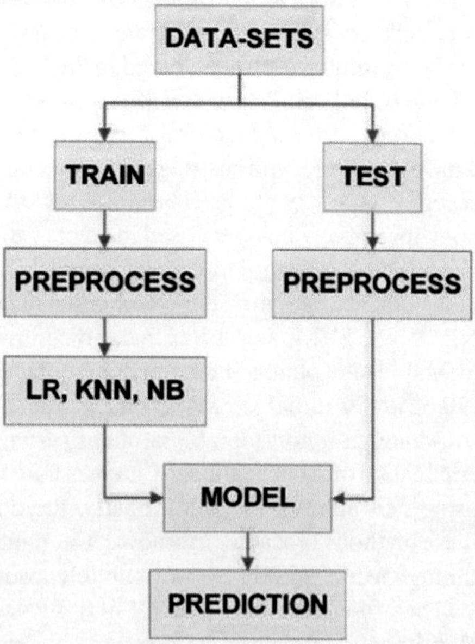

Fig. 1 Framework of the proposed research methodology

purpose, whereas 70% is used for training purpose. In the current study, patient data is utilized for the analysis purpose. The dataset contains 1000 cases, 5 attributes, and class (0 or 1). Out of 1000, 744 are grouped into class (1) having cataract, and 256 are classified as class (0) who do not have cataract. The attributes considered for data collection are patient's age, blood pressure, heart rate, vision, and intra-ocular pressure.

3.2 Preprocessing of Data

The dataset should be preprocessed to increase forecast accuracy. Data normalization is an effective way to enhance the precision of ML algorithms. In this study, the data is normalized using the min–max scalar (MMS) technique. The factors are rescaled using MMS. Using the data's maximum and minimum values, the variables in the dataset are rescaled to be between [− 1 and 1] or [0–1]. Equation (1) is used to follow MMS normalization.

$$M_{\text{new}} = \frac{M_i - \min(M)}{\max(M) - \min} \quad (1)$$

3.3 Application of ML Algorithms

After putting together, the data for modeling and three frequently applied ML algorithms are employed for cataract prediction. The three ML algorithms employed in this study are summarized in the next section.

3.3.1 *K*-Nearest Neighbor (*K*NN)

The *K*NN algorithm is a supervised learning technique that is commonly used in pattern categorization. Because of its simplicity, the *K*NN principle has become a widely used method in a variety of applications. The approach, for example, uses the feature vectors and the specified distance to identify the *K*-nearest neighbors in the factor space to categorize a sample *Di*. Following that, the system counts the votes of these neighbors based on their labels. The object sample will be assigned to the group with the most neighbors who have the same label. The *K*NN classifier requires the use of a variety of distance metrics, the best of which is the equivalent distance, as illustrated in Eq. (2).

$$Z(m, p) = \sqrt{\sum_{i=1}^{n} (m_i - p)} \tag{2}$$

where Z represents the Euclidean distance, m represents data from the dataset, p represents new data to be predicted, and n represents the number of dimensions.

3.3.2 Naive Bayes (NB)

NB is a supervised learning technique based on the use of Bayes' theorem. It is a straightforward, powerful, and widely used machine learning technique that assumes all features are independent. According to NB, the classification of one characteristic within a class has no bearing on the classification of another. It is a supervised learning algorithm that is effective and is based on conditional probability. Equation (3) shows how the classification demonstrates this likelihood.

$$p\left(\frac{m}{n}\right) = \frac{p\left(\frac{n}{m}\right)p}{p(n)} \tag{3}$$

where $p(m)$ represents the prior probability of class, $p(n/m)$ represents the likelihood probability of (m) conditional on Ni, $p(m/n)$ represents the posterior probability of class, and $P(n)$ represents the prior probability of (n).

3.3.3 Logistic Regression (LR)

The linear classification model is known as LR. It depicts the likelihood of a group creating a z function vector. A logistic algorithm is used to establish a relationship between the class and the feature. It is based on the $z(M/N)$ distribution, where N is the feature vector and M is the class, and it is on a boundary shape, as illustrated in the training data. The sigmoid function in class M defines the likelihood $z(M/N)$ of N, which is explained in Eqs. (4) and (5).

$$d(M, N) = \sum_{i=1}^{n} w_i f_i(M, N) \tag{4}$$

$$z(M/N) = \frac{1}{1 + \exp(-d(M, N))} \tag{5}$$

where z denotes the probability (M/N), w is the word's weight, i denotes picked at random, f denotes the frequency, M denotes the class, N denotes the feature vector, and exp defines the exception.

4 Metric for Evaluation

All of the evaluation metrics utilized in this cataract prediction model are listed in this section.

4.1 Precision

Precision is defined as the number of true positives divided by the number of false positives that are truly in the positive category. Equation (6) is used for expressing precision.

$$\text{Precision} = \frac{\text{TP}}{\text{TP} + \text{FP}} \tag{6}$$

where FP: false positives and TP: true positives.

4.2 Recall

Recall measures the capacity to determine the number of positive class predictions made of all positive cases in the collection, as presented in Eq. (7).

$$\text{Recall} = \frac{\text{TP}}{\text{TP} + \text{FN}} \tag{7}$$

where TP: true positives and FN: false negatives

4.3 F1-Measure

$F1$-measure: The harmonic mean of the model's precision and recall is also known as the F score or F measure. It is very useful when it comes to false negatives and positives. $F1$ is calculated using Eq. (8).

$$F\text{-measure} = \frac{2 + \text{Recall} + \text{Precision}}{\text{Recall} + \text{Precision}} \tag{8}$$

5 Results and Discussion

After separating the dataset into test and train groups, the findings are presented and examined in this section. In order to get overall prediction, the dataset is divided into 70% for train and 30% for test. Table 1 shows the output obtained from Jupyter Notebook. As depicted from Table 1, considering the precision aspect *K*NN model emerges out as best (in which precision decreases with increasing *K*), followed by NB and then LR model. It can also be interpreted from Table 1 that for *F*1-measure, all the models are giving almost similar performance with given hyperparameter settings. *K*NN and NB model have only slightly higher *F*1-score than LR model. Within the *K*NN model, $K = 3$, 5, and 10 are best. However, this ranking of the models might vary if precision metric is used for sorting; as mostly in disease prediction cases, precision is slightly more important than recall [16, 18]. Table 2 summarizes the results obtained from different models.

Figures 2, 3 and 4 display the precision, recall, and *F*1-measure trend for all the models employed in this work in the form of bar graphs. The behavior of each classification is shown using the various measures.

Table 1 Results for cataract dataset on evaluation metrics

Mode type	Model parameter	Precision	Recall	*F*1	Precision ranking	Recall ranking	*F*1 ranking
*K*NN	$K = 3$	0.995708	0.775920	0.872180	3	3	1
*K*NN	$K = 5$	0.995708	0.775920	0.872180	4	4	2
*K*NN	$K = 10$	0.995708	0.775920	0.872180	5	5	3
*K*NN	$K = 1$	1.000000	0.773333	0.872180	1	6	4
*K*NN	$K = 2$	1.000000	0.773333	0.872180	2	7	5
Naïve Bayes	n/a	0.773333	1.000000	0.872180	6	1	6
Logistic regression	LR = 0.0001, Epochs = 500	0.772881	0.995633	0.870229	7	2	7

Table 2 Results for cataract prediction dataset on three parameters

Algorithms	Precision	Recall	*F*1-measure
*K*NN	99	77	87
NB	77	100	87
LR	77	96	87

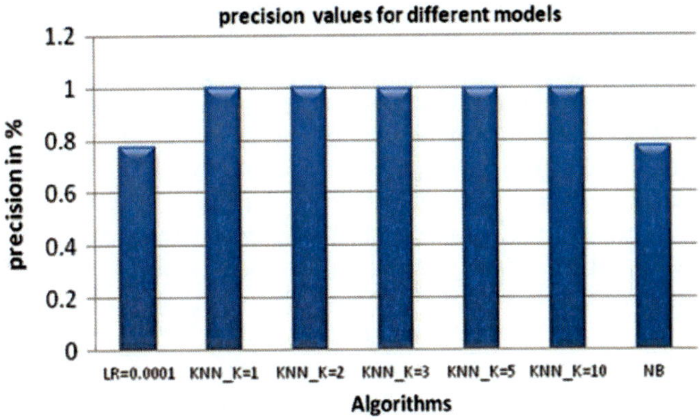

Fig. 2 Precision measurement for different models

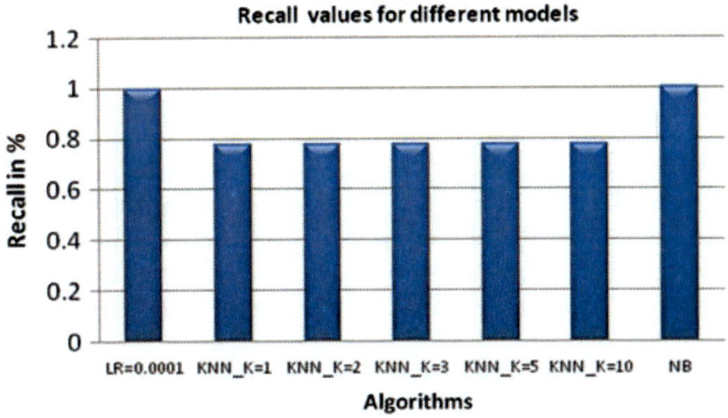

Fig. 3 Recall measurement for different models

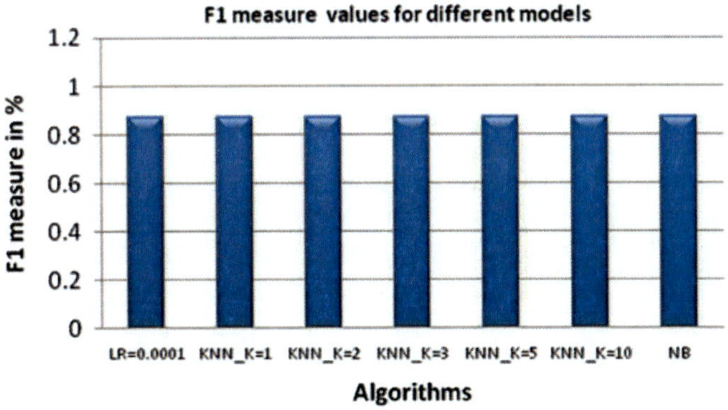

Fig. 4 *F*1 measurement for different models

6 Conclusions

Cataract is one of the leading causes of blindness worldwide each year. The severity and risk factor can also be avoided if a keeping accurate forecast is attainable. The suggested ML-based approach can predict if a patient has cataracts or not. Various metrics are used to compare three machine learning algorithms. The cataract dataset obtained from a private eyecare center is used in the experiments. As observed earlier from the training summary data frame, the $F1$-measure is approximately the same for all the models. This can be due to the fact that model parameters like learning rate, value of K, threshold probability for NB, etc., have been selected by trial-and-error method. The performance might differ (and even improve) if we search across different hyperparameters and find optimal values for which the models perform best. The results showed that the KNN algorithm was more effective than other algorithms in predicting diabetes.

A cataract prediction model based on ML algorithms is developed in this work. This represents a technological basis for development of a medical device for patient-oriented cataract surgery. The identification model has the potential to serve as a complementary screening procedure for the early detection of Cataract, which could be especially useful in underdeveloped and remote areas. A centralized classification database may improve classification accuracy as the classification algorithm learns from the growing database. However, collecting such a large volume of hospital-based medical data from patients with cataract diseases is very difficult. In this study, 1000 cases based on 5 attributes of patients including young to old aged were analyzed and trained identification models. This exhibited satisfactory discrimination and stability, indicating that medical AI techniques achieve satisfactory performance in identifying cataract even with limited data.

Future Scope

More recently, other applications of AI in cataract surgery have included OR scheduling, intra operative support, and postoperative management of complications. With advances in technology, there has been an emergence of several AI-based imaging tools and smart phone applications to provide support for clinical decisions. Our study may provide a reference for the development of AI-based preventive strategies for other congenital diseases. Such application can help deliver ophthalmology care across screening, diagnosis, and monitoring.

Acknowledgements We would like to express our special thanks to the eyecare center for their interest in providing data and encouragement for conducting this study.

References

1. Ray A, Chaudhuri AK (2020) Smart healthcare disease diagnosis and patient management innovation, improvement and skill development. Mach Learn Appl 3:100011
2. Saha P, Sircar R, Bose A (2021) Using hospital admission, discharge & transfer (ADT) data for predicting readmissions. Mach Learn Appl 5:100055
3. Sunarti S, Rahmana FF, Naufal M, Riskya M, Febriyantoa K, Masnina R (2021) Artificial intelligence in healthcare: opportunities and risk for future. Gac Sanit 35(S1):S67–S70
4. Abu Daqar MAM, Smoudy AKA (2019) The role of artificial intelligence on enhancing customer experience. Int Rev Manag Mark 9(4):22–31
5. Dogan A, Birant D (2021) Machine learning and data mining in manufacturing. Expert Syst Appl 166:114060
6. Paturi UMR, Cheruku S (2020) Application and performance of machine learning techniques in manufacturing sector from the past two decades: a review. Mater Today Proc
7. Haleema A, Javaida M, Singhb RP, Suman R (2021) Applications of artificial intelligence (AI) for cardiology during COVID-19 pandemic. Sustain Oper Comput 2:71–78
8. Rawat S, Rawat A, Kumar D, Sai Sabitha A (2021) Application of machine learning and data visualization techniques for decision support in the insurance sector. Int J Inf Manag Data Insights 1:100012
9. Tognetto D, Giglio R, Vinciguerra AL, Milan S, Rejdak R, Rejdak M, Zaluska-Ogryzek K, Zweifel S, Toro MD (2021) Artificial intelligence applications and cataract management: a systematic review. Surv Ophthalmol
10. Cecula P, Yu J, Dawoodbhoy FM, Delaney J, Tan J, Peacock I, Cox B (2021) Applications of artificial intelligence to improve patient flow on mental health inpatient units—narrative literature review. Heliyon 7:e06626
11. Haleem A, Javaid M, Khan IH (2019) Current status and applications of artificial intelligence (AI) in medical field: an overview. Curr Med Res Pract 19:S2352-0817(19)30193-X
12. Long GJ, Lin BH, Cai HX, Nong GZ (2019) Developing an artificial intelligence (AI) management system to improve product quality and production efficiency in furniture manufacture. In: 3rd international conference on mechatronics and intelligent robotics (ICMIR-2019)
13. Mojjada RK, Yadav A, Prabhu AV, Natarajan Y (2020) Machine learning models for covid-19 future forecasting. Mater Today Proc
14. Dharani NP, Bojja P, Kumari PR (2021) Evaluation of performance of an LR and SVR models to predict COVID-19 pandemic. Mater Today Proc
15. Gothai E, Thamilselvan R, Rajalaxmi RR, Sadana RM, Ragavi A, Sakthivel R (2021) Prediction of COVID-19 growth and trend using machine learning approach. Mater Today Proc
16. González-Carrasco I, Jiménez-Márquez JL, López-Cuadrado JL, Ruiz-Mezcua B (2019) Automatic detection of relationships between banking operations using machine learning. Inf Sci 485:319–346
17. Pratap T, Kokil P (2019) Computer-aided diagnosis of cataract using deep transfer learning. Biomed Signal Process Control 53:101533
18. Lin D, Chen J, Lin Z, Li X, Zhang K, Wu X, Liu Z, Huang J, Li J, Zhu Y, Chen C, Zhao L, Xiang Y, Guo C, Wang L, Liu Y, Chen W, Lin H (2020) A practical model for the identification of congenital cataracts using machine learning. EBioMedicine 51:102621
19. Caixinha M, Velte E, Santos M, Perdigao F, Amaro J, Gomes M, Santos J (2015) Automatic cataract classification based on ultrasound technique using machine learning: a comparative study. Phys Proc 70:1221–1224

Strokes-Related Disease Prediction Using Machine Learning Classifiers and Deep Belief Network Model

M. Anand Kumar, Kamlesh Chandra Purohit, and Anuj Singh

Abstract The concept of bringing fundamental theories and analytical procedures to medicine and biology is known as biomedical engineering. Bio-artificial intelligence medical engineering is one of the newest and fastest-growing branches of research. From the adoption of medical devices to diagnostic expert systems, it can be profitable in the field of healthcare. These gadgets and expert systems generate data that are multidimensional and erratic. The use of deep learning (DL) algorithms in those devices will be useful for signal analysis and disease detection. Multiple levels of neural networks are used in DL, which are a subset of machine learning. It can learn features on its own. Strokes are one of the deadly diseases worldwide, impacting a huge percentage of the population. Stroke-related diseases require immediate medical attention, which requires emergency treatments. If it was treated in the initial stages, then brain injury and other consequences can be avoided. The main objective of this work was to evaluate different machine learning algorithms to find a suitable approach for stroke-related diseases prediction applications. This research work evaluates different machine learning algorithms, namely logistic regression, random forest, support vector machine (SVM), KNN, and the decision tree for predicting and identifying stroke and its related diseases. Finally, deep belief networks are applied for the symptom dataset to get results that are more accurate.

Keywords Bio-artificial intelligence · Dataset · Deep belief networks · Disease · Engineering · Strokes · Machine learning

M. Anand Kumar (✉)
Department of Computer Applications, Graphic Era Deemed to be University, Dehradun, India
e-mail: anandkumar@geu.ac.in

K. C. Purohit
Department of Computer Science, Graphic Era Deemed to be University, Dehradun, India

A. Singh
Department of Computer Science and Engineering, Graphic Era Deemed to be University, Dehradun, India

P. Chatterjee et al. (eds.), *Computational Intelligence for Engineering and Management Applications*, Lecture Notes in Electrical Engineering 984,
https://doi.org/10.1007/978-981-19-8493-8_11

1 Introduction

ICT plays a vital role for digital firms which may boost their working efficiency and competitiveness. In the current digital era, most of the industries are using the latest digital equipment and methods for their latest innovations especially in biomedical-related firms [1]. Similarly, there are no exclusions for ICT in health-related businesses. In developed economies, hospitals and care providers are aggressively deploying digital technologies such as artificial intelligence (AI), machine learning, smart sensors, robots, big data, and the Internet of things (IoT) in order to improve the quality of patient care [2]. Artificial neural networks (ANNs) and deep learning (DL) are the main machine learning approaches in a variety of areas, including image analysis and defect diagnostics [3]. The DL uses in biomedical fields span the entire medical spectrum, from genomic applications like gene expression to public health applications like predicting demographic data or infectious disease epidemics.

Most recently, artificial intelligence-based methodologies are widely implemented in healthcare industries for improvements in quality treatments and performance of medical-related data and equipment [4]. Machine learning technologies provide tremendous prospects for the latest healthcare firms since they comprise MI-based smart robotics with natural language processing which provides a better solution for medical datasets [5]. Machine learning is widely used in biomedical applications in three ways: (1) as a machine-aided diagnosis to assist health professionals in making more accurate and timely diagnoses with better harmonization and less contradictory diagnoses; (2) to improve patient healthcare with much improved personalized treatments; and (3) to improve overall human wellness with the latest AI-based ML algorithms which use the massive volume of medical data to learn and predict diseases from the patients' historical data which enables healthcare workers for better diagnosis and treatment measures.

It was very necessary to identify the nature of stroke patients in the initial stage before there was a blockage of blood flow in the brain; otherwise, the supply of oxygen and nutrients to the tissue cells will be blocked which will result in serious conditions for a stroke patient. In such circumstances, brain cells will start to die in a fraction of time [6]. A blocked artery is one of the leading causes of stroke. There are two types of stroke risk factors: modifiable and non-modifiable. Lifestyle risk factors and medical risk factors are two types of modifiable danger issues. Lifestyle habits such as alcohol use, smoking, physical idleness, and fatness can all be reduced, whereas clinical issues such as high blood pressure, diabetes mellitus, atrial fibrillation, and high cholesterol can typically be preserved. Non-modifiable risk issues, on the other hand, cannot be modified, although they can help identify people who are at high risk of having a stroke [7].

The rest of the paper was structured as follows. Section 2 presents a review of the relevant works related to MI algorithms for strokes. Section 3 presents the research methodology used for this research. Section 5 presents the research design followed by experimental results in Sect. 6. Section 7 concludes the work by summarizing the results and the findings.

2 Literature Review

This section presents the latest research works that are carried out on machine learning algorithms for stroke-related diseases and the new advancements. A few of them are presented in this section.

The recent work presented a model which had been developed with artificial neural network (ANN) for stroke detection [8]. In this work, the cardiovascular health study (CHS) database was used to compile their findings. Three datasets were created, each with 212 strokes (all three) and 52, 69, and 79 for non-stroke patients. The final data includes 357 characteristics and 1824 units, as well as 212 stroke rates. The C4.5 decision tree approach had been utilized for feature selection, and principle component analysis (PCA) was applied for length reduction. They employed the backpropagation learning approach to implementing ANN. For the three datasets, the accuracy was 95%, 95.2%, and 97.7%, respectively.

For stroke classification, decision trees, Bayesian classifiers, and neural networks are employed [9]. They used thousand patient data in their dataset. For dimensionality reduction, the PCA technique was applied. They achieved the highest accuracy in ten rounds of each algorithm, with 92%, for the neural network model, 91% for the Naive Bayes classifier, and 94% for decision tree algorithms, respectively.

Several deep learning approaches for the segmentation and prediction of stroke-related diseases are based on the imaging concept [10]. This research also presented the importance of 2 deep neural network approaches for stroke diseases, particularly image processing such as conventional neural network (CNN) and fully convolutional network (FCN). It suggests that various deep learning models could be developed for improved outcomes in detecting stroke patients. This assessment also includes information on upcoming trends and advancements in stroke detection [9]. Early encouraging results suggest that machine learning approaches could be effective as decision support tools in AIS treatment decisions. However, several constraints in currently available designs must be addressed to improve the generalizability of the findings presented above. The sample size is the first constraint. Deep learning methods that use medical imaging frequently demand massive datasets that are not always readily available [11].

Another model suggested a deep learning-based model for predicting stroke illness that was trained using data from real-time EEG sensors [12]. To construct the classification model, a backpropagation neural network classification algorithm was used, along with a decision tree algorithm for feature selection, a principal component analysis algorithm for reducing the dimension, and a decision tree algorithm for feature selection. Several conventional models for the prediction of stroke detection mechanisms were presented based on CT scans or MRI scan images [13]. It also provides information regarding the feature extraction and classification methods that are used in stroke detection and prevention mechanisms. This work is also helpful for radiologists to accurately predict the stroke regions affected by the diseases and to take measures for future treatments for the patients. Different analysis has been done to know error rates and percussive measures that are to be taken to avoid serious issues related to the disease. The experimental results were presented in this work with different parameters for future research works in strokes-related problems.

3 Research Methodology

3.1 Data Collection

This research work collected the dataset for the stroke disease prediction mechanism from Kaggle, a subsidiary of Google, an online community for data scientists and machine learning practitioners. The dataset contains patient who is concerned with stroke events that were collected at several locations around the world. It contains a total of 5111 records with 11 attributes. Out of 5111 records, male patients are 41%, and female patients are 59%. The data are divided such that 75% are for training, and 25% are for testing.

3.2 Data Parameters

Table 1 presents the data parameters that are used in this research work for stroke-related diseases. These parameters are selected from the Kaggle dataset based on the criteria that are exactly matching for the disease predictions.

3.3 Data Pre-processing

There is a considerable chance that data gathered from any archive will have missing numbers or perhaps contain outliers. For the medical dataset, the likelihood of missing data values increases [10]. Some values in the Kaggle dataset are also missing. The values of FBS and AGL, for example, are absent. The missing data can be handled in a variety of ways. Here for imputation, the strategy of most frequent was used. There are both categorical and numerical values in the dataset. Six of the twelve traits are categorized, while the remaining six are numerical. Data are typically transformed linearly using the min-max normalization technique. It was used to convert the data on a scale of 0–1. Ensuring that all of the features were scaled the same. According to the min-max normalization technique:

Table 1 Data parameters

Attributes	Description	Values
Gender	Male or female	1 = Male 0 = Female
Age	Age denoted in years	Incessant
Hypertension	Yes or No	1 = Yes 0 = No
Heart disease	Yes or No	1 = Yes 0 = No
RBP	Resting blood pressure	Incessant value in mm hg
Heart rate	heart rate achieved	Incessant value
FBS	Fasting blood sugar	$1 \geq 120$ mg/dl $0 \leq 120$ mg/dl
Avg. glucose level	The average glucose level in blood	Floating point number
BMI	Patient's body mass index	Floating point number
Smoking status	Smoking status of the patient	String literal
DEP	Exercise-induced depression when compared to rest	Incessant value
Stoke	Output	1 = Stroke 0 = No stroke

$$V^{/} = \frac{V - \text{Min}}{\text{Max} - \text{Min}} \tag{1}$$

4 Machine Learning Algorithms

This research work applied logistic regression, random forest, support vector machine (SVM), KNN, and the decision tree. Here is a brief overview of the algorithms:

4.1 Logistic Regression (LR)

LR is a continuous/discrete predictor supervised machine learning technique. For binary classification, LR is commonly employed. It is a statistical model that determines the best fitted linear model to depict the relationship between the binary dependent variable's logical transformation and independent variables (one or more than one). When compared to other non-parametric machine learning models, this model is a straightforward prediction strategy using baseline accuracy values provided by the model [14].

4.2 Random Forest (RF)

This approach can be used to classify and solve regression problems. During the process of creating distinct decision trees for qualities, this technique corrects the overfitting of their training. It is easy to use because of its flexibility.

4.3 Support Vector Machine (SVM)

It is a supervised learning test that can be used to diagnose and prevent relapse, but it is most typically employed for characterization. SVM functions perform brilliantly and are capable of handling both linear and nonlinear situations [15].

$$f(x) = B_0 + \sum_{I}^{1}(a_i * (x, xi))) \tag{2}$$

x denotes the new input vector, $B0$ is the bias, and a_i is the weight that must be obtained from the training data

4.4 K-Nearest Neighbors (KNNs)

It can be used for classification as well as regression. The KNN algorithm [16] is mostly used to find values for critical disease factors utilizing K values, which allows for the discovery of new values for critical disease criteria which defines boundaries for each class of attributes. Using the Euclidean formula presented below, this approach simply calculates the distances between training samples.

$$\text{Dist}(x, y) = \sum_{i=1}^{k}\left(xi \cos\frac{n\pi x}{L} + b_n \sin\frac{n\pi x}{L}\right) \tag{3}$$

4.5 Decision Tree (DT)

A decision tree (DT) [16] is a data extraction model that can be used to solve alliancing and categorization problems. In the most continuous reckoning implementations, the DT is the most essential recursive block. DT includes both nodes (parent) and leaves (child) [17]. Three max-leaf nodes and three DT criteria were utilized in this investigation, with the random state set to 0. The random state determines the

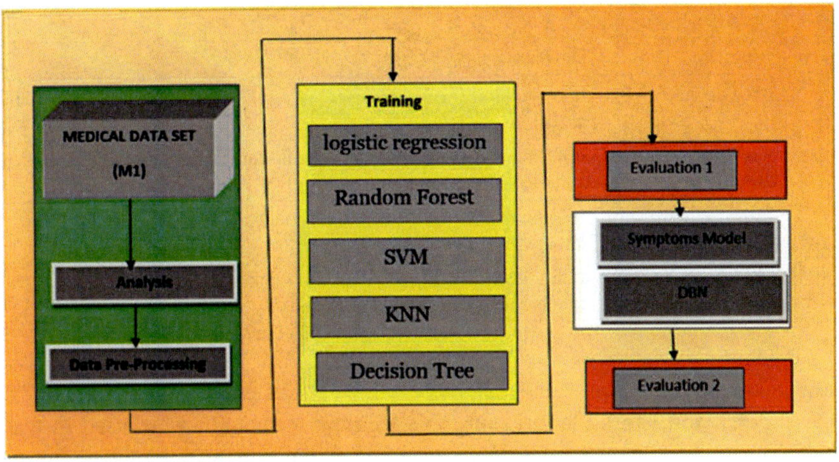

Fig. 1 Proposed model

unpredictability of the estimator, and a value of 0 indicates that the randomness was untrue in this case.

5 Research Design

Step 1: Input: patient records from medical dataset
Step 2: The dataset is subjected to analysis, which aids in the explanation of the data.
Step 3: The dataset is pre-processed after analysis. The dataset was scaled using the min-max normalization approach for this investigation.
Step 4: On the scaled dataset, various algorithms are trained using various k fold cross-validation methods.
Step 5: Evaluate the result from the above steps (1–4)
Step 6: Deep belief network was employed with the symptom set (S1)
Step 7: Evaluate the result from the above step (6)
Step 8: Compare the results between step (5) and step (7) (Fig. 1).

6 Experiments

6.1 Experimental Setup

Computer simulation is a way for examining a wide range of machine learning models that represent real-world systems using specific simulation application software

Table 2 Performance metrics

S. No.	Metrics	Description
1	True positive (TP)	It is the number of occurrences of stroke disease that are classified as true and are true
2	True negative (TN)	It is the number of cases of stroke disease that are classified as fake and are, in fact, false
3	False negative (FN)	It is the number of cases of stroke disease that are labeled as false but are true
4	False positive (FP)	It refers to the number of cases of stroke disease that are labeled as true but are false

designed to emulate some of the system's essential attributes. Microsoft's ML.NET Open-Source and Cross-Platform machine learning techniques were used in this study. The machine learning algorithms were implemented using this method. It also adds artificial inelegancy to any application by creating trained machine learning models for medical applications of any kind.

6.2 Evaluation Metrics

Machine learning algorithms' overall performance is measured using performance metrics. It was also used to track how well machine learning models are implemented and performed on a given dataset under different conditions. Selecting the appropriate metric is crucial for comprehending the model's behavior and making the necessary changes to improve it. Table 2 indicates the metrics.

6.3 Mathematical Model

When it comes to medical data, a mathematical model is one of the most significant aspects to consider because these are vital data that must be properly assessed for better disease detection [18]. This section explains the mathematical model that was used to evaluate this outcome.

- Accuracy: It is the number of right forecasts made by the model over all types of forecasts given in classification problems [19]. The proper forecasts (true positives and true negatives) are in the numerator, while the type of all algorithmic predictions (right and wrong) is in the denominator, and it is stated as

$$\text{Accuracy} = \frac{\text{TP} + \text{TN}}{(\text{TP} + \text{FP} + \text{FN} + \text{TN})} \tag{4}$$

- Precision: Precision [20, 21] is a measure that determines what % of stroke patients have actually had a stroke. The projected positives (those who are expected to have a stroke, abbreviated as TP and FP) and the ones who have had a stroke, abbreviated as TP, are as follows:

$$\text{Precision} = \frac{\text{TP}}{\text{TP} + \text{FP}} \tag{5}$$

- Recall or Sensitivity: Recall [20, 22] is a measure that shows what percentage of stroke patients the algorithm correctly diagnosed. The true positives (those who have strokes are TP and FN) and the patient diagnosed with strokes by the model are both TP. FN is included since the person experienced a stroke despite the model's prediction, and it is given as

$$\text{Sensitivity} = \frac{\text{TP}}{\text{TP} + \text{FN}} \tag{6}$$

- $F1$-Score: When creating a model to solve a classification problem, it is not necessary to carry both precision and recall. So if it can receive a single score that represents both precision (P) and recall (R) that would be ideal (R). Taking their arithmetic mean is one approach to do this. $(P + R)/2$, with P denoting precision and R denoting recall. However, in other circumstances, this is not a good thing. It can be written as [23–25]:

$$F1\text{Score} = \frac{2 * \text{Precision} * \text{Recall}}{\text{Precision} + \text{Recall}} \tag{7}$$

7 Results and Discussion

7.1 Accuracy Results Based on Algorithms

This section presents the various results obtained during the experiments based on the mathematical model that was presented in the previous section. Figure 2 shows the results and accuracies of the implemented machine learning algorithms such as logistic regression, random forest, support vector machine (SVM), KNN, and the decision tree. The experimental results clearly show that the random forest algorithm gives a better result with an accuracy of 87.43% when compared to other algorithms. It also indicated that the random forest algorithm was better suitable for stroke-related disease prediction mechanisms. It was also identified that logical regression with an accuracy of 82.16%, support vector machine with 81.23%, KNN with 81.02, and decision tree with 79.37% accuracy.

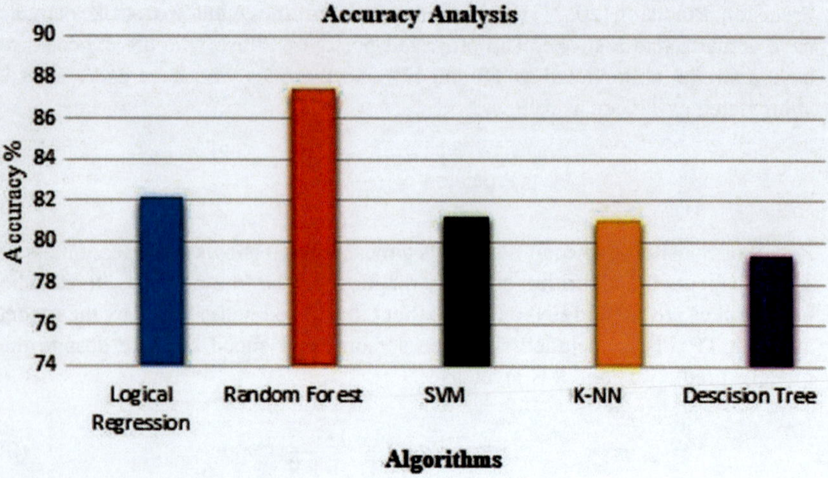

Fig. 2 Accuracy analysis of algorithms

7.2 *Accuracy Results Based Training Dataset*

Figure 3 shows the results based on the selection of different dataset sizes ranging from 45 to 90. The highest value is attained for all the used algorithms in the 75% to an 80% range of data training, according to the analysis. Only SVM exhibits an increase in accuracy as the training size grows. The remaining algorithms, on the other hand, provide outstanding results and peak accuracy values of 80% in the training size realm. The best size range for forecasting the disease and getting the accuracy, precision, recall, and $F1$-score is 75–80% train data size, according to this research. It was also indicated that there is a larger possibility of under-fitting in the 50% train dataset. Overfitting is possible for 90% of the train dataset. As a result, a dataset with a 75–80% training rate is excellent for any disease diagnosis application.

Fig. 3 Accuracy based on dataset size

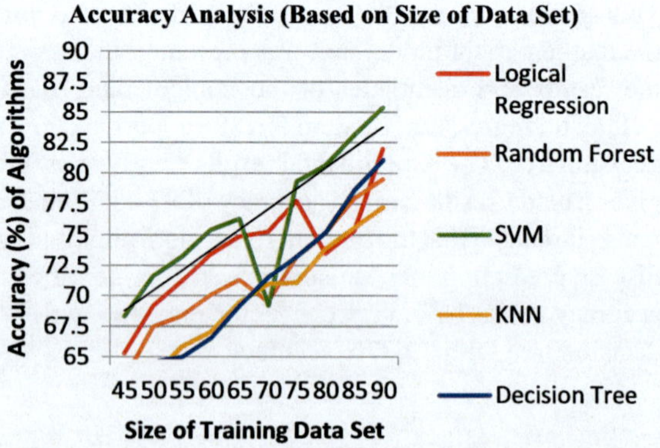

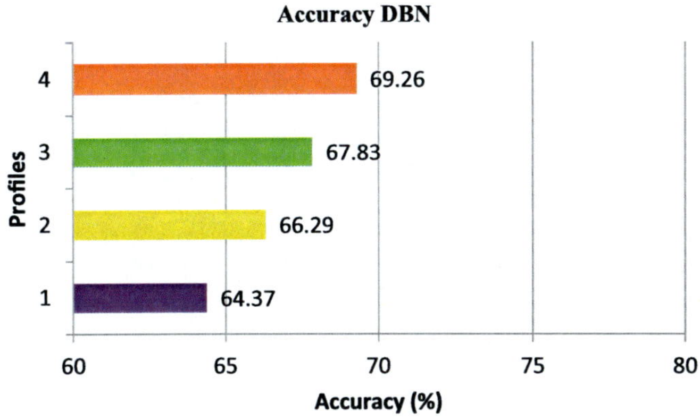

Fig. 4 Accuracy based on DBN

7.3 *Accuracy Results-Based Deep Belief Networks (DBN) with Symptoms*

According to the number of layers, nodes, and hyperparameters, the accuracy of the stroke prediction model can be seen in Fig. 4. "Layers" refer to the number of layers in the DBN model, whereas "nodes" refer to the number of nodes per layer. This model was run in four separate profiles, each with its own set of parameters. Profile 4 was discovered to have a higher accuracy of 69.26%. The training set includes several symptoms as a parameter.

8 Conclusion

Strokes are one of the major health problems worldwide. So there is a need for early intervention and the timely diagnosis of the disease to reduce risk factors. It was observed from the literature survey, there is a lot of research work has been done for the early detection of strokes using various machine learning algorithm techniques. However, there is still an utmost need for the identification of relevant attributes that could detect strokes at a very early stage. This research work employed well-known machine learning algorithms. The experimental results show that the random forest algorithm is better suited for disease diagnosis, especially for stroke-related issues. Another main observation from this research work was that 80% training dataset is ideal for accurate results as well as for good performances.

References

1. Yang F, Gu S (2021) Industry 4.0, a revolution that requires technology and national strategies. Complex Intell Syst 7:1311–1325
2. Lakkamraju P, Anumukonda M, Chowdhury SR (2020) Improvements in accurate detection of cardiac abnormalities and prognostic health diagnosis using artificial intelligence in medical systems. IEEE Access 8:32776–32782
3. Xu D, Sheng JQ, Hu PJ-H, Huang T-S, Hsu C-C (2021) A deep learning-based unsupervised method to impute missing values in patient records for improved management of cardiovascular patients. IEEE J Biomed Health Inf 25(6):2260–2272
4. Kaur S et al (2020) Medical diagnostic systems using artificial intelligence (AI) algorithms: principles and perspectives. IEEE Access 8:228049–228069
5. Ravì D et al (2017) Deep Learning for Health Informatics. IEEE J Biomed Health Inform 21(1):4–21
6. Sarmento RM, Vasconcelos FFX, Filho PPR, Wu W, de Albuquerque VHC (2020) Automatic neuroimage processing and analysis in stroke—a systematic review. IEEE Rev Biomed Eng 13:130–155
7. Kwon S, Yu J, Park S, Jun J-A, Pyo C-S (2021) Stroke medical ontology for supporting AI-based stroke prediction system using bio-signals. In: 2021 Twelfth international conference on ubiquitous and future networks (ICUFN), pp 53–59
8. Ufumaka I (2021) Comparative analysis of machine learning algorithms for heart disease prediction. Int J Sci Res Publ 11(1)
9. Potdar V, Santhosh L, Yashu Raj Gowda CY (2021) A survey on stroke disease classification and prediction using machine learning algorithms. Int J Eng Res Technol (IJERT) 10(08)
10. Karthik R, Menaka R Neuroimaging and deep learning for brain stroke detection—a review of recent advancements and prospects. Comput Methods Programs Biomed 197:105728
11. Kamal H, Lopez V, Sheth SA (2018) Machine learning in acute ischemic stroke neuroimaging. Front Neurol 9(1):945–952
12. Choi Y-A, Park S-J, Jun J-A, Pyo C-S, Cho K-H, Lee H-S, Yu J-H (2021) Deep learning-based stroke disease prediction system using real-time bio signals. Sensors 21:4269
13. Surya S, Yamini B, Rajendran T, Narayanan KE (2021) A comprehensive method for identification of stroke using deep learning. Turk J Comput Math Educ 12(7):647–652
14. Sangari N, Qu Y (2020) A comparative study on machine learning algorithms for predicting breast cancer prognosis in improving clinical trials. In: 2020 International conference on computational science and computational intelligence (CSCI), pp 813–818
15. Ketpupong P, Piromsopa K (2018) Applying text mining for classifying disease from symptoms. In: 2018 18th International symposium on communications and information technologies (ISCIT), pp 467–472
16. Enriko KA, Suryanegara M, Gunawan D (2018) Heart disease prediction system using k-nearest neighbor algorithm with simplified patient's health parameters. J Telecommun Electron Comput Eng 8(12)
17. Yarasuri VK, Indukuri GK, Nair AK (2019) Prediction of hepatitis disease using machine learning technique. In: 2019 Third international conference on I-SMAC (IoT in social, mobile, analytics, and cloud) (I-SMAC), pp 265–269
18. Ghosh M, Raihan M, Sarker M, Raihan M, Akter L, Bairagi AK (2021) A comparative analysis of machine learning algorithms to predict liver disease, Intell Autom Soft Comput 30(3):917–924
19. Liu N, Kumara S, Reich E (2021) Gaining insights into patient satisfaction through interpretable machine learning. IEEE J Biomed Health Inf 25(6):2215–2226
20. Kumar P, Chauhan R, Stephan T, Shankar A, Thakur S (2021) A machine learning implementation for mental health care. Application: smart watch for depression detection. In: 2021 11th International conference on cloud computing, data science and engineering (confluence)
21. Shamout F, Zhu T, Clifton DA (2021) Machine learning for clinical outcome prediction. IEEE Rev Biomed Eng 14:116–126

22. Pande A, Manchanda M, Bhat HR, Bairy PS, Kumar N, Gahtori P (2021) Molecular insights into a mechanism of resveratrol action using hybrid computational docking/CoMFA and machine learning approach. J Biomol Struct Dyn 1–15
23. Gupta A, Lohani MC, Manchanda M (2021) Financial fraud detection using naive bayes algorithm in highly imbalance data set. J Discrete Math Sci Crypt 24(5):1559–1572
24. Singh N, Singh DP, Pant B (2019) ACOCA: ant colony optimization based clustering algorithm for big data preprocessing. Int J Math Eng Manag Sci 4:1239–1250
25. Singh N, Singh DP, Pant B, Tiwari UK (2021) μBIGMSA-Microservice-based model for big data knowledge discovery: thinking beyond the monoliths. Wireless Pers Commun 116(4):2819–2833
26. Kabiraj S et al. (2020) Breast cancer risk prediction using XGBoost and random forest algorithm. In: 2020 11th International conference on computing, communication and networking technologies (ICCCNT), pp 1–4

Analysis and Detection of Fraudulence Using Machine Learning Practices in Healthcare Using Digital Twin

B. J. D. Kalyani, Kopparthi Bhanu Prashanth, Kopparthi Praneeth Sai, V. Sitharamulu, and Srihari Babu Gole

Abstract Traditional IT with advent of cloud computing and industry 4.0 leads a trend set for manufacturing industry, smart cities, and healthcare. An essential upsurge during pandemic is virtualization of hospital operational strategies, recruitment, and care models. Virtual models can assist in bed shortages, spreading of germs, staff schedules, operating rooms, and monitor systems. These will help to enhance patient care and performance. Digital twin technology consists of creating virtual replicas of objects or processes that simulate the behavior of their real counter parts. This is enormously important in healthcare as it enables informed strategic decisions to take place in a highly complex and sensitive environment. Digital twins can virtualize the hospital in order to create a safe environment, which tests the influences of changes on system performance without risks. The digital twin model can reveal information from the historical, optimize the present, and even predict the future performance of the different areas analyzed. The motive behind this work is to implement the increased fraud detection mechanism in the field of health record system and to obviously reduce the error ratios using machine learning algorithms, especially supervised restricted Boltzmann machine based on multilevel key catching and low latency classification model. The work also exposures the prediction analysis on the wide variety of fraud data in medical databases of digital twin.

Keywords IoT · Digital twin · Deep learning · Software defined approach (SDA)

B. J. D. Kalyani (✉) · V. Sitharamulu
Department of Computer Science and Engineering, Institute of Aeronautical Engineering, Hyderabad, India
e-mail: kjd_kalyani@yahoo.co.in

K. B. Prashanth
Department of Computer Science and Engineering, Vardhaman College of Engineering, Hyderabad, India

K. P. Sai
Wipro Ltd., Hyderabad, India

S. B. Gole
Department of Computer Science and Engineering, Koneru Lakshmaiah Education Foundation, Vijayawada, India

P. Chatterjee et al. (eds.), *Computational Intelligence for Engineering and Management Applications*, Lecture Notes in Electrical Engineering 984,
https://doi.org/10.1007/978-981-19-8493-8_12

1 Introduction

Digital twin is a mechanism to analyze, observe, and predict the consequences of the quality and performance in the business processes. Fuller et al. [1] stated that these digital twins resemble various process to formulate various business outcomes in the current market. The close observation of digital twin describes the three components to formulate the performance aspects. The three components comprise a data model consisting of data, set of analytics with various algorithms, and deep knowledge database. The newest technologies may be sought to prevent the cloud accessibility frauds across the medical domain. The fraud detection system in the cloud is based on the usage of multilevel clustering in the feature space using static resemblances. The proposed methodology enhances the protection mechanism using dynamic feature extraction in order to enhance additional protection to data in the cloud. Bose and Mahapatra during [2] surveyed an increased protection mechanism for the cloud. The proposed mechanism is cleared explained in Fig. 1 for better analysis.

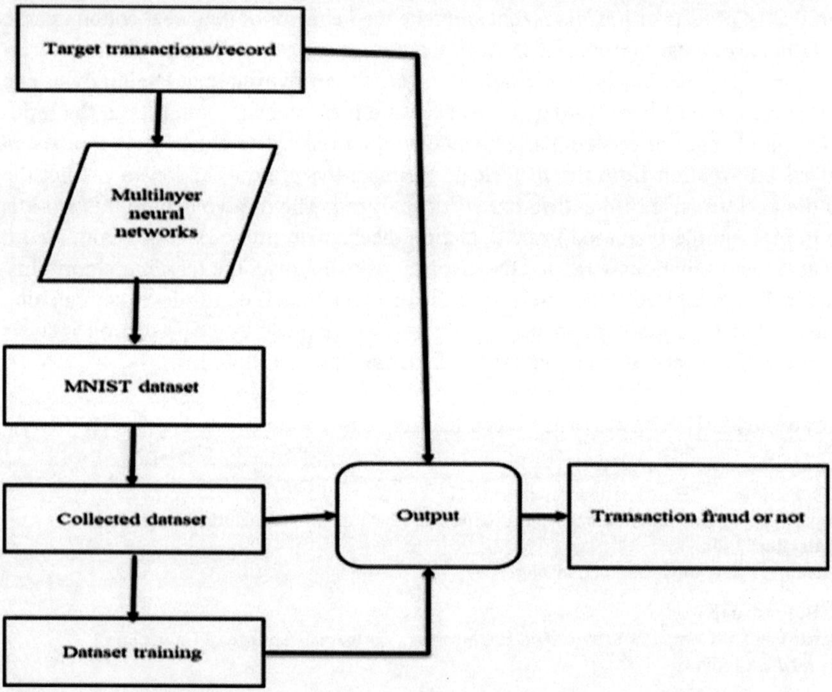

Fig. 1 Block diagram of proposed system

1.1 Literature Review

US FDA [3] concentrates on digital twins with bidirectional mapping with different level of sophistications by considering whole body as on twin or a separate twin for part of the body. Additionally focused on how to create digital twin instances and digital twin thread and processes. Boulos and Al-Shorbaji [5] illustrate that how IoT driven services can improve quality of life and focus also on smart city implementation plans with IoT sensors. Grieves, M [4] encourages cloud-based fraud detection with machine learning techniques. The objectives of this research are as follows:

- Deep and machine learning are considered as a key function to minimize the error rate and to improve the over cloud frauds.
- The proposed methodology provides the number of multilevel clustering which can be varied according to cloud space.
- A dynamic feature extraction is used to identify the frauds in the available cloud space.

2 Proposed System

The central cloud frauds [6] appearance at high traffic generating proximal theft from application. Note that the lack of mass effect and lack of high technical models facing frauds in their environment. Therefore, an efficient classification model [7] for predicting the severity level of the heterogeneous feature types is essential with high true positivity and low error rate. For an improved accessibility and clarity in the approach for real-time information, the proposed system followed implementing the John stanko Database which was duly used worked by Jiang et al.[8]. The considered dataset comprises various controlled buttons to control the incoming data and process the data and producing the right information for the transaction. The proposed and selected model is a novel approach which can be apply used for real-time applications to detect and analyze the fraud mechanism in the cloud-based transactions, especially in the medical domain [9]. The network uses various machine learning mechanisms like auto encoder to further analyze, classify, categorize, and segregate the transaction producing the real-time decisions. This mechanism automatically generates fraud detection ailments and produces them as the output.

The current section of the paper entails the usage of the auto encoder mechanism and navigated by RBM machine learning model so as to identify the fraud. The main purpose of choosing this kind of approach using machine learning is to quickly detect the fraud being raised, and mechanism is very suitable to solve many complications other than fraud detection. The present approach clearly provides optimal solutions to solve many complications arising on the cloud. The proposed approach is purely developed on the Python development environment. The datasets being used in this methodology are purely depend upon the records collected from Europe, MNIST,

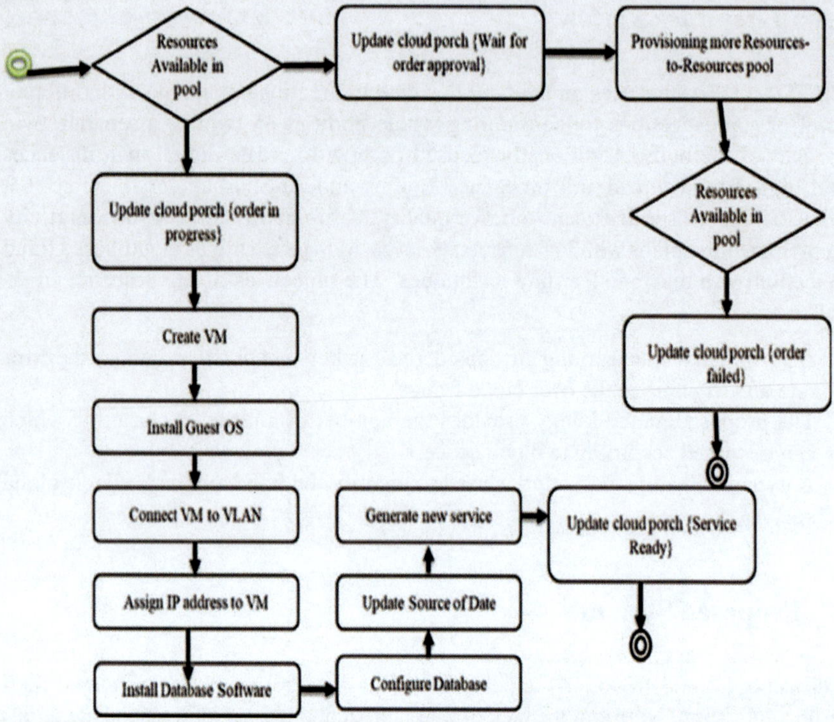

Fig. 2 Flow of software defined approach

Australia, and Germany. These datasets are further compared with SVM, KNN, and LR. The current studies clearly denote the importance of SVM's because of its intensive performance upon the classification datasets being selected. However, some of the relations of ANN's have also achieved tremendous result sets compared to the nonlinear models.

The flow diagram of software defined approach is described in Fig. 2.

3 Implementation

The proposed methodology inherited the concept of hidden and pooling layers from FGNN [10] classification. As part of implementation process, the training is carried out on a daily basis credit card fraud transaction as an input dataset and used FCNN mathematical operations to verify the accuracy ratio. As deep learning is platform independent and can work on any type of technologies like data mining cloud and big analytics, it is aptly considered to be part of the work for the training process. Rectified linear units (ReLUs) function is abundantly used by the FCNN network to normalize the occurrence of dropouts [11]. The 3 * 3 sigmoid activation layer is obtained with the usage of ReLU function in the FCNN network.

The proposed mechanism serves more benefits during usage of cloud-based drop down network issues. This is made possible because of the activation of sublayers present in the ReLU from the FCNN network for fraud detection [12]. The proposed mechanism duly used $N = 56$ FCNN CON2 dataset for computing the available frauds in the clouds. If datasets found to be null values, classifiers return non-fraud prevalence objects. The general architecture of the layered model is demonstrated with the help of Fig. 3. The complete operation is based upon 300 * 300 * 56 confuse matrix. Merely, 500 training weights are trained using FCNN Keras library. The FCNN on the other hand processed the classification datasets duly by activating the ReLU function within the filter layers. The normalization is mandate to convolute the process and hence, ReLU.

1. The normalized dropouts along with the crucial parameters of the cloud are precisely depicted in the first residual.
2. The last layer may be replaced with a set of alternative layers such as 1, 2, 4, 8, and 16.
3. Validating the test data is done on the basis of the combination of second and third layers. The training is done on FCNN CON 2 which is an input to detect the fraud detection.

Figure 4 entails the description of both the encoder and decoder models. The main functionality of the encoder is to detect the total number of frames per the

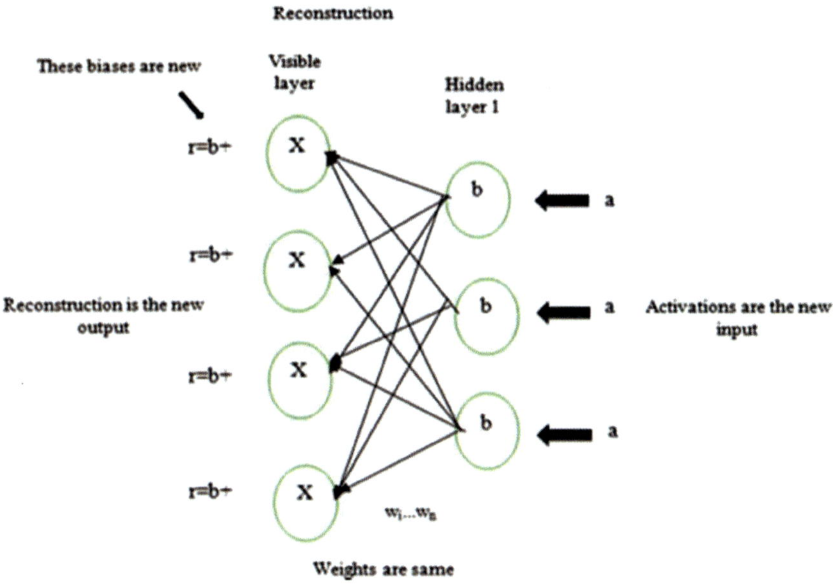

Fig. 3 Multiple input RBL

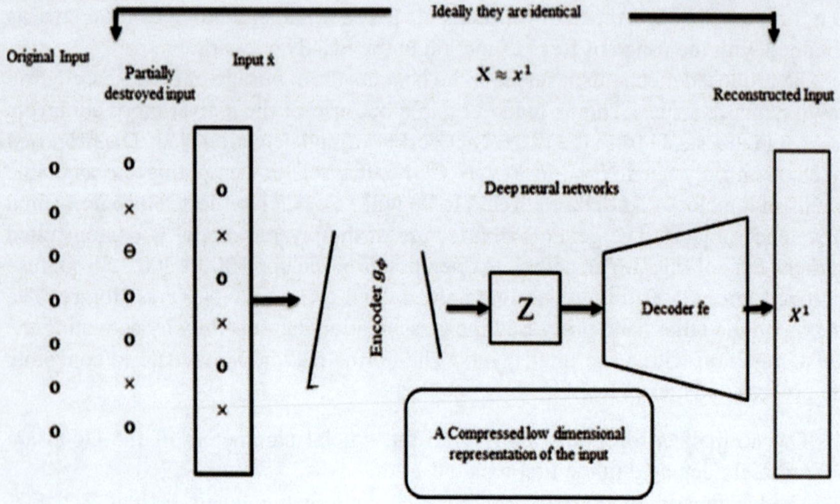

Fig. 4 Auto encoder method

maximum number of frames observed for utmost maximum operations. These are the key circumstances where the clouds prone to lose their security perspectives paving for increase in the frauds during the original inputs [13]. The data which is given as input now decodes the reconstructed data and fed back to the encoder stage [14].

4 Mathematical Modeling of SDA

$$\text{Encoding} = (h) = g(a(\text{x})) = \text{Sigm}(Wx)\text{or} = \tanh((Wx) \quad (1)$$

$$\text{Decoding} \quad X^\wedge = o(\hat{a}\ (\text{x})) = \text{Sigm}(W * h(x))\text{or} = \tanh((W * h(x)) \quad (2)$$

$$= J(W, b; x) = \frac{1}{2}||x - x^\wedge||^2 \quad (3)$$

$$z^1 = W^1 x + b^1 \\ a = f(z^1) \quad (4)$$

$$\begin{aligned} z^2 &= w^2 a + b^2 \\ x^{\wedge} &= f\left(z^2\right) \end{aligned} \tag{5}$$

Equations (1) and (2) analyze event driven approaches for encoding and decoding. Here, "*h*" is the encoding element, and sigma and tanh function identify the variant weights from average pooling layer. Equation (2) is the decoding function which is easily attained with convolution operator [15]. Equation (3) explains about event extraction from available event in transaction dataset. Equation (4) is the query message; here, *Z* is query element, *W* is weight of transaction, *x* is input transaction, and *b* is the number of parameters affecting the information and transaction. Equation (5) is the consecutive transaction information in the dataset. These all parameters generate the feature extracted data.

4.1 Mathematical Modeling of RBM

$$s(x) = \frac{1}{1+e^{-x}} = \frac{e^x}{1+e^x} \tag{6}$$

$$h^1 = S\left(V^{(O)T} W + a\right) \tag{7}$$

$$V^{(1)} = S\left(h^{(1)} w^T + a\right) \tag{8}$$

$$= p\left(h^{(1)} | v^o; w\right) \tag{9}$$

$$= p\left(v^{(1)} | h^{(1)}; w\right) \tag{10}$$

$$p(v, h) \tag{11}$$

$$E(v, h) = -\sum_{i \in \text{visible}} a_i v_i - \sum_{i \in \text{hidden}} b_j h_j - \sum_{ij} v_i h_j w_{ij} \tag{12}$$

$$\Delta w_{ij} = \propto \left(\langle v_i h_j \rangle_{\text{data}} - \langle v_i h_j \rangle_{\text{model}}\right. \tag{13}$$

$$\text{CD}_k\left(w, v^{(o)}\right) = -\sum_h p(h|v_k) \frac{\partial E(v_k, h)}{\partial w} + \sum_h p(h|v_k) \frac{\partial E(v_k, h)}{\partial w} \tag{14}$$

In Eq. (6): "*s*" is the sigmoid function; it calculates corresponding weight from inputs, and *h* and *V* are the vectors for hidden and visible layers. In Eq. (7): *V* (0) = input provide from the database, *h*1 = hidden mean value, *a* is the bias weights, *W* = weight of each vector, this equation can produce the hidden data from the selected database. In Eq. (8): "p" is the phase learning value, which can estimate from visible vectors and weights. In Eq. (9): complete phase learning value from hidden and visible layers in Eqs. (10) and (11).

5 Results and Discussion

The current section of the paper illustrates the usage of DN-CON2 dataset for detection of fraud in the cloud. The availability of datasets for fraud detection and ULB group on the big data is available over online with the name Kaggle. Further, the Kaggle dataset comprises of several attributes, variants, integers, and many more integrated functions. These datasets act as classifiers to act upon the data and have around 285,000 instances which are being trained. The comparative and feasibility of the proposed system with precision levels are clearly depicted as described in the Fig. 5.

5.1 Confusion Matrix

An error matrix or a confusion matrix is either pronounced in most of the areas which contains the matrix values that specify the performance analysis on the selected classification model [16]. The number of correct and incorrect predictions are summarized with count values and broken down by each class as in Fig. 6, and we have 492 fraud data points and 284,315 regular data points of training set size: (199364, 30).

The performance measures have shown significant improvement compared to existing methods. With an increase in accuracy to 98%, the use of cloud computing for feature applications is a remarkable achievement.

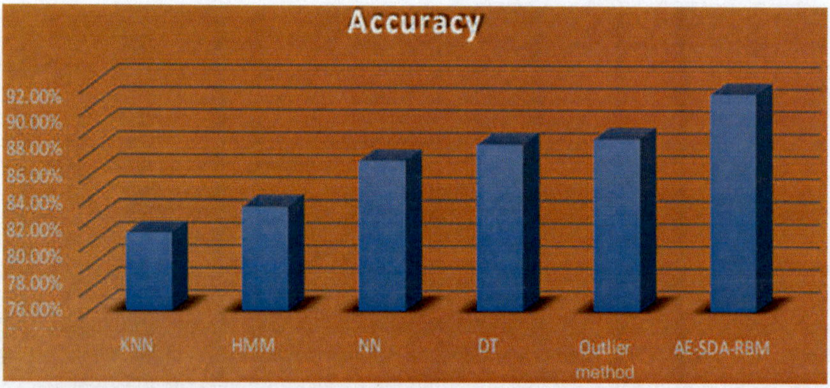

Fig. 5 Comparative study

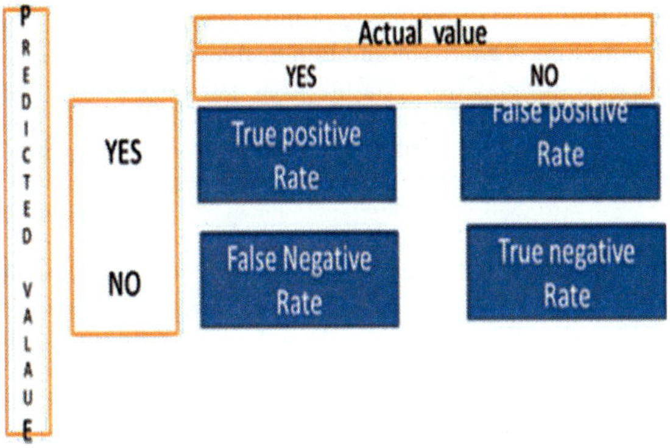

Fig. 6 Confusion matrix

6 Conclusion

The online medical records can be accessed by various customers via cloud servers. During usage of cloud services, users may be hacked by fraudsters, and further, result set may detect the total amount of debita. Due to this reason, clouds are necessary to update regarding reverse of fraud. In this paper, auto encoder and software-defined approach-based RBM machine learning model implemented for fraud identification. This examination solves the many issues on clouds. The entire design is developed on Python software with various packages. This implemented technique applied on various datasets like European, German, Australian, and MNIST records. The accuracy of proposed system is observed to be 98% which is more precision that older methods.

References

1. Fuller A, Fan Z, Day C (2020) Digital twin: enabling technologies, challenges and open research
2. Bose I, Mahapatra RK (2001) Business data mining—a machine learning perspective. Inf Manag 39(3):211–225
3. U.S. FDA (2013) Paving the way for personalized medicine-FDA'S role in a new era of medical product development. U.S. Food and Drug Administration: Silver Spring, MD, USA. Retrieved from https://www.fdanews.com/ext/resources/files/10/10-28-13-Personalized-Medicine.pdf. Accessed on 2 July 2021
4. Grieves M (2015) Digital Twin: manufacturing excellence through virtual factory replication (Digital Twin White Paper-2004). Retrieved from https://www.researchgate.net/publication/275211047_Digital_Twin_Manufacturing_Excellence_through_Virtual_Factory_Replication. Accessed on 2 July 2021
5. Boulos KMN, Al-Shorbaji NM (2014) On the internet of things, smart cities and the WHO healthy cities. Int J Health Geog 13:10
6. Randhawa K et al. (2018) Credit card fraud detection using AdaBoost and majority voting. IEEE Access 6:14277–14284. https://doi.org/10.1109/access.2018.2806420
7. Phyu TN (2009) Survey of classification techniques in data mining. In: Proceedings of the international multi conference of engineers and computer scientists, vol 1

8. Jiang C et al. (2018) Credit card fraud detection: a novel approach using aggregation strategy and feedback mechanism. IEEE Internet Things J 5:3637–3647
9. Melo-Acosta GE et al. (2017) Fraud detection in big data using supervised and semi-supervised learning techniques. In: 2017 IEEE Colombian conference on communications and computing (COLCOM). https://doi.org/10.1109/colcomcon.2017.8088206
10. Roy A et al. (2018) Deep learning detecting fraud in credit card transactions. In: 2018 systems and information engineering design symposium (SIEDS). https://doi.org/10.1109/sieds.2018.8374722
11. Zareapoor M, Seeja KR, Alam MA (2012) Analysis of credit card fraud detection techniques: based on certain design criteria. Int J Comput Appl (09758887) 52(3)
12. Ghosh S, Reilly DL (1994) Credit card fraud detection with a neural-network. In: Proceedings of the twenty-seventh Hawaii international conference on system sciences, vol 3. IEEE
13. Xuan S et al. (2018) Random forest for credit card fraud detection. In: 2018 IEEE 15th international conference on networking, sensing and control (ICNSC). https://doi.org/10.1109/icnsc.2018.8361343
14. Zhou X et al. (2018) A state of the art survey of data mining-based fraud detection and credit scoring. MATEC Web Conf 189. EDP Sciences
15. Seeja KR, Zareapoor M (2014) Fraudminer: a novel credit card fraud detection model based on frequent itemset mining. Sci World J 2014
16. Wang Z, Wang N, Su X, Ge S (2020) An empirical study on business analytics affordances enhancing the management of cloud computing data security. Int J Inf Manag 50:387–394

Prediction and Analysis of Polycystic Ovary Syndrome Using Machine Learning

Shivangi Raghav, Muskan Rathore, Aastha Suri, Rachna Jain, Preeti Nagrath, and Ashish Kumar

Abstract Polycystic ovary syndrome (PCOS) is a widespread pathology that affects many aspects of women's health, with long-term consequences beyond the reproductive age. The wide variety of clinical referrals, as well as the lack of internationally accepted diagnostic procedures, had a significant impact on making it difficult to determine the exact aetiology of the disease. The exact cause of PCOS is not yet clear. Therefore, multiple features are responsible for it. The aim of this project is to analyze simple factors (height, weight, lifestyle changes, etc.) and complex (imbalances of bio hormones and chemicals such as insulin and vitamin D) factors that contribute to the development of the disease. The data we used for our project was published in Kaggle, written by Prasoon Kottarathil, called polycystic ovary syndrome (PCOS) in 2020. This database contains records of 543 PCOS patients tested on the basis of 40 parameters. For this, we have used machine learning techniques such as logistic regression, decision trees, SVMs, and random forests. A detailed analysis of all the items made using graphs and programmes and prediction using machine learning models helped us to identify the most important indicators for the same.

Keywords Random forest · Machine learning · PCOS · Chi-square test · Feature selection · Information gain

S. Raghav · M. Rathore (✉) · A. Suri · R. Jain · P. Nagrath · A. Kumar
Computer Science and Engineering, Bharati Vidyapeeth's College of Engineering, New Delhi, India
e-mail: muskanrathore01@gmail.com

P. Nagrath
e-mail: preeti.nagrath@bharatividyapeeth.edu

A. Kumar
e-mail: ashish.kumar@bharatividaypeeth.edu

P. Chatterjee et al. (eds.), *Computational Intelligence for Engineering and Management Applications*, Lecture Notes in Electrical Engineering 984,
https://doi.org/10.1007/978-981-19-8493-8_13

1 Introduction

Polycystic ovary syndrome (PCOS) was first observed by Stein and Leventhalm [1] who in 1935 described seven girls with amenorrhoea, hirsutism and increased volume of the ovaries characterized by the presence of several cysts. It is one of the most relevant and prevalent hormonal disorders seen amongst the girls of childbearing age. The exact histology of PCOS is not yet clear. It is, therefore, a multifaceted study, which shares genetic and environmental factors [2]. This is a heterogeneous endocrine disorder that is quite prone to infertility, anovulation, cardiovascular disease, type 2 diabetes, obesity, etc. PCOS is diagnosed based on the presence of a combination of clinical signs of menstrual irregularities or anovulation, clinical or biochemical hyperandrogenism, and polycystic ovaries [3]. It is generally diagnosed in about 12–21% women of reproductive age. They are confronted with infertility, signs, and symptoms of hyperandrogenism, inclusive of acne, alopecia androgenica, and hirsutism [4]. Approximately, half of the women with PCOS are obese [5]. However, there is no proof that PCOS is because of obesity [6], but it could increase insulin levels and make PCOS symptoms worse as many women with PCOS have insulin resistance. This means the body cannot use insulin well. Insulin levels building up in the body and can cause higher androgen levels. A biochemical and clinical hyperandrogenism of ovarian origin and to some point adrenal are evident in approximately 60–80% of PCOS patients, resulting in one of the most important features of the syndrome [7]. Extra-ovarian factors, such as excessive levels of LH, low levels of FSH, and intraovarian factors, such as anti-Müllerian hormone (AMH) and inhibin, may enhance the hyperandrogenism state. PCOS might also run-in families. It is common for sisters or a mother and daughter to have PCOS.

The condition can be treated to some extent by controlled medication and bringing alterations in lifestyle. The first therapeutic method in obese patients with PCOS is to obtain weight loss. In addition to an improvement of metabolic comorbidities related to obesity, weight loss reduces hyperinsulinaemia with a consequent increase of insulin sensitivity, reduced LH and androgen levels, and improvement of both menstrual cycle and fertility [8]. Numerous molecules take part in the insulin signalling pathway. Many of them come from natural sources, and their concentration depends on every day consumption through food. Thus, correct dietary habits additionally help maintain physiological ovarian functions [9]. Other techniques include the treatment methods with pills for birth-control, diabetes, fertility, anti-androgen medicines, and scanning procedures like ultrasound scan. When such interventions fail, invasive treatment procedures like surgical drilling of ovaries are also used for improving the ovulation ability of the ovary by reducing the male hormone level.

In this project, we have analyzed the extent of various factors in contributing to PCOS. Around 70% of PCOS cases go undetected, this is a major issue in making it such a widespread problem. So, we have predicted PCOS on the basis of various simple and complex parameters using machine learning techniques. Along with that, we have analyzed diverse implications of PCOS that may happen in the long term. The main contributions of our work are as follows:

(i) Analyzing the previous work done on PCOS.
(ii) Selection of the most important features of the given dataset for PCOS prediction using feature selection method.
(iii) Applying machine learning algorithms (prediction and visualization) on "all features", "some selected important features", "physical features", "hormonal features", "breathing and blood parameters", and "parameters showing the functioning of internal organs" of PCOS dataset.
(iv) Analyzing our work and concluding the results we found.
(v) Comparing the results, we got with others mentioned in literature review.

The organization of paper is as follows. Section 1 provides a brief literature review on the application of statistical tools and machine learning on the diagnosis of PCOS. Section 2 describes the overall methodology applied to conduct this research work. The process of finding the most significant features of PCOS is described in Sect. 3. The results obtained using the application of machine learning on PCOS dataset are described in Sect. 4. Finally, Sect. 5 provides the concluding remarks.

Literature Review

The prevalence of PCOS is increasing in the modern world and is displaying a galloping growth in parallel with the rising prevalence of type 2 diabetes mellitus (T2DM). PCOS has also been stated to affect 28% of unselected obese and 5% of lean women. Chauhan et al. [10] convey into the picture that an estimated one in 5 women suffers from PCOS. It is seen that most girls neglect the common indication of PCOS and visit the physician only when they face issues conceiving. If not diagnosed in time, the circumstance can cause serious health issues.

Bhosale et al. [11] proposed an approach that offers a basis for the automated quality evaluation of PCOS data using a deep convolutional neural network. Their work emphasized that for enhanced performance, machine learning algorithms as well as improved feature selection can be done.

Adla et al. [12] investigated the use of classification algorithms to identify patients with polycystic ovary syndrome (PCOS). In order to enhance the performance of their model, sequential forward floating selection (SFFS) was used to select the best features. The best algorithm was found to be linear support vector machine with 24 features. Both precision and accuracy of their work were about 90%, whereas its recall was around 80%.

Works of Mehr et al. [9] have highlighted that recently, machine learning methods have acquired promising outcomes in medical diagnosis. Kaggle PCOS dataset has been used to diagnose polycystic ovary syndrome for their analysis. Furthermore, the performance of various classifiers (i.e., ensemble random forest, extra tree, adaptive boosting (AdaBoost), and multi-layer perceptron (MLP)) was investigated using the dataset with all features which inspired us to pursue our work in a similar fashion. It has also been deduced that feature selection techniques that produced the most significant subset of features can minimize the computational time and maximize the performance of classifiers.

Bharati et al. [13] applied a number of classifiers such as gradient boosting, random forest, logistic regression, and hybrid random forest and logistic regression (RFLR) to the dataset. Their results demonstrated that RFLR exhibits the best testing accuracy of 91.01% and is concluded as suitable method for reliably classifying PCOS patients. FSH/LH has been observed as the most important feature for the detection of PCOS in their works. PCOS women have an elevated level of LH and a decreased degree of FSH, which leads to disorders in the regulation of the menstrual cycle. The increased degree of LH leads to the development of a surplus of man's sexual hormones (androgens) and oestrogen in the female organism.

Franks et al. [14] inferred that either hypothalamic/pituitary dysfunction or primary ovarian abnormalities (or both) can set up a cycle of events that lead to anovulation. Chronic anovulation seen in PCOS implies prolonged oestrogen excess or lack of progesterone. The main conclusion is that calorie restriction and weight loss in case of obesity reduced hyperinsulinemia by restoring physiological levels of FSH, thereby breaking the cycle that led to arrest of follicle development i.e., anovulation.

Li et al. [15] suggested endometrial progesterone resistance in PCOS. Progesterone resistance implies a decreased responsiveness of target tissue to bioavailable progesterone, and such an impaired progesterone response is seen within the endometrium of the females with PCOS. Along with these hormonal imbalances, several different factors also impact PCOS. Higher BMI and weight gain are amongst the prime features [13].

Legro et al. [6] highlight that there appears to be an epidemic of both obesity and polycystic ovary syndrome (PCOS) in the world today. Although most treatments of obesity, aside from bariatric surgery, obtain modest reductions in weight and improvements in the PCOS phenotype, encouraging weight loss in the obese patient remains one of the most significant therapies. However, further studies are needed to identify the best treatments, and the role of lifestyle therapies in women of normal weight with PCOS is uncertain. This fact is further mentioned in the works of Sirmans et al. [16] which brings out that weight loss improves menstrual irregularities, symptoms of androgen excess, and infertility. Considering the baseline defects in insulin sensitivity and secretion in PCOS and the deleterious impact of obesity on these measures, women with this condition are expected to have a high occurrence of impaired glucose tolerance.

2 Methodology

For the development of an appropriate machine learning model-based diagnostic aid for PCOS and a proper analysis of it, a comparison of performance of various existing algorithms in our dataset is presented as well as various visualization techniques have been used for the analysis. Preparation of the model is the most crucial step that provides the outline of the research. Steps that are included in the development of

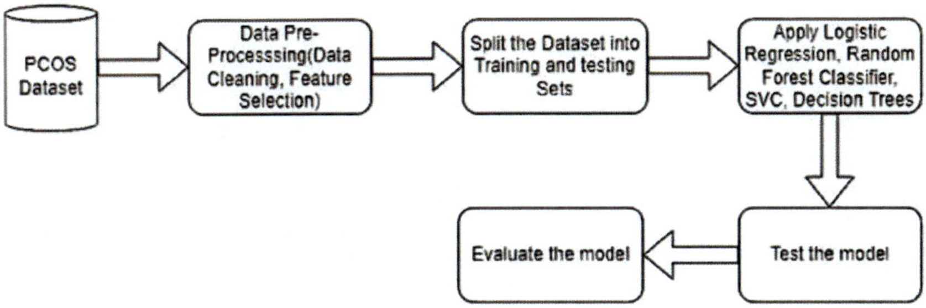

Fig. 1 Flowchart of PCOS prediction

an appropriate model and tuning it for obtaining possibly the best result are detailed above with the help of a workflow diagram, (Fig. 1).

The following section describes these aspects.

2.1 *Dataset Used*

PCOS dataset is a dataset taken from the Kaggle Website [16]. This is in CSV format. This dataset consists of 543 rows and 40 columns. These columns are ('Age (yrs)', 'Weight (Kg)', 'Height (Cm)', 'BMI', 'Blood Group', 'Pulse rate (bpm)', 'RR (breaths/min)', 'Hb (g/dl)', 'Cycle (*R*/*I*)', 'Cycle length (days)', 'Marriage Status (Yrs)', 'Pregnant (*Y*/*N*)', 'No. of aborptions', 'I beta-HCG (mIU/mL)', 'II beta-HCG (mIU/mL)', 'FSH (mIU/mL)', 'LH (mIU/mL)', 'FSH/LH', 'Hip (inch)', 'Waist (inch)', 'Waist:Hip Ratio' 'TSH (mIU/L)', 'AMH (ng/mL)', 'PRL (ng/mL)', 'Vit D3 (ng/mL)', 'PRG (ng/mL)', 'RBS (mg/dl)', 'Weight gain (*Y*/*N*)', 'hair growth(Y/N)', 'Skin darkening (*Y*/*N*)', 'Hair loss (*Y*/*N*)', 'Pimples (*Y*/*N*)', 'Fast food (*Y*/*N*)', 'Reg.Exercise (*Y*/*N*)', 'BP _Systolic (mmHg)', 'BP _Diastolic (mmHg)', 'Follicle No. (*L*)', 'Follicle No. (*R*)', 'Avg. *F* size (*L*) (mm)', 'Avg. *F* size (*R*) (mm)', 'Endometrium (mm)', representing the various parameters affecting the PCOS patients.

2.2 *Data Pre-Processing*

2.2.1 Data Cleaning

In data cleaning, we removed the null values and irrelevant columns. Along with that, we changed the categorical values of object data type to numerical values of float data type which is easy to use. This helped us in improving the standard of the training data for analytics and enabled accurate decision-making.

2.2.2 Feature Selection

We used filter methods, importance gain (Fig. 2), and chi-square test for feature selection. The chi-square test is used for categorical features in a dataset. We calculated chi-square between each feature and the target and selected the desired number of features with the best chi-square scores. Information gain calculated the reduction in entropy from the transformation of a dataset. It can be used for feature selection by evaluating the information gain of each variable in the context of the target variable. It helped us identify the most important parameters responsible for PCOS, and with the chi-square test ($k = 30$), we selected those parameters. We also used correlation matrix to cross verify the results of importance gain.

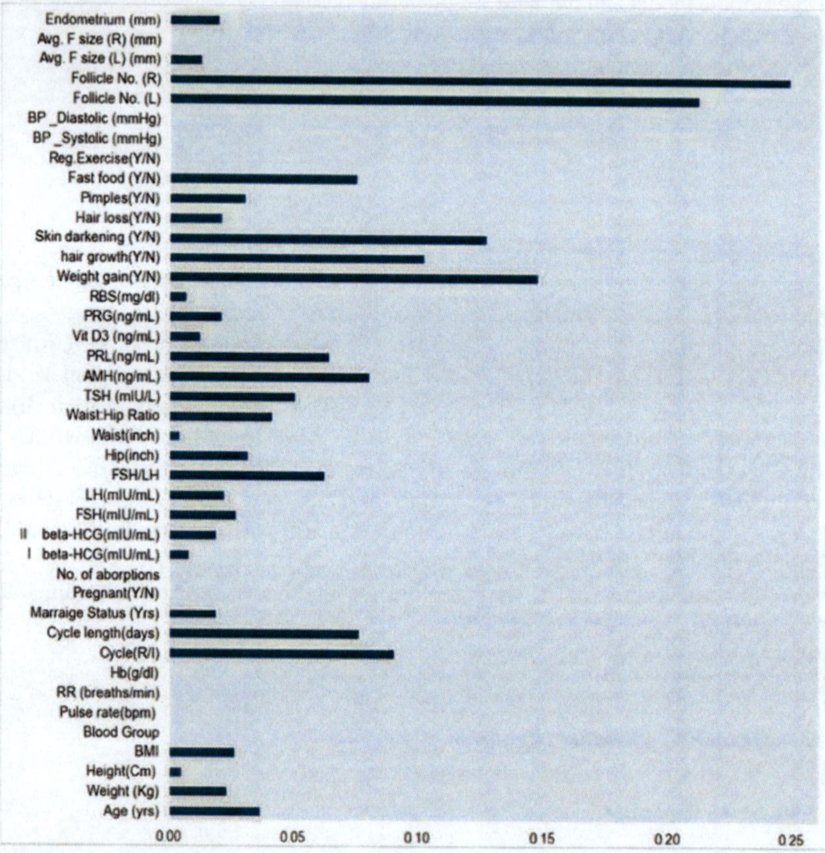

Fig. 2 Importance gain result for all the parameters

2.3 Splitting into Training and Testing

The train-test split is a method for evaluating the performance of a machine learning algorithm. It can be used for classification or regression problems and may be used for any supervised learning algorithm. The method involves taking a dataset and dividing it into two subsets. The first subset is used to fit the model and is directed as the training dataset. The second subset is not used to train the model; instead, the input element of the dataset is fed to the model, then predictions are made and compared with the expected values. This second dataset is referred to as the test dataset [8]. Our ratio of train and test datasets is 4:1, and we applied it on all the prediction models to train them and get their accuracy score.

2.4 Modelling Data

We have used various ML classifiers for making our model for the prediction of PCOS. The classifiers we used are random forest, decision tree, SVC, and logistic regression. Out of these, random forest (depth = 40) performed the best when all the parameters were considered. When applied on the features selected after the chi-square test, the accuracy got further improved (Fig. 3).

We categorized the features in four sets of parameters: "physical features", "hormones", "breathing and blood parameters", and "internal organs functioning" and applied various classifiers on them to predict the occurrence of PCOS depending on these parameters.

3 Data Analysis

3.1 Using Correlation Matrix

We used the correlation matrix to view the correlation between various parameters. It helped us conclude that the number of follicles, weight gain, skin darkening, hair growth, fast food intake, pimples, AMH levels, and cycles are the most important factors in predicting PCOS.

3.2 Using Pair Plot and KDE Plot

We plotted four pair plot merged with the KDE plot for each set of parameters: "physical features", "hormones", "breathing and blood parameters", and "internal organs functioning". It helped us to identify the relation between various attributes.

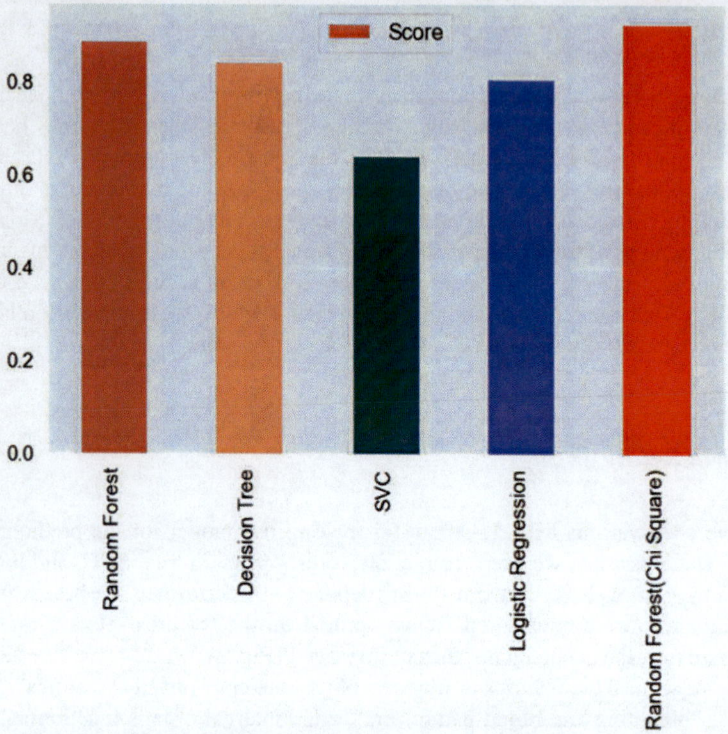

Fig. 3 Comparing accuracy of various models for PCOS prediction

The plot of physical features in Fig. 4 shows that weight gain is associated with hair growth, skin darkening, hair loss, pimples, fast food, and lack of regular exercise; fast food is associated with hair growth, skin darkening, hair loss, and pimples; and pimples are in turn associated with skin darkening and hair loss, in most of the PCOS patients. Also, we see that pimples, fast food intake, and lack of regular exercise are more prominent in PCOS patients.

The plot of internal organs concluded that a greater number of follicles are associated with high BP systolic, BP diastolic, and larger average follicle size in PCOS patients. So if a symptom of PCOS occurs, it further aggravate other symptoms and the cycle continues.

3.3 Using Swarm Plot and Boxen Plot

Using swarm plot and boxen plot, we got to know the degree of impact various individual parameters have on PCOS. Figure 5 shows that high blood sugar and AMH levels are more pronounced in PCOS patients.

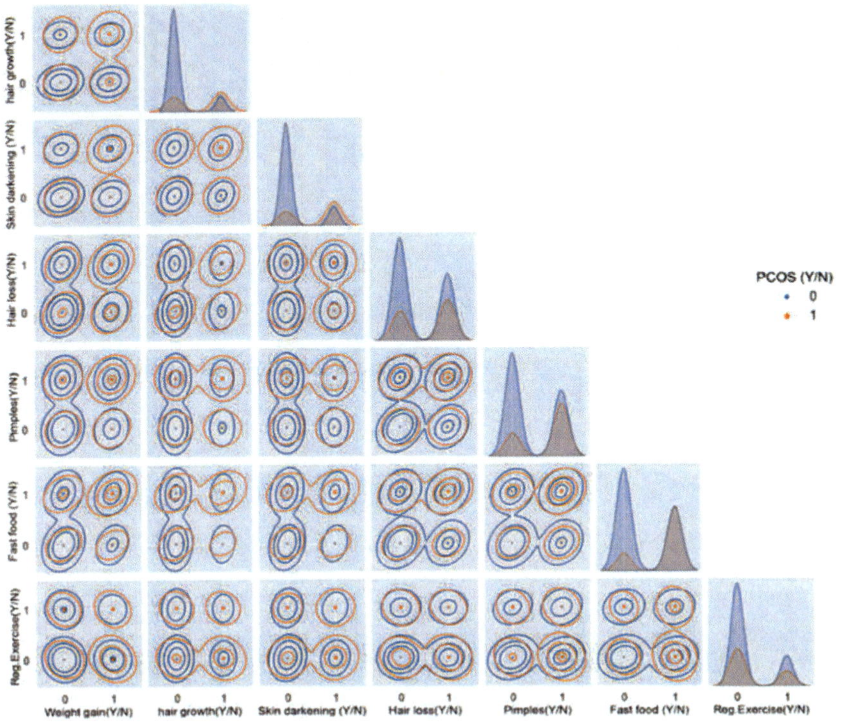

Fig. 4 Pair plot and KDE plot of physical features affecting PCOS

Other plots showed higher weight, BMI, cycle length, follicle number, and reduced endometrium lining in PCOS patients. Also, higher PCOS occurrence is found between 25 and 33 years of age.

4 Result and Discussion

Polycystic ovary syndrome (PCOS) is one of the most common types of endocrine disorder in the reproductive age of women. This may result in infertility and anovulation. The database we used for our project was published in Kaggle, written by Prasoon Kottarathil, called polycystic ovary syndrome (PCOS) in 2020. A database of 543 PCOS patients obtained from Kaggle repository is used. The data comprised the women's reproductive age group i.e., in between 20 and 48 years, most of them being diagnosed around their thirties. The probability of PCOS is predicted using various models and feature selection, and the highest amongst them was 92.66%. Table 1 shows the accuracies of various algorithms used.

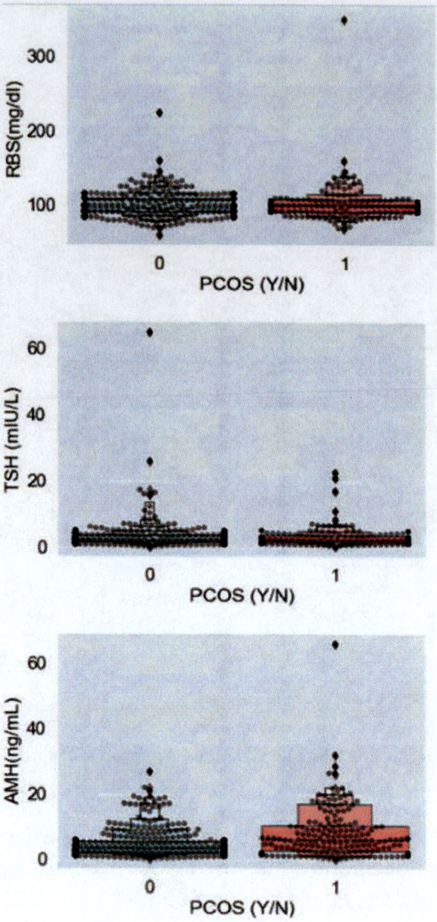

Fig. 5 Swarm plot and boxen plot of various hormones

Table 1 Accuracy of various algorithms used

S. No.	Algorithms used	Accuracy (%)
1	SVC	64
2	Logistic regression	80.7
3	Decision trees	84.4
4	Random forest	88.9
5	Random forest (after applying feature selection)	92.6

In the analysis of our data, PCOS patients are found with a lesser rate of haemoglobin (Hb) than usual which can cause anaemia. Also, people tend to gain weight resulting in higher BMI, leading to obesity. Infrequent, irregular, or prolonged menstrual cycles are the foremost common signs of PCOS due to unbalanced hormones. For example, you would possibly have fewer than nine periods a year, over 35 days between periods and abnormally heavy periods. In PCOS affected women, the ovaries may be enlarged and contain follicles that surround the eggs. In our research, more follicles tend to be found in PCOS patients. As a result, the ovaries might fail to function regularly leading to irregular ovulation. Unpredictable menstrual cycles can even make it difficult to induce pregnancy.

As per our research, PCOS reduces endometrium lining in women, leading to decrease in the rate of fertility. The prolactin level remains unchanged, which means it has no effect on motherhood. Also, I beta-HCG levels are reduced, which is produced after 10 days of conception resulting in a decreased chance of conceiving.

Higher blood sugar level was found in women with PCOS. Insulin is the hormone produced in the pancreas that allows cells to use sugar, our body's primary energy supply. If our cells become resistant to the action of insulin, then our blood sugar levels can rise, and our body might produce more insulin. Excess insulin might increase androgen production, causing difficulty with ovulation, excess facial and body hair (hirsutism), and occasionally severe acne and male-pattern baldness.

Despite our efforts, there are some limitations in the proposed study. The results we obtained have not been validated by various data sets. A wider range of studies can be carried out, including more parameters, resulting in more reliable analyzes in many respects. Along with that ultrasound images can also be used for prediction. In addition, several hybrid methods can be developed to improve the classification accuracy of machine learning algorithms. Additionally, the early effects of PCOS, such as dysmenorrhea and infertility, can be used as an opportunity to prevent long-term consequences by raising awareness of the importance of healthy living and helping to optimize modifiable lifestyle factors such as obesity.

5 Conclusion

A number of different classifiers are used on the 40 features. The diagnostic criterion includes the simple and complex parameters which are biomarkers for the disease. Our methodology involves the concepts of machine learning algorithms. Amongst the various algorithms used, random forest algorithm is found superior in performance with 88.99%. After removing redundant attributes using feature selection, the accuracy rose up to 92.66%.

As a future work, the results obtained from this paper can be validated with a number of different datasets from different parts of world so that an inclusive conclusion can be drawn. The research can be done on a broader scale including more parameters and patients, so that a more reliable analysis can be produced on various aspects. Ultrasound images can also be used for prediction with the help of advanced deep learning algorithms. Moreover, a number of hybrid methods can be developed to increase the classification accuracy of the machine learning algorithms. Furthermore, early consequences of PCOS, such as menstrual cycle disturbances and infertility, could be used as a window of opportunity to prevent long-term consequences by increasing awareness about the importance of a healthy lifestyle and providing support to optimize modifiable lifestyle factors such as obesity.

References

1. Stein IF, Leventhalm ML (1935) Amenorrhea associated with bilateral polycystic ovaries. Am J Obstet Gynecol 29:181–191
2. Jahanfar S, Seppala M, Eden JA, Nguyen TV, Warren P (1995) A twin study of polycystic-ovary-syndrome. Fertil Steril 63(3):478–486
3. Wekker V, van Dammen L, Koning A, Heida KY, Painter RC, Limpens J, Laven JSE, Roeters van Lennep JE, Roseboom TJ, Hoek A (2020) Long-term cardiometabolic disease risk in women with PCOS: a systematic review and meta-analysis. Hum Reprod Update 26(6):942–960. https://doi.org/10.1093/humupd/dmaa029
4. McLuskie I, Newth A (2017) New diagnosis of polycystic ovary syndrome. BMJ 356:i6456
5. Glueck CJ, Dharashivkar S, Wang P, Zhu B, Gartside PS, Tracy T, Sieve L (2005) Obesity and extreme obesity, manifest by ages 20–24 years, continuing through 32–41 years in women, should alert physicians to the diagnostic likelihood of polycystic ovary syndrome as a reversible underlying endocrinopathy. Eur J Obstet Gynecol Reprod Biol 122:206–212
6. Legro RS (2012) Obesity and PCOS: implications for diagnosis and treatment. Semin Reprod Med 30:496–506
7. Franks S (2006) Diagnosis of polycystic ovarian syndrome: in defense of the Rotterdam criteria. J Clin Endocrinol Metab 91(3):786–789
8. Jensterle MJA, Mlinar B, Marc J, Prezelj J, Pfeifer M (2008) Impact of metformin and rosiglitazone treatment on glucose transporter 4 mRNA expression in women with polycystic ovary syndrome. Eur J Endocrinol 158:793–801
9. Danaei Mehr H, Polat H (2021) Diagnosis of polycystic ovary syndrome through different machine learning and feature selection techniques. Heal Technol. https://doi.org/10.1007/s12553-021-00613-y
10. Chauhan P, Patil P, Rane N, Raundale P, Kanakia H (2021) Comparative analysis of machine learning algorithms for prediction of PCOS. In: 2021 International conference on communication information and computing technology (ICCICT). https://doi.org/10.1109/iccict50803.2021.9510128
11. Bhosale S, Joshi L, Shivsharan A (2022) PCOS (polycystic ovarian syndrome) detection using deep learning
12. Adla YAA, Raydan DG, Charaf MZJ, Saad RA, Nasreddine J, Diab MO (2021) Automated detection of polycystic ovary syndrome using machine learning techniques. In: 2021 Sixth international conference on advances in biomedical engineering (ICABME). IEEE, pp 208–212
13. Bharati S, Podder P, Hossain Mondal MR (2020) Diagnosis of polycystic ovary syndrome using machine learning algorithms. In: 2020 IEEE region 10 symposium (TENSYMP). https://doi.org/10.1109/tensymp50017.2020.923
14. Franks S, Hardy K (2020) What causes anovulation in PCOS? Curr Opin Endocr Metab Res. https://doi.org/10.1016/j.coemr.2020.03.001
15. Li X, Feng Y, Lin J-F, Billig H, Shao R (2014) Endometrial progesterone resistance and PCOS. J Biomed Sci 21(1). https://doi.org/10.1186/1423-0127-21-2
16. Sirmans SM, Pate KA (2013) Epidemiology, diagnosis, and management of polycystic ovary syndrome. Clin Epidemiol 6:1–13

Feature Selection for Medical Diagnosis Using Machine Learning: A Review

Kushagra, Rajneesh Kumar, and Shaveta Jain

Abstract Feature selection is a dimensionality reduction approach that aims to choose a small subset of the key characteristics from the original features by deleting redundant, superfluous, or noisy attributes. Improved learning performance, such as increased learning accuracy, lower processing costs, and better model interpretability, can all be attributed to feature selection. Without the proper utilization of FS approach, machine learning models can give inappropriate results. Researchers in computer vision, text mining, and other domains recently presented a variety of feature selection algorithms and shown their usefulness via theory and experiments. In this paper, work done by feature selection researchers to improve the performance of machine learning algorithms is presented which will be useful to budding researchers in this field.

Keywords Machine learning · Feature selection · Unsupervised · Filter · Wrapper · Embedded

1 Introduction

The quantity of data available in many ML applications, such as text mining, computer vision, and biomedicine, has lately increased dramatically in terms of sample size and complexity. It is critical to learn how to use vast amounts of data to obtain information. Our main focus is on high-dimensional data. The huge chunk of high-dimensional data has greatly hampered traditional machine learning methodologies. Learning algorithms may become exceedingly slow, if not degenerate, in their execution of learning tasks as a result of the addition of noisy, redundant, and irrelevant aspects, creating concerns regarding model interpretability. Feature selection can pick a subset of acceptable traits from the original collection by deleting noisy, irrelevant, or redundant information.

Kushagra (✉) · R. Kumar · S. Jain
MMEC, MM (Deemed to be University), Mullana, Ambala, India
e-mail: Agrawal_kushagra@rediffmail.com

P. Chatterjee et al. (eds.), *Computational Intelligence for Engineering and Management Applications*, Lecture Notes in Electrical Engineering 984,
https://doi.org/10.1007/978-981-19-8493-8_14

Depending on the availability of label processed data, feature selection strategies are divided into three categories: supervised techniques [1], semi-supervised techniques [2], and unsupervised techniques [3]. Because label data is readily available, supervised feature selection methods may be successful in discovering discriminative and significant characteristics that differentiate samples from distinct classes. Several supervised learning algorithms have been developed and tested [1]. We may utilize semi-supervised feature selection, which works with both labeled and unlabeled data, when just a fraction of the data is tagged. The bulk of existing semi-supervised feature selection methods [4] is focused with constructing the similarity matrix and selecting features that top match the similarity matrix. Because there are no labels to help in the search for distinguishing characteristics, unsupervised FS is deemed a lot more challenging approach [3].

FS can also be divided into three groups based on different search methods, which include filtering, folding, and embedding methods, respectively. Filter algorithms identify the top most discriminatory features based on data attributes. Sorting techniques, in general, select pre-segmentation features and merging functions, and usually work in two categories. First, all attributes are sorted using specific terms. After that, high-quality features are picked. Many filtering approaches have been utilized, which includes "reliefF" [5], "*F*-statistic" [6], "mRMR" [7], and information retrieval [5]. Wrapper approach analysis of features using a structured learning algorithm. The study [8] used support vector machine methods based on the Elimination of Repetitive Tragedy to identify a gene closely related genetic disease (RFE) gene. Embedded models select features as they construct a model. Figure 1 shows the division of methods for selecting features.

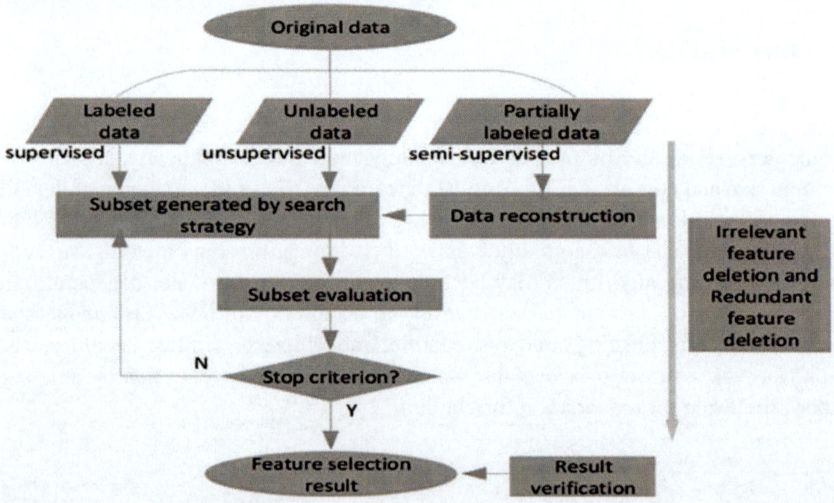

Fig. 1 Feature selection strategy

2 FS in the Prediction of Medical Diseases

FS alias variable selection [9] is a prominent data pre-processing approach in knowledge discovery in data that is primarily used for data depletion by removing extraneous and unwanted features from any data file [10]. Furthermore, this approach improves prediction performance by boosting data comprehension, allowing for improved data visualization, lowering learning algorithm training time, and saving time.

Relevant feature identification algorithms have a big range of applications in the field of health care. Filter, wrapper, ensemble, and embedding techniques are some of the most frequent variable selection approaches. Recently, the vast majority of writers have concentrated on hybrid feature selection tactics.

In order to obtain accurate results quickly, it is usually preferable to decrease noisy and inconsistent data before applying any model to the data. In real-world applications, bringing down the dimensionality of a dataset is critical. Furthermore, by selecting only the most critical criteria, the complexity is greatly reduced.

Many FS algorithms have been used on medical datasets in recent years to get meaningful information. On clinical datasets, FS algorithms are used to predict a variety of illnesses such as diabetes, hypertension, heart disease, thalassemia, and strokes. Different learning algorithms perform better and even more accurate results when the data is more relevant and non-redundant. Because healthcare datasets include a substantial amount of repetitive and irrelevant information, an effective feature selection approach is required to uncover fascinating disease-related components.

The authors presented a greatly accurate diagnostic technique for detecting knee-joint problems using VAG signals in [11]. The method was created by utilizing a one-of-a-kind feature selection and classification technique. The apriori method and the genetic algorithm were used to determine the most significant and stable traits. Random forest and LS-SVM classifiers were employed to assess their performance. Wavelet decomposition was also utilized to distinguish between normal and pathological VAG signals. When the results were compared using assessment metrics, the LS-SVM utilizing the apriori approach is the best performer, with 94.31% accuracy. Proposed method might be extremely useful for early detection of knee-joint disorders, allowing patients to undergo therapy at an early stage.

Ang et al. [12] have established a basic tax selective trait and several genetic selection strategies. These strategies are divided into three types by authors: supervised, sub-surveyed, and selection of unchecked features. It also investigated various barriers and challenges in generating information from gene expression data. The key issues discussed were (1) diminishing data size by hundreds of myriad of factors, (2) dealing with inaccurate and inaccurate data, (3) dealing with very uneven data, and (4) determining genetic/explicit relationships and extract relevant biological information from genetic engineering. According to genetic testing comparisons, the accuracy of the categories of under-monitored and supervised strategies was equally promising as that of genetically modified selection.

Balakrishnan et al. [13] describe a unique feature selection method based on SVM level and a retrospective search strategy for selecting the best subset of traits in the Type-2 diabetes database. The presented method greatly improved the expected accuracy of the Naive Bayes divider. The method used was simple but effective and would no doubt help doctors and medical professionals in diagnosing Type-2 diabetes. Ang et al. created ModifiedFAST, a speedy and efficient feature detection method. Using symmetric uncertainty, the maximum value (SU) was obtained. Using symmetric uncertainty, the shortest tree was formed (SU) [12].

3 Approaches to Feature Selection

Figure 1 depicts a feature selection strategy for reducing an input dataset before giving it to the learning algorithm.

Filter Method—One of the most often utilized FS strategies. The use of a filter method for variable selection requires filtering characteristics prior to executing any learning algorithm. It assigns rankings to traits based on a set of assessment criteria [10, 14]. Its prediction performance varies since it is not affected by the classifier used. When used correctly, these strategies produce quick and efficient outcomes. As a result, they outperform wrapper approaches for large datasets. The disadvantage of these methods is that they overlook feature reliance and classifier interaction, thus missing out on finding the most "useful" traits [12].

Wrapper Method—Wrapper strategies choose features with the learning process in mind. The main benefit of this technique over filter alternatives is that it detects the most "useful" traits and picks material for the learning process correctly [15]. Furthermore, it observes feature dependencies and produces more accurate results than filter approaches [12]. It does have the disadvantage, though, that if another learning algorithm is used, this method must be re-executed. Furthermore, these strategies are challenging and prone to overfitting with limited training datasets.

Embedded Method—Search is often controlled by the learning procedure in the integrated feature selection strategy. This technique, also called the method of nested subset [16], regularly evaluates the "utility" of feature subsets and incorporates FS into the training process [9]. They typically operate in accordance with a certain learning algorithm, which aids in the optimization of the performance of a learning algorithm. Because it does not need partitioning training data into training and validation sets, this strategy makes greater use of current data and gives a speedier solution. They are less costly to calculate than wrapper approaches and have a smaller danger of overfitting. Furthermore, embedded approaches are more computationally demanding than wrapper methods [17]. The fundamental disadvantage of these strategies is that they rely on the classifier to make choices. As a result, the classifier's hypothesis may have an influence on feature selection, which may or may not be compatible with another classifier [18].

4 Related Work

FS is a useful way of defining a set of classification features linked to data selection. The following sections of the study examined the specific work performed in the application and use of Selection Feature in the medical field. Komeili et al. [19] have developed a novel selection method for coping the ECG and temporary data output otoacoustic emission (TEOAE). The effectiveness of this strategy is compared with the efficiency of the seven main FS algorithms and the six ECG and TEOAE biometric detection methods. According to research, the proposed method surpasses competing algorithms with broad genes. The fully automatic tube scanning system analyzes incoming chest X-rays (CXRs) by image processing approaches to improve image quality as well as flexibility depends on model collections to define the pulmonary sections obtained using a number of image metrics [20]. These attributes are then modified using a feature selection method. Results in developing the definition of a class divider ultimately determine whether the processed image is normal or abnormal. Then select the best feature subset from the larger set of image quality—initially used for tasks like tracking an object, image detection, and so on. To test performance, measures such as AUC and ACC were investigated.

The neural network division of the two datasets ("Montgomery and Shenzhen") was 0.99 and 97.03%, respectively, with a central curve and accuracy. Compare their findings with advanced algorithms and radiologists' recommendations for diagnosing asthma (tuberculosis). Pereira et al. [21] also visited and completed a comprehensive survey, as well as a new classification of a collection of relevant field features such as literature classification, biomolecular reviews, spatial classification, and clinical diagnostics, all of which were developed by multiple labels. He completed this research by combining key elements from his classification and exploring research that could be a functional group with multiple labels. Chatterjee et al. [22] indicated how many different types of measures were used to select the top feature set and compare it to their proposed FDM method. In their experiments, they used both a capture method and 10 times the authentication to use the algorithm suggested for various datasets. Accuracy was used to calculate the impacts of selected subsets, calculated using Assembly Classifier variations and Support Vector Machine (SVM). The strongest discovery on this work is reliable and surpasses other ancient T tests, the Kullback–Leibler Divergence (KLD). For 12 and 24 real AAR datasets, our method of selecting a role based on FDM capture achieves 80% accuracy and 78.57%, respectively. The holdout method overrides the original feature sets even when using the best unequal function component (without using any FS approach).

To reduce ambiguity in the set of values, Rostami et al. [23] proposed an Alternative Multiplication Limit of Functional Selection (PCFS). It is applied to eight databases, in which he selects the ones with the smallest repetition and most closely related to the subset set of available attributes. The efficiency of a given system was compared with modern performance with less controlled options in eight databases. According to the numerical findings, the proposed method increased the accuracy of the separation by about 3% while reducing the required features by 1%. As a result, while improving partition accuracy, the proposed method reduced the computer complexity of the machine learning algorithm. In addition, Tubishat et al. [24] propose the "Dynamic Butterfly Optimization Algorithm (DBOA)" addressing FS concerns as an improved version of the "Butterfly Optimization Algorithm (BOA)." BOA is one of the recently proposed development algorithms. In comparison with other development strategies, the BOA has demonstrated its ability to address many problems with comparable results. The BOA, on the other hand, strives to maximize problems with high magnitude. These concerns include a lack of options throughout the development process and the complete suspension of the site. To address these issues, the first BOA made two important improvements: create a local search algorithm (LSAM) operator based on location modification to minimize site beauty and use LSAM to expand various BOA solutions. Using twenty sets of benchmark datasets from the UCI collection, the reliability and height of the DBOA algorithm were verified. The accuracy of the categories, number of selected jobs, health values, statistical findings, and meeting turns of DBOA, and its competitors are all disclosed. According to these findings, DBOA exceeds comparable algorithms with broad genes in the majority of the process studied.

In addition, Ghaddar et al. [25] suggested a repetitive adjustment approach that would allow the number of selected features to be converted to a much greater limit with the $l1$-norm vector separator. In the definition of real life, you have investigated two complexities with high-intensity traits. Its first use was in vegetative testing using microarray data, which suggested a way to differentiate genetic cancer. The second sample included online comments from Yelp, Amazon, and IMDb. The discovery shows that the presented method of categorization and selection is easy, measurable and provides low error rates, all of which are essential for the creation of complex decision-making systems. Lee et al. [26] develop new strategies. The contemporary classifier or predicament that is most involved in feature selection enhances classification and speculation significantly. The new wrapper-based C4.5 algorithm is intended to assist in making informed clinical decisions in the medical and healthcare industries. In addition to dealing with data distortions, the newly published S-C4.5-SMOTE sample optimization device improves device performance by trying to reduce data size while maintaining balanced and efficient datasets. This functionality directly allows Wrapper's approach to choosing a profitable job, which does not require large amounts of data. Jain et al. [27] discussed the various methods of selecting features, as well as their natural benefits and limitations. He then investigated flexible separation systems and simultaneous separation systems to predict incurable diseases. Pashaei et al. [28] suggested using the Binary version of the "Black Hole Algorithm (BBHA)" to address career selection issues in biological knowledge. BBHA is an extension of an existing BHA banner. In addition, the six groups of important cutting trees ("Random Forest, Bagging, C5.0, C4.5, Boosted C5.0, and CART") were compared significantly better as the recommended method tester in his study. C4.5 is a differentiated method of producing DT. This is achieved by pressing the continuous features and the missing data cycle. C4.5 was used to construct DT, also known as analytical separator and can also be used for integration. The nodes and arms are integrated into the DT. Each node collects concerns based on one or more factors, such as comparing adjective value and consistency or comparing fewer features using special functions. The result tree is another name for the training dataset by the popular tree. C4.5 is a collection of machine learning and algorithms for mining data sharing. Hassan et al. [29] "Black Hole Algorithm (BHA)" is a transformation and successful global search method based on Black Holes behavior. It is used to develop a wide range of scenarios. However, it has not yet investigated BHA's job selection capabilities. Its findings showed that the efficiency of RF exceeds the other decision tree algorithms in all measures, and that BBHA's preferred wrap option strategy surpasses BPSO, GA, SA, and CFS. BBHA exceeds BPSO and GA in terms of processing time, number of model configuration parameters, and number of selected configuration functions. In addition, BBHA exceeds competitors in other ways mentioned in the literature. Tuba et al. [30] presented a practical strategy for developing cognitive thinking in the collection of clinical data elements. Separation is done with the help of a vector support machine, and the parameters of which are selected using the process of developing a brainstorm. Using sets of standard, transparent medical data, the suggested method is compared to existing methods. By analyzing the data collected, a robust approach was demonstrated, and the number of required signals was reduced. Sakri et al. [31] investigated the effect of incorporating a FS algorithm within the breast cancer prediction phase algorithm. They argue that by using selection processes to limit the number of features, we may improve the performance of multiple classification systems. Some factors have a greater correlation and impact on the effects of segregation than others. The results of our test of three common classification algorithms using and without the selective particle development feature have been published (PSO). In the end, with the PSO and beyond, the Naive Bayes performed much better than the other two alternatives.

5 Conclusions

In order to pick only useful data from the dataset, feature selection is an essential strategy that is employed before applying classifiers to a data collection. A solid FS approach that selects key features aids in modeling successfully and meaningfully while maintaining a cheap computing cost and excellent classification accuracy. In this paper, a sketch of the FS techniques is presented in the literature. FS is an important element in categorizing diverse difficulties. To begin, the concept may be simplified, and computing costs can be decreased by giving less inputs, while a model is employed for practical implementation. Second, reducing superfluous variables from data gathering would improve model transparency and interpretation, aiding in the explanation of potential diagnoses, which eventually is important in medical applications. The feature selection approaches can assist in reducing unwanted data and therefore improving classification accuracy. As presented in earlier sections, this study investigated many recent publications on feature selection in health care that were published recently. This study gives a briefing about the work already done in this field which will help budding researchers to get a clear picture about the importance of feature selection and how to investigate with better accuracy in the field of medical sector. Summary of the related work is given in Table 1.

In the future, we will implement devising an approach which will perform best in selecting best possible features for pulmonary diseases dataset, then performance of classification algorithm will be analyzed.

Table 1 Examine the algorithms for feature selection

Ref	Dataset	FS/algorithm	Problems	Accuracy
Komeili et al. [19]	ECG, TEOAE, auxiliary, synthetic	EECG and TEOAE outperformed seven cutting-edge FS algorithms. Biometric identification as six alternate approaches	ECG address signals and transient otoacoustic emissions elicited address (TEOAE)	75, 85, 95 and 99%
Vajda et al. [20]	Montgomery Shenzhen	Lung segmentation, features description, FS, and classification	Detection of pulmonary anomalies including TB (TB)	97.03%
Chatterje et al. [22]	• Brain–Computer Interface (BCI) • Competition on III motor imagery • Electroencephalog ram (EEG)	FS and DM and PCA	To identify the most suitable feature subset	Accuracy rates were 80% and 78.57% for the 12 and 24 feature AAR datasets, respectively
Rostami et al. [23]	SPECTF SpamBase sonar arrhythmia madelon colon	Novel pairwise constraint FS method (PCFS)	When dealing with very high-dimensional datasets in medical applications, the classifier output drops significantly	79.66%
Tubishat et al. [24]	UCI	There are two types of FS techniques: filter-based methods and wrapper-based methods	Typically, datasets contain irrelevant properties that might impact the classifier's result	DBOA surpassed all other baseline algorithms with greater average accuracy (7.83, 4.71, 8.09, 3.00, 8.94, and 7.18%) than (BOA, GA, GOA, POS, ALO, and SCA)

(continued)

Table 1 (continued)

Ref	Dataset	FS/algorithm	Problems	Accuracy
Ghaddar et al. [25]	Real-world datasets	FS and SVM	High dimensionality microarray data is a difficulty due to the presence of various noisy qualities that do not contribute to the decrease of classification mistakes	Achieves low error rates
Lee et al. [26]	ECG	SMOTE ((SMOTE)) is a unique bagging C4.5 method based on wrapper feature selection	How to deal with the multidimensionality and large volume of data generated by IoT medical systems	100%
Pashaei et al. [28]	Eight medical–biological dataset	BBHA for solving FS	To choose a small or significant number of relevant characteristics in order to improve classification efficiency	Best performance
Tuba et al. [30]	Standard publicly available medical datasets	SVM and FS	Reduce the feature set	91.46%
Sakri et al. [31]	Wisconsin Prognosis Breast Cancer dataset	• FS with Naive Bayes • FS with REP Tree • FS with IBK	Fear of recurrence of breast cancer and early illness prediction may aid patients in receiving timely treatment	• 81.3% • 80% • 75%

References

1. Nie F, Huang H, Cai X, Ding CH (2010) Efficient and robust feature selection via joint ℓ2, 1-norms minimization. In: Advances in neural information processing systems, pp 1813–1821
2. Wang P, Li Y, Chen B, Hu X, Yan J, Xia Y, Yang J (2013) Proportional hybrid mechanism for population based feature selection algorithm. Int J Inf Technol Decis Making 1–30
3. Liu R, Rallo R, Cohen Y (2011) Unsupervised feature selection using incremental least squares. Int J Inf Technol Decis Mak 10(06):967–987
4. Cheng Q, Zhou H, Cheng J (2011) The fisher-markov selector: fast selecting maximally separable feature subset for multiclass classification with applications to high-dimensional data. IEEE Trans Pattern Anal Mach Intell 33(6):1217–1233
5. Raileanu LE, Stoffel K (2004) Theoretical comparison between the gini index and information gain criteria. Ann Math Artif Intell 41(1):77–93
6. Ding C, Peng H (2005) Minimum redundancy feature selection from microarray gene expression data. J Bioinf Comput Biol 3(02):185–205

7. Peng H, Long F, Ding C (2005) Feature selection based on mutual information criteria of max-dependency, max-relevance, and min-redundancy. IEEE Trans Pattern Anal Mach Intell 27(8):1226–1238
8. Guyon I, Weston J, Barnhill S, Vapnik V (2002) Gene selection for cancer classification using support vector machines. Mach Learn 46(1–3):389–422
9. Guyon I, Elisseeff A (2003) An introduction to variable and feature selection. J Mach Learn Res 3(Mar):1157–1182
10. Tang J, Alelyani S, Liu H (2014) Feature selection for classification: a review. In: Data classification: algorithms and applications, p 37
11. Nalband S, Sundar A, Prince AA, Agarwal A (2016) Feature selection and classification methodology for the detection of knee-joint disorders. Comput Methods Programs Biomed 127:94–104
12. Ang JC, Mirzal A, Haron H, Hamed H (2015) Supervised, unsupervised and semi- supervised feature selection: a review on gene selection. IEEE/ACM Trans Comput Biol Bioinf 99
13. Balakrishnan S, Narayanaswamy R, Savarimuthu N, Samikannu R (2008) SVM ranking with backward search for feature selection in type II diabetes databases. In: IEEE international conference on systems, man and cybernetics, SMC 2008. IEEE, pp 2628–2633
14. Dash M, Liu H (1997) Feature selection for classification. Intell Data Anal 1(3):131–156
15. Kumar V, Minz S (2014) Feature selection. SmartCR 4(3):211–229
16. Shahana AH, Preeja V (2016) Survey on feature subset selection for high dimensional data. In: 2016 International conference on circuit, power and computing technologies (ICCPCT). IEEE, pp 1–4
17. Wikipedia (2016) Feature selection. https://en.wikipedia.org/wiki/Feature_selection. Accessed 24 Oct 2016
18. Hira ZM, Gillies DF (2015) A review of feature selection and feature extraction methods applied on microarray data. Adv Bioinf
19. Komeili M, Louis W, Armanfard N, Hatzinakos D (2018) Feature selection for nonstationary data: application to human recognition using medical biometrics. IEEE Trans Cybern 48(5):1446–1459. https://doi.org/10.1109/TCYB.2017.2702059
20. Vajda S, Karargyris A, Jaeger S, Santosh KC, Candemir S, Xue Z, Antani S, Thoma G (2018) Feature selection for automatic tuberculosis screening in frontal chest radiographs. J Med Syst 42(8). https://doi.org/10.1007/s10916-018-0991-9
21. Pereira RB, Plastino A, Zadrozny B, Merschmann LHC (2018) Categorizing feature selection methods for multi-label classification. Artif Intell Rev 49(1):57–78. https://doi.org/10.1007/s10462-016-9516-4
22. Chatterjee R, Maitra T, Hafizul Islam SK, Hassan MM, Alamri A, Fortino G (2019) A novel machine learning based feature selection for motor imagery EEG signal classification in Internet of medical things environment. Futur Gener Comput Syst 98:419–434. https://doi.org/10.1016/j.future.2019.01.048
23. Rostami M, Berahmand K, Forouzandeh S (2020) A novel method of constrained feature selection by the measurement of pairwise constraints uncertainty. https://doi.org/10.1186/s40537-020-00352-3
24. Tubishat M, Alswaitti M, Mirjalili S, Al-Garadi MA, Alrashdan MT, Rana TA (2020) Dynamic butterfly optimization algorithm for feature selection. IEEE Access 8:194303–194314. https://doi.org/10.1109/access.2020.3033757
25. Ghaddar B, Naoum-Sawaya J (2018) High dimensional data classification and feature selection using support vector machines. Eur J Oper Res 265(3):993–1004. https://doi.org/10.1016/j.ejor.2017.08.040
26. Lee SJ, Xu Z, Li T, Yang Y (2018) A novel bagging C4.5 algorithm based on wrapper feature selection for supporting wise clinical decision making. J Biomed Inf 78:144–155. https://doi.org/10.1016/j.jbi.2017.11.005
27. Jain D, Singh V (2018) Feature selection and classification systems for chronic disease prediction: a review. Egypt Inf J 19(3):179–189. https://doi.org/10.1016/j.eij.2018.03.002

28. Pashaei E, Aydin N (2017) Binary black hole algorithm for feature selection and classification on biological data. Appl Soft Comput 56:94–106. https://doi.org/10.1016/j.asoc.2017.03.002
29. Hassan OMS, Abdulazeez AM, Tiryaki VM (2018) Gait-based human gender classification using lifting 5/3 wavelet and principal component analysis. In: ICOASE 2018—International conference on advanced science and engineering, pp 173–178. https://doi.org/10.1109/ICOASE.2018.8548909
30. Tuba E, Strumberger I, Bezdan T, Bacanin N, Tuba M (2019) Classification and feature selection method for medical datasets by brain storm optimization algorithm and support vector machine. Procedia Comput Sci 162(3):307–315. https://doi.org/10.1016/j.procs.2019.11.289
31. Sakri SB, Abdul Rashid NB, Muhammad Zain Z (2018) Particle swarm optimization feature selection for breast cancer recurrence prediction. IEEE Access 6(c):29637–29647. https://doi.org/10.1109/ACCESS.2018.2843443

Computational Intelligence in Image/Gesture Processing

Convolutional Neural Network Architectures Comparison for X-Ray Image Classification for Disease Identification

Prince Anand, Pradeep, and Aman Saini

Abstract The coronavirus (COVID-19) disease was caused by the SARS-CoV-2 virus. Due to the worldwide spread of the disease, the World Health Organization declared it as a global pandemic. The disease attacks the respiratory system and human lungs. Computer vision can be a viable option for detecting diseases from X-ray images. In our study, we are classifying X-ray images into four predefined classes for disease identification using some CNN models. Here, we are comparing VGG-16, VGG-19, Inception V3, and DenseNet201 architectures based on their accuracy, precision, recall, and *F*-Score. We trained and validated all CNN models with the dataset available on Kaggle containing X-ray images.

Keywords X-ray · Convolutional neural network · Deep learning · VGG-16 · VGG-19 · Inception V3 · DenseNet201

1 Introduction

The coronavirus (COVID-19) disease was caused by the SARS-CoV-2 virus. The initial known case was occurred in Wuhan, China, in early December 2019 and identified [1]. The main symptoms of this disease are fever [2], cough, headache, [3] fatigue, breathing difficulties, and a lack of scent and taste [4–6].

The result of a decrease in the ratio of gas to soft tissue (blood, lung parenchyma, and stroma) in the lung is known as pulmonary opacification or lung opacity [7, 8].

Viral pneumonia is pneumonia caused by a virus. Pneumonia is an infection that causes inflammation in one or both of the lungs [9]. The pulmonary alveoli fill with fluid or pus, making it difficult to breathe [10].

P. Anand (✉) · Pradeep · A. Saini
Department of ECE, G.B. Pant Government Engineering College Delhi, Delhi 110020, India
e-mail: prince32anand@gmail.com

A. Saini
e-mail: amansaini@gbpec.edu.in

P. Chatterjee et al. (eds.), *Computational Intelligence for Engineering and Management Applications*, Lecture Notes in Electrical Engineering 984,
https://doi.org/10.1007/978-981-19-8493-8_15

A convolutional neural network (CNN or ConvNet) [11] is a class of artificial neural network, most commonly used to analyze visual imagery.

The VGG-16 is a type of convolutional neural network. It is a deep network having 16 layers and is used for image classification. This network is pre-trained with over a million photos to categorize images into 1000 classes or categories. The input to the network is images of size which is 224 * 224.

Similarly, the VGG-19 is a deep network having 19 layers and is used for image classification. The Inception is a type of convolutional neural network. It is a deep neural network having 48 layers and is used for image classification. This network is pre-trained to categorize photos into 1000 classes or categories.

So, the network has clearly learnt feature-rich representations for a variety of picture genres. The image having a resolution of 299 * 299 pixels is given as parameter for the network. The Inception was initially named "GoogleLeNet".

The DenseNet201 is a type of convolutional neural network. It is a 201-layer deep network used for image classification. This network is pre-trained with over a million images to categorize images into 1000 classes more than a million pictures to organize into 1000 classifications or variety of categories. The image of size is 224 * 224 which is given as parameter for the network.

In our research, we are comparing the existing pre-built CNN architectures based on accuracy, recall, *F*-Score. We trained all the four CNN architectures using the identical dataset consisting of X-ray images belonging to four classes, namely COVID, viral pneumonia, lung opacity, and normal. The dataset is available on Kaggle.

The dataset is partitioned into two subsets which contains the 80–20% ratio of images of total images for training and validation of CNN architectures. The CNN architectures should classify the X-ray pictures into four classes as a person diagnosed with COVID, a person diagnosed with lung opacity, a person diagnosed with viral pneumonia, or a person with good health, i.e., person with no disease.

2 Literature Survey

1. Deep convolutional neural network (DCNN) such as VGG-16 (VGG with 16-weights layers) as well as DenseNet121, with transfer learning methods, was applied in to enhance the generalization ability and perform the better classification accuracy based on the pediatric chest X-ray dataset [12].
2. Automated image analysis based on artificial intelligence is being developed to detect, quantify, and monitor COVID-19 infections, as well as to separate healthy lungs from diseased ones by using transfer learning [13].
3. Two CNN architectures: (1) channel-shuffled dual-branched (CSDB) CNN and (2) CSDB CNN augmented with distinctive filter learning (DFL) paradigm are used [14].
4. Deep transfer learning algorithm focuses on binary classification of images in order to classify the various CXR and CT scan images for fast and accurate detection of COVID-19. The algorithm considers the pre-trained weights to

extract simple features and then learn the pattern of COVID-19 cases obtained from the patient's CXR and CT scan images. The main feature of the technique is that it is fully connected layers with five extra layers in VGG-19 model [15].
5. COVID-19 chest images dataset was preprocessed before being trained with deep learning models. In this preprocessing, the dataset was reconstructed using the fuzzy technique and stacking technique and then trained the three datasets using the MobileNetV2 and SqueezeNet deep learning models and classified the models by the SVM method [16].
6. An automatic COVID screening (ACoS) system employs hierarchical classification using conventional ML algorithms and radiomic texture descriptors to segregate normal, pneumonia, and nCOVID-19-infected patients. The major advantage of the system is that it can be easily modeled using the limited number of annotated images and can be deployed even in a resource-constrained environment [17].
7. Automatic classification of the images was carried out through the two different convolutional neural networks (CNNs). In the study, experiments were carried out for the use of images directly, using local binary pattern (LBP) as a preprocess and dual-tree complex wavelet transform (DT-CWT) as a secondary operation, and the results of the automatic classification were calculated separately [18].
8. A method of prostate ultrasound image segmentation based on the improved mask R-CNN includes the following steps: first, the improved S-Mask R-CNN network model was established. Then, the prostate ultrasound images to be segmented were used as input into the network for segmentation. Finally, the ultrasound images of prostate after segmentation were used as output [19].
9. A hybrid bi-branched CNN model based on dilated convolution is to classify CXR images into three infection types: COVID-19, non-COVID-19, and normal. The first ten layers of the VGG-16 architecture were used as the front end of the proposed model, VGG-COVIDNet. The back end was based on dilated kernels. The main idea was the deployment of a deeper CNN to produce high-level features without losing detailed information in the images [20].
10. A binary encoding method is to represent the structure of the model in a binary string, where "1" and "0" indicate whether the feature is allowed to pass into a layer or not. Three GA operations, selection, cross-over, and mutation, are employed to evolve the structure. After conducting the selection on each generation, poorly performing structures are discarded [21].

3 Methodology

3.1 Dataset

For our study, the dataset is what we are working with is freely and publicly available on Kaggle. The dataset is named as COVID-19 radiography database that contains

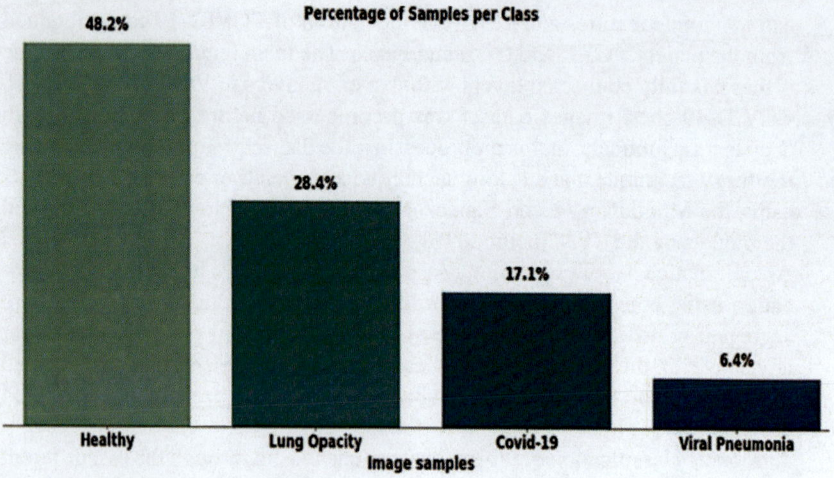

Fig. 1 Percentage of samples per class

X-ray images in four classes, namely patients diagnosed with coronavirus having 3616 images, patients diagnosed with viral pneumonia having 1345 images, patients diagnosed with lung opacity having 6012 images, and healthy person having 10.2 k images [22].

Every X-ray image has resolution of 299 × 299 pixels and is in portable network graphics (PNG) file format [22]. The percentage of X-ray image samples per class is shown in Fig. 1.

3.2 VGG-16 Architecture

An image with dimensions (224, 224, 3) is fed into the network as input. The first two layers each have a 3 * 3 filter with 64 channels and the same padding, followed by a max pool layer of stride (2, 2), two convolution layers, each of which is 256 times the filter's size (3, 3)*;* after which, there is a max pooling layer of stride (2, 2) similar to the previous layer. There are convolution layers of filter size (3, 3*)* and *256* filters which are two in number. Following that, there are two sets of three convolution layers and a max pooling layer. Each layer has 512 filters of the same size (3, 3) and the same padding, and then, the image is passed to a convolution layer stack consisting of two layers. In these max pooling and convolution layers, instead of AlexNet's 11 * 11 size and ZF-Net 7 * 7 size, the filters used are of size 3 * 3. A 1 * 1 pixel is used to change the number of input channels in some layers. After each convolution layer to prevent the spatial feature of the image, padding of 1-pixel (same padding) is present [23].

Then, after the stack of convolution and max pooling layer, we got a (7, 7, 512) feature map. The operation of flattening was performed on this result to make it a one-column (1, 25088) feature vector. Following that, there are three fully connected layers: the initial layer obtains input from the last feature vector and provides an output a vector of size (1, 4096), the second layer also gives a vector of size (1, 4096) as an output, but the third layer outputs 1000 channels, and then after, the output of fully connected third layer is passed to Softmax layer to normalize the classification vector. The rectified linear unit (ReLU) activation function is applied to all the hidden layers [24]. The ReLU results in faster learning and also lowers the chances of vanishing gradient problems; thus, it is more computationally. The architecture is shown in Fig. 2 [23].

ConvNet Configuration					
A	A-LRN	B	C	D	E
11 weight layers	11 weight layers	13 weight layers	16 weight layers	16 weight layers	19 weight layers
input (224 × 224 RGB image)					
conv3-64	conv3-64 LRN	conv3-64 conv3-64	conv3-64 conv3-64	conv3-64 conv3-64	conv3-64 conv3-64
maxpool					
conv3-128	conv3-128	conv3-128 conv3-128	conv3-128 conv3-128	conv3-128 conv3-128	conv3-128 conv3-128
maxpool					
conv3-256 conv3-256	conv3-256 conv3-256	conv3-256 conv3-256	conv3-256 conv3-256 conv1-256	conv3-256 conv3-256 conv3-256	conv3-256 conv3-256 conv3-256 conv3-256
maxpool					
conv3-512 conv3-512	conv3-512 conv3-512	conv3-512 conv3-512	conv3-512 conv3-512 conv1-512	conv3-512 conv3-512 conv3-512	conv3-512 conv3-512 conv3-512 conv3-512
maxpool					
conv3-512 conv3-512	conv3-512 conv3-512	conv3-512 conv3-512	conv3-512 conv3-512 conv1-512	conv3-512 conv3-512 conv3-512	conv3-512 conv3-512 conv3-512 conv3-512
maxpool					
FC-4096					
FC-4096					
FC-1000					
soft-max					

Fig. 2 Different VGG configurations [23]

3.3 VGG-19 Architecture

The architecture of VGG-19 is completely the same as VGG-16 but with some additions. The VGG-19 has 19 layers; hence, three convolutional layers are added in this architecture. The architecture is shown in Fig. 2 [23].

3.4 Inception V3 Architecture

The Inception architecture is made up of special types of layers called inception layers. In these layers, three convolution operations and one max pooling operation are done on a single layer, and then, the result of each is concatenated to make the final output of the layer. Initially, Inception V1 was made where 1 * 1, 3 * 3, and 5 * 5 convolution operation and one max pooling are performed on the same level also; to reduce computation, an extra 1 * 1 convolutional layer is added before the 3×3 and 5×5 convolutions [25].

The inception V1 was good, but to further reduce computations, Inception V2 was designed; here, the 5 * 5 convolution layer was replaced by two 3 * 3 convolutional layers, to reduce the computation. Moreover, the $n * n$ convolutional layer was decomposed as $1 * n$ and $n * 1$ to reduce further computation [26].

Finally, at last, Inception V3 was made having all features of Inception V2 with some changes, and the optimizer was changed as RMSprop optimizer; also, the convolution factor was changed to 7 * 7. Batch normalization in the fully connected layer of the auxiliary classifier can help improve the performance of the model by reducing internal covariate shift and improving the generalization ability of the network [26].

3.5 DenseNet201 Architecture

The DenseNet contains all the fundamental layers of a CNN, i.e., convolutional layer, pooling layer, activation layer, fully connected layer, Softmax layer.

In addition to all layers mention above, the DenseNet has direct connections from any layer to all subsequent layers [24]. For L layers, there are $L(L + 1)/2$ direct connections.

Consequently, the lth layer gets all previous layers' feature maps, $x_0,\ldots, x_{(l-1)}$ layer, as input:

$$x_l = H_l([x_0, x_1, \ldots, x_{(l-1)}]),$$

where $[x_0, x_1,\ldots, x_{(l-1)}]$ refers to the joining of the feature maps produced in layers $0,\ldots, l-1$. Since the network is densely connected, thus this network architecture is known as the dense convolutional network (DenseNet) [27]. The architecture is shown in Fig. 3.

Layers	Output Size	DenseNet-121	DenseNet-169	DenseNet-201	DenseNet-264
Convolution	112 × 112	7 × 7 conv, stride 2			
Pooling	56 × 56	3 × 3 max pool, stride 2			
Dense Block (1)	56 × 56	[1 × 1 conv, 3 × 3 conv] × 6	[1 × 1 conv, 3 × 3 conv] × 6	[1 × 1 conv, 3 × 3 conv] × 6	[1 × 1 conv, 3 × 3 conv] × 6
Transition Layer (1)	56 × 56	1 × 1 conv			
	28 × 28	2 × 2 average pool, stride 2			
Dense Block (2)	28 × 28	[1 × 1 conv, 3 × 3 conv] × 12	[1 × 1 conv, 3 × 3 conv] × 12	[1 × 1 conv, 3 × 3 conv] × 12	[1 × 1 conv, 3 × 3 conv] × 12
Transition Layer (2)	28 × 28	1 × 1 conv			
	14 × 14	2 × 2 average pool, stride 2			
Dense Block (3)	14 × 14	[1 × 1 conv, 3 × 3 conv] × 24	[1 × 1 conv, 3 × 3 conv] × 32	[1 × 1 conv, 3 × 3 conv] × 48	[1 × 1 conv, 3 × 3 conv] × 64
Transition Layer (3)	14 × 14	1 × 1 conv			
	7 × 7	2 × 2 average pool, stride 2			
Dense Block (4)	7 × 7	[1 × 1 conv, 3 × 3 conv] × 16	[1 × 1 conv, 3 × 3 conv] × 32	[1 × 1 conv, 3 × 3 conv] × 32	[1 × 1 conv, 3 × 3 conv] × 48
Classification Layer	1 × 1	7 × 7 global average pool			
		1000D fully-connected, softmax			

Fig. 3 Different DenseNet configurations [27]

Growth Rate

If every function H_l generates k feature maps, following that, the lth layer has $k_0 + k\times(l-1)$ input feature maps, where number of count of channels in the input layer is k_0 [27].

Bottleneck Layer

A 1 * 1 convolutional layer is added as a bottleneck layer before the 3 * 3 convolutional layer to decrease the number of input feature maps and thus to improve computational efficiency [27].

4 Results and Discussions

Our model is trained to 500 epochs. The input images are X-ray images belonging to four classes COVID, viral pneumonia, lung opacity, and normal, all having a resolution of 299 × 299 pixels.

In our research, we have observed some CNN architectures VGG-16, VGG-19, Inception V3, and DenseNet201. All the architectures were primarily used for image classification.

Here, we have compared all the architectures by classifying X-ray images into four classes, namely a person diagnosed with COVID, a person diagnosed with lung opacity, a person diagnosed with viral pneumonia, or a person with good health, i.e., person with no disease.

All architectures are compared based on accuracy, recall, *F*-Score. These are determined as follows:

TP—True positive
TN—True negative
FP—False positive
FN—False negative.

$$\text{Accuracy} = (\text{TP} + \text{TN})/(\text{TP} + \text{TN} + \text{FP} + \text{FN}) \quad (1)$$

$$\text{Precision} = \text{TP}/(\text{TP} + \text{FP}) \quad (2)$$

$$\text{Recall} = \text{TP}/(\text{TP} + \text{FN}) \quad (3)$$

$$F1\ \text{Score} = 2 \times (\text{Precision} \times \text{Recall})/(\text{Precision} + \text{Recall}) \quad (4)$$

The VGG-16 has an accuracy of 93.50%, precision of 93.74%, recall of 92.55%, and *F*-Score of 93.12%. The VGG-19 has an accuracy of 93.10%, precision of 93.14%, recall of 93.17%, and *F*-Score of 93.11%. The Inception *V*3 has an accuracy of 93.90%, precision of 94.66%, recall of 91.21%, and *F*-Score of 92.77%. The DenseNet201 has an accuracy of 82.28%, precision of 84.57%, recall of 79.72%, and *F*-Score of 80.05%. Hence, Inception V3 possesses the greatest precision and accuracy, DenseNet201 has the lowest all factors, VGG-16 has the highest *F*-Score, and VGG-19 has the highest recall (Table 1).

On observing the confusion matrix of all the architectures, i.e., Figs. 4, 5, 6 and 7, the viral pneumonia is the class having minimum misclassifications and the lung opacity is the class having maximum misclassifications. The Densenet201 has the highest number of misclassifications, and Inception V3 has minimum misclassifications.

Table 1 Results of different CNN models

CNN Models	Accuracy (%)	Precision (%)	Recall (%)	*F*-Score (%)
VGG-16	93.50	93.74	92.55	93.12
VGG-19	93.10	93.14	93.17	93.11
Inception V3	93.90	94.66	91.21	92.77
DenseNet201	82.28	84.57	79.72	80.05

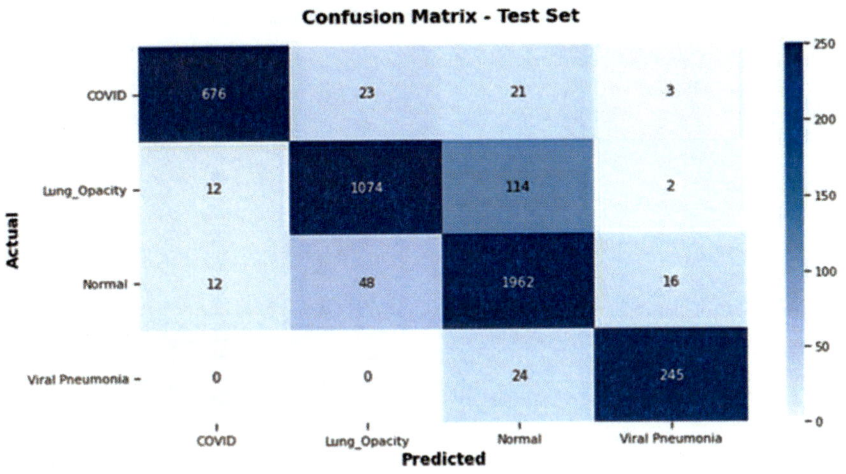

Fig. 4 Confusion matrix of VGG-16 architecture

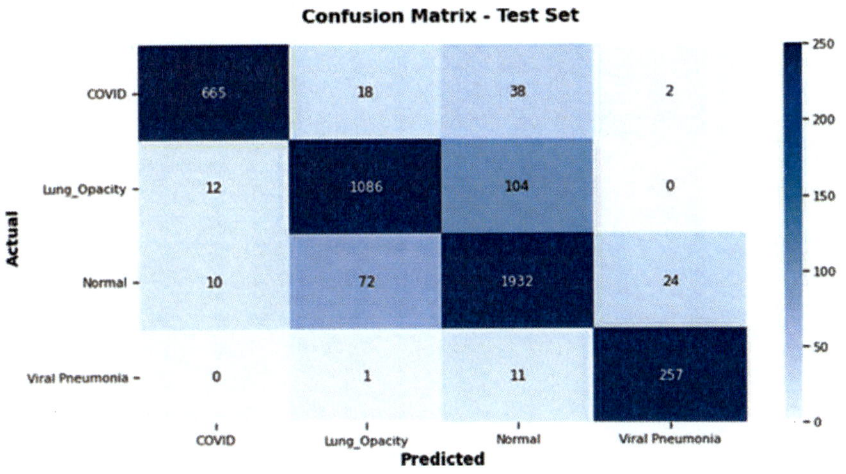

Fig. 5 Confusion matrix of VGG-19 architecture

5 Conclusion

After a comparison of VGG-16, VGG-19, Inception V3, DenseNet201 architectures, it is clear that all are extremely complex architectures and all have high accuracy above 90%; thus, all can classify X-ray images efficiently.

The DenseNet201 has the highest number of misclassifications with the smallest accuracy, and Inception V3 has minimum misclassifications and the highest accuracy.

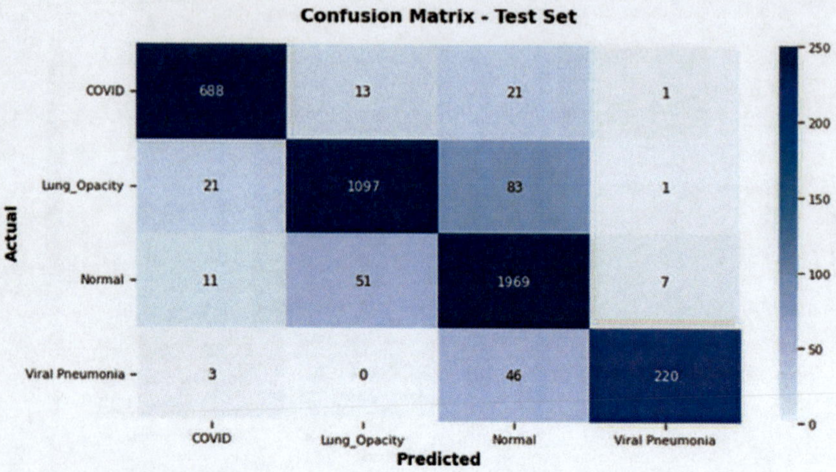

Fig. 6 Confusion matrix of Inception V3 architecture

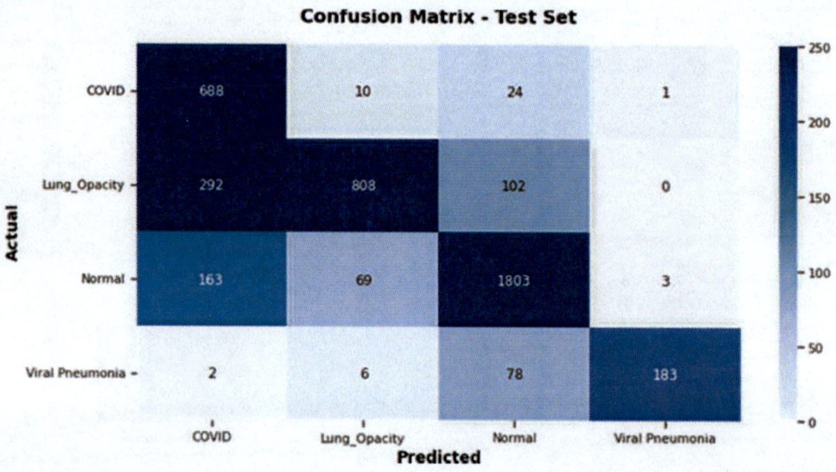

Fig. 7 Confusion matrix of DenseNet201 architecture

So, these models are quite effective in classifying X-ray images and hence can be utilized for disease identification with high accuracy. Moreover, these models can be used at places where a lot of identification needs to be performed. Hence, computer vision can help us to solve this issue very easily and efficiently.

As X-ray machines are available in many hospitals, the CNN models can help in identifying diseases and can replace testing kits used to identify disease. We believe that this method can be used to identify disease with better accuracy and precision than testing kits and other methods used for identify diseases. This CNN models can also be used to identify other diseases like plant diseases, brain diseases, and many more.

References

1. Khan Page J, Hinshaw D, McKay B (2021) In hunt for Covid-19 origin, patient zero points to second wuhan market—the man with the first confirmed infection of the new coronavirus told the WHO team that his parents had shopped there. Wall Street J
2. Islam M, Kundu S, Alam S, Hossan T, Kamal M, Hassan R (2021) Prevalence and characteristics of fever in adult and paediatric patients with coronavirus disease 2019 (COVID-19): a systematic review and meta-analysis of 17515 patients. PLoS ONE 16:e0249788
3. Islam M, Alam S, Kundu S, Hossan T, Kamal M, Cavestro C (2020) Prevalence of headache in patients with coronavirus disease 2019 (COVID-19): a systematic review and meta-analysis of 14,275 patients. Front Neurol. https://doi.org/10.3389/fneur.2020.562634
4. Saniasiaya J, Islam M (2021) Prevalence of olfactory dysfunction in coronavirus disease 2019 (COVID-19): a meta-analysis of 27,492 Patients. https://doi.org/10.1002/lary.29286
5. Anand R, Sindhwani N, Saini A (2021) Emerging technologies for COVID-19. In: Enabling healthcare 4.0 for pandemics: a roadmap using AI, machine learning, IoT and cognitive technologies. https://doi.org/10.1002/9781119769088.ch9
6. Agyeman A, Chin K, Landersdorfer C, Liew D, Ofori-Asenso R (2020) Smell and taste dysfunction in patients with COVID-19: a systematic review and meta-analysis. Mayo Clin Proc 95:1621–1631
7. Jones J Pulmonary opacification | radiology reference article | Radiopaedia.org. In: Radiopaedia.org, https://radiopaedia.org/articles/pulmonary-opacification. Accessed on 6 Apr 2022
8. What are interstitial opacities? In: Askinglot.com, https://askinglot.com/what-are-interstitial-opacities?cf_chl_f_tk=.hN5s8D444OG.Sk10xPXyKVt.xAB_2pwkni5JzR.o2k-1640172196-0-gaNycGzNBqU. Accessed on 6 Apr 2022
9. Pneumonia. In: Hopkinsmedicine.org, https://www.hopkinsmedicine.org/health/conditions-and-diseases/pneumonia. Accessed on 6 Apr 2022
10. Pneumonia—what is pneumonia? | NHLBI, NIH. In: Nhlbi.nih.gov, https://www.nhlbi.nih.gov/health/pneumonia. Accessed on 6 Apr 2022
11. Singh H, Rehman T, Gangadhar C, Anand R, Sindhwani N, Babu M (2021) Accuracy detection of coronary artery disease using machine learning algorithms. Appl Nanosci. https://doi.org/10.1007/s13204-021-02036-7
12. Deng X, Shao H, Shi L, Wang X, Xie T (2020) An classification–detection approach of COVID-19 based on chest X-ray and CT by using keras pre-trained deep learning models. Comput Model Eng Sci 125:579–596. https://doi.org/10.32604/cmes.2020.011920
13. Ohata E, Bezerra G, Chagas J, Lira Neto A, Albuquerque A, Albuquerque V, Reboucas Filho P (2021) Automatic detection of COVID-19 infection using chest X-ray images through transfer learning. IEEE/CAA J Automatica Sinica 8:239–248. https://doi.org/10.1109/jas.2020.1003393
14. Karthik R, Menaka R, Hariharan M (2021) Learning distinctive filters for COVID-19 detection from chest X-ray using shuffled residual CNN. Appl Soft Comput 99:106744. https://doi.org/10.1016/j.asoc.2020.106744
15. Panwar H, Gupta P, Siddiqui M, Morales-Menendez R, Bhardwaj P, Singh V (2020) A deep learning and grad-CAM based color visualization approach for fast detection of COVID-19 cases using chest X-ray and CT-Scan images. Chaos Solitons Fractals 140:110190
16. Toğaçar M, Ergen B, Cömert Z (2020) COVID-19 detection using deep learning models to exploit social mimic optimization and structured chest X-ray images using fuzzy color and stacking approaches. Comput Biol Med 121:103805

17. Chandra T, Verma K, Singh B, Jain D, Netam S (2021) Coronavirus disease (COVID-19) detection in chest X-ray images using majority voting based classifier ensemble. Expert Syst Appl 165:113909
18. Yasar H, Ceylan M (2020) A new deep learning pipeline to detect covid-19 on chest X-ray images using local binary pattern, dual tree complex wavelet transform and convolutional neural networks. Appl Intell 51:2740–2763
19. Liu Z, Yang C, Huang J, Liu S, Zhuo Y, Lu X (2021) Deep learning framework based on integration of S-mask R-CNN and Inception-v3 for ultrasound image-aided diagnosis of prostate cancer. Futur Gener Comput Syst 114:358–367
20. Binsawad M, Albahar M, Bin Sawad A (2021) VGG-CovidNet: Bi-branched dilated convolutional neural network for chest X-ray-based COVID-19 predictions. Comput Mater Continua 68:2791–2806
21. Fang Z, Ren J, Marshall S, Zhao H, Wang S, Li X (2021) Topological optimization of the DenseNet with pretrained-weights inheritance and genetic channel selection. Pattern Recogn 109:107608
22. COVID-19 radiography database. In: Kaggle.com, https://www.kaggle.com/tawsifurrahman/covid19-radiography-database. Accessed on 6 Apr 2022
23. Simonyan K, Zisserman A (2015) Very deep convolutional networks for large-scale image recognition. https://doi.org/10.48550/arXiv.1409.155624
24. Krizhevsky A, Sutskever I, Hinton G (2017) ImageNet classification with deep convolutional neural networks. Commun ACM 60:84–90
25. Szegedy C, Liu W, Jia Y, Sermanet P, Reed S, Anguelov D, Erhan D, Vanhoucke V, Rabinovich A (2014) Going deeper with convolutions. https://doi.org/10.48550/arXiv.1409.4842
26. Szegedy C, Vanhoucke V, Ioffe S, Shlens J, Wojna Z (2015) Rethinking the inception architecture for computer vision. https://doi.org/10.48550/arXiv.1512.00567
27. Huang G, Liu Z, van der Maaten L, Weinberger K (2018) Densely connected convolutional networks. https://doi.org/10.48550/arXiv.1608.06993

Secure Shift-Invariant ED Mask-Based Encrypted Medical Image Watermarking

Paresh Rawat, Jyoti Dange, Prashant Kumar Shukla, and Piyush Kumar Shukla

Abstract Security is a major concern for huge medical imaging data captured on regular basis by medical authorities. In this regard, many watermark algorithms have been proposed in past using edge detection, but improving robustness is an open challenge. Thus, in this paper, a novel two-pass image watermarking algorithm is designed. The proposed algorithm takes the advantage of shift invariance (SI) mask for generating asymmetric edge detection (ED) in wavelet transform domain. Then, in order to enhance the security and robustness, the simple and fast random key-based encryption is combined to securely store the watermarked images in cipher form. The image is first decrypted using the private key, and then, watermark reconstruction is performed. The performance is evaluated under various noisy attacks for qualitative evaluation. The parameter quantitative evaluation is presented in terms of MSE and SNR.

Keywords Watermarking · Encryption · Key · Image security · Medical images

P. Rawat (✉)
SN Technology, Bhopal, Madhya Pradesh 462032, India
e-mail: parrawat@gmail.com

J. Dange
Atharva College of Engineering, Mumbai, Maharashtra 400095, India

P. K. Shukla
Department of Computer Science and Engineering, Koneru Lakshmaiah Education Foundation Vaddeswaram, Guntur, Andhra Pradesh 522302, India

P. K. Shukla
University Institute of Technology, Rajiv Gandhi Proudyogiki Vishwavidyalaya, Bhopal, Madhya Pradesh 462033, India

P. Chatterjee et al. (eds.), *Computational Intelligence for Engineering and Management Applications*, Lecture Notes in Electrical Engineering 984,
https://doi.org/10.1007/978-981-19-8493-8_16

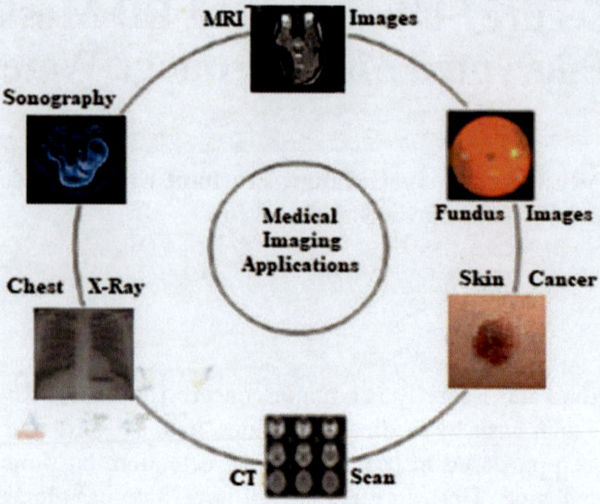

Fig. 1 Frequent applications of the medical imaging

1 Introduction

The watermarking is a very frequent approach for securing the digital images. An huge amount of medical imaging data such as CT [1], MRI [2], X-ray [3], sonography images are captured on daily basis. The authenticity of the patient data and the storage of data have become the critical issues. This paper is aimed to design the robust and secure medical image watermarking algorithm. There are huge imaging applications in the medical fields. Major medical imaging applications are shown in Fig. 1.

All these applications require authentication. Since medical images are directly related to health of humans, thus it is highly required to secure or encrypt imaging data. Using random key encryption improves robustness of watermarking.

2 Related Works

Many methods were designed for securing the medical images [1–5]; for digital watermarking, edge detection (ED) is widely being used [3, 4, 7]. Conventional EDs are slightly sensitive to the added noisy attacks. The edge detector also follows the shift invariance properties as suggested by [4]. The classification of watermarking methods is given in Fig. 2. This paper is focused to design the hybrid watermarking method as highlighted.

ED is common for image watermarking, since it improves watermark invisibility. Ellinas [3] proposed watermarking based on wavelet ED using Sobel mask with two-level DWT. They demonstrated using image dilation of 3×3 matrix mask along with detected edge image. They also aided Gaussian noise as watermark. Tiwari et al. [4] have earlier proposed to use the shift-invariant edge detection mask for scale invariance property. They demonstrated the effectiveness of the SI-ED mask.

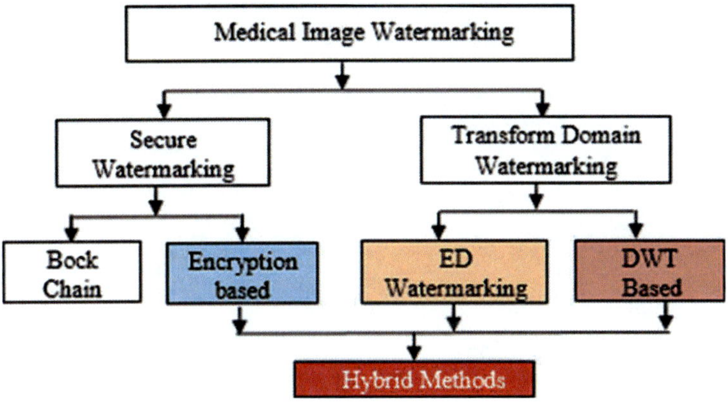

Fig. 2 Classification of the watermarking algorithms

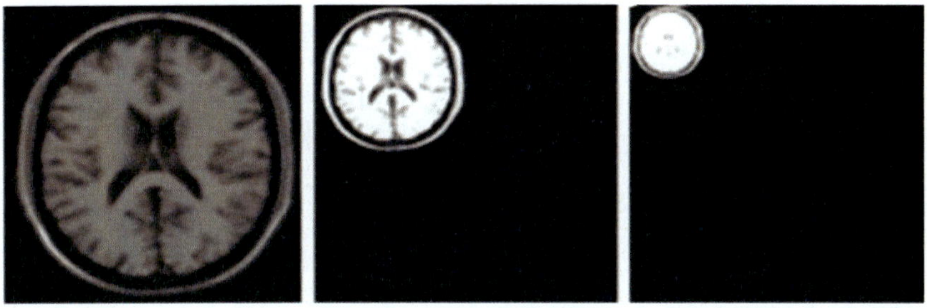

Fig. 3 Wavelet decompositions up to second level for CT image

Paper by Kumari et al. [6] used advance encryption standard (AES) for securing DWT-SVD-based watermarking. Good use of encryption based watermarking is presented by Singh et al. [7] have used Sobel ED with dilation for the watermarking. The watermark in ED domain is invisible and secure with encryption standards. Block-based edge detection is used in embedding process for robustness. There are certain image security methods too. So, this research is proposed to merge secure encryption with method of Tiwari et al. [4] and Singh et al. [7] for improving the robustness. Bharathi et al. [8] have used AES-based security [9, 10]. Singh et al. [11] have presented SWT-based watermarking but was sensitive to noise. Figure 3 present example of DWT decompositions.

3 Random Key Image Encryption

The encryption is method of hiding the private key to the imaging data. This paper presented random simple and fast key generation algorithm for image size of $M \times N$.

$$x_N = 1 - 2 * \text{bin}_{x_{N_{\text{Minus}_1}}} * \text{bin}_{x_{N_{\text{Minus}_1}}} \tag{1}$$

$$\text{key(ind1)} = \sum_{i=1}^{M} \sum_{j=1}^{N} \text{key(ind1)} + \text{bin}_{x(\text{ind2}*\text{ind1})} * 2 \wedge (\text{ind2} - 1) \tag{2}$$

4 Proposed Methodology

This paper proposed a two-pass watermarking methodology designed for medical image security is sequentially presented. The proposed ED mask is asymmetric in nature and thus robust too. Section first presents the basic block diagram of methodology and then describes watermark embedding and extraction algorithms. This paper proposed a combination of wavelet based ED + dilation-based watermarking along with the advantage of crypto encryption based security. The ED is representing the local features of images and is sparse in nature. Method uses the shift-invariant ED filter mask called SI-ED for generating the robust medical image watermarking. Sequential block diagram of proposed secure watermarking is presented in Fig. 4.

4.1 Linear Shift-Invariant Edge Detector

Tiwari et al. [4] have presented ED using the shift-invariant high-pass filter for image watermarking [4]. The edge is formed by using the convolution filter mask and second-level DWT HH component. Then, sum up the multiplied values. The discreet W dimensional ED filter of SI-ED operation is mathematically represented

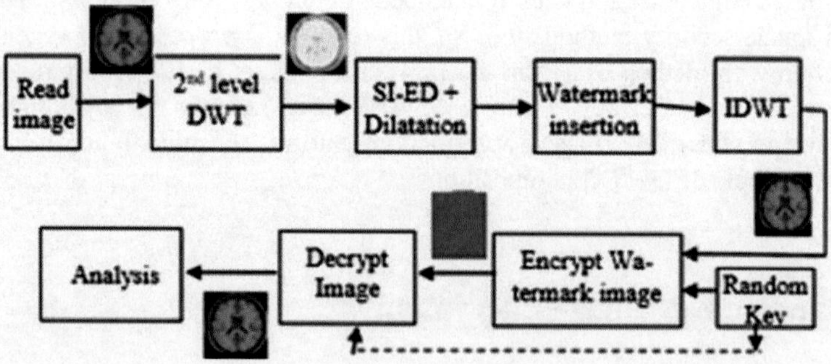

Fig. 4 Proposed block diagram of SI-ED-based secure encrypted watermarking

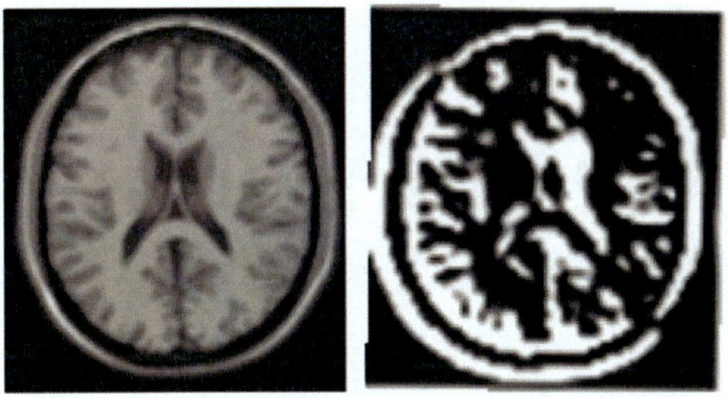

a) Input CT scan image b) Shift invariant mask based ED image

Fig. 5 Example of the ED using proposed asymmetric mask for second-level DWT

as simplified vector form as

$$g'_n = \sum_{n'=-R}^{R} h_{-n'} g_{n+n'} \tag{3}$$

Linear shift/scale-invariant ED used in paper in HH wavelet component is shown below:

$$\text{HSI} = \begin{bmatrix} 0 & -1 & -2; & 1 & 0 & -1; & 2 & 1 & 0 \end{bmatrix} \tag{4}$$

This linear SI-ED mask is an asymmetric matrix and is a high-pass derivative-based ED filter. Main advantages of SI-ED are properties of superposition and shift invariance. Thus, proposed mask provides robustness as compared to standard Sobel and Prewitt masks. An example of edge detected image on the second-level DWT is presented in Fig. 5 for the CT scan image. The HH coefficients of the second wavelet decompositions are taken and SI-ED is applied over them.

5 Results and Discussions

The watermarking results of the proposed method are sequentially presented in this section. The input medical images used for performance evaluation are presented in Fig. 6. The images are considered for CT scan, MRI, chest X-ray, sonography, and ILD image. The results are presented in two pass: in first pass, the wavelet-domain ED-based watermarking using scale-invariant mask is presented. Then, in second phase, the encryption-based image decryption and reconstruction results are presented.

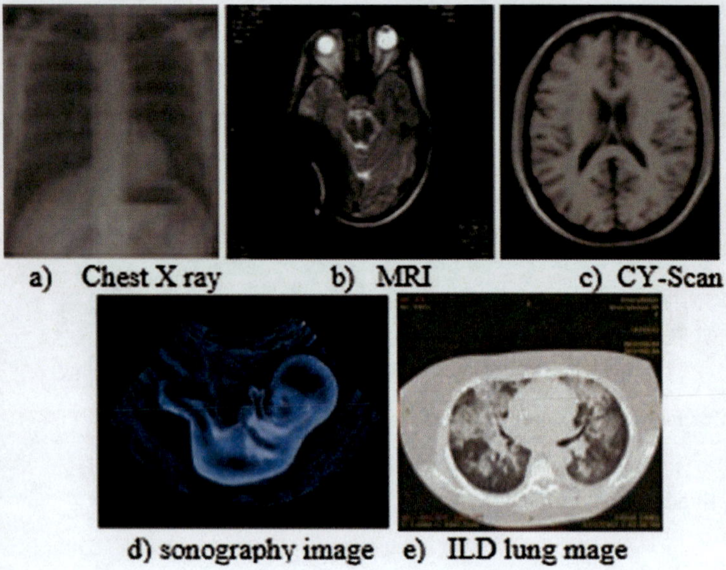

Fig. 6 Input image data used for performance evaluation in the paper

The dilated edge image and there difference are taken as the watermark in this paper. The dilated coefficient is generated as difference of

$$\mathrm{di} = D_{\mathrm{Mark}} = D((X)_{\mathrm{ED}})_{\mathrm{LL}}) - ((X)_{\mathrm{ED}})_{\mathrm{LL}})) \tag{5}$$

where $\mathrm{di} = D_{\mathrm{Mark}}$ is the dilated difference, and $(X)_{\mathrm{ED}}$ is the edge coefficient of the LL component. The watermark is generated using the following law:

$$W_{\mathrm{Mark}} = (1 - \alpha) * \mathrm{di} + \alpha * N \tag{6}$$

where α is the scaling parameter set to 0.7 to maintain the invisibility. The result of the stepwise watermark generation process is shown in Fig. 7 for CT scan image. It is clear that watermark is added to LL component using acted coefficient di.

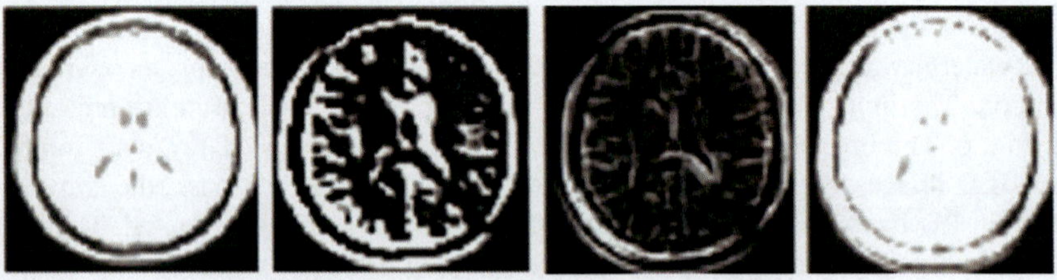

Fig. 7 Process of watermark generation based on ED and dilation

The initial results of the complete encrypted secure watermarking are presented in Fig. 8. The cipher key is encrypted and generated as the random private key to match patient. The encryption efficiency is shown as comparison of the restive decrypted image histogram representing cipher weights in Fig. 9 (Tables 1 and 2).

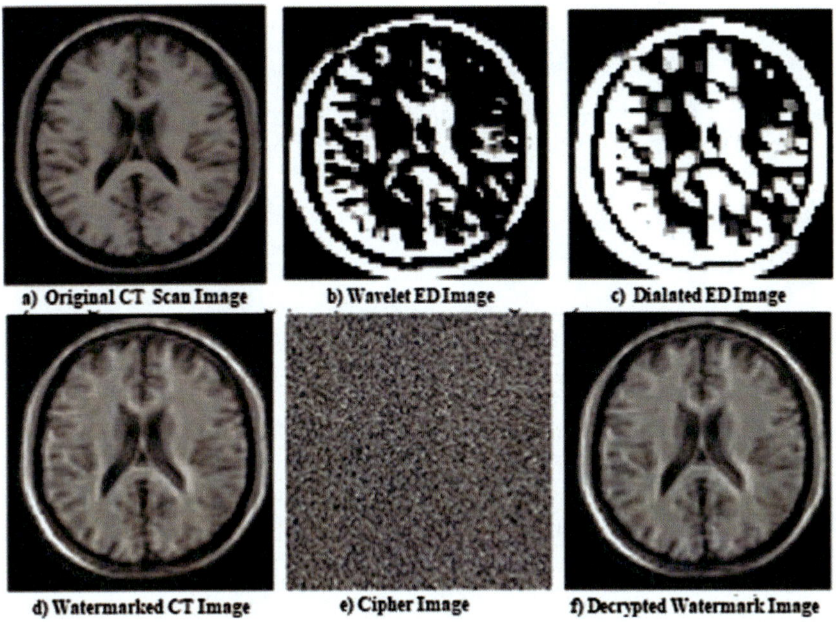

Fig. 8 Sequential results of proposed two-pass secure watermarking for CT image

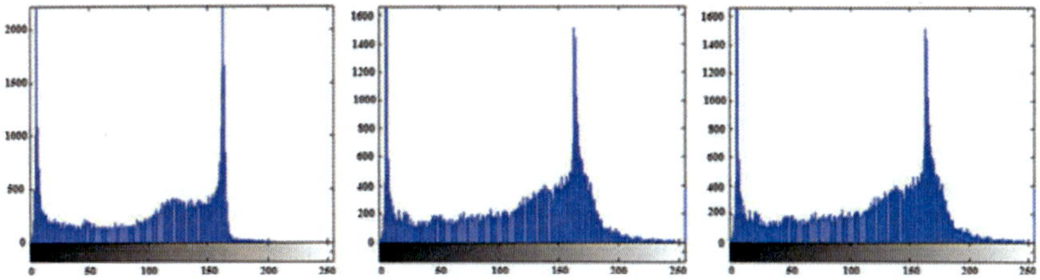

Fig. 9 Comparison of the histogram of watermarked and decrypted images

Table 1 Comparison of entropy of the different stages of encrypted watermark

Images	Original	Watermark	Encrypted	Decrypted
CT scan	6.3996	6.954	7.997	6.954
MRI	5.8538	6.3816	7.9974	6.3816
Chest X-ray	6.5725	6.5711	7.9971	6.5711

Table 2 Comparison of SNR of calculated between original and watermark images

SNR			
CT scan	MRI	X-ray	ILD
25.91251	2.36757	17.799	6.86662

The SNR is defined as*

$$SNR = \sqrt{\left(\frac{MAX_1^2}{MSE}\right)} \tag{7}$$

6 Conclusions

This paper presents a two-pass secure watermarking method for the medical images. The algorithms use asymmetrical invariant ED mask and simple and efficient random key generation-based encryption method for security enhancement. It is found that after decryption, the watermarked image is truly recovered. The entropy analysis justifies the statement well. The watermark images offer high SNR ranges. The cipher image is generated to same size as of input. Using the multilevel security improves the robustness of the algorithm and also can be used authentication purpose. The watermark reconstruction is future scope of work.

References

1. Xiao S, Zhang Z, Zhang Y, Yu C (2020) Multipurpose watermarking algorithm for medical images. Hindawi Scientific Programming
2. Khan PW, Byun Y (2020) A blockchain-based secure image encryption scheme for the industrial internet of things. MDPI J Entropy 22:175
3. Ellinas JN (2008) A robust wavelet-based watermarking algorithm using edge detection. IEEE J Image Process 197–208
4. Tiwari A, Singh V (2013) Digital image watermarking using DWT and shift invariant edge detection. Int J Comput Technol Electron Eng (IJCTEE) 3(6)
5. Singh R, Shukla P, Rawat P (2019) Efficient Ed3wt method for robust medical image watermarking. Int J Sci Technol Res 8(12)
6. Kumari N, Chauhan DS (2019) Robust image watermarking algorithm DWT-SVD-AES for telemedicine application. Int J Comput Sci Mob Comput 8(5):268–279
7. Singh R, Rawat P, Shukla P, Shukla PK (2019) Invisible medical image watermarking using edge detection and discrete wavelet transform coefficients. Int J Innov Technol Exploring Eng (IJITEE) 9(1)
8. Bharathi C, Amudha A (2018) Secure transmission of tele medical information using DWT watermarking and AES encryption algorithm. Int J Sci Res Dev 6(04)
9. Lalani S, Doye DD (2016) A novel DWT-SVD canny-based watermarking using a modified torus technique. J Inf Process Syst 12(4):681–687

10. Rathi N, Holi G (2014) Securing medical images by watermarking using DWT-DCT-SVD. Int J Comput Trends Technol (IJCTT) X(Y)
11. Singh R, Rawat P, Shukla P (2017) Robust medical image authentication using 2-D stationary wavelet transform and edge detection. In: 2nd IET international conference on biomedical image and signal processing (ICBISP 2017), pp 1–8

Fusion-Based Feature Extraction Technique Using Representation Learning for Content-Based Image Classification

Khushbu Kumari, Chandrani Singh, Archana Nair, Pankaj Kumar Manjhi, Rik Das, and Debajyoti Mukhopadhyay

Abstract The upsurge in advancements of digital imaging technology has led to archival of high-quality images embedded with rich content. This has facilitated content-based image classification (CBIC) as one of the most popular techniques to accurately classify these images by extracting robust low-level features like color, shape, texture, and so on. However, designing a robust feature descriptor to cater diversified set of images is a challenging task. Traditional handcrafted feature extraction techniques have taken a back sit after the advent of deep neural network-based image feature extraction. However, generalization of neural network-based feature vectors for smaller datasets has posed significant challenges due to overfitting caused by high variance of pre-trained networks. Hence, this research has primarily investigated the possibilities of designing generalized features using final layer of different pre-trained neural networks individually vis-a-vis handcrafted feature extraction. The process is followed by concatenation of final layers from diversified pre-trained networks along with handcrafted feature. In the process of experimentation, the paper aims at using representation learning using VGG19 and ResNet50. It has also considered handcrafted technique of feature extraction named color histogram (CH). Each of the techniques are separately tested for feature generalization efficiency using assorted classifiers. Further, the research has proposed fusion of features in several

K. Kumari (✉)
Yogoda Satsanga Mahavidyalaya, Ranchi, India
e-mail: mishra.khushbu09@gmail.com

C. Singh
Sinhgad Institute of Management, Pune, India

A. Nair
Sinhgad Institute of Business Administration and Research, Pune, India

P. K. Manjhi
VinobaBhave University, Hazaribag, India

R. Das
Siemens Technology and Services Pvt. Ltd., Bengaluru, India

D. Mukhopadhyay
WIDiCoReL Research Lab, Bennett University, Greater Noida, India

P. Chatterjee et al. (eds.), *Computational Intelligence for Engineering and Management Applications*, Lecture Notes in Electrical Engineering 984,
https://doi.org/10.1007/978-981-19-8493-8_17

combinations using each of these descriptors to observe any improvement caused in feature generalization while tested for classification results. The outcomes using the proposed fusion approach have resulted in improved generalization by outshining classification accuracy of individual methods.

Keywords Content-based image classification · Handcrafted feature · Deep neural networks · Pre-trained convolution neural networks · Feature fusion · Representation learning

1 Introduction

Image classification using low-level features extracted from image content has embraced immense significance with affinity of the mass toward communicating with image data [1]. Extraction of robust descriptors from image content is the baseline for achieving good results for content-based image classification. There are various traditional handcrafted methods such as histograms of oriented gradient (HOG) [2], color histogram (CH) [3], image binarization [4], image transformation [5], as well as automated descriptor definition techniques using pre-trained convolutional neural networks (CNNs). The pre-trained CNN-based feature extraction is a representation learning approach to detect the systematic pattern automatically from the image databases. Although due to advancement of tools and technologies, the quality and size of image databases improved drastically which affect the classification performance, thus the representation learning approaches perform markable in comparison to the handcrafted methods due to its extensive learning approaches for feature detection [6]. However, it is often found that either of the techniques are unable to perform impressively due to lack of feature generalization [7, 8]. Extraction of handcrafted features requires domain expertise, whereas designing representation learning-based descriptors with pre-trained CNNs requires huge amount of training data. The authors have identified this problem, especially in case of smaller datasets, for which no relevant pre-trained neural network exists and the features are extracted using handcrafted techniques.

The extracted features are evaluated using diverse classification environments such as Support Vector Machine (SVM), Neural Network (NN), Logistic Regression (LR), Random Forest (RF) [9, 10]. Hence, this paper attempts to investigate and compare the effectiveness of features extracted by representation learning to the handcrafted features to ensure feature generalization. Further, a fusion approach of the handcrafted feature named color histogram is carried out with two different sets of features extracted using representation learning models, namely, VGG19 and ResNet50. The fusion approach is attempted to evaluate improvement in generalization of features.

Therefore, the objectives of this paper are as given below:

- Designing robust descriptors to test feature generalization efficiency.

- Carrying out comparative study of feature generalization between handcrafted technique and automated deep learning-based technique of feature extraction.
- Evaluate the efficiency of feature fusion in leveraging feature generalization.
- Comparing content-based image classification results with diverse set of feature vectors under various classification environments to evaluate feature generalization efficiency.

We have used Oliva and Torralba (OT Scene) dataset with 2688 images of eight different categories and Corel dataset with 5000 images. The results of the experimentations with the proposed techniques are encouraging while compared to state of the art.

2 Literature Survey

Content-based image classification is considered as an important research field in the area of image classification. Research based on various feature extraction techniques has helped accelerate advancements in CBIC and has contributed to the improvement of classification models. In a study by Rao et al. [11], on the spatial color histogram for content-based image retrieval, the traditional color histogram was modified to capture the spatial layout of different colors and three new types of spatial color histograms were introduced named annular, angular, and hybrid color histograms. Sergyan [12] discussed the color histogram features-based image classification in content-based image retrieval (CBIR) system, and the study described a new approach which was based on low-level image histogram features. The result of their study was the quick generation and comparison of an applied feature vector. Agrawal et al. [13] studied content-based classification of color images using Support Vector Machine (SVM). They used a dataset of 500 images divided into four different categories and further compared to the histogram for RGB, CMYK, YUV, HSV, HVC, and YJQ color spaces. Kumar and Saravanan [14] described the CBIR using color histogram. The grid code of an image was used to create a color histogram by quantizing the colors of the images and counting pixel numbers for each color. Yang and Loog [15] created a benchmark and compared active learning for Logistic Regression (LR). Three synthetic datasets and 44 real-world datasets were used for experimenting. These datasets provided insight into the behavior of these active learning methods in comparison with the area of the learning curve and their costs for computation. Kadhim and Abed [16] applied convolutional neural networks on multimedia approaches and was successful in creating a system that was capable of classification without any human interference. Effective methods were produced for satellite image classification based on deep learning. [17] studied transfer learning with ResNet50 for malaria cell image classification. The paper focused on the implementation of transfer learning-based cell image classification of the malaria-infected cells for improving diagnostic accuracy. Reddy and Juliet [18] discussed image classification using SVM and Artificial Neural Network (ANN). The images were at first separated into various sub-images, and then after, it was further classified into responsive class using ANN. [19] have introduced a fusion framework for image classification which incorporated better and accurate results rather than a single feature extraction technique. Das et al. [20] have emphasized on defining the hybrid descriptor for obtaining robust feature vector using fusion architecture. The proposed fusion model used neural network and handcrafted method for feature extraction in content-based image classification system. The specified domain of CBIC needs more expansion in research and thus created more scope for exploration.

3 Proposed Techniques and Tools

The proposed research is trying to embrace feature engineering with two distinct techniques with their amalgamation. Here, for extracting the feature content from images, the following three techniques are proposed in this research paper which are as follows:

(a) Handcrafted technique using color histogram (CH).
(b) Representation learning method based on VGG19 and ResNet50 architecture.
(c) Feature fusion of both handcrafted and representation learning methods for feature generalization.

The above three methods are being tested in the proposed research for obtaining the features from image database with the help of different classifiers, namely Support Vector Machine (SVM), Logistic Regression (LR), and Artificial Neural Network (ANN) using Orange data mining tool.

3.1 Color Histogram (CH)

Color histogram is a feature extraction technique for extracting features from images based on the distribution of colors in the images. This technique mainly defines the frequency dispensation of colors in an image. Color histograms have been implemented in two different ways, global color histogram and local color histogram [21], where the global color histogram focuses on the entire color frequency distribution of images, and the local color histogram focuses on the specific part of the images.

Color histogram feature extraction techniques have mainly worked on the intensity of images, where images (I) are represented with number (N) of color bins and each color bin represents the collection of the identical color. The histogram (H) consists of several pixels present in the image (I) along with its intensity value (i), which is clearly shown in the given Eq. (1). Figure 1 shows the graphical representation of the color histogram for the sample image taken from the OT Scene dataset.

$$CH(i) = I * i \tag{1}$$

where

I = Image, and

i = Intensity value of an image.

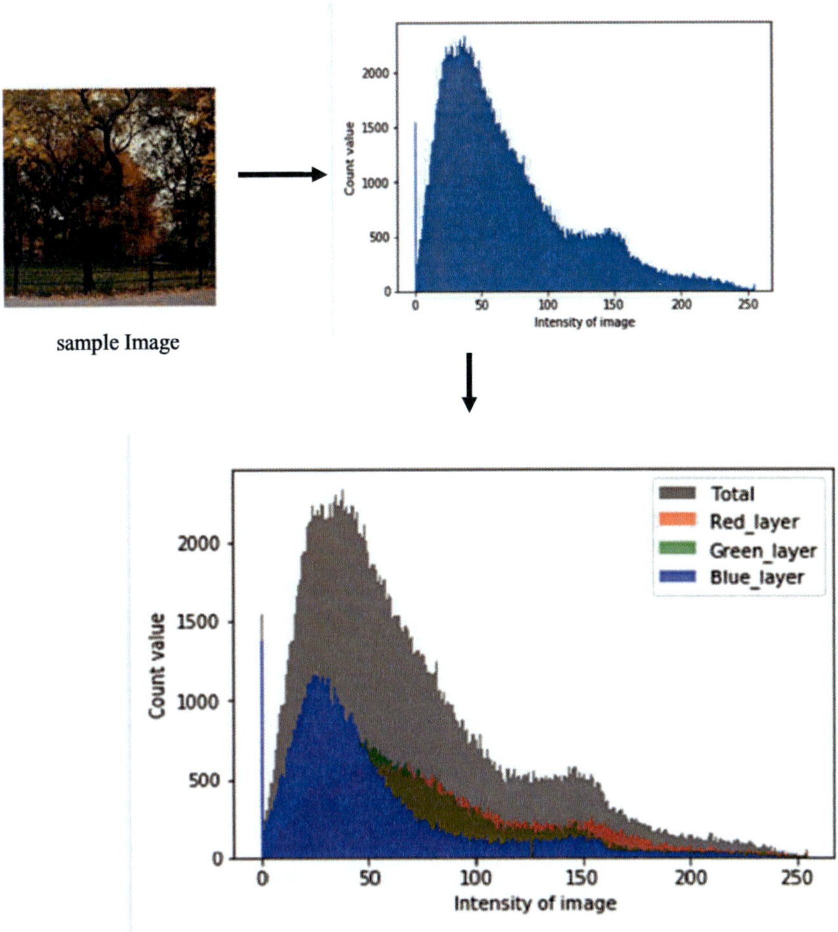

Fig. 1 Graphical representation of color histogram for sample image

3.2 Representation Learning

Here, two different representation learning models are used for performing the test: the first one is VGG19 and the other is ResNet50. VGG19 [22] is a pre-trained convolutional neural network (CNN) comprising 19 layers architecture from visual geometry group, and ResNet50 is another pre-trained model with 50 layers comprising 48 convolution layers, 1 max pool, and an average pool layer [23] which is shown in Fig. 2.

The authors have used the above models for representation learning and extracted the features from the Corel 5000 dataset and OT Scene dataset using these models.The evaluation metrics for the experimentation process are given in Eqs. (2–5).

$$\text{Accuracy} = \frac{\text{TP} + \text{TN}}{\text{TP} + \text{FN} + \text{FP} + \text{FN}} \tag{2}$$

Fig. 2 ResNet50 architecture

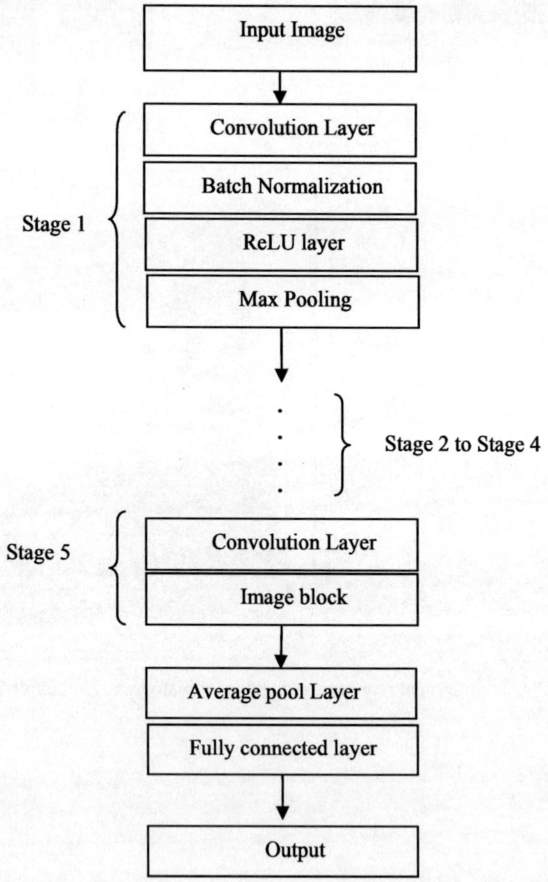

Fig. 3 **a** Sample of OT Scene dataset, **b** sample of Corel 5 K dataset

$$\text{Precision} = \frac{\text{TP}}{\text{TP} + \text{FP}} \tag{3}$$

$$F1 - \text{Score} = \frac{2 * \text{Precision} * \text{Recall}}{\text{Precision} + \text{Recall}} \tag{4}$$

$$\text{Recall/TPR} = \frac{\text{TP}}{\text{TP} + \text{FN}} \tag{5}$$

Here, TP = True positive
TN = True negative
FP = False positive
FN = False negative
TPR = True-positive rate.

4 Datasets

Two different datasets named Oliva and Torralba (OT Scene) and Corel dataset are used in this research. OT Scene dataset consists of 2688 images of eight different categories, and the Corel dataset consists of 5000 images with 50 different categories. These datasets are open-access datasets. Features were extracted from the image datasets using the proposed techniques. Figure 3a, b show the sample dataset taken from OT Scene and Corel datasets, respectively.

5 Results and Discussion

The features were extracted from the datasets using color histogram, VGG19, and ResNet50. These individual feature vectors were classified using three different

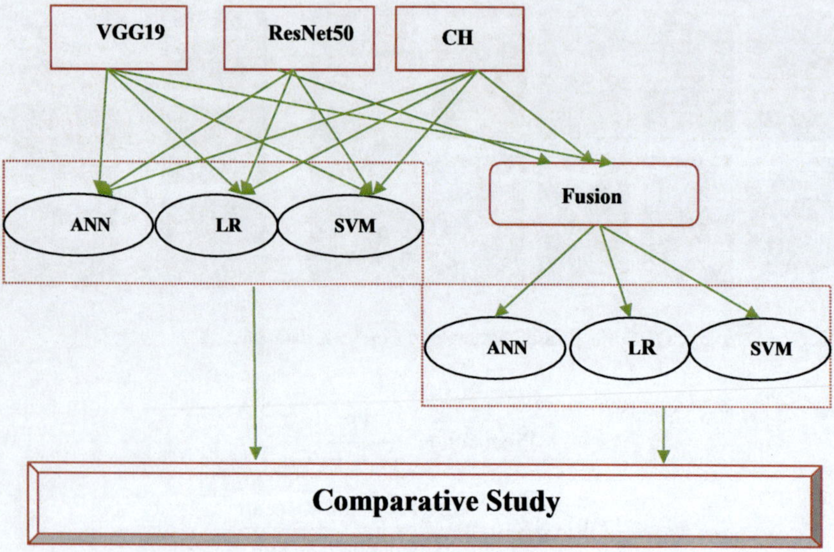

Fig. 4 Model design for evaluation of the normalized features

classifier environments, namely Support Vector Machine (SVM), Artificial Neural Network (ANN), and Logistic Regression (LR). As a second step, the feature vectors from all three techniques were fused together and were classified using the same set of classifiers. Since the individual feature extraction method is not much robust to capture results with generalized feature, the fusion approach has been proposed here for feature generalization. The designed model for feature extraction is shown below in Fig. 4.

Results of classification are shown in Tables 1 and 2, respectively, for OT Scenes and Corel datasets. Here, the table consists of classification performance using metrics such as accuracy, $F1$-score, precision, and recall matrix for all three classifiers called SVM, Artificial Neural Network, and Logistic Regression as mentioned in the designed model shown in Fig. 4. The results are estimated and analyzed with the help of Orange data mining tool in the presence of three different classifier environments such as Support Vector Machine, Artificial Neural Network, and Logistic Regression. Further, the analysis has been recorded in the mentioned Tables 1 and 2 for both the dataset OT Scenes and Corel, respectively.

From Tables 1 and 2, it is clearly alluded that the result recorded for fused feature extraction techniques is having better classification outcomes over the individual feature extraction method with all the three mentioned classifiers, namely Support vector machine (SVM), Neural Network, and Random Forest (RF), whereas the outcome recorded for both the datasets for all three classifiers has also demonstrated distinctly that the results for Logistic Regression are more precise as compared to other two mentioned classifiers. However, the individual feature extraction techniques are not much capable to capture the multidimensional aspects of the image

Table 1 Comparison table of classification performance based on accuracy recorded with three different classifiers for OT Scenes dataset

Techniques	SVM				Neural network				Logistic regression			
	ACC	*F*1-score	Precision	Recall	ACC	*F*1-score	Precision	Recall	ACC	*F*1-score	Precision	Recall
CH	0.802	0.397	0.455	0.407	0.86	0.57	0.571	0.571	0.83	0.476	0.481	0.48
VGG19_fc2	0.992	0.894	0.896	0.893	0.993	0.905	0.905	0.905	0.99	0.915	0.915	0.914
ResNet50	0.994	0.915	0.917	0.913	0.995	0.922	0.923	0.922	1	0.928	0.929	0.928
CH + VGG	0.992	0.892	0.894	0.891	0.993	0.909	0.909	0.908	0.99	0.914	0.915	0.914
CH + ResNet	0.994	0.913	0.916	0.912	0.993	0.917	0.918	0.917	1	0.29	0.93	0.929
ResNet50 + VGG19	0.994	0.912	0.914	0.911	0.995	0.924	0.925	0.924	1	0.932	0.933	0.932
CH + ResNet + VGG	0.994	0.907	0.909	0.906	0.994	0.927	0.927	0.926	1	0.935	0.935	0.935

Table 2 Comparison table of classification performance based on accuracy recorded with three different classifiers for Corel dataset

Techniques	SVM				Neural network				Logistic regression			
	ACC	*F*1-score	Precision	Recall	ACC	*F*1-score	Precision	Recall	ACC	*F*1-score	Precision	Recall
CH	0.952	0.475	0.491	0.49	0.952	0.598	0.597	0.601	0.934	0.418	0.435	0.448
VGG19_fc2	0.999	0.949	0.951	0.948	0.999	0.951	0.951	0.951	1	0.961	0.961	0.961
ResNet50	0.952	0.475	0.491	0.49	0.952	0.598	0.597	0.601	0.934	0.418	0.435	0.448
CH + VGG	0.999	0.949	0.951	0.948	0.999	0.952	0.952	0.952	1	0.964	0.964	0.964
CH + ResNet50	0.999	0.964	0.966	0.964	1	0.966	0.966	0.966	1	0.977	0.977	0.977
ResNet50 + VGG19	0.999	0.949	0.951	0.948	0.999	0.952	0.952	0.952	1	0.964	0.964	0.964
CH + Resnet + VGG19	0.999	0.966	0.967	0.965	1	0.967	0.968	0.967	1	0.977	0.977	0.977

dataset for achieving feature generalization. But contrarily, the fusion model is potentially viable to capture significant aspects of feature content from the large database to obtain feature generalization. The graphical representation of the results is shown in Figs. 5 and 6 for OT Scenes and Corel datasets, respectively, which technically underline that the outcome recorded with Logistic Regression is counted more considerable rather than the other two mentioned classifier environments.

The graph in Figs. 5 and 6 reveals that the results with Logistic Regression have outclassed the other two classifiers. The difference is significant while classifying with fusion-based features.

The sample Confusion Matrix for fused features of CH, VGG19, and ResNet50 was recorded for OT Scene dataset with all eight mentioned categories are shown in Fig. 7a–c for Support Vector Machine, Artificial Neural Network, and Logistic Regression classifiers, respectively. The mentioned figure compares the calculated

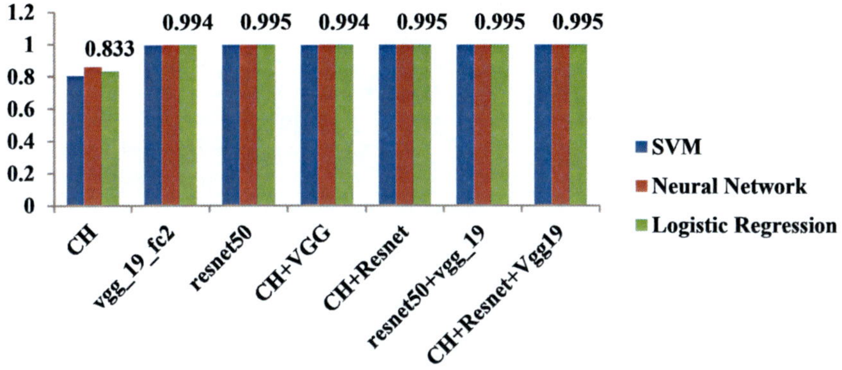

Fig. 5 Graphical representation of performance accuracy recorded for OT Scenes dataset with three different classifiers

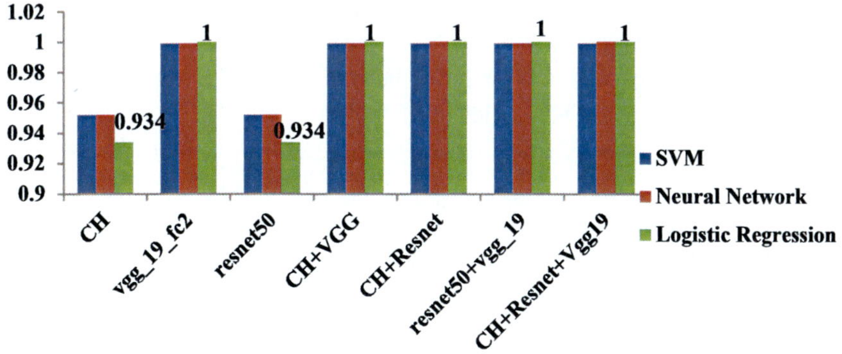

Fig. 6 Graphical representation of performance accuracy recorded for Corel dataset with three different classifiers

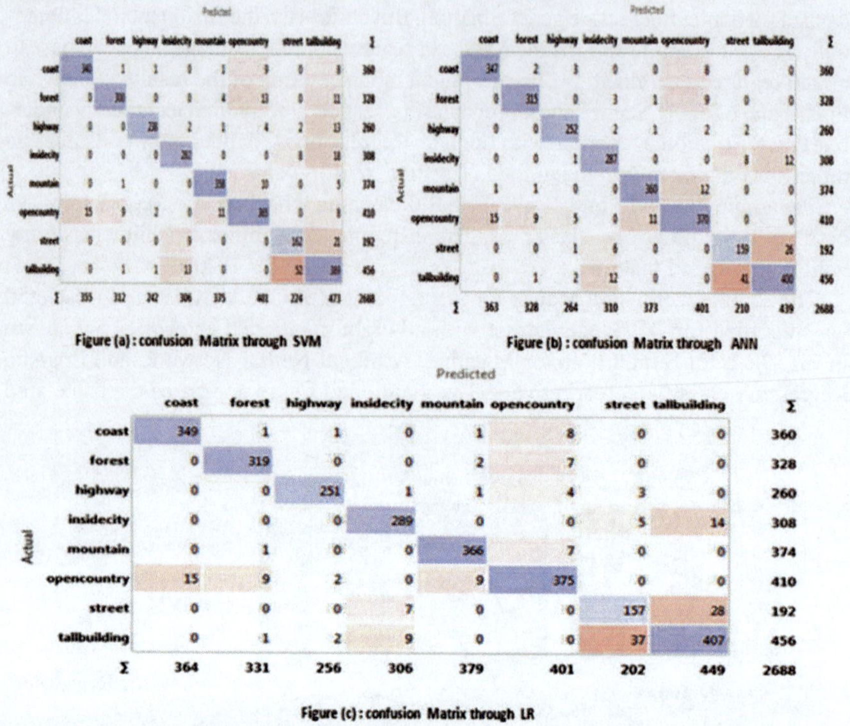

Fig. 7 Confusion Matrix of fusion feature of a color histogram, VGG19, and ResNet50 model through SVM, ANN, and LR

results of all three classifiers, and it was concluded that better accuracy has been recorded with logical regression classifier in comparison to the other two classifiers. For example, the category coast has been identified as 349 true images out of 360 images, while the result with SVM recorded 340 out of 360 and ANN recorded 347 out of 360 images of OT Scenes dataset.

The results clearly show that the fusion-based technique for feature generalization performs better in capturing significant aspects of the images as compared to individual techniques.

6 Conclusion

The proposed research emphasizes into the outcome from representation learning-based feature extraction techniques, namely VGG19 and ResNet50, over traditional handcrafted method named color histogram. However, the handcrafted method for extracting features has its own challenges for designing robust feature vector with multitudinous set of image databases. At the same time, representation learning methods emerge with drastic outgrowth in the classification model. The result enlisted the consequences from deep neural network-based architecture like VGG19 and ResNet50 reveals better classification results instead of the handcrafted technique of feature extraction. The experimentation reveals that the process of representation learning helps to avoid overfitting cost due to transfer learning over smaller dataset and ensures extraction of robust features which in turn provides better feature generalization. Additionally, the hybrid architecture designed in this work to promote feature generalization with fusion-based approach has resulted in better classification outcomes establishing the efficiency of feature fusion for designing robust descriptors for content-based image classification. The future work is aimed at designing automated computer-aided diagnosis using fusion-based approach.

References

1. Yadav BAM, Sengar BPS (2014) A survey on: 'content based image retrieval systems.' Int J Emerg Technol Adv Eng 4(6):22–26
2. Giveki D, Soltanshahi MA, Montazer GA (2017) A new image feature descriptor for content based image retrieval using scale invariant feature transform and local derivative pattern. Optik 131:242–254
3. Nazir A, Ashraf R, Hamdani T, Ali N (2018) Content based image retrieval system by using HSV color histogram, discrete wavelet transform and edge histogram descriptor. In: 2018 International conference on computing, mathematics and engineering technologies (iCoMET). IEEE, pp 1–6
4. Thepade S, Das R, Ghosh S (2014) A novel feature extraction technique using binarization of bit planes for content based image classification. J Eng
5. Rehman M, Iqbal M, Sharif M, Raza M (2012) Content based image retrieval: survey. World Appl Sci J 19(3):404–412
6. Rukmangadha PV, Das R (2022) Representation-learning-based fusion model for scene classification using convolutional neural network (CNN) and pre-trained CNNs as feature extractors. In: Computational intelligence in pattern recognition. Springer, Singapore, pp 631–643
7. PerumalSankar S, Vishwanath N, Jer Lang H (2017) An effective content based medical image retrieval by using ABC based artificial neural network (ANN). Curr Med Imaging 13(3):223–230
8. Wan J, Wang D, Hoi SCH, Wu P, Zhu J, Zhang Y, Li J (2014) Deep learning for content-based image retrieval: a comprehensive study. In: Proceedings of the 22nd ACM international conference on multimedia, pp 157–166
9. Kamavisdar P, Saluja S, Agrawal S (2013) A survey on image classification approaches and techniques. Int J Adv Res Comput Commun Eng 2(1):1005–1009
10. Xu B, Ye Y, Nie L (2012) An improved random forest classifier for image classification. In: 2012 IEEE international conference on information and automation. IEEE, pp 795–800
11. Rao A, Srihari RK, Zhang Z (1999) Spatial color histograms for content-based image retrieval. In: Proceedings 11th international conference on tools with artificial intelligence. IEEE, pp 183–186
12. Sergyan S (2008) Color histogram features based image classification in content-based image retrieval systems. In: 2008 6th International symposium on applied machine intelligence and informatics. IEEE, pp 221–224

13. Agrawal S, Verma NK, Tamrakar P, Sircar P (2011) Content based color image classification using SVM. In: 2011 Eighth international conference on information technology: new generations. IEEE, pp 1090–1094
14. Kumar AR, Saravanan D (2013) Content based image retrieval using color histogram. Int J Comput Sci Inf Technol 4(2):242–245
15. Yang Y, Loog M (2018) A benchmark and comparison of active learning for logistic regression. Pattern Recogn 83:401–415
16. Kadhim MA, Abed MH (2019) Convolutional neural network for satellite image classification. In: Asian conference on intelligent information and database systems. Springer, Cham, pp 165–178
17. Reddy ASB, Juliet DS (2019) Transfer learning with ResNet-50 for malaria cell-image classification. In: 2019 International conference on communication and signal processing (ICCSP). IEEE, pp 0945–0949
18. Thai LH, Hai TS, Thuy NT (2012) Image classification using support vector machine and artificial neural network. Int J Inf Technol Comput Sci 4(5):32–38
19. Das R, Kumari K, Manjhi PK, Thepade SD (2019) Ensembling handcarafted features to representation learning for content based image classification. In: 2019 IEEE Pune section international conference (PuneCon). IEEE, pp 1–4
20. Das R, Kumari K, De S, Manjhi PK, Thepade S (2021) Hybrid descriptor definition for content based image classification using fusion of handcrafted features to convolutional neural network features. Int J Inf Technol 1–10
21. Chakravarti R, Meng X (2009) A study of color histogram based image retrieval. In: 2009 Sixth international conference on information technology: new generations. IEEE, pp 1323–1328
22. Shaha M, Pawar M (2018) Transfer learning for image classification. In: 2018 Second international conference on electronics, communication and aerospace technology (ICECA). IEEE, pp 656–660
23. Mahmood A, Bennamoun M, An S, Sohel F (2017) Resfeats: residual network-based features for image classification. In: 2017 IEEE international conference on image processing (ICIP). IEEE, pp 1597–1601

A Comparative Study on Challenges and Solutions on Hand Gesture Recognition

Jogi John and Shrinivas P. Deshpande

Abstract Any human–computer interaction application needs to be able to recognize gestures. Hand gesture detection systems that recognize gestures in real time can improve human–computer interaction by making it more intuitive and natural. Color gloves and skin color detection are two prominent hand segmentation and detection technologies, but each has its own set of benefits and drawbacks. For physically challenged people, gesture identification is a crucial technique of sharing information. Support Vector Machine algorithm with Principal Component Analysis, hidden Markov model, superposed network with multiple restricted Boltzmann machines, Growing Neural Gas algorithm, convolutional neural network, double-channel convolutional neural network, artificial neural network, and linear Support Vector Machine with gesture dataset match the interaction of gestures for various postures in real time. Although this method can recognize a huge number of gestures, it does have certain downsides, such as missing movements due to the accuracy of the categorization algorithms. Furthermore, matching the vast dataset takes a longer time. The main contribution of this work lies in a conceptual framework based on the findings of a systematic literature review that provides fruitful implications based on recent research findings and insights that can be used to direct and initiate future research initiatives in the field of hand gesture recognition and artificial intelligence. As a result, a novel study based on a hybrid Recurrent Neural Network (RNN) with Chaos game optimization may reduce classification mistakes, increase stability, maximize resilience, and efficiently use the hand gestures' recognition system.

Keywords Artificial intelligence · Hand gesture (HG) · Algorithm · ANN

J. John (✉)
Priyadarshini College of Engineering, Nagpur, India
e-mail: jogi.john@pcenagpur.edu.in

S. P. Deshpande
Hanuman Vyayam Prasarak Mandal, Amravati, India

P. Chatterjee et al. (eds.), *Computational Intelligence for Engineering and Management Applications*, Lecture Notes in Electrical Engineering 984,
https://doi.org/10.1007/978-981-19-8493-8_18

1 Introduction

In a variety of computer–human interaction applications, gesture identification is an important task. As a result, establishing a reliable gesture identification system is a crucial challenge. The risk factor of HG identification is further refined by the hand movement experienced due to imaging quality, viewpoint, and crowded backdrops [1, 2]. "Theoretical examination of palm gestures for correct and efficient identification is difficult and fascinating. Real-time hand gesture detection apps are making computer–human interaction more natural and intuitive. Virtual reality and computer games both benefit from this recognition mechanism" [3]. Color gloves [4] and skin color detection [5] are two common strategies for hand segmentation and detection, but both have few advantages and disadvantages. Applications that recognize hand gestures play an essential role in people's daily lives. This gesture recognition approach not only increases human-to-human communication, but it also serves as a primary source of information.

Because hand gesture recognition is considered simple due to its widespread use in HCI applications, research in this topic has gotten a lot of interest in recent years. Telemedicine [6], interactive augmented reality [7], human–computer interaction interfaces [8], and other uses of this hand gesture recognition are a few examples. "HCI interfaces are divided into two types: hardware-based and vision-based techniques" [9]. With the use of markers, gloves, magnetic sensors, or other hardware solutions, the hand and feature points from that hand may be easily located. Although these methods produce high-point palm identification results, they are not generally used in real-time applications due to the increased hardware requirements and need of spontaneousness for end users. As a result of its benefits, such as being contact-free and requiring no additional hardware, the image analysis approach has gotten a lot of attention. "Furthermore, this image analysis develops a reliable technique for dealing with and normalizing variable environmental conditions, which is the most difficult issue for real-time applications" [10].

Because of the development of passive vision-based hand gesture identification systems, images captured by a single camera attain an accurate gesture recognition rate [11]. Hand gestures are divided into two categories: static and motional. When it comes to gesture identification, the features extracted by elastic graph matching are beneficial for distinguishing hand postures from complex backgrounds, with an accuracy of 85%. For recognition, a disjunctive normal-based learning technique is used, which achieves a 93% recognition rate. In learning approach, compactness and normalized hand moments are also applied. The recognition of finger spell is carried out using the CamShift algorithm, which yields a processing rate of 125 ms for single image frames [12]. Principal Component Analysis (PCA) is also employed in the palm gesture detection process. For motion-based hand gesture recognition, three fundamental techniques are used: HMM, optical flow, and model-based approaches. The construction of a hand gesture model uses a hybrid AdaBoost classifier and Haar features, as well as the benefit of a Kalman filter to reduce false detection. A model-based hand gesture tracking was considered for gesture recognition in [13], and the HMM for hand gesture was handled in [14].

"A multimodal deep learning approach was studied in order to study the cross-modality representation in an audio-visual speech recognition setting" [15]. The deep speech system was proposed lately [16], which merges deftly optimized RNN with this system to obtain a minimum error rate over a noisy speaking dataset. Deep architectures were used in the recognition process in the methodologies outlined above, resulting in better performance on high-level feature extraction and recognition. However, neural networks may be found in a variety of machine learning-based fields, including prediction of student performance, face recognition, and blood cell identification [17, 18].

The most current and prominent topic in machine learning is deep learning, which involves neural networks but contains a large number of hidden layers (i.e., > 1). "In speech recognition, natural language processing, and facial recognition applications, deep learning techniques have a high rate of success" [19]. Deep learning-based networks complement the benefits of learning algorithms and biologically inspired frameworks by obviating the need for a standard feedforward network, resulting in a higher overall recognition rate. In general, deep learning training is done layer-by-layer, much like the human visual cortex, and it relies on a more hierarchical and dispersed feature learning process [20]. Due to this dependence, more interesting features from the highly nonlinear functions are discovered during the training phase, and the complex issues are modeled in a perfect manner.

Research Gaps

Conventional machine learning classifiers appear superior in acknowledgment execution, but they give more precision with tall computational fetches. Standard acknowledgment calculations with routine classification methods experienced tall mistakes with less exactness. Whereas signal acknowledgment on any framework cannot be completely distinguished until the end of time, real-time acknowledgment of the sign dialects is basic to protect the frameworks in an viable way. Constrained investigation exists on effective signal acknowledgment models. Errors are more common in writing than they appear on the surface. Some well-known proof models are unable to recognize the sign dialect because the classifier easily falls into nearby ideals. In this manner, in this proposed work, it is basic to create a precise classification technique.

Objective

The main contribution of this work lies in a conceptual framework based on the findings of a systematic literature review that provides fruitful implications based on recent research findings and insights that can be used to direct and initiate future research initiatives in the field of hand gesture recognition and artificial intelligence.

2 Related Work

Tan et al. [21] presented an enhanced hand gesture (HG) recognition called enhanced densely connected CNN (EDenseNet) for the recognition of image-based hand gestures. High-density blocks were used to support gradient flow and feature propagation in the network. The dense network reduced the number of parameters required for network training while improving parameter efficacy. TensorFlow was used to carry out the implementation. Multiple HG datasets were used in the experiments, including one NUS dataset and two ASL (American Sign Language) datasets. "The EDenseNet has obtained 98.50% accuracy without data augmentation and 99.64% accuracy with data augmentation, outperforming other deep learning approaches" [21]. The major limitations faced while recognizing the HG in images were illumination variations and background noise which affects the recognition accuracy.

Tsinganos et al. [22] developed an HG recognition using the deep learning model named CNN-based Hilbert surface electromyography (EMG). An image of the sEMG signal was created using the space-filled Hilbert curve. Performance was evaluated using a single-fold neural network and a multi-fold neural network architecture. The MSHilbNet network architecture uses multiple scales of an initial Hilbert curve representation to achieve the same level of performance with fewer convolutional layers [22]. Ninapro was the dataset that was used. The HilbTime and HilbElect techniques were tested, and the results revealed that they outperformed the window segmentation approach across a variety of networks [22]. Higher computational time and the use of a single dataset for testing were the major challenges.

Gao et al. [23] proposed multimodal fusion of data and parallel multi-scale CNN-based method for HG recognition. "Initially, the data fusion was performed using sEMG signals, RGB images, and depth images of hand gestures. Then, the fused images are sent via parallel CNNs in two different sizes. The image was reduced to two HG detection results" [23]. After that, the results were combined together to get the last HG identification result. Finally, testing was conducted using a self-built database including ten typical hand identifications, demonstrating the method's efficacy and benefit for HG identification. The overall accuracy was 92.45%, and speed 32 ms proves the effectiveness of the fusion model. The challenge with this model was only static hand gestures which were identified.

Tan et al. [24] to detect the movement of the palm of a photograph, we have developed a deep learning model based on a convolutional neural network and spatial pyramid pooling. To overcome the difficulties of traditional pooling, the SPP concept was combined with convolutional neural networks. This multi-level linking is built together to increase the quality being fed into a completely connected layer. Provided with an intake of varying measurements, SPP also gives a set-length factor presentation. "Extensive experiments were conducted to examine the convolutional neural network–spatial pyramid pooling performance on three datasets such as American Sign Language (ASL), ASL digits, and NUS hand gesture dataset" [24]. The results reveal that CNN–SPP won over other deep learning-driven instances. The average accuracy was reported as 98.35% without DA and 99.34% with DA, respectively. The processing of vision-based HG recognition was performed with CNN-SPP, not the dynamic model.

Mujahid et al. [25] a deep learning model was used to show HG recognition in real time. A lightweight model YOLOv3 CNN was used for palm identification without any extra preprocessing, image filtering, and improvement of images. On a labeled dataset of palm motions in both Pascal VOC and YOLO formats, the YOLOv3 model was estimated. The experimentation was done on Python 3.7. The outcomes were achieved by extracting features from the palm and recognized palm gestures with the metrics such as precision, recall, accuracy, and an $F1$-score of 97.68, 94.88, 98.66, and 96.70%, respectively. With Single-Shot Detector and Visual Geometry Group, which achieved lesser accuracy between 82 and 85%, YOLOv3's overall performance was different. When compared to other DL models, the negative was the lower accuracy value.

Anazco et al. [26] developed HG recognition with the six-axis single patchable inertial measurement unit (IMU) attached to the wrist using Recurrent Neural Networks (RNNs). The IMU is made up of electronic pieces based on integrated circuits that are adhered to a stretchy substrate and have very long organized interconnections. The signal distortion (that is, motion artifacts) caused by vibrations during motion has been reduced. This patchable IMU uses a Bluetooth connectivity module to send the detected signal to each processing device. Cognitive performance was calculated by placing the current six-axis patchable IMU on the correct wrists of five participants and interpreting three hand gestures using two RNN-based models. In the network training procedure, the 6DMG public database was used. RNN-GRU had a classification accuracy of 95.34%, whereas RNN-BiLSTM had a classification accuracy of 94.12%. The most significant flaw was the intricate design.

Yuan et al. [27] presented the recognition of HG with wearable sensors using Deep Fusion Network. Initially, a specially integrated three-dimensional flex sensor and a data glove with two arm rings were designed to capture fine grain motion from the all knuckles and full arm. Deep Fusion Network was used to detect long distance dependency in complex HG. The CNN-based fusion task was performed to track detailed motion features from multi-sensors by extracting both shallow and deep features. We classified complex hand movements using a long short-term memory model with fused feature vectors. The datasets used for implementation were ASL and CSL datasets on the Python environment. Results of experiments are acquired especially in Chinese Sign Language (with 96.1% precision) and American Sign Language (with the precision of 99.93%). The difference between batch size and learning rate has a big impact on training time.

Abdulhussein and Firas [28] proposed a static HG recognition using deep learning model. The recognition process is characterized using dual phases. The scaling of bicubic static ASL binary pictures was first attempted. The border was discovered using the RE detection method. The CNN model was used to categorize the 24 alphabet letters in the second phase. The ASL hand motions' dataset was utilized to assess the testing performance. There are 2425 photos in this collection, each containing five persons. The classification accuracy achieved was 99.3%, with a loss function error of 0.0002 in 36-minutes with a 15-second elapsed duration with hundred repetitions. In compared to other similar CNN, SVM, and ANN for training efforts, the training time was short and the outcomes were excellent. The tuning of weights was done with SGD optimization which takes longer time to achieve convergence to the minima of the loss function which was one of the major limits.

Al-Hammadi et al. [29] introduced a deep learning model for recognizing gestures with the efficient HG representation. The significance of hand motion recognition has increased due to the rapid growth of the hearing-impaired population and the prevalence of touchless applications. The dynamic HG recognition is presented using multiple deep learning architectures for local and global feature representations, hand segmentation, and sequence feature globalization and recognition. The current method is based on a difficult dynamic hand gesture dataset that includes forty dynamic HGs performed by forty people in an uncontrolled setting. The system using MLP fusion attained the maximum accuracy of 87.69% in the signer-independent scenario. The drawback of the system shows that it was not strong enough to capture the long-term temporal dependence of the HG of the video data.

Sharma et al. [30] introduced an HG recognition using the integration of feature extraction (FE) and image processing (IP) techniques. The major objective was to recognize and categorize the hand gestures with its appropriate meaning and much possible accuracy. The preprocessing approaches used were Principal Component Analysis, histogram of gradients, and local binary pattern. The FE techniques used were ORB, bag-of-word, and canny edge detection methods. The identification of images was characterized with an ASL dataset. To obtain successful results, the preprocessed data were run through multiple classifiers (K-nearest neighbors, Random Forests, Support Vector Machines, Logistic Regression, naïve–Bayes, Multilayer Perceptron). ASL-KNN-ORB had a classification accuracy of 95.81%, whereas ASL-MLP-ORB had a classification accuracy of 96.96%. The new models' accuracy has been shown to be significantly higher than that of older ones. The technology had only been tested on static gesture pictures, which was a flaw.

Ameur et al. [31] introduced a hybrid network model using leap motion and dynamic HG recognition. Initially, HG identification was performed on continuous time series data collected from jump motion using a long short-term memory model network. Both bidirectional and unidirectional long short-term memory model architectures were used separately. The final prediction was made using a hybrid bidirectional unidirectional long short-term memory model. This significantly improves the performance of the model by taking into account the spatial and temporal relationships between the network layer and the jumping motion data during forward and backward paths. The recognition models were tested on the LeapGestureDB dataset and the RIT dataset, which are both publicly available benchmark datasets. The study demonstrates the HBU long short-term memory network's capability for dynamic hand gesture detection, with average recognition rates of 73.95% and 89.98% on both datasets, respectively. The increasing time consumption posed a significant barrier.

Mirehi et al. [32] developed a meaningful set of shape features using Growing Neural Gas algorithm-based graph construction. The graph properties had improved the stability against different scale, noise, and deformations. "This approach has been tested using the latest methods on NTU's manually numbered datasets". In addition, a thorough dataset (SBU-1) for various hand movements was created, which comprises 2170 photos" [32]. Many conceivable deformations and variants, as well as certain articulations, were included in this dataset. "The mean accuracy was calculated using several experiments such as utilizing half of the data for training and the other half for testing (*h-h*), leaving one subject out (*l-o-o*), and leaving nine subjects out (*l-9-o*)" [32]. With NTU dataset, the accuracy obtained was 98.68%, 98.6%, and 98% respectively. The mean accuracy of about 90% was obtained with the SBU-1 dataset. The system model is characterized with challenges such as enhanced time consumption and sensitivity to noise.

Li et al. [33] introduced CNN for gesture recognition. The feature extraction was done within the CNN, so no additional parameters had to be learned. During the recognition procedure, this CNN performed flawlessly in terms of unsupervised learning. The error backpropagation method was loaded alongside this CNN, and the weight and threshold of CNN were adjusted to reduce the error created by this model. Finally, the Support Vector Machine (SVM) was combined with this CNN for classification purposes, maximizing the resilience, and validity of the whole model. Experimental datasets were employed in the picture information obtained by Kinect, datasets consisting of five persons in the light of eight different motions (*G*1, *G*2, *G*3, *G*4, *G*5, *G*6, *G*7, and *G*8), image samples, totaling roughly 2000. MATLAB was used to integrate the gathered color and depth pictures. Eight species are identified in a semi-surveillance scenario [33]. However, the challenge is noticed with the fixed size value as CNN is not as good as the long short-term memory model in the long dependence.

Wu [34] developed a novel algorithm named double-channel CNN (DC-CNN) to enhance the recognition rate. Initially, the denoizing, preprocessing, and edge detection were performed on input images to spot the hand edge images. The hand edge and hand motion photos were then sent into CNN as input. CNNs have the same number of parameters and convolutional layers as LSTMs, but each layer has a different weights. The entire link layer of CNN was used for functional fusion, and CNN's Softmax classifier was used for classification. The experiment was performed using the NAO camera hand posture database (NCD) and Jochen Triesch database (JTD). The implementation was performed using MATLAB 2017a. The detection rate achieved was 98.02%. The issue represents a half image with a redundant background.

Chen et al. [35] introduced Kinect sensor-based gesture recognition. Some problems associated with recognizing gestures include poor robustness and accuracy. To address these issues, the planned Kinect sensor was primarily utilized to acquire gesture samples such as depth and color, which were subsequently analyzed. In addition, a network that combines CNN and RBM has been proposed for gesture recognition. This method uses an overlay network with a large number of RBMs to integrate both unsupervised and supervised CNN feature extractions for classification purposes. In basic gesture recognition, simulation analysis with a combined network yields a high recognition rate and a low error rate of only 3.9%. Because RBM requires precise data dispersion, the joint network, complicated sample, and other centralized networks function poorly.

Sharma introduced an gesture recognition system in et al. [36] Indian Sign Language using Fine-tuned Deep Learning Model accuracy performance is (99.92%) better than existing approaches [36] like Fine-Tuned Deep Transfer learning model (FTDT) 100% (Table 1).

3 Conclusion

In summary, HG offers promising research because it can ease communication and provide a means of interaction that can be used across different real-time applications. Most challenging issues for vision-based gesture identification systems are hand size, variation in skin color, and viewpoints. Other promises include similarity of gestures, mixed illuminations, and noisy background present in the images. Moreover, the use of wearable devices for HG recognition is significantly limited as signers are required to wear relevant devices beforehand, which entails cost and inconvenience. Due to the existence of skin-colored objects in the background which is considered as the challenging issue for the feature extraction process, along with these issues, few other issues also encountered by this system are lighting changes, efficiency, complex circumstances, speed, system inactivity, occlusion, etc. Above all, identification of powerful modeling techniques for capturing the specific sign language is found difficult. Traditional classifiers show better recognition performance, but they provide low accuracy with high computational cost. Ordinary recognition algorithms that used traditional classification approaches had a lot of errors and were less accurate. The majority of the research publications focus on improving hand gesture detection frameworks or building new algorithms. The greatest challenge faced by the researcher is to develop a robust framework that controls the most common issues with fewer limitations and accuracy and reliability. Hence, a new study based on hybrid Recurrent Neural Network with Chaos game optimization is needed which aims to perform effective HG recognition.

Table 1 Hand gesture recognition: a review of existing approaches

Author name	Method used	Objective	Advantages	Disadvantages
Tan et al. [21]	EDenseNet	Recognition of image-based hand gestures	Minimized the number of parameters for network training and enhanced the parameter efficacy	The image was a fluctuation of lighting and background noise, which impaired the recognition accuracy
Tsinganos et al. [22]	CNN-based Hilbert surface EMG	HG recognition using the deep learning model	When compared to the window segmentation approach, the performance across multiple networks is superior	Higher computational time and the use of a single dataset for testing were the major challenges
Gao et al. [23]	Multimodal fusion of data and parallel multi-scale CNN	HG recognition	Self-build database containing ten common hand identifications	Only static hand gestures were identified
Tan et al. [24]	Convolutional neural network–spatial pyramid pooling	Identification of palm gestures in images	CNN–SPP won over other deep learning-driven instances	Not a dynamic model
Mujahid et al. [25]	YOLOv3 CNN model	Palm identification without any extra preprocessing, image filtering, and improvement of images	Based on PascalVOC and YOLO format tagged palm gesture datasets	Lower precision value
Anazco et al. [26]	HG recognition with the six-axis single patchable inertial measurement unit (IMU) attached to the wrist using recurrent neural networks (RNNs)	HG recognition	The network training method used the public database 6DMG	Complicated design
Yuan et al. [27]	Deep fusion network is used to recognize HG utilizing wearable sensors	Recognition of HG	We classified complex hand movements using a long short-term memory model with fused feature vectors	Training time is greatly affected by the difference between batch size and learning rate

(continued)

Table 1 (continued)

Author name	Method used	Objective	Advantages	Disadvantages
Abdulhussein and Firas [28]	Bicubic static ASL, RE detection approach, CNN model	Static HG recognition using deep learning models	The training was completed in a short amount of time and with excellent outcomes	Takes longer time to achieve convergence
Al-Hammadi et al. [29]	Deep learning model for recognizing gestures	Recognition of HG	The system using MLP fusion attained the highest accuracy in the signer-independent scenario	It was not powerful enough to capture the video data's HG's long-term temporal dependency
Sharma et al. [30]	ASL-MLP-ORB	Recognize and classify hand gestures with the proper which means and as a great deal precision as feasible	It turns out that the accuracy of the new model is significantly higher than that of the previous model	Only static gesture images were used to test the system
Ameur et al. [31]	Hybrid network model using leap motion and dynamic HG recognition	HG recognition	Improves the model performance	Higher time consumption
Mirehi et al. [32]	Growing Neural Gas algorithm-based graph construction	HG recognition	Enhanced stability in the face of varying scales, noise, and deformations	Enhanced time consumption and sensitivity to noise
Li et al. [33]	CNN	Gesture recognition	Minimize the error produced	Not as good as the long short-term memory model in the long dependence
Wu [34]	DC-CNN	To enhance the HG recognition rate	The rate of recognition achieved is pretty good	The dataset posture represents a half image with a background that is redundant

(continued)

Table 1 (continued)

Author name	Method used	Objective	Advantages	Disadvantages
Chen et al. [35]	CNN and RBM joint network	Kinect sensor-based gesture recognition	In basic gesture recognition, simulation analysis with a combined network yields a high recognition rate and a low error rate of only 3.9%	The performance of joint networks, complicated samples, and other centralized networks is poor

References

1. Li SZ, Yu B, Wu W, Su SZ, Ji RR (2015) Feature learning based on SAE–PCA network for human gesture recognition in RGBD images. Neurocomputing 5(151):565–573
2. Pugeault N, Bowden R (2011) Spelling it out: real-time ASL fingerspelling recognition. In: IEEE international conference on computer vision workshops (ICCV Workshops), pp 1114–1119
3. Wachs JP, Kölsch M, Stern H, Edan Y (2011) Vision-based hand-gesture applications. Commun ACM 54(2):60–71
4. Wang RY, Popović J (2009) Real-time hand-tracking with a color glove. ACM Trans Graph (TOG) 28(63):1–8
5. Lee T, Hollerer T (2009) Multithreaded hybrid feature tracking for markerless augmented reality. IEEE Trans Visual Comput Graph 15(3):355–368
6. Wachs J, Stern H, Edan Y, Gillam M, Feied C, Smith M, Handler J (2006) A real-time hand gesture interface for medical visualization applications. Applications of soft computing AISC, vol 36. Springer, Heidelberg, pp 153–162
7. Shen Y, Ong SK, Nee AYC (2011) Vision-based hand interaction in augmented reality environment. Int J Hum Comput Interact 27(6):523–544
8. Lee DH, Hong KS (2010) Game interface using hand gesture recognition. In: Proceedings of the 5th international conference on computer sciences and convergence information technology (ICCIT 2010). IEEE, pp 1092–1097
9. Czupryna M, Kawulok M (2012) Real-time vision pointer interface. In: Proceedings of the 54th international symposium ELMAR (ELMAR 2012). IEEE, pp 49–52
10. Huang Y, Monekosso D, Wang H, Augusto JC (2011) A concept grounding approach for glove-based gesture recognition. In: Proceedings of the 7th international conference on intelligent environments (IE 2011). IEEE, pp 358–361
11. Rodriguez S, Picon A, Villodas A (2010) Robust vision-based hand tracking using single camera for ubiquitous 3D gesture interaction. In: Proceedings of IEEE symposium on 3D user interfaces. Waltham, pp 135–136
12. Park A, Yun S, Kim J, Min S, Jung K (2009) Real-time vision based Korean finger spelling recognition system. Int J Electr Comput Eng 4:110–115
13. Ren Y, Zhang F (2009) Hand gesture recognition based on MEB-SVM. In: Proceedings of international conference on embedded software and systems. Hangzhou, pp 344–349
14. Bradski G, Davis J (2000) Motion segmentation and pose recognition with motion history gradients. In: Proceedings of IEEE workshop on applications of computer vision. Palm Springs, pp 238–244
15. Ngiam J, Khosla A, Kim M, Nam J, Lee H, Ng AY (2011) Multimodal deep learning. In: Proceedings of the 28th international conference on machine learning, pp 689–696
16. Hannun A, Case C, Casper J, Catanzaro B, Diamos G, Elsen E, Prenger R, Satheesh S, Sengupta S, Coates A et al. (2014) Deepspeech: scaling up end-to-end speech recognition. arXiv preprint arXiv:1412.5567

17. Khashman A (2012) A investigation of different neural models for blood cell type identification. Neural Comput Appl 21(6):1177–1183
18. Oyedotun OK, Tackie SN, Olaniyi EO, Khashman A (2015) A data mining of students' performance: Turkish students as a case study. Int J Intell Syst Appl 7(9):20–27
19. Oyedotun OK, Khashman A (2017) Deep learning in vision-based static hand gesture recognition. Neural Comput Appl 28(12):3941–3951
20. Kruger N, Janssen P, Kalkan S, Lappe M, Leonardis A, Piater J, Rodriguez-Sanchez AJ, Wiskott L (2012) Deep hierarchies in the primate visual cortex: what can we learn for computer vision? IEEE Trans Pattern Anal Mach Intell 35(8):1847–1871
21. Tan YS, Lim KM, Lee CP (2021) Hand gesture recognition via enhanced densely connected convolutional neural network. Expert Syst Appl 175:114797
22. Tsinganos P, Cornelis B, Cornelis J, Jansen B, Skodras A (2021) Hilbert sEMG data scanning for hand gesture recognition based on deep learning. Neural Comput Appl 33(7):2645–2666
23. Gao Q, Liu J, Ju Z (2021) Hand gesture recognition using multimodal data fusion and multiscale parallel convolutional neural network for human–robot interaction. Expert Syst 38(5):e12490
24. Tan YS, Lim KM, Tee C, Lee CP, Low CY (2021) Convolutional neural network with spatial pyramid pooling for hand gesture recognition. Neural Comput Appl 33(10):5339–5351
25. Mujahid A, Awan MJ, Yasin A, Mohammed MA, Damaševičius R, Maskeliūnas R, Abdulkareem KH (2021) Real-time hand gesture recognition based on deep learning YOLOv3 Model. Appl Sci 11(9):4164
26. Añazco EV, Han SJ, Kim K, Lopez PR, Kim T-S, Lee S (2021) Hand gesture recognition using single patchable six-axis inertial measurement unit via recurrent neural networks. Sensors 21(4):1404
27. Yuan G, Liu X, Yan Q, Qiao S, Wang Z, Yuan L (2020) Hand gesture recognition using deep feature fusion network based on wearable sensors. IEEE Sens J 21(1):539–547
28. Abdulhussein AA, Raheem FA (2020) Hand gesture recognition of static letters American sign language (ASL) using deep learning. Eng Technol J 38(6A):926–937
29. Al-Hammadi M, Muhammad G, Abdul W, Alsulaiman M, Bencherif MA, Alrayes TS, Mathkour H, Mekhtiche MA (2020) Deep learning-based approach for sign language gesture recognition with efficient hand gesture representation. IEEE Access 8:192527–192542
30. Sharma A, Mittal A, Singh S, Awatramani V (2020) Hand gesture recognition using image processing and feature extraction techniques. Procedia Comput Sci 173:181–190
31. Ameur S, Khalifa AB, Bouhlel MS (2020) A novel hybrid bidirectional unidirectional LSTM network for dynamic hand gesture recognition with leap motion. Entertainment Comput 35:100373
32. Mirehi N, Tahmasbi M, Targhi AT (2019) Hand gesture recognition using topological features. Multimedia Tools Appl 78(10):13361–13386
33. Li G, Tang H, Sun Y, Kong J, Jiang G, Jiang D, Tao B, Xu S, Liu H (2019) Hand gesture recognition based on convolution neural network. Clust Comput 22(2):2719–2729
34. Wu XY (2019) A hand gesture recognition algorithm based on DC-CNN. Multimedia Tools Appl 1–3
35. Cheng W, Sun Y, Li G, Jiang G, Liu H (2019) Jointly network: a network based on CNN and RBM for gesture recognition. Neural Comput Appl 31(1):309–23. Li SZ, Yu B, Wu W, Su SZ, Ji RR (2015) Feature learning based on SAE–PCA network for human gesture recognition in RGBD images. Neurocomputing 151:565–73
36. Sharma CM, Tomar K, Mishra RK, Chariar VM (2021) Indian sign language recognition using fine-tuned deep transfer learning model. In: Proceedings of international conference on innovations in computer and information science (ICICIS), pp 62–67

Design a Computer-Aided Diagnosis System to Find Out Tumor Portion in Mammogram Image with Classification Technique

Rashmi Ratnakar Bhale and Ratnadeep R. Deshmukh

Abstract For the original, mammogram is a one type of X-ray image which is used to create detailed image of breast. Nowadays, medical images are playing important role in diagnosis and plays important role in prognosis. Images captured by machines may not be directly diagnosed the problem or find out tumor, so it is very important to process image and analyze on the basis of its features. Manual reading of an image may cause misdiagnosis. Computer vision supports scholars to analyze images to get required information, understand information or descriptions, and various pattern. In this paper, we are designing a CAD system which finds region of interest, calculates GLCM features, and compares classification technique to decide the one which gives highest accuracy with this algorithm. For the experimental work, images are taken from mini-MIAS database, and for experimental work, MATLAB has been used.

Keywords CAD · MIAS · Mammogram · GLCM · Insert

1 Introduction

Detection of breast cancer at early stage is the way to improve breast cancer prognosis meanwhile the reasons of the disease are still unidentified to us. Nowadays, breast cancer has become a significant health problem in the world [1]. From last two to three decades, breast cancer is the type of cancer which is common and also the key cause of mortality because of cancer in female everywhere in the world. If we saw the historical data, near about 1.38 million new breast cancer cases were diagnosed in 2008 with almost 50% of all breast cancer patients. In case of developed nations alike the USA, near about 232,340 cases of breast cancer in female will be diagnosed. In an American female, 12.38% is the lifetime risk of developing breast cancer [2].

There is enormous variance in breast cancer survival rates over the world, which will be with an assessed 5-year survival of 80% in developed countries which would

R. R. Bhale · R. R. Deshmukh (✉)
Department of Computer Science and Information Technology, Dr. Babasaheb Ambedkar Marathwada University, Aurangabad, India
e-mail: rrdeshmukh.csit@bamu.ac.in

P. Chatterjee et al. (eds.), *Computational Intelligence for Engineering and Management Applications*, Lecture Notes in Electrical Engineering 984,
https://doi.org/10.1007/978-981-19-8493-8_19

be less than 40% for developing countries. Resource and infrastructure constraints are the challenges faced by developing countries, so this needs to improve for breast cancer outcomes of timely recognition and diagnosis [3].

In last few years, in Taiwan and other developed countries, breast cancer has become one of the top reasons of death. The prime screening tool used is mammogram and the secondary tool is breast ultrasound which is commonly used to recompense the inadequacies of mammography; then, its sensitivity is lesser in higher breast density [4].

In human body, normal cells grow and divide to produce new cells in order to fulfill the body requirements. When these normal cells grow old, then they die. So in that case, the newly formed cells take their place. Cancer arises when there is no need of newly formed cells and these new cells are generated; in this case, the old and destroyed cells are not died as they should be, so these additional cells are built as a massive portion of tissues which is known as a lump, abnormal growth, or a tumor [5].

It has been observed the best examination technique for breast cancer detection at an early stage reducing mortality rates by up to 25%, and their analysis requires skill as well as experience by a well-trained radiologist. At first, radiologist screens the mammograms to find abnormalities. Then, if any suspicious abnormality is noticed, then the supplementary diagnostic examination is accomplished to assess the possibility that the abnormality is malignant. The main signs of breast cancer in mammographic include cluster by microcalcifications and masses. A mass can be roughly reflected as a circle with a luminance that grows from its border to its center. Identify a mass in a mammogram is tough because of their low contrast [6].

Mammograms are signified as very correct and complex images to be interpreted. Many cases of cancers are not detected in the screen with human eyes. So, there is a need of advance computer-aided diagnosis systems which will help a radiologist in his diagnosis for the cancer [7]. For these reasons, our aim in this paper is to work on GLCM features and compare various classification techniques which will give highest accuracy. A mass detection approach was referred by Kok et al. [8] and Cerneaz [9] which was given by Guissin and Brady [10] and that was based on iso-intensity contours, which stated that the measures that they used to reduce FPs are not appropriate for mammogram images.

Digital mammography is a technique for recording X-ray images in computer code instead of on X-ray film, as with conventional mammography. The first digital mammography system received the US Food and Drug Administration (FDA) approval in 2000 [11].

Digital mammogram is one of the important methods to identify the breast cancer at an early stage at some extend. The advantages of digital mammography include the lack of ionizing radiation, its non-invasiveness, the relatively compact instrumentation, and its cost-effectiveness.

In order to reduce the increasing workload and improve the accuracy of interpreting mammograms, a variety of CAD systems that perform computerized mammographic analysis have been proposed, as stated by Shan et al. [12].

Mudigonda and Rangayyan performed classification, and they used shape and texture features for performing classification of mammogram images into two classes malignant and benign. For classification, SVM classifier was applied on it [13].

Chithra and team have studied various images to remove unwanted noise and performed enhancement techniques like contrast limited adaptive histogram equalization, Laplacian and Harr filtering and color models, and so on; then, the clustering algorithms are used for data logically, and then, patterns were extracted for pattern analysis, grouping, decision-making, and machine learning techniques and segment the regions using binary, K-means, and OTSU segmentation algorithm. They classify the images with SVM and K-nearest neighbor classifier to produce good results for those images [14].

Sreedevi and Sherly proposed an algorithm which may help to remove Gaussian as well as impulse noise very effectively without any loss of chosen data. Author combines robust outlyingness ratio mechanism with extended NL-Means filter based on discrete cosine transform to detect and remove of noise. For segmenting and removing pectoral muscles, they used global thresholding to identify pectoral muscles and to identify the edge of the full breast and connected component labeling to identify and remove the connected pixels outside the breast region edge detection processes. This algorithm removes Gaussian and impulse noise effectively without any loss of desired data, and they obtained 90.06% accuracy [15].

2 Breast Anatomy

A mammogram image has two regions: those are the exposed breast region and the other is unexposed non-breast region. Background region which is non-breast area in a mammogram image usually looks like as a black color section, and it moreover has high-intensity portions like bright rectangular labels or numbering, solid markers, and artifacts. Breast portion can be divided into four parts as (Fig. 1):

1. Near-skin tissue region, which covers uncompressed fatty tissue which is positioned at the edge of the breast, close to the skin–air interface where the breast is poorly compressed.
2. Fatty region, which contains the fatty tissue that is located next to the uncompressed fatty tissues adjoining the denser region of fibroglandular tissue.
3. Glandular regions, which consist of non-uniform breast density tissue with assorted texture that is surrounded by the hyperdense region of the fibroglandular tissue.
4. Hyperdense region, which represents the high-density portions of the fibroglandular tissue, or it may be a tumor [16].

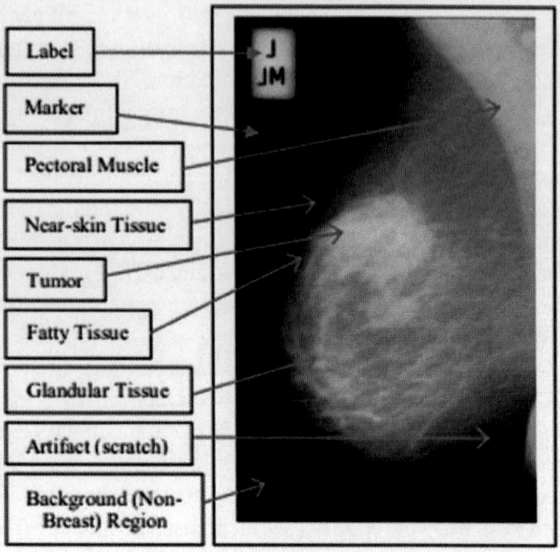

Fig. 1 Mammogram image composes of the image background, label, marker, artifact (scratch), near-skin tissue, fatty tissue, pectoral muscle, and denser glandular tissue

3 Different Forms of Breast Abnormalities

There are some types of irregularity that may affect breast tissues. These irregularities are frequently categorized into three categories that are the opacities, microcalcifications, and architectural distortions:

A. Masses are space conquering lesions which are seen on two different impressions. They are categorized by their shape, their contour, and density either it is high, medium, or low fat. Breast cancers are certainly not made of fat though they may trap lubricant. Lesions containing fat are: oil cysts, lymphomas, the galactocele, and mixed lesions. Mass that contains fat is always benign.
B. Microcalcifications are distributed into three classes: typically benign, suspicious, and with high probability of malignancy.
C. Architectural distortions are not the normal architecture of breast tissue with invisible mass. They can be dense center or clear center [17].

4 Materials and Methods

The mini-MIAS database has been used for the experimental work; each image is of size 1024 × 1024 pixels at a resolution of 200 m. The dataset includes circumscribed and speculated cases in both benign and malignant categories. The cases in each category are further divided into dense glandular, fatty, and fatty glandular types. The center of anomaly and an approximate radius of mass of each image are given in the database [18]. Mini-MIAS database contains a total of 322 mammogram images of both left and right breasts of 161 patients. Out of them, 61 are benign, 54 are malignant, and 207 are normal mammogram [19]. The whole implementation is done in MATLAB.

5 Work Flow

Work flow is the diagrammatic representation of process which shows a diagram or pictorial representation of the steps performed during experiment. In image processing, there are certain phases which play an important role in processing an image. These phases include taking input from database, then enhancement of an image, finding the part which is to be consider for further operation which is considered as region of interest, then perform operation, and find the results and analyze them (Fig. 2).

5.1 Input Image

The image is taken from mini-Mammographic Image Analysis Society (MIAS) database which is of size 1024 × 1024. The image is taken as it is from database which is of .PGM format (portable gray map format file).

Fig. 2 Work flow diagram

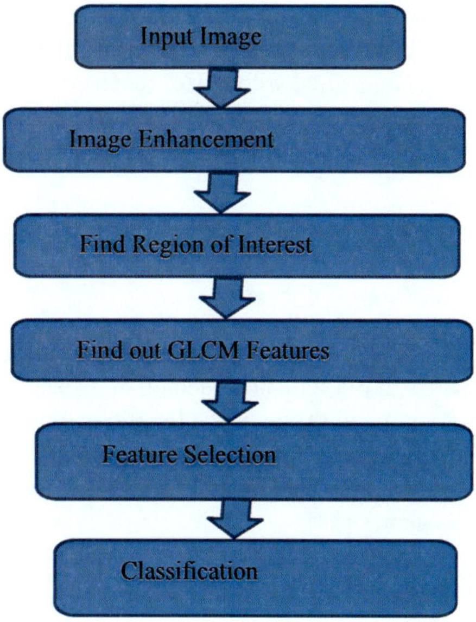

5.2 Image Enhancement

It is known to all that the process of image enhancement is a problem-oriented process and the aim of the image enhancement is to increase the optical exterior of the image or to deliver better transform depiction for future automated image processing [20].

Image enhancement is considered as preprocessing step for better observation, analysis, demonstration, and diagnosis. The definition of image contrast is the difference between intensity at highest and lowest levels of an image [21].

The simplest form of discrete wavelet transformation is Haar wavelet transformation; it operates by calculating the sums and differences between the intensity values of the pixels. The region of interest is considered as the input for the further operation, and at each level of it, the algorithm finds out four coefficients, and those are approximation, vertical, horizontal, and diagonal. This procedure is repeated yet again for the generation of approximation coefficients.

5.3 Region of Interest

As the successive calculations and investigation ominously depend upon the ROI size, therefore the size and the location of the region of interest are very important [22].

To find the region which contains the massive portion or tumor and to separate that much part of an image from the background, there are different image processing techniques which have been used by researchers. But, in mini-MIAS database, the coordinates of regions of interest (ROI) are given by expert radiologist, in the form of chain code allied with MIAS database as *.overlay files. Here, the ROI is extracted by reading the chain code *.overlay files.

5.4 GLCM Features

Features' extraction is one of the key steps in breast cancer detection. Some of the most commonly used texture measures are derived from the gray-level co-occurrence matrix. It is a statistical method of inspecting texture that reflects the spatial relationship of pixels. The GLCM functions illustrate the texture of an image by calculating how often pairs of pixel with explicit values and in a specified spatial relationship occur in an image, creating a GLCM, and then extracting statistical measures from this matrix [23].

First-order statistical features and second-order statistical features can be extracted from ROI.

First-order statistics features: FOS is a bundle of various statistical properties of the image histogram. It depends on the gray value of a distinct pixel without caring about the relation with the neighborhood pixels.

Second-order statistics features: It is a statistical tool used for mining second-order texture features of a mammogram image. GLCM defines the relation between two gray-level pixels in an image. An element in GLCM Pd, $\theta(i, j)$, gives the probability of the existence of the pair of gray level (i, j), parted by distance d at direction θ [24].

5.5 *Feature Selection*

Here, we are using first-order and second-order gray-level co-occurrence matrix. With first and second order, twenty-six features have been calculated, and for now in this experimental work, we are using all features for classification.

5.6 *Classification*

The mini-MIAS database has been used for the feature extraction and the statistical data, i.e., GLCM features were given as input to classification. For classification, we used the classification learner application provided by MATLAB. After applying various classification methods, we can say that as per result of classification with cubic SVM, we got 71.6% accuracy, and with fine and weighted KNN, we get 78.4% accuracy. So, according to the scenario, fine and cubic KNN get high accuracy than any support vector machine.

6 Results and Discussion

The highest accuracy result is obtained by classification when applied various classification method. Here, we have used benign and malignant images. Fine KNN finely detailed distinctions between classes. The number of neighbors is set to 1, and weighted KNN medium distinctions between classes, using a distance weight. The number of neighbors is set to 10. For future work, we are working to improve accuracy of the algorithm.

7 Conclusion

The aim of designed algorithm is to develop a CAD system which will give accurate diagnosis of mammogram image to avoid delay in the treatment of breast cancer. It helps to reduce mortality rate. For implementation of developed algorithm, mini-MIAS database is used. For statistical features, GLCM features have been calculated, and then, classification is carried out with SVM and KNN classification technique. As per implementation, fine and weighted KNN gives better result as compared to other classifiers.

References

1. Bandyopadhyay SK, Bandyopadhyay SK (2010) Pre-processing of mammogram images. Int J Eng Sci Technol 2(11):6753–6758
2. Akram M, Iqbal M, Daniyal M, Khan AU (2017) Awareness and current knowledge of breast cancer. Biol Res 50:33. https://doi.org/10.1186/s40659-017-0140-9
3. Coleman M, Quaresma M, Berrino F, Lutz JM, Angelis R, Capocaccia R et al (2008) Cancer survival in five continents: a worldwide population-based study (CONCORD). Lancet Oncol 9:730–756
4. Hsu C-Y, Chou Y-H, Chen C-M (2014) A tumor detection algorithm for whole breast ultrasound images incorporating breast anatomy information. In: 2014 international conference on computational science and computational intelligence
5. Ponraj D, Jenifer M, Poongodi P, Manoharan J (2011) A survey on the preprocessing techniques of mammogram for the detection of breast cancer. J Eng Trends Comput Inf Sci 2(12):656–664. http://www.csijournal.org
6. Maitra IK (2011) Identification of abnormal masses in digital mammography images. Int J Comput Graph 2(1)
7. Al-Bayati M, El-Zaart A (2013) Mammogram images thresholding for breast cancer detection using different thresholding methods. Adv Breast Cancer Res 2:72–77
8. Kok SL, Brady M, Highnam R (1998) Comparing mammogram pairs for the detection of lesions. In: Karssemeijer N, Thijssen M, Hendriks J, van Erning L (eds) Proceedings of 4th international workshop digital mammography, Nijmegen, June 1998, pp 103–110
9. Cerneaz NJ (1994) Model-based analysis of mammograms. Ph.D. dissertation, Department of Engineering Science, University of Oxford, Oxford
10. Guissin R, Brady JM (1992) Iso-intensity contours for edge detection. Technical report OUEL 1935/92, Department of Engineering Science, University of Oxford, Oxford
11. Rangayyan R, Ayres F, Desautels JL (2007) A review of computer-aided diagnosis of breast cancer: toward the detection of subtle signs. J Franklin Inst 344(3–4):312–348
12. Shan J, Ju W, Guo Y, Zhang L, Cheng HD (2010) Automated breast cancer detection and classification using ultrasound images—a survey. Pattern Recogn 43:299–317
13. Mudigonda NR, Rangayyan RM (2001) Detection of breast masses in mammograms by density slicing and texture flow-field analysis. IEEE Trans Med Imaging 20(12)
14. Chithra PL, Bhavani P (2019) A study on various image processing techniques. Int J Emerg Technol Innov Eng 5(5):316. ISSN: 2394-6598
15. Sreedevi S, Sherly E (2014) A novel approach for removal of pectoral muscles in digital mammogram. In: International conference on information and communication technologies (ICICT 2014)
16. Radovic M, Djokovic M, Peulic A, Filipovic N (2013) Application of data mining algorithms for mammogram classification. ISBN: 978-1-4799-3163-7/13/$31.00 © 2013 IEEE

17. Bahaalddin BM, Ahmed HO (2020) Breast mass classification based on hybrid discrete cosine transformation, Haar wavelet transformation. UKH J Sci Eng 4(2). E-ISSN: 2520-7792
18. Saidin N, Sakim HAM, Ngah UK, Shuaib IL (2012) Segmentation of breast regions in mammogram based on density: a review
19. Hela B, Hela M, Kamel H, Sana B, Najla M (2013) Breast cancer detection: a review on mammograms analysis techniques. In: 2013 10th international multi-conference on systems, signals & devices (SSD), Hammamet, 18–21 Mar 2013. ISBN: 978-1-4673-6457-7/13/$31.00 © 2013 IEEE
20. Zakeri FS, Behnam H, Ahmadinejad N (2012) Classification of benign and malignant breast masses based on shape and texture features in sonography images. J Med Syst 36:1621–1627
21. Suckling J, Parker J, Astley S, Dance D (1994) The mammographic images analysis society digital mammogram database. Excerpta Med Int Congr Ser 1069:375–378
22. Agaian SS, Lentz KP, Grigoryan AM (2000) A new measure of image enhancement, Jan 2000. www.researchgate.net/publication/244268659
23. Padma Priya G, Venkateswarlu T (2017) Image contrast enhancement in wavelet domain. Adv Comput Sci Technol 10(6):1915–1922. ISSN: 0973-6107
24. Sharma V (2017) Comparative analysis of region of interest of different sizes for breast density classification. Int J Med Res Health Sci 6(3):76–84. ISSN: 2319-5886

Performance Analysis of Panoramic Dental X-Ray Images Using Discrete Wavelet Transform and Unbiased Risk Estimation

J. Jeslin Libisha, S. Harishma, D. Jaisurya, and R. Bharani

Abstract The images acquired in any modality will tend to have imposed noise in it. Denoising has to be done as a major step to perform further analysis. The denoising technique employed should be capable of removing all the noises present in the image. The dental panoramic X-ray images are denoised using discrete wavelet transform (DWT) and orthonormal wavelet transform (OWT) with Stein's unbiased risk estimator (SURE). The performance analysis is done based on mean squared error (MSE) and peak signal-to-noise ratio (PSNR) values obtained from both the techniques. The denoising process preserves the information present in the signal like edges of the diseased signal. The denoised image is subjected to feature extraction and thresholding process to extract the diseased portion of the teeth. Artificial neural network is built to classify the images as normal or diseased teeth based on the extracted features. The performance analysis shows that orthonormal wavelet transform with Stein's unbiased risk estimator performs well with good PSNR and MSE value.

Keywords Panoramic X-ray images · Discrete wavelet transform · Orthonormal wavelet transform · Mean squared error

1 Introduction

An image is represented by two-dimensional function $f(x, y)$, where x and y are its spatial coordinates. When the amplitude of image f and its spatial coordinates are given in discrete form, then the source image becomes digital. The digital images are usually represented in bits (0 or 1). Processing the transformed digital images through best algorithms to enhance the quality to improve understandability of image

J. Jeslin Libisha (✉) · D. Jaisurya · R. Bharani
Department of Biomedical Engineering, Dr. N.G.P Institute of Technology, Coimbatore, India
e-mail: jeslinlibisha@gmail.com

S. Harishma
Department of Biomedical Engineering, Kalaignarkarunanidhi Institute of Technology, Coimbatore, India

P. Chatterjee et al. (eds.), *Computational Intelligence for Engineering and Management Applications*, Lecture Notes in Electrical Engineering 984,
https://doi.org/10.1007/978-981-19-8493-8_20

and extract useful information from it is referred to as digital image processing. For the past few years, many researchers have been endeavored in developing algorithm for the improvement in medical image analysis. We chose dental panoramic X-ray image as input for denoising. The purpose of selecting this image is that many are suffering from dental diseases such as cyst, oral cancer, etc., which should be diagnosed effectively. Hence, the goal of our project is to provide the best denoising algorithm which will be appropriate for better diagnosis. Denoising is an important factor to be considered in image processing. This process is usually done before the data are analyzed. It is done so that the essential information is retained by removing all the noises present in the data. The noise is usually of high-frequency content and it retains the low-frequency content. Henceforth, denoising is performed to recuperate the vital data. In this work, we focus much on how well the artifact granularity has been taken out and how well the data are obtained in dental panoramic images. In mathematics, wavelets are simply the functions that exhibit some type of oscillatory behavior. For few signals, when wavelet transform is applied to the data, it gives exact data about the sign information than some other sign investigation procedures. For few signals, the wavelet analysis provides accurate information about the signal data than any other signal analysis techniques. Wavelet transform is widely used in areas of digital image processing, biomedical imaging, image compression, etc. Wavelet analysis can be performed in MATLAB—wavelet toolbox to compute the wavelet transform coefficients.

Yadav et al. [1] discussed about the importance of edge preservation and the amount of noise that has been removed in the image. This process is accomplished by simulating different thresholding techniques and by comparison of those PSNR values. Donoho [2] has proposed about threshold function which is developed for image denoising algorithms. In this process, the removal of noise is done by using wavelet transform along with threshold functions. Finally, Universal, Visu Shrink, Sure Shrink, Bayes Shrink, and normal are compared with their threshold functions that improve signal-to-noise ratio efficiently. Vetterli et al. [3] introduce a new technique to denoise images using orthonormal wavelet technology. In this paper [4–6], the process of denoising is directly parameterized as a sum of nonlinear processes with unknown weights. Later, the estimated mean square error of the output between the denoised image and the clean image is minimized. Luisier et al. [7] have discussed about denoising of images containing Gaussian and salt-pepper noise. The workflow is structured as follows. The image is denoised by filtering method and by wavelet based techniques using threshold. Simultaneously, hard threshold and filtering method are applied on noisy image. The results of peak signal-to-noise ratio and mean squared error are calculated by comparing with all the cases. After inferring through all these literature survey papers, we got an idea that denoising the image by using Stein's unbiased risk estimator (SURE) [8, 9] will provide better results. We proved it by comparing the PSNR values of SURE and DWT technique. We also found that the SURE algorithm is found to be the best technique for medical image analysis.

The wavelet limit denoising strategy [10–12] is to handle the disintegrated wavelet coefficients by tracking down suitable edges in the fitting technique and to eliminate the wavelet coefficients having a place with the commotion, leaving the wavelet coefficients having a place with the sign, and afterward reproduce the handled wavelet coefficients to get the denoised signal. Old style edge handling wavelet denoising has the accompanying sorts: wavelet hard edge denoising, delicate edge denoising, and delicate and hard edge denoising [13]. Strategies proposed in the past are predominantly arranged into spatial space and change-based methods. The choice is subjected to the application and the idea of commotion present in the picture. Denoising in the spatial area is performed utilizing direct techniques dependent on an averaging channel, e.g., Gabor proposed a model dependent on Gaussian smoothing [14] and edge recognition-based denoising utilizing anisotropic sifting [15–17]. Yaroslavsky and Manduchi utilized an area sifting strategy [18, 19]. Picture denoising in the spatial area is perplexing and additional tedious particularly for ongoing issues. A few scientists worked in the recurrence area to denoise pictures, e.g., Wiener channels [18]. Chatterjee and Milanfar worked on the presentation of denoising by exploiting patch excess from the Weiner channel-based technique [20]. The extent of these direct procedures is restricted to fixed information, as the Fourier change is not appropriate to perform examination on non-fixed and nonlinear information. Subsequently, denoising techniques dependent on multi-scale denoising utilizing nonlinear activities in the change space came up as another option. These numerous scales give scanty portrayal of the sign in the change area.

2 Methodology

The block diagram in Fig. 1 depicts the steps carried out in this work. The images may be affected due to artifacts, which have to be removed in order to obtain the useful information. The image denoising is performed using discrete wavelet transform and orthonormal wavelet transform, and the performance of these algorithms is compared later. After denoising, the image is set to thresholding procedure to differentiate normal and malformed regions of teeth. Finally, decision-making is done with artificial neural network to confirm the region of abnormality.

The principle of an image denoising calculation is then to diminish the artifact level, while protecting the image highlights. The multi-goal investigation performed by the wavelet change has been demonstrated to be an integral asset for denoising. In wavelet area, the commotion is consistently spread all through the coefficients, while the greater part of the picture data is packed in the few largest coefficients. The most direct method of recognizing data from noise in the wavelet area comprises thresholding the wavelet coefficients [21].

However, the disadvantage of the thresholding methods dependent on wavelet transform is that they need an earlier information on the noise/artifact power present in the image for the ideal threshold figuring. The downside of the wavelet coefficients thresholding methods is just the picked limit which may not match the particular appropriation of sign and commotion parts in various scales [22, 23]. Therefore, metaheuristic calculations can give great arrangements of observing ideal limit an

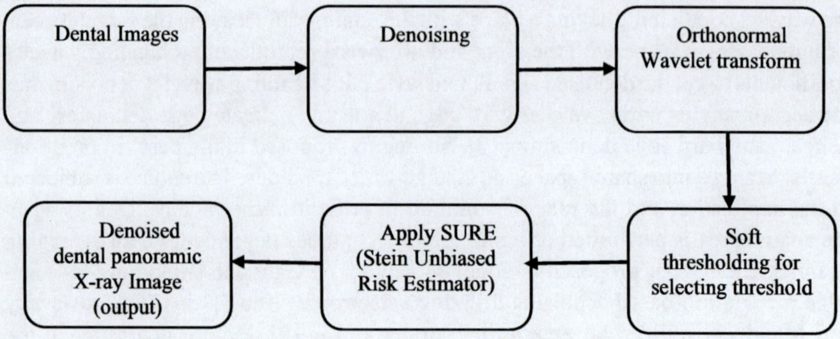

Fig. 1 Block diagram

incentive for each information picture without the need of any earlier information on the clamor and indeed, even the commotion assessment. We have proposed the utilization of the SSO and CSA advancement calculations to view as the ideal limit that limits the PSNR rules. These calculations are picked on the grounds that they have low intricacy and quick assembly dissimilar to different techniques [24].

In this paper, three assessment index of image displays, SNR (signal-to-noise ratio), and MSE (mean square error) are considered in the correlation of different denoising strategies. Every marker confirms one of the benefits and detriments of the denoising strategy. Among them, the image display shows that the denoising image can be outwardly noticed; the signal-to-noise ratio is pointed at retaining the sign while smothering all noises; the mean square error is reasonable for displaying the sharpness of an image. Bnou proposed a strategy dependent on an unaided learning model. They introduced versatile word reference learning-based denoising of guess. The wavelet coefficients were denoised by utilizing a versatile word reference gained over the arrangement of removed patches from the wavelet portrayal of the adulterated picture [25].

2.1 Discrete Wavelet Transform

The discrete signal in time function is transformed to a wavelet representation in discrete format, which is achieved using discrete wavelet transform. In wavelet transform, a signal is divided into small wavelets. The analysis of non-stationary signals is possible by DWT.

2.2 Unbiased Risk Estimator

Stein's unbiased risk estimator is a tool that is mostly used in statistical field, which is upcoming technique in image processing. It is an unbiased estimator of mean squared error. It is given by,

$$E\mu\,\{\text{SURE}\,(H)\} = \text{MSE}\,(H)$$

where

$$\text{MSE}\,(H) = E\mu\|H(x) - \mu\|2$$

2.2.1 PSNR

Peak signal-to-noise ratio is the important parameter in analyzing the performance (accuracy) of denoising.

$$\text{PSNR} = 10\log 10(\text{Maxi})2/\text{MSE}$$

where
 Maxi is the maximum intensity (255);
 MSE is the mean squared error.

2.2.2 MSE

Mean square error is found by taking the difference between actual value and the predicted value. The MSE is a measure of the quantity of the estimator. The value of MSE is always positive. For better results, the value lies closer to zero.

$$\text{MSE} = 1/n\sum_{i=1}^{n}\left(Y_i - \widehat{Y_i}\right)2$$

3 Results and Discussion

3.1 PSNR Comparisons

The images are denoised using DWT and OWT with SURE. It is important to know the performance level of each denoising algorithm, by calculating PSNR and elapsed time for processing. The PSNR comparison between the denoised image by discrete wavelet transform and unbiased risk estimator is given in Figs. 2 and 3.

The denoising process has been performed for about three dental panoramic images using DWT and unbiased risk estimator method. Of these two methods, unbiased risk estimator is found to be the best technique for denoising medical images by inferring the graph and PSNR values of both the techniques. The computation time is also found to be effective in unbiased risk estimator method (Figs. 4, 5; Table 1).

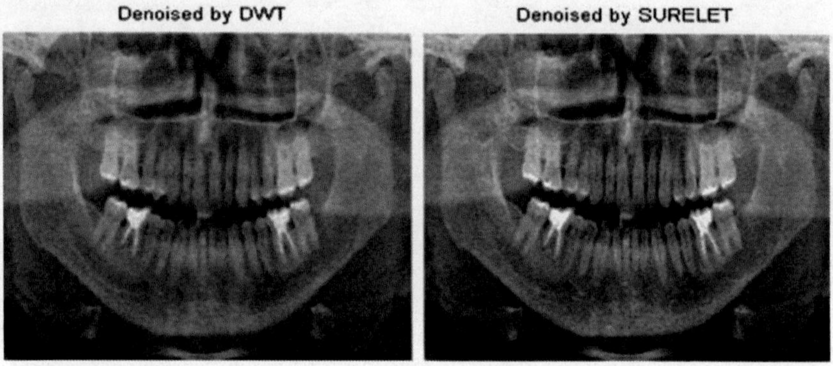

Fig. 2 Output 1

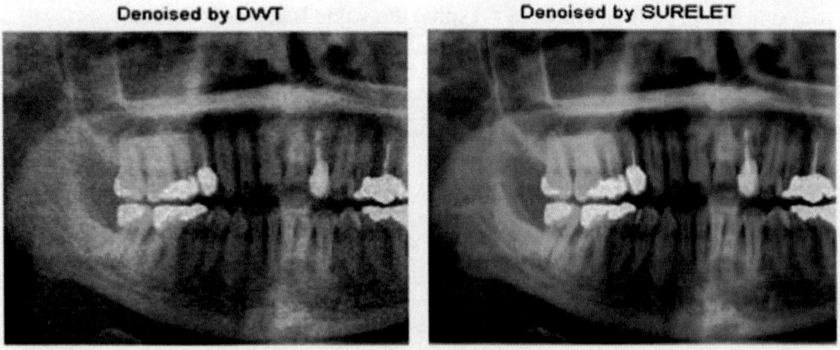

Fig. 3 Output 2

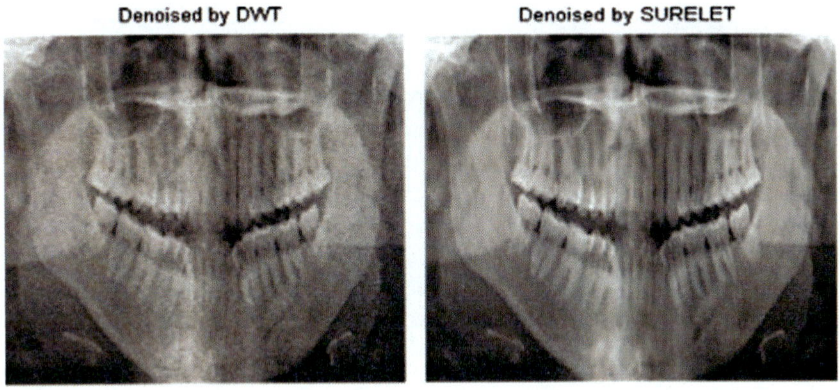

Fig. 4 Output 3

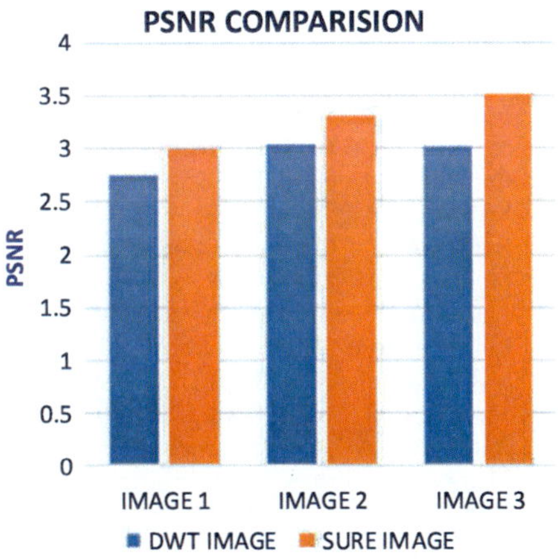

Fig. 5 Bar chart comparison of PSNR values

Table 1 Tabulation of PSNR values obtained from DWT and OWT method

Images	DWT_PSNR	OWT_PSNR	Elapsed time (s)
Image 1	2.747	2.995	0.0313
Image 2	3.0208	3.2999	0.0340
Image 3	3.01330	3.5134	0.028781

As per the evaluation protocol of denoising performance, the signal-to-noise ratio of the denoised image by orthonormal wavelet transform unbiased risk estimator is higher than that of the denoised image by discrete wavelet transform, and the elapsed time is lower. The data show that the denoised image by orthonormal wavelet transform unbiased risk estimator has better similarity with the original image, and the denoised signal by discrete wavelet transform retains more energy of the original image.

4 Conclusion

Performance analysis of panoramic dental images is done by computing the peak signal-to-noise ratio (PSNR). Many of the current methods such as DWT for denoising the medical image is not as better as unbiased risk estimator. We first preprocessed the dental panoramic image and then applied DWT and unbiased risk estimator. We then computed the PSNR values of both. We found that unbiased risk estimator provides better quality images. The unbiased risk estimator algorithm can be applied for denoising all type of medical images.

References

1. Yadav M et al (2014) Image denoising using orthonormal wavelet transform with stein unbiased risk estimator. In: 2014 IEEE students' conference on electrical, electronics and computer science
2. Donoho DL (1995) De-noising by soft thresholding. IEEE Trans Inf Theory 41(3):613–627
3. Vetterli M, Chang SG, Yu B (2000) Adaptive wavelet thresholding for image denoising and compression. IEEE Trans Image Process 9(9):1532–1546
4. Coifman R, Donoho D (1995) Translation-invariant de-noising. In: Wavelets and statistics. Springer Verlag, pp 125–150
5. Buades A et al (2005) A review of image denoising algorithms, with a new one. Multiscale Model Simul 4(2):490–530
6. Jansen M (2000) Wavelet thresholding and noise reduction. Ph.D. thesis
7. Luisier F et al (2007) A new SURE approach to image denoising: interscale orthonormal wavelet thresholding. IEEE Trans Image Process 16(3)
8. Muramatsu S, Han D, Kikuchi H (2011) SURE-LET image denoising with directional LOTs
9. Raja MA et al (2016) Performance comparison of adaptive algorithms with improved adaptive filter based algorithm for speech signals. Asian J Inf Technol 15(11):1706–1712
10. Wang FG, Zhang TG (2018) Real-time data processing of automatic settlement sensor based on wavelet method. Geomat Spat Inf Technol 41(11):168–170
11. Chen B-Q, Cui J-G, Xu Q, Shu T, Liu H-L (2019) Coupling denoising algorithm based on discrete wavelet transform and modified median filter for medical image. J Cent South Univ 26(1):124–135
12. Crouse MS, Nowak RD, Baraniuki RG (1998) Wavelet-based signal processing using hidden Markov models. IEEE Trans Signal Process 46(4):886–902
13. Xu LM (2018) Application of full frequency wavelet denoising method in quadrupole logging while drilling. Prog Geophys 33(5):274–280
14. Bruckstein A, Lindenbaum M, Fischer M (1994) On Gabor contribution to image enhancement. Comput Methods Programs Biomed 27:1–8
15. Malik J, Perona P (1990) Scale space and edge detection using anisotropic diffusion. IEEE Trans Pattern Anal 12:629–639
16. Catté F, Lions PL, Morel JM, Coll T (1992) Image selective smoothing and edge detection by nonlinear diffusion. J Numer Anal 29:845–866
17. Hosotani F, Inuzuka Y, Hasegawa M, Hirobayashi S, Misawa T (2015) Image denoising with edge-preserving and segmentation based on mask NHA. IEEE Trans Image Process 24:6025–6033
18. Yaroslavsky L (1985) Digital picture processing—an introduction. Springer, Berlin, Heidelberg

19. Manduchi R, Tomasi C (1998) Bilateral filtering for gray and color images. In: Proceedings of the sixth international conference on computer vision, Bombay, 7 Jan 1998, pp 839–846
20. Chatterjee P, Milanfar P (2012) Patch-based near optimal image denoising. IEEE Trans Image Process 21:1635–1649
21. Jansen M (2001) Noise reduction by wavelet thresholding, vol 161, 1st edn. Springer Verlag, United States of America
22. Sagheer SVM, George SN (2020) A review on medical image denoising algorithms. Biomed Signal Process Control 61:102036. https://doi.org/10.1016/j.bspc.2020.102036
23. Pesquet J-C, Leporini D (1997) A new wavelet estimator for image denoising. In: Proceedings of 6th international conference on image processing and its applications, 14–17 July 1997, vol 1, pp 249–253
24. Benhassine NE, Boukaache A, Boudjehem D (2021) Medical image denoising using optimal thresholding of wavelet coefficients with selection of the best decomposition level and mother wavelet. Int J Imaging Syst Technol
25. Bnou K, Raghay S, Hakim A (2020) A wavelet denoising approach based on unsupervised learning model. EURASIP J Adv Signal Process 2020
26. Jaiswal A et al (2014) Image denoising and quality measurement by using filtering and wavelet based techniques. Int J Electron Commun (AEU) 68(8):699–705

Efficient Image Retrieval Technique with Local Edge Binary Pattern Using Combined Color and Texture Features

G. Sucharitha, B. J. D. Kalyani, G. Chandra Sekhar, and Ch. Srividya

Abstract Image proceeding is foundation for pattern recognition and machine learning. The recent research focuses on image retrieval techniques with machine and deep learning methodologies. This paper recommends an other approach for extracting images based on combined characteristics of color and texture with edge binary pattern technique. The suggested descriptor first translates an RGB image into HSV color space. In this approach, the HSV color model makes use of the fundamental properties of color image as color, intensity, and brightness of an image. The hue (H) and saturation (S) components support to extract the color features, and the value component is more feasible to obtain the texture features; upon the value constituent of each image, local maximum edge binary patterns (LMEBPs) are used to find the relations among the pixels for every 3×3 matrix in order to extract texture features. Finally, all three histograms were used to extract the feature vector. The presented algorithm is experimented on two well-known color databases, Corel-10k and MIT-Visex. The comparative analysis of the proposed approach in terms of retrieval performance, with existing methods like CS-LBP, LEPSEG, LEPINV, has shown a substantial perfection in terms of precision and recall.

Keywords Histogram · Image retrieval · Local edge binary pattern · HSV color space · Color database

1 Introduction

The image retrieval has become active research area since 90's. Generally, image retrieval system depend two techniques. The first method uses the annotation technique. Because manually keywording photos is a tedious operation, annotating all images for huge databases is extremely challenging. In this, the great challenging is to provide the same keyword to the similar images for unlike users based on

G. Sucharitha (✉) · B. J. D. Kalyani · G. Chandra Sekhar · Ch. Srividya
Computer Science Engineering, Institute of Aeronautical Engineering, Hyderabad, Telangana, India
e-mail: sucharithasu@gmail.com

P. Chatterjee et al. (eds.), *Computational Intelligence for Engineering and Management Applications*, Lecture Notes in Electrical Engineering 984,
https://doi.org/10.1007/978-981-19-8493-8_21

their perception with respect to the objects involved. Content-based image retrieval (CBIR) may produce clarifications to these challenges by the multimedia. The CBIR represents and indexes an image in the database upon the consideration and evaluation of optical contents, such as color, texture, shape, and spatial layout. Color, texture, and shape are examples of general features that can be found in images. In CBIR, feature extraction is critical since the efficiency of the system is determined by the methods used to extract the features. CBIR [1–4] presents a comprehensive and broad literature review.

Color and texture examination has invited a challenge of attention because of its conceivable value in computer visualization, virtual reality, pattern detection, and pattern identification. In color texture analysis, the application of mixed color texture in texture feature extraction has yielded positive results. The texture feature is dependent on an image's local intensity. As a result, texture pattern is exposed for neighborhood and statistical aspects. Color histogram, color correlogram, color coherence vector, and other color feature descriptors are used to represent the distribution of intensity in different color channels. Palm [5] developed and implemented the correlation between textures of different color channels for a content-based picture retrieval system. For texture categorization, Ahmadian and Mostafa [6] employed the wavelet transform using extended Gaussian density and Kullback–Leibler (K–L) distance. Do and Vetterli [7] discussed an efficient utilization of discrete wavelets for texture features' observations of an image, but DWT can only extract features in three directions. Rotated wavelet filters [8] and Gabor transform [9] are also used to extract the texture features. Some of the significant transforms like Gabor filters [10] and wavelet packets [11] are used for image retrieval.

The color features of an object play a vital role in CBIR; if it is conserved semantic influence, the feature will decide the perpetual performance of the system. In addition, the size, position, and resolution of the visual scenery change by this concept. The color histogram was proposed by Swain and Ballard [12] for picture retrieval, and it is fairly simple to implement. To extract the color feature for picture retrieval, color coherence vector, color distribution feature, and cumulative histograms were studied in [13, 14].

Since a decade local patterns are observed as one of the dominant texture feature descriptors, the local patterns are designed to deal with pixel intensities directly in the regular fashion. The primary local pattern named local binary pattern (LBP) was projected for rotational and scale invariant texture characteristics of an image by Ojala et al. [15]. This local pattern got famous with its properties and forced to utilize for facial feature analysis and recognition [1, 16]. For shape localization, Huang et al. [17] developed a texture properties based on segmentation using extended LBP, and Li and Staunton [18] proposed a combination of LBP and Gabor filter. Aimed at face identification, Zhang et al. [19] introduced a local derivative patterns (LDPs). Takala et al. [20] debuted the block-based texture property analyzer, which makes usage of LBP descriptor to describe texture patterns in the images. For texture feature extraction, the authors used a mix of center symmetric local binary pattern (CS-LBP) and scale invariant feature transform (SIFT) [21, 22]. LEPSEG for image segmentation and LEPINV for image retrieval are two forms of local edge pattern histograms introduced by Yao and Chen [23], and in [24–26], the authors provide few more feature extraction methods and [27] proposes a multi-joint histogram for texture feature extraction in image retrieval. The combination of local edge patterns and lower-order Zernike moments has been proposed by Sucharitha and Senapati [28] for retrieval of biomedical images using texture and shape attributes. She also proposed a combination of local quantized edge binary patterns and color features for efficient image retrieval [29].

The proposed methodology introduces an efficient feature description for retrieval of similar images and image indexing based on recent methodologies on spatial patterns LMEBP and the significance of color feature. The following is a list of the proposed methodology main contributions:

(1) To improve picture retrieval, the proposed technique separates the image into R, G, and B color channels.
(2) Extracts the LMEBP and RGB features.
(3) Constructs the feature vector that comprises the color and texture features.

The remaining sections of the paper is as follows, Sect. 2 involves the introduction of color space and various texture descriptors LBP, CS-LBP, and LMEBP. Section 3 proposed method, similarity measurements, and structure of proposed method. Section 4 gives experimental results and discussions on two different databases. As a final point, the conclusion is achieved in Sect. 5.

2 Low Level Feature Descriptors

2.1 Color Model

Generally, images are accessible in three different categories: binary images, grayscale images, and color images. For black and white pixels, binary images have only two intensity levels. In a single band, grayscale images have a range of intensities [30]. The last color images include many bands, each with a different intensity range. The RGB images, which have three color bands named red, green, and blue, were commonly utilized in color images. As a result, it is known as RGB color space. These three bands hold information about an image's red, green, and blue colors. HSV stands for hue, saturation, and value in the other color space.

Color and hue are inextricably linked, and hue is expressed as an angle. The lightness and brightness of a color segment are represented by saturation, whereas the intensity of a color component is represented by value. Hue provides angle information ranging from 0 to 360°, with each degree occupying a different color. The intensity of an image is characterized by saturation, which extends from 0 to 1, with the intensity of color increasing as of low to high as the concentration of color enhances. In addition, the value scales from 0 to 1. Numerous studies have shown that individual RGB components are rarely recommended and that the HSV color model is more suited than the RGB model. The RGB image is transformed to HSV color space in the suggested method.

2.2 Local Binary Patterns

One of the dominant texture descriptor-known LBPs proposed by Ojala et al. [15] has rotation and scale-invariant characteristics. LBP is well known in many study fields due its specific qualities like discriminative power and simplicity. Face recognition and analysis, object tracking, texture classification, fingerprint identification, and picture retrieval have all been seen as a result of its performance.

Assume a grayscale image I with the dimensions $m \times n$ pixels, where $I(g)$ denotes the gray level of the image's gth pixel. For every 3×3 array with spatial organization, a pixel in the center turns out to be the threshold for determining the local binary pattern. The following mathematical Expressions (1) and (2) give the details about the implementation of LBP.

$$\mathrm{LBP}_{P,R} = \sum_{i=1}^{P} 2^{(i-1)} f\left(g^{p} - g^{c}\right) \tag{1}$$

$$f(x) = \begin{cases} 1, & x \geq 0 \\ 0, & x < 0 \end{cases} \tag{2}$$

where g^c and g^p are the intensities of center pixel and neighborhood pixel intensities, respectively, and P indicates the number of neighbors, and R represents the neighborhood length.

After the derivation of the LBP structure for the entire image, a histogram is constructed to describe the image according to Eqs. (3) and (4).

$$\mathrm{Hi}_{\mathrm{LBP}} = \sum_{i=1}^{m}\sum_{j=1}^{n} f_1(\mathrm{LBP}(i, j), \mathrm{bins}); \quad n \in \left[0, 2^{P} - 1\right] \tag{3}$$

$$f_1(u, v) = \begin{cases} 1, & u = v \\ 0, & \text{else} \end{cases} \tag{4}$$

where m and n are the dimensions of an image.

Figure 1 demonstrates calculate the LBP pattern for a 3×3 matrix. The circulation of edges in an image is represented by the histogram of these patterns.

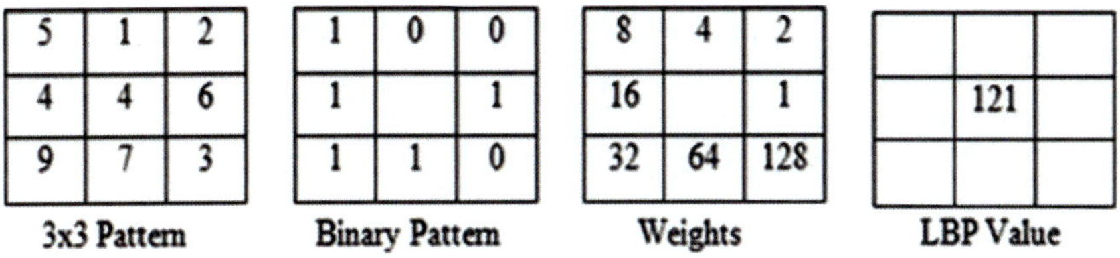

Fig. 1 LBP design and pattern construction procedure for a 3 × 3 pattern

2.3 *Center Symmetric Local Binary Pattern (CS-LBP)*

Instead of comparing each pixel to the center pixel, Heikkilä et al. [21] suggested the CS-LBP method, which compares the center symmetric pair of pixels as given in Eq. (5).

$$\text{CS\_LBP}_{P,R} = \sum_{p=1}^{P} 2^{(p-1)} X f\left(I\left(g_p - g_{p+\left(\frac{P}{2}\right)}\right)\right) \tag{5}$$

A histogram will be constructed to the whole image, after calculating each pixel's (x, y) CS-LBP pattern.

2.4 *Local Edge Patterns*

By comparing intensity values, local binary patterns link the center pixel with reference nearby pixels [15]. Subrahmanyam et al. [26] presented local max edge binary patterns, which is the continuation of LBP in that they extract information based on the distribution of edges in the image. The LMEBP can easily capture the edges' information of the objects involved in the image with the differences between the center pixel and its eight neighbors. It does not consider any magnitude of edges. For every center pixel I_c in every 3 × 3 pattern and it's consistent eight neighbor pixels I_i, the LMEBP is calculated as follows:

$$I_m(d_i) = I_m(I_i) - I_m(I_c) \quad i = 1, 2, 3 \ldots 8 \tag{6}$$

$$i_s = \text{sort}(\max(|I_m(d_1)|, |I_m(d_2)| \ldots |I_m(d_8)|)) \tag{7}$$

In the above expression, max value in the array I can be identified by max(I) function and Sort is a function that sorts an array in descending order, regardless of I's magnitude.

$$I^n(d_c) = f(I_m(d_c)) \tag{8}$$

where $f(x)$ is a function, and it is defined to produce the edge value as 1 or 0 for positive and negative values, respectively.

$$f(x) = \begin{cases} 1, & x \geq 0 \\ 0, & \text{else} \end{cases} \tag{9}$$

LMEBP is defined as:

$$\text{LMEBP}(I(d_c)) = \left\{ I^n(d_c), I^n(d_1), I^n(d_2), \ldots I^n(d_8) \right\} \tag{10}$$

The whole image is characterized by generating a histogram based on LMEBP after it has been calculated.

$$H_{\text{LMEBP}}(j) = \sum_{k=1}^{m} \sum_{l=1}^{n} f_2(\text{LMEBP}(k, l), j); \quad j \in [0\ 255] \tag{11}$$

where image size is $m \times n$.

Figure 2 demonstrates an sample of LMEBP calculation in a 5×5 matrix for a 3×3 window. Figure 2 shows a 3×3 matrix with the 3×3 pattern rotated around the center pixel as an example. The eight neighbors of the center pixel of Fig. 2a have become the center of their respective 3×3 pattern. For each 3×3 pattern, differences have been calculated using Eq. (6). After determining the variances, all of the values are sorted in descending order, regardless of their signs using, and assign the '0' and '1' to the patterns according to their signs using Eqs. (7)–(9). Finally, eight maximum edge binary patterns and their equivalent values have been calculated as shown in Fig. 2d.

3 Proposed Method

With the help of the previously described methods, more information from the image was attempted to be extracted in this present method. A new picture retrieval approach based on image color and texture information is proposed here. Color and texture are both prominent components of an image, as previously stated. The color image is initially transformed into HSV color space in the proposed method [31]. Color information is represented by hue, with each angle corresponding to a different color. Two distinct quantizations of the hue component, namely 18 and 36 bins, were used in this technique, and the accomplishment of the proposed work was assessed. The greater the quantization, the larger the feature vector, and the image sizes in the Corel-10k database are modest in comparison to the Corel-1k database. As a result, with this method, the lower quantization is applied. The two quantization approaches separate all colors into various portions in order to obtain the most useful color information. For reasonable information extraction, saturation is quantized into 20 bins. For the

3	1	4	6	7
4	$16_{(4)}$	$8_{(3)}$	$2_{(2)}$	5
5	$8_{(5)}$	$9_{(0)}$	$14_{(1)}$	2
6	$1_{(6)}$	$10_{(7)}$	$7_{(8)}$	3
10	5	9	2	4

(a)

$7_{(i3)}$	$-1_{(i6)}$	$-7_{(i2)}$
$-1_{(i7)}$		$5_{(i4)}$
$-8_{(i1)}$	$1_{(i8)}$	$-2_{(i5)}$

(b)

i1	*i2*	*i3*	*i4*	*i5*	*i6*	*i7*	*i8*
-8	-7	7	5	-2	-1	-1	1
0	0	1	1	0	0	0	1

(c)

	0	*1*	*2*	*3*	*4*	*5*	*6*	*7*	*8*	*LMEBP*	
i1	0	0	1	1	0	1	0	0	1	105	1st Edge
i2	0	0	1	0	0	0	0	0	0	64	2nd Edge
i3	1	0	1	0	0	0	0	0	0	320	3rd Edge
i4	1	0	1	1	0	0	0	1	0	354	4th Edge
i5	0	0	1	0	0	0	0	0	0	64	5th Edge
i6	0	0	1	0	0	1	0	0	1	73	6th Edge
i7	0	0	1	1	0	1	0	0	1	105	7th Edge
i8	1	0	1	1	0	1	0	0	1	361	8th Edge

(d)

Fig. 2 LMEBP calculation example

optimal combination, the histogram is built for both hue and saturation. Because the value component is closely identical to the grayscale component of a color (RGB) image, it is utilized to take out texture information. Edge binary patterns can be used to extract the local information of each pixel. As illustrated in the LMEBP map in Fig. 1, it gives each pixel eight edge values (2). The next step is to create three histograms, as shown in Fig. 3. The feature vector is obtained by joining these three histograms together. The database image is interpolated to a size of 256×256 to minimize the size of the feature vector. As a result, the feature vector is $3 \times 8 \times 256$ in size.

3.1 Similarity Measure and Query Matching

All images in database including query have gone through the proposed algorithm to create the feature vector database. Following the successful construction of the database, the following similarity metrics are applied to find the similar images for each query. Here, three types of similarity metrics are defined, and among those, $d1$

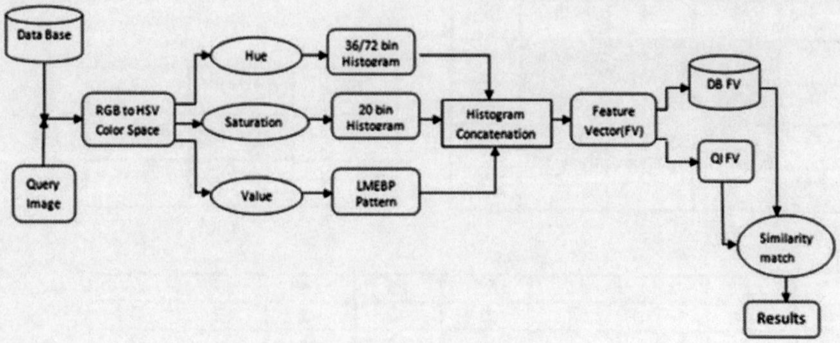

Fig. 3 Block diagram representation for the proposed algorithm

distance has been identified with satisfied results.

$$d1 \text{ distance: } d(q, b) = \sum_{i=1}^{\text{flen}} \left| \frac{f_b(i) - f_{q(i)}}{1 + f_b(i) + f_q(i)} \right| \quad (12)$$

$$\text{Canberra distance: } d(q, b) = \sum_{i=1}^{\text{flen}} \left| \frac{f_b(i) - f_{q(i)}}{f_b(i) + f_q(i)} \right| \quad (13)$$

$$\text{Manhattan distance: } d(q, b) = \sum_{i=1}^{\text{flen}} \left| f_b(i) - f_{q(i)} \right| \quad (14)$$

q and b are the query and database image, respectively.

3.2 *Working Procedure of the Proposed Algorithm*

Figure 3 is a block diagram representation of proposed algorithm, and the frame work is explained as steps from 1 to 6.

1. Change the color space of the RGB image to HSV.
2. Construct the histograms for hue and saturation in the quantized bins.
3. Calculate the edge binary patterns for value.
4. Construct feature vector.
5. Using Eq. (12) find the like images for query image from the database.
6. Choose the best matches to retrieve the images.

3.3 Benefits of Proposed Algorithm

1. The RGB color space contains only gray level information, whereas HSV consists of three different information. As a result, it can be argued that HSV outperforms RGB in terms of feature extraction. Color features were extracted using the H and S color spaces, while texture features were extracted using the LMEBP method on the V color space.
2. The image size is lowered to 256 × 256 to reduce the proposed descriptor's feature vector length.
3. Two large experiments were conducted on the Corel-10k and MIT-Vistex datasets, to validate the retrieval results of the designed algorithm.

4 Analysis of Results on Different Databases

Diverse databases are used in image retrieval for various purposes, including Corel database, MIT color database, Brodtz texture database, and so on. The Corel dataset (1k, 5k, and 10k) is a well-known and widely used database for verifying retrieval results. Corel is accessible in three volumes: 1k, 5k, and 10k. For color and texture feature analysis, the MIT-Color dataset was used, and the Brodtz dataset was used for texture analysis.

Existing color and texture algorithms are compared to the proposed technique and renowned metrics precision and recall for all database images, and many approaches are determined. To validate the results, largest number of images from the database has grabbed as query image to analyze the results. Based on these results, the all retrieval analysis parameters are calculated. The retrieval results are quantified in terms of precision, recall, average retrieval (ARR), and average precision (ARP) using Eqs. (15)–(19) [25].

The mathematical expression for precision for a query image I_q is:

$$P(Q_i, n) = \frac{1}{n}\sum_{k=1}^{|DB|}\delta\left(f(I_k), f\left(I_q\right)\right)\left|\text{Rank}\left(I_k, I_q\right) < n\right| \tag{15}$$

The rank for each retrieved image will be provided based on the similarity for top n matches from the database |DB|, and $f(x)$ is the category of 'x'.

$$\delta\left(f(I_k), f\left(I_q\right)\right) = \begin{cases} 1 & f(I_k) = f\left(I_q\right) \\ 0 & \text{Otherwise} \end{cases} \tag{16}$$

Similarly, recall is defined as

$$R\left(I_q, n\right) = \frac{1}{N_R}\sum_{k=1}^{|DB|}\delta\left(f(I_k), f\left(I_q\right)\right)\left|\text{Rank}\left(I_k, I_q\right) \le n\right| \tag{17}$$

The following Eqs. (18) and (19) are used to calculate the ARR and ARP, respectively, in which N_R represents relevant or similar images for a query from the database.

$$\text{ARR} = \frac{1}{|\text{DB}|}\sum_{k=1}^{|\text{DB}|} R(I_k, n)|_{n \le N_R} \tag{18}$$

and

$$\text{ARP} = \frac{1}{|\text{DB}|}\sum_{k=1}^{|\text{DB}|} P(I_k, n) \tag{19}$$

In which, |DB| is the count of images of the dataset. Finally, the entire precision and recall can be calculated for the entire database using Eqs. (20) and (21).

$$P_{\text{total}} = \frac{1}{N_{\text{Ca}}}\sum_{i=1}^{N_{\text{Ca}}} \text{ARP}(i) \tag{20}$$

$$R_{\text{total}} = \frac{1}{N_{\text{Ca}}}\sum_{i=1}^{N_{\text{Ca}}} \text{ARR}(i) \tag{21}$$

where i represents the number of categories and N_{Ca} denotes the complete number of types in the database. The outcomes with respect to precision and recall are observed as more significant when compared to the other existing methods. One further metric, the F-measure, has been added to all of them. A relationship between precision and recall is specified as an F-measure [32].

Equation (22) shows how it is defined:

$$F\text{-measure} = \frac{2 \times \text{precision} \times \text{recall}}{\text{precision} + \text{recall}} \tag{22}$$

Experiments were conducted on two benchmark databases, the main database is being Corel-10k and the second is being the MIT-color database, in order to assess the suggested method's capacity.

4.1 Corel-10k Database

This database [32] has ten thousand images which includes hundred categories, and each category has hundred and many similar images. This database is more adoptable for all image retrieval systems due its capacity and more similar images as compared with Corel-1k and 5k. Images of animals, such as foxes, tigers, and deer, as well as humans, natural settings, ships, food, buses, the army, ocean, cats, and airplanes, are included. The suggested work's database retrieval performance is measured in terms of precision, recall, average precision, average retrieval, and *F*-measure.

The significance of the proposed algorithm is declared by comparing the state-of-the-art techniques, which include colorwavelets (color histogram + Wavelets), colorCS-LBP (color histogram + CS-LBP), colorLEPINV (color histogram + LEPINV), colorLEPSEG (color histogram + LEPSEG), and colorLECoP (color histogram + LECoP) algorithms along with color histograms. Figure 4a, b are the graphical representations for precision and recall with respect to various methods. The precision and recall values for existing methods along with proposed method for each category have represented.

Average precision (ARP) and average retrieval (ARR) are depicted in Fig. 4c, d. The figures are evidently representing the substantial progress in the retrieval nearby 4.5% as compared to colorLEPSEG, 6.5% as compared to colorLEPINV, and 8.2% as compared to colorCS_LBP in terms of ARP. Figure 4e is presenting the relation between *F*-measure and top ranked images, and the metric *F*-measure is computed using Eq. (22).

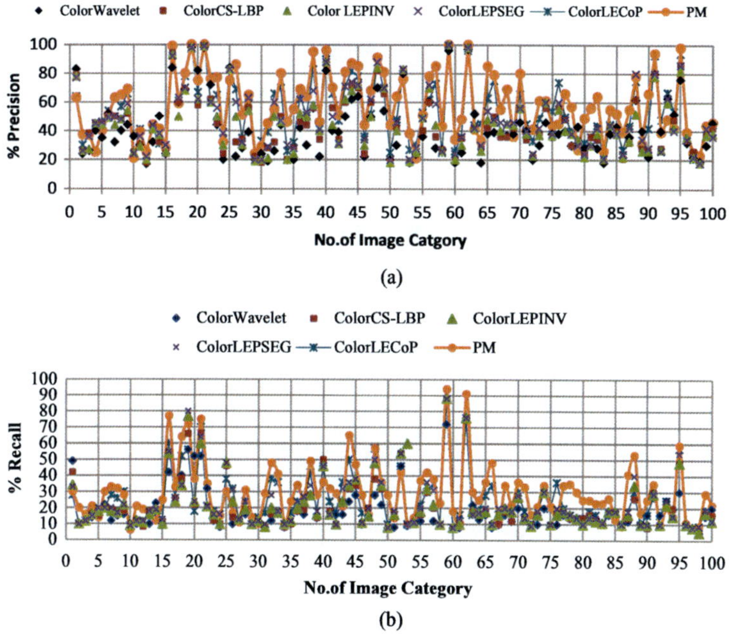

Fig. 4 Results on Corel-10k. **a** Category-wise precision, **b** category-wise recall, **c** average precision (ARP) and images retrieved, **d** average retrieval (ARR) and number of images retrieved, **e** *F*-measure versus top ranked images

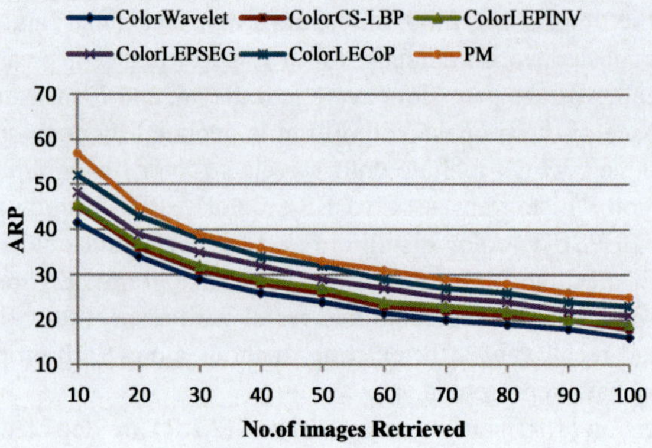

(c)

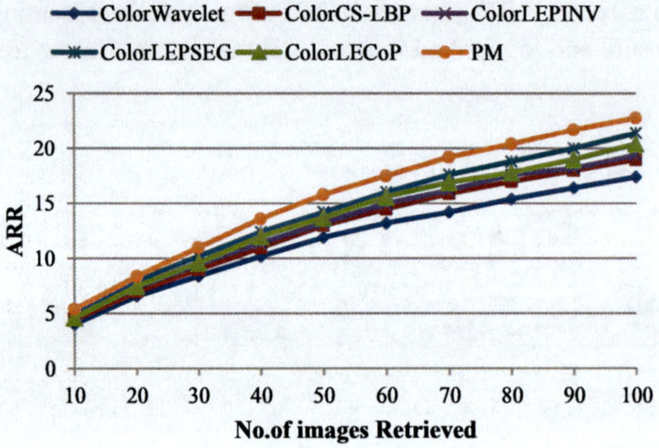

(d)

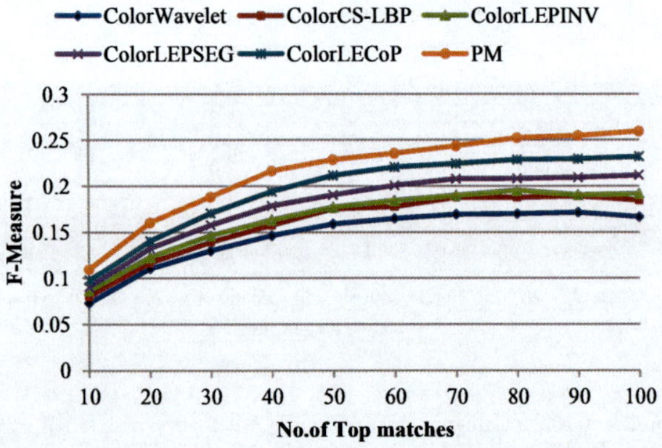

(e)

Fig. 4 (continued)

Fig. 5 Samples from MIT-Vistex database

4.2 MIT-Vistex Database

This dataset has an enormous number of colorful and regular texture-based images [33]. This collection has 40 different colored texture images, each of which is 512 × 512 pixels in size. These images were separated into 16 blocks for image retrieval, with each block having a size of 128 × 128, resulting in a 640 (40 × 16) image database. Figure 5 shows some of the sample images. The results of the presented algorithm are related to state-of-the-art techniques in Fig. 6a, b. In terms of average retrieval (ARR), it is obvious that the proposed descriptor has improved by 7.5% when compared to colorLEPSEG and by 9% when compared to colorLEPINV. The similarity measurement $d1$ distance is used for all the experiment calculations.

5 Conclusion

For picture retrieval, an integrated color texture-featured descriptor is proposed. Using the HSV color space, it extracts texture and color properties using local

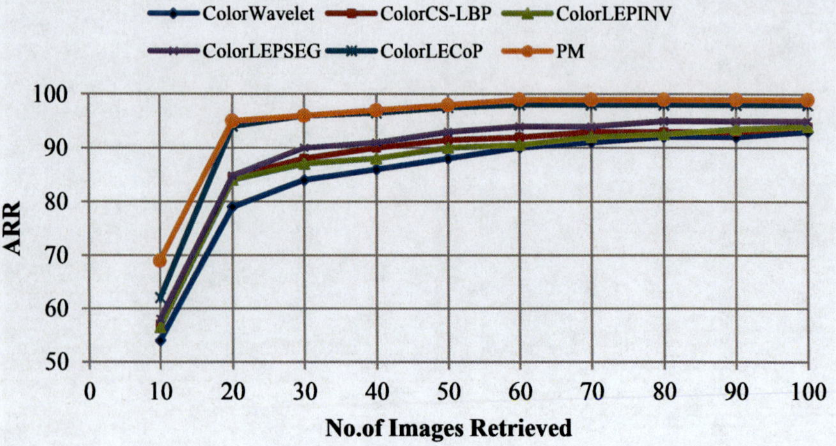

(a)

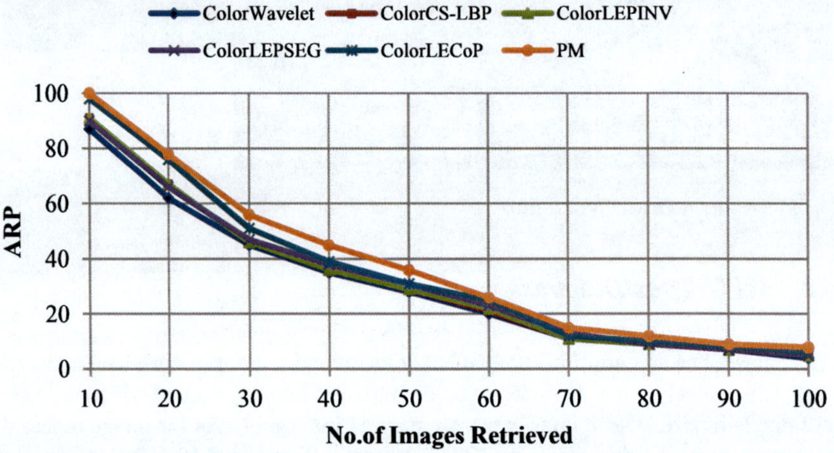

(b)

Fig. 6 MIT-Vistex: **a**, **b** retrieved images versus ARP and ARR

patterns. The LMEBP retrieves each pixel's edge information as well as color attributes from the HSV color space. Histograms were used to extract color attributes from hue and saturation components in the HSV color space. On two big databases, Corel-10k and MIT-Vistex, LEBP was used on a value component for texture features and a composite feature vector was applied. To justify the proposed work in resemblance to state-of-the-art color and texture characteristics approaches, the suggested descriptor and previous techniques' experimental results are discussed using graphs with suitable evolution metrics, which indicate a considerable improvement.

Acknowledgements Here I am declaring that, there is no conflict of interest in publishing this paper.

References

1. Ahonen T, Hadid A, Pietikainen M (2006) Face description with local binary patterns: application to face recognition. IEEE Trans Pattern Anal Mach Intell 28(12):2037–2041
2. Rui Y, Huang TS, Chang S-F (1999) Image retrieval: current techniques, promising directions, and open issues. J Vis Commun Image Represent 10(1):39–62
3. Smeulders AWM et al (2000) Content-based image retrieval at the end of the early years. IEEE Trans Pattern Anal Mach Intell 22(12):1349–1380
4. Liu Y et al (2007) A survey of content-based image retrieval with high-level semantics. Pattern Recogn 40(1):262–282
5. Palm C (2004) Color texture classification by integrative co-occurrence matrices. Pattern Recogn 37(5):965–976
6. Ahmadian A, Mostafa A (2003) An efficient texture classification algorithm using Gabor wavelet. In: Proceedings of the 25th annual international conference of the IEEE engineering in medicine and biology society, 2003, vol 1. IEEE
7. Do MN, Vetterli M (2002) Wavelet-based texture retrieval using generalized Gaussian density and Kullback–Leibler distance. IEEE Trans Image Process 11(2):146–158
8. Kokare M, Biswas PK, Chatterji BN (2007) Texture image retrieval using rotated wavelet filters. Pattern Recogn Lett 28(10):1240–1249
9. Manjunath BS, Ma W-Y (1996) Texture features for browsing and retrieval of image data. IEEE Trans Pattern Anal Mach Intell 18(8):837–842
10. Idrissa M, Acheroy M (2002) Texture classification using Gabor filters. Pattern Recogn Lett 23(9):1095–1102
11. Laine A, Fan J (1993) Texture classification by wavelet packet signatures. IEEE Trans Pattern Anal Mach Intell 15(11):1186–1191
12. Swain MJ, Ballard DH (1992) Indexing via color histograms. In: Active perception and robot vision. Springer, Berlin, Heidelberg, pp 261–273
13. Pass G, Zabih R, Miller J (1997) Comparing images using color coherence vectors. In: Proceedings of the fourth ACM international conference on multimedia. ACM
14. Stricker MA, Orengo M (1995) Similarity of color images. In: IS&T/SPIE's symposium on electronic imaging: science & technology. International Society for Optics and Photonics
15. Ojala T, Pietikäinen M, Harwood D (1996) A comparative study of texture measures with classification based on featured distributions. Pattern Recogn 29(1):51–59
16. Zhao G, Pietikainen M (2007) Dynamic texture recognition using local binary patterns with an application to facial expressions. IEEE Trans Pattern Anal Mach Intell 29(6)
17. Huang X, Li SZ, Wang Y (2004) Shape localization based on statistical method using extended local binary pattern. In: Third international conference on image and graphics (ICIG'04). IEEE
18. Li M, Staunton RC (2008) Optimum Gabor filter design and local binary patterns for texture segmentation. Pattern Recogn Lett 29(5):664–672
19. Zhang B et al (2010) Local derivative pattern versus local binary pattern: face recognition with high-order local pattern descriptor. IEEE Trans Image Process 19(2):533–544
20. Takala V, Ahonen T, Pietikäinen M (2005) Block-based methods for image retrieval using local binary patterns. In: Scandinavian conference on image analysis. Springer, Berlin, Heidelberg

21. Heikkilä M, Pietikäinen M, Schmid C (2009) Description of interest regions with local binary patterns. Pattern Recogn 42(3):425–436
22. Tan X, Triggs B (2010) Enhanced local texture feature sets for face recognition under difficult lighting conditions. IEEE Trans Image Process 19(6):1635–1650
23. Yao C-H, Chen S-Y (2003) Retrieval of translated, rotated and scaled color textures. Pattern Recogn 36(4):913–929
24. Murala S, Maheshwari RP, Balasubramanian R (2012) Local tetra patterns: a new feature descriptor for content-based image retrieval. IEEE Trans Image Process 21(5):2874–2886
25. Liao S, Law MWK, Chung ACS (2009) Dominant local binary patterns for texture classification. IEEE Trans Image Process 18(5):1107–1118
26. Subrahmanyam M, Maheshwari RP, Balasubramanian R (2012) Local maximum edge binary patterns: a new descriptor for image retrieval and object tracking. Signal Process 92(6):1467–1479
27. Vipparthi SK, Nagar SK (2014) Multi-joint histogram based modelling for image indexing and retrieval. Comput Electr Eng 40(8):163–173
28. Sucharitha G, Senapati RK (2020) Biomedical image retrieval by using local directional edge binary patterns and Zernike moments. Multimed Tools Appl 79(3):1847–1864
29. Sucharitha G, Senapati RK (2018) Local quantized edge binary patterns for colour texture image retrieval. J Theoret Appl Inf Technol 96(2)
30. Smith AR (1978) Color gamut transform pairs. ACM Siggraph Comput Graph 12(3):12–19
31. Verma M, Raman B, Murala S (2015) Local extrema co-occurrence pattern for color and texture image retrieval. Neurocomputing 165:255–269
32. Corel 10k database. Available online: http://www.ci.gxnu.edu.in/cbir/
33. MIT Vision and Modeling Group, Cambridge. Vision texture. Available online: http://vismod.media.mit.edu/pub/

Texture and Deep Feature Extraction in Brain Tumor Segmentation Using Hybrid Ensemble Classifier

Divya Mohan, V. Ulagamuthalvi, and Nisha Joseph

Abstract The abnormal cell development in the brain is referred to as a tumor. Brain tumors are treated by physicians using radiation and surgery. The brain tumor is categorized as benign or malignant. The benign tumor can be treated and cured using the appropriate medication suggested. A malignant tumor is an abnormal tissue that affects the nearby tissues, and it can be cured only through proper surgery by a physician. Manual identification process of malignant and benign tumors is very time taking and has a good chance of occurrence error in the process. To overcome this limitation, automatic brain tumor classification technique is proposed. An efficient methodology for the detection of brain tumors is done. Initially, the brain MRI image is smoothed and enhanced by a Gaussian filter. Then, deep and texture features are extracted. In the proposed work, an ensemble technique using three different classifiers based on majority voting method is used. The research work proposed in this paper is tested using BRATS 2017, 2018 datasets, and the obtained results are compared with recent methods and proved its efficacy.

Keywords Tumor · Texture · Deep features · Classifier

1 Introduction

Brain tumor must be detected earlier since it is dangerous. After detecting a brain tumor as benign or malignant, proper treatment can be given by the physician. The brain tumor is referred to as dangerous and serious based on the extending capacity of

D. Mohan (✉) · V. Ulagamuthalvi
Department of Computer Science and Engineering, Sathyabama University, Chennai, India
e-mail: divyamohanresearch@gmail.com

V. Ulagamuthalvi
e-mail: ulagamuthalvi.cse@sathyabama.ac.in

N. Joseph
Computer Science and Engineering, SAINTGITS College of Engineering, Kottayam, Kerala, India

P. Chatterjee et al. (eds.), *Computational Intelligence for Engineering and Management Applications*, Lecture Notes in Electrical Engineering 984,
https://doi.org/10.1007/978-981-19-8493-8_22

the brain. Benign tumors are not spread over other tissues but malignant tumors widen from their normal area. In the case of a benign tumor, medications are sufficient, and they can be cured easily. But surgery is needed for malignant tumors.

To identify the type of tumor in brain, image processing can be used effectively. MRI image is used to identify the inner structure details of the brain and the variations in the brain cell. There are various researches in the field of brain tumor segmentation that uses pre-trained network models for feature extraction and classification. Similarly, handcrafted features are also used for brain tumor segmentation. The objective of this paper is to identify brain tumor with the combination of handcrafted and deep features. This paper also uses hybrid classifier approach to further improve the accuracy.

The steps associated with the proposed work are removal of noise, image segmentation, extraction of characteristics, and using the features perform classification. Basic procedure in the first phase is to develop the brain MRI image quality using reducing the noise and by enhancing the brightness of the image. Then in feature extraction, residual network (ResNet) features, Local Derivative Pattern (LDP) features, and characteristics obtained through Gray Level Co-occurrence Matrix (GLCM) are used. Finally, an ensemble technique using three classifiers—Random Forest, Support Vector Machine (SVM), and Naïve Bayes (NB) classifier—is used to find the type of tumor.

The rest of the paper is organized in the following manner: Sect. 2 discusses some papers related to brain tumor segmentation. Section 3 elaborates the projected work with its experiments in Sect. 4. Section 5 briefs the work with its future improvements.

2 Related Work

A segmentation method is used for segmenting the tissues in magnetic resonance imaging (MRI) images as tumor and non-tumor [1]. The non-tumor region of the brain is executed along the tumor region. A modification was recorded as it occurs in the non-tumor region. This is mainly helpful for the physicians to identify the tumor.

A Computer-Aided Design (CAD)-based segmentation method [2] is proved as one of the efficient techniques for brain tumor segmentation. Local Independent Projection-Based Classification (LIPC) was used to separate the voxel into its respective labeled group.

A brain tumor segmentation technique has been implemented using SVM classifier [3] that uses GLCM for feature extraction. Viswa Priya introduced a clustering technique for identifying the tumor [4]. In the preprocessing stage, the input image is smoothened, and the noise is eliminated using the adaptive mean filter. Some morphological processing is also done to identify the tumor.

A genetic algorithm technique has been developed to identify the brain tumor [5]. This method has achieved precision of 94%. The thresholding technique is applied for segmenting the tumor regions [6]. Then, the basic preprocessing, morphological processing, and thresholding are done to identify the normal and abnormal tissues.

A novel algorithm has been developed for identifying brain metastases [7]. Shape features and energy contrast features are used in this technique for identifying brain metastases.

A Fully Convolutional Network (FCN) architecture using has been implemented to detect tumor [8] which performs two dimensional convolution operations on input. It was faster when compared to others because of less resource utilization. A deep learning model was introduced which isolates the tumor more accurately in BRATS 2013 dataset [9]. Patch-wise Convolutional Neural Network (CNN) [10] was used to improve the efficiency of the model [9]. But these models take more time because it consumed 3D image at the input stage.

Kansas has introduced the Ensembles of Multiple Models and Architectures (EMMA) method [11]. The advantage is that it did not depend on a particular database. Texture and abnormality features were considered for isolating the tumor region in the input which is followed by RF classifier to provide proper annotation at the output. These two methods suffer from less accuracy and efficiency.

In [12], for accurately segmenting the tumor growth in the input image, a combination of k-means and Fuzzy C-Means was implemented. Another method [13] used histograms to extract features from 2D slice of the 3D input [2] which is followed by thresholding combined with median filtering. Then, connectivity was identified on each 2D slice. The biggest cluster was taken as the region having tumor growth. The collection of all the 2D slices produced the required output.

Chen et al. have developed a method using super-pixel segmentation for tumor classification [14]. Features were taken out of the super pixels which was followed by SVM. Another technique has been employed in [15] cascading Random Decision Forest (RDF) classifier in multiple levels.

3 Methodology

The architecture of the proposed framework is shown in Fig. 1.

The proposed work constitutes two major processes: selection of discriminating features known as feature extraction and assigning labels to similar groups of such features called classification. In the first process, features such as ResNet, GLCM, and LDP are extracted. These features are concatenated to form a single array which is fed as input to the second phase. The second process uses an ensemble technique which combines RF, NB, and SVM to identify the region having tumor in the input.

3.1 Initial Processing

It improves the visualization of the image. Noises are present in the brain image input. To clean the input, the adaptive median filter is used which retains the fine details of the image. This will increase the accuracy of classification.

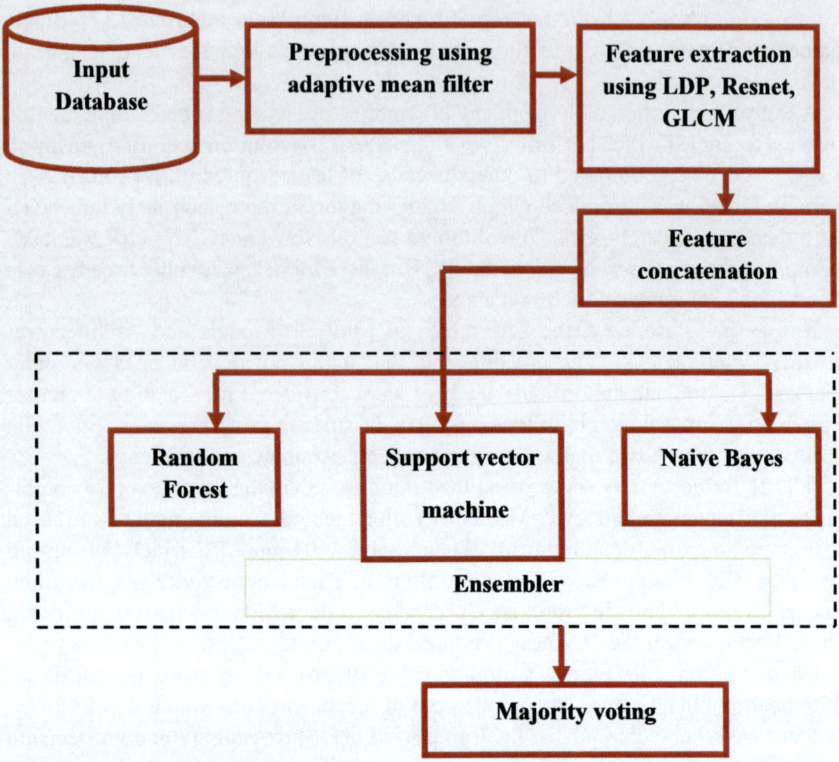

Fig. 1 Proposed system architecture

3.2 *Extraction of Features*

Here, features are extracted using different techniques: ResNet, GLCM, and LDP. All these features are explained in this section.

3.2.1 Local Derivative Pattern (LDP)

Local Derivative Pattern is used to extract the deep features using CNN. These features enhance the accuracy of classification efficiently. In the deep learning-based LDP, neighboring values are taken initially, and then, Eq. 1 is applied to find the difference.

$$\mathrm{DLDP} = \mathrm{Cp} - \mathrm{Np} \tag{1}$$

where Cp is the value of center pixel and Np is the value of neighboring pixel. Then, direction is estimated using the formulas mentioned below.

$$\text{if DLDP1} > 0 \text{ and DLDP2} > 0 \ \text{IV} = 1 \tag{2}$$

$$\text{DLDP1} < 0 \text{ and DLDP2} < 0 \ \text{IV} = 2 \tag{3}$$

$$\text{DLDP1} > 0 \text{ and DLDP2} < 0 \ \text{IV} = 3 \tag{4}$$

$$\text{DLDP1} < 0 \text{ and DLDP2} > 0 \ \text{IV} = 4 \tag{5}$$

$$\text{PD} = \sum_{i=1}^{8} \text{IV}_i \tag{6}$$

3.2.2 ResNet Feature Extraction

This system uses 194 layers of the residual network. It has three convolutional layers. ResFeatures are extracted from the residual units. The entire shape of the brain MRI contained within the ROI is extracted as the residual network features. The deep filter bank produces the residual network outcome. The outcome is of the form $w \times h \times c$ where w denotes width and h represents height of the resultant feature descriptor and c represents the number of convolutional layer channels. It is expressed as

$$\begin{aligned} Z_i &= g(y_i) + Fn(y_i, w_i) \\ y_{i+1} &= fn(Z_i) \end{aligned} \tag{7}$$

where the residual function is represented as *Fn*, ReLU function is represented as *fn*, the weight matrix is represented by w_i and ith layer input is represented as y_i, and ith layer output is represented as Z_i. The identity value of mapping h is represented as

$$h(y_i) = y_i \tag{8}$$

The residual function F is defined in as

$$F(y_i, w_i) = w_i \cdot \sigma\big(B\big(w_i'\big) \cdot \sigma(B(x_i))\big) \tag{9}$$

3.2.3 Gray Level Co-occurrence Matrix

The spatial relationship of a pixel is represented using the statistical measure GLCM. The texture of the Brain MRI is calculated using corresponding pixels frequency and spatial relationships among those pixels. Contrast, correlation, energy, homogeneity, kurtosis, and skewness measures are used as features. The detailed information on these features are given below.

$$\text{Contrast } (\mathbf{C}) = \sum_{t,r=1}^{T,R} |t - r|^2 \mathbf{q}(t, r) \tag{10}$$

where $q(t, r)$ is GLCM, t and r are row and column, T is total rows, and R is total columns.

$$\text{Correlation (Corr)} = \sum_{t,r=1}^{T,R} ((t - \mu)(r - \mu)q(t, r))/(\sigma(t) * \sigma(r)) \tag{11}$$

where mean is μ, and the standard deviation is σ

$$(\mathbf{E}) = \sum_{t,r=1}^{T,R} \mathbf{q}(t, r)^2 \tag{12}$$

$$\text{Homogeneity } (\mathbf{H}) = \frac{\sum_{t,r=1}^{T,R} \mathbf{q}(t, r)}{1 + |t - r|} \tag{13}$$

$$\text{Kurtosis } (\mathbf{K}) = \left\{ \frac{1}{T * R} * \sum_{t=1}^{T} \sum_{r=1}^{R} \left(\frac{\mathbf{q}(t, r) - \mu}{\sigma} \right)^4 \right\} - 3 \tag{14}$$

$$\text{Skewness } (\sigma) = \sqrt[2]{\frac{1}{T * R} * \sum_{t=1}^{T} \sum_{r=1}^{R} (\mathbf{q}(t, r) - \mu)} \tag{15}$$

The proposed feature selection algorithm is given in Algorithm 1.

Table 1 Details of matrices used in the feature mining algorithm

Matrix	Size
Input MRI, A	(240, 240, 155)
Intermediate feature vectors (F_1, F_2, and F_3)	(240, 240) each
Final feature descriptor F	(3, 240, 240)

Algorithm 1 For mining features from input

Input: MRI image sequence
Output: Feature descriptor
MRI image with size (r, c, l)
Output: Feature vector size $(3, m, n)$
Steps:
1. For every input MRI image
   1.1 For each input Brain MRI
      1.1.1 Assume $A = I(x, y, z)$
      1.1.2 Compute $F_1 = \text{Resnet}(A_i)$
      1.1.3 $F_2 = \text{DLLDP}(A_i)$
      1.1.4 $F_3 = \text{GLCM}(A_i)$
   1.2 End
2. End
3. Compute F as a 3D array of 3 subbands

The characteristics are accepted based on their performance. The size of features matrices used in the property selection algorithm is represented in Table 1. The size of matrix A is the size of input image. In BRATS dataset, the size of the image is 240 × 240 × 155. The accepted properties for each feature descriptor are appended one after other for all subbands, i.e., the 3D image is converted to 2D features. Hence, the size of each feature matrix (F_1, F_2, and F_3) is 240 × 240. The concatenation of matrices F_1, F_2, and F_3 produces the final feature descriptor for classification operation which is named as F. Thus, the size of the final feature matrix F is 3 × 240 × 240.

The concatenated features are classified using the ensemble classifier explained in the next section.

3.3 *Classification*

The concatenated feature vector of the MRI is fed as input in the classification. For the classification process, RF, SVM, and NB classifiers are used. These classifiers' results are combined to form a hybrid ensemble classifier. The advantages of using these three classifiers are as follows:

- RF algorithm is also used to measure the importance of each feature on the prediction. It computes a score for the given features automatically after training and then scales the results. By feature score, it can be decided to possibly drop the less scored features. This is done because the features having low scores don't contribute to classifying the results.

- SVM is the most widely used classifier in many applications.
- NB classifier works on the principle of the maximum likelihood called the Bayes Theorem. To minimize computational cost, class conditional independence was assigned by Naive. The attributes within the class are independent. The execution, classification, estimation, and prediction steps are performed sequentially. NB overcomes various limitations including iteration, computational time, and computational cost.

3.3.1 Hybrid Ensemble Classifier

The proposed hybrid ensemble classifier technique was RF–SVM–NB. The identification of brain tumors is based on voting. Among three classifiers at least two ratios one voting is identified as the corresponding tumor type as benign or malignant.

Algorithm 2 Hybrid ensemble classifier technique

Input: Id from three classifiers Id1, Id2, Id3
Output: Brain tumor type
Steps:

1. right = 0
2. left = 0
3. If Id1 = malignant then
   right = right + 1
4. Else
   left = left + 1
5. End
6. If Id2 = malignant then
   right = right + 1
7. Else
   left = left + 1
8. End
9. If Id3 = malignant then
   right = right + 1
10. Else
    left = left + 1
11. End
12. If right > left then
    Type = malignant
13. Else
    Type = Benign
14. End

4 Results and Discussion

The projected work is executed using two challenging BRATS datasets such as 2017 and 2018. The BRATS dataset consists of two gliomas Low (LGG) and High Grade (HGG). They also divided the dataset into training, testing, and leaderboard datasets. In BRATS 2017, there are 431 cases (both HGG and LGG), among which 285 are used for training and 146 cases in testing [16]. BRATS 2018 dataset contains 285 training and 191 testing cases. Some examples of BRATS 2017 and 2018 are shown in Fig. 2.

4.1 Performance Measures

For analyzing the performance of the projected work, accuracy, sensitivity, and specificity are used. If T_p represents the True Positive, F_p corresponds to False Positive, T_n is the True Negative, and F_n delegates False Negative, then the above-mentioned performance parameters can be expressed as

$$\mathrm{Ac}_r = \frac{T_p + T_n}{T_p + T_n + F_p + F_n} \times 100 \tag{16}$$

$$\text{Sensitivity} = \frac{\mathrm{TP}}{\mathrm{TP} + \mathrm{FN}} \tag{17}$$

$$\text{Dice Score} = \frac{2 \times \mathrm{TP}}{\mathrm{FP} + (2 \times \mathrm{TP}) + \mathrm{FN}} \tag{18}$$

4.2 Results

The performance evaluation of the projected work tested using datasets BRATS 2017 and 2018 are listed in Table 2. The projected method is also compared with

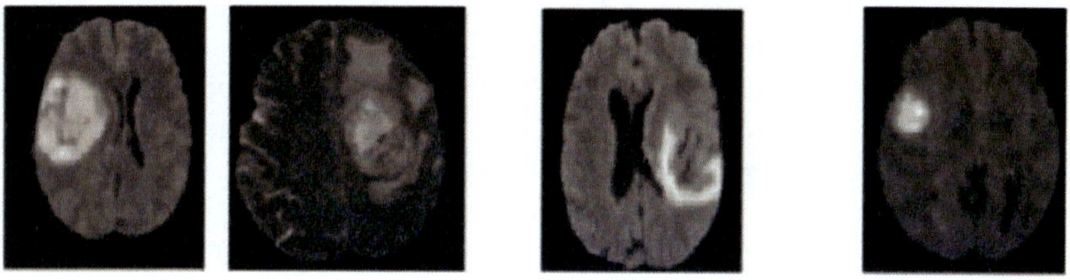

Fig. 2 Examples of BRATS dataset

Table 2 Results provided by the projected work using datasets BRATS 2017 and 2018

Measure	Classifier	BRATS 2017	BRATS 2018
Accuracy	RF	96	97
	NB	94.8	95
	SVM	96	97
	Hybrid ensemble	98.3	99
Sensitivity	RF	97	98
	NB	98	98.5
	SVM	98	99
	Hybrid ensemble	99	99
Dice score	RF	98	98
	NB	98	98.6
	SVM	98	99
	Hybrid ensemble	99	99

the performance of individual classifiers such as RF, NB, and SVM without using majority voting method.

From Table 2, it is observed that the accuracy, sensitivity, and dice score obtained by the proposed method using SVM classifier is better than other individual classifiers. The accuracy obtained by individual classifiers ranges 94–97%. But the accuracy of hybrid ensemble classifier ranges 98–99%. The sensitivity and dice score obtained by hybrid ensemble classifier are 99% for both the datasets which are also greater than the sensitivity and dice score obtained by individual classifiers. It is evident that the hybrid ensemble classifier outperforms all the individual classifiers in all the measured metrics.

4.3 Comparison of Proposed Method with Recent Methods

From Sect. 4.2, it is analyzed that hybrid ensemble classifier obtained satisfied results when compared to all other classifiers. Hence, the results obtained by hybrid ensemble classifier are compared with some recent methods [17–22]. Tables 3, 4, and 5 give the performance comparison of the projected work on datasets BRATS 2017 and 2018 using parameters accuracy, dice score, and sensitivity, respectively.

From Table 3, it is observed that the accuracy obtained by all other methods on both datasets are less than 98%. But the proposed method achieved a higher accuracy of above 98% on both datasets. Table 4 implies that Saba et al.'s and Amin et al.'s method achieve highest dice score of 99%, whereas other methods achieve very less dice score. The proposed method reaches this maximum on both datasets. From Table 5, it is inferred that the sensitivity obtained by the proposed method also reaches its maximum obtained by other methods.

Table 3 Accuracy comparison of projected work with recent methods on BRATS 2017 and 2018 datasets

Method	BRATS 2017	BRATS 2018
Rehman et al. [17]	96.97	92.67
Khan et al. [18]	96.9	92.5
Sharif et al. [19]	96.9	92.5
Saba et al. [20]	–	99
Amin et al. [21]	–	98
Aboussaleh et al. [22]	98	–
Proposed method	98.3	99

Table 4 Dice score comparison of proposed method with recent methods on BRATS 2017 and 2018 datasets

Method	BRATS 2017	BRATS 2018
Saba et al. [20]	99	–
Amin et al. [21]	–	99
Ranjbarzadeh et al. [23]	–	92.03
Liu et al. [24]	89.28	–
Wang et al. [25]	87	–
Myronenko [26]	–	81
Nema et al. [16]	–	94
Proposed method	99	99

Table 5 Comparison of projected work with recent methods on BRATS 2017 and 2018 datasets in terms of parameter sensitivity

Method	BRATS 2017	BRATS 2018
Saba et al. [20]	99	
Amin et al. [21]		98
Aboussaleh et al. [22]	99	
Ranjbarzadeh et al. [23]		97.12
Proposed method	99	99

5 Conclusion

Healthcare applications are the most widely used applications in all parts of the world. One among them is brain tumor identification and segmentation. This paper proposed to use handcrafted and deep features for feature extraction. Then, three different classifiers are combined to form a hybrid ensemble classifier. The proposed method is tested on BRATS 2017 and 2018 datasets. The proposed method achieves a higher accuracy of 99% with 99% sensitivity and dice score. This method can also be tested on recent datasets such as BRATS 2019 and 2020.

References

1. Demirhan A, Törü M, Güler İ (2015) Segmentation of tumor and edema along with healthy tissues of brain using wavelets and neural networks. IEEE J Biomed Health Inform 19(4):1451–1458
2. Meier R, Bauer S, Slotboom J, Wiest R, Reyes M et al (2014) Patient-specific semi-supervised learning for postoperative brain tumor segmentation. In: Medical image computing and computer assisted intervention—MICCAI. Springer, pp 714–721
3. Rajaguru H, Ganesan K, Bojan VK (2016) Earlier detection of cancer regions from MR image features and SVM classifiers. Int J Imaging Syst Technol 26(3):196–208
4. Viswa Priya V (2016) Segmentation in MRI. Indian J Sci Technol 9(19)
5. Rajesh Chandra G, Rao KRH (2016) Tumor detection in brain using genetic algorithm. Procedia Comput Sci 79:449–457
6. Isselmou A, Zhang S, Xu G (2016) A novel approach for brain tumor detection using MRI images. J Biomed Sci Eng 9:44–52
7. Perez U, Arana E, Moratal D (2016) Brain metastases detection algorithms in magnetic resonance imaging. IEEE Lat Am Trans 14(3):1109–1114
8. Shreyas V, Pankajakshan V (2017) A deep learning architecture for brain tumor segmentation in MRI images. IEEE
9. Havaei M, Davy A, Warde-Farley D, Biard A, Courvill A, Bengio Y, Pal C, Jodoin P-M, Larochelle H (2017) Brain tumor segmentation with deep neural networks. Med Image Anal 35:18–31
10. Pereira S, Pinto Pereira A, Pinto A, Alves V, Silva CA (2016) Brain tumor segmentation using convolutional neural networks in MRI images. IEEE Trans Med Imaging 35(5):1240–1251
11. Kamnitsas K, Bai W, Ferrante E, McDonagh S, Sinclair M, Pawlowski N, Rajchl M, Lee M, Kainz B, Rueckert D, Glocker B (2017) Ensembles of multiple models and architectures for robust brain tumour segmentation. In: Brainlesion: glioma, multiple sclerosis, stroke and traumatic brain injuries. BrainLes 2017. Lecture notes in computer science, vol 10670. Springer, Cham
12. Sharma H, Bhadauria HS (2017) An effective approach on brain tumor segmentation with polynomial hybrid technique. IEEE
13. Akter MK, Khan SM, Azad S, Fattah SA (2017) Automated brain tumor segmentation from MRI data based on exploration of histogram characteristics of the cancerous hemisphere. In: 2017 IEEE region 10 humanitarian technology conference (R10-HTC), pp 815–818
14. Chen W, Qiao X, Liu B, Qi X, Wang R, Wang X (2017) Automatic brain tumor segmentation based on features of separated local square. IEEE, pp 6489–6493
15. Shah N, Ziauddin S, Shahid AR (2017) Brain tumor segmentation and classification using cascaded random decision forests. In: 2017 14th international conference on electrical engineering/electronics, computer, telecommunications and information technology
16. Nema S, Dudhane A, Murala S, Naidu S (2020) RescueNet: an unpaired GAN for brain tumor segmentation. Biomed Signal Process Control 55:101641
17. Rehman A, Khan MA, Saba T, Mehmood Z, Tariq U, Ayesha N (2021) Microscopic brain tumor detection and classification using 3D CNN and feature selection architecture. Microsc Res Tech 84(1):133–149
18. Khan MA, Ashraf I, Alhaisoni M, Damaševičius R, Scherer R, Rehman A, Bukhari SAC (2020) Multimodal brain tumor classification using deep learning and robust feature selection: a machine learning application for radiologists. Diagnostics 10:565
19. Sharif MI, Li JP, Khan MA, Saleem MA (2020) Active deep neural network features selection for segmentation and recognition of brain tumors using MRI images. Pattern Recogn Lett 129:181–189
20. Saba T, Mohamed AS, El-Affendi M, Amin J, Sharif M (2020) Brain tumor detection using fusion of hand crafted and deep learning features. Cogn Syst Res 59:221–230
21. Amin J, Sharif M, Yasmin M, Fernandes SL (2020) A distinctive approach in brain tumor detection and classification using MRI. Pattern Recogn Lett 139:118–127

22. Aboussaleh I, Riffi J, Mahraz AM, Tairi H (2021) Brain tumor segmentation based on deep learning's feature representation. J Imaging 7(12):269
23. Ranjbarzadeh R, Kasgari AB, Ghoushchi SJ, Anari S, Naseri M, Bendechache M (2021) Brain tumor segmentation based on deep learning and an attention mechanism using MRI multi-modalities brain images. Sci Rep 11(1):1–17
24. Liu P, Dou Q, Wang Q, Heng PA (2020) An encoder-decoder neural network with 3D squeeze-and-excitation and deep supervision for brain tumor segmentation. IEEE Access 8:34029–34037
25. Wang G, Li W, Ourselin S, Vercauteren T (2017) Automatic brain tumor segmentation using cascaded anisotropic convolutional neural networks. In: International MICCAI brainlesion workshop. Springer, Cham, pp 178–190
26. Myronenko A (2018) 3D MRI brain tumor segmentation using autoencoder regularization. In: International MICCAI brainlesion workshop. Springer, Cham, pp 311–320

Reviews in Computational Intelligence

A Systematic Review on Sentiment Analysis for the Depression Detection During COVID-19 Pandemic

Sofia Arora and Arun Malik

Abstract Various social media platforms like Facebook, Twitter, etc., are the powerful tools to express sentiments and emotions across the globe. Sentiment analysis and its evaluation are used to reveal the positive or negative opinions associated with an individual. In this paper, we have tried to study the sentiment analysis based on their count, year wise, country wise, university wise, and keyword wise progression to understand the depth of the study in the field of sentiment analysis. Results show that sentiment analysis is not a new field, and authors are contributing to this field since 2008. Collaboration among different countries and universities has also seen during our study. In 2019, maximum contributions are received. Further, this study shows that 200 keywords with 149 unique and 18 as repeated are used by different author. Authors from 65 universities with 40 universities listed in Times Higher Education ranking 2020 are observed with highest number of authors from India.

Keywords Sentiment analysis · Depression · Naïve Bayes · COVID-19

1 Introduction

Health is an important part of well-being of every country. For the economy, the occurrence of the COVID-19 pandemic is an unforeseen shock. Before COVID-19 struck, the economy was already in a parlous condition. A sudden outbreak of a disease may trigger panic among individuals. Early identification of a disease is an incredibly challenging task. It is necessary to stop it from spreading in an area. By posting on social networking sites, people share their thoughts, emotions and feelings. Analysis of sentiment is an analysis to determine their thoughts, attitudes, views, etc. Depression is also part of this study to classify the person via their posts on Facebook, Instagram, Twitter, Blog, and Forums as being in depression or not.

S. Arora · A. Malik (✉)
Lovely Professional University, Phagwara, India
e-mail: arun.17442@lpu.co.in

P. Chatterjee et al. (eds.), *Computational Intelligence for Engineering and Management Applications*, Lecture Notes in Electrical Engineering 984,
https://doi.org/10.1007/978-981-19-8493-8_23

In these days, electronic information is evolving rapidly in every phase of life, which produces the large amount of data. As an outcome the large amount of data are generated in the fields of business, health care, tourism, eMarketing, technology etc. Automated analysis systems are related for analyzing, summarization and classification of data and number of efficient methods to store large amount of data. Sentiment analysis is an approach used in different fields such as machine learning, information retrieval, statistics, and computational linguistics for opinion mining. Social networking has become a community these days and ranked as one of the most common kinds of online activity. It offers the people the place to connect online. With the aid of social networking sites such as Twitter, Facebook, Blogs, and Forums, people share their views, emotions, perceptions, and thoughts with each other. The people's decision can be evaluated or defined with the aid of such posts, tweets, and forums. Analyzing these perceptions, ideas, feelings, and emotions is called analyzing sentiment. The aim of an analysis of sentiments is to identify and categorize emotions that people express on social networking sites.

Analysis of sentiments used natural language processing for text exploration and statistics to identify sentiment among people. This method includes the study of sentimental data in five separate steps:

1. The first move toward analyzing the emotional data is data collection from social networking sites. This data is conveyed in various ways by the use of vocabulary, writing material, etc. Text mining and the analysis of natural languages are also used for extraction and classification.
2. The second step consists the cleaning of extracted data before analysis. This process is called text preparation.
3. The feeling is identified from data in the third phase of an analysis of sentiment. In this, the comments and opinions from the extracted sentences are analyzed. The sentences which are subjective words such as thoughts, values, and views are engaged, and on the other hand, sentences which are objective communications such as facts are rejected and reliable knowledge.
4. The next step of evaluating sentiments is classification of sentiments. In this, classification of the subjective sentences is marked as positive, negative, and neutral in polarity.
5. The last step of this analysis is presenting the output. The main goal of an examination of sentiments is to turn amorphous text into useful details. After the analysis, the results are displayed in the form of graphs such as pie chart, bar chart, and line graphs.

The text classification is based on the criteria as shown in Figs. 1 and 2.

Fig. 1 Criteria of text classification

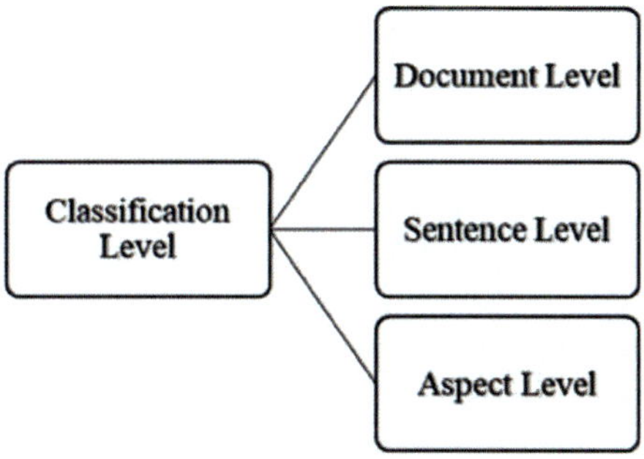

Fig. 2 Classification levels

2 Literature Review

Another form of medical aid for identifying symptoms of mental illness, such as depression, is social networking. This paper summarized the results of the identification of depressive mood disorders, using methods and techniques of emotion analysis. The author concentrated on research which automatically identifies irregular patterns of activity on social networks. In order to evaluate the knowledge available for lexicon use, the selected studies used classic off-the-shelf classifiers [1]. This work uses two techniques for machine learning on Twitter info, such as the Naïve Bayes classifier and K-Nearest Neighbor. According to the report, LinkAja and Go-Pay have positive feelings as opposed to other providers of services [2].

An individual who is depressed will often feel sad, helpless, and sometimes loses interest in activities on a daily basis and faces physical symptoms as well. There is a lack of work in this application, according to the source, which is used to recognize depression on social media such as Twitter, etc. The web application that performs sentiment analysis with the assistance of the classification feature that recognizes the proportion of depressive and non-depressive thoughts is built to solve this problem. The different phases of text classification were mainly based on supervised machine learning techniques. It is noted that, according to the source, k-NN performs well in various classification techniques [3, 4]. The connection between social media and stock comments made by individuals has been clarified. The Thai stock exchange is Pantip.com. This site contains a tool for tagging. All posts on this website are labeled stock. This model provides a 74% accuracy result. Through this analysis, ADVANC and CPALL stock volumes have been found to contribute to social media feelings. In this paper, a systematic approach to machine learning and lexicon-based methods is carried out, and then, the hybrid approach proposes the problem of rating-prediction in Persian [5, 6].

A methodology known as Customer Experience Management is used to create and improve customer service. In this paper, this work uses sentiment analysis to assess consumer feedback experiences. The data available on social media is therefore massive and formless, using the Naïve Bayes classification classifier applied to contribute to the process of sentiment analysis. The author offered healthcare Twitter analysis which deals with health-related tweets through sentiment analysis. In terms of accuracy, the author also reveals the contrast of current work with the method proposed [7, 8].

The author suggested a new show called Bag of Sub-Emotions (BoSE). The new representation generates close-grained emotions that are automatically generated using a lexical tool of emotions and subwords that is embedded in Fast-Text. It produces better performance than the baselines suggested. The whole approach also aims to describe the contextual polarity of an individual's Internet communications and also helps to segregate the user's most common words [9, 10].

An emoji expression detection system was developed by the author that finds feelings such as smile, anger, crazy, funny, sick, sleepy, cool, and sadness. It extracts the emoji image from the feature and uses the KNN algorithm for detection. Work was based on image classification, and using emoji gave us a successful result in the area of emotion. No local information will be used when a given image does not contain suitable objects; otherwise, sub-images will be collected based on the important object detection window [11, 12]. The Internet is a sea of raw data these days, and it can only be seen as information after processing, establishing, and organizing raw data. A novel method, called partial textual entailment, was introduced to solve this problem of effectively evaluating feelings. It was used to test the semantic similarity between tweets shared, so that it could be easier to group related tweets. According to the scientist, this technique was first used in this paper to reduce the overhead of computing. The main result of this paper is that it offers a system where signs of cyberbullying are automatically identified on social media [13, 14].

The research is being done in this Twitter paper in which the author examined the views, emotions, expectations, and behaviors of people on an outdoor game of 'Lawn Tennis.' How many people really like this game, and how popular this game is in different countries, has been investigated. Because of the vast amount of data handling, Hadoop was selected because of its dispersed architecture and ease of handling. The multi-modality blend concept was created by the author to combine audio, video, and text modalities to recognize biomarkers that predict depression [15, 16].

A medical data science system based on the measurement of different emotions and emotional processing methodologies has been proposed by the author. In terms of patient medical information, the structure of this analysis was considered in the context of data mining, data analytics, and data visualization. The main goal was to construct a self-serving psychometric analyzer capable of conducting rapid computational linguistics, providing a mental well-being summary based on previous studies, patient medications, and treatments [17, 18].

The author presented a method in this paper that uses tweets as a database. It uses SentiStrength sentiment analysis to construct training dataset and to categorize the given tweets into depressed and non-depressed individuals, using Back Propagation Neural Networks. With the aid of this hybrid model, it is easy to identify the social activity, thought, and mental level of the depressed patient. The aim of this paper is to offer an overview of the state of the art in the study of machine learning techniques [19, 20]. To demonstrate depression-oriented emotional analysis, this paper focused on the application of natural language processing on Twitter. To detect depression on Twitter info, text dependent emotion AI is used. In the class calculation process, the help vector machine and the Naïve-Bayes classifier were used. In this paper, the author proposed a new method to categorize the user by using SNS as a data source and using artificial intelligence through a selection tool [21, 22]. The model is generated by the author, who uses two separate classifiers to organize the UGC: Vector Machine Support and Naïve Bayes Support. It ultimately categorizes patients into one out of four categories of minimal, medium, moderate, or extreme depression [23]. The aim of an examination of feelings is to acknowledge opinions or emotions. Methodologies for sensitivity analysis may provide useful methods and frameworks for tracking mental illness and depression. SVM is larger than Naïve Bayes and Maximum Entropy classifiers. Technological advances in the processing of natural languages and techniques for machine learning are helpful [24–26].

The author uses four state-of-the-art machine learning classifiers, such as Naïve Bayes, J48, BFTree, and OneR, in this paper to optimize sentiment analysis. This type of approach also offers social workers the ability to reach distressed people who need care in the early stages. With the aid of this technique, various fields such as marketing, politics, and sociology are being explored [27–29]. Analysis of emotions is a method of automatic representation of characteristics through the thoughts of other people, with regard to specific products, services, or experiences. The objective of sentiment analysis is to create an automated mechanism capable of recognizing and categorizing emotions. The machine learning classification technique used gives 90% accuracy to classify sentiment tweets into positive, negative, and neutral [30, 31]. The internal model identifies the polarization of the messages in this paper, which reduces the need for normalization. With this internal model, the algorithm's efficiency increases. Together with the results, the algorithmic approach also demonstrates the functioning of the application in a well-organized manner, and this approach is also useful for holding suicide attempts due to cyber-depression. Facebook was used as a responsible source for detecting individual depression trends by this built-up method. The tool is designed to classify distressed Twitter users and to analyze spatial designs using GIS technology. The strategies for treating depression can be improved by this method [32–34]. In machine learning, lexicon-based, and hybrid technologies, there are a number of classifiers. It is important for the classifier to be able to quantify the risk associated with each classification decision. An arithmetical approach to classifying patterns is Bayesian decision theory. The primary aim is to explain why some NB classifiers perform better than others and to what extent the best solution to the standard decision method is. In comparison with other algorithms such as CART, DT, and MLP's, the experimental results have proved the efficacy of this algorithm [35–37]. Coronavirus disease 2019 (COVID-19) impacted worldwide psychological health in contrast to being a public physical health emergency, as illustrated by panic buying globally as cases soared. Changes in levels of psychological effects, stress, anxiety, and depression during this pandemic are little understood. This longitudinal research surveyed the general population twice, surveying demographics, symptoms, awareness, fears, and precautionary measures against COVID-19 during the initial outbreak and the peak epidemic four weeks later [38–40]. This research aimed to identify factors correlated with symptoms of depression, anxiety, and PTSD in young adults in the USA during the COVID-19 era. Research on social networks will help us understand the intractable issues that are fundamental to today's healthcare system's challenges: silo work, bottlenecks, discrepancies, inadequate coordination, professional alienation, and other social structures that are vulnerable to undermining patient safety and quality care [41–44]. In view of the dynamism and complexities of health care and the need for more longitudinal, mixed-methods network research, and design decisions are debated [45, 46].

We conclude by addressing some significant remaining restrictions of the Facebook strategy, as well as highlighting some specific strengths of the Facebook targeted advertising strategy, including the ability to rapidly collect data in response to research opportunities, rich and scalable sample targeting capabilities, and low cost, and suggesting wider applications of this technique [47–50].

3 Discussion and Analysis

In this paper, we have studied the papers which are related to sentiment analysis. We have taken the paper from 2008 to 2020. All the papers define the idea of sentiment analysis, tools and applications, and depression detection techniques, etc. In all the papers, supervised learning techniques are used to identify the depression. It was observed that from year 2008 to 2013, very few contributions were there, but from 2013 to 2020, this field has shown enormous growth.

On the basis of papers, we have studied the keyword trend used by author. From the papers, 200 keywords are recognized. From these 200 keywords, 149 keywords are unique in these papers and 18 keywords are repeated. Further, we have explored the most used top 10 keywords with frequency of 3 or more (Figs. 3 and 4).

We have also studied the countries with respect to the author affiliations. In all papers, authors were from 29 different countries. We have listed top 5 countries with maximum contributions. 12 authors were from India who have worked on sentiment analysis. Similarly, 3 authors from China, 3 from Italy, 4 from Australia, and 7 authors from USA have done work on sentiment analysis on detecting depression (Figs. 5 and 6).

In university wise analysis, we have analyzed that 64 universities were involved with the contribution of their authors in sentiment analysis. We have found the ranks of these universities from time higher ranking website. While checking the ranking of universities, it has been observed that there are 6 universities who have ranked in top 100. Seven universities are in 101–200 rank, two universities in 201–300 and 301–400 rank simultaneously and 20 universities with rank in top 500. Hence, we can conclude that number of authors have affiliations from world level universities.

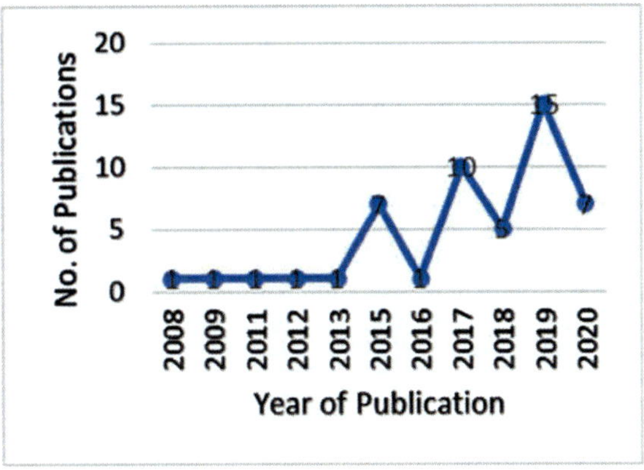

Fig. 3 Shows the year wise publication

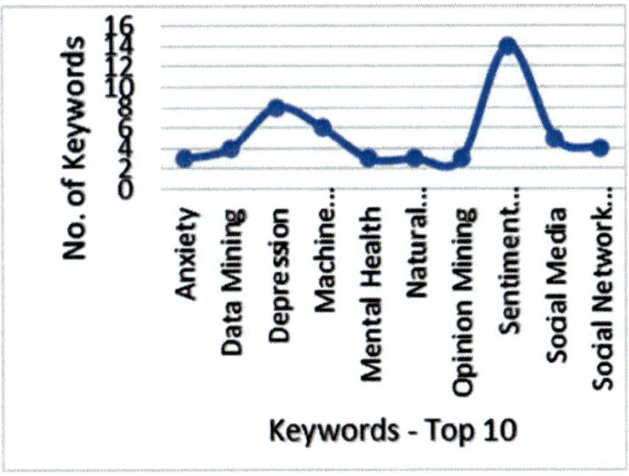

Fig. 4 Shows the keywords trend

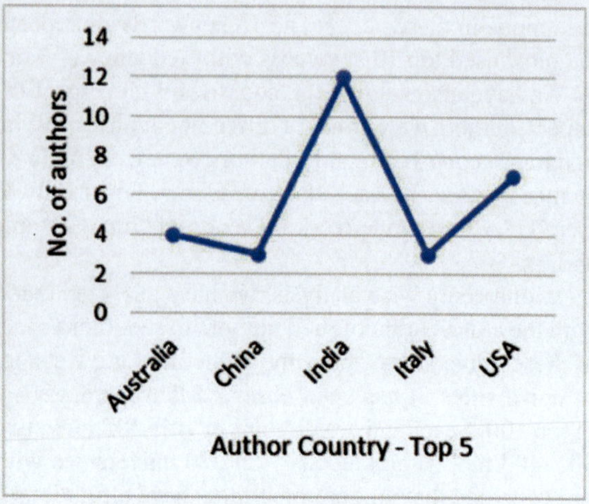

Fig. 5 Shows the country wise variation of sentiment analysis

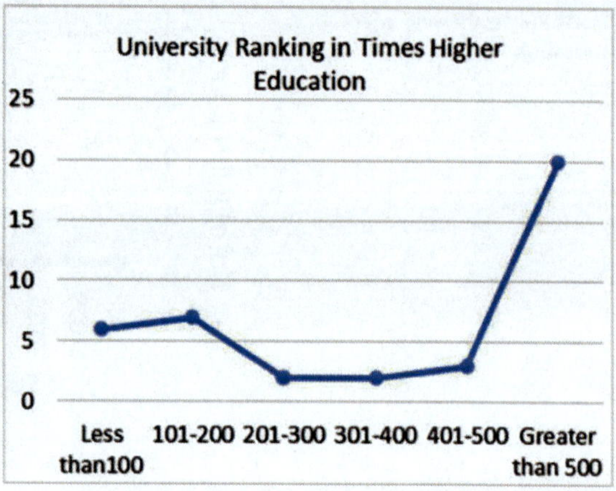

Fig. 6 Shows the university ranks from Times Higher Education 2020 ranking

4 Conclusion

We have studied papers of different publications, and authors from all over the world have completed their work on sentiment analysis. It has been observed that most probably all authors worked on supervised learning techniques. We have also studied supervised learning techniques such as Naïve Bayes, support vector machines, Maximum Entropy to detect the depression. In these techniques, the predefined datasets are used by the authors to provide more accurate results. We have tried to propose the system based on unsupervised learning techniques for the detection of depression in COVID-19 pandemic. Data will be collected from surveys, interviews, and questionnaires. Proposed study has some limitations as it gives better results only on predefined dataset that is supervised learning techniques as compared to unsupervised learning techniques.

References

1. Giuntini FT et al (2020) A review on recognizing depression in social networks: challenges and opportunities. J Ambient Intell Humaniz Comput 11(11):4713–4729
2. Wisnu H, Afif M, Ruldevyani Y (2020) Sentiment analysis on customer satisfaction of digital payment in Indonesia: a comparative study using KNN and Naïve Bayes. J Phys Conf Ser 1444(1). IOP Publishing
3. Ziwei BY, Chua HN (2019) An application for classifying depression in tweets. In: Proceedings of the 2nd international conference on computing and big data
4. Kadhim AI (2019) Survey on supervised machine learning techniques for automatic text classification. Artif Intell Rev 52(1):273–292
5. Padhanarath P, Aunhathaweesup Y, Kiattisin S (2019) Sentiment analysis and relationship between social media and stock market: pantip.com and SET. IOP Conf Ser Mater Sci Eng 620(1). IOP Publishing
6. Basiri ME, Kabiri A (2020) HOMPer: a new hybrid system for opinion mining in the Persian language. J Inf Sci 46(1):101–117
7. Alamsyah A, Bernatapi EA (2019) Evolving customer experience management in internet service provider company using text analytics. In: 2019 international conference on ICT for smart society (ICISS), vol 7. IEEE
8. Arora P, Arora P (2019) Mining Twitter data for depression detection. In: 2019 international conference on signal processing and communication (ICSC). IEEE
9. Aragón ME et al (2019) Detecting depression in social media using fine-grained emotions. In: Proceedings of the 2019 conference of the North American chapter of the association for computational linguistics: human language technologies, vol 1 (long and short papers)
10. Nagarajan SM, Gandhi UD (2019) Classifying streaming of Twitter data based on sentiment analysis using hybridization. Neural Comput Appl 31(5):1425–1433
11. Wongkoblap A, Vadillo MA, Curcin V (2017) Researching mental health disorders in the era of social media: systematic review. J Med Internet Res 19(6):e228
12. Priya BG (2019) Emoji based sentiment analysis using KNN. Int J Sci Res Rev 7(4):859–865
13. Wu L et al (2019) Visual sentiment analysis by combining global and local information. Neural Process Lett 1–13
14. Gupta S, Lakra S, Kaur M (2019) Sentiment analysis using partial textual entailment. In: 2019 international conference on machine learning, big data, cloud and parallel computing (COMITCon). IEEE
15. Van Hee C et al (2018) Automatic detection of cyberbullying in social media text. PLoS One 13(10):e0203794
16. Malik M, Naaz S, Ansari IR (2018) Sentiment analysis of Twitter data using big data tools and Hadoop ecosystem. In: International conference on ISMAC in computational vision and bio-engineering. Springer, Cham
17. Samareh A et al (2018) Detect depression from communication: how computer vision, signal processing, and sentiment analysis join forces. IISE Trans Healthc Syst Eng 8(3):196–208
18. Vij A, Pruthi J (2018) An automated psychometric analyzer based on sentiment analysis and emotion recognition for healthcare. Procedia Comput Sci 132:1184–1191
19. Zhang W, Xu M, Jiang Q (2018) Opinion mining and sentiment analysis in social media: challenges and applications. In: International conference on HCI in business, government, and organizations. Springer, Cham
20. Biradar A, Totad SG (2018) Detecting depression in social media posts using machine learning. In: International conference on recent trends in image processing and pattern recognition. Springer, Singapore

21. Fatima I et al (2018) Analysis of user-generated content from online social communities to characterise and predict depression degree. J Inf Sci 44(5):683–695
22. Deshpande M, Rao V (2017) Depression detection using emotion artificial intelligence. In: 2017 international conference on intelligent sustainable systems (ICISS). IEEE
23. Aldarwish MM, Ahmad HF (2017) Predicting depression levels using social media posts. In: 2017 IEEE 13th international symposium on autonomous decentralized system (ISADS). IEEE
24. Zucco C, Calabrese B, Cannataro M (2017) Sentiment analysis and affective computing for depression monitoring. In: 2017 IEEE international conference on bioinformatics and biomedicine (BIBM). IEEE
25. Hassan AU et al (2017) Sentiment analysis of social networking sites (SNS) data using machine learning approach for the measurement of depression. In: 2017 international conference on information and communication technology convergence (ICTC). IEEE
26. Guntuku SC et al (2017) Detecting depression and mental illness on social media: an integrative review. Curr Opin Behav Sci 18:43–49
27. Singh J, Singh G, Singh R (2017) Optimization of sentiment analysis using machine learning classifiers. HCIS 7(1):1–12
28. Tao X et al (2016) Sentiment analysis for depression detection on social networks. In: International conference on advanced data mining and applications. Springer, Cham
29. Alessia D et al (2015) Approaches, tools and applications for sentiment analysis implementation. Int J Comput Appl 125(3)
30. Kaushik A, Naithani S (2015) A study on sentiment analysis: methods and tools. Int J Sci Res (IJSR) 4(12)
31. Dinakar S, Andhale P, Rege M (2015) Sentiment analysis of social network content. In: 2015 IEEE international conference on information reuse and integration. IEEE
32. Gupta E et al (2015) Mood swing analyser: a dynamic sentiment detection approach. Proc Natl Acad Sci India Sect A Phys Sci 85(1):149–157
33. Hussain J et al (2015) SNS based predictive model for depression. In: International conference on smart homes and health telematics. Springer, Cham
34. Yang W, Mu L (2015) GIS analysis of depression among Twitter users. Appl Geogr 60:217–223
35. Nunzio D, Maria G (2014) A new decision to take for cost-sensitive Naïve Bayes classifiers. Inf Process Manage 50(5):653–674
36. Wang X et al (2013) A depression detection model based on sentiment analysis in micro-blog social network. In: Pacific-Asia conference on knowledge discovery and data mining. Springer, Berlin, Heidelberg
37. Hsu C-C, Huang Y-P, Chang K-W (2008) Extended Naive Bayes classifier for mixed data. Expert Syst Appl 35(3):1080–1083
38. Wang C et al (2020) A longitudinal study on the mental health of general population during the COVID-19 epidemic in China. Brain Behav Immunity 87:40–48
39. Elbay RY et al (2020) Depression, anxiety, stress levels of physicians and associated factors in COVID-19 pandemics. Psychiatry Res 290:113130
40. Liu CH et al (2020) Factors associated with depression, anxiety, and PTSD symptomatology during the COVID-19 pandemic: clinical implications for US young adult mental health. Psychiatry Res 290:113172
41. Pomare C et al (2019) Social network research in health care settings: design and data collection. Soc Netw
42. Schneider D, Harknett K (2019) What's to like? Facebook as a tool for survey data collection. Sociol Methods Res 0049124119882477
43. Pak A, Paroubek P (2010) Twitter as a corpus for sentiment analysis and opinion mining. In: LREc, vol 10, no 2010
44. Archibald MM et al (2019) Using zoom videoconferencing for qualitative data collection: perceptions and experiences of researchers and participants. Int J Qual Methods 18:1609406919874596
45. Morrell-Scott NE (2018) Using diaries to collect data in phenomenological research. Nurse Res 25(4):26–29

46. Doody O, Noonan M (2013) Preparing and conducting interviews to collect data. Nurse Res 20(5)
47. Carr EM et al (2019) Qualitative research: an overview of emerging approaches for data collection. Australas Psychiatry 27(3):307–309
48. Roh Y, Heo G, Whang SE (2019) A survey on data collection for machine learning: a big data—AI integration perspective. IEEE Trans Knowl Data Eng
49. Agarwal A et al (2011) Sentiment analysis of Twitter data. In: Proceedings of the workshop on language in social media (LSM 2011)
50. Wendt DE, Starr RM (2009) Collaborative research: an effective way to collect data for stock assessments and evaluate marine protected areas in California. Mar Coast Fish Dyn Manage Ecosyst Sci 1(1):315–324

Vehicular Adhoc Networks: A Review

Gagan Preet Kour Marwah and Anuj Jain

Abstract Vehicle ad hoc networks (VANETs) have a lot of potential for improving road safety and passenger comfort in cars. On the other hand, they are exposed to many risks that affect the reliability of these features because they use an open medium for communication. Our mission is to provide a lightweight security model based on privacy-based applications that work in VANET environments. To ensure that a specific safety message is focused on actual events, the messages sent on the network involve trusted software components not from a hostile vehicle and not injected.

Keywords VANET · Channel state identification · Machine learning

1 Introduction

In recent years, there has been a lot of focus on wireless networks. It is also garnering a lot of attention, and it is currently the most popular field in both industry and academia. Due to Vehicular Ad-hoc Networking (VANET), the vehicle manufacturing industry in the automotive sector is receiving lot of awareness due to self-driving and autonomous cars [1].

Intelligent Transport System (ITS) assists self-driving vehicles in reducing traffic congestion on roads while also improving road safety. Vehicle communication in VANET [2] can be accomplished by sending data using Vehicles to Infrastructure (V-2-I), Vehicles to a House (V-2-H), Vehicles to Vehicles (V-2-V), and Vehicles to Everything (V-2-X) interactions as shown in Figs. 1, 2, and 3, respectively. Two units where VANET generally focuses:

(a) On-Board Unit (OBU)—It is a device present in the vehicle with all communication competencies.

G. P. K. Marwah (✉) · A. Jain
School of Electronics and Electrical Engineering, Lovely Professional University, Phagwara, Punjab, India
e-mail: gaganmarwaha.marwaha@gmail.com

P. Chatterjee et al. (eds.), *Computational Intelligence for Engineering and Management Applications*, Lecture Notes in Electrical Engineering 984,
https://doi.org/10.1007/978-981-19-8493-8_24

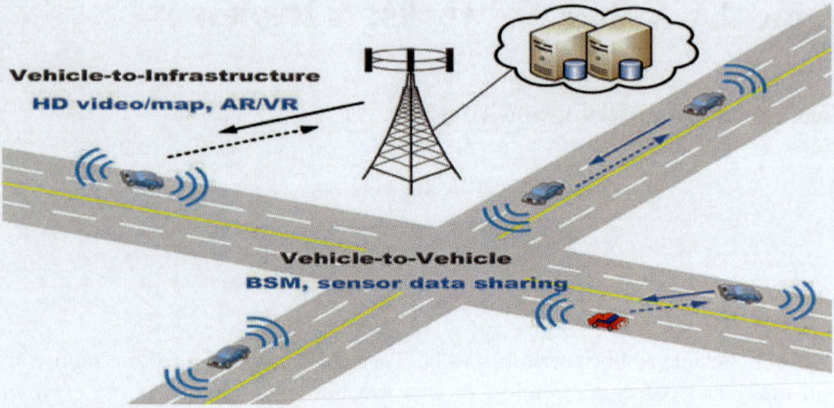

Fig. 1 Vehicular communication: a quick overview

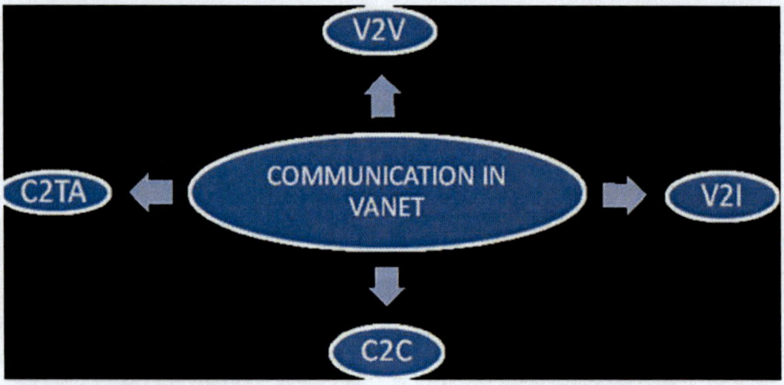

Fig. 2 Communication in VANET [4]

(b) Application Unit (AU)—It is a device that use OBU's communication competencies and executes one or series of an application [3].

AU is always working while connected with OBU, and it can be movable and portable like a laptop. Apart from this, VANET [4] architecture can have multiple networks like:

(a) Cellular/WLAN: In this type of network, vehicle information is transferred with road-side-unit (RSU) whose other name is base station (BS).
(b) Ad-Hoc: Under this, the information among vehicles will be transferred directly irrespective of intermediary bodies like (V-2-V communication).
(c) Hybrid: This is a type of network which can have the combination of last two networks.

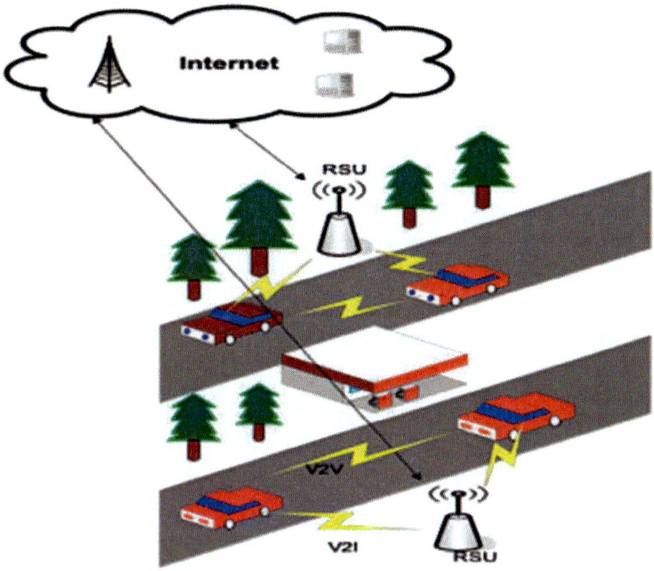

Fig. 3 Architecture of dedicated short range communication. RSU: road side unit, V2I: vehicle to infrastructure

In VANET vehicles, information such as speed, distance, direction, traffic situation, position, and so on can be immediately communicated to fixed remote nodes.

In both developed and under developing countries, especially in automotive sector, vehicles face congestion and heavy road traffic. This is a major problem. In order to address the difficulty with congestion, a new benchmark dedicated short range communication (DSRC) [5] has emerged to carry IEEE-802.11 while communicating in vehicles. The federal communication commission (FCC) set aside 75 MHz of DSRC spectrum at 5.9 GHz for V-2-V communication in 1999. Vehicle safety consortium (VSC), Car-to-Car (C-2-C) Communications Consortium Europe, advanced safety vehicles (ASV) Japan, and different countries are working on other consortiums to create a secure environment for self-driving and autonomous vehicles, with emerging trends in vehicle networking across countries such as the US [6].

VANET has a number of distinguishing characteristics that set it distinct from many other kinds of networks.

a. Extreme Mobility: VANET nodes are moving at rapid speed. Only if the location of these moving nodes can be predicted can they be safeguarded from attacks and other security concerns. Other VANET difficulties are exacerbated by high mobility [7, 8].
b. Rapid Topology Changes: High-speed automobiles create rapid topology changes in a VANET [7–9].

c. No Power Constraints: In VANET, high-speed vehicles lead to rapid changes in network topology in different networks, but in VANET vehicles, the long-life battery will supply power to the onboard unit (OBU) [7–9]. Thus, energy restriction, as in MANETs, is not always an important obstacle.
d. Unbounded Network Size: Because VANET can be constructed for a single town or area, its network size is geographically infinite [8, 9].
e. Time Critical: Data delivery must be done in a timely manner. Only if data is available in the required format actions may be taken [10–12].

1.1 Importance of the Work to Be Carried Out

It is anticipated to open up a slew of new possibilities from enhancing safety on road to improve efficiency in traffic, from automotive driving to universal vehicle Internet access [13, 14]. This new generation of networks will, in the end, have far-reaching implications for society and the daily lives of millions of individuals all over the world. Because of the tight and complicated quality of service (QoS) regulations, in addition to the underlying dynamics in vehicle surroundings, e.g., rapidly changing radio transmission mediums and constantly modifying the topology of the network, the vehicle network poses unique challenges not seen in traditional wireless communication systems, despite its enormous transformational potential daily vehicle experiences.

Meanwhile, cars would be more than just a basic means of transport, with facilities for high-performance computing, storage, and numerous sophisticated on-board sensors fitted, like light detection and ranging (LIDAR), radars, and cameras. They generate, represent, reserve, process, and transfer vast quantities of data to make driving more convenient and safe. Usually, applications in vehicle environments can enhance road safety, increase traffic quality, and provide passengers with entertainment. Accident avoidance is typically the purpose of safety applications, and therefore this form of application is also the key impetus for the creation of ad hoc vehicle networks.

There is a great need for vehicle-to-vehicle contact or vehicle-to-infrastructure contact for such solicitations as crash avoidance [15]. Vehicles outfitted with numerous sensors continuously monitor traffic data and track the surroundings, and afterwards using V2I or V2V communication, cooperative vehicle safety applications can modify real-time traffic information and send/receive warning messages to upgrade road safety and prevent accidents. In 2006, eight safety applications were reported by several transport departments in the US, which are thought to provide the most benefits, i.e., traffic signal violation, curve speed alert, emergency brake lights, pre-crash sensing, collision warning, left turn assist, lane change warning, and stop sign assist [16].

1.2 Literature Review

To support a scalable VANET, a large amount of research had been undertaken. Despite all of the study, there is still a disconnect between industry standards and research community norms. As 4G and 5G technologies mature, future VANETs will most likely converge with these technologies.

In [17], the authors introduced a data collection mechanism that is perceived to be an intrinsic obstacle to Vehicular Ad-Hoc networks. For VANETs, an adaptive data collection protocol using reinforcement learning (ADOPEL) is suggested. Its foundation is built on a distributed Q-learning approach that makes the collecting process more sensitive to changes in node mobility and topology. The simulation results confirm the efficacy of our technique in comparison to a non-learning version, as well as the trade-off between time and selection ratio.

By transferring data to a new carrier (vehicle) before the present data carrier leaves a specific area, the authors in [18] suggested a convention where data could be stored in VANETs. The convention uses fuzzy logic for the choice of the another carrier data node and to determine instant compensation by taking multiple metrics into consideration. Besides, to understand the potential reward of a decision, a reinforcement learning-based algorithm is used. They employed theoretical studies and computer simulations to test the proposed protocol.

Accordingly, the authors in [19] Vehicular fog networks (VeFNs) have appeared to allow the dividing of computing resources through computer task discharge, presuming a broad variety of fog applications. However, due to the increased mobility of vehicles, ensuring the latency that accounts for the entire job offloading process for both communication and computation is problematic. In this paper, authors first discuss the state of the art of offloading tasks in VeFNs and state that mobility is not just a hindrance in VeFNs for timely computation, but it can also benefit from delay efficiency.

The authors in [20] concluded that an intensive analysis requires low-delay output in VANET. They suggested a novel priority approach in this article to reduce the one-hop transmission delay in VANET. Each message was labeled with a priority based on static variables, dynamic variables, and message size. Based on the importance of messages, the scheduling of messages was introduced. The findings of the simulation are consistent with theoretical derivation. Therefore, low delays and more efficient contact scenarios can be provided in the Vehicular Ad-hoc Network by using the new system.

A study of vehicle network research topics, such as allocation of resources, discharge of data, placement of cache, ultra-reliable low-latency communication (URLLC), and higher mobility, was provided by the authors in [21]. In addition, they showed MARL's potential applications that allow for decentralized and flexible decision-making in scenarios of vehicle-to-everything (V2X).

The authors in [22] proposed a Q-learning-based technique where large number of vehicles may be linked together and data packets can be shared efficiently. This work, in particular, focuses on a contention-based medium access control (MAC) technique compatible with IEEE 802.11p to efficiently share the wireless channel between numerous vehicle stations. A collective contention estimation (CCE) technique that they implemented into the Q-learning agent also helped them achieved faster convergence, higher throughput, and short-term fairness.

The authors characterized the out loading choice as a resource scheduling issue with one or more objective functions and limits in [23], in which few customized heuristics algorithms were studied. Moreover, it is a difficult decision to unload many data dependence tasks in a complex operation, as an ideal approach must consider the requirement of resources, the access network, user mobility, and, most importantly, data dependency.

The authors proposed a software-defined trust-based deep reinforcement learning system (TDRL-RP) in [24, 25], stationing a deep Q-learning algorithm into a logically centralized software-defined networking controller (SDN). In particular, the SDN controller is employed as an agent to determine the routing path's greatest trust value in a VANET system via convolution neural network, in which the trust model is programmed to determine the conduct of forwarding packets by neighbors. To demonstrate the efficacy of the proposed TDRLRP system, simulation results are presented.

The authors of [26] tackled one of the most fundamental challenges in wireless communication systems: channel state information (CSI) estimation. Till date, numerous techniques have been invented to perform CSI estimation, which typically involves elevated computational complexity. Moreover, because of numerous approaches (e.g., massive MIMO, OFDM, and millimeter-Wave (mm-Wave)) to be used in 5G wireless communication systems, these techniques are not ideal for 5G wireless communications. In this paper, the authors proposed a successful CSI projection method on the internet which is known as OCEAN to forecast CSI from data from the past wireless communication solutions that are 5G.

Channel state information (CSI) and signal strength obtained are reliable assessments of the wireless channel, which have been accessed by the authors in [27], thus increasing throughput and transmission efficiency in vehicular communications by allowing for quick customization of transmission settings. The authors presented their efforts to allow accurate, updated estimated channel in vehicular communications in this paper. They begin by gathering IQ (in phase and quadrature) samples of IEEE 802.11p transmission and implementing CSI extraction techniques to build and perform a measurement campaign to get and assess wireless channel estimations from various real-world scenarios.

This is a literature survey of work done by different researchers. Now, we will try to find out the drawbacks of works done by the authors which are explained in the next section.

2 Research Issues

Based on the above literature survey, we identified the number of open issues for future work. These are the main drawbacks of various existing works, which motivate us to do this research on:

(a) Learning vehicular network dynamics: Vehicle networks are dynamic in a variety of ways, such as wireless transmission channels, network topologies, and traffic dynamics. How such dynamics may be efficiently taught and robustly predicted based on past data supplied by various onboard sensors or prior transmission is still an outstanding subject.
(b) Identification of relevant parameters affecting vehicular communications: Parameters are very critical factors that normally attack vehicle-to-vehicle networking. So, it is also a big job to pick the parameters.
(c) Accurate and fast Prediction of Channel State Information (CSI) for vehicular communications. One of the most important parts of wireless communication is channel state information (CSI). This refers to a radio link's established channel characteristics. CSI will characterize the cumulative influence of route loss, scattering, diffraction, fading, shadowing, and other factors as a signal propagates from a transmitter to its corresponding receiver. As a result, this data can be used to determine if a radio link is in good or terrible condition.
(d) Improving the efficiency of channel prediction in vehicular communications, which is also one of the most significant outstanding concerns due to the high speed of vehicle communication mobility.
(e) Minimum latency communication of basic safety messages (BSM) will have a short safety message and that message will be transmitted between vehicles in a minimum of time.
(f) Efficient communication within vehicle networks, which is one of the essential tasks of communication between vehicles.

3 Proposed Methodology

We will first focus on features (such as frequency band, location, time, temperature, humidity, and weather) that affect Channel State Information (CSI), which is one of the most important challenges in wireless communication systems [26]. To date, a variety of ways to CSI estimation have been developed, which often requires a high level of computer complexity and flowchart of proposed methodology as mentioned in Fig. 4.

(a) After reviewing the features that affect CSI, the next step is to collect a data sample that includes information from both these features and the CSI.
(b) After collecting the data samples for CSI features, now pre-processing of the dataset according to our requirement using machine learning (ML) techniques [3].

Fig. 4 Mechanism of congestion control

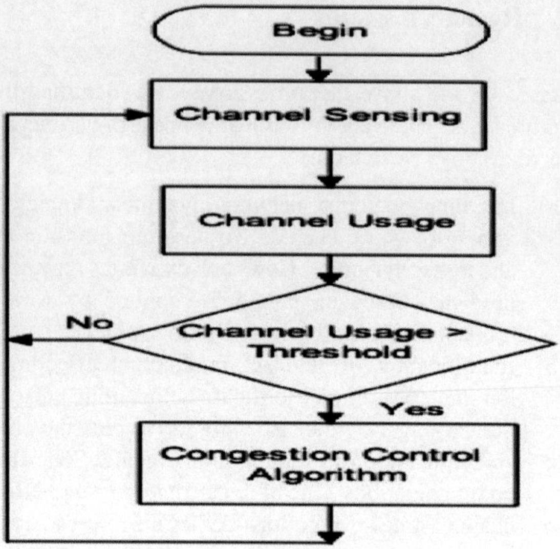

(c) The prediction of CSI will be done after selecting appropriate learning techniques.
(d) Next, we will validate the predicted data by comparing it with actual information.
(e) After validation of predicted data, the impact of channel state information (CSI) on the performance of vehicular communications will be investigated.
(f) The predicted information will be exploited to make communication fast and reliable.

4 Conclusion

We examined various research articles on VANET applications, security, and routing protocols in this work and compare their performance as mentioned in Table 1. In terms of security, VANET is falling behind. Authentication for VANET was developed by a team of researchers. However, there is not much effort done in terms of secrecy and availability. As a result, additional effort related to VANET security is required, as it has become the primary priority of users.

Table 1 Comparison of various techniques

Methodology	Accuracy	Efficiency	Complexity	Latency
ADOPEL	Low	Moderate	High	High
Vehicular fog networks	High	Low	Moderate	High
Q learning	High	High	High	Moderate
TDRL-RP	Moderate	High	Moderate	Moderate
OCEAN	High	Moderate	Very high	Low
Hybrid meta-heuristic	High	Moderate	Low	Low

References

1. Liang W, Li Z, Zhang H, Sun Y, Bie R (2014) Vehicular ad hoc networks: architectures, research issues, challenges and trends
2. Khan UA, Lee SS (2019) Multi-layer problems and solutions in VANETs: a review. Electronics (Switzerland)
3. Ye H, Liang L, Li GY, Kim J, Lu L, Wu M (2017) Machine learning for vehicular networks
4. Chadha D, Reena (2007) Vehicular ad hoc network (VANETs): a review. Int J Innov Res Comput Commun Eng
5. Tong W, Hussain A, Bo WX, Maharjan S (2019) Artificial intelligence for vehicle-to-everything: a survey. IEEE Access 7:10823–10843
6. Liu Y (2019) Intelligent processing in vehicular ad hoc networks: a survey
7. Jakubiak J, Koucheryavy Y (2008) State of the art and research challenges for VANETs
8. Vehicle safety communications project task 3 final report identify intelligent vehicle safety applications enabled by DSRC (2005)
9. Yousefi S, Siadat Mousavi M, Fathy M (2006) Vehicular ad hoc networks (VANETs): challenges and perspectives
10. Raw RS, Kumar M, Singh N (2013) Security issues and solutions in vehicular adhoc network: a review approach. In: ICCSEA, SPPR, CSIA, WimoA—2013. Academy and Industry Research Collaboration Center (AIRCC), pp 339–347
11. Nekovee M (2005) Sensor networks on the road: the promises and challenges of vehicular ad hoc networks and grids
12. Olariu S, Weigle M (2009) Vehicular networks: from theory to practice. Chapman and Hall/CRC
13. Liang L, Peng H, Li GY, Shen X (2017) Vehicular communications: a physical layer perspective. IEEE Trans Veh Technol 66:10647–10659
14. Peng H, Liang L, Shen X, Li GY (2019) Vehicular communications: a network layer perspective. IEEE Trans Veh Technol 68:1064–1078
15. Al-Sultan S, Al-Doori MM, Al-Bayatti AH, Zedan H (2014) A comprehensive survey on vehicular ad hoc network. J Netw Comput Appl 37:380–392
16. Hartenstein H, Laberteaux K (2009) VANET: vehicular applications and inter-networking technologies
17. Soua A, Afifi H (2013) Adaptive data collection protocol using reinforcement learning for VANETs, pp 1041–1045
18. Wu C, Yoshinaga T, Ji Y, Murase T, Zhang Y (2016) A reinforcement learning-based data storage scheme for vehicular ad hoc networks. IEEE Trans Veh Technol 66:6336–6348
19. Zhou S, Sun Y, Jiang Z, Niu Z (2019) Exploiting moving intelligence: delay-optimized computation offloading in vehicular fog networks. IEEE Commun Mag 57:49–55
20. Tian J, Han Q, Lin S (2019) Improved delay performance in VANET by the priority assignment. IOP Conf Ser Earth Environ Sci 234. Institute of Physics Publishing
21. Althamary I, Huang C-W, Lin P (2019) A survey on multi-agent reinforcement learning methods for vehicular networks, pp 1154–1159

22. Pressas A, Sheng Z, Ali F, Tian D (2019) A Q-learning approach with collective contention estimation for bandwidth-efficient and fair access control in IEEE 802.11p vehicular networks. IEEE Trans Veh Technol 68:9136–9150
23. Qi Q, Wang J, Ma Z, Sun H, Cao Y, Zhang L, Liao J (2019) Knowledge-driven service offloading decision for vehicular edge computing: a deep reinforcement learning approach. IEEE Trans Veh Technol 68:4192–4203
24. Zhang D, Yu FR, Yang R (2018) A machine learning approach for software-defined vehicular ad hoc networks with trust management
25. Liang L, Ye H, Li GY (2019) Toward intelligent vehicular networks: a machine learning framework
26. Luo C, Ji J, Wang Q, Chen X, Li P (2018) Channel state information prediction for 5G wireless communications: a deep learning approach. IEEE Trans Netw Sci Eng
27. Joo J, Park MC, Han DS, Pejovic V (2019) Deep learning-based channel prediction in realistic vehicular communications. IEEE Access 7:27846–27858

The Data Vortex Switch Architectures—A Review

Amrita Soni and Neha Sharma

Abstract All optical packet switching is the most desirable and faster switching technology that enhances the speed and data transmission rate. In order to provide high performance communication networks with high throughput, low latency, high terminal reliability, high component reliability, low cross talk, low ASE noise, low bit error rate, and high fault tolerance capability for all optical packet switched networks, a novel interconnection architecture, namely, the Data Vortex (DV) switch was explored. This paper presents a thorough review of various Data Vortex architectures and a comparison of these architectures on the basis of different performance parameters. It may help to provide an effective solution for proper choice of Data Vortex architecture as a sub-network for next-generation Data Center Networks (DCNs) or High Performance Computer Systems (HPCs) architectures.

Keywords Augmented Data Vortex (ADV) switch · Data Vortex switch · High-performance computing · Modular DV · Original Data Vortex

---

A. Soni (✉) · N. Sharma
Department of Electronics and Communication, Ujjain Engineering College, Ujjain, Madhya Pradesh 456010, India
e-mail: amuujn@uecu.ac.in

N. Sharma
e-mail: nehatripathi@uecu.ac.in

P. Chatterjee et al. (eds.), *Computational Intelligence for Engineering and Management Applications*, Lecture Notes in Electrical Engineering 984,
https://doi.org/10.1007/978-981-19-8493-8_25

1 Introduction

Nearly in all large scale high performance workloads, in large DCNs, core routers (used in telecommunication), supercomputers, and in parallel and distributed HPC systems, interconnection network plays very important role to achieve the desired goals such as high bandwidth, high throughput, and low latency [1]. To provide higher bandwidth in optical packet switching (OPS) networks, buffering, scalability, and latency are the key designing parameters. If we want to take maximum advantage of fiber optic technology, then the use of optical buffering, slow switching algorithms, and complex routing should be avoided.

The optical multistage interconnection networks (OMINs) based on dense wavelength division multiplexing (DWDM) technique support transmission of Tb/s of bandwidth and have reduced limitations of electronic communication [2]. However, some new challenges are introduced by the constraints of optical component technique such as path distance losses, inefficient buffering, optical crosstalk, ASE, and a bit level processing in optical domain. Therefore, in a single all OPS network, it is tough to have at the same time all the qualities, i.e., scalability, low bit error rate, low latency, low cross talk, and high capacity due to above stated limitations [3]. With the drastic growth in technology-based applications such as cloud computing, mobile consumption and multimedia information, the large DCN have become very important computing infrastructure [4–6]. However, the main challenge for these applications is to build an interconnecting network which is based on fast optical switching technique with the advantage of high bandwidth, is lack of optical buffers for packet congestion [7]. The all optical interconnection network, namely, the Data Vortex (DV) can provide a viable solution to overcome the above stated challenges by offering high throughput, ultra-low latency, ultra-high capacity, low optical crosstalk, low BER, and reduced power consumption and also by providing buffer-less operation, i.e., deflection method for routing of packets to avoid packet contention.

The main objectives of this review paper are as follows:

- A detailed investigation on traditional DV switch architectures.
- A detailed qualitative comparison for various DV switch architectures on different performance parameters which may help to provide an effective solution for proper choice of DV architecture as a sub network in the future next generation applications.

This paper presents a detailed review of important architectures of DV switch that have been recently presented in the research literature, as suitable all optical interconnection networks. Since it is a very broad area, we have tried to study those Data Vortex architectures which have had or expected to have, in our knowledge, a significant role in all optical packet switched networks. Section 2 presents the Data Vortex Switch architecture and its various hierarchical models. Section 3 discusses the comparison of various architectures, and is presented in Table 1. And finally Sect. 4 concludes the findings of the review and directions for future work.

2 Data Vortex Switch Architecture and Its Various Hierarchical Models

A step-by-step evolution of the successive models is reviewed in this section.

2.1 *Original DV Switch Architecture*

DV is a multistage routing network made of 2×2 routing nodes (switching elements). There are multiple interconnecting links provided between source and destination node pair. Thus, the DV architecture was in a way that it has inherent fault tolerant capability and reliability. Yang et al. explored this new switching architecture by excluding the usage of internal buffering and reducing the count of switching and logic operations. This switching architecture is a hierarchical model, and to avoid contention, it employs synchronous packet clocking and distributed control signaling arrangements. Therefore, as compared to old topologies, data flow and node design were greatly improved. It was assumed that the packets were having similar size and alignment in timing when they were being injected at the input port. For detailed architecture refer [8, 9].

In DV, routing nodes reclines on collection of concentric cylinders. The architecture was described by some important designing parameters like height parameter 'H' which corresponds to the count of nodes existing alongside the cylinder height and angle parameter 'A' which corresponds to the count of nodes along the periphery, with value of 'A' is kept a tiny odd count (< 10). For each concentric cylinder, total count of nodes is $A \times H$. Total count of cylinders is $C = \log_2 H + 1$, as shown in Fig. 1. Thus, total numbers of nodes in the architecture are $A \times H \times C$. Each routing node is identified on cylinder using (a, c, h) parameters, where $0 \leq a < A$, $0 \leq c < C$, and $0 \leq h < H$. All the packets get in from $c = 0$ cylinder level and emerge at $c = \log_2 H$ cylinder level. Thus, every packet is automatic routed in binary tree decoding fashion toward its output port [8–12].

The main advantage of Original Data Vortex architecture is that it uses deflection method for routing of packets and adopts virtual buffering mechanism to get hardware simplicity, high injection rate, and scalability. Additionally, wavelength division multiplexing (WDM) is used to encode the packet payload and header bits to enhance the throughput and reduce the switching latency, respectively. It also simplifies the routing strategy.

The drawback of Original Data Vortex (ODV) was that it used $LiNbO_3$ switching-based routing node [9] which was not an economical application due to problem of integration and huge insertion loss for practical large-scale switching fabrics. Therefore, to simplify the problem, semiconductor optical amplifier (SOA) gate-based node with a same switching speed as used in place of $LiNbO_3$ switching-based routing node. It provided higher ON–OFF ratio in addition to integration advantages within the similar type of device.

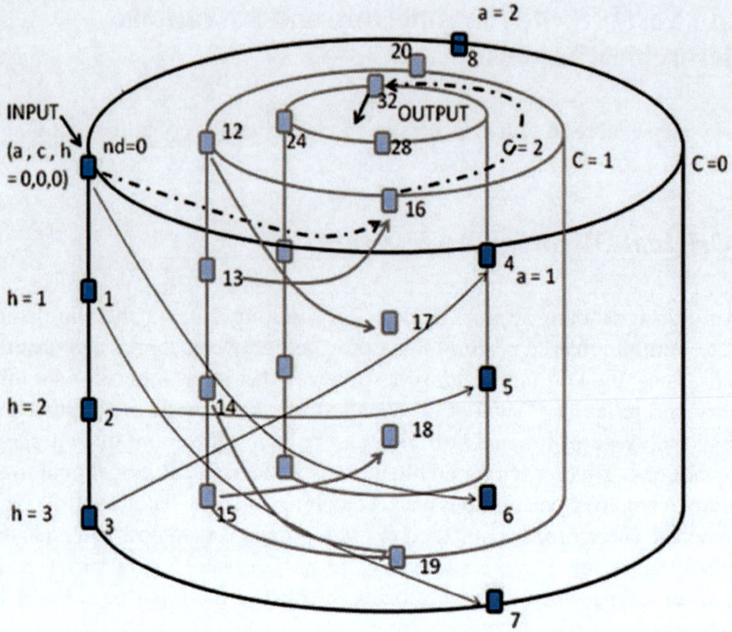

Fig. 1 Interconnection between neighboring cylinders of DV switch for $A = 3$, $C = 3$, $H = 4$ [8, 9]

2.2 *Equivalent Chained (EC) Planar (MIN) Model of the Multiplexed DV Architecture*

For any interconnection network, the main issue is fault tolerance analysis, assessment of which lacked attention for the ODV switch. To analyze FT, a simple model was required, which was proposed in Ref. [13] and is discussed in this section.

In ODV network, each of external input (or output) device is interconnected with its respective input/output link at individual input (or output) node. If among these links any link fails, then external input (or output) device disconnects from the network in spite of fault tolerance introduced within the network. To resolve this problem, a new architecture was proposed with multiplexed input (output) devices through extra switches. The proposed idea of multiplexing at input and output has improved fault tolerance. The converted 3D architecture of DV into its equivalent chained MIN (planar) model was introduced for the first time in Ref. [13] and is shown in Fig. 2.

The main advantage of this conversion is simplification of performance analysis of the network, due to which its topological comparison is easy with different MINs.

The main drawback is that it has analyzed the fault tolerance and reliability only by analytical method. It could be analyzed by simulation method and experimental test bed to compare the results.

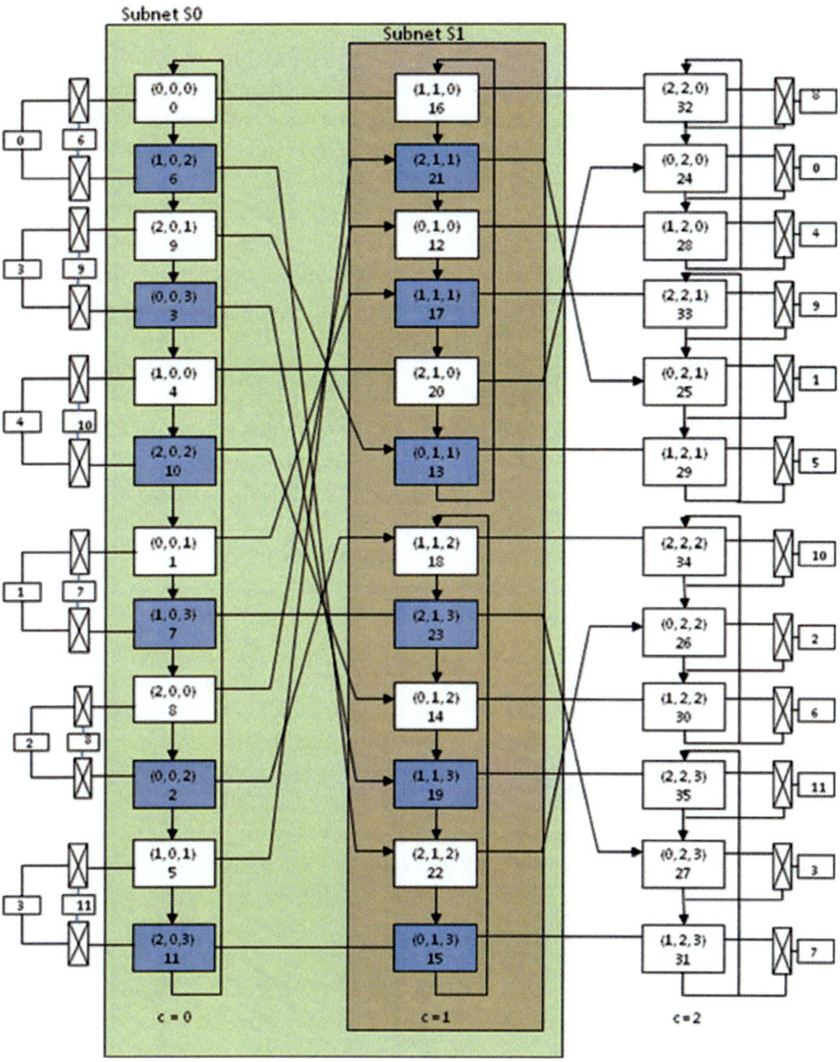

Fig. 2 EC planar model for the multiplexed DV network for $A = 3$, $C = 3$, $H = 4$ [13]

2.3 ADV Switch Architecture

Again, the challenge for researchers was to further improve the fault tolerance capability of the ODV switch. It was resolved by modification in the architecture of DV by introducing an augmented link among the inter-cylinder stages of DV, and is explained in this section.

A new architecture was proposed named as Augmented Data Vortex (ADV) architecture which was an enhancement of the design and architecture of the original DV architecture [3, 14]. In this architecture, extra (augmented) links are provided between the cylinder stages. In this way, the number of paths to route the packet between distinct cylindrical levels also increases hence fault tolerance and reliability both are improved. As compared to original DV switch architecture, due to extra augmented link at each switching node, it works even after failure of two output links. Apart from extra links, multiplexing method at input and output (I/O) ports is also provided which again enriches the FT capability of the network. In addition, a novel self-routing method and priority method are also proposed for distributed control signaling. The equivalent planar diagram of the 3D switch is also presented. In Ref. [3], it is observed that with the additional routes, the throughput is also increased and average latency is reduced.

Labeling scheme for original DV switch and ADV switch is same for comparison purpose. In ADV switch, as shown in Fig. 3, collection of 3×3 (i.e., three inputs and three outputs) routing nodes are arranged on multilayer fiber cylinders. Here, 'A' and 'H' are determining the switch fabric size. The value of 'A' is always kept to small odd count (< 10), and it does not depend on choice of 'H'. In this architecture, available count of I/O ports can be represented by the formula $A \times H$, where $C = \log_2 H + 1$, represents the total number of cylinder levels which are numbered from 0 to $\log_2 H$. Here, 0 represents the ingress (input) level, and $\log_2 H$ is an exit (output) level. Each routing node is represented by labeling the coordinates as (a, c, h), where $0 \le a < A$, $0 \le c < C$, and $0 \le h < H$. Also, a new numbering scheme is proposed to identify each node with its single decimal equivalent address in which each node address (a, c, h) is altered to its decimal equivalent presentation 'nd' (ndNode). Here, ndNode $= H \times a + (A \times H) \times c + h$.

In ADV switch, each node has three inputs and three outputs, and to get without buffer functioning and easy routing logic, only one incoming packet is allowed to enter through a node in a particular clock cycle. It may be possible, all the three inputs of a node are competing for common targeted node accessibility, and in that case, conflict may occur.

Thus, to avoid this situation, a modified priority scheme is adopted to convey message by using separate control bits, apart from header and the payload bits. By implementing this scheme, more than two incoming packets never dispute for common outgoing terminal. Thus, it also eliminates the need of buffer at every node. It oversimplifies the logic operation at every node of the architecture. Detail of priority scheme, new routing scheme, and new equivalent chained MIN Model are given in [3, 14]. In an Original Data Vortex switch, the blocked packet has minimal deflection penalty of 2 hops to reach its destination output port, while in ADV switch, this is only one hop.

The advantages of proposed ADV architecture are that it shows low latency, improved fault tolerance and reliability. In Data Vortex, if one path is faulty or not working due to any reason, then after 2 hops the required target pathway is accessible, but for similar case in Augmented Data Vortex, next desired path can be

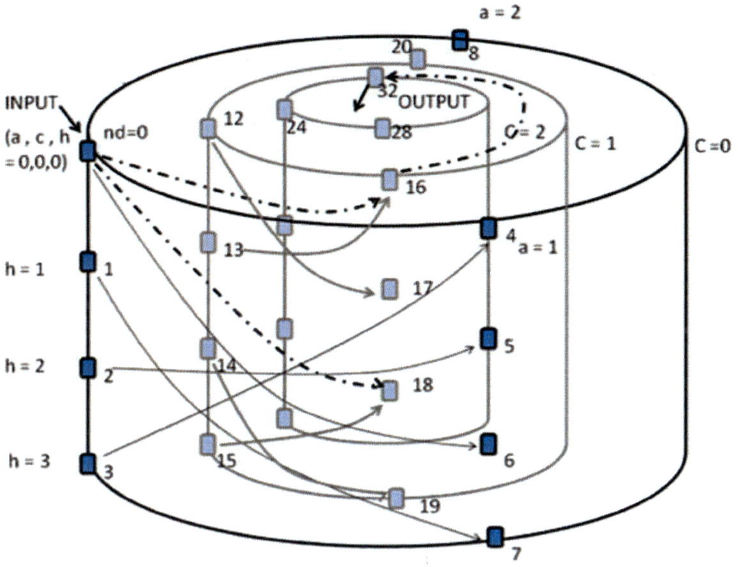

Fig. 3 Interconnection arrangement of ADV architecture for $A = 3$, $C = 3$, $H = 4$ [3]

found itself after only one hop. ADV also shows improvement in average latency, latency distribution, and injection rate which were computed by simulation method.

The drawback of ADV is that due to extra link provided between inter-cylinder pathways, the cost of the switching element is greater than original DV architecture. Either we choose good performance parameters or the cost, depends on the requirement of the application in which the architecture would be used.

2.4 4 x 4 Optical Data Vortex Switch Architecture

To further improve the performance parameters of ADV such as fault tolerance, reliability, injection rate, and average latency the enhancement of ADV was proposed. This is included in this section.

Further enhancement of 3 × 3 ADV was suggested in the form of 4 × 4 Optical Data Vortex (4 × 4 ODV) switch [15]. Figure 4 is showing the 4 × 4 ODV framework. Similar to DV switch, packets enter from outmost cylinder level (CL) ($c = 0$) and exit from inmost CL ($c = \log_2 H$). The node arrangement along with the periphery of the cylinder is represented as Height = H, Angle = A, and total count of Cylinder levels = C. The remaining architecture, numbering scheme and labeling scheme used is same as DV and ADV with one extra input and output link as compared to ADV.

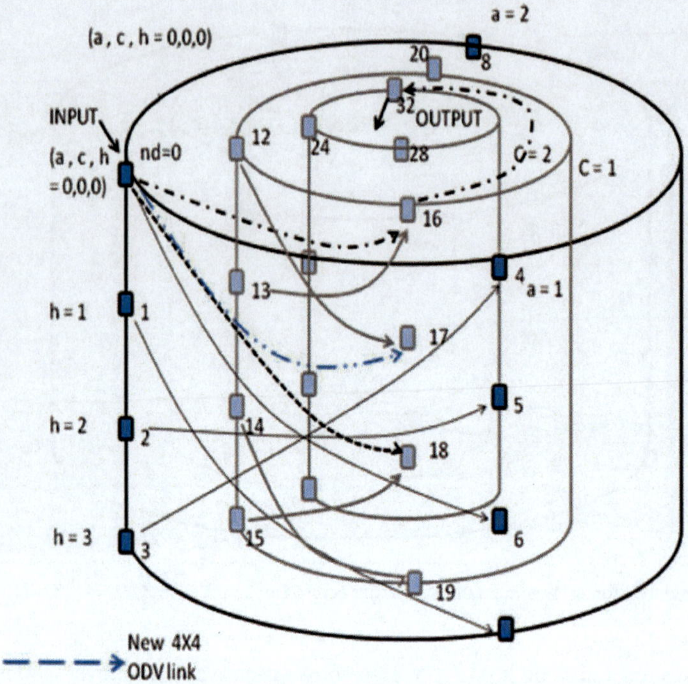

Fig. 4 4 × 4 Optical Data Vortex switch architecture [15]

In 4 × 4 ODV architecture, there are four input links and four output links; one pathway link is interconnected between them to same cylinder level, and three pathways are connected to next adjacent cylinder level toward destined output ports. Thus, in this architecture, number of input and output links is increased which increases the availability of the output terminals also.

In 4 × 4 ODV, packets follow self routing scheme and proper synchronization. As investigated in original DV, in 4 × 4 ODV architecture each node allows only one packet at a time. Similar to DV, each packet contains header bits, frame bit, and payload. Header bits uses WDM technique, in which each bit is decoded with distinct wavelengths. These different wavelengths represent different addresses. The frame bit is utilized to recognize the existence of the packet. In each time slot, packet moves in advance either in current CL or in next or adjacent CL, with one angle forward, i.e., from 'a' to '$a + 1$'.

When packet deflects on current cylinder level, it proceeds from node address (a, c, h) to node address ($a + 1$, c, $h$nxt). In 4 × 4 Optical Data Vortex switch architecture, three interconnection links are provided to route the packets for next cylinder level towards its destination node at three different heights.

The main advantage of 4 × 4 ODV is that it provides increased path to all the nodes thereby enhances the accessibility to the entire destinations. With the multiplexing at I/O ports the FT has also been further improved. Due to the increased count of components in each node, it has improved the node reliability as compared to original DV (ODV) and augmented DV.

The drawback of 4×4 ODV is that extra links between inter-cylinder stages increases the hardware cost also.

2.5 k-ary Data Vortex Switch Architecture

k-ary Data Vortex (k-ary DV) architecture was designed from original DV by using k-ary decoding at each stage [16]. In original DV, binary tree decoding is used which means the value of k is 2. The binary tree decoding uses one header bit at each node for decoding and when routing on any specific cylinder it selects one out of two groups, i.e., either upper or lower group. In this method, deflection latency penalty (DLP) is 2 hops while in case of k-ary DV architecture, k-ary decoding scheme is used, in which $\log_2 k$ bits are decoded at each stage. In this scheme, each hop allows the packet to select one out of k groups along the same cylinder level as shown in Fig. 5. When the targeted destination address ($\log_2 k$) header bits and routing node height address ($\log_2 k$) bits are matched, the packet will enter to the inner CL if the correspondent traffic having non-blocking pathway at the similar time; otherwise, packet deflects on the same CL. In k-ary DV architecture, total count of cylinders is $C = \log_k H + 1$. Hence with large value of k, forward latency is still much smaller. The value of $k = 4$ is taken for comparison with original DV switch, where each hop decodes two header bits in an architecture of $A = 4$, $H = 16$. The count of cylinders is $C = \log_4 H + 1 = 3$, as shown in Fig. 5. To design the 4-ary decoding network, the interconnection arrangement at various cylinders in the ODV network are incorporated and conversed at appropriate angle. It is found that for larger value of k ($k > 4$) there is large deflection latency and traffic backpressure. The DLP increases to k hops in place of two hops in the original DV architecture [16].

The advantage of k-ary Data Vortex is that it requires less number of cylinders for same switch fabric when compared with Original Data Vortex. For example, for $k = 4, A = 4$ and $H = 16$ the value of $C = 3$. While for $A = 4$ and $H = 16$ the value of $C = 5$ in case of Original Data Vortex architecture.

The k-ary network shows slight reduction in average latency, but its main drawback is deflection-induced traffic backpressure which limits the throughput and makes it less desirable.

2.6 Reverse Data Vortex (RDV) Switch Architecture

In proposed [17] architecture of Reverse Data Vortex (RDV), interconnection pattern between routing nodes is similar to DV, but the orientation of data flow has been reversed. In RDV architecture (as shown in Fig. 6), packets intervene from $c = \log_2 H$

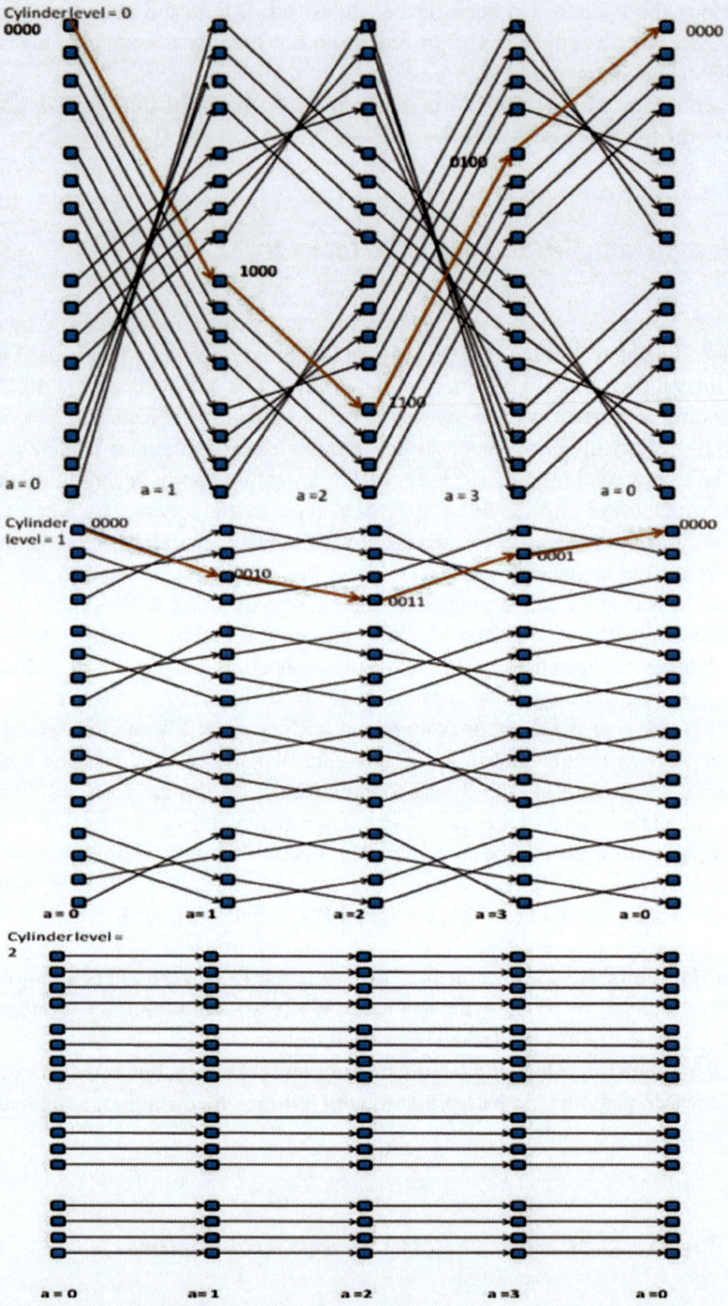

Fig. 5 Routing arrangements at each level of cylinder in a 4-ary decoding DV network. $A = 4$, $H = 16$, $C = \log_4 H + 1 = 3$ [16]

cylinder level and egress from $c = 0$ cylinder level. The count of subgroups is 2^c, and total count of nodes in every subgroup is 2^N c, respectively. In DV switch, each packet enters from $c = 0$ (outmost) cylinder level and through the intermediate nodes exits from innermost cylinder level. When packet is routed, if some intermediate nodes are not working due to fault, then certain nodes from $c = \log_2 H$ level cannot have reachability to desired destination output port. Thus, to interconnect all the target nodes from their corresponding root node, RDV switch architecture was proposed. Here, the total number of cylinders is $C = \log_2 H + 1$ and the total count of nodes is $A \times C \times H$, respectively.

The node counts are labeled as (a, c, h) designing dimensions, and their decimal equivalence presentation is $d = H \times a + (A \times H) \times c + h$. In RDV, header bit decoding and control signaling are done in reverse direction. Here, priority scheme is also modified as compared to DV. In DV, if directing pathway of next cylinder is not accessible due to some reason, then packet is deflected to the current pathway of cylinder level, whereas in RDV, if routing path is not available, then packet is deflected to adjacent cylinders.

Each node in the next (adjacent) CL contains a specific pathway to connect its desired node as all the nodes are linked in outmost cylinder level. Hence, here, higher priority is given to adjacent cylinder level unlike DV. Routing decision is done by using the control signaling from the neighboring cylinder level.

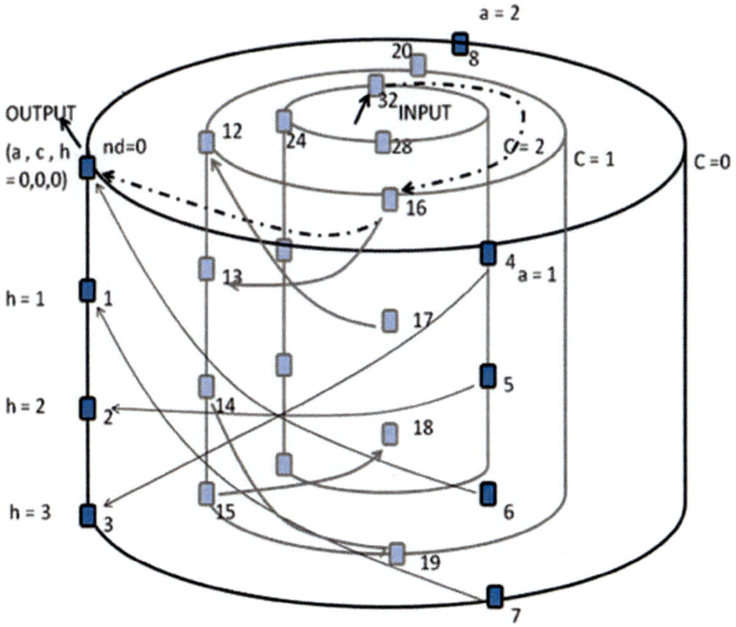

Fig. 6 Reverse Data Vortex architecture [17]

The advantage of proposed RDV architecture is that it shows improved fault tolerance and reachability. In Data Vortex, if one path is faulty or not working due to any reason then after 2 hops the required target pathway is accessible, whereas in RDV for similar case only 1 hop is required. Therefore in RDV, only one hop latency penalty exists which reduces the requirement of buffering and hence backpressure also reduces. Here, average latency reduces, and comparable throughput is obtained. In case of any fixed architecture like $A = 3$, and $H = 4$ for both Data Vortex and Reverse Data Vortex, the count of nodes and count of pathways remain similar and thus reliability of the node, terminal reliability, and path reliability also remains the same [17].

The disadvantage of RDV is that it shows slightly greater routing complexity as compared to original DV.

2.7 Planar Layout (PL) of Data Vortex Switch Architecture

In PL of Data Vortex switch [18], a new planar structure is proposed in which 3D topology is converted to multiple planes of routing levels to support construction and integration for physical implementations. In planar structure, same semi-twisted pattern is used to minimize deflection probability with advantage of easy layered integration for parallel planes. Here, unlike original DV switch architecture, paths are added along the same angle at the first and last angle as shown in Fig. 7, and half of the routing paths traveling in opposite directions. Due to this arrangement (traveling in same plane), it forms a looping structure same as original DV switch architecture.

In this new architecture design, different routing level parallel planes correspond to the different cylinder levels of original DV architecture. Only parallel routing paths are required between different planes. For more clarity, paths of control signals are not shown. In planar system, output nodes are located on same angle and control signals are applied on edge angle. For more detail Ref. [18] can be referred. With the help of planar optical waveguide technique, all nodes which have same plane are manufactured on an identical board or if angles are connected with a flexible waveguide, then identical angle nodes can be manufactured on an identical board. Logically, planar layout of DV architecture contains similar connections as that of original DV architecture. Parallel inter stage forwarding path, control signaling system, and additional last cylinder are also same, hence similar routing performance also expected in planar layout when compared to original DV architecture; but in planar system, new same angle path affects the routing performance. The reason is that the nodes which have different angles carry different load, and hence, there is different traffic distribution in the entire network. Thus, this fact should be investigated in detail also for comparison in architectures.

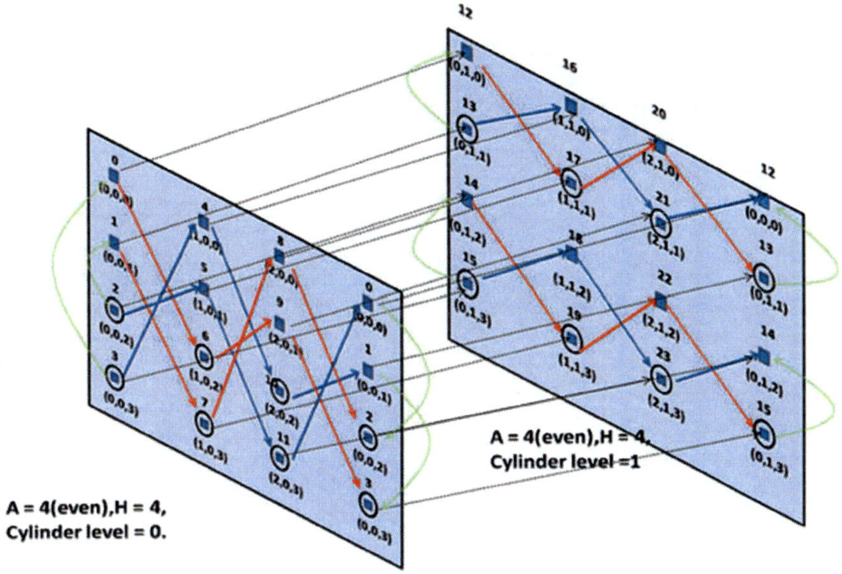

Fig. 7 Planar layout (PL) of Data Vortex switch architecture [18]

2.8 *Modular Data Vortex Switch Architecture*

In proposed architecture of Modular DV switch architecture [19], clos network and DV switch topology are successfully combined. Modular design focuses on use of single angle design, in which each sub network is defined by $H \times H$. In original DV architecture, $\log_2 H$ header bits are used for decoding, whereas in Modular DV architecture, $\log_2 H$ bits break into multiple stages. In each stage, as shown in Fig. 8, small DVs are used which decodes subset of the header bits in each stage. Stages are connected with each other similar to clos network. The cost of modular switch architecture may differ from original DV switch because it depends on use of final buffering cylinder in each sub network. The cost also depends on total counts of nodes requirement and links to cover the entire architecture. Modular DV switch architecture is shown in Fig. 8.

In addition, effective hybrid switching is also introduced in Modular DV architecture to obtain better quality of service (QoS) within the OPS network. In hybrid switching network architecture, during the transmission of packets the circuit request is established and route is reserved. It means that circuit traffic is sharing the resources of regular packet traffic. Advantage of using this architecture is versatility and easy adaptation for various HPC applications requirements [20].

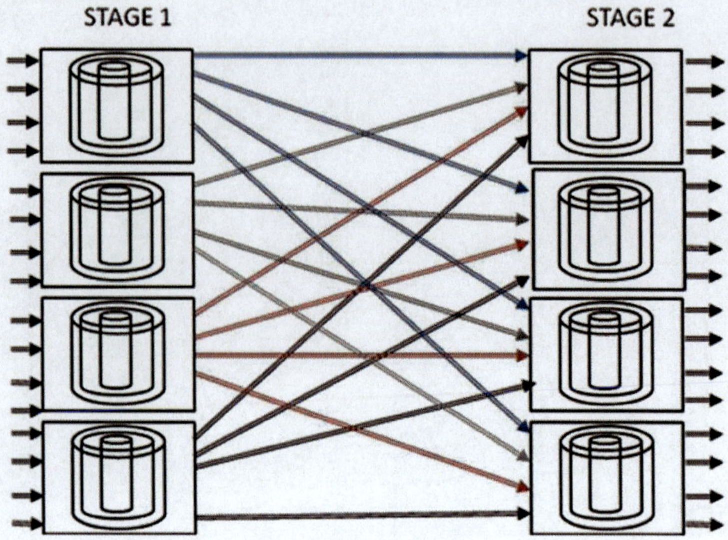

Fig. 8 Modular Data Vortex switch architecture (16 × 16 designed from two 4 × 4 DVs in 2 stages) [19]

3 Comparison of Various Architectures

Comparison for various architectures of DV switch is given in Table 1. Based on the study from earlier sections, we have prepared a comprehensive comparison chart for various architectures of Data Vortex compared on the basis of different performance parameters. This chart represents the characteristics of various DV architectures in a nutshell.

Original DV switch architecture uses 2 × 2 switching node design. The traffic is self-routed in binary tree fashion toward its output port. Its hardware complexity is normal. As compared to other topologies like butterfly, banyan, shufflenet, and debruijn, the fault tolerance capacity is large. If a packet is unable to go at destined output port, it is routed to another node by deflection routing mechanism and after 2 hops it reaches to its targeted output port. Thus latency penalty is of 2 hops.

The equivalent chained MIN model of DV is 2D planar conversion from 3D architecture of ODV. It also uses 2 × 2 switching node design. Hardware requirement is slightly greater than ODV due to multiplexing at I/O ports with improved FT. The latency penalty is same as of ODV. The main advantage of this type of conversion is simplification in performance analysis of the network due to which its topological comparison is easy with other MINs.

The ADV architecture uses 3 × 3 switching node design. In the equivalent chained planar prototype of ADV architecture, multiplexing and de-multiplexing are also provided; thus, on the first layer, it has four inputs at every switching node, and the number of output per node is three. The use of multiplexing and de-multiplexing at

Table 1 Comparison of different Data Vortex architectures

S. No.	Architecture	Switch type	I/P elements per root	O/P elements per target	Routing complexity	Hardware complexity	Latency penalty	References
1	Original DV	2 × 2	2	2	Simple	Normal	02 hops	[8–12]
2	Equivalent chained MIN model of DV	2 × 2	2	2	Simple	Not complex	02 hops	[13]
3	ADV	3 × 3	4	3	Comparable	Slightly greater than DV	01 hop	[3, 14]
4	4 × 4 ODV	4 × 4	5	4	Complex than DV	Greater than DV and ADV	01 hop	[15]
5	k-ary DV	$k \times k$	k	k	Same as DV	Greater than DV	k hops	[16]
6	RDV	2 × 2	2	2	Slightly greater than DV	Normal	01 hop	[17]
7	Planar DV	2 × 2	2	2	Same as DV	Slightly greater than DV	Same as DV	[18]
8	Modular DV	16 × 16	16	16	Same as DV	Greater than other architectures	–	[19, 20]

input and output ports, respectively, with an extra augmented link between the inter cylinder stages increases the fault tolerance capacity, but with penalty of hardware requirements as compared to ODV. The traffic is routed with a modified priority scheme. Always three inputs are competing for a destined common node this makes the routing comparable with ODV. When output port is not available due to blocking of node the packet deflects in the same cylinder and after 1 hop again.

The 4 × 4 ODV is an enhanced version of ADV. It uses 4 × 4 switching element. In comparison with original DV, two extra inputs and outputs are added at each switching node. Thus fault tolerance increases, but it also increase the hardware and routing complexity. The latency penalty is 1 hop which is same as ADV. Due to extra inter cylinder links and multiplexing and de-multiplexing at input and output ports respectively fault tolerance improves, but hardware requirement is also increased and results in increased cost.

The k-ary DV uses $k \times k$ switching node, with the variable value of k. The k-ary DV architecture allows the packet to select one out of k groups along the same cylinder level. If targeted output port is unavailable due to some reason then it deflects the packet with k-hops in the current cylinder. Hence, latency penalty increases to k hops. The rest of the routing mechanism is same as ODV. The routing process is slightly difficult than ODV. Hardware requirements are also increased with the increased value of k.

The RDV architecture again uses 2×2 switching element. In RDV, header bit decoding and control signaling are done in reverse direction. Hardware requirement is same as ODV. At the time of traffic flow if directing path of next cylinder is not accessible due to blocking of node then packet deflects to the adjacent cylinder. This process is just reverse of routing mechanism of ODV; therefore, we can say that routing complexity is slightly greater than ODV. Here, due to reverse direction of traffic and reverse routing mechanism the latency penalty is 1 hop. Thus, average latency reduces and fault tolerance improves as compared to ODV. The count of nodes and pathways are same as ODV therefore reliability of node, terminal reliability and path reliability are also same as ODV.

The planar DV architecture also uses 2×2 switching element. The routing mechanism is same as ODV. In planar DV, new same angle path affects the routing performance because the node which contains different angles, carry different load, and hence, it leads to different traffic distribution in the entire network. Thus, we can say that due to the arrangement of 3D topology into parallel planes, the hardware requirement is slightly greater as compared to ODV. Here, the latency penalty is same as ODV.

The Modular architecture of DV is a combination of clos network and DV topology. Here, 16×16 DV network is built from 4×4 Data Vortex in two stages, each stage decodes half of the header bits. If the buffering cylinder is removed from each sub DV then the cost of the non-buffered modular DV and non-buffered DV remains the same. The routing behavior is same as that of ODV. The combination of clos network and DV switch requires more nodes and links which are necessary to fully connect all the nodes. Thus, in this architecture, hardware complexity is greater than other tabulated architectures.

4 Conclusion

This paper presents a review which gives an upgraded and detailed investigation on traditional Data Vortex switch architectures. Apart from this, other critical issues which are related to proper choice of Data Vortex switching architectures are examined such as injection rate, average latency, latency penalty, fault tolerance capacity, hardware complexity, routing complexity, cost, input/output requirements, and implementation technology. For different types of architectures, modified routing mechanism and performance evaluation of parameters are also compared and discussed. Additionally, a detailed comparison for various Data Vortex switch architectures on different performance parameters is also presented in tabular form which may help to provide an effective solution for proper choice of DV architecture as a sub network in future next generation large DCNs and HPC applications.

References

1. Nezhadi A, Koohi S (2019) OMUX: optical multicast and unicast capable interconnection network for data centers. J Opt Switch Netw 33:01–12. https://doi.org/10.1016/j.osn.2019.01.002
2. Meyer H, Sancho JC, Mrdakovic M, Miao W, Calabretta N (2017) Optical packet switching in HPC. An analysis of applications performance. Future Gener Comput Syst
3. Sharma N, Chadha D, Chandra V (2007) The augmented data vortex switch fabric: an all-optical packet switched interconnection network with enhanced fault tolerance. J Opt Switch Netw 4:92–105
4. Minakhmetor A, Ware C, Iannone L (2020) Hybrid and optical packet switching supporting different service classes in data centre networks. Photon Netw Commun 40:293–302
5. Soto KI (2018) Realization and application of large scale fast optical circuit switch for data centre networking. J Lightwave Technol 36(7)
6. Lallas EN (2019) A survey on key roles of optical switching and labelling technologies on data traffic of data centers and HPC environment. AIMS Electron Electr Eng 3:233–256. https://doi.org/10.39341/ElectrEng.2019.3.233
7. Yan F, Yuan C, Li C, Deng X (2021) FOSquar: a novel optical HPC interconnect network architecture based on fast optical switched with distributed optical flow control. Photonics. https://doi.org/10.3390/photonics8010011
8. Yang Q et al (2000) Optics in computing, vol 4089. SPIE, p 555
9. Yang Q, Bergman K, Hughes GD, Johnson FG (2001) WDM packet routing for high-capacity data networks. J Lightwave Technol 19:1420–1426
10. Reed C (1999) Multiple level minimum logic network. U.S. patent 5,996,020, 30 Nov 1999
11. Yang Q, Bergman K (2002) Traffic control and WDM routing in the data vortex packet switch. IEEE Photon Technol Lett 14:236–238
12. Yang Q, Bergman K (2002) Performance of the data vortex switch architecture under nonuniform and bursty traffic. J Lightwave Technol 20:1242–1247
13. Sharma N, Chadha D, Chandra V (2006) Fault tolerance in a multiplexed data vortex all-optical interconnection network. In: National conference on communications, NCC 2006. IIT Delhi, India, pp 27–29
14. Sharma N, Chadha D, Chandra V (2007) Performance evaluation of the augmented data vortex switch fabric: an all optical packet switching interconnection network. J Opt Switch Netw 4:213–224
15. Sangeetha RG, Sharma N, Chadha D, Chandra V (2008) 4 × 4 optical data vortex switch fabric: a fault tolerant study. In: International conference on photonics, New Delhi, India, 13–17 Dec 2008
16. Yang Q (2009) Performance evaluation of K-ary data vortex networks with bufferless and buffered routing nodes. In: Asia photonic and communication conference (ACP), Shanghai, pp 1–2
17. Sangeetha RG, Chandra V, Chadha D (2010) Optical interconnection reverse data vortex network. In: International conference signal processing and communications, Bangalore, pp 1–5
18. Yang Q (2012) Performance evaluation of a planar layout of data vortex optical interconnection network. In: The second international conference on advanced communications and computation, IARIA
19. Burley B, Yang Q (2012) Modular design of data vortex switch network. In: Proceeding of the ACM/IEEE symposium on architectures for networking and communications systems
20. Yang Q (2014) Hybrid switching in modular data vortex optical interconnection network. In: 16th international conference on transparent optical networks (ICTON), pp 1–4. https://doi.org/10.1109/ICTON.2014.6876471

Survey on Genomic Prediction in Biomedical Using Artificial Intelligence

Shifana Rayesha and W. Aisha Banu

Abstract The diverse variety of diseases creates a threat to human lives, which increases the mortality rate. The antibiotics are practiced in the medical profession to find an antidote for these diseases. The evolution of antibiotic genes prediction is necessary to find the comparable antimicrobial substance. From this antibiotic gene, we can obtain antibiotic resistance medicine to minimize the emergent global disease. The current tools to predict the genome by using big data analytics is performed by collecting the genetic information of antibiotics and match the sequence. The obstacle in the existing method is the genome which is available for the restricted family, and the obtained antibiotic resistant genome is challenging to discover that it has the immunity to struggle the antibiotic. The approaches used to predict the antibiotic gene include imaging freely moving bacterial cells and evaluating the recordings using a deep learning system. The second method is by utilizing the bacterial genetic information to predict the genetic sequence the different diverse array of microorganisms. In the next model, they have enhanced the existing drug by learning about the correct genomic chemical structure that presents in microorganisms. Final techniques are to forming frame sequencing facilitates linear access and perform DNA pooling with all metagenomic data, where antibiotic gene resistance is regularly distinguished or anticipated dependent, on the "fittest scores" of grouping look into the actual databases. From the above model, we reviewed the papers to bring the optimized solution, alternative solutions, or improve the efficiency of the algorithm to predict the antimicrobial genetic substance.

Keywords Phenotypic antimicrobial susceptibility · Metagenomic · Variant mapping · Antibiotic resistances · Antibiotic gene resistance · KEGG orthology database · Next generation DNA sequencing · Card · ARDB · UniProtKB

S. Rayesha (✉) · W. Aisha Banu
B S Abdur Rahman Crescent Institute of Science and Technology, Chennai, India
e-mail: shifana.nt@crescent.education

P. Chatterjee et al. (eds.), *Computational Intelligence for Engineering and Management Applications*, Lecture Notes in Electrical Engineering 984,
https://doi.org/10.1007/978-981-19-8493-8_26

1 Elucidation

In the most recent decade, we have seen a dramatic increment both in the extent and supreme number of bacterial pathogens introducing multidrug protection from antibacterial agents [1]. Yearly, more than 50,000 new born are assessing to pass on from sepsis because of microorganisms impervious to first-line antimicrobials. Antibiotics are medicines used to treat or prevent infection caused by bacteria [2]. Anti-infection opposition is the progressions or changes in the DNA of the microscopic organisms or the securing of antimicrobial obstruction qualities from other bacterial species through the level of quality exchange [3]. Antibiotic resistance occurs when microscopic organisms build up the capacity to endure presentation to anti-infection agents are intend to slaughter them or stop bacterial contamination development [4]. The application has emerged in current years and addresses the drug discovery problem [5]. Using deep learning techniques, we predict the antibiotic gene of microorganisms with the help of genetic information of antibodies, the model able to determine the character, and to identify the specific bacteria among different classes that are used in the field of medicine. This model used to reduce the minimal time and real-time analysis to find out the antimicrobial classes [6]. Deep learning techniques to predict the antibiotic genome is categorized into four different types (Fig. 1).

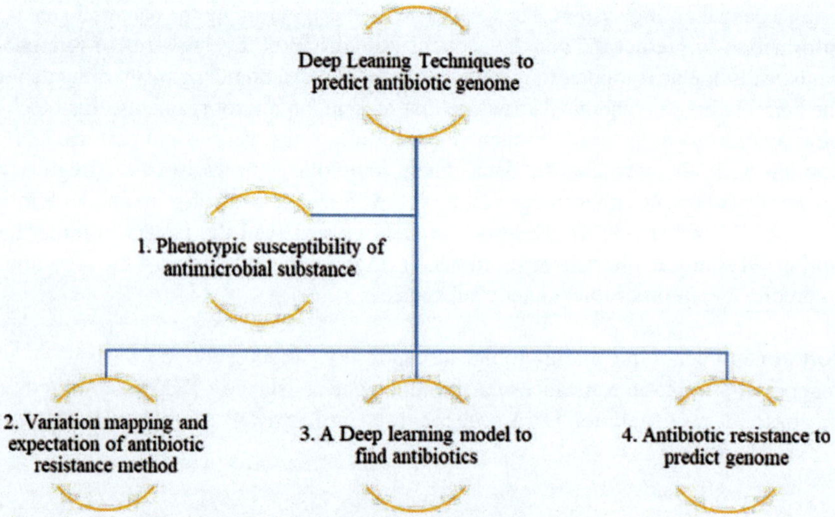

Fig. 1 Deep learning classification method to predict antibiotics

1.1 Phenotypic Susceptibility of Antimicrobial Substance

Using the real-time devices, the image of freely moving bacteria is recorded and analyzed. With deep learning algorithm, the determination of microorganisms inhibits a bacterial cell by learning the phenotypic features of the cell requirement for characterizing and evaluating each component [7]. In this approach, the antibiotic is predicted by the real-time image or recorded video of freely moving bacteria [8].

1.2 Variation Mapping and Expectation of Antibiotic Resistance Method

This method builds another bio-information device to encourage comprehension of the associations between DNA variety and phenotypic antimicrobial resistance, to (1) infer quality ortholog-based grouping highlights for protein variations [9]; (2) examine these reasonable quality level variations for their known or novel relationship with resistance to antibiotics [10]; and (3) predict the antibiotic resistance genomic sequence using this technique [11]. The optimized antimicrobial-resistant genome was obtained from extensive data analysis and machine learning procedures [12].

1.3 A Deep Learning Model to Find Antibiotics

In this exemplar, one molecule is selected and is predicted to determine robust antibacterial action, which consist of the different antibiotic chemical structures from an existing structure [13]. Using the various machine learning pattern, the researchers also showed that this molecule would likely produce low infections to human cells [14]. The deep learning algorithm derives the chemical structure of the antimicrobial element.

1.4 Antibiotic Resistance to Predict Genome

The profound learning approach consists of the different frameworks made utilizing a familiar gene that is resistant to antibiotics, which belongs to a single class [15]. Two profound learning models are deep learning model, using antibiotic resistance-SS and antibiotic resistance-LS, developed for a short read and comprehensively long gene sequence length [16]. DNA sequencing method mapping the antimicrobial substance's short sequence and long sequencing length is explored.

Every model is examined; from the above observation, there is specific kind of advantages and shortcomings. Therefore, by the proposed four models, they can be able to foresight the antibiotic genome for the drugs and medicine using artificial intelligence of deep learning models [14]. Future work and development of this model carried out either improve the algorithm or strategy to solve the problem that can be optimized elegantly.

2 Research Objective

To identify potential antimicrobial-resistant microorganisms based on the current chemical composition. Various Algorithm and implementation to discover the targeted antibiotic resistance genome. To identify the specific bacteria among different classes used in the field of medicine and to improve the algorithm's efficiency to predict the antimicrobial genetic substance.

3 Existing Methods to Detect the Presence of Antibiotic Elements

Over the decades, the methods to obtain the antibiotic substances are either from self-medication with natural resources or propelling techniques for checking of ecological media (e.g., wastewater, horticultural waste, nourishment, and water) [17] to identify the potential antibiotics among the different antimicrobial elements to derive new antibiotic resistance substance [18]. The actual method to discover and predict microorganisms is by laboratory-based in-depth testing of high intensities [19]. The three stages of antibiotic detection are i) natural fragmentation, (ii) semi-synthetic, (iii), and (iii) synthetic. Existing algorithms that predict the antibiotic gene are used in these three techniques.

3.1 The Antibiotic Genomic Recognition Method

Cutting edge next-generation sequencing [20] allows immediate access and profiling of the DNA pool with all metagenomic data, where antibiotic gene resistance is regularly distinguished or anticipated dependent on the "most desirable fits" of grouping look toward subsisting databases.

This approach, unfortunately, produces a high false rate of negative values [21]. There will be in mismatching of DNA sequence with the existing DNA sequence [22]. The experimental setup is expensive because of the usage of a detector and a laser.

3.2 Decision Tree-Based Model

Selection tree-based models have demonstrated significant for anticipating opposition and microbe obtrusiveness from genomic groupings [23]. Although, these investigations were restricted in both the hereditary highlights utilized and the techniques applied.

This method is limited to investigate antimicrobial class [24]. Therefore, if a new genome sequence is derived, then it is necessary to construct a new antimicrobial classification.

3.3 Whole-Cell Biosensors

Another technique utilizes manufactured science to outfit the inside hardware of bacterial cells to decide neighborhood convergences of anti-toxins [25]. As exposed to observing the reaction to anti-toxin nearness, bacterial cells hereditary is composed to react typically to varying concentrations of anti-toxins. Cultivating the embedded plasmid into bacterial cells will predict the properties of antibiotics [26]. The item is known all in all cell biosensor (Table 1).

From the above two models, they concluded that each model has its state of drawbacks and defects. The first technique is time consuming, and it has minimal possibility to find out the correct DNA sequence from the existing system [29]. In the second model, the new antibiotic gene prediction is difficult. Hence, this model is limited to looming antimicrobial elements. The results of Whole-Cell Biosensors yield poor responsiveness and semi-quantitative characteristics [30].

Table 1 Drawbacks in existing schema

S. No.	Existing schema	Implementation tools	Drawbacks
1.	The antibiotic genomic recognition method	Florescence and ion semiconductor [27]	High false rate of negative values Experimental setup is expensive
2.	Decision tree-based model	Spectrometer [28]	Limited to antimicrobial class
3.	Whole-cell biosensors	Genetically modified plasmid [25]	Poor sensitivity and semi-quantitative results

4 Algorithm and Implementation

4.1 Algorithm to Implement the Phenotypic Susceptibility

To deploy the phenotypic [7] convolutional neural system model, the single cell compacted picture is first reshaped the real-time image or video into a pixel. Two convolutional layers are incorporated different dataset of minimized pixels separately, trailed by subsampling layers to scale back the pictures by max pooling [31]. A completely associated layer of neurons, with every neuron associated with all neurons in the past layer, was put preceding the yield layer. A back-engendering calculation is prepared by numerous cycles to take place in this model [32]. For every emphasis, a set of compacted single cell pictures are arbitrarily chosen from the preparation dataset and took care of into the model as information. The model updates every neuron to limit the distinction between results. The final result will compare the images or real-time video of antimicrobial substances, and the device will predict the family of microorganism [33].

The dataset of images or video is evaluated and calculated with the guidance of convolution neural networks; the intensity and pictures of each image of a single cell are selected to foretell the class of antibiotics [34].

4.2 Algorithm to Implement Variation Mapping and Antibiotic Resistance Prediction Method

The author [11] made complete investigation for the genotypic prediction of anti-infection resistance and expanded to created novel techniques for using Next Generation Sequence information to better (1) initially, to characterize the features of amino acid-based variant [35], (2) develop the information base on genetic affiliations with resistance to the antibiotic, and (3) build precise forecast models for deciding resistance to phenotypic.

It was constructed based on the enormous dataset of bacterial genomes from the Biotechnology Information is in for National Center (NCBI) [36]. Read the sequence length of the genome present in Archive (SRA) [37], besides the susceptibility data of paired antibiotic genome from NCBI BioSample. Downloaded the publicly available bacterial genomes from the NCBI Short Read Archive (SRA) and matched antibiotic sensitivity data from the NCBI BioSample Antibiograms study [38]. To distinguish bacterial genome variations, they performed all over again gathering and adjusted the amassed frameworks to a curated Anti-infection Resistance (AIR) KEGG orthology database (KO) [39]. Through this procedure, we recognize the variation in KO-based group variation in bacterial genomes [40]. At last, they figured both the variations in genome and anti-infection resistance phenotypes, affiliation, and expectation models were built based on this phenomenon.

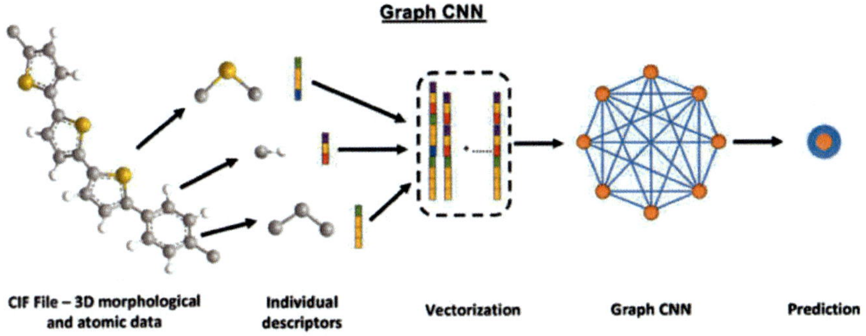

Fig. 2 Graph convolution neural network to predict chemical composition in drugs [44]

4.3 Algorithm to Implement to Find Antibiotic Using Deep Learning Model

The main objective of this proposed method is to predict the antibiotic using antibacterial molecules [14]. Preparing an introductory model and separating halicin from other antibacterial compounds. The particle named halicin was accounted for to have a development inhibitory movement against *Mycobacterium tuberculosis*. The machine model, which can screen more than a hundred million substance blends rapidly, is intended to choose potential antimicrobials that eliminate microscopic organisms utilizing unexpected components in comparison with those of existing medications [41]. Using prescient computer models for "in silico" screening isn't new, however as of not long ago, these models were not adequately precise to change sedate revelation [42]. Certain chemical groups reflect the presence or absence of vectors group is the traditional method to represent the molecules. However, the representation of graph convolution neural network, we can learn about the molecules are outlined in a continuous vector, by this method predict the characteristics of antibiotics [43] (Fig. 2).

4.4 Algorithm to Predict the Resistant Antibiotic Genome

In this proposed [16], two techniques are commonly used in this terminology to recognize Antibiotic Resistance Gene (ARG) from metagenomic information. Short read and open section frames (i.e., a gene of full-length progression) are the two methods to predict the antibiotic resistance gene from accumulated contiguous to prognosticate ARGs [45]. Two profound learning models are deployed to consider both comment techniques, Deep Antibiotic Resistance Gene-SS and Deep Antibiotic Resistance Gene-LS, were created to prepare exact peruses and full feature-length groupings, separately.

4.4.1 Database Amalgamates

The underlying assortment of Antibiotic Resistance Genes was gotten from three significant databases: Comprehensive Antibiotic Resistance Database, Antibiotic Resistance Genes Database [10], and UniProtKB [46]. For UniProtKB, all qualities that contained the Antibiotic Resistance catchphrase (KW-0046) were recovered [47], synchronically with their archive depictions when accessible. By grouping all the sequences (ARDB, CARD, and UniProt) using the Cluster Database at High Identity with Tolerance (CD-HIT), which eliminates all the 100% identical sequences that were the same length.

4.4.2 ARGs Notation of Comprehensive Antibiotic Resistance Database and Antibiotic Resistance Genes Database

The ARDB and CARD databases contain both the data which help in the characterization of Antibiotic Resistance Genes, counting the anti-infection class to which a quality gives resistance [48] (macrolides, beta-lactamases, or aminoglycosides) and the anti-infection gathering to which the class it belongs to (e.g., tetA, sul1, macB, oxa, mir, or dha). The standard investigation uncovered a few qualities of explicit arrangements of antimicrobials rather than anti-infection resistant classes or classifications [49].

4.4.3 Notation for UniProtKB Gene

Contrasted with the Antibiotic Resistance Gene in CARD and ARDB, the UNIPROT qualities consist of antimicrobial opposition catchphrases are less well curated [48]. In particular, given the Cluster Database at High Identity with Tolerance [50] grouping conclusions, bunches consist of just UniProtKB qualities were characterized into two classes: (1) with no explanation were labeled as "obscure" and (2) with portrayals were content mined to distinguish conceivable relationship with anti-toxin opposition.

The above four proposed techniques are deployed with the deep learning algorithm [51] and derived the solution for the existing systems and overcome the drawbacks and difficulties in predicting the antibiotic genome. The major drawback that the extant system is clinical setup and regressive experimentation. Deep learning methods are used to mitigate this disadvantage.

5 Research Gap

The antibiotic genetic association and prediction model is a tool for the fundamental scientist to identify potential antibiotics, which is used to cure bacterial infections.

Table 2 Summarize the algorithm and implementation techniques

S. No.	Method to predict antibiotic gene	Input dataset	Algorithm deployed	Advantages
1.	Phenotypic susceptibility	Real-time image and videos	Convolutional neural network model [7]	Detect the antibiotics from non-antibiotic substance
2.	Variation mapping	Bacterial genome sequence in NCBI	KEGG orthology database (KO) [11]	Comparing the dataset in the database to predict the model for all antimicrobial classes
3.	Deep learning model	Antibacterial molecules	Graph convolution neural networks [14]	To discover property of antibiotic by chemical structure
4.	Resistant antibiotic genome	Database merging of CARD, ARDB, and UNIPROT	Deep ARG-SS and deep ARG-LS [16]	Datasets are compared with existing dataset to find the antibiotic gene resistance for all family sets

From the review of the proposed prototype, the drawbacks in the working model to predict the antibiotic-resistant gene class applies to the restricted class of antimicrobial substances. The implementation of the algorithm is different for the datasets. The collection of new bacterial resistance genome from various factor is hectic task. The working prototype might not adopt to the flexible environment. The challenging task of the working prototype is to identify the potential antibiotic resistance with microorganisms.

6 Tabulation Method to Summarize the Algorithm and Implementation Techniques

See Table 2.

7 Conclusion

The corresponding studies are made, in the deep learning models to deploy and predict the antibiotic genome. From the survey, we can conclude that the four different techniques yield two main methods to found out the gene of antibiotic resistance: (i) evaluation of the dataset in the database and (ii) with the algorithms in deep learning working model (real-time image processing), they can predict the antibiotic resistance. The predicted four models obtained high-level accuracy that lies between in the range of 80–90%. The deep learning algorithm interprets the antibiotic gene using the computational model without human interaction.

In the future work, the model has to cover the maximum number of antibiotic resistance gene classes and algorithms are composed together to increase the efficiency of existing system. The accuracy yields by this algorithm are comparatively more prominent than the previous model. These methods will overcome the barrier of laboratory setup and reduce the error rate in determining antimicrobial substances. Identify potential antibodies among the microorganisms, which enables us to reduce the side effects of anti-toxic elements in drug discovery.

References

1. Van Boeckel TP, Gandra S, Ashok A, Caudron Q, Grenfell BT, Levin SA, Laxminarayan R (2014) Global antibiotic consumption 2000 to 2010: an analysis of national pharmaceutical sales data. Lancet Infect Dis 14:742–750
2. Dixit A, Kumar N, Kumar S, Trigun V (2020) Progress in the decade since emergence of New Delhi metallo-B-lactamase in India. Indian J Community Med
3. Forslund K, Sunagawa S, Kultima JR et al (2013) Country-specific antibiotic use practices impact the human gut
4. Reddy GS, Pranavi S, Srimoukthika B, Reddy VV (2017) Isolation and characterization of bacteria from compost for municipal solid waste from Guntur and Vijayawada. J Pharm Sci Res 9:1490
5. Ekins S (2016) The next era: deep learning in pharmaceutical research. Pharm Res
6. Coates ARM, Halls G, Hu Y (2011) Novel classes of antibiotics or more of the same? Br J Pharmacol 163:184–194
7. Yu H, Jing W, Iriya R, Yang Y, Syal K, Mo M, Grys TE, Haydel SE, Wang S, Tao N (2018) Phenotypic antimicrobial susceptibility testing with deep learning video microscopy. Anal Chem
8. Kaczorek E, Małaczewska J, Wójcik R, Rękawek W, Siwicki AK (2017) Phenotypic and genotypic antimicrobial susceptibility pattern of *Streptococcus* spp. isolated from cases of clinical mastitis in dairy cattle in Poland. J Dairy Sci 100:6442–6453
9. Natale DA, Arighi CN, Barker WC, Blake JA, Bult CJ, Caudy M, Drabkin HJ, D'Eustachio P, Evsikov AV, Huang H et al (2010) The protein ontology: a structured representation of protein forms and complexes. Nucleic Acids Res 39:D539–D545
10. Alcock BP, Raphenya AR, Lau TTY, Tsang KK, Bouchard M, Edalatmand A, Huynh W, Nguyen A-LV, Cheng AA, Liu S et al (2020) CARD 2020: antibiotic resistome surveillance with the comprehensive antibiotic resistance database. Nucleic Acids Res 48:D517–D525
11. Kim J, Greenberg DE, Pifer R, Jiang S, Xiao G, Shelburne SA, Koh A, Xie Y, Zhan X (2020) VAMPr: VAriant Mapping and Prediction of antibiotic resistance via explainable features and machine learning. Int J PLoS Comput Biol
12. Hunt M, Bradley P, Lapierre SG, Heys S, Thomsit M, Hall MB, Malone KM, Wintringer P, Walker TM, Cirillo DM et al (2019) Antibiotic resistance prediction for *Mycobacterium tuberculosis* from genome sequence data with Mykrobe. Wellcome Open Res 4
13. Béahdy J (1974) Recent developments of antibiotic research and classification of antibiotics according to chemical structure. Adv Appl Microbiol 18:309–406
14. Stokes JM, Yang K, Swanson K, Jin W, Cubillos-Ruiz A, Donghia NM, MacNair CR, Carfrae LA, French S, Bloom-Ackermann Z, Anush C-P, Tran VM, Chiappino-Pepe A, Badran AH,

Andrews IW, Chory EJ, Church GM, Brown ED, Jaakkola TS, Barzilay R, Collins JJ (2020) A deep learning approach to antibiotic discovery. Cell
15. Bagely MC, Dale JW, Merritt EA, Xiong X (2005) Thiopeptide antibiotics. Chem Rev
16. Arango-Argoty G, Garner E, Pruden A, Heath LS, Vikesland P, Zhang L (2018) DeepARG: a deep learning approach for predicting antibiotic resistance genes from metagenomic data. Microbiome
17. Linder JA, Doctor JN, Friedberg MW, Nieva HR, Birks C, Meeker D, Fox CR (2014) Time of day and the decision to prescribe antibiotics. JAMA Intern Med 174:2029–2031
18. Rosamond J, Allsop A (2000) Harnessing the power of the genome in the search for new antibiotics. Science 287:1973–1976
19. Pal C, Bengtsson-Palme J, Kristiansson E, Larsson DJ (2015) Co-occurrence of resistance genes to antibiotics, biocides and metals reveals novel insights into their co-selection potential. BMC Genom 16:1–14
20. Mardis ER (2008) Next-generation DNA sequencing methods. Annu Rev Genom Hum Genet 9:387–402
21. Kel AE, Gossling E, Reuter I, Cheremushkin E, Kel-Margoulis OV, Wingender E (2003) MATCHTM: a tool for searching transcription factor binding sites in DNA sequences. Nucleic Acids Res 31:3576–3579
22. Davis NM, Proctor DM, Holmes SP, Relman DA, Callahan BJ (2018) Simple statistical identification and removal of contaminant sequences in marker-gene and metagenomics data. Microbiome 6:1–14
23. Maier L, Pruteanu M, Kuhn M, Zeller G, Telzerow A, Anderson EE, Brochado AR, Fernandez KC, Dose H, Mori H et al (2018) Extensive impact of non-antibiotic drugs on human gut bacteria. Nature 555:623–628
24. Roca I, Akova M, Baquero F, Carlet J, Cavaleri M, Coenen S, Cohen J, Findlay D, Gyssens I, Heure OE et al (2015) The global threat of antimicrobial resistance: science for intervention. New Microbes New Infect 6:22–29
25. Wu Y, Wang C-W, Wang D, Wei N (2021) A whole-cell biosensor for point-of-care detection of waterborne bacterial pathogens. ACS Synth Biol 10:333–344
26. Thouand G, Belkin S, Daunert S, Freemont P, Hermans J, Karube I, Martel S, Michelini E, Roda A. Handbook of cell biosensors
27. Panek M, Paljetak HČ, Barešić A, Perić M, Matijašić M, Lojkić I, Bender DV, Krznarić Ž, Verbanac D (2018) Methodology challenges in studying human gut microbiota—effects of collection, storage, DNA extraction and next generation sequencing technologies. Sci Rep 8:1–13
28. Moradigaravand D, Palm M, Farewell A, Mustonen V, Warringer J, Parts L (2018) Prediction of antibiotic resistance in *Escherichia coli* from large-scale pan-genome data. PLoS Comput Biol 14:e1006258
29. Klein DC, Hainer SJ (2020) Genomic methods in profiling DNA accessibility and factor localization. Chromosome Res 28:69–85
30. Bueno J (2020) Antimicrobial screening: foundations and interpretation. In: Preclinical evaluation of antimicrobial nanodrugs. Springer, pp 1–14
31. Weng W, Zhu X (2021) INet: convolutional networks for biomedical image segmentation. IEEE Access 9:16591–16603
32. Krizhevsky A, Sutskever I, Hinton GE (2017) ImageNet classification with deep convolutional neural networks. Commun ACM 60:84–90
33. Camuñas-Mesa LA, Domínguez-Cordero YL, Linares-Barranco A, Serrano-Gotarredona T, Linares-Barranco B (2018) A configurable event-driven convolutional node with rate saturation mechanism for modular ConvNet systems implementation. Front Neurosci 12:63
34. Ding L, Fang W, Luo H, Love PED, Zhong B, Ouyang X (2018) A deep hybrid learning model to detect unsafe behavior: integrating convolution neural networks and long short-term memory. Autom Constr 86:118–124
35. Maienschein-Cline M, Dinner AR, Hlavacek WS, Mu F (2012) Improved predictions of transcription factor binding sites using physicochemical features of DNA. Nucleic Acids Res 40:e175

36. Sayers EW, Barrett T, Benson DA, Bolton E, Bryant SH, Canese K, Chetvernin V, Church DM, DiCuccio M, Federhen S et al (2010) Database resources of the national center for biotechnology information. Nucleic Acids Res 39:D38–D51
37. Kitsou K, Kotanidou A, Paraskevis D, Karamitros T, Katzourakis A, Tedder R, Hurst T, Sapounas S, Kotsinas A, Gorgoulis V et al (2020) Upregulation of human endogenous retroviruses in bronchoalveolar lavage fluid of COVID-19 patients. medRxiv
38. Shumway M, Cochrane G, Sugawara H (2010) Archiving next generation sequencing data. Nucleic Acids Res 38:D870–D871
39. Kanehisa M, Furumichi M, Tanabe M, Sato Y, Morishima K (2017) KEGG: new perspectives on genomes, pathways, diseases and drugs. Nucleic Acids Res 45:D353–D361
40. Humphries RM, Abbott AN, Hindler JA (2019) Understanding and addressing CLSI breakpoint revisions: a primer for clinical laboratories. J Clin Microbiol 57:e00203–e00219
41. Yang JH, Wright SN, Hamblin M, McCloskey D, Alcantar MA, Schrübbers L, Lopatkin AJ, Satish S, Nili A, Palsson BO et al (2019) A white-box machine learning approach for revealing antibiotic mechanisms of action. Cell 177:1649–1661
42. Viceconti M, Juárez MA, Curreli C, Pennisi M, Russo G, Pappalardo F (2019) Credibility of in silico trial technologies—a theoretical framing. IEEE J Biomed Health Inform 24:4–13
43. Winter R, Montanari F, Noé F, Clevert D-A (2019) Learning continuous and data-driven molecular descriptors by translating equivalent chemical representations. Chem Sci 10:1692–1701
44. Bronstein M (2020) Do we need deep graph neural networks. Towards Data Science, 20 July 2020
45. Pruden A, Arabi M, Storteboom HN (2012) Correlation between upstream human activities and riverine antibiotic resistance genes. Environ Sci Technol 46:11541–11549
46. Magrane M et al (2011) UniProt Knowledgebase: a hub of integrated protein data. Database 2011
47. MacDougall A, Volynkin V, Saidi R, Poggioli D, Zellner H, Hatton-Ellis E, Joshi V, O'Donovan C, Orchard S, Auchincloss AH et al (2020) UniRule: a unified rule resource for automatic annotation in the UniProt Knowledgebase. Bioinformatics 36:4643–4648
48. Yang Y, Jiang X, Chai B, Ma L, Li B, Zhang A, Cole JR, Tiedje JM, Zhang T (2016) ARGs-OAP: online analysis pipeline for antibiotic resistance genes detection from metagenomic data using an integrated structured ARG-database. Bioinformatics 32:2346–2351
49. Medema MH, Blin K, Cimermancic P, De Jager V, Zakrzewski P, Fischbach MA, Weber T, Takano E, Breitling R (2011) antiSMASH: rapid identification, annotation and analysis of secondary metabolite biosynthesis gene clusters in bacterial and fungal genome sequences. Nucleic Acids Res 39:W339–W346
50. Huang Y, Niu B, Gao Y, Fu L, Li W (2010) CD-HIT Suite: a web server for clustering and comparing biological sequences. Bioinformatics 26:680–682
51. LeCun Y, Bengio Y, Hinton G (2015) Deep learning. Nature 521:436–444

A Brief Review on Right to Recall Voting System Based on Performance Using Machine Learning and Blockchain Technology

Vivek R. Pandey and Krishnendu Rarhi

Abstract Voting is an important part of a democracy's political life cycle. A democracy is defined by citizens voting freely, fairly, and after due consideration. Citizens of the country vote based on candidates' pledges or incumbents' performance, as well as a perception formed from information provided by political parties, media houses, and social media and election manifesto. The electorate, who returned a midway result, had placed their trust in the candidate in vain. In this case, the electorate has no choice but to wait until the next election to see if the incumbent can be ousted from office. To address this inadequacy, remedies have been provided in some of the democracies in the form of the right to recall and No Confidence Motion (NCM), by a re-election or replacement of one elected representative with other for the same position in the government house. To address these issues, we propose a stable right to recall e-voting system in place of our traditional signature campaign based on blockchain and machine learning concepts. The measure intends to hold elected governments more accountable for pledges made to the country's voters. The power to recall gives the electorates of an elected representative the ability to withdraw or replace their mandate before the elected person's customary term ends. Here mean while you can say that the voters have a power to de-elect their representatives from the government through a direct vote. For the purpose of protecting the integrity and security of votes, we deploy blockchain technology, as well as a model of machine learning, to detect infiltration into voting data centers and electronic polling locations. Using the principles of personal blockchain and public blockchain, we offer a new paradigm for blockchain technology. With the proposed blockchain-based electronic voting system, users will have greater transparency, better treasury management, and greater confidence. The system will also prohibit unauthorized access to the data exchange network.

Keywords Blockchain · Right to recall e-voting · NCM · Machine learning

V. R. Pandey · K. Rarhi (✉)
Department of Computer Science and Engineering, Chandigarh University, Mohali, Punjab, India
e-mail: rarhik@gmail.com

P. Chatterjee et al. (eds.), *Computational Intelligence for Engineering and Management Applications*, Lecture Notes in Electrical Engineering 984,
https://doi.org/10.1007/978-981-19-8493-8_27

1 Introduction

In present days, because of the potential benefits of electronic voting, such as dependability, speed, transparency, performance, and accountability, some countries have switched from traditional to electronic voting. This is possible when we use a blockchain technology and machine learning.

India's Varun Gandhi submits a bill in the Lok Sabha that would allow MPs and MLAs to be recalled for non-performance: According to a measure submitted by Bharatiya Janata Party MP Varun Gandhi, members of Parliament and members of the Legislative Assembly (MLAs) should be recalled within two years of being elected if 75% of those who voted for them are unhappy with their performance. "Logic and justice necessitate that if the people have the power to elect their representatives, they should also have the power to remove these representatives when they engage in misdeeds or fail to fulfill the duties whatever they given promises before election." The Lok Sabha Member of Parliament suggested a change to the Representation of the People Act 1951 in his Representation of the People (Amendment) Bill, 2016, arguing that countries all over the world have experimented with the notion of right to recall in their electoral processes. Any voter in the constituency has the authority to initiate the recall process by presenting to the Speaker of the House of Representatives a petition, it is necessary that at least one-fourth of all voters in a given electoral district sign this document.

As soon as the Speaker determines that the application is legally valid, he will transmit it to the Election Commission (EC) for Verify and certification of the signatures of the registered voters. After validating petition signatures, the Commission will hold polling at ten locations within each MP/constituency, MLA's according to the release. If three-fourths of the votes cast in a member's election vote in favor of recall, the member is re-elected, according to the statute. The Speaker will tell the general public of the result within 24 h of receiving it. The Commission can call a by-election in that constituency after the seat has been properly vacated.

The right to vote in a free and fair election is one that all citizens of the country have, and if their elected representatives do not continue to inspire confidence in them, as Varun Gandhi has stated, the people must have the ability to remove them in order to ensure that the country's democracy is preserved. He went on to say that the genuine idea of democracy can only be realized if politicians are held accountable [1]. The electorate currently has no redress if they are dissatisfied with their elected representative.

1.1 Use of Direct Democracy

Representative democracy, which is more extensively utilized, is usually contrasted with direct democracy. Voters elect politicians and parties to make decisions on their behalf in a representational democracy. Direct democracy, on the other hand, allows citizens to decide for themselves on individual topics rather than delegating the decision-making process to their representatives. In referendums, voters make decisions on constitutional or policy issues rather than their elected representatives; Voters may actively strive to introduce constitutional or legislative changes through citizen initiatives. Finally, if people are displeased with the performance of their elected officials, they can use the recall power to replace them (i.e., with the choices that have been made in their name).

1.2 Recall in India

Recall has been operational only as NCM at the national and state levels in India and as recall at municipal level in some of the states. Both these differ a great deal from recall as a direct democracy measure provided in the USA. Recall as a direct democracy measure entails the government to be accountable to the voter seven after being elected for a particular tenure. The right in the USA, for instance, allows the voters to demand recall of an elected representative at the city and state levels for certain reasons, which included dissatisfaction with their functioning. A decision on recall is made at a special election where the voters of a constituency decide whether an elected representative should continue in office. In case the elected representative loses this recall election, the position is deemed vacant for which fresh elections.

As Atal Bihari Vajpayee Say in "Right to Recall Elected Representative." The "Right to Recall," like the "Right to Party Platform," finds its validity in the universal democratic system. When a candidate is elected to power by the people on the basis of his party's platform, that platform takes on the status of a contract, and the elected person is bound to uphold it. A default on the part of an elected representative vests in the electorate an inherent and non-negotiable right to recall that representative under a universal democracy. "As a result, right to recall is a democratic instrument that ensures greater accountability in the democratic system by allowing people to maintain control over legislators who are under performing or misusing their positions for personal gain." Right to recall is currently in use in several Indian states, including Madhya Pradesh, Rajasthan, Chhattisgarh, and Bihar. In Punjab, however, it is referred to as No Confidence Motion.

The right to recall (RTR) gives the electorate the power to recall their representatives. Every elector in a constituency who brings a recall petition signed by at least one-fourth of the total number of voters has this right. It can be found in many modern constitutions. The power to recall is also allowed in Canada and the USA for misbehavior and misconduct.

1.3 Objectives

The right to recall e-voting method, according to this paper, is essentially a mechanism in which the voter has the authority to recall elected individuals before the end of their regular term based on their promises before election and performance after election in a democratic society. Because of the above system that comes into play, the progress of the country will increase, and also the trust of voters will increase.

2 Related Works

As we know that blockchain technology and machine learning increases day by day, the research society perspective on security, privacy, transparency and authenticity using BCT has shifted. Blockchain technology, which provides a distributed and unchangeable ledger for storing transactions, makes use of cryptographic proofs, digital signatures, and p2p networking to accomplish this [2–4]. Researchers are drawn to blockchain for the development of distributed applications because of these qualities [5].

The potential of blockchain for electronic voting was recently discussed in a paper. The developers of used blind signature technology in conjunction with the blockchain to create an e-voting procedure [6]. This method used a bitcoin transaction with an additional 80 bytes of information for voting. The researchers created an e-voting blockchain technology-based solution to ensure the system's security and reliability [7]. Furthermore, there are some constraints, as it is expected that voters will vote using secure system and a secure network, but in practice, a hacker can damage the system by putting a harmful application on a utilized system or through a thorough network hole.

In a similar vein, the authors of Ref. [8] described a BCT-based electronic voting system that relied on commitments and blind signature technology to achieve success. The research tackles the essential challenge of a blockchain-based electronic voting system's verifiability; nevertheless, it ignores the concerns about e-voting privacy and security. The authors of Khan's research have merged the blockchain network with an electronic voting mechanism to create a novel voting system. In order to develop a long-lasting BCT-based electronic democracy framework that would be broadly accepted, the authors focused on the blockchain's adaptability concerns [9].

Salman's authors described a shift in voting systems from traditional to electronic voting, citing possible benefits such as performance and reliability. The author examines the current system in Iraq and the major issues that it faces in order to better understand how it works. The current voting system contains security issues, according to the findings, and must be replaced utilizing information and communication technology in accordance with industry standards [10]. None of the exiting methods fully supports voters or election participants in verifying the correctness and authenticity of cast votes at every election phase (i.e., various levels of verifiability); they do not protect against all conceivable threats by additionally keeping vote verification challenges. As a result, the goal of this work was to offer a vote verification technique that could verify votes against major potential threats while also allowing all election participants to verify votes [11].

The authors of Ref. [12] described a smart contract-based blockchain-based electronic voting system that ensures a secure and cost-effective election while maintaining voter anonymity. The authors suggest that the blockchain offers a new means to get over electronic voting systems' limitations and adoption barriers, maintaining election security and integrity while also paving the road for transparency. To reduce the burden on the network, hundreds of transactions per second can be transferred onto an Ethereum private blockchain using every component of the smart contract.

The authors of Panja's paper developed a cryptographic technique for conducting a secret ballot election that is verified, end-to-end verifiable, and secure. Except for the direct-recording electronic-i and direct-recording electronic-ip systems, nearly all verifiable e-voting systems now require trusted authorities to complete the tallying procedure. The author presents a method for voter registration and authentication that is both secure and dependable. The proposed approach by the author prevents a ballot stuffing attack. The author of this paper updated the direct-recording electronic-ip system such that no one may generate and submit a legitimate ballot on the public bulletin board without being detected [13]. The authors of Ref. [14] detailed the components and architecture of a fully functional and widely used online voting system, as well as some of the potential threats that such a deployment might face, and, most importantly, a description of the integration with the SMESEC framework and how it benefited the particular online voting solution.

3 Methodology

Our proposed system's right to recall e-voting method is presented in Fig. 2 along with the system paradigm for it. As can be observed, the system includes a large number of e-voting stations, all of which are connected to the public blockchain technology, which is a significant benefit. "Aside from that, we have a database that has all of the citizens of the country's information for the entire city, allowing us to assess whether or not a voter is eligible to vote at a certain polling station." There are servers, voters, and voting tools at every electronic voting station that can retrieve information from the primary database if necessary. As illustrated in Fig. 1, we employ both public as well as private blockchain in our system. The public blockchain, on the other side, is used to reveal the Merkle tree's root hash in order to assure data integrity and to publish the polling station's final conclusions for all to see.

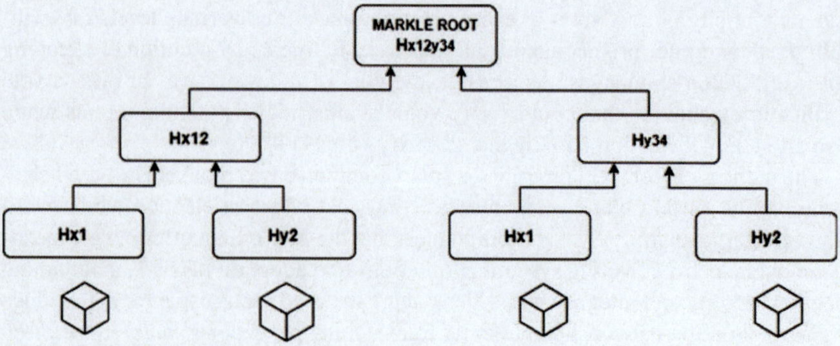

Fig. 1 Merkle tree

3.1 Merkle Root

Our hashing algorithm of choice is the Merkle tree hashing algorithm. We use the public blockchain to broadcast the Merkle tree's root hash in order to offer an additional layer of information integrity to our citizen's record that is stored in the primary data center (where the citizen's record is saved). Merkle tree is a crucial technique for storing data securely and reliably. Each Merkle tree leaf has a cryptographic hash associated with it. As shown in Fig. 1, this hash takes the voter's information as input and generates a distinct hash. The Merkle tree continues to merge it from the first step all the way up to the hash at the bottom. The root hash must therefore be protected in order to retain the data's integrity. When a voter's trustworthiness (citizen) information is stored in our data center, it is verified using the Merkle tree, which employs the SHA-256 hashing method. Only validated voters are then allowed to vote at a later time [15].

4 Proposed System

This suggested system is a secure right to recall e-voting system that leverages voter identification, also known as UID, as its back-end database. This document enables individual user identification by matching UIDs calculated by estimating the voter's age, obviating the need for existing voting cards. There are two databases in our suggested system. One is the central database, and the other is the polling booth's local database. The central database, known as the central identities data repository (CIDR), is the system's backbone. Every person in the country's voter information is stored in this database [16]. To lessen the stress on the central database, local databases will be installed beside the servers, including cached copies of data from the residents who live within its zone. These zones are determined by characteristics such as population density and other variables. Only those people who fall under the

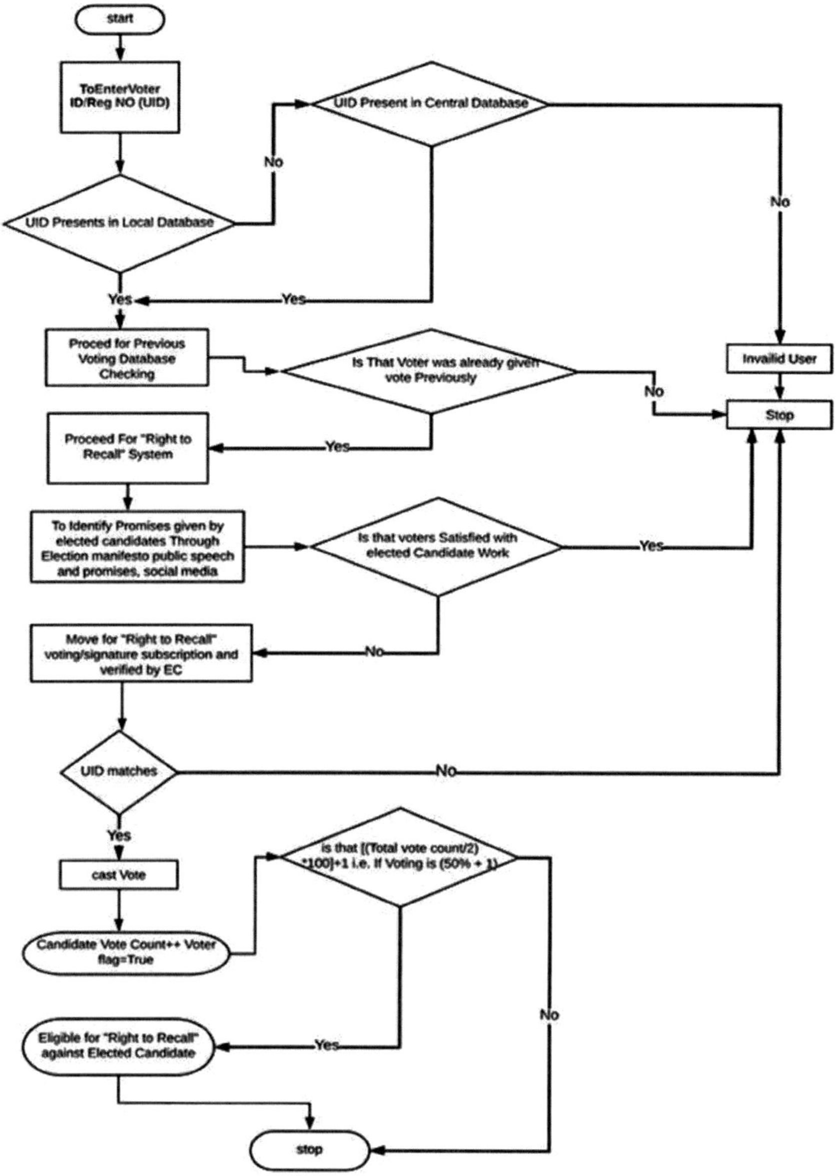

Fig. 2 Flowchart of proposed right to recall e-voting system

scope of CIDR are retrieved by all local databases [17]. This information is updated on a regular basis and saved in a volatile format so that it can be deleted if and when necessary, such as during security breaches, natural disasters, or routine maintenance.

Only data relevant to the voting process will be retrieved from the local database, with all other data being excluded. These databases will be utilized to generate election-related statistics and outcomes. These databases allow voters to vote from anywhere as long as they are within the electoral circle. The flowchart of the proposed system is shown in Fig. 2.

4.1 Authentication and Verification of the Right to Recall Voters

Authentication is the process of determining whether or not someone or something is who or what they claim to be in the realm of information security. We demand that a voter have a valid user identification number in order to authenticate them (UID). First, the number will be validated against the records in the local database. If that isn't found, it will go to the central repository to look for it. It's a one-to-many match. If a person's number cannot be discovered in the central database, voters will naturally avoid participating in the right to recall voting procedure. The data of the voters will be cached in the local database if the number is found in the central database [17]. This record is fetched from the local database and forwarded to the authentication server to be processed further. On the client side, the voters' user identification will be scanned and compared one by one with data collected from the local database on the server side for verification. The above procedure reduces the load on the local database and increases data traffic [9, 18].

4.2 Generating Report of Voting

When a voter casts a recall vote in support of their preferred candidate, the candidate's vote total in the local database is increased. The total number of votes received by the candidate is calculated by adding the votes from all of the local databases [17]. As a result, this technique produces immediate results while avoiding the waste of personnel and time. This technology has various advantages because it is electronic and uses digital data. With the aid of machine learning and blockchain technology, we might answer questions like how many people moved for right to recall voted from a given region, how many females moved for right to recall voted, which age group voted the most, largest turnouts, past comparisons, and so on. All the above mention terms not possible through traditional way of right to recall process [17]. It would provide valuable information into the election results and aid in the future improvement of the system.

4.3 Blockchain Technology

As previously stated, we employ the concepts of public and private blockchains for a variety of purposes. The Ethereum platform, which connects to and runs the Ethereum blockchain, is responsible for establishing a blockchain connection and deploying smart contracts defined in the solidity programming language. Each smart contract had its own set of features and functions, and they were used in conjunction with one another. For example, some of the most important activities of the initial smart contract include sharing of the root hash and communicating results of e-voting stations to the public blockchain from each e-voting station (when the voting period is over). Using the second smart contract, which communicates with the personal blockchain, authorized voters can register to vote and cast their ballots for various candidates [5]. If the voter gives accurate information, he will be registered to vote. Once a voter is registered with the correct address, he can vote for the political party of his choice.

When the voting procedure is finished, we may utilize the winning functionality to determine the winner, which is then shared together with the vote total on the public blockchain. The smart contract is installed on a personal blockchain and its capabilities are tested using web3 libraries, a python script, and a solidity remix [5]. We employ the Ganache technology developed by the truffle industry to interface with the blockchain [19]. Setting up ten accounts with hundred Ethers apiece for testing, with auto mining enabled by default, can be utilized for various blockchain scenarios (because in the test environment, no real Ethers will be lost). An interactive real-time environment for interacting with the blockchain, as well as a graphical user interface (GUI) for viewing the smart contract formation block and many other smart contract transactions, is also included in the package [5, 20].

However, even if we propose to use the blockchain in our proposed architecture to ensure that votes are anonymized, that their integrity is maintained, and that they are secure, electronic voting can be impacted by external and internal cyber-attacks. In order to prevent this, we will create an intrusion detection system that employs machine learning to detect e-voting network invasions and add the IP addresses of the perpetrators to a blacklist of known hackers. We're also using the machine learning model to predict a denial-of-service attack on our data center, which houses residents' personal information and could jeopardize data access at a critical time. It is vital to investigate how non-transactions have no effect on any of the voters in order to disregard the proposed e-voting system's scalability. As a result of the blockchain's gas block limit, each block in Ethereum can only hold a certain number of transactions. It increases the amount of time and blocks required to transfer data between nodes [5, 18].

5 Conclusions

The purpose of this research project is to create a reliable and effective right to recall electronic voting architecture by providing security, transparency, and integrity to the right to recall electronic voting system. As an e-voting station network, this suggested system not only protects the Electoral integrity, but also the information of inhabitants of the country. This proposed system gives power to proceed for right to recall against the elected candidate if their performance is not supposed to as per manifesto submitted to election commission or in promises given to voter in public speech. Here mean while you can say that the voters have a power to de-elect their representatives from the government through a direct vote. The proposed blockchain-based right to recall electronic voting system that provides the transparency while also preventing intrusion into the system of exchanging data. Here we use BCT and machine learning platform to detect infiltration in voting data centers to defend the integrity and security of each and every votes. In the coming year a large amount of in this field, work is to be expected.

Acknowledgements I'd like to express my gratitude to everyone who has helped, motivated, or supported me in carrying out this job, whether directly or indirectly. The authors would like to thanks his guide Dr. Krishnendu Rarhi for their supportive and constructive comments of this article.

References

1. Li Y, Susilo W, Yang G, Yu Y, Liu D, Guizani M (2019) A blockchain based self-tallying voting scheme in decentralized IoT. arXiv preprint arXiv:1902.03710
2. Gao J, Wu T, Li X (2020) Secure, fair and instant data trading scheme based on bitcoin. J Inf Secur Appl 53:102511. [Online]. Available: http://www.sciencedirect.com/science/article/pii/S2214212619309688
3. Lafourcade P, Lombard-Platet M (2020) About blockchain interoperability. Inf Process Lett 161:105976. [Online]. Available: http://www.sciencedirect.com/science/article/pii/S0020019020300636
4. Hassan MU, Rehmani MH, Chen J (2020) Differential privacy in blockchain technology: a futuristic approach. J Parallel Distrib Comput. [Online]. Available: http://www.sciencedirect.com/science/article/pii/S0743731520303105
5. Cheema MA, Ashraf N, Aftab A, Qureshi HK, Kazim M, Azar AT (2020) Machine learning with blockchain for secure e-voting system. In: 2020 first international conference of smart systems and emerging technologies (SMARTTECH), pp 177–182. https://doi.org/10.1109/SMART-TECH49988.2020.00050
6. Cruz JP, Kaji Y (2017) E-voting system based on the bitcoin protocol and blind signatures. IPSJ Trans Math Model Appl 10(1):14–22
7. Ayed AB (2017) A conceptual secure blockchain-based electronic voting system. Int J Netw Secur Appl 9(3):01–09
8. McCorry P, Shahandashti SF, Hao F (2017) A smart contract for boardroom voting with maximum voter privacy. In: International conference on financial cryptography and data security. Springer, pp 357–375
9. Khan KM, Arshad J, Khan MM (2020) Investigating performance constraints for blockchain based secure e-voting system. Future Gener Comput Syst 105:13–26

10. Salman W, Yakovlev V, Alani S (2021) Analysis of the traditional voting system and transition to the online voting system in the Republic of Iraq. In: 2021 3rd international congress on human–computer interaction, optimization and robotic applications (HORA), pp 1–5. https://doi.org/10.1109/HORA52670.2021.9461387
11. Al-Shammari AFN, Weldemariam K, Villafiorita A, Tessaris S (2011) Vote verification through open standard: a roadmap. In: 2011 international workshop on requirements engineering for electronic voting systems, pp 22–26. https://doi.org/10.1109/REVOTE.2011.6045912
12. Hjálmarsson FÞ, Hreiðarsson GK, Hamdaqa M, Hjálmtýsson G (2018) Blockchain-based e-voting system. In: 2018 IEEE 11th international conference on cloud computing (CLOUD), pp 983–986. https://doi.org/10.1109/CLOUD.2018.00151
13. Panja S, Roy B (2021) A secure end-to-end verifiable e-voting system using blockchain and cloud server. J Inf Secur Appl 59. https://doi.org/10.1016/j.jisa.2021.102815
14. Cucurull J et al (2020) Integration of an online voting solution with the SMESEC security framework. In: 2020 IEEE international systems conference (SysCon), pp 1–8. https://doi.org/10.1109/SysCon47679.2020.9275838
15. Adiputra CK, Hjort R, Sato H (2018) A proposal of blockchain-based electronic voting system. In: Second world conference on smart trends in systems, security and sustainability (WorldS4). IEEE, pp 22–27
16. Gupta S, Gupta A, Pandya IY, Bhatt A, Mehta K (2021) End to end secure e-voting using blockchain & quantum key distribution. Mater Today Proc. ISSN: 2214-7853. https://doi.org/10.1016/j.matpr.2021.07.254
17. Lakshmi CJ, Kalpana S (2018) Secured and transparent voting system using biometrics. In: 2018 2nd international conference on inventive systems and control (ICISC), pp 343–350. https://doi.org/10.1109/ICISC.2018.8399092
18. Mishra A, Mishra A, Bajpai A, Mishra A (2020) Implementation of blockchain for fair polling system. In: 2020 international conference on smart electronics and communication (ICOSEC), pp 638–644. https://doi.org/10.1109/ICOSEC49089.2020.9215354
19. Lee W-M (2019) Testing smart contracts using ganache. In: Beginning Ethereum smart contracts programming. Springer, pp 147–167
20. Usmani ZA, Patanwala K, Panigrahi M, Nair A (2017) Multi-purpose platform independent online voting system. In: 2017 international conference on innovations in information, embedded and communication systems (ICIIECS), pp 1–5. https://doi.org/10.1109/ICIIECS.2017.8276077

Sentiment Analysis Techniques: A Review

Divyanshi Sood, Nitika Kapoor, and Dishant Sharma

Abstract Sentiment can be described in the form of any type of approach, thought or verdict which results because of the occurrence of certain emotions. This approach is also known as opinion extraction. In this approach, emotions of different peoples with respect to meticulous rudiments are investigated. For the attainment of opinion related data, social media platforms are the best origins. Twitter may be recognized as a social media platform which is socially accessible to numerous followers. When these followers post some message on twitter, then this is recognized as tweet. The sentiment of twitter data can be analyzed with the feature extraction and classification approach. In this paper, various sentiment analysis methods are reviewed and analyzed.

Keywords Sentiment analysis · Machine learning · Twitter data · Lexical analysis

1 Introduction

Sentiments refer to the thoughts, beliefs, or feelings of an author for something, which may include a person, thing, corporation, or position. An opinion of a writer toward some subject or on the entire relative polarity of the manuscript is summarized in the process of analyzing the sentiments. The outlook can be defined as the point of view or assessment or the emotional message of an individual [1]. Opinions are critical influencer of a person's behavior. The opinions and insights for the reality are based on the way of others for perceiving the world. The fundamental objective

D. Sood (✉) · D. Sharma
Computer Science and Engineering, Chandigarh University, Gharuan, Mohali, Punjab, India
e-mail: 20MCS1010@cuchd.in

D. Sharma
e-mail: 20MCS1003@cuchd.in

N. Kapoor
Department of CSE, Chandigarh University, Gharuan, Mohali, Punjab, India
e-mail: nitika.cse@cumail.in

P. Chatterjee et al. (eds.), *Computational Intelligence for Engineering and Management Applications*, Lecture Notes in Electrical Engineering 984,
https://doi.org/10.1007/978-981-19-8493-8_28

of the opinion mining (OM) is to deduce the inclusive polarity of the document on some particular topic [2]. SA is a major research sector in which various regions related to technology as well as social disciplines, such as sociology, psychology, and ethics are included. Mining the opinions is a skill in which the perception of people toward something or some particular concept is followed from an enormous judgments or reviews which are openly obtainable in web. OM plays a significant role when the decision is made after searching out other opinions [3, 4]. To illustrate, to buy a camera or any gadget, we can check reviews or comments or take the view of others before buying the product. Opinion mining called as SA is a technique using which judgment of an individual is revealed for a topic or a product. This technique is utilized to classify the scrutiny of a user toward a region, event and object, etc., in three classes: positive, negative, or neutral. Subjective understanding related to any topic is contained in an opinion text. The weblog, reviews, reaction, etc., are involved in opinion text. These reviews are recognized as positive and negative reviews [5].

OM and recapitulation process is executed in three phases in which the opinion is retrieved, classified, and summarized.

A. **Opinion Retrieval**

This process focuses on selecting the review text from several review sites. People are posted review on various places, news, objects, and movies on the review websites. These reviews help other buyers to attain an idea about the quality and services related to that particular place or thing [6]. The data of review text is gathered from diverse sources and stored in a database using a number of methods. This process contains a stage to retrieve the reviews, microblogs, and comments of different users.

B. **Opinion Classification**

This process is executed for classifying the review text at first. To illustrate, in a given document $M = \{M_1 \ldots M_i\}$ and a predefined category set $K = \{\text{positive, negative}\}$, the main intend is to classify each point in M. This process classifies the review in two parts such as positive and negative [7]. These kinds of tasks are accomplished using dictionary-based techniques and machine learning (ML) techniques.

C. **Opinion Summarization**

The process to summarize the opinions is the major stage of OM [8]. The reviews are summarized on the basis of sub concepts or attributes present in reviews.

1.1 Techniques of Sentiment Classification

Various prime data mining (DM) techniques are adopted for extracting the facts and information. The methods of OM are represented in Fig. 1.

The entire procedure is consisted of various stages, in which text is cleaned online, white spaces are eliminated, acronym is amplified, stemming is done, stop words are removed elimination, refusal is managed, and the attributes are selected [9, 10].

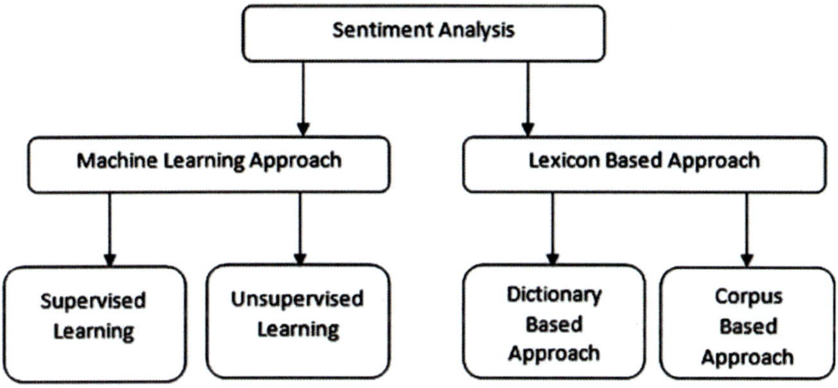

Fig. 1 Techniques of sentiment analysis

Thereafter, the classifiers are exploited to classify the opinions as positive, negative, and neutral.

1.1.1 Machine Learning Approaches

ML is planned on the basis of its diverse techniques. The issue related to classify the sentence level is resolved using these techniques. Moreover, a judgment is made on the syntactic attribute. ML is categorized in two kinds: supervised learning and unsupervised learning [11]. ML assists the machines in adjusting their interior configuration in such a way for predicting the upcoming performance boost.

A. *Supervised Learning*

Supervised learning is effective for dealing with the classification issues. This category is emphasized on acquiring the workstation so that a classifier can be investigated. The major instance of classification learning is the digit recognition [12, 13]. Any issue, in which classification learning is significant and the classification can be detected easily, is tackled using the classification learning. In some scenarios, the execution of programmed classifications is not required for each occurrence of a problem in case the technique is capable of performing the classification itself.

B. *Unsupervised Learning*

Such algorithms are utilized to assume the patterns of any dataset for which no labeled outcomes are considered [14].

Unlike the initial category, these methods are ineffective for dealing with the problem of regression or a classification. Due to this, it becomes difficult to train the algorithm normally. The unsupervised learning is useful to discover the underlying data structure. The prior unknown data patterns are exhibited using this approach.

1.1.2 Lexicon-Based Approach

The method of classifying the sentiment is utilized to estimate the opinion lexicons from text. Opinion lexicons are classified as: positive and negative. The most expected phase is confirmed using positive outlook terms and the negative outlooks are assisted in representing the most redundant phase [15, 16]. Opinion lexicons are called opinion clauses or idioms.

A. *Dictionary-Based Approach*

This technique is a suitable technique to collect the sentiment words due to the involvement of synonyms and opposites in the majority dictionary list for each and every word [17]. Therefore, some seed sentiment words are generated with the help of simple and easy technique for reset in accordance with antonym and synonym organization of a lexicon.

B. *Corpus-Based Approach*

The above-mentioned technique has limitation that it is incapable of discovering the view or sentiment words according to the application domains [18]. Thus, this technique is suggested for tackling the issue of dictionary-based system. This technique is able to acquire the opinion with the region-based orientations. This technique becomes popular as it provides opinion words with domain-based systems.

1.2 Applications

Various applications utilized to analyze the sentiments are discussed as:

1. **Aid in Decision Making**—This application is become significant in daily life. The sentiment analysis (SA) is useful to select a particular product and decide from the accessible options in accordance with the views of other people.
2. **Designing and Building Innovative Products**—The SA is implemented to analyze the products based on the public reviews and opinions. The utilization and adaptive nature of a product are considered while analyzing the products.
3. **Recommendation System**—Various applications make the implementation of the recommendation system. This system assists the users related to books, online-media, entertainment, music, film industry, and other art types [19]. This system deploys the information regarding an individual, history, likes, and dislikes to make proposals.
4. **Products Analysis**—SA is capable of examining several goods and to make selections. This analysis is efficient to select a product with regard to its specified attributes.
5. **Business Strategies**—The reactions of the public are considered to plan any business. The major goal of industries is to meet the demands of consumers. Hence, the strategies of companies are decided on the basis of public opinions and remarks.

6. **User Modeling**—This application provides a mechanism of interfaces and utilizes to produce more interactive design. This mechanism emphasizes on establishing a communication amid human and computer [20].
7. **Information Diffusion**—This procedure is deployed to transmit and spread the information by the means of interactions. This theory is utilized to make same decisions in a sequential manner. The behavior of individuals acts significantly to analyze the sentiments.

2 Literature Review

Gupta et al. [21] performed a case study of many unexplored fields of emotion analysis. The deployment of right knowledge is required to enhance the earlier methods. Summarize the text is a suitable method that was used to extract only valuable information for users from massive volume of gathered textual data. An intelligent system was planned using ML methods for extracting the data and analyzing the sentiments. With regard to summarize the text and analyze the review, a survey was conducted in this work. The merits and demerits of existing technologies were ascertained in this work.

Zirpe and Joglekar [22] discussed that many schemes to analyze the sentiments were reviewed for polarity shift discovery. Reviews have shown that all types of polarity shifts can be noticed and removed through polarity shift discovery, removal, and hybrid models. Therefore, the polarity shift was capable of detecting and removing a variety of issues related to detect the polarity shift. The performance of ML algorithms was outstanding in this work.

Bouazizi and Ohtsuki [23] discussed that the multi-class classification techniques were utilized to classify the Twitter users' online posts. This work analyzed the advantages and disadvantages of this approach. This research suggested a novel framework for representing diverse thoughts and presenting the potential of this model in understanding the relationship between emotions. The accuracy of multi-class classification gets improved. Also, this model resolved the existing issues.

Bouazizi and Ohtsuki [24] presented a novel approach to sarcasm on Twitter. The predictive technique employs different components of the Tweet. His plan uses part-of-speech-tags for exposing the blueprints showing the level of disrespect for tweets. Although the outcomes obtained are found good, a huge training set is deployed to enhance these outcomes. This implies that the dataset assisted in extracting the tweets probably unable of covering all the possible samples of sarcasm. The authors also envisioned a more efficient way to grow their set having an initial training set of 6000 tweets and a more robust prototype with the hash tag "#sarcasm".

Alshari et al. [25] suggested a novel technique so that the distinction between SentiWordNet and corpus terminology was extended to improve the feature set for analyzing the sentiments. Learning from SentiWordNet was done by assigning polarity scores to available non-opinion words in the corpus vocabulary. A labeled dataset from Film Review was used to gage this approach. The outcomes depicted the efficiency of the suggested technique over the standard SentiWordNet.

John et al. [26] reviewed that semantic gap was assessed to measure the polarity difference through making up hybrid lexicons. The acronyms were employed to remove the sentiment scores present in the text to assign the scores in a precise way. The hybrid lexicon was implemented to deal with this issue. This problem could be avoided by the hybrid lexicon, however, could not remove all the issues. With the aim to enhance the output, a hybrid technique was devised that integrated two techniques with other contextual sentiment alteration schemes that provided very accurate outcomes.

Taj et al. [27] projected a lexicon dependent scheme in order to analyze the sentiment on news articles. BBC News dataset was executed to implement the projected scheme. The experiments were conducted to validate the projected scheme. The results indicated that the business and sports categories had a higher number of positive articles, and the entertainment and technology-based categories had negative articles.

Ahmed and Danti [28] presented a novel method that was efficient for punters to decide on online reviews available on the web. The focus was on generating an effective technique to web reviews through mixed rule-based ML algorithms. The exploratory results demonstrated the efficiency of the new approach with maximal accuracy. The comprehensive experiments have been conducted on a variety of rule-based ML algorithms for classifying emotions.

Tsytsarau and Palpanas [29] discussed the issue of dissenting perceptions of feelings and opinions and their recognition in terms of a single or time factor for each case. He proposed procedures for the information preservation emporium of several emotions and for understanding negativity for vast and coarse data reasons, which was the first comprehensive and systematic explanation for FIX. A probe estimates together with the model, and actual data demonstrates the appropriateness and effectiveness of the proposed interpretation.

Ankit and Saleena [30] presented a new ensemble classification system to improve efficacy to classify the sentiments in tweet. A single superior classification algorithm was developed by integrating the base learning classification algorithms. The experiments were carried out to assess the working of the presented approach. A comparative analysis was conducted on the results obtained from the presented system against existing algorithms, and the results exhibited the effectiveness of the presented classifier over the traditional classifier. Customers were able to choose the best products according to public opinion by the means of this innovative technique. The work ahead can expand this investigation to study neutral tweets present in the dataset.

Sun et al. [31] conducted a study which focused on computing several techniques adopted for classifying the Tibetan micro-blog sentiments with regard to accuracy. For this, DL algorithms were implemented. The hybrid DL model was analyzed on the basis of several evaluation parameters and the results revealed an improvement of 1.22% in accuracy by using the presented technique.

Park and Seo [32] presented a novel framework in order to analyze tweets and select an artificial intelligence approach to make better decisions. Users' opinion was taken to understand the tweets better. The purpose of sentiment analysis was to improve the sentiment of products or services by actively researching them and developing NLP.

Yadav and Bhojane [33] presented three approaches based on which Hindi multi-domain review were utilized to analyze the sentiments. The Devanagari script available in the UTF-8 encoding system was utilized for the input in this research. In the first approach, neural network for pre-classified words was used for classifying the data. The second technique made the implementation of IIT Bombay Hindi Center to classify the data. The third technique employed NN prediction in pre-classified sentences as labeled data with the objective of classifying the data. Finally, the accuracy of each approach was obtained. In the results, the first technique provided the accuracy up to 52%, second around 71%, and third approach offered accuracy up to 70%.

Gupta and Joshi [34] introduced feature-based TSA system along with enhanced negation accounting for which a variety of attributes were employed. This system made the deployment of three classification algorithms namely support vector machine (SVM), Naïve Bayes (NB), and decision tree (DT). These algorithms were computed with diverse feature group in the experimentation. SemEval-2013 Task 2 dataset was applied in order to quantify the introduced system. The experimental outcomes depicted the supremacy of SVM algorithms over other algorithms. In the end, effect of every module to pre-process the data was presented on classification efficacy.

Yang et al. [35] developed a novel model to analyze the sentiment called SLCABG planned on the basis of the sentiment lexicon. In his approach, the convolutional neural network (CNN) was integrated with attention-based bidirectional gated recurrent unit (BiGRU). The sentiment attributes were improved using the sentiment lexicon. Thereafter, CNN and gated recurrent unit (GRU) were adopted for extracting the major sentiment attributes and context features in the reviews. In the end, the weighted sentiments were classified. The experimental outcomes exhibited that the developed model was adaptable for enhancing the efficiency of analyzing the sentiments.

Wisnu et al. [36] focused on deploying Naïve Bayes (NB) and support vector machine (SVM) for constructing a framework to classify the sentiment. The patterns were recognized, and a topic from the association among sentiment data of twitter was discovered using latent Dirichlet allocation (LDA). This dataset contained Indonesian tweets for the classifier having1600 tweets. The results of analyzing the sentiment indicated that SVM attained superior accuracy of 92.9% in comparison with NB.

Biradar et al. [37] developed an algorithm on the basis of analyzing the sentiments with the help of technique of classifying the customers review. The data was pre-possessed, the specific domains were considered to cluster the data, the synonyms were extracted using TF-IDF vectors, and the sentiments were classified. The outcomes indicated that the developed algorithm was 1.5 times faster in contrast to conventional database to Hadoop cluster and yielded the accuracy around 80% (Table 1).

Table 1 Table of comparison

Author name	Year	Description	Outcomes
Yadav and Bhojane [33]	2019	Hindi multi-domain review was utilized to analyze the sentiments. The Devanagari script available in the UTF-8 encoding system was utilized for the input in this research. In the first approach, neural network for pre-classified words was used for classifying the data. The second technique made the implementation of IIT Bombay Hindi Center to classify the data. The third technique employed NN prediction in pre-classified sentences as labeled data with the objective of classifying the data	Finally, the accuracy of each approach was obtained. In the results, the first technique provided the accuracy up to 52%, second around 71%, and third approach offered accuracy up to 70%
Bouazzi and Ohtsuki [23]	2019	It discussed that the multi-class classification techniques were utilized to classify the Twitter users' online posts. This work analyzed the advantages and disadvantages of this approach. This research suggested a novel framework for representing diverse thoughts and presenting the potential of this model in understanding the relationship between emotions	The accuracy of multi-class classification gets improved. Also, this model resolved the existing issues
John et al. [26]	2019	Semantic gap was assessed to measure the polarity difference through making up hybrid lexicons. The acronyms were employed to remove the sentiment scores present in the text to assign the scores in a precise way	A hybrid technique was devised that integrated two techniques with other contextual sentiment alteration schemes that provided very accurate outcomes

(continued)

Table 1 (continued)

Author name	Year	Description	Outcomes
Taj et al. [27]	2019	A lexicon dependent scheme in order to analyze the sentiment on news articles. BBC News dataset was executed to implement the projected scheme	The experiments were conducted to validate the projected scheme. The results indicated that the business and sports categories had a higher number of positive articles, and the entertainment and technology-based categories had negative articles
Alshari et al. [25]	2018	A novel technique so that the distinction between SentiWordNet and corpus terminology was extended to improve the feature set for analyzing the sentiments	A labeled dataset from Film Review was used to gage this approach. The outcomes depicted the efficiency of the suggested technique over the standard SentiWordNet
Ankit and Saleena [30]	2018	A new ensemble classification system to improve efficacy to classify the sentiments in tweet. A single superior classification algorithm was developed by integrating the base learning classification algorithms	The experiments were carried out to assess the working of the presented approach. A comparative analysis was conducted on the results obtained from the presented system against existing algorithms, and the results exhibited the effectiveness of the presented classifier over the traditional classifier
Sun et al. [31]	2018	A study which focused on computing several techniques adopted for classifying the Tibetan micro-blog sentiments with regard to accuracy	The hybrid DL model was analyzed on the basis of several evaluation parameters, and the results revealed an improvement of 1.22% in accuracy by using the presented technique
Park and Seo [32]	2018	A novel framework in order to analyze tweets and select an artificial intelligence approach to make better decisions. Users' opinion was taken to understand the tweets better	The purpose of sentiment analysis was to improve the sentiment of products or services by actively researching them and developing NLP

(continued)

Table 1 (continued)

Author name	Year	Description	Outcomes
Zirpe and Joglekar [22]	2017	It discussed that many schemes to analyze the sentiments were reviewed for polarity shift discovery. Reviews have shown that all types of polarity shifts can be noticed and removed through polarity shift discovery, removal, and hybrid models	The polarity shift was capable of detecting and removing a variety of issues related to detect the polarity shift. The performance of ML algorithms was outstanding in this work
Gupta et al. [21]	2016	The deployment of right knowledge is required to enhance the earlier methods. Summarize the text is a suitable method that was used to extract only valuable information for users from massive volume of gathered textual data. An intelligent system was planned using ML methods for extracting the data and analyzing the sentiments	With regard to summarize the text and analyze the review, a survey was conducted in this work. The merits and demerits of existing technologies were ascertained in this work
Bouazzi and Ohtsuki [24]	2016	It presented a novel approach to sarcasm on Twitter. The predictive technique employs different components of the Tweet. His plan uses part-of-speech-tags for exposing the blueprints showing the level of disrespect for tweets	The authors also envisioned a more efficient way to grow their set having an initial training set of 6000 tweets and a more robust prototype with the hash tag "#sarcasm"
Ahmed and Danti [28]	2016	A novel method that was efficient for punters to decide on online reviews available on the web. The focus was on generating an effective technique to web reviews through mixed rule-based ML algorithms	The exploratory results demonstrated the efficiency of the new approach with maximal accuracy. The comprehensive experiments have been conducted on a variety of rule-based ML algorithms for classifying emotions

(continued)

Table 1 (continued)

Author name	Year	Description	Outcomes
Tsytsarau and Palpanas [29]	2016	He proposed procedures for the information-preservation emporium of several emotions and for understanding negativity for vast and coarse data reasons, which was the first comprehensive and systematic explanation	The model and actual data demonstrates the appropriateness and effectiveness of the proposed interpretation
Gupta and Joshi [34]	2021	Introduced feature-based TSA system along with enhanced negation accounting for which a variety of attributes were employed. This system made the deployment of three classification algorithms namely support vector machine (SVM), Naïve Bayes (NB), and decision tree (DT)	The experimental outcomes depicted the supremacy of SVM algorithms over other algorithms
Yang et al. [35]	2020	Developed a novel model to analyze the sentiment called SLCABG planned on the basis of the sentiment lexicon. In his approach, the convolutional neural network (CNN) was integrated with attention-based bidirectional gated recurrent unit (BiGRU)	The experimental outcomes exhibited that the developed model was adaptable for enhancing the efficiency of analyzing the sentiments
Wisnu et al. [36]	2020	Focused on deploying Naïve Bayes (NB) and support vector machine (SVM) for constructing a framework to classify the sentiment	The results of analyzing the sentiment indicated that SVM attained superior accuracy of 92.9% in comparison with NB
Biradar et al. [37]	2021	Developed an algorithm on the basis of analyzing the sentiments with the help of technique of classifying the customers review	The outcomes indicated that the developed algorithm was 1.5 times faster in contrast to conventional database to Hadoop cluster and yielded the accuracy around 80%

3 Conclusion

Sentiment analysis is an umbrella term that covers many diverse fields, related to computer science, as well as social disciplines such as sociology, psychology, and ethics. There are various stages of sentiment analysis methods proposed so far. The pre-processing step involves removing missing and unnecessary values from the dataset. The technique of extracting attributes establishes the relationship amid the attribute and the target set. The final stage applies a classifier to classify the data into certain classes such as positive, negative, and neural. The various techniques for the sentiment analysis are reviewed in terms of certain parameters. It is analyzed that machine learning is best performing technique for the sentiment analysis.

References

1. Raut VB, Londhe AD (2014) Survey on opinion mining and summarization of user reviews on web. Int J Comput Sci Inf Technol 5:1026–1030
2. Li J, Chiu B, Shang S, Shao L (2022) Neural text segmentation and its application to sentiment analysis. IEEE Trans Knowl Data Eng 34:828–842
3. Alowisheq A, Al-Twairesh N, Altuwaijri M (2021) MARSA: multi-domain Arabic resources for sentiment analysis. IEEE Access 9:142718–142728
4. Long Y, Xiang R, Lu Q, Huang C-R, Li M (2021) Improving attention model based on cognition grounded data for sentiment analysis. IEEE Trans Affect Comput 12:900–912
5. Wu Z, Li Y, Liao J, Li D, Li X, Wang S (2020) Aspect-context interactive attention representation for aspect-level sentiment classification. IEEE Access 8:29238–29248
6. Wrobel MR (2020) The impact of lexicon adaptation on the emotion mining from software engineering artifacts. IEEE Access 8:48742–48751
7. Studiawan H, Sohel F, Payne C (2021) Anomaly detection in operating system logs with deep learning-based sentiment analysis. IEEE Trans Dependable Secure Comput 18:2136–2148
8. Xuanyuan M, Xiao L, Duan M (2021) Sentiment classification algorithm based on multi-modal social media text information. IEEE Access 9:33410–33418
9. Boumhidi A, Benlahbib A, Nfaoui EH (2022) Cross-platform reputation generation system based on aspect-based sentiment analysis. IEEE Access 10:2515–2531
10. Singh PK, Paul S (2021) Deep learning approach for negation handling in sentiment analysis. IEEE Access 9:102579–102592
11. Lin P, Yang M, Lai J (2021) Deep selective memory network with selective attention and inter-aspect modeling for aspect level sentiment classification. IEEE/ACM Trans Audio Speech Lang Process 29:1093–1106
12. Shyamasundar LB, Jhansi Rani P (2020) A multiple-layer machine learning architecture for improved accuracy in sentiment analysis. Comput J 63:395–409
13. Zhu Q, Jiang X, Ye R (2021) Sentiment analysis of review text based on BiGRU-attention and hybrid CNN. IEEE Access 9:149077–149088
14. Liu G, Huang X, Liu X, Yang A (2020) A novel aspect-based sentiment analysis network model based on multilingual hierarchy in online social network. Comput J 63:410–424
15. Yang M, Yin W, Qu Q, Tu W, Shen Y, Chen X (2021) Neural attentive network for cross-domain aspect-level sentiment classification. IEEE Trans Affect Comput 12:761–775
16. Bhatia S (2021) A comparative study of opinion summarization techniques. IEEE Trans Comput Soc Syst 8:110–117
17. Aurangzeb K, Ayub N, Alhussein M (2021) Aspect based multi-labeling using SVM based ensembler. IEEE Access 9:26026–26040

18. Zhang H, Sun S, Hu Y, Liu J, Guo Y (2020) Sentiment classification for Chinese text based on interactive multitask learning. IEEE Access 8:129626–129635
19. Abas AR, El-Henawy I, Mohamed H, Abdellatif A (2020) Deep learning model for fine-grained aspect-based opinion mining. IEEE Access 8:128845–128855
20. Mabrouk A, Díaz Redondo RP, Kayed M (2020) Deep learning-based sentiment classification: a comparative survey. IEEE Access 8:85616–85638
21. Gupta P, Tiwari R, Robert N (2016) Sentiment analysis and text summarization of online reviews: a survey. In: International conference on communication and signal processing (ICCSP)
22. Zirpe S, Joglekar B (2017) Polarity shift detection approaches in sentiment analysis: a survey. In: International conference on inventive systems and control (ICISC)
23. Bouazizi M, Ohtsuki T (2019) Multi-class sentiment analysis on Twitter: classification performance and challenges. Big Data Min Anal 6:3232–3240
24. Bouazizi M, Ohtsuki T (2016) A pattern-based approach for sarcasm detection on Twitter. IEEE Access 4:5477–5488
25. Alshari EM, Azman A, Doraisamy S, Mustapha N, Alkeshr M (2018) Effective method for sentiment lexical dictionary enrichment based on Word2Vec for sentiment analysis. In: Fourth international conference on information retrieval and knowledge management (CAMP)
26. John A, John A, Sheik R (2019) Context deployed sentiment analysis using hybrid lexicon. In: 1st international conference on innovations in information and communication technology (ICIICT)
27. Taj S, Shaikh BB, Meghji AF (2019) Sentiment analysis of news articles: a lexicon based approach. In: 2nd international conference on computing, mathematics and engineering technologies (iCoMET)
28. Ahmed S, Danti A (2016) Effective sentimental analysis and opinion mining of web reviews using rule based classifiers. Adv Intell Syst Comput 41:171–179
29. Tsytsarau M, Palpanas T (2016) Managing diverse sentiments at large scale. IEEE Trans Knowl Data Eng 28:3028–3040
30. Ankit, Saleena N (2018) An ensemble classification system for Twitter sentiment analysis. Procedia Comput Sci 4:24254–24262
31. Sun B, Tian F, Liang L (2018) Tibetan micro-blog sentiment analysis based on mixed deep learning. In: International conference on audio, language and image processing (ICALIP)
32. Park CW, Seo DR (2018) Sentiment analysis of Twitter corpus related to artificial intelligence assistants. In: 5th international conference on industrial engineering and applications (ICIEA)
33. Yadav M, Bhojane V (2019) Semi-supervised mix-Hindi sentiment analysis using neural network. In: 9th international conference on cloud computing, data science & engineering (confluence)
34. Gupta I, Joshi N (2021) Feature-based Twitter sentiment analysis with improved negation handling. IEEE Trans Comput Soc Syst 67:661–679
35. Yang L, Li Y, Wang J, Sherratt RS (2020) Sentiment analysis for e-commerce product reviews in Chinese based on sentiment lexicon and deep learning. IEEE Access 4:117–124
36. Wisnu GRG, Ahmadi, Muttaqi AR, Santoso AB, Putra PK, Budi I (2020) Sentiment analysis and topic modelling of 2018 Central Java gubernatorial election using Twitter data. In: International workshop on big data and information security (IWBIS), vol 23, pp 1202–1210
37. Biradar SH, Gorabal JV, Gupta G (2021) Machine learning tool for exploring sentiment analysis on Twitter data. Mater Today Proc 6:868–875

Network Traffic Classification Techniques: A Review

Nidhi Bhatla and Meena Malik

Abstract The network traffic classification task is focused on recognizing diverse kinds of applications or traffic data for which the received data packets are analysed that is essential in communication networks in these days. A network controller must have efficient understanding of applications and protocols in the network traffic to deploy the suitable security solutions. In addition, the name or kind of application is recognized and classified in the network for treating some aspects in advance. The process to classify the network traffic has become popular among research community along with the industrial field. A number of schemes have been put forward and constructed over the last two decades. The network traffic can be classified in several stages, in which pre-processing is done, attributes are extracted and classification is performed. The various machine learning models are reviewed in this paper for the network traffic classification.

Keywords Network traffic classification · Machine learning · Classification · KDD

1 Introduction

The escalating growth of network technology has pushed the expansion of the forms and amounts of traffic data in the virtual world. Identifying and classifying the traffic flow within a network is an important research work in the field of network protection and management. It is the basis for dynamic access control, network resource planning, content-based invoicing, interference and malware finding among others [1]. Service quality guarantee, dynamic access control and anomalous network behaviour recognition are some of the tasks in which classifying traffic efficiently and accurately is of great realistic significance. Network traffic classification (NTC) innovations based on port and deep packet inspection are steadily deprecated with full traffic encryption. Machine learning (ML) technology has grown into the best performing and prevailing approach. A great deal of work on efficient traffic feature mining

N. Bhatla · M. Malik (✉)
Chandigarh University, Gharuan, Mohali, India
e-mail: meena.e9536@cumail.in

P. Chatterjee et al. (eds.), *Computational Intelligence for Engineering and Management Applications*, Lecture Notes in Electrical Engineering 984,
https://doi.org/10.1007/978-981-19-8493-8_29

and the search for optimum classification networks have appeared in the education sector and have yielded promising outcomes [2]. Nevertheless, the majority of works ignored two big challenges. First, network traffic leads to a usual imbalance distribution. There is broad difference in the amount of traffic produced by the various protocols and applications. Next, the design objective of almost all machine learning algorithms is to maintain the maximum generalized accuracy irrespective of class imbalance, which switches the classifier's training to the majority class. Typically, the class with the greatest and the smallest sample dimension is respectively known as the majority and minority class. As a result, the productivity of existent network traffic classification-based machine learning approaches deteriorates greatly in practical unbalanced traffic classification operations [3]. In some cases, for example network control, intrusion finding, etc., where high-quality traffic occupies only a small portion, the performance deterioration on the minority class is disastrous. Hence, it is required to give sufficient consideration to the imbalance problem in NTC.

1.1 Description of Class Imbalance

There are several applications that give rise to natural allocation of skewed data. In these applications, positive class arises with low frequency, such as data originated in disease diagnostics, fraud detection, computer security and image recognition. At one side, internal imbalance occurs due to normally existing data frequencies, for example [4], clinical diagnosis where most patients are physically fit. On the other side, external imbalance is the result of external factors, for example, accumulation or storage processes. The representative of minority and majority classes must be considered when learning from unbalanced data. It is possible to obtain high-quality results without considering class inequality through good representation of both groups coming from non-overlapping distributions. To study the impacts of class imbalance, few investigators have created artificial datasets with different blends of complexity, training set dimension and imbalance levels [5].

According to results, imbalance sensitivity increases with the increase in issue complexity and that simple, linearly separable problems remain untouched by all degrees of class imbalance. There is really a lack of data in some areas owing to the low frequency with which events happen, such as oil leak detection. It is highly important to learn from acute class imbalanced data, where the minority class has the proportion up to 0.1% of the training data, as it is usually these rare events that are of utmost interest. The overall existing minority samples is of more interest than the proportion or ratio of the minority. Assume a minority group that accounts for just 1% of a dataset covering 10 lakh samples. Several positive samples (10,000) still occur for a model training despite the imbalance of high degree [6]. In contrast, an imbalanced dataset where the minority class exhibits infrequency or under-representations has higher degree of probability to undermine the classifier's performance.

$$\rho = \frac{\max_i\{|C_i|\}}{\min_i\{|C_i|\}}$$

In the above example, a ratio ρ depicts the greatest between-class imbalance level. C_i specifies an example set in class i, and $\max_i\{|C_i|\}$ and $\min_i\{|C_i|\}$, respectively, return the maximal and minimal class size over all i classes.

1.2 Existing Solutions to Class Imbalance

The solutions to class imbalance in network traffic classification are divided into three major levels: data-level, algorithm level and cost-sensitive level [7] (Fig. 1).

1.2.1 Data-Level Methods

Data-level techniques deal with class imbalances by over-sampling and under-sampling, respectively, by enlarging the sample size of minority classes or reducing the sample number of majority classes. These techniques reduce the imbalance degree or noise by modifying the training distributions for example, mislabelled samples or irregularities [8]. In its most non-complex types, random under-sampling (RUS)

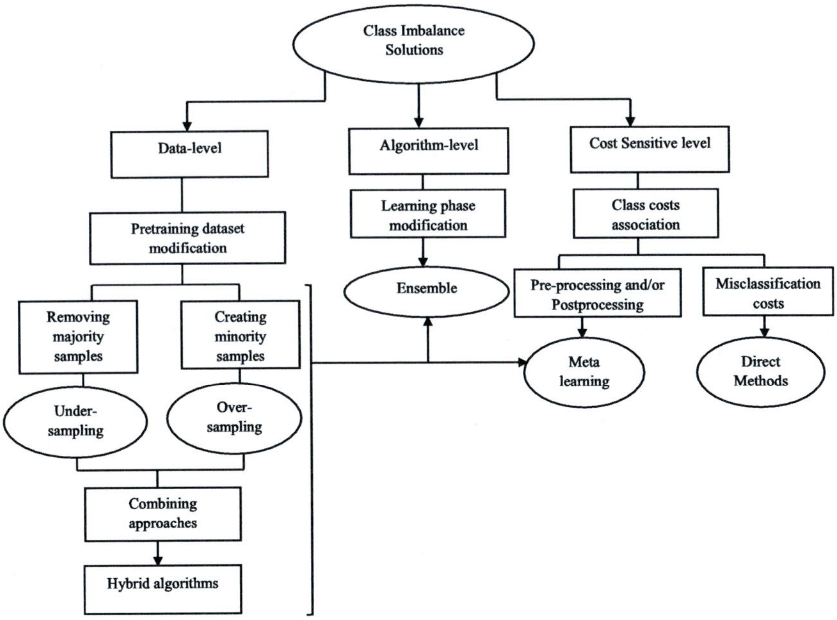

Fig. 1 Existing solutions to class imbalance

rejects random samples from the majority class, whereas random over-sampling (ROS) replicates random samples from the minority class.

(a) Under-sampling methods: Under-sampling willingly rejects the data in order to decrease the overall information the model needs to learn. Various smart sampling approaches have been devised as a means to balance these shifts [9]. The purpose of smart under-sampling schemes is to maintain important information for learning. Some researchers have obtained under-sampling using the K-nearest neighbours (KNN) classifier. The features of the provided data distribution have contributed in the development of four KNN under-sampling techniques called near miss-1, near miss-2, near miss-3 and 'most distant' technique. Rather than using the whole set of over-represented majority training samples, a small subset of these samples is chosen so that the resultant training data is less heterogeneous [10].

- The near miss-1 approach chooses the majority of instances that have the shortest average distance from the three nearest minority instances.
- The near miss-2 approach chooses the majority class instances whose mean distance is the shortest from the three most distant minority class instances.
- Near miss-3 chooses a provided number of nearest majority instances to each minority instance to guarantee that each minority instance is encircled by some majority instance [11].
- At last, the 'most distant' approach selects the majority class instances that have the greatest average distance of the three nearest minority class instances.

(b) Over-sampling methods: Random over-sampling attempts to balance the class distribution by randomly repeating the minority class examples. However, many authors come to a conclusion that this approach may improve the probability of overfitting, as it creates same replicas of existent instances. Several reported over-sampling techniques have also been developed to toughen class boundaries [12], decrease overfitting and increase differentiation.

- Synthetic Minority Over-sampling Technique (SMOTE): It is well-known over-sampling technique. The synthetic minority over-sampling technique (SMOTE) is implemented for generating the novel minority class examples for which various minority class instances which are put together are interrupted. This technique assists in avoiding the overfitting problem. But, its procedure is naturally dangerous as the minority class is often generalized blindly by it without considering any majority class [13] and this strategy becomes problematic during highly skewed class distributions due to the sparse nature of the minority class in terms of the majority class in some scenarios. Consequently, a greater chance of class mixture is obtained. Various enhanced over-sampling algorithms are implemented for retaining the benefits of SMOTE and lessening its drawbacks.
- Modified SMOTE (MSMOTE): This algorithm is an enhancement of synthetic minority over-sampling technique (SMOTE). This algorithm is

adopted for partitioning the instances of the minority class into three groups: safe, border and latent noise instances after computing the distances among all examples. The new instances are created by modified synthetic minority over-sampling technique. The strategy of selecting the nearest neighbours is changed in accordance with earlier technique and the group previously assigned to the instance [14]. This algorithm emphasizes on selecting a data point at random from the KNNs for safe instances, choosing the nearest neighbour for border instances and selecting nothing for latent noise instances. This technique efficiently decreases the risk of introducing artificially mislabelled instances. Thus, this technique is useful for performing more accurate classification in contrast to earlier technique.

1.2.2 Algorithm-Level Methods

Different from other techniques, these techniques have not changed the training data distribution to handle the class imbalance. Its learning or decision process is adjusted to maximize the efficacy of the positive class [15]. The algorithms which assist in awarding the minority classes and punishing the majority under the training phase are deployed in these techniques. A class penalty or weight is considered by modifying the algorithms or the shifting of the decision threshold for mitigating the bias towards the negative class. The major intend of the algorithm-level category is to modify the existing learner for eliminating its bias towards the majority classes. The ensemble methods are well-known algorithm-level techniques in which a re-sampling stage is included during the development of ensembles [16]. Ensemble classification algorithms are utilized for enhancing the accuracy of single classifiers with the integration of various models and can be implemented on the imbalanced data sets.

1.2.3 Cost-Sensitive Methods

In these techniques, the data-level transformations are integrated with the algorithm-level modifications. The costs related to the misclassifying samples are taken in account using the cost matrix. The learner is pressurized for classifying the minority class samples in exact manner for which a high cost is set to the misclassified samples of minority class. There is not any penalty for correct classification samples. However, the cost of misclassification is greater as compare to the majority samples [17]. The cost-sensitive techniques concentrate on diminishing the total cost of the training dataset, but it is challenging task to determine the cost values due to their dependency on multi factors having trade-off relationships. These techniques have two categories namely direct methods and meta-learning methods.

(a) Direct methods: These techniques have cost-sensitive potentials which are attained by enhancing the learner's underlying algorithm such as the costs are considered under the learning process. The optimization process is changed amid one of minimizing total error and cost.

(b) Meta-learning: A wrapper is utilized for converting the cost-insensitive learners into cost-sensitive systems in this technique. In case of generation of a cost-insensitive classification algorithm [18], a new threshold p^* is defined using the cost matrix as

$$p^* = \frac{c_{10}}{c_{10} + c_{01}}$$

In general, p^* is utilized in the thresholding techniques for redefining the output decision threshold when the samples are classified. The above-mentioned equation is utilized to perform the threshold moving or post-process the output class probabilities and it is a meta-learning approach using which the cost-insensitive learner is transformed into a cost-sensitive system [19].

1.2.4 Hybrid Methods

The integration of data-level is done with the algorithm-level techniques in several ways and deployed for dealing with the class imbalance issues. This strategy has sampled the data for mitigating the class noise and imbalance. Thereafter, the cost-sensitive learning or thresholding is utilized for further alleviating the bias towards the majority group. A number of methods in which the ensemble techniques are put together with the sampling and cost-sensitive learning have introduced. Two well-known hybrid algorithms are EasyEnsemble and BalanceCascade utilized for learning multiple classifiers [20]. For this purpose, the subsets of the majority group are combined with the minority group and pseudo-balanced training sets are generated for every individual classification algorithm.

(a) EasyEnsemble: The chief objective of this technique is that a high efficiency of under-sampling has to be maintained and the risk of avoidance of potentially useful information contained in majority class examples must be diminished. A simple strategy is utilized in this technique. Initially, this technique emphasizes on creating the multiple subsamples $S_{\mathrm{maj}1}, S_{\mathrm{maj}2}, \ldots, S_{\mathrm{maj}n}$ from the majority class sample [21]. The size of each subsample is similar to the minority class sample S_{min}, that is, $|S_{\mathrm{maj}i}| = |S_{\mathrm{min}}|, 1 \leq i \leq n$. Subsequently, an adaboost ensemble is trained using the union of each possible pair $(S_{\mathrm{maj}i}, S_{\mathrm{min}})$. All the base learners are integrated in all the adaboost ensembles to develop the final ensemble. The technique provides optimal outcomes in comparison with adaboost, bagging, random forest (RF), SMOTEBoost and BRF while tackling the issues of binary imbalance [22].

(b) BalanceCascade: This technique focuses on deploying the guided instead of random deletion of majority class examples. Unlike the EasyEnsemble, this method performs in a supervised way. The ith round is executed to generate a subsample $S_{\mathrm{maj}i}$ at random from the current majority class dataset S_{maj} with sample size $|S_{maji}| = |S_{min}|$. Therefore, adaboost is adopted to train an ensemble Hi from the union of S_{maji} and S_{min}. Later on, the majority class data examples that the Hi algorithm has classified are eliminated from the S_{maj} [23]. The BalanceCascade assists in eliminating the correctly classified majority class examples in every iteration. Thus, its efficacy must be enhanced for highly imbalanced datasets.

2 Literature Review

2.1 Class Imbalance for Network Traffic Classification Using Machine Learning

Gómez et al. [24] emphasized on tackling the issue related to class imbalance while classifying the network traffic. The presence of this phenomenon was analysed, and various solutions were examined in two diverse Internet environments. Twenty-one data-level algorithms, six ensemble techniques and a cost-level method were employed in the experimentation. The issue related to imbalance was resolved considering the methodological aspects such as DOB-SCV validation technique. Moreover, for this, the binary techniques, in which two ensemble techniques included, were also adopted in machine learning (ML). The experimental outcomes depicted that some methods led to diminish the class imbalance and boost the accuracy by 8% under diverse scenarios.

Guo et al. [25] suggested a focal loss-based adaptive gradient boosting (FLAGB) model to classify the imbalanced traffic. This model was adaptable in classifying the network traffic at diverse imbalance levels and tackling the imbalance without any prior knowledge regarding process to distribute the data. BOT and KDD99' datasets were applied to conduct the experiments in which binary and multiple classes were covered. The experimental outcomes demonstrated the supremacy of the suggested model over the traditional methods. This model consumed least time in the training phase; thus, it becomes an effective tool to classify the highly imbalanced traffic.

Dong [26] introduced a modified SVM algorithm known as cost-sensitive support vector machine (CMSVM) for addressing the imbalance issue while classifying the network traffic. A multi-class SVM algorithm was implemented with active learning due to its proficiency of assigning a weight for applications in dynamic way. The MOORE_SET and NOC_SET datasets were utilized for the quantification of introduced algorithm with regard to accuracy and efficiency. The experimental outcomes confirmed that the introduced algorithm was applicable for lessening the computation cost, increasing the accuracy and tackling the imbalance problem in comparison with other machine learning (ML) methods.

Peng et al. [27] constructed an imbalanced data gravitation-based classification (IDGC) system for dealing with the issue related to classify the Internet traffic.

Initially, the generation of six imbalanced traffic datasets was done from three original traffic datasets. Subsequently, their packet sizes were considered to extract the attributes. A comparative analysis was conducted on the constructed system against various algorithms in the experimentation. The results obtained in experiments revealed the effectiveness and stability of the constructed system with regard to diverse parameters while classifying the imbalanced traffic.

Saber et al. [28] suggested a new correlation-based algorithm in which a cost-sensitive technique was put together with a bagged random forest (BRF) ensemble algorithm for dealing with the issue related to the inter-class imbalance and fulfilling the time requirements in a data centre environment. Moreover, an innovative technique was also presented on the basis of reverse k-nearest neighbours (RkNN) for capturing the rebalancing weights that expressed inter-flow correlations. A comparative analysis was conducted on the suggested algorithm against traditional techniques. The outcomes acquired on datasets depicted the supremacy of suggested algorithm over others concerning precision, recall and $F1$ measure (Table 1).

2.2 *Class Imbalance for Network Traffic Classification Using Deep Learning*

Wang et al. [29] developed a new technique known as PacketCGAN using CGAN with the objective of controlling the modes of data before their construction. The efficiency of conditional generative adversarial network (CGAN) was adopted in this technique for creating the specified samples with the input of types of applications as conditional. Hence, the data was balanced. Four kinds of encrypted traffic datasets were classified using three traditional deep learning (DL) models. The random over sampling (ROS), synthetic minority over-sampling technique (SMOTE), Vanilla GAN and PacketCGAN were deployed to augment these datasets. The results of experiments indicated that the developed technique performed more successfully against other algorithms for classifying the encrypted traffic.

Song et al. [30] presented a technique on the basis of text convolution neural network (CNN) algorithm was established in which the traffic data was represented as vectors. The key attributes were extracted using this algorithm to classify the traffic. ISCX VPN-non-VPN dataset was applied to evaluate the established algorithm. The results proved that the established algorithm outperformed the earlier technique concerning $F1$-score. Moreover, the issue regarding class imbalance was resolved with the implementation of a novel loss function and a suitable technique of allocating the class weight to perform multi-class classification. The adaptability of these techniques was proved.

Guo et al. [31] formulated an innovative generative adversarial network (GAN) algorithm for generating the traffic samples. Furthermore, the stability and efficiency were offered to this procedure using the classification algorithm and the pre-training module. An end-to-end (E2E) model recognized as ITCGAN was put

Table 1 Machine learning techniques comparison

Author	Year	Technique used	Dataset	Results
Gómez et al. [24]	2019	DOB-SCV validation approach	HOST datasets	Some methods were assisted in diminishing the class imbalance and boosting the accuracy by 8% under diverse scenarios
Guo et al. [25]	2020	Focal loss-based adaptive gradient boosting framework (FLAGB)	BOT and KDD99' dataset	This model consumed least time in the training phase; thus, it becomes an effective tool to classify the highly imbalanced traffic
Dong [26]	2021	A cost-sensitive SVM (CMSVM)	MOORE_SET and NOC_SET datasets	The introduced algorithm was applicable for lessening the computation cost, increasing the accuracy and tackling the imbalance problem
Peng et al. [27]	2017	Imbalanced data gravitation-based classification (IDGC)-based model	UNIBS-SKYPE and UJN-CD datasets	The results obtained in experiments revealed the effectiveness and stability of the constructed system with regard to diverse parameters while classifying the imbalanced traffic
Saber et al. [28]	2020	Bagged random forest, reverse k-nearest neighbours	CAIDA dataset, UNI1 dataset, UNI2 dataset, UNIBs dataset	The outcomes acquired on datasets depicted the supremacy of suggested algorithm over others concerning precision, recall and $F1$ measure

forward for generating traffic samples for minority classes so that the original traffic was rebalanced in adaptive way. A publicly available dataset named ISCXVPN2016 was utilized to validate the formulated algorithm considering the global and individual parameters. The experimental outcomes exhibited that the formulated algorithm was effective to classify the imbalanced network traffic subsequent to decrease the performance degradation.

Wang et al. [32] designed a generative adversarial network (GAN) algorithm named FlowGAN was for addressing the class imbalance issue during classifying the traffic. The designed algorithm was computed by training the multilayer perceptron (MLP)-based classification algorithm. The experiments were conducted on ISCX dataset. The results of experiments validated the supremacy of the designed algorithm over the existing techniques and assisted in enhancing the precision up to 13.2%, recall around 17.0% and $F1$-score up to 15.6%.

Lopez-Martin et al. [33] focused on expanding the traditional radial basis function (RBF) for which it was incorporated as a policy network in an offline reinforcement learning (RL) algorithm. The results were enhanced using additional dense hidden-layers and the number of radial basis kernels. Five datasets were utilized to evaluate the presented approach against other algorithms. The outcomes revealed that the presented approach was efficient to develop classification algorithms using the constraints imposed by detecting the network intrusion. The significance of dataset imbalance was also discussed.

Pan et al. [34] recommended a zero-shot fault recognition technique which had two stages. Initially, a new network used to generate feature was constructed on the basis of conditional generative adversarial network (CGAN) in which a feature extractor, a discriminator and a generator were included for capturing the distribution of normal samples. The synthetic pseudo fault attributes were created via a generator. Subsequently, the training of an enhanced deep neural network (DNN) was done using those attributes as the classification algorithm. Eventually, three datasets were applied for quantifying the recommended technique. The results proved the effectiveness of the recommended technique for detecting the crucial faults even under the absence of the fault data during training (Table 2).

2.3 Class Imbalance for Network Traffic Classification Using Ensemble Technique

Oeung and Shen [35] projected a mechanism for developing an effectual classification algorithm from NetFlow to classify the traffic. First of all, the C4.5 decision tree (DT) algorithm was utilized with attributes of NetFlow records for analysing the application. In addition, an ensemble feature selection (FS) technique was suggested with the objective of boosting the accuracy and mitigating the computational complexity. In the end, an integration of clustering-based under-sampling with synthetic minority over-sampling technique (SMOTE) was implemented so the issue related to data

Table 2 Comparison of deep learning techniques

Author	Year	Technique used	Dataset	Results
Wang et al. [29]	2020	PacketCGAN using conditional GAN	ISCX2012 and USTC-TFC2016	The developed technique performed more successfully against other algorithms for classifying the encrypted traffic
Song et al. [30]	2019	Text convolution neural network	ISCX VPN-nonVPN dataset	The issue regarding class imbalance was resolved with the implementation of a novel loss function and a suitable technique of allocating the class weight to perform multi-class classification
Guo et al. [31]	2021	Generative adversarial network (GAN)	ISCXVPN2016 dataset	The formulated algorithm was effective to classify the imbalanced network traffic subsequent to decrease the performance degradation
Wang et al. [32]	2019	FlowGAN	ISCX dataset	The designed algorithm was superior to the existing techniques and assisted in enhancing the precision up to 13.2%, recall around 17.0% and $F1$-score up to 15.6%
Lopez-Martin et al. [33]	2021	Offline reinforcement learning algorithm	NSL-KDD, UNSW-NB15, AWID, CICIDS2017 and CICDDOS2019	The outcomes revealed that the presented approach was efficient to develop classification algorithms using the constraints imposed by detecting the network intrusion

(continued)

Table 2 (continued)

Author	Year	Technique used	Dataset	Results
Pan et al. [34]	2021	Two-stage zero-shot fault recognition, generative adversarial network	CWRU dataset, SQ dataset	The results proved the effectiveness of the recommended technique for detecting the crucial faults even under the absence of the fault data during training

imbalance was resolved. The experimental outcomes indicated that the projected mechanism offered higher F-measure and least computational complexity.

Zaki and Chin [36] presented a hybrid algorithm known as filter-wrapper feature selection (FWFS) in order to classify the network traffic. The robust attributes were selected using this algorithm for providing the resistance against concept drift. The wrapper function was utilized to discard the redundant attributes. The results confirmed the reliability and stability of the presented algorithm to classify the new data, and the acquired accuracy was calculated 98.7% while classifying the new data and the F-measure was found 0.8 above in every class.

Xu et al. [37] suggested an innovative algorithm to classify the traffic on the basis of packet transport layer payload for which an ensemble learning (EL) was implemented. Three types of neural networks (NNs) were deployed for generating an effective classification algorithm. The training of every model was done at individual level. The weight voting was implemented to decide the predictive outcome. The experimental outcomes demonstrated that the suggested algorithm yielded the accuracy around 96.38% and proved superior over the traditional techniques.

Wang et al. [38] designed an end-to-end (E2E) technique of classifying the encrypted traffic with one-dimensional convolution neural networks (1D-CNN). The techniques of extracting the attributes, selecting the attributes and a classification algorithm were integrated into unified E2E model to learn the nonlinear relationship amid raw input and expected output. ISCX VPN-nonVPN traffic dataset was executed to authenticate the intended algorithm. The experimental outcomes revealed that the intended technique was more adaptable in comparison with the traditional technique.

Liu et al. [39] projected a new difficult set sampling technique (DSSTE) algorithm using which a classifier attained capability to support the process executed to learn the imbalanced network data. The imbalance of network traffic was mitigated when the number of minority samples of learning was maximized. There were six traditional classifiers implemented in machine learning (ML) and deep learning (DL), and their integration was done with other sampling methods. The experimental results indicated that the projected algorithm was applicable for determining the samples whose modification was required in the imbalanced network traffic and enhancing the efficiency to recognize the attack in more successful way.

Shang et al. [40] presented a hybrid AID method by employing random forest-recursive feature elimination (RF-RFE) algorithm and a long-short term memory (LSTM) network augmented through a Bayesian optimization algorithm (BOA). First, underlying traffic variables and their composites were used to build a comparatively broad set of initial variables. Next, the RF-RFE algorithm was used to choose feature variables from the initial variables. Thirdly, the feature variables were adopted to train the LSTM network, and use of BOA augmented the hyper-metrics of the LSTM network. The output of experimentation revealed that the methodology presented had performed better in terms of the majority of the evaluation criteria. It also demonstrated that the presented methodology was outstanding for addressing imbalance issues and small sample dimensions of traffic occurrence data (Table 3).

3 Review Methodology

This section presents the state-of-the-art review layout, a step-by-step method for the literature discussed in the previous sections. This research focuses on categorizing the current literature on network traffic classification assessing the current trends. This evaluation finds relevant research articles from reputable electronic databases and the top conferences in the field. After then, inclusion and exclusion criteria were used to reduce the number of papers that were considered. Following that, final research studies were chosen based on a variety of variables. The information given here is the product of a thorough investigation. For this review study, various electronic database sources were investigated; some of the popular electronic databases used in this search like google scholar, Elsevier, Science direct, etc. Using the inclusion criterion, which mainly depends on the techniques, the relevant work of network traffic classification is retrieved from the enormous collection of data given by search engines. The data shows that journals account for most of the work in this study (51%), with conferences accounting for 40% of the work and book chapters accounting for 9%. In addition, the data depicts a year-by-year study of work relevant to network traffic classification. The major data is available on the google scholar as compared to Elsevier and Science direct. The google scholar has 60% data, Elsevier has approx. 10% and Science direct has approx. 30 data on network traffic classification. The data division has been presented in Fig. 2.

In Fig. 2, the percentage of data sharing is shown in figure approx. 60% data is available on Google scholar, 30% is available on science direct and very less amount of data that is 10% is available on Elsevier.

As shown in Fig. 3, the data is available which is available through conference, journal and books. The conference has approx. 51% of total data, 40% data has available through journal and 9% is available through books.

Table 3 Comparison of ensemble techniques

Author	Year	Technique used	Dataset	Results
Oeung and Shen [35]	2019	Ensemble feature selection (FS) method	UNIBS and Auckland dataset	The experimental outcomes indicated that the projected mechanism offered higher *F*-measure and least computational complexity
Zaki and Chin [36]	2019	Filter and wrapper feature selection (FWFS)	ISCX dataset	The presented algorithm offered the accuracy of 98.7% to classify the new data, and the *F*-measure was found 0.8 above in every class
Xu et al. [37]	2019	A novel traffic classification approach	UNB ISCX VPN-nonVPN dataset	The suggested algorithm yielded the accuracy around 96.38% and proved superior over the traditional techniques
Wang et al. [38]	2017	An end-to-end encrypted traffic classification method with one-dimensional convolution neural networks	ISCX VPN-nonVPN traffic dataset	The experimental outcomes revealed that the intended technique was more adaptable in comparison with the traditional technique
Liu et al. [39]	2021	Difficult set sampling technique (DSSTE) algorithm, machine learning and deep learning	NSL-KDD, CSE-CIC-IDS2018	The experimental results indicated that the projected algorithm was applicable for determining the samples whose modification was required in the imbalanced network traffic and enhancing the efficiency to recognize the attack in more successful way

(continued)

Table 3 (continued)

Author	Year	Technique used	Dataset	Results
Shang et al. [40]	2021	Random forest-recursive feature elimination and long short-term memory network with Bayesian optimization algorithm	UNB ISCX VPN-nonVPN dataset	The output of experimentation revealed that the methodology presented had performed better in terms of the majority of the evaluation criteria. It also demonstrated that the presented methodology was outstanding for addressing imbalance issues and small sample dimensions of traffic occurrence data

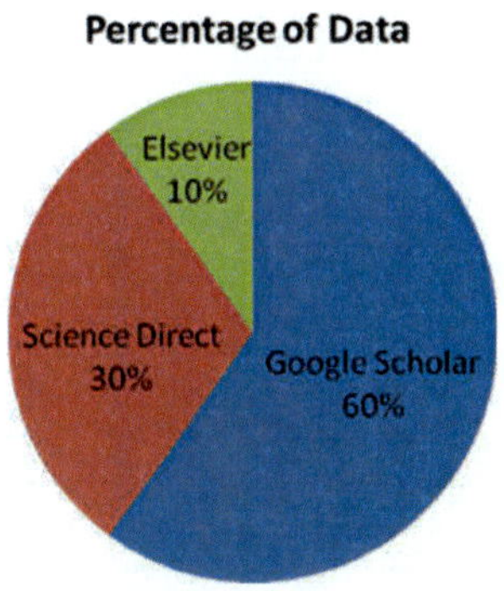

Fig. 2 Percentage of data sharing

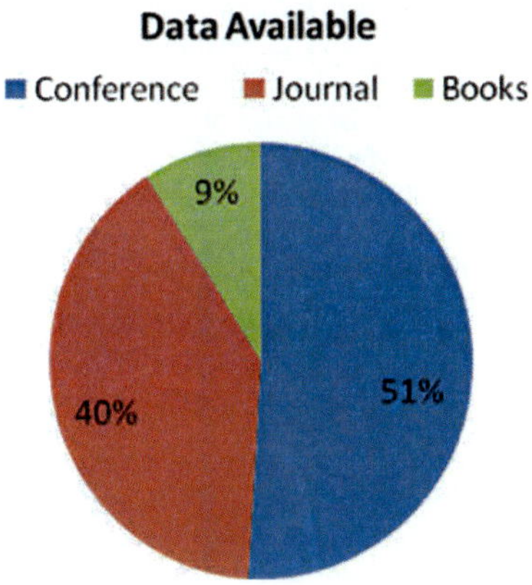

Fig. 3 Data available

4 Conclusion and Future Aspects

The network traffic classification techniques have three categories namely port-based, payload-based and flow statistics-based. The process to classify the network traffic is deal with recognizing distinct kinds of applications or traffic data for which the received data packets are investigated that is essential in the communication networks of real world. The new network management functions such as to ensure the network quality of service (QoS) and for detecting the network anomaly on the basis of accuracy obtained so as the network traffic can be classified. The network is capable of various relevant applications and services. Thus, the differentiation of the network packets or flows is done on the basis of the applications or services offered through the network. It is analysed that machine learning algorithms performance best for the network traffic classification in terms of accuracy, precision and recall. In future, method of hybrid machine learning will be implemented for network traffic classification.

References

1. Hasibi R, Shokri M, Fooladi MDT (2019) Augmentation scheme for dealing with imbalanced network traffic classification using deep learning. arXiv:1901.00204v1
2. Shafiq M, Yu X, Bashir AK, Chaudhry HN, Wang D (2018) A machine learning approach for feature selection traffic classification using security analysis. J Supercomput 74:4867–4892
3. Peng L, Zhang H, Chen Y, Yang Bo (2017) Imbalanced traffic identification using an imbalanced data gravitation-based classification model. Comput Commun 102:177–189
4. Vu L, Bui CT, Nguyen QU (2018) A deep learning based method for handling imbalanced problem in network traffic classification. In: The eighth international symposium, vol 15, pp 3478–3485
5. Tanha J, Abdi Y, Samadi N, Razzaghi N, Asadpour M (2020) Boosting methods for multi class imbalanced data classification: an experimental review. J Big Data 4:6754–6762
6. Liu Q, Liu Z (2014) A comparison of improving multi-class imbalance for internet traffic classification. Inf Syst Front 8:5432–5440
7. Zhena L, Qiong L (2016) A new feature selection method for internet traffic classification using ML. In: International conference on medical physics and biomedical engineering, vol 9, pp 9654–9663
8. Yang J, Wang Y, Dong C, Cheng G (2012) The evaluation measure study in network traffic multi-class classification based on AUC. In: International conference on ICT convergence (ICTC), vol 21, pp 362–367
9. Dhote Y, Agrawal S, Deen AJ (2015) A survey on feature selection techniques for internet traffic classification. In: International conference on computational intelligence and communication networks (CICN), vol 5, pp 1375–1380
10. Wang Z, Wang P, Zhou X, Li S, Zhang M (2019) FLOWGAN: unbalanced network encrypted traffic identification method based on GAN. In: IEEE international conference on parallel and distributed processing with applications, big data and cloud computing, sustainable computing and communications, vol 11, pp 975–983
11. Sharif MS, Moein M (2021) An effective cost-sensitive convolutional neural network for network traffic classification. In: International conference on innovation and intelligence for informatics, computing, and technologies (3ICT), vol 21, pp 40–45

12. Jiang K, Wang W, Wang A, Wu H (2020) Network intrusion detection combined hybrid sampling with deep hierarchical network. IEEE Access 8:32464–32476
13. Sadeghzadeh AM, Shiravi S, Jalili R (2021) Adversarial network traffic: towards evaluating the robustness of deep-learning-based network traffic classification. IEEE Trans Netw Serv Manag 18:1962–1976
14. Bu Z, Zhou B, Cheng P, Zhang K, Ling Z-H (2020) Encrypted network traffic classification using deep and parallel network-in-network models. IEEE Access 8:132950–132959
15. Chen W, Lyu F, Fan Wu, Yang P, Xue G, Li M (2021) Sequential message characterization for early classification of encrypted internet traffic. IEEE Trans Veh Technol 70:3746–3760
16. Wang X, Wang X, Jin L, Lv R, Dai B, He M, Lv T (2021) Evolutionary algorithm-based and network architecture search-enabled multiobjective traffic classification. IEEE Access 9:52310–52325
17. Alizadeh H, Vranken H, Zúquete A, Miri A (2020) Timely classification and verification of network traffic using Gaussian mixture models. IEEE Access 8:91287–91302
18. Iliyasu AS, Deng H (2020) Semi-supervised encrypted traffic classification with deep convolutional generative adversarial networks. IEEE Access 8:118–126
19. Shapira T, Shavitt Y (2021) FlowPic: a generic representation for encrypted traffic classification and applications identification. IEEE Trans Netw Serv Manage 18:1218–1232
20. Yoo J, Min B, Kim S, Shin D, Shin D (2021) Study on network intrusion detection method using discrete pre-processing method and convolution neural network. IEEE Access 9:142348–142361
21. Mezina A, Burget R, Travieso-González CM (2021) Network anomaly detection with temporal convolutional network and U-Net model. IEEE Access 9:143608–143622
22. Hu X, Gu C, Wei F (2021) CLD-Net: a network combining CNN and LSTM for internet encrypted traffic classification. Secur Commun Netw 12:138502–138510
23. Bei Lu, Luktarhan N, Ding C, Zhang W (2021) ICLSTM: encrypted traffic service identification based on inception-LSTM neural network. Symmetry 13:1080–1087
24. Gómez SE, Hernández-Callejo L, Sánchez-Esguevillas AJ (2019) Exploratory study on class imbalance and solutions for network traffic classification. Neurocomputing 343:100–119
25. Guo Y, Li Z, Li Z, Xiong G, Jiang M, Gou G (2020) FLAGB: focal loss based adaptive gradient boosting for imbalanced traffic classification. In: International joint conference on neural networks (IJCNN), pp 1–8
26. Dong S (2021) Multi class SVM algorithm with active learning for network traffic classification. Expert Syst Appl 176
27. Peng L, Zhang H, Yang Bo (2017) Imbalanced traffic identification using an imbalanced data gravitation-based classification model. Comput Commun 102:177–189
28. Saber MAS, Ghorbani M, Bayati A, Nguyen K-K, Cheriet M (2020) Online data center traffic classification based on inter-flow correlations. IEEE Access 8:60401–60416
29. Wang P, Li S, Ye F, Wang Z, Zhang M (2020) PacketCGAN: exploratory study of class imbalance for encrypted traffic classification using CGAN. In: IEEE international conference on communications (ICC), pp 1–7
30. Song M, Ran J, Li S (2019) Encrypted traffic classification based on text convolution neural networks. In: IEEE 7th international conference on computer science and network technology (ICCSNT), pp 432–436
31. Guo Y, Xiong G, Li Z, Shi J, Cui M, Gou G (2021) Combating imbalance in network traffic classification using GAN based oversampling. In: IFIP networking conference (IFIP networking), pp 1–9
32. Wang ZX, Wang P, Zhou X, Li SH, Zhang M (2019) FLOWGAN: unbalanced network encrypted traffic identification method based on GAN. In: IEEE international conference on parallel and distributed processing with applications, big data and cloud computing, sustainable computing and communications, social computing and networking (ISPA/BDCloud/SocialCom/SustainCom), pp 18–25
33. Lopez-Martin M, Sanchez-Esguevillas A, Arribas JI, Carro B (2021) Network intrusion detection based on extended RBF neural network with offline reinforcement learning. IEEE Access 9:153153–153170

34. Pan T, Chen J, Xie J, Zhou Z, He S (2021) Deep feature generating network: a new method for intelligent fault detection of mechanical systems under class imbalance. IEEE Trans Ind Inf 17:6282–6293
35. Oeung P, Shen F (2019) Imbalanced internet traffic classification using ensemble framework. In: International conference on information networking (ICOIN), pp 37–42
36. Zaki FAM, Chin TS (2019) FWFS: selecting robust features towards reliable and stable traffic classifier in SDN. IEEE Access 7:166011–166020
37. Xu L, Zhou X, Ren Y, Qin Y (2019) A traffic classification method based on packet transport layer payload by ensemble learning. In: IEEE symposium on computers and communications (ISCC), pp 1–6
38. Wang W, Zhu M, Zeng X, Ye X, Sheng Y (2017) Malware traffic classification using convolutional neural network for representation learning. In: International conference on information networking (ICOIN), pp 712–717
39. Liu L, Wang P, Lin J, Liu L (2021) Intrusion detection of imbalanced network traffic based on machine learning and deep learning. IEEE Access 9:7550–7563
40. Shang Q, Feng L, Gao S (2021) A hybrid method for traffic incident detection using random forest-recursive feature elimination and long short-term memory network with Bayesian optimization algorithm. IEEE Access 9:1219–1232

Hand Gesture Identification Using Deep Learning and Artificial Neural Networks: A Review

Jogi John and Shrinivas P. Deshpande

Abstract Any human–computer interaction application needs to be able to recognize gestures. Hand gesture detection systems that recognize gestures in real time can improve human–computer interaction by making it more intuitive and natural. Colour gloves and skin colour detection are two prominent hand segmentation and detection technologies, but each has its own set of benefits and drawbacks. For physically challenged people, gesture identification is a crucial technique of sharing information. Support vector machine algorithm with principal component analysis, hidden Markov model, superposed network with multiple restricted Boltzmann machines, growing neural gas algorithm, convolutional neural network, double channel convolutional neural network, artificial neural network, and linear support vector machine with gesture dataset matches the interaction of gestures for various postures in real time. Although this method can recognize a huge number of gestures, it does have certain downsides, such as missing movements due to the accuracy of the categorization algorithms. Furthermore, matching the vast dataset takes a longer time. The main contribution of this work lies in a conceptual framework based on the findings of a systematic literature review that provides fruitful implications based on recent research findings and insights that can be used to direct and initiate future research initiatives in the field of hand gesture recognition and artificial intelligence. As a result, a novel study based on a hybrid recurrent neural network (RNN) with chaos game optimization may reduce classification mistakes, increase stability, maximize resilience, and efficiently use the hand gestures recognition system.

Keywords Artificial intelligence · Hand gesture (HG) · Algorithm · ANN

J. John (✉) · S. P. Deshpande
P.G. Department of Computer Science & Technology, D.C.P.E, Hanuman Vyayam Prasarak Mandal, Amravati University, Amravati, India
e-mail: jogi.john@pcenagpur.edu.in

P. Chatterjee et al. (eds.), *Computational Intelligence for Engineering and Management Applications*, Lecture Notes in Electrical Engineering 984,
https://doi.org/10.1007/978-981-19-8493-8_30

1 Introduction

In a variety of computer–human interaction applications, gesture identification is an important task. As a result, establishing a reliable gesture identification system is a crucial challenge. The risk factor of HG identification is further refined by the hand movement experienced due to imaging quality, viewpoint, and crowded backdrops [1, 2]. Theoretical examination of palm gestures for correct and efficient identification is difficult and fascinating. Real-time hand gesture detection apps are making computer–human interaction more natural and intuitive. Virtual reality and computer games both benefit from this recognition mechanism [3]. Colour gloves [4] and skin colour detection [5] are two common strategies for hand segmentation and detection, but both have few advantages and disadvantages. Applications that recognize hand gestures play an essential role in people's daily lives. This gesture recognition approach not only increases human-to-human communication, but it also serves as a primary source of information.

Because hand gesture recognition is considered simple due to its widespread use in HCI applications, research in this topic has gotten a lot of interest in recent years. Telemedicine [6], interactive augmented reality [7], human–computer interaction interfaces [8], and other uses of this hand gesture recognition are a few examples. HCI interfaces are divided into two types: hardware-based and vision-based techniques [9]. With the use of markers, gloves, magnetic sensors, or other hardware solutions, the hand and feature points from that hand may be easily located. Although these methods produce high-point palm identification results, they are not generally used in real-time applications due to the increased hardware requirements and need of spontaneousness for end users. As a result of its benefits, such as being contact-free and requiring no additional hardware, the image analysis approach has gotten a lot of attention. Furthermore, this image analysis develops a reliable technique for dealing with and normalizing variable environmental conditions, which is the most difficult issue for real-time applications [10].

Because of the development of passive vision-based hand gesture identification systems, images captured by a single camera attain an accurate gesture recognition rate [11]. Hand gestures are divided into two categories: static and motional. When it comes to gesture identification, the features extracted by elastic graph matching are beneficial for distinguishing hand postures from complex backgrounds, with an accuracy of 85%. For recognition, a disjunctive normal-based learning technique is used, which achieves a 93% recognition rate. In learning approach, compactness and normalized hand moments are also applied. The recognition of finger spell is carried out using the CamShift algorithm, which yields a processing rate of 125 ms for single image frames [12]. Principal component analysis (PCA) is also employed in the palm gesture detection process. For motion-based hand gesture recognition, three fundamental techniques are used: HMM, optical flow, and model-based approaches. The construction of a hand gesture model uses a hybrid Adaboost classifier and Haar features, as well as the benefit of a Kalman filter to reduce false detection. A model-based hand gesture tracking was considered for gesture recognition in [13], and the HMM for hand gesture was handled in [14].

A multimodal deep learning approach was studied in order to study the cross-modality representation in an audio-visual speech recognition setting [15]. The deep speech system was proposed lately [16], which merge deftly-optimized RNN with this system to obtain a minimum error rate over a noisy speaking dataset. Deep architectures were used in the recognition process in the methodologies outlined above, resulting in better performance on high-level feature extraction and recognition. However, neural networks may be found in a variety of machine learning-based fields, including prediction of student performance, face recognition, and blood cell identification [17, 18].

The most current and prominent topic in machine learning is deep learning, which involves neural networks but contains a large number of hidden layers (i.e. > 1). In speech recognition, natural language processing, and facial recognition applications, deep learning techniques have a high rate of success [19]. Deep learning-based networks complement the benefits of learning algorithms and biologically inspired frameworks by obviating the need for a standard feed-forward network, resulting in a higher overall recognition rate. In general, deep learning training is done layer-by-layer, much like the human visual cortex, and it relies on a more hierarchical and dispersed feature learning process [20]. Due to this dependence, more interesting features from the highly nonlinear functions are discovered during the training phase and the complex issues are modelled in a perfect manner.

Research Gaps

Conventional classifiers appear superior in acknowledgement execution, but they give moo precision with tall computational fetches. Standard acknowledgement calculations with routine classification methods experienced tall mistakes with less exactness. Whereas signal acknowledgement on any framework cannot be completely distinguished until the end of time, real-time acknowledgement of the sign dialects is basic to protect the frameworks in a viable way. Constrained investigation exists on effective signal acknowledgement models. Errors are more common in writing than they appear on the surface. Some well-known proof models are unable to recognize the sign dialect because the classifier easily falls into nearby ideals in this manner; in this proposed work, it is basic to create a precise classification technique.

Objective

The main contribution of this work lies in a conceptual framework based on the findings of a systematic literature review that provides fruitful implications based on recent research findings and insights that can be used to direct and initiate future research initiatives in the field of hand gesture recognition and artificial intelligence.

2 Related Work

Tan et al. [21] presented an enhanced hand gesture (HG) recognition called enhanced densely connected CNN (EDenseNet) for the recognition of image-based hand gestures. High density blocks were used to support gradient flow and feature propagation in the network. The dense network reduced the number of parameters required for network training while improving parameter efficacy. Tensorflow was used to carry out the implementation. Multiple HG datasets were used in the experiments, including one NUS dataset and two American Sign Language (ASL) datasets. "The proposed EDenseNet has obtained 98.50% accuracy without data augmentation and 99.64% accuracy with data augmentation, outperforming other deep learning approaches" [21]. The major limitations faced while recognizing the HG in images were illumination variations and background noise which affects the recognition accuracy.

Tsinganos et al. [22] developed an HG recognition using the deep learning model named CNN-based Hilbert surface electromyography (EMG). An image of the sEMG signal was created using the space-filled Hilbert curve. Performance was evaluated using a single-fold neural network and a multi-fold neural network architecture. The MSHilbNet network architecture uses multiple scales of an initial Hilbert curve representation to achieve the same level of performance with fewer convolutional layers [22]. Ninapro was the dataset that was used. The HilbTime and HilbElect techniques were tested, and the results revealed that they outperformed the window segmentation approach across a variety of networks [22]. Higher computational time and the use of a single dataset for testing were the major challenges.

Gao et al. [23] proposed multimodal fusion of data and parallel multi-scale CNN-based method for HG recognition. Initially, the data fusion was performed using sEMG signals, RGB images, and depth images of hand gestures. Then, the fused images are sent via parallel CNNs in two different sizes. The image was reduced to two HG detection results [23]. After that, the results were combined together to get the last HG identification result. Finally, testing was conducted using a self-built database including ten typical hand identifications, demonstrating the method's efficacy, and benefit for HG identification. The overall accuracy was 92.45%, and speed 32 ms proves the effectiveness of the fusion model. The limitation of this model was that it could only recognise static hand gesture.

Tan et al. [24] to detect the movement of the palm of a photograph, we have developed a deep learning model based on a convolutional neural network and spatial pyramid pooling. To overcome the difficulties of traditional pooling, the SPP concept was combined with convolutional neural networks. This multi-level linking is built together to increase the quality being fed into a completely connected layer. Provided with an intake of varying measurements, SPP also gives a set-length factor presentation. Extensive experiments were conducted to examine the convolutional neural network-spatial pyramid pooling performance on three datasets such as American Sign Language (ASL), ASL digits, and NUS hand gesture dataset [24]. The results reveal that CNN–SPP won over other deep learning-driven instances. The average accuracy was reported as 98.35% without DA and 99.34% with DA, respectively. The processing of vision-based HG recognition was performed with CNN-SPP, not the dynamic model.

Mujahid et al. [25] a deep learning model was used to show HG recognition in real time. A lightweight model YOLO v3 CNN model was used for palm identification without any extra pre-processing, image filtering, and improvement of images. On a labelled dataset of palm motions in both Pascal VOC and YOLO format, the YOLO v3 model was estimated. The experimentation was done on Python 3.7. The outcomes were achieved by extracting features from the palm and recognized palm gestures with the metrics such as precision, recall, accuracy, and an F-1 score of 97.68, 94.88, 98.66, and 96.70%, respectively. With single shot detector and visual geometry group, which achieved lesser accuracy between 82 and 85%, YOLO v3's overall performance was different. When compared to other DL models, the negative was the lower accuracy value.

Añazco et al. [26] developed HG recognition with the six-axis single patchable inertial measurement unit (IMU) attached to the wrist using recurrent neural networks (RNNs). The IMU is made up of electronic pieces based on integrated circuits that are adhered to a stretchy substrate and have very long organized interconnections. The signal distortion (i.e. motion artefacts) caused by vibrations during motion has been reduced. This patchable IMU uses a Bluetooth connectivity module to send the detected signal to each processing device. Cognitive performance was calculated by placing the current six-axis patchable IMU on the correct wrists of five participants and interpreting three hand gestures using two RNN-based models. In the network training procedure, the 6DMG public database was used. RNN-GRU had a classification accuracy of 95.34%, whereas RNN-BiLSTM had a classification accuracy of 94.12%. The most significant flaw was the intricate design.

Yuan et al. [27] presented the recognition of HG with wearable sensors using deep fusion network. Initially, a specially integrated three-dimensional flex sensor and a data glove with two arm rings were designed to capture fine grain motion from the all knuckles and full arm. Deep fusion network used to detect long distance dependency in complex HG. The CNN-based fusion task was performed to track detailed motion features from multi-sensors by extracting both shallow and deep features. We classified complex hand movements using a long short-term memory model with fused feature vectors. The datasets used for implementation were ASL and CSL datasets on the Python environment. Results of experiments acquired especially in Chinese Sign Language (with 96.1% precision) and American Sign Language (with the precision of 99.93%). The difference between batch size and learning rate has a big impact on training time.

Abdulhussein and Raheem [28] proposed a static HG recognition using deep learning model. The recognition process is characterized using dual phases. The scaling of bicubic static ASL binary pictures was first attempted. The border was discovered using the RE detection method. The CNN model was used to categories the 24 alphabet letters in the second phase. The ASL hand motions dataset was utilized to assess the testing performance. There are 2425 photos in this collection, each containing five persons. The classification accuracy achieved was 99.3%, with a loss function error of 0.0002. In compared to other similar CNN, SVM, and ANN for training efforts, the training time was 36-minute with 100 repeats at 15 seconds and the outcomes were excellent. The tuning of weights was done with SGD optimization which takes longer time to achieve convergence to the minima of the loss function was one of the major limits.

Al-Hammadi et al. [29] introduced a deep learning model for recognizing gestures with the efficient HG representation. The significance of hand motion recognition has increased due to the rapid growth of the hearing-impaired population and the prevalence of touchless applications. The dynamic HG recognition is presented using multiple deep learning architectures for local and global feature representations, hand segmentation, and sequence feature globalization and recognition. The current method is based on a difficult dynamic hand gesture dataset that includes forty dynamic HGs performed by forty people in an uncontrolled setting. The system using MLP fusion attained the maximum accuracy of 87.69% in the signer-independent scenario. The drawback of the system shows that it was not strong enough to capture the long-term temporal dependence of the HG of the video data.

Sharma et al. [30] introduced an HG recognition using the integration of feature extraction (FE) and image processing (IP) techniques. The major objective was to recognize and categorize the hand gestures with its appropriate meaning and much possible accuracy. The pre-processing approaches used were principal component analysis, histogram of gradients, and local binary pattern. The FE techniques used were ORB, bag of word, and canny edge detection methods. The identification of images was characterized with an ASL dataset. To obtain successful results, the preprocessed data was run through multiple classifiers (K-nearest neighbours, random forests, support vector machines, logistic regression, Nave Bayes, and multilayer perceptron). ASL-KNN-ORB had a classification accuracy of 95.81%, whereas ASL-MLP-ORB had a classification accuracy of 96.96%. The new models' accuracy has been shown to be significantly higher than that of older ones. The technology had only been tested on static gesture pictures, which was a flaw.

Ameur et al. [31] introduced a hybrid network model using leap motion and dynamic HG recognition. Initially, HG identification was performed on continuous time series data collected from jump motion using a long short-term memory model network. Both bidirectional and unidirectional long short-term memory model architectures were used separately. The final prediction was made using a hybrid bidirectional unidirectional long short-term memory model. This significantly improves the performance of the model by taking into account the spatial and temporal relationships between the network layer and the jumping motion data during forward and backward paths. The recognition models were tested on the Leap Gesture DB dataset and the RIT dataset, which are both publicly available benchmark datasets. The study demonstrates the HBU long short-term memory network's capability for dynamic hand gesture detection, with average recognition rates of 73.95% and 89.98% on both datasets, respectively. The increasing time consumption posed a significant barrier.

Mirehi et al. [32] developed a meaningful set of shape features using growing neural gas algorithm-based graph construction. The graph properties had improved the stability against different scale, noise, and deformations. This approach has been tested using the latest methods on NTU's manually numbered datasets. In addition, a thorough dataset (SBU-1) for various hand movements was created, which comprises 2170 photos [32]. Many conceivable deformations and variants, as well as certain articulations, were included in this dataset. "The mean accuracy was calculated using several experiments such as utilizing half of the data for training and the other half for testing (h-h), leaving one subject out (l-o-o), and leaving nine subjects out (l-9-o)" [32]. With NTU dataset, the accuracy obtained were 98.68%, 98.6%, and 98%, respectively. The mean accuracy of about 90% was obtained with the SBU-1 dataset. The system model is characterized with challenges such as enhanced time consumption and sensitivity to noise.

Li et al. [33] introduced CNN for gesture recognition. The feature extraction was done within the CNN, so no additional parameters had to be learned. During the recognition procedure, this CNN performed flawlessly in terms of unsupervised learning. The error back propagation method was loaded alongside this CNN, and the weight and threshold of CNN were adjusted to reduce the error created by this model. Finally, the support vector machine (SVM) was combined with this CNN for classification purposes, maximizing the resilience and validity of the whole model. Experimental datasets were employed in the picture information obtained by Kinect, datasets consisting of five persons in the light of eight different motions (G1, G2, G3, G4, G5, G6, G7, and G8), image samples, totalling roughly 2000. MATLAB was used to integrate the gathered colour and depth pictures. Eight species identified in a semi-surveillance scenario [33]. However, the challenge is noticed with the fixed size value as CNN is not as good as the long short-term memory model in the long dependence.

Wu [34] developed a novel algorithm named double channel CNN (DC-CNN) to enhance the recognition rate. Initially, the denoising, pre-processing, and edge detection were performed on input images to spot the hand edge images. The hand edge and hand motion photos were then sent into CNN as input. CNNs have the same number of parameters and convolutional layers as LSTMs, but each layer has a different weight. The entire link layer of CNN was used for functional fusion, and CNN's Softmax classifier was used for classification. The experiment was performed using the NAO camera hand posture database (NCD) and Jochen Triesch database (JTD). The implementation was performed using MATLAB 2017a. The detection rate achieved was 98.02%. The issue represents a half image with a redundant background.

Cheng et al. [35] introduced Kinect sensor-based gesture recognition. Some problems associated with recognizing gestures include poor robustness and accuracy. To address these issues, the planned Kinect sensor was primarily utilized to acquire gesture samples such as depth and colour, which were subsequently analysed. In addition, a network that combines CNN and RBM have been proposed for gesture recognition. This method uses an overlay network with a large number of RBMs to integrate both unsupervised and supervised CNN feature extractions for classification purposes. In basic gesture recognition, simulation analysis with a combined network yields a high recognition rate and a low error rate of only 3.9%. Because RBM requires precise data dispersion, the joint network, complicated sample, and other centralized networks function poorly.

Sharma [36] introduced gesture recognition in Indian Sign Language using Fine-tuned Deep Learning Model accuracy performance is (99.92%) better than existing approaches [36] like Fine-Tuned Deep Transfer learning model (FTDT) 100% (Table 1).

3 Conclusion

In summary, HG offers promising research because it can ease communication and provide a means of interaction that can be used across different real-time applications. Most challenging issues for vision-based gesture identification systems are hand size, variation in skin colour, and viewpoints. Other promises include similarity of gestures, mixed illuminations, and noisy background present in the images. Moreover, the use of wearable devices for HG recognition is significantly limited as signers are required to wear relevant devices beforehand, which entails cost and inconvenience. Due to the presence of background items which are skin-colored, which are regarded as difficult problem for feature extraction. Along with these issues, few other issues also encountered by this system are lighting changes, efficiency, complex circumstances, speed, system inactivity, occlusion, etc. Above all, identification of powerful modelling techniques for capturing the specific sign language is found difficult. Traditional classifiers show better recognition performance, but they provide low accuracy with high computational cost. Ordinary recognition algorithms that used traditional classification approaches had a lot of errors and were less accurate. The majority of the research publications focus on improving hand gesture detection frameworks or building new algorithms. The greatest challenge faced by the

Table 1 Hand gesture recognition: a review of existing approaches

Author name	Method used	Objective	Advantages	Disadvantages
Tan et al. [21]	EDenseNet	Recognition of image-based hand gestures	Minimized the number of parameters for network training and enhanced the parameter efficacy	The image was a fluctuation of lighting and background noise, which impaired the recognition accuracy
Tsinganos et al. [22]	CNN-based Hilbert surface EMG	HG recognition using the deep learning model	When compared to the window segmentation approach, the performance across multiple networks is superior	Higher computational time and the use of a single dataset for testing were the major challenges

(continued)

Table 1 (continued)

Author name	Method used	Objective	Advantages	Disadvantages
Gao et al. [23]	Multimodal fusion of data and parallel multi-scale CNN	HG recognition	Self-build database containing ten common hand identification	Only static hand gestures were identified
Tan et al. [24]	Convolutional neural network–spatial pyramid pooling	Identification of palm gestures in images	CNN–SPP won over other deep learning-driven instances	Not a dynamic model
Mujahid et al. [25]	YOLO v3 CNN model	Palm identification without any extra pre-processing, image filtering, and improvement of images	Based on Pascal VOC and YOLO format tagged palm gesture datasets	Lower precision value
Añazco et al. [26]	HG recognition with the six-axis single patchable inertial measurement unit (IMU) attached to the wrist using recurrent neural networks (RNN)	HG recognition	The network training method used public database 6DMG	Complicated design
Yuan et al. [27]	Deep fusion network is used to recognize HG utilizing wearable sensors	Recognition of HG	We classified complex hand movements using a long short-term memory model with fused feature vectors	Training time is greatly affected by the difference between batch size and learning rate
Abdulhussein and Raheem [28]	Bicubic static ASL, RE detection approach, and CNN model	Static HG recognition using deep learning models	The training was completed in a short amount of time and with excellent outcomes	Takes longer time to achieve convergence
Al-Hammadi et al. [29]	Deep learning model for recognizing gestures	Recognition of HG	The system using MLP fusion attained the highest accuracy in the signer-independent scenario	It was not powerful enough to capture the video data's HG's long-term temporal dependency

(continued)

Table 1 (continued)

Author name	Method used	Objective	Advantages	Disadvantages
Sharma et al. [30]	ASL-MLP-ORB	Recognize and classify hand gestures with the proper which means and as a great deal precision as feasible	It turns out that the accuracy of the new model is significantly higher than that of the previous model	Only static gesture images were used to test the system
Ameur et al. [31]	Hybrid network model using leap motion and dynamic HG recognition	HG recognition	Improves the model performance	Higher time consumption
Mirehi et al. [32]	Growing neural gas algorithm-based graph construction	HG recognition	Enhanced stability in the face of varying scales, noise, and deformations	Enhanced time consumption and sensitivity to noise
Li et al. [33]	CNN	Gesture recognition	Minimize the error produced	Not as good as the long short-term memory model in the long dependence
Wu [34]	DC-CNN	To enhance the HG recognition rate	The rate of recognition achieved is pretty good	The dataset posture represents a half image with a background that is redundant
Cheng et al. [35]	CNN and RBM joint network	Kinect sensor-based gesture recognition	In basic gesture recognition, simulation analysis with a combined network yields a high recognition rate and a low error rate of only 3.9%	The performance of joint networks, complicated samples, and other centralized networks is poor

researcher is to develop a robust framework that controls the most common issues with fewer limitations and accuracy and reliability. Hence, a new study based on hybrid recurrent neural network with chaos game optimization is needed which aims to perform effective HG recognition.

References

1. Li SZ, Yu B, Wu W, Su SZ, Ji RR (2015) Feature learning based on SAE–PCA network for human gesture recognition in RGBD images. Neurocomputing 151:565–573
2. Pugeault N, Bowden R (2011) Spelling it out: real-time ASL fingerspelling recognition. In: IEEE international conference on computer vision workshops (ICCV workshops), pp 1114–1119
3. Wachs JP, Kölsch M, Stern H, Edan Y (2011) Vision-based hand-gesture applications. Commun ACM 54(2):60–71
4. Wang RY, Popović J (2009) Real-time hand-tracking with a color glove. ACM Trans Graphics (TOG) 28(63):1–8
5. Lee T, Hollerer T (2009) Multithreaded hybrid feature tracking for markerless augmented reality. IEEE Trans Visual Comput Graphics 15(3):355–368
6. Wachs J, Stern H, Edan Y, Gillam M, Feied C, Smith M, Handler J (2006) A real-time hand gesture interface for medical visualization applications. In: Applications of soft computing. AISC, vol 36. Springer, Heidelberg, pp 153–162
7. Shen Y, Ong SK, Nee AYC (2011) Vision-based hand interaction in augmented reality environment. Int J Human-Comput Interact 27(6):523–544
8. Lee DH, Hong KS (2010) Game interface using hand gesture recognition. In: Proceedings of the 5th international conference on computer sciences and convergence information technology (ICCIT 2010). IEEE, pp 1092–1097
9. Czupryna M, Kawulok M (2012) Real-time vision pointer interface. In: Proceedings of the 54th international symposium ELMAR (ELMAR 2012). IEEE, pp 49–52
10. Huang Y, Monekosso D, Wang H, Augusto JC (2011) A concept grounding approach for glove-based gesture recognition. In Proceedings of the 7th international conference on intelligent environments (IE 2011). IEEE, pp 358–361
11. Rodriguez S, Picon A, Villodas A (2010) Robust vision-based hand tracking using single camera for ubiquitous 3D gesture interaction. In: Proceedings of IEEE symposium on 3D user interfaces. Waltham, MA, pp 135–136
12. Park A, Yun S, Kim J, Min S, Jung K (2009) Real-time vision based Korean finger spelling recognition system. Int J Electr Comput Eng 4:110–115
13. Ren Y, Zhang F (2009) Hand gesture recognition based on MEB-SVM. In: Proceedings of international conference on embedded software and systems, Hangzhou, China, pp 344–349
14. Bradski G, Davis J (2000) Motion segmentation and pose recognition with motion history gradients. In: Proceedings of IEEE workshop on applications of computer vision, Palm Springs, CA, pp 238–244
15. Ngiam J, Khosla A, Kim M, Nam J, Lee H, Ng AY (2011) Multimodal deep learning. In: Proceedings of 28th international conference on machine learning, pp 689–696
16. Hannun A, Case C, Casper J, Catanzaro B, Diamos G, Elsen E, Prenger R, Satheesh S, Sengupta S, Coates A (2014) Deepspeech: scaling up end-to-end speech recognition. arXiv preprint arXiv:1412.5567
17. Khashman (2012) A investigation of different neural models for blood cell type identification. Neural Comput Appl 21(6):1177–1183
18. Oyedotun OK, Tackie SN, Olaniyi EO, Khashman A (2015) Data mining of students' performance: Turkish students as a case study. Int J Intell Syst Appl 7(9):20–27
19. Oyedotun OK, Khashman A (2017) Deep learning in vision-based static hand gesture recognition. Neural Comput Appl 28(12):3941–3951
20. Kruger N, Janssen P, Kalkan S, Lappe M, Leonardis A, Piater J, Rodriguez-Sanchez AJ, Wiskott L (2012) Deep hierarchies in the primate visual cortex: what can we learn for computer vision? IEEE Trans Pattern Anal Mach Intell 5(8):1847–1871
21. Tan YS, Lim KM, Lee CP (2021) Hand gesture recognition via enhanced densely connected convolutional neural network. Expert Syst Appl 175:114797
22. Tsinganos P, Cornelis B, Cornelis J, Jansen B, Skodras A (2021) Hilbert sEMG data scanning for hand gesture recognition based on deep learning. Neural Comput Appl 33(7):2645–2666

23. Gao Q, Liu J, Ju Z (2021) Hand gesture recognition using multimodal data fusion and multiscale parallel convolutional neural network for human–robot interaction. Expert Syst 38(5):e12490
24. Tan YS, Lim KM, Tee C, Lee CP, Low CY (2021) Convolutional neural network with spatial pyramid pooling for hand gesture recognition. Neural Comput Appl 33(10):5339–5351
25. Mujahid A, Awan MJ, Yasin A, Mohammed MA, Damaševičius R, Maskeliūnas R, Abdulkareem KH (2021) Real-time hand gesture recognition based on deep learning YOLOv3 model. Appl Sci 11(9):4164
26. Añazco EV, Han SJ, Kim K, Lopez PR, Kim T-S, Lee S (2021) Hand gesture recognition using single patchable six-axis inertial measurement unit via recurrent neural networks. Sensors 21(4):1404
27. Yuan G, Liu X, Yan Q, Qiao S, Wang Z, Yuan Li (2020) Hand gesture recognition using deep feature fusion network based on wearable sensors. IEEE Sens J 21(1):539–547
28. Abdulhussein AA, Raheem FA (2020) Hand gesture recognition of static letters American sign language (ASL) using deep learning. Eng Technol J 38(6A):926–937
29. Al-Hammadi M, Muhammad G, Abdul W, Alsulaiman M, Bencherif MA, Alrayes TS, Mathkour H, Mekhtiche MA (2020) Deep learning-based approach for sign language gesture recognition with efficient hand gesture representation. IEEE Access 8:192527–192542
30. Sharma A, Mittal A, Singh S, Awatramani V (2020) Hand gesture recognition using image processing and feature extraction techniques. Procedia Comput Sci 173:181–190
31. Ameur S, Khalifa AB, Bouhlel MS (2020) A novel hybrid bidirectional unidirectional LSTM network for dynamic hand gesture recognition with leap motion. Entertainment Comput 35:100373
32. Mirehi N, Tahmasbi M, Targhi AT (2019) Hand gesture recognition using topological features. Multimedia Tools Appl 78(10):13361–13386
33. Li G, Tang H, Sun Y, Kong J, Jiang G, Jiang D, Tao B, Xu S, Liu H (2019) Hand gesture recognition based on convolution neural network. Cluster Comput 22(2):2719–2729
34. Wu XY (2019) A hand gesture recognition algorithm based on DC-CNN. Multimedia Tools Appl 1–3
35. Cheng W, Sun Y, Li G, Jiang G, Liu H (2019) Jointly network: a network based on CNN and RBM for gesture recognition. Neural Comput Appl 31(1):309–323. Li SZ, Yu B, Wu W, Su SZ, Ji RR (2019) Feature learning based on SAE–PCA network for human gesture recognition in RGBD images. Neurocomputing 151:565–573
36. Sharma CM, Tomar K, Mishra RK, Chariar VM (2021) Indian sign language recognition using fine-tuned deep transfer learning model. In: Proceedings of international conference on innovations in computer and information science (ICICIS), pp 62–67

Various Aspects of IOT, Machine Learning and Cyber-Network Security

IoT-Assisted Solutions for Monitoring Cancer Patients

Rohit Tanwar and Keshav Kaushik

Abstract The emergence of the Internet of Things (IoT) has significantly influenced and shaped the world of innovation in terms of network, interconnection, and compatibility with smartly embedded devices, devices, electronics, data, and services. IoT has had a significant influence on global production and social experience, ranging from company to company and across various applications, especially medical services. It is necessary to foster a constant connection and communication of items (gadgets) with individuals on the planet using the Internet of Things. As a result, it is critical to comprehend the potential and benefits of IoT innovation in delivering medical solutions to ensure that lives are saved and that individual happiness is improved via smart connected devices. In this article, we have concentrated on an IoT-assisted solution for cancer patient care and monitoring. In addition, the decision and implementation of IoT/WSN technology to extend present treatment options and deliver pharmaceutical services arrangements were emphasized. In this case, business investigation/cloud services set up the enabling components for significant information examination, interactivity, information conveyance, and disclosing for improving illness medications. As a result, we emphasized a range of configurations and layouts to demonstrate and support the helpful IoT-based organization that is being considered or employed in our targeted smart emergency clinics that provide solutions for cancer care. Lastly, we have included the IoT-related findings in the cancer-care area.

Keywords Internet of things (IoT) · Cancer care · Healthcare system · Patient monitoring · Smart devices

R. Tanwar (✉) · K. Kaushik
School of Computer Science, University of Petroleum and Energy Studies, Dehradun, India
e-mail: rohit.tanwar.cse@gmail.com

K. Kaushik
e-mail: keshavkaushik@ddn.upes.ac.in

P. Chatterjee et al. (eds.), *Computational Intelligence for Engineering and Management Applications*, Lecture Notes in Electrical Engineering 984,
https://doi.org/10.1007/978-981-19-8493-8_31

1 Introduction

Internet of things (IoT) has established one of the latest perceptive ideas of the advanced age, whose effects have been seen in each part of human undertaking with incredible possibilities for smarter living. Through IoT, gadgets/items, systems, and applications are being associated with encourage a trade of information and data subsequently prompting the enablement and initiation of an assortment of administrations. Fundamentally, it assists with conveying and coordinate a few system advances (conventions, topologies, wired/remote network, structures, frequency bands, and so on.) and correspondence arrangements utilizing the associated gadgets. Curiously, the IoT innovation has bolstered an assortment of administrations and applications. These incorporate shrewd city, keen home mechanization, brilliant traffic control, perceptive leaving, keen lighting, keen office, and brilliant vehicle control. Google, for example, has recently used the Internet of Things to guide self-driving automobiles by connecting them to the transportation framework and road constraints. In this case, the cars may drive and enjoy the world and highway on their own, with both human and autonomous cars being integrated into the entire transport system to ensure secure and safe driving. This investigation looks at two of the many agencies that are acceptable for smart human services delivery regarding unavoidable social security and patient monitoring. Cancer development solutions and company evaluation establishments are two of them.

This is left upon to upgrade and improve living, and furthermore to build the personal satisfaction, which we see will elevate and prompt a proactive instead of a responsive social insurance conveyance and experience. In particular, we have proposed the usage of IoT innovation in malignancy care conveyance alongside the consolidation of business investigation/cloud benefits as empowering influences for disease care administrations. More or less, the blend of these administrations is proposed to offer a system, engineering, and arrangement using sensor systems, brilliant associated gadgets, and information investigation apparatuses/strategies to improve, screen, and upgrade malignancy medicines. Likewise, the IoT execution will improve the personal satisfaction while simultaneously helping social insurance suppliers to transform a surge of information into a knowledge, significant experiences, and proof-based medicinal services choices. In light of this, it is foreseen that the advanced IoT-empowered solution will improve the effectiveness and nature of social insurance conveyance. Wearable IoT devices are designed with certain qualities that make them a suitable component of the human body. Edge computing and other cutting-edge technologies are aiding wearables in delivering the anticipated performance. This article Singh et al. [18] aims to emphasize the importance of wearables in IoT healthcare, as well as the design principles of the newest mobile sensors and different IoT networking devices. In particular, the features of the various wearables utilized in IoT healthcare are highlighted in this article.

The article's goal is to implement a Smart Integrated IoT Healthcare System for Cancer Treatment. As observed, it tends to be viewed as an action plan or, best yet, notwithstanding all structure utilization, a decision afterwards on is to accomplish it on a certain scale. However, depending on the three stages of construction: pre-usage, real-world usage (go-live), and post-usage, it might be perceived or contrasted to pre-execution. Consequently, we foresee directing an approval by then, as it will be exceptionally basic to guarantee a structure usage that characterizes the elements of different foundations inside the general system in meeting the crucial system plan objectives of versatility, accessibility, execution, excess, practicality, security, resilience, and sensibility that will guarantee expanded or upgraded interconnectivity, interoperability, and intercommunication as contended. In spite of the fact that the idea of IoT has been concentrated in the writing with numerous human services structures and systems, generally centered around general medicinal services related encounters with none on the utilization of IoT to disease care benefits as we have completed.

2 Related Work

The literature of past 10 years is considered for deep understanding the use of IoT in cancer. The keyword used for searching the relevant publication on google scholar is ("iot" OR "internet of things") AND ("Cancer"). The methodology that has been followed for literature review has been shown in Fig. 1.

The results retrieved from the Google Scholar as a result of the query are further used to derive various types of information. One of the information derived is the trend of year-wise publication on IoT in Cancer as shown in Fig. 2.

We can observe that it took almost five years to publish the first work on IoT in cancer. Even after that, the growth was not significant for the next three years. After 2018, the researches in this domain started increasing. However, the work done was still very less.

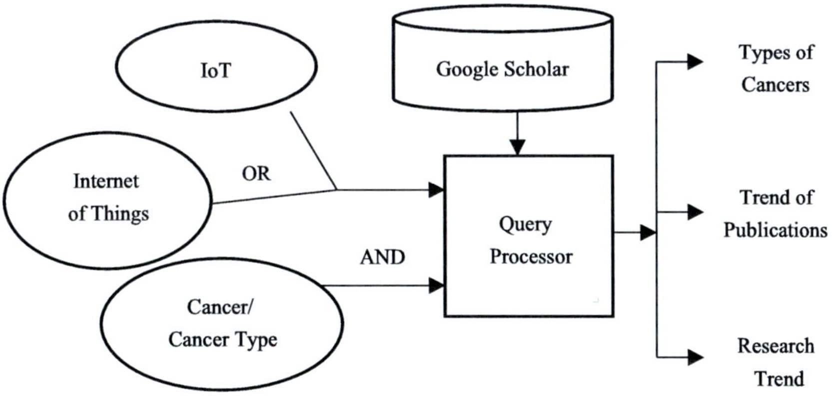

Fig. 1 Methodology of literature review

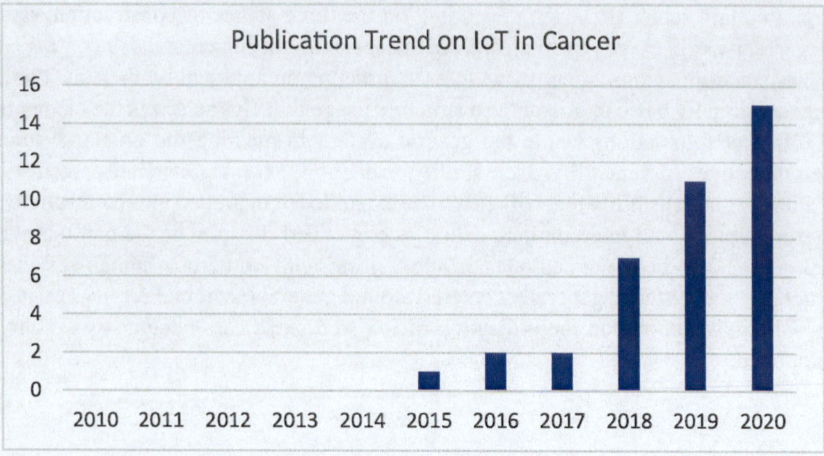

Fig. 2 Publications on IoT in cancer

In order to investigate the work done on specific type of cancer, the keyword used for searching the relevant literature on Google Scholar was modified appropriately by prefixing type of cancer. For example ("iot" OR "internet of things") AND ("Breast Cancer") to search for the publications on "Breast Cancer". Similarly, the number of publications in each year was find out for thirteen types of the cancer, respectively. Figure 3 shows the number of publications in each year for different types of cancer.

From the analysis, it can be observed that, initially the researchers focused on cervical cancer, prostate cancer, and thyroid cancer. That too started after 2014. Later on, the other types of cancers were considered for research. There are certain kind cancers that are not explored yet.

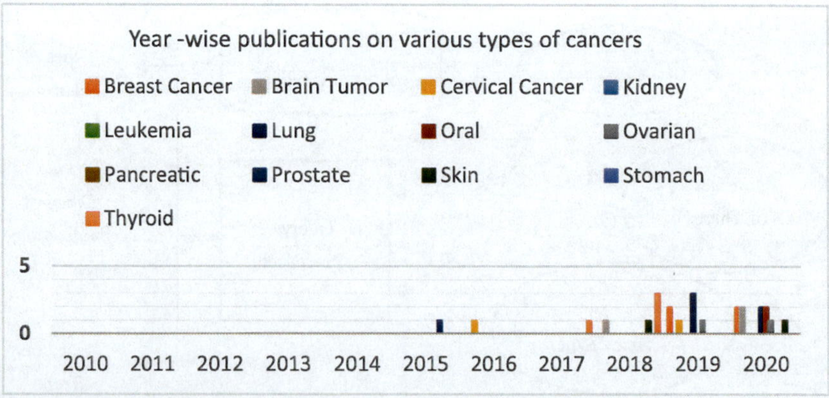

Fig. 3 Year-wise publications on different cancer types

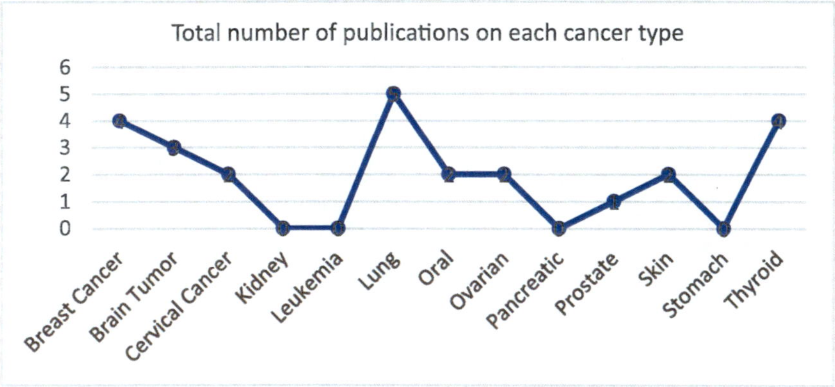

Fig. 4 Number of publications on each cancer type

Figure 4 represents the number of publications on each type of cancer in the last 10 years. It helps in identifying the unexplored areas in this domain to attract researchers. It can be observed that the average number of publications on the of use IoT in treating various types of cancers is '2' which is very less. Kidney cancer, leukemia, pancreatic cancer, and stomach cancer are among those types where no publication is done till now.

Figure 5 gives a proportionate view of research that has been published in the last one decade on IoT and Cancer. Lung cancer, breast cancer, and thyroid cancer are dominating as a preferred choice of researchers. Their contributions are 20% and 16%, respectively. While most of the other types of cancers are in initial stage of research with 4% and 8% contributions generally, some of the areas are still not explored. Table 1 summarizes the relevant literature studied.

3 IoT Applied on Cancer Types

In this section, the existing research work on IoT applied on different types of cancers has been discussed.

3.1 Internet of Things and Lung Cancer

The IoT is extremely essential to everyone for its numerous applications in various fields. It may be used in a variety of settings, including healthcare, smart buildings, etc. These applications are used to reduce emergency department long waits, monitor patient wellbeing, supervise personnel practices, manage inventory, and control essential equipment like the hearable, which enables persons with hearing

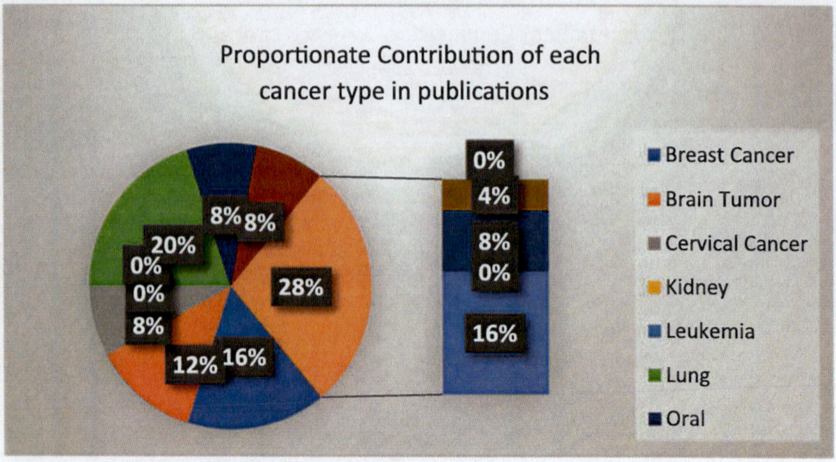

Fig. 5 Research proportion of each type of cancer

impairment interact with their surroundings. In Palani and Venkatalakshmi [11], the Single Internet of Things (IoT) is implemented for aided modeling through continuous monitoring and improvements in health care by providing diagnostic advice, on a classification for the prediction of lung cancer and fumigational cluster-based increase. The fuzzy clustering method, based on transitional zone extraction, for effective image segmentation is used. The right edge image and anatomical dilution are both used to increase the differentiation precision. To acquire the entity regions a lung cancer image is performed by morphological cleaning and image area filling. Figure 6 shows the process that is discussed in paper [11].

Smart medicine has recently evolved as a cross-disciplinary field as a result of the integration of multiple computing methods into medicine. Smart medicine's key purpose is to provide them with ubiquitous and customized treatment and medical facilities. Examples of both are computer-aided forecasts and decisions for a personalized care plan. Specifically, intelligent Chinese medicine aimed at providing a differentiation in intelligent syndrome by introducing artificial intelligence techniques into traditional Chinese clinical practice, with edge/cloud computing, is a promising case of intelligent medicine. Easy detection of lung cancer and care can enhance the efficacy of treatment and prolong survival. The representative, profound strengthening learning models for the diagnosis of lung cancer is discussed. In Liu et al. [6], the recent problems and potential research aspects in the future using deep reinforcement learning to diagnose lung cancer are discussed, which is supposed to accelerate the advancement of smart medicine across the medical Internet of Things. An analysis of nearly 65 articles using machine learning algorithms to forecast various diseases was performed in Pradhan and Chawla [13]. The research focuses on a variety of machine learning techniques that are used to predict a variety of diseases in order to identify a need that can be filled in the future for diagnosing the lung cancer in IoT healthcare. Each methodology was examined systematically, and

Table 1 Summary of related work

Citation	Technology used	Type of cancer	Remarks
Memon et al. [7]	Machine learning algorithms like SVM	Breast cancer	To distinguish benign and malignant persons, a sequential backward evaluation method was employed to pick highly appropriate characteristics from a breast cancer dataset
Palani and Venkatalakshmi [11]	Fuzzy cluster algorithm, Association rule mining to existing decision tree	Lung cancer	An IoT-based prediction model with fuzzy cluster enhancement and constant monitoring for detecting lung cancer
Rahman et al. [15]	DApps, AI model, Blockchain and off-chain technologies	Post cancer treatment	A cancer patient's in-home efficiency monitoring system that enabled for day-to-day surveillance. A variety of performance detectors, as well as ambient sensing systems, is employed
Valluru et al. [19]	Optimal SVM, GWO-GA, MATLAB2014b	Lung cancer	Usage of optimized support vector machine for the detection of lung cancer assisted by IoT-based cloud system
Han et al. [5]	CNN algorithm, BAS algorithm	Cancer rehabilitation	This model is very useful for aiding doctors in selecting the personalized nutrition and rehabilitation program for the patient
Onasanya and Elshakankiri [9]	WSN, Hadoop cluster		Internet of things healthcare solutions with protected cancer care and cloud solutions. Additional advantages of utilizing this include its quick reaction time and real-time capabilities

(continued)

Table 1 (continued)

Citation	Technology used	Type of cancer	Remarks
Savitha et al. [17]	DK-AES algorithm, OKM-ANFIS algorithm	Breast cancer	On the Internet of Things, the authors recommended using a decentralized key authentication system in conjunction with a breast cancer predictive model
Elouerghi et al. [4]	CMS type heat microsensor, OpenCV-python, PHP script	Breast cancer	The authors proposed a new way of diagnosis based on an ultrasensitive micro-bio-heat sensor. A matrix design of 3 × 3 bio-microsensors was proposed in this paper
Prachumrasee et al. [12]	RFID-based system, Google data sheets, and Google app scripts	Breast cancer	Authors used an IoT strategy toward the pre-analytical process of moving breast cancer samples from OR to the laboratory
Premavathi et al. [14]	M4, M5, M7 subtype tumor cells image preprocessing, segmentation, classification, and transferring the information through IoT module	Leukemia	Image segmentation and classifications are explained with the help of algorithms, but the IoT module isn't explained properly. They would have improved the implementation of a proper IoT architecture
Zaminpira et al. [20]	Intravenous Ozone Therapy (IOT) and Hyperbaric Oxygen Therapy (HBOT) with a Ketogenic Diet (SKD) (HBO2T)	Brain, breast, kidney, liver, lung and colorectral cancer	This diet is mainly to weak the cancer tumor cells by providing less amount of protein and carbohydrate. 180 days of this diet have resulted improvement of lifestyles in cancer patients

(continued)

Table 1 (continued)

Citation	Technology used	Type of cancer	Remarks
Reethu et al. [16]	Temperature sensor, flow sensor, and pH sensor, followed by fetching these information to MATLAB and sense via SVM (Machine learning algorithm)	Oral cancer	A smart device which monitors the damaged tissues of the tumor cells in order to analyze the data using SVM algorithm for early detection of oral cancer
Onasanya and Elshakankiri [8]	IoT architecture for WSN's and the lab equipment of cancer treatments. Hadoop cluster framework for business analytics	Cancer care system	WSN's are used for monitoring the changes in the patient. IoT framework not only consists of the WSN's but also the servers of the hospital equipment. The data from pathology, radiology systems sends all the information through cloud and this data is used for the business analytics so that cancer prediction will be more accurate
Onasanya and Elshakankiri [10]	Mesh technology along with WSN and also Hadoop	Cancer care system	Smart IoT integrated health care system for cancer care. Different layers are transferring the information to the top layers in the network for an efficient cancer treatment monitoring

the overall disadvantages were identified. Furthermore, it examines the form of data used to forecast the disease in question, whether it is metric or manually gathered data. Healthcare's Digital Twin, which will be a significant component of the next 6G network, is a virtual subjective experience that uses IoT and artificial intelligence models to anticipate wellness and respond to a variety of clinical queries. To enable healthcare smart sensors, the necessary cyber resilience techniques and regulations should be established and preserved [21]. Vulnerability detection is vital information management infrastructure in digital twins in healthcare. Deep learning (DL) was recently used in the discovery of bugs to solve the weaknesses of classical machine learning.

It is critical to comprehend code contextual relationships and pay attention to weakness phrases while searching for an IoT vulnerability in medical digital twins. A completely automated technique for aiding cyber readiness tests in genuine situations

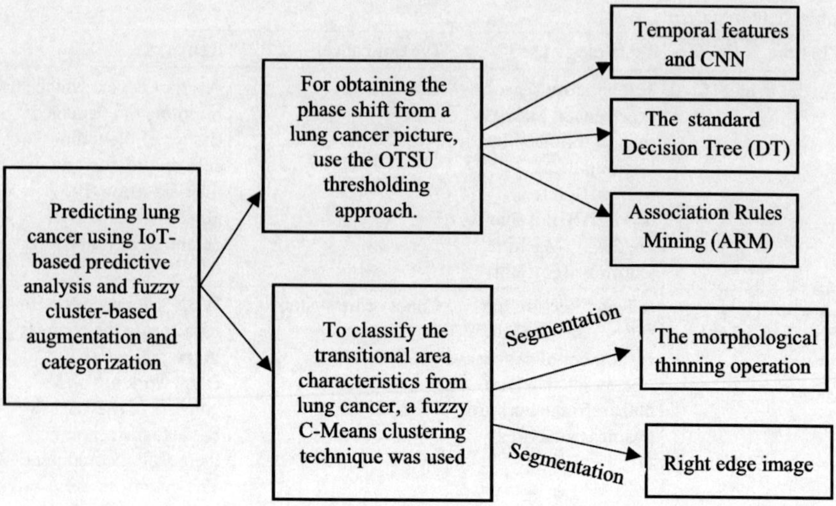

Fig. 6 IoT-assisted cluster-based prediction model for lung cancer

is important due to the vast programs and complexity of the medical smart city. The Internet of Things (IoT) and cloud hosting are closely linked, and they may be used to follow patients in far-flung places. It is made possible by the cloud's near-infinite capabilities and tools, which allow it to overcome technological limitations such as memory and computational capacity. The IoT-centric-cloud architecture will be utilized to develop new products and services in the healthcare industry. The mortality rates have been increasing when cancer is detected at an earlier age. Early detection, on the other hand, is a challenging task. In the intermediate or late stages of cancer, around a third of individuals [19] are accurately categorized. Figure 7 shows an IoT-assisted lung cancer detection utilizing an optimum SVM classifier.

Diagnostic-assisted computer models are found to be useful in early and fast detecting and diagnosing abnormal. In order to predict the prevalence of lung cancer by computer tomography, images (CT) machine learning (ML) models have been used. In controlled and unattended methods of classification, current image classification methods may be categorized. The purpose of the Onasanya and Elshakankiri [9] is specifically to incorporate IoT technologies in the provision of carcinogenic health care, as well as market intelligence and cloud computing for treatment and diagnostics of cancer. The combination of these services offers a solution and platform for the analysis of IoT health data from various smart connected devices and other sensor networks, which lets healthcare providers transform the data stream into actionable observations and evidentiary decision-making regarding patients' health using the appropriate analytical resources to enhance and improve the situation.

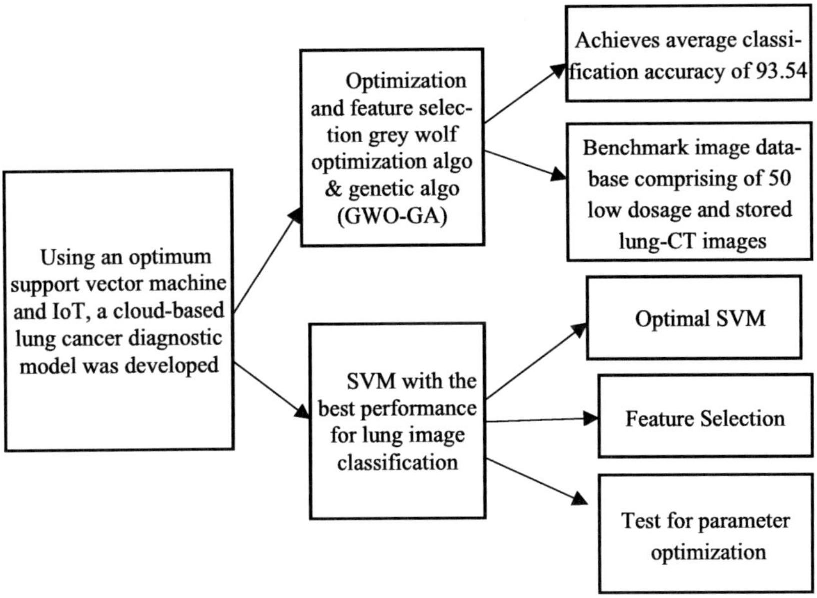

Fig. 7 IoT-assisted lung cancer diagnosis using optimal support vector machine

3.2 *Internet of Things and Ovarian Cancer*

Analytical methods are now being developed to develop innovative biosensor methods for effective point-of-care screening of biomarkers for diseases, such as the newly discovered early diagnostic epididymium [1] protein 4 (HE4) serum marker. This is an initial use of enzyme-labeled magnetic characteristics to read differentiated pulse voltammetry with a home built IoT Wi-Fi cloud-based portable potentiostat. The electrochemical method has been designed to allow independent measurement and collection of information, alternating between calibration and measurements: first, a base analysis approach is applied for the proper top calibration; then a calibration function consists of the interpellant data with a 4-parameter logistic feature. In order to assess concentration of unknown samples, the calibration parameters are processed in the cloud for reverse prediction. Calibration of the interpolation mechanism and concentration assessment was carried out directly on board reducing the energy consumption.

Ovarian cancer (OC) is a form of cancer that influences women's ovaries and that is not easily found in the initial phase, leading to an increasing death rate. Distinguishing OCs can be used with data created from the Internet of Medical Things (IoMT). We use self-maps and optimal recurrent neural networks to characterize OC to accomplish this objective. SOM algorithms were used for improved sub-set collection of functions and were used to separate profitable, comprehensive, and fascinating data from large medical data measurements. An optimal classifier is also

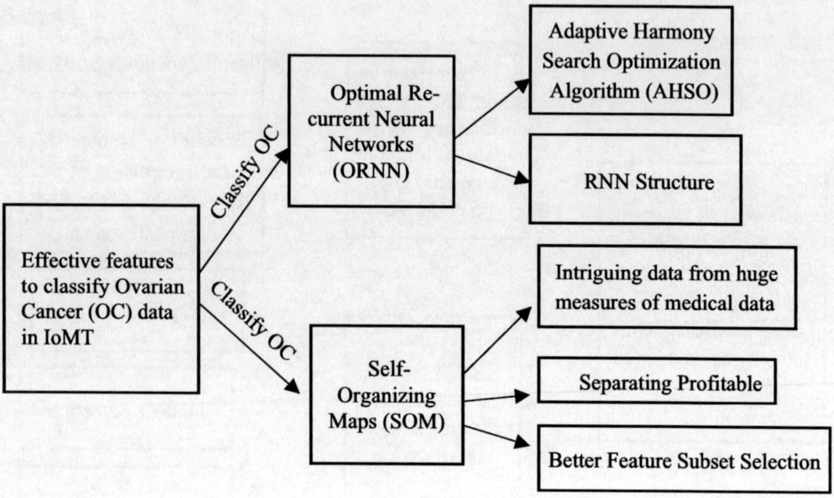

Fig. 8 Effective features to classify ovarian cancer

used, called optimal recurrent network (ORNN). Effective features to classify ovarian cancer is shown in Fig. 8 using the Adaptive Harmony Search Optimization (AHSO) [3] algorithm for optimizing the classification rate of the OC detection process. A series of experiments were performed using data obtained from women who are at great risk of OC due to their family or particular cancer records.

3.3 Internet of Things in Thyroid Cancer

It is described as the development of cancerous cells in the thyroid gland which is presented at the base of the throat. Generally, it is difficult to feel this gland through the skin. Generally, the thyroid cancer is classified into two categories: (i) differentiated thyroid cancer, and (ii) medullary thyroid cancer.

It is usually simple to treat differentiated thyroid cancer. Different procedures are used to test thyroid cancer like, laryngoscopy, blood hormone studies, ultrasound, CT Scan, etc. Cui et al. [2] developed an IoT-based medical system for thyroid care risk as shown in Fig. 9. The estimation of thyroid cancer is done by monitoring the basal body temperature. Based on the temperature variations, the outcome is determined.

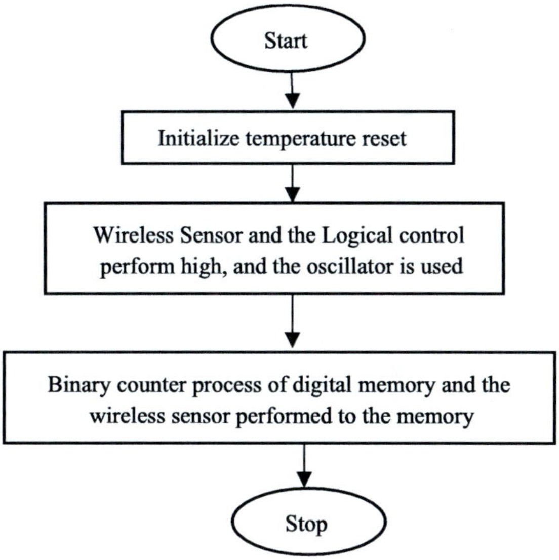

Fig. 9 IoT-based medical system for thyroid care risk [2]

4 IoT in Cancer Care

IoT is one of the emerging technology that has impacted the human life to a great extent in recent days. The topping of IoT in various applications ranging from education to healthcare has shown great improvements. The cancer patients demand an utmost care in terms of drugs, real-time monitoring, and timely reporting of various body parameters. Foreseeing the impact of IoT in other applications, various frameworks have been suggested by researchers for cancer care. A general IoT-enabled framework for cancer care is shown in Fig. 10 [9].

The working of various layers are described as:

1. **Service Layer**: This layer provides various smart services. These services are highly impacted by the observations and results of different IoT devices used. It supports the decision-making process in terms of diagnosis, detection, and treatment.
2. **Data Center Layer**: This layer is responsible for movement of data from smart devices to a repository. The information can be used in many clinical applications from the repository.
3. **Hospital Layer**: This layer is responsible for communication and interaction with different healthcare centers, like pathology, various scans, and physicians and medical practitioners. It will help all to be on same path of treatment.
4. **Cancer Care Layer**: This layer provides the required interaction among patient, the equipment being used in treatment, etc. By the use of smart devices (IoT), it provides various types of visualizations. The required smart services are ensured even through remote access using suitable communication.

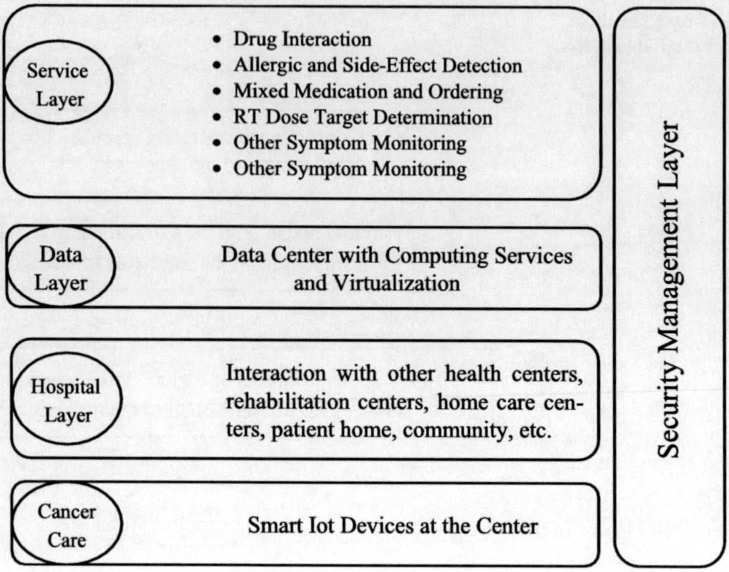

Fig. 10 IoT-enabled cancer care system

5. **Security Management Layer**: The required security of patient data, reports, security of the devices, communications, etc., is taken care by this layer.

5 Observations

The use of technology is benefiting the healthcare professionals as well as patients these but still many challenges are there that need resolutions. Some of the challenges are listed below:

- Standard government policies are not there.
- Lack of regulations for compatible interfaces.
- Interoperability issues.
- Vulnerability of security and devices.
- Device diversity.
- Ownership of data collected and stored.
- Power capacity of connected devices.
- Strength and security of communication channels.
- Reliability of network communication.

A lot of research is expected from the science community in this domain to exploit the benefit of IoT in healthcare in greater extent.

6 Conclusion

Due to various numbers of connected devices, IoT is a game-changing technology. In this paper, we have done a comprehensive literature survey of IoT-assisted solutions for monitoring the cancer patients. We have targeted the lung, ovarian, and thyroid cancer primarily and have highlighted the IoT solutions for the same. The cancer is a deadly lifestyle disease and is fatal in nature. IoT-assisted solutions in the cancer treatment are a supporting factor for both patients and treating doctors. The available literature in this domain is relatively less and is expected to grow with the advancements in IoT and supporting technologies. Our future work will focus on artificial intelligence and blockchain-assisted solutions for cancer patients.

References

1. Bianchi V, Mattarozzi M, Giannetto M, Boni A, De Munari I, Careri M (2020) A self-calibrating IoT portable electrochemical immunosensor for serum human epididymis protein 4 as a tumor biomarker for ovarian cancer. Sensors (Switzerland) 20(7). http://doi.org/10.3390/s20072016
2. Cui J, Zhang Y, Cao M, Wang S, Xu Y (2021) Thyroid tumour care risk based on medical IoT system. Microprocess Microsyst 82:103845. http://doi.org/10.1016/j.micpro.2021.103845
3. Elhoseny M, Bian GB, Lakshmanaprabu SK, Shankar K, Singh AK, Wu W (2019) Effective features to classify ovarian cancer data in internet of medical things. Comput Netw 159:147–156. https://doi.org/10.1016/j.comnet.2019.04.016
4. Elouerghi A, Bellarbi L, Afyf A, Talbi T (2020) A novel approach for early breast cancer detection based on embedded micro-bioheat ultrasensitive sensors: IoT technology. In: 2020 international conference on electrical and information technologies, ICEIT 2020, 2–5. http://doi.org/10.1109/ICEIT48248.2020.9113180
5. Han Y, Han Z, Wu J, Yu Y, Gao S, Hua D, Yang A (2020) Artificial intelligence recommendation system of cancer rehabilitation scheme based on IoT technology. IEEE Access 8:44924–44935. https://doi.org/10.1109/ACCESS.2020.2978078
6. Liu Z, Yao C, Yu H, Wu T (2019) Deep reinforcement learning with its application for lung cancer detection in medical internet of things. Futur Gener Comput Syst 97:1–9. https://doi.org/10.1016/j.future.2019.02.068
7. Memon MH, Li JP, Haq AU, Memon MH, Zhou W, Lacuesta R (2019) Breast cancer detection in the IOT health environment using modified recursive feature selection. Wirel Commun Mobile Comput. https://doi.org/10.1155/2019/5176705
8. Onasanya A, Elshakankiri M (2017) IoT implementation for cancer care and business analytics/cloud services in healthcare systems. In: UCC 2017—proceedings of the 10th international conference on utility and cloud computing, pp 203–204. http://doi.org/10.1145/3147213.3149217
9. Onasanya A, Elshakankiri M (2019) Secured cancer care and cloud services in IoT/WSN based medical systems. In: Lecture notes of the institute for computer sciences, social-informatics and telecommunications engineering, LNICST, vol 256. Springer International Publishing. http://doi.org/10.1007/978-3-030-05928-6_3
10. Onasanya A, Elshakankiri M (2019) Smart integrated IoT healthcare system for cancer care. Wireless Netw 1. https://doi.org/10.1007/s11276-018-01932-1
11. Palani D, Venkatalakshmi K (2019) An IoT based predictive modelling for predicting lung cancer using fuzzy cluster based segmentation and classification. J Med Syst 43(2). http://doi.org/10.1007/s10916-018-1139-7

12. Prachumrasee K, Juthong N, Waisopha B, Suthiporn W, Manerutanaporn J, Koonmee S (2019) IoT in pre-analytical phase of breast cancer specimens handling in Thailand hospitals, pp 23–26
13. Pradhan K, Chawla P (2020) Medical Internet of things using machine learning algorithms for lung cancer detection. J Manag Analytics 7(4):591–623. http://doi.org/10.1080/23270012.2020.1811789
14. Premavathi M, Rajbrindha M, Srimathi P, Swetha A (2019) IOT based automated detection of WBC cancer diseases. IJIRAE 6(03):148–154
15. Rahman MA, Rashid M, Barnes S, Shamim Hossain M, Hassanain E, Guizani M (2019) An IoT and blockchain-based multi-sensory in-home quality of life framework for cancer patients. In: 2019 15th international wireless communications and mobile computing conference, IWCMC 2019, pp 2116–2121. https://doi.org/10.1109/IWCMC.2019.8766496
16. Reethu R, Preetha D, Parameshwaran P, Sivaparthipan CB, Kalaikumaran T (2020) A design of smart device for detection of oral cancer using IoT, vol 3, pp 3–6
17. Savitha V, Karthikeyan N, Karthik S, Sabitha R (2020) A distributed key authentication and OKM-ANFIS scheme based breast cancer prediction system in the IoT environment. J Ambient Intell Humanized Comput 0123456789. https://doi.org/10.1007/s12652-020-02249-8
18. Singh K, Kaushik K, Ahatsham, Shahare V (2020) Role and impact of wearables in IoT healthcare. Adv Intell Syst Comput 1090:735–742. https://doi.org/10.1007/978-981-15-1480-7_67
19. Valluru D, Jasmine Selvakumari Jeya I (n.d.) IoT with cloud based lung cancer diagnosis model using optimal support vector machine. http://doi.org/10.1007/s10729-019-09489-x
20. Zaminpira S, Niknamian S (n.d.) Miraculous effect of specific ketogenic diet (SKD) plus intravenous ozone therapy (IOT) and hyperbaric oxygen therapy (HBO2T) in the treatment of several cancer types in human models
21. Zhang J, Li L, Lin G, Fang D, Tai Y, Huang J (2020) Cyber resilience in healthcare digital twin on lung cancer. IEEE Access 8:201900–201913. https://doi.org/10.1109/ACCESS.2020.3034324

Application of User and Entity Behavioral Analytics (UEBA) in the Detection of Cyber Threats and Vulnerabilities Management

Rahma Olaniyan, Sandip Rakshit, and Narasimha Rao Vajjhala

Abstract Technological advancements such as the Internet of Things, mobile technology, and cloud computing are embraced by organizations, individuals, and society. The world is becoming more reliant on open networks, which fosters global communication and cloud technologies like Amazon Web Services to store sensitive data and personal information. This changes the danger landscape and opens new opportunities. As the number of people who use the Internet grows, so does the number of cyber risks and data security challenges that hackers pose. A cybersecurity threat is an action that aims to destroy or damage data, steal data, or otherwise disrupt digital life. Computer viruses, data breaches, and Denial of Service (DoS) assaults are all examples of cyber dangers. We'd be witnessing a notion of scanning enormous volumes of data across the internet if we described how AI systems might discover where hacks came from and recommend solutions to decision-makers within the corporation.

Keywords Artificial intelligence · Machine learning · UEBA · Cybersecurity threats · Vulnerability · Machine learning · Security

R. Olaniyan · S. Rakshit
American University of Nigeria, Yola, Nigeria
e-mail: rahma.olaniyan@aun.edu.ng

S. Rakshit
e-mail: sandip.rakshit@aun.edu.ng

N. R. Vajjhala (✉)
University of New York Tirana, Tirana, Albania
e-mail: narasimharao@unyt.edu.al

P. Chatterjee et al. (eds.), *Computational Intelligence for Engineering and Management Applications*, Lecture Notes in Electrical Engineering 984,
https://doi.org/10.1007/978-981-19-8493-8_32

1 Introduction

Because cyber-attacks are so common these days, businesses are taking precautions to intercept potentially suspicious and dangerous data to avert a security breach [1]. Because advanced threats can be carried out months or even years in advance of the actual attack, security professionals must act ahead of time to prevent such attacks. Cybercrime can be controlled or avoided in several ways. A traditional healthy technique [2] includes incorporating proper password hygiene, investing in strong anti-virus, employing a firewall for Internet connection, using a VPN, and encrypting equipment. Despite these efforts, cyber criminals continue to develop and carry out increasingly sophisticated and multi-dimensional attacks that put data at risk. Attacks are growing increasingly complex, incorporating a mix of network-based methods, malware, and web application attacks [3]. Millions of new malware variants are discovered every year, adding to the complexity of cyber threats [4]. Traditional anti-virus software can't detect these types of malwares, and some don't even use binary files. Insider threats are difficult to distinguish from legitimate user activity because of these factors. This is where artificial intelligence (AI) can help to bridge the gap by using AI to the management of cyber threats.

Every company faces a variety of risks, the majority of which originate in cyberspace. Data for both the business and the target market or consumers is the most crucial factor to most firms and organizations. Technology has accelerated the growth of businesses and organizations in the modern era. Cyber dangers are becoming more widespread, and data is becoming more vulnerable. According to the Comcast research community, 53,308 security breaches were reported in 2020, with 2316 data breaches. For every 1 million records damaged, companies pay around $40 million [5]. Hackers are updating and improving their techniques as organizations expand and improve their businesses [6]. Many IT security departments rely on AI technology, which is essential. By 2024, the AI cybersecurity market is expected to reach $35 billion, and enterprises and organizations have become completely reliant on AI security technologies for their operations. AI systems provide a variety of services for security structures, including network surveillance, risk assessment, and control, as well as a variety of other functions [7].

Companies can process massive volumes of threat data and successfully avert and respond to breaches and cyberattacks using artificial intelligence technologies. Many of these emerging hazards can be identified and mitigated using artificial intelligence (AI) systems based on machine learning techniques. They can intelligently evaluate a far bigger volume of data than humans, discover anomalies and suspicious activity, and investigate dangers by comparing several data points. AI security systems aren't flawless and require human monitoring, setup, and tuning, but they're becoming an important part of the 21st-century cybersecurity armory [8].

Machine learning (ML) is the brain of artificial intelligence. Machine learning is defined as an application of artificial intelligence (AI) that provides systems the ability to learn and develop without being explicitly programmed [9]. Machine learning is concerned with creating computer algorithms that can access data and learn on their own. Machine learning is a method of extracting information from data by evaluating it and monitoring previous experiences that approximates human behavior. Machine learning in cybersecurity saves time by recognizing security risks and vulnerable areas rapidly and, in some situations, automatically responding to them.

2 Traditional Ways of Preventing Cyber Attacks Using Artificial Intelligence

Passwords have always been a delicate matter when it comes to security. They are also frequently the only line of defense between cybercriminals and our accounts. Biometric authentication has been tried as a password replacement; however, it is inconvenient and vulnerable to hacking. For example, when a facial recognition system does not recognize the user because of a new hairstyle or because they are wearing a hat, it can be aggravating to use. Attackers can also get around it by using the user's Facebook or Instagram photos. Developers use AI to improve biometric authentication and eliminate flaws to make it a more trustworthy system.

Take, for example, the facial recognition technology of Apple, which is featured on the latest iPhones starting from the iPhone X. The system is known as 'Face ID.' It analyzes the user's face features using inbuilt neural engines and infrared sensors. The AI software generates a detailed model of the user's face by finding significant connections and patterns. The AI software design can also perform in different lighting settings and correct for changes such as a new hairdo, facial hair growth, or wearing a hat, among other things [10]. Apple states that fooling the AI and accessing the users iPhone with a different face is a one-in-a-million probability with this technology.

Creating a security policy and determining an organization's network topography are crucial aspects of network security. Both activities are typically time-consuming [11]. We can now utilize AI to speed up these procedures by analyzing and learning network traffic patterns and recommending security measures. This saves time and a lot of work and resources that can be put to better use in areas like technical growth [1, 4, 12].

3 User and Entity Behavioral Analysis (UEBA)

The strong integration of user and entity behavioral analytics (UEBA) with endpoint monitoring, detection, and response is one way to find insider and undiscovered security risks [13]. This is where behavioral analytics, which combines artificial intelligence, machine learning, big data, and data analytics, comes into play. Gartner is the most widely utilized cybersecurity procedure for detecting insider risks, targeted assaults, and financial fraud. User and entity behavioral analytics (UEBA) use advanced data analytics to study user activity on networks and other systems to detect suspicious actions [13]. It involves behavioral analysis of things other than users, such as routers, servers, and endpoints. These can be used to detect security threats such as malicious insiders and privileged account compromise that standard security technologies cannot detect [14]. Because it can evaluate behavior across various individuals, IT devices, and IP addresses, UEBA is far more potent in detecting complex attacks [13]. When the activity score reaches a certain risk threshold, an alert is sent out to the security team.

4 UEBA—Applications of Cybersecurity

4.1 Vectra's Cognito Detect

Vectra is a well-known company that specializes in providing effective preventive and remedial methods for network vulnerabilities and cyber-attacks. Thanks to their offering of the "Cognito Detect" platform, they have been able to secure the safety of data generated and stored by more than seven million hosts in 35 countries worldwide. The fact that they have a retention rate of over 95% also attests to this [14].

How it works:

- Rich Metadata: Cognito Detect gathers metadata from network packets and provides real-time visibility into network traffic. This fetch operation is carried out on every device in the network that has an IP address.
- Identify Attacker Behaviors: The extracted metadata is then evaluated using pre-programmed "behavioral detection algorithms" to identify concealed attackers. Backdoors, tools that allow for remote access, credential misuse, and lateral movement are examples of significant attacker activities revealed by such assessments. Devices that have already been compromised are exposed and labeled as dangers during these analyses. Furthermore, Cognito Detect detects any suspicious activity carried out by personnel in firms who attempt to access data they are not authorized to access or who violate agreed-upon data usage and movement policies. When risks are detected, alerts are issued to initiate necessary response procedures.
- Automated analysis: The "Threat Certainty Index," an important platform aspect, keeps track of prior events to identify dangers with the greatest potential for destruction. This is to prevent the addition of new occurrences that must be examined from the beginning. It links and exposes attacker activities and compares them to previous similar events to further automate its threat hunting process. In this approach, it learns from previous events in a constant manner. This is consistent with any artificially intelligent platform's operating principles.

- Drive response: As a result of all the automated analyzes performed, Cognito Detect can respond to recognized risks quickly by ensuring that all required information to address such threats effectively is made available to the relevant party. This type of data contains the dangers with the greatest potential for harm. These dangers are given top emphasis. Cognito Detect integrates with endpoint security, firewalls, and network access control (NAC) to respond to unknown threats automatically.

4.2 Darktraces Enterprise Immune System

Over the years, Darktrace has established itself as a reliable cybersecurity service. They are the creators of the "Enterprise Immune System," a self-learning defense platform. This technique is used by more than 3500 firms to protect their digital data and requires no extra setup.

How it works:

- Both businesses and industries can use Darktrace's Immune System. The Enterprise Immune System is utilized in the cloud, email, the Internet of Things, and organization networks, whereas the Industrial Immune System is utilized on a larger scale in industries.
- The Enterprise Immune System is self-learning because it does not rely on any pre-existing rules or assumptions. It's like how the human immune system was created. It can learn and comprehend how 'self' operates and use what it learns to pinpoint the location of vulnerabilities in a network while they are still in the early stages of the attack.
- On the other hand, the Industry Immune System learns what the normal working standard is and uses real-time comparisons to discover any irregularities in the system. Antigena Network, built with Darktrace's world-class AI, automatically calculates the optimal procedures to take in the shortest amount of time in response to these discovered threats and get rid of vulnerabilities in real time in a cyber network. Antigena Email also works using the principles of AI to fight advanced threats which may cause harm to employees through their email inboxes. Instead of basing its working principles on the use of past events, it studies the everyday patterns of life to be able to identify and deal with malicious emails precisely.

4.3 Paladon's AI-Based Managed Detection and Response Service (MDR)

Paladon's managed detection and response is a service that gives businesses the tools they need to understand risks better and detect and respond to threats. The Managed Detection and Response service includes threat intelligence, threat hunting, security monitoring, incident analysis, and incident response.

How it works:

- The provider's tools and technologies are used to deliver managed detection and response services, but they are installed on the user's premises. These tools will be managed and monitored by the service provider. These tools work together to protect the Internet's gateways. The gadgets are also strategically placed to detect or detect threats. The supplier employs these methods to keep the network of individuals safe.
- In addition, the service handles incident validation and remote response. It aids in the detection of indicators of a breach in this case. They offer consultation services via which individuals or businesses can get assistance on fixing problems and mitigating security risks.
- Finally, managed detection and response employs a combination of analytics and human experience to eliminate network risks. It also relies on humans to monitor the network. It scans the network for threats, and once one is detected, human operators take control and mitigate the damage caused by successful attacks and breaches.

5 User and Entity Behavior Analytics (UEBA) Tools of 2021

Exabeam: Exabeam has a SIEM platform that incorporates its own standalone technologies into a single system. Some of these technologies include UEBA, log management, incident response, and querying. The solution that Exabeam uses as a tool is called "Exabeam Advanced Analytics" and is designed to tackle and detect advanced threats, quickly analyze incidents, and efficiently hunt down threats. "Exabeam's analytics dashboard provides a snapshot of the environment's dangers, including open cases and associated high-risk individuals and assets." [1]

Forcepoint: Forcepoint's tool for monitoring user behavior has been in operation for over fifteen years and has up to 20,000 clients using the solution. This tool allows security experts analyze and learn from existing data to proactively tackle and monitor high-risk behaviors. An article explains that "to give context for varied user actions, the platform collects and analyzes data from a variety of sources, including communication platforms and security devices." [1] A sequence of events leading to a particular high-risk score can be traced by using the entity timeline that Forcepoint provides.

Microsoft: Microsoft Azure Advanced Threat Analytics (ATA) is a security system that works by using cloud-based technologies to detect and identify advanced threats and compromised accounts and users. Following its acquisition of Aorato, "Microsoft integrated Advanced Threat Analytics into its Enterprise Mobility Suite and released it as a standalone product in 2015". [1] Microsoft ATA is only available for implementation on-premises.

LogRhythm: LogRhythm UEBA works by applying machine learning solutions and concepts into the detection of known and unknown user-based threats. After the solution has detected the threats, it then weighs it on a scale to figure out the gravity of the threat. An article by Kyle Guercio explains that "It provides a broad range of security analytics, spanning scenario-based and behavior-based approaches. LogRhythm works as a standalone UEBA product or as an add-on to existing SIEM or log management systems to enhance corporate security environments" [1].

6 Conclusion

Many cybersecurity problems can be detected immediately by artificial intelligence, which can then be escalated to human experts. AI can save human analysts a lot of time and identify threats that they would not be able to otherwise. Simultaneously, it will not be able to totally replace dedicated IT personnel. As a first line of defense, AI is increasingly being integrated into next-generation cybersecurity solutions. Artificial intelligence will become more effective as it grows more robust.

Companies like Vectra, Cyr3con, and Darktrace, as evidenced in the situations described in this article, have successfully used artificial intelligence to efficiently carry out threat hunting and proper vulnerability management practices. Spark Cognitions, for example, has been able to integrate AI concepts with traditional methodologies to build powerful cybersecurity solutions. Artificial Intelligence isn't going away anytime soon. Its applications in guaranteeing cyber platform security are still being researched. Though it has flaws, it has a lot of applications and is a subject that deserves a lot of research.

References

1. Salem MB et al (2008) A survey of insider attack detection research. In: Stolfo SJ et al (eds) (2008) Insider attack and cyber security: beyond the hacker. Springer US, New York, pp 69–90
2. Karjalainen M, Kokkonen T (2020) Comprehensive cyber arena; the next generation cyber range. In: Proceedings of 2020 IEEE European symposium on security and privacy workshops (EuroS&PW), pp 11–16
3. Al-Mhiqani MN et al (2018) A new taxonomy of insider threats: an initial step in understanding authorised attack. Int J Inf Syst Manag 1(4):343–359. https://doi.org/10.1504/IJISAM.2018.094777
4. Livshitz II et al (2020) The effects of cyber-security risks on added value of consulting services for IT-security management systems in holding companies. In: Proceedings 2020 international conference quality management, transport and information security, information technologies (IT&QM&IS), pp 119–122
5. Mendsaikhan O et al (2020) Quantifying the significance and relevance of cyber-security text through textual similarity and cyber-security knowledge graph. IEEE Access 8:177041–177052. http://doi.org/10.1109/ACCESS.2020.3027321

6. Al-Turkistani HF, Ali H (2021) Enhancing users' wireless network cyber security and privacy concerns during COVID-19. In: Proceedings of 2021 1st international conference on artificial intelligence and data analytics (CAIDA), pp 284–285
7. Thuraisingham B (2020) Cyber security and artificial intelligence for cloud-based internet of transportation systems. In: Proceedings of 2020 7th IEEE international conference on cyber security and cloud computing (CSCloud)/2020 6th IEEE international conference on edge computing and scalable cloud (EdgeCom), pp 8–10
8. Shu F et al (2020) Research and implementation of network attack and defense countermeasure technology based on artificial intelligence technology. In: Proceedings of 2020 IEEE 5th information technology and mechatronics engineering conference (ITOEC), pp 475–478
9. Vajjhala NR et al (2021) Novel user preference recommender system based on Twitter profile analysis. In: Proceedings of soft computing techniques and applications. Springer, Singapore, pp 85–93
10. Basallo YA et al (2018) Artificial intelligence techniques for information security risk assessment. IEEE Lat Am Trans 16(3):897–901. https://doi.org/10.1109/TLA.2018.8358671
11. Ho TY et al (2020) The burden of artificial intelligence on internal security detection. In: Proceedings of 2020 IEEE 17th international conference on smart communities: improving quality of life using ICT, IoT and AI (HONET), pp 148–150
12. Saxena N et al (2020) Impact and key challenges of insider threats on organizations and critical businesses. Electronics (Basel) 9:1DW+
13. Khaliq S et al (2020) Role of user and entity behavior analytics in detecting insider attacks. In: Proceedings of 2020 international conference on cyber warfare and security (ICCWS), pp 1–6
14. Shashanka M et al (2016) User and entity behavior analytics for enterprise security. In: Proceedings of 2016 IEEE international conference on big data (Big Data), pp 1867–1874

Review of Software-Defined Network-Enabled Security

Neelam Gupta, Sarvesh Tanwar, and Sumit Badotra

Abstract In order to exploit any of the system's security holes attackers will always target software-defined networking (SDN) security. As a game-changer in modern internetworking technologies, SDN appears to be ubiquitous these days, and it has been widely accepted by the scientific community, organizations, and industry. These days, enterprise networks are increasingly using it because of its flexibility in network management and lower operational expenses. No flexibility is available in a dynamic network environment while using traditional networks. Software-defined networking is recognized as it's thought of as an effective way to manage the entire network and it's also thought to simplify the complex network design into something more manageable. Research on security challenges and concerns in software-defined networks has been conducted up to this point, as described in this paper. SDN creates substantial security risks and increases the complexity of the system as its approach replaces traditional networking in order to provide remote and centralized control features. This article presents methodologies and techniques used for the security enhancements of SDN, as well as the security issues that are explored, as well as a categorization of the research literature over the last ten years. The results present outline the information on the most active research areas, security technologies and methodologies, pros and cons in security in SDN. A detailed review like this will immensely benefit the research community in designing more consistent and effective security solutions for SDN networks.

Keywords Software defined networking · Interfaces · Security · Application plane · Data plane · Control plane

N. Gupta · S. Tanwar (✉)
Amity Institute of Information Technology, Amity University Noida, Noida, Uttar Pradesh, India
e-mail: s.tanwar1521@gmail.com

N. Gupta
e-mail: neelam.gupta@s.amity.edu

S. Badotra
School of Computer Science and Engineering, Bennett University, Noida, UP, India
e-mail: Sumit.badotra@bennett.edu.in

P. Chatterjee et al. (eds.), *Computational Intelligence for Engineering and Management Applications*, Lecture Notes in Electrical Engineering 984,
https://doi.org/10.1007/978-981-19-8493-8_33

1 Introduction

Software-defined networking has emerged as a strong new paradigm for enabling networking research and development as well as a low-cost network technology capable of supporting the dynamic nature of future network functions and intelligent applications while lowering operating costs through simplified hardware, software, and management [1]. Separating the network's control plane from the information plane is the fundamental concept. As seen in Fig. 1, the open networking foundation has provided a reference model for SDN. This paradigm comprises of three levels that overlay on top of one another, namely, an infrastructure layer, an impression layer, and an application layer, limiting the flexibility and creativity of networking infrastructure.

Energy efficiency and security are getting increasingly important in networking [2]. To effectively address the increasing demands of this evolving network landscape, network operators and repair and products providers require a replacement network solution. The open networking foundation (ONF) has formed a security working party. There are several challenges raised here that show the requirement for more research and development of security solutions. Network forensics, security change, and Military Intelligence [38] are all things that SDN may help with in terms of security. SDN security is barely nearly as good because the security policy that has been defined. Implementing current authentication and authorization procedures can typically solve certain aspects of the protection difficulty. In the meantime, techniques for identifying and protecting against threats will continue to improve. The network function virtualization (NFV) [3] Industry Specification Group of the European Telecommunications Standards Institute (ETSI), the IETF's Forwarding and Control Element Separation (ForCES) section, ONF, and other industry working groups must collaborate, and there's a desire to deal with the subsequent limitations in traditional network as shown in Fig. 2.

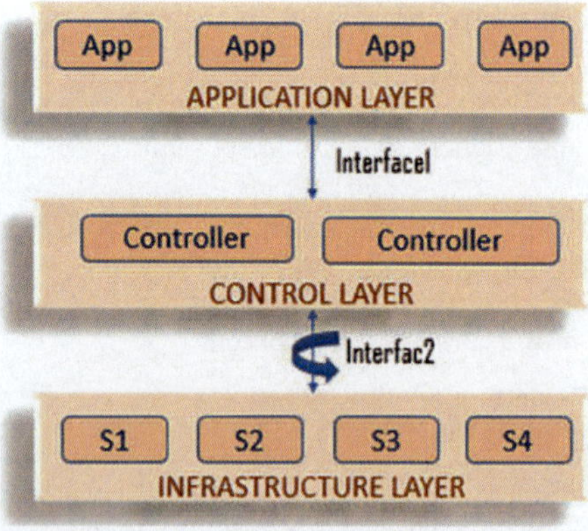

Fig. 1 SDN architecture model

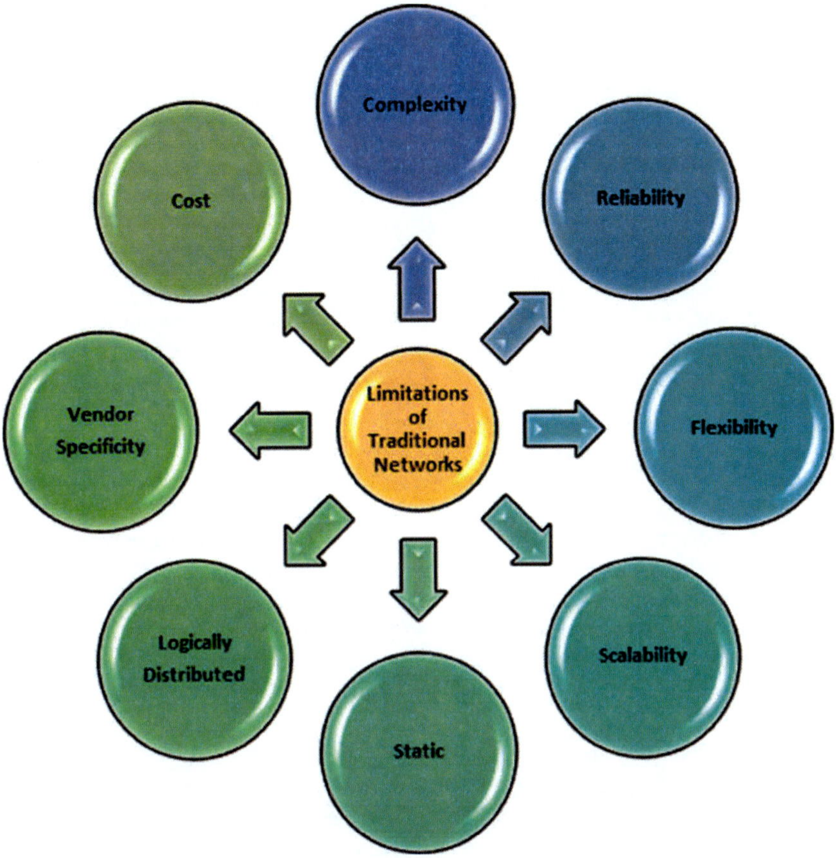

Fig. 2 Limitations of traditional network

SDN takes the idea of a centralized network control plane and adds programmability, making network management easier and allowing security methods to be organized in real time. It can then react quickly to network irregularities and malicious traffic. The three basic functional levels or SDN planes will help you better understand the SDN architecture such as application, control, and data planes. The human and organizational components of SDN must be thoroughly examined in order to determine who oversees what and why, as well as to anticipate any conflicts that could compromise the system's security. A DDoS [4] assault is one in which a group of compromised computers known as bots or zombies attack one system. The seven most common types of DDoS attacks include flood attacks, core melt attacks, land attacks, TCP SYN attacks, CGI request attacks, and authentication server attacks. SDN could be a target of an attack, and we analyze previous research on introducing SDN attacks and how to deal with the problem.

1.1 Motivation

We review the SDN literature in this article with the goal of presenting security in SDN and its principles, offering an overview of recent SDN achievements, and reviewing research difficulties during the previous 10 years, as well as future SDN advancements. We give the results of a literature evaluation and responses to research questions in Sect. 1 of this article, as well as a set of recommendations for future research areas on SDN security. The technique utilized in the literature review is discussed in Sect. 2, as well as the replies to the research questions. Section 4 contains the conclusion, which will be published in its entirety later this year in the Journal of Network and Data Security.

2 Literature Review

SDN technologies have emerged as viable alternatives for reducing network costs and increasing network agility. It separates network control logic from the underlying hardware, providing network managers more control over network performance and a more comprehensive picture of the network [5]. During the previous two decades, the use of Internet-based services and applications has skyrocketed. Concerns regarding internet security have risen dramatically as an estimated 59% of the world's population now accesses the Internet. By leveraging centralized controllers, global network awareness, and on-demand generation of traffic forwarding rules, the SDN architecture can improve network security. Common Internet abnormalities include worms, denial-of-service assaults, and Trojans. Network security, scalability, and supportability, on the other hand, remain issues and concerns for SDN [6]. Security [7, 8] is at the Center of all these issues. Since the centralized controller oversees network administration, its failure would have a negative impact on the entire network. Sophisticated assaults could target centralized control and communication between controllers and switches. It addresses network programmability, the rise of virtualization, device setup and debugging, OpenFlow, OpenFlow architecture, Flow and group tables, and the OpenFlow protocol, among other issues faced by older network designs. The benefits of SDN in terms of security as well as security challenges are discussed, as well as a categorization of the academic literature over the previous 10 years.

From 2017 through 2021, various researchers [9–12] employed in their simulation situations use a single controller (e.g., POX, NOX, Floodlight, and OpenDaylight, among others). This single controller, however, could end up becoming the network's single point of failure due to a lack of a secure and dependable controller. Implementing a large number of controllers and distributed war implements, on the other hand, may be a preferable alternative for future defense methods because it distributes overhead across multiple machines and allows for load balancing as needed. Many controller topologies and distributed DDoS defense systems can be created in the future with little synchronization and communication overheads. There are numerous OpenFlow software projects available now [13]. The NOX controller and Mininet simulator are the key tools in SDN research. Commercial controllers including NEC's Programmable Flow Controller and Big Switch's Big Network Controller are accessible, as are open-source controllers like OMNI, Trema, Floodlight, NOX, and OpenDaylight, as illustrated in Table 1. Other networking-related initiatives, such as OpenStack Quantum for cloud computing networks, have begun to use OpenFlow.

Many researchers (from 2013 to 2020) [14–19], switch intelligence can increase device complexity and cost, so it's critical to make sure devices are secure before they reach the control plane. It's still a challenge to properly install security modules in switches while reducing device communication complexity. Most fellow researchers (from 2015 to 2018) [20–23]. The researchers built their recommended experimental settings on a single computer, which is a major hurdle in the evaluation of Internet security measures. They also exploited virtualization of network devices and connections to arrive at their conclusions. For anomaly detection in systems based on information theory measurements, predetermined threshold values (depending on baseline network behavior) are used (from 2014 to 2018) [20, 21, 23–26]. The absence of benchmark data sets that reflect both regular and attack traffic is a significant problem. Researchers utilized several techniques to imitate regular traffic, which did not correctly represent today's high-speed network. These tools are incapable of mixing background, regular, and attacker traffic in a balanced manner. Most security solutions (from 2010 to 2016) [16, 24, 27–31], some approaches have been developed

Table 1 OpenFlow controllers

Organization	Platform	Controller's name	Key points
Stanford	Java	Beacon	Both event-based and threaded operation are supported, provide Web UI
Stanford	C++/Python	NOX/POX	First OpenFlow controller
Linux Foundation	Java	OpenDaylight controller	Provide REST API and web GUI
Yale University	Haskell	McNettle	High-level declarative expressive language, multi-core optimization
Big Switch	Java	Floodlight	Developer-friendly with plenty documents
Rice University	Java	Maestro	Multi-threaded support
NTT Laboratories OSRG Group	Python	Ryu	With OpenStack support
Universidade Federal do Rio de Janeiro	Python/Java	OMNI	Web interface on NOX

to reduce processing overhead for the centralized control plane in large networks. These approaches use native OpenFlow statistics gathering mechanisms to capture network properties from the data plane. The sFlow technique has been employed by certain writers to decouple the network statistics activity from the switch forwarding mechanism. However, because sFlow only collects a part of the data, the solution's accuracy suffers. As a result, there is a research gap in the current work in terms of low-cost network statistics collection.

In addition, we've included a literature review of recent SDN research in the infrastructure, control, and application layers, as summarized in Table 2.

Many authors in 2021 work on the controller's usage of the link layer discovery protocol (LLDP) to discover network topology has been proved to be unable to ensure the integrity of its communications. This weakness could be used by attackers to transmit phoney LLDP packets and establish a false link between two switches. SDN network counter-attack detection is based on path latencies and offers effective protection. In order to establish an in-band covert connection between the two cheating switches, it injects a relay host and three ways into the networks. This allows the recovery of data packet address fields and forwarding to their intended

Table 2 SDN-related research

S. No.	Layer	Related area	Issue
1	Application	Routing, traffic engineering, NDN, wireless network	Adaptive routing
		Computer security, function outsourcing, network	Boundless mobility
		Virtualization, web, caching, green networking	Consolidate security
			Network virtualization
			Ease maintenance
			Green networking
2	Control	Programming language, formal methods, compilers	High level language
		Distributed system	Rule update
		Formal methods	Policy and rule validation
		Network measurement and distributed system, database	Network status collection and synchronization
		Algorithm analysis, software engineering	Increase passing ability
3	Data	Integrated circuits, embedded systems, hardware testing SDR, mobile ad hoc GMPLS, ROADMs	Storage
			Processing
			Hardware platform
			Performance evaluation
			Wireless radio
			Optical fibers

destinations [32]. Figure 3 shows which plane is vulnerable to a particular attack in SDN architecture.

SDN's efficiency and performance in detecting other ransomware variants such as NotPetya were evaluated. Five modules of the SDN system were introduced last year to tackle the threat. In recent years, DDoS attacks are increasingly being countered with machine learning techniques. Several machine learning algorithms have been used to detect DDoS assaults in recent years. DDoS attack mitigation leveraging ML/DL techniques in SDN settings was also explored in this study, as virtualized systems are fast growing and becoming far more extensively used due to the numerous advantages they provide over traditional setups.

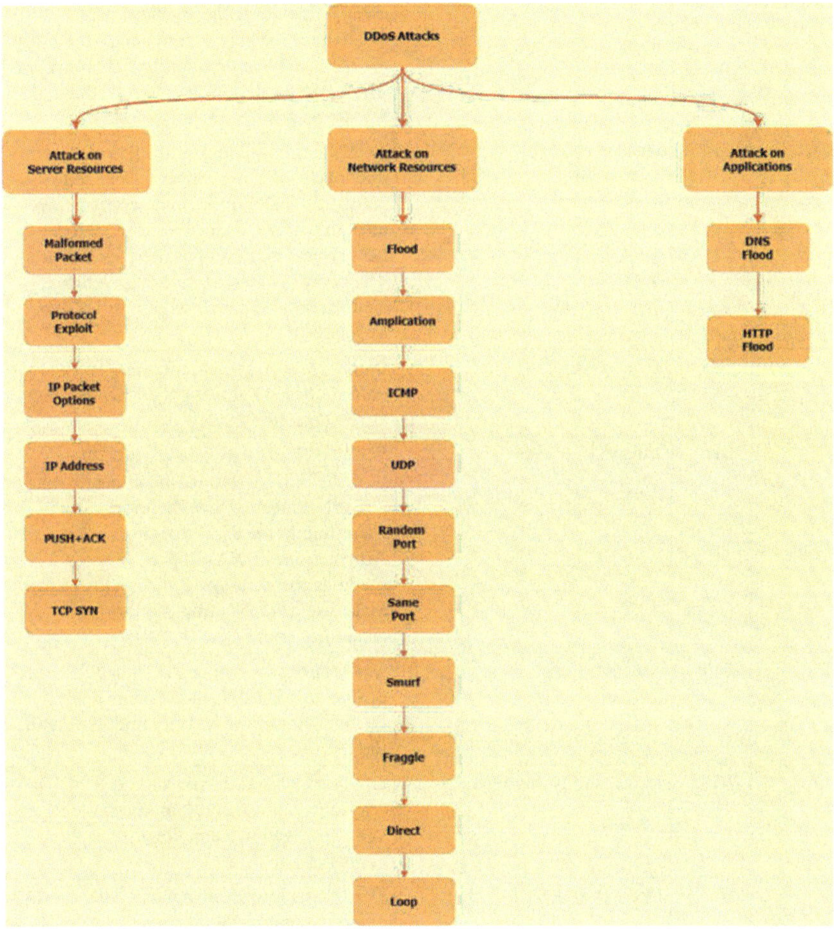

Fig. 3 Types of DDoS attacks at different layers of SDN

In the below table, we present concepts or techniques used by authors of the academic literature over the previous ten years, as summarized in Table 3.

Table 3 Authors have worked on various SDN concepts for the last ten years

S. No.	Year	Authors	Concepts/techniques
1	2011	Mehdi et al. [33]	Four prominent traffic anomaly detection techniques are substantially more accurate in detecting hostile actions in home networks as compared to ISPs
2	2013	Sezer et al. [1]	A new study looks at how difficult it is to establish an effective carrier-grade network using software-defined networking. Challenges include performance, scalability, security, and interoperability. The study was conducted by the US National Institute of Standards and Technology (USIST)
3	2014	Jammal et al. [3]	In this article, we look at the advantages and drawbacks of using Software-Defined Networks (SDN) in various scenarios. We also go over the challenges that SDN deployment on a campus network can bring. The essay is based on research on data centers, data center networks, and network as a service
4	2015	Tri and Kim [34]	In the OF edge switch, offer an entropy-based lightweight DDoS flooding attack detection model
5	2016	Xu and Liu [29]	SDN operators may quickly identify potential DDoS victims and attackers using a small number of flow monitoring rules, according to the SDN Operators' Association (SDNA) Report. Propose methods for identifying DDoS attacks that make use of SDN's flow monitoring capability
6	2017	Xu et al. [35]	The security mechanism designed to protect against this attack is based on an analysis of the behavior of the security middleware in the IoT
7	2018	Lawal and Nuray [36]	DDoS attacks on the SDN are detected in real-time, and a control approach based on sFlow mitigation technology is used to combat them
8	2019	Dong et al. [37]	DDoS attack state-of-the-art in SDN and cloud computing situations

(continued)

Table 3 (continued)

S. No.	Year	Authors	Concepts/techniques
9	2020	Hua et al. [38]	A worm-hole attack in SDN could be the first of its kind to achieve packet transfer across a falsified link without using any out-of-band channels. Flows routed to the fake connection will result in 100% packet loss, and so will be readily detected
10	2021	Alotaibi and Vassilakis [39]	A review of the use of SDN to detect and mitigate the risk of self-propagating ransomware. Analyzes research on DDoS detection methods in modern networking systems that use single and hybrid machine learning algorithms

2.1 Ques: How to Solve the Research Problem Through Literature Review?

To tackle the research topic, the literature was thoroughly examined in order to have a thorough understanding of the problem domain and relevant technologies. In this paper, we illustrate layout of SDN security analysis shown in Fig. 4.

Table 4 shows a survey of numerous papers from the last ten years, as well as their goals. Following research approach has been devised to provide an effective answer for the given problem by examining the current literature. The paper's objective is to:

- Clarify definitional ambiguities and explain the topic's scope.
- Provide a comprehensive, synthesized summary of current information.
- Look for contradictions in previous findings and possible reasons (e.g., moderators, mediators, measures, and approaches).
- Examine current methodological techniques as well as unique ideas.

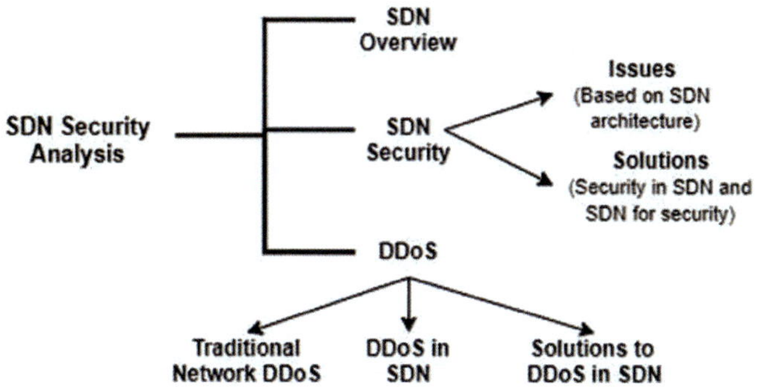

Fig. 4 Layout of SDN security analysis

Table 4 A literature review on various papers from the last ten years and their objectives

Authors	Year	Objective	Advantages	Limitations
Mehdi et al. [33]	2011	Using Openflow compatible switches and NOX as a controller, four well-known traffic anomaly detection algorithms can be deployed in an SDN context	The standardized programmability of SDN enables these solutions to reside inside a bigger framework while also improving threat mitigation opportunities	This study focused on small office/home office networks rather than large-scale networks
Sezer et al. [1]	2013	Several performance, scalability, security, and interoperability issues are discussed. The model's purpose is to improve network traffic processing performance, scalability, security, and interoperability issues are discussed. The model's purpose is to improve network traffic processing	Network performance, scalability, security, and interoperability issues are all examined in-depth, as well as possible solutions	Some of these issues may be solved by existing research and industry solutions, and a few working groups are currently discussing possible answers
Jammal et al. [3]	2014	It describes the challenges that network operators face while adopting SDN in a campus network, as well as fresh ideas and strategies for those considering building their own solution utilizing existing technologies	A variety of issues can be avoided by just avoiding using typical networking strategies to solve the problem	The SDN architecture that exists today is suitable for use, but it will gain in the future from more experimentation, deployment skills, and targeted research
Tri and Kim [34]	2015	An SDN network's response to a resource attack	The impact of a resource attack on the SDN is thoroughly explored using Mininet and OpenDaylight in terms of delay and bandwidth, as well as the need to manage flow tables while keeping in mind their size constraints	Furthermore, the controller's size constraint issue may result in unanticipated actions, posing further security concerns

(continued)

Table 4 (continued)

Authors	Year	Objective	Advantages	Limitations
Xu and Liu [29]	2016	Propose ways for detecting DDoS assaults using the flow monitoring capacity of SDN	Using a limited number of flow monitoring criteria, quickly discover probable DDoS victims and attackers	This method was not tested using packet traces from real DDoS attacks
Xu et al. [35]	2017	Propose a sensible security technique to combat the new- flow attack (SSM)	By studying the behavior of the security middleware in the IoT, the SSM proposes a dynamic access control solution to prevent the new-flow attack	It just explains how to fight against a new-flow assault in general; it doesn't explain how to generate normal baselines or evaluate filtering results
Lawal and Nuray [36]	2018	Based on the sFlow mitigation technology, this paper presents a real-time detection and control technique for distributed denial of service (DDoS) attacks on the SDN	A real-time security system for detecting and mitigating DDoS flood attacks was simulated on the SDN network	Only physical interfaces are supported by sFlow. Two sFlow collectors are supported by the switch. When the device is in stack mode, sFlow is not supported
Dong et al. [37]	2019	The state of the art in DDoS attacks in SDN and cloud computing scenarios and architecture	Examine the research on deploying DDoS attacks on SDN and how to address the issue	Detecting and mitigating DDoS assaults in a virtualized SDN configuration is a major research topic
Hua et al. [38]	2020	Present and evaluate the proposed attack's countermeasures	Address this problem by presenting the first genuine wormhole attack in SDN, which could allow packet transmission through a forged link without using out-of-band channels	It is only utilized within a reasonable range
Alotaibi and Vassilakis [39]	2021	The use of SDN to detect and reduce the risk of self-propagating ransomware is investigated	We evaluated our system's efficiency and performance in terms of detection time, CPU utilization, and TCP and ping latency	The IDPS's performance and efficiency are not suitable for use in a live network

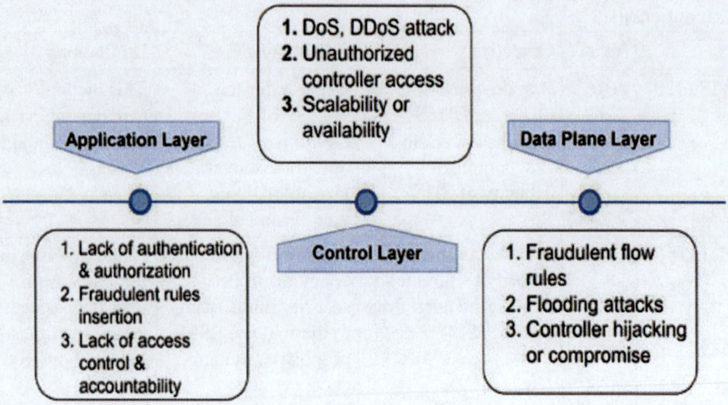

Fig. 5 Types of threats in different layers in SDN architecture

- Create conceptual frameworks that tie together and expand previous research.
- Outline key findings, research gaps, and future research directions.

3 Technology and Research Opportunities: Future Directions

The northbound and southbound SDN/OpenFlow interfaces require some attention because they are empirical means to a broader technological approach. More complex, yet simple to use and administrate solutions are almost certain to emerge in the future, according to experts at Google and Microsoft. Figure 5 shows the different sorts of threats in different tiers of the SDN architecture. We believe that these interfaces will need to be further investigated in order to gain a better knowledge of networks and optimize resource allocation.

We envisage the following research opportunities in the future: (1) SDN controller virtualization for multi-controller sensor clusters; (2) Improved global network experience: northbound interface communication enhancements; (3) Southbound interface optimization for efficient device access and controller communication; and (4) SDN strategies for sensor cluster runtime and computational overhead.

4 Conclusion

As discussed in this paper, research on security difficulties in software-defined networks has been undertaken up to this point. As attackers have developed new ways to use modern technologies, DDoS attacks have become much more difficult to mitigate. Despite the fact that academics have devised a variety of mitigation measures, DDoS attacks continue to pose a significant threat to service providers [36]. Many studies attempted to mitigate specific types of such attacks, but the methods remained vulnerable to other types of SDN security breaches. While designing an effective and realistic defense system, there are some research hurdles to address. Several works did not take advanced attacker strategies into account when solving the type of attack, they were attempting to address. After a review and extensive examination of the possible uses of many techniques to combat attacks, further work is needed to create and construct a reliable mitigation system based on several tactics.

References

1. Sezer S, Scott-Hayward S, Chouhan PK, Fraser B, Lake D, Finnegan J et al (2013) Are we ready for SDN? Implementation challenges for software-defined networks. IEEE Commun Mag 51(7):36–43
2. Chourishi D, Miri A, Milić M, Ismaeel S (2015) Role-based multiple controllers for load balancing and security in SDN. In: 2015 IEEE Canada international humanitarian technology conference (IHTC2015). IEEE, pp 1–4
3. Jammal M, Singh T, Shami A, Asal R, Li Y (2014) Software defined networking: state of the art and research challenges. Comput Netw 72:74–98
4. Wang R, Jia Z, Ju L (2015) An entropy-based distributed DDoS detection mechanism in software-defined networking. In: 2015 IEEE Trustcom/BigDataSE/ISPA, vol 1. IEEE, pp 310–317
5. Dacier MC, König H, Cwalinski R, Kargl F, Dietrich S (2017) Security challenges and opportunities of software-defined networking. IEEE Secur Priv 15(2):96–100
6. Singh J, Behal S (2020) Detection and mitigation of DDoS attacks in SDN: a comprehensive review, research challenges and future directions. Comput Sci Rev 37:100279
7. Badotra S, Nagpal D, Panda SN, Tanwar S, Bajaj S (2020) IoT-enabled healthcare network with SDN. In: 2020 8th international conference on reliability, infocom technologies and optimization (trends and future directions) (ICRITO). IEEE, pp 38–42
8. Kakkar L, Gupta D, Saxena S, Tanwar S (2021) IoT architectures and its security: a review. In: Proceedings of the second international conference on information management and machine intelligence. Springer, Singapore, pp 87–94
9. Zanna P, Hosseini S, Radcliffe P, O'Neill B (2014) The challenges of deploying a software defined network. In: 2014 Australasian telecommunication networks and applications conference (ATNAC). IEEE, pp 111–116
10. Ahalawat A, Dash SS, Panda A, Babu KS (2019) Entropy based DDoS detection and mitigation in OpenFlow enabled SDN. In: 2019 international conference on vision towards emerging trends in communication and networking (ViTECoN). IEEE, pp 1–5
11. Li R, Wu B (2020) Early detection of DDoS based on φ-entropy in SDN networks. In: 2020 IEEE 4th information technology, networking, electronic and automation control conference (ITNEC), vol 1. IEEE, pp 731–735
12. Deepa V, Sudar KM, Deepalakshmi P (2019) Design of ensemble learning methods for DDOS detection in SDN environment. In: 2019 international conference on vision towards emerging trends in communication and networking (ViTECoN). IEEE, pp 1–6
13. Xia W, Wen Y, Foh CH, Niyato D, Xie H (2014) A survey on software-defined networking. IEEE Commun Surv Tutorials 17(1):27–51
14. Shin S, Yegneswaran V, Porras P, Gu G (2013) Avant-guard: scalable and vigilant switch flow management in software-defined networks. In: Proceedings of the 2013 ACM SIGSAC conference on computer and communications security, pp 413–424

15. Wang B, Zheng Y, Lou W, Hou YT (2015) DDoS attack protection in the era of cloud computing and software-defined networking. Comput Netw 81:308–319
16. Piedrahita AFM, Rueda S, Mattos DM, Duarte OCM (2015) FlowFence: a denial of service defense system for software defined networking. In: 2015 global information infrastructure and networking symposium (GIIS). IEEE, pp 1–6
17. Kalkan K, Gür G, Alagöz F (2017) SDNScore: a statistical defense mechanism against DDoS attacks in SDN environment. In: 2017 IEEE symposium on computers and communications (ISCC). IEEE, pp 669–675
18. Li C, Wu Y, Yuan X, Sun Z, Wang W, Li X, Gong L (2018) Detection and defense of DDoS attack–based on deep learning in OpenFlow-based SDN. Int J Commun Syst 31(5):e3497
19. Conti M, Lal C, Mohammadi R, Rawat U (2019) Lightweight solutions to counter DDoS attacks in software defined networking. Wireless Netw 25(5):2751–2768
20. Mousavi SM, St-Hilaire M (2015) Early detection of DDoS attacks against SDN controllers. In: 2015 international conference on computing, networking and communications (ICNC). IEEE, pp 77–81
21. Boite J, Nardin PA, Rebecchi F, Bouet M, Conan V (2017) Statesec: stateful monitoring for DDoS protection in software defined networks. In: 2017 IEEE conference on network softwarization (NetSoft). IEEE, pp 1–9
22. Sahoo KS, Puthal D, Tiwary M, Rodrigues JJ, Sahoo B, Dash R (2018) An early detection of low rate DDoS attack to SDN based data center networks using information distance metrics. Futur Gener Comput Syst 89:685–697
23. Jiang Y, Zhang X, Zhou Q, Cheng Z (2016) An entropy-based DDoS defense mechanism in software defined networks. In: International conference on communications and networking in China. Springer, Cham, pp 169–178
24. Giotis K, Argyropoulos C, Androulidakis G, Kalogeras D, Maglaris V (2014) Combining OpenFlow and sFlow for an effective and scalable anomaly detection and mitigation mechanism on SDN environments. Comput Netw 62:122–136
25. Kalkan K, Altay L, Gür G, Alagöz F (2018) JESS: joint entropy-based DDoS defense scheme in SDN. IEEE J Sel Areas Commun 36(10):2358–2372
26. Hong GC, Lee CN, Lee MF (2019) Dynamic threshold for DDoS mitigation in SDN environment. In: 2019 Asia-Pacific signal and information processing association annual summit and conference (APSIPA ASC). IEEE, pp 1–7
27. Braga R, Mota E, Passito A (2010) Lightweight DDoS flooding attack detection using NOX/OpenFlow. In: IEEE local computer network conference. IEEE, pp 408–415
28. Dotcenko S, Vladyko A, Letenko I (2014) A fuzzy logic-based information security management for software-defined networks. In: 16th international conference on advanced communication technology. IEEE, pp 167–171
29. Xu Y, Liu Y (2016) DDoS attack detection under SDN context. In: IEEE INFOCOM 2016-the 35th annual IEEE international conference on computer communications. IEEE, pp 1–9
30. Hu D, Hong P, Chen Y (2017) FADM: DDoS flooding attack detection and mitigation system in software-defined networking. In: GLOBECOM 2017–2017 IEEE global communications conference. IEEE, pp 1–7
31. Wang P, Chao KM, Lin HC, Lin WH, Lo CC (2016) An efficient flow control approach for SDN-based network threat detection and migration using support vector machine. In: 2016 IEEE 13th international conference on e-business engineering (ICEBE). IEEE, pp 56–63
32. Aljuhani A (2021) Machine learning approaches for combating distributed denial of service attacks in modern networking environments. IEEE Access 9:42236–42264
33. Mehdi SA, Khalid J, Khayam SA (2011) Revisiting traffic anomaly detection using software defined networking. In: International workshop on recent advances in intrusion detection. Springer, Berlin, pp 161–180
34. Tri HTN, Kim K (2015) Assessing the impact of resource attack in software defined network. In: 2015 international conference on information networking (ICOIN). IEEE, pp 420–425
35. Xu T, Gao D, Dong P, Zhang H, Foh CH, Chao HC (2017) Defending against new-flow attack in sdn-based internet of things. IEEE Access 5:3431–3443

36. Lawal BH, Nuray AT (2018) Real-time detection and mitigation of distributed denial of service (DDoS) attacks in software defined networking (SDN). In: 2018 26th signal processing and communications applications conference (SIU). IEEE, pp 1–4
37. Dong S, Abbas K, Jain R (2019) A survey on distributed denial of service (DDoS) attacks in SDN and cloud computing environments. IEEE Access 7:80813–80828
38. Hua J, Zhou Z, Zhong S (2020) Flow misleading: worm-hole attack in software-defined networking via building in-band covert channel. IEEE Trans Inf Forensics Secur 16:1029–1043
39. Alotaibi FM, Vassilakis VG (2021) Sdn-based detection of self-propagating ransomware: the case of badrabbit. IEEE Access 9:28039–28058

A Concise Review on Internet of Things: Architecture and Its Enabling Technologies

Vandana Choudhary and Sarvesh Tanwar

Abstract Internet of Things (IoT) is an amalgamation of diverse technologies including big data analytics, machine learning, wireless sensor networks, artificial intelligence, etc. IoT is contemplated to alchemize the way Internet works. It has become the next communication era. It enables machine-to-machine (M2M) communication viable. Its prime objective is to furnish seamless communication among heterogeneous physical devices/objects connected through Internet to achieve smart application goals without human mediation. Communication technology plays a substantial role in enabling the universal connectivity of physical devices/objects in view of IoT. For being able to instrument a smart world, it is not only vital to inspect the discipline of IoT, but also crucial to make the required amendments in the architecture of the IoT for realizing reliable end-to-end communication. In this paper, we have outlined the basics of IoT, IoT architectures, and its enabling technologies.

Keywords Internet of things (IoT) · Connected devices · Architecture · Enabling technologies and protocols · Smart world

1 Introduction

IoT is anticipated to grow expeditiously due to escalation in the sum total of physical devices being connected to the Internet provided the expansion of communication technology [1]. According to Gartner, Inc. [2], the number of personal computers and mobile devices employed worldwide will accumulate to 6.2 billion units in 2021. As per their report, additional 125 million devices such as tablets and laptops are forecasted to be operational in 2021 as compared to 2020. Figure 1 depicts the rise in IoT-connected devices with population in billions worldwide.

V. Choudhary (✉) · S. Tanwar
Department of Information Technology, Amity Institute of Information Technology, Amity University, Noida, Uttar Pradesh 201301, India
e-mail: vandana.choudhary@s.amity.edu

P. Chatterjee et al. (eds.), *Computational Intelligence for Engineering and Management Applications*, Lecture Notes in Electrical Engineering 984,
https://doi.org/10.1007/978-981-19-8493-8_34

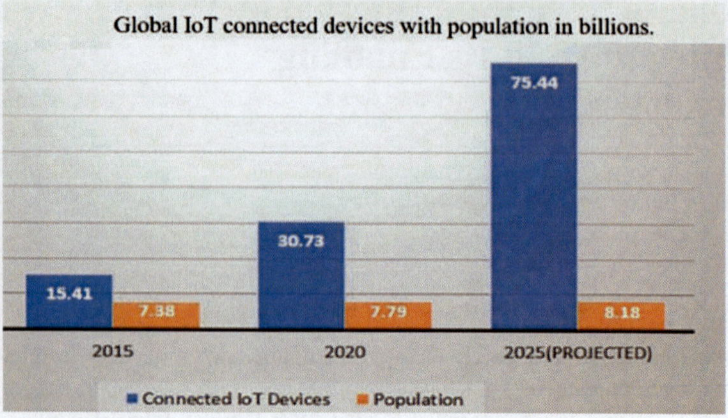

Fig. 1 Global IoT-connected devices with population in billions

The enactment of IoT applications is increasing worldwide. According to another survey [3] from Gartner Inc., in spite of disruptive effects of COVID-19 made in year 2020, 47% of organizations intent to raise their investments in IoT to reduce costs, while on the other hand, 35% of organizations decreased their investment in IoT as shown in Fig. 2. According to another survey [4], the worldwide repercussions of COVID-19 on IoT market is predicted to raise to USD 243 billion in 2021 from USD 150 billion in 2019, at a CAGR of 13.7%.

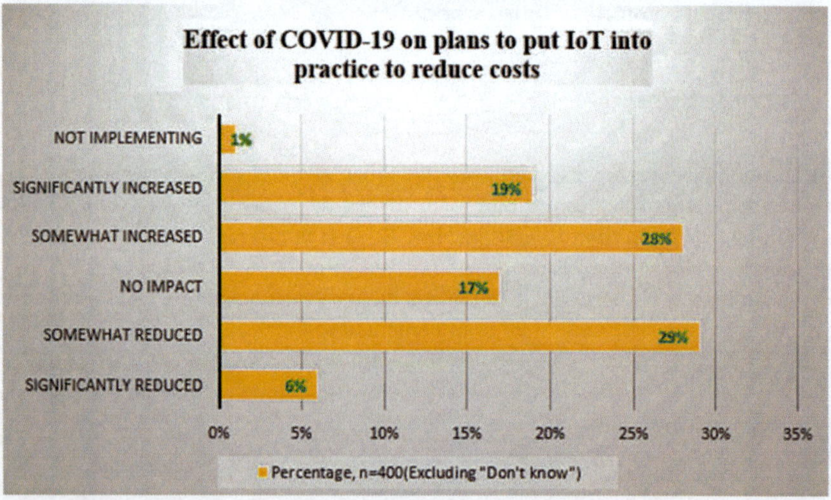

Fig. 2 Effect of COVID-19 on ideas to put IoT into practice to bring down the costs in organizations. *Source* Gartner, October 2020

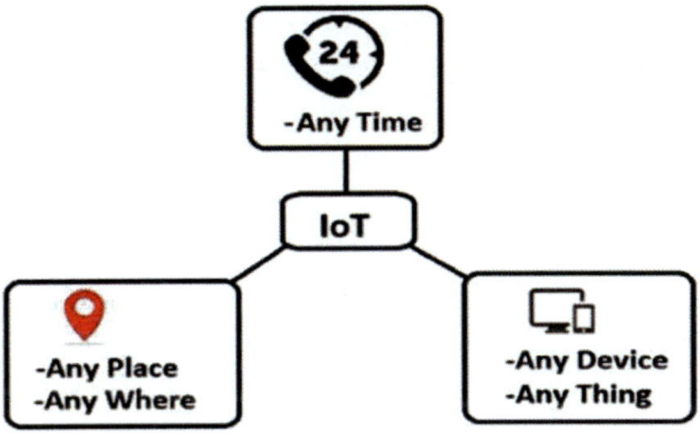

Fig. 3 IoT facets

The surveys clearly reveal that IoT will be rudimentary in the time to come as the concept empowers new services, applications, and new innovations. With all this gamut of IoT applications, the matter of security [5] and privacy show up and are an area of utmost concern. Without a reliable and coherent IoT environment, emerging IoT applications cannot be exploited comprehensively and thereby may lose all their potential.

At present, a standard architecture of IoT is much needed that clearly describes how various technologies should be implemented and how to enable secure interaction among IoT devices. As per the authors of [6], IoT implementation is comprised of four main building blocks, namely, things (Sensors/Actuators), network infrastructure (Repeaters, Aggregators, and Routers), gateways, and cloud infrastructure. Things are used to gather information out of encompassing environment without any human interaction. Network infrastructure manages the flow of data in a secure and smooth manner from things to cloud infrastructure. Gateways are used for connectivity purposes. And finally, cloud infrastructure, implemented by data storage units or virtualized servers is empowered with information depositories and computing potentiality. Figure 3 depicts various facet of IoT.

2 Motivation

With the proliferation of devices connected to Internet, almost every device has become potentially vulnerable to attacks. In addition, the way these devices are connected in a network to create an IoT ecosystem and the way information is gathered, transmitted, and processed through them is equally important since whole IoT ecosystem is thereupon susceptible to attacks. It is, therefore, fundamental to first understand the underlying architecture of IoT, its enabling technologies, and protocols applicable at each layer.

Hence, through this paper, we intend to present existing fundamental layered architectures of IoT, taking which as basis a smart world could be realized by selecting most efficient enabling technology and protocols at each layer in IoT architecture.

3 IoT Architecture

Many architectures of IoT have been realized in the literature by several researchers. But so far, there is not a single architecture which has been adopted globally in the research community. In this section, we have described the most commonly referenced layered architectures of IoT.

3.1 *Three-Layer Architecture*

Three-layer architecture of IoT is the primitive architecture of all the available architectures [7, 8]. It was proposed during the earliest stages of research and comprised of three layers, specifically, perception, network, and application layer as shown in Fig. 4.

- **Perception Layer**

Perception layer is the bottommost layer. It encompasses devices such as sensors and actuators for sensing and gathering information regarding the surrounding environment.

- **Network Layer**

Next is the middle layer, i.e., network layer. It is responsible for disseminating and processing input provided by the bottommost layer to the application layer.

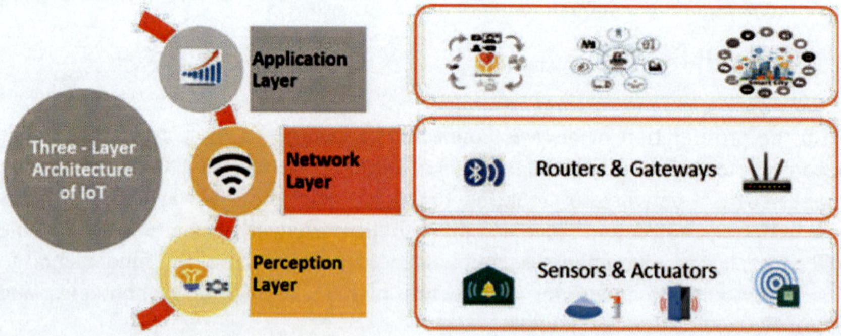

Fig. 4 Three-layer architecture of IoT

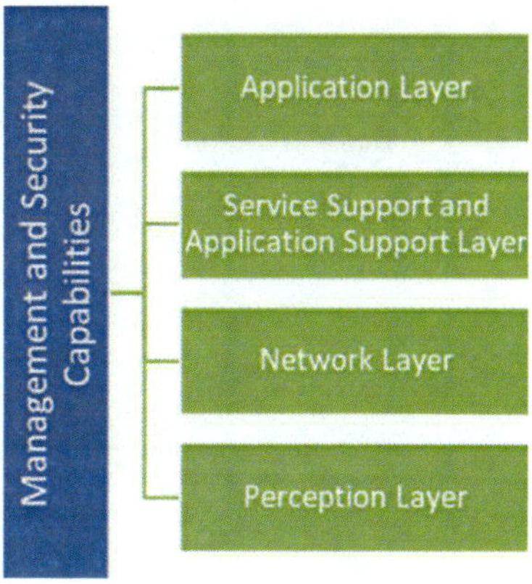

Fig. 5 Four-layer architecture of IoT. *Source* Rec. ITU-T Y.2060 (06/2012)

- **Application Layer**

The application layer is the topmost layer. It is accountable for providing application specific services to the end user.

3.2 *Four-Layer Architecture*

Due to inability of three-layer architecture to provide well-grounded solution taking into account the various aspects IoT, inability to provide proper insights in researching, and due to continuous advancement in the field of IoT, a four-layer architecture of IoT was proposed. Of four layers, it has three layers with same functionality as indicated for the preceding architecture. In addition, it has a layer called service support and application support layer/processing layer which consists of generic support capabilities which involve common capabilities such as data processing or data storage and specific support capabilities which involve particular capabilities which serve the obligation of a lot many different applications and may invoke generic support capabilities as well. It has management capabilities and security capabilities as well which are linked with all the layers [9]. Figure 5 depicts a four-layer architecture of IoT as per the recommendation by the ITU-T (International Telecommunications Union-Telecommunication Standardization Sector).

3.3 *Five-Layer Architecture*

Five-layer architecture can be considered as an extension of preceding architectures. It is comprised of five layers, namely, perception, network, processing, application, and business in order [10, 11]. The purpose of perception, network, processing, and

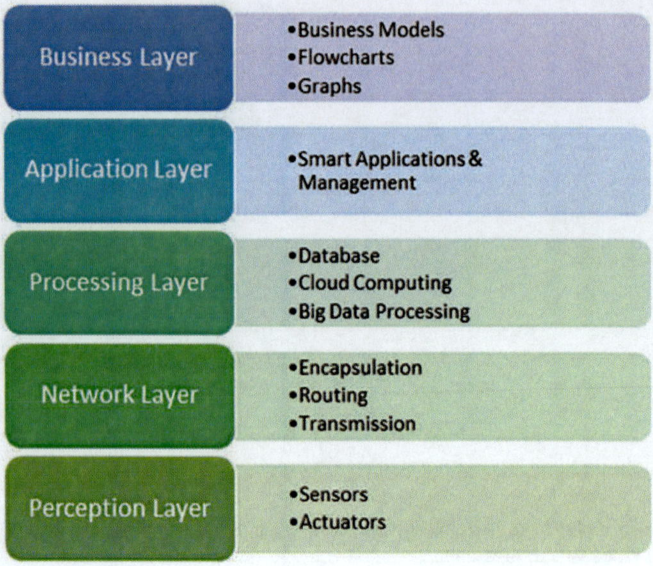

Fig. 6 Five-layer architecture of IoT

application layers in five-layer architecture of IoT is same as defined for preceding architectures of IoT. Here, processing layer is termed as middleware layer or service support layer as it links network layer and application layer together in a bidirectional way. It makes use of many technologies such as big data processing modules, databases, and cloud computing. to accomplish its functionality. Figure 6 depicts a five-layer architecture of IoT. Next, the functionality of business layer has been considered.

- **Business Layer**

It is accountable for administrating the entire IoT system. It has the capacity to build business models, flowcharts, etc. depending on data attained from application layer. This layer enables prediction of future actions and business strategies for businesses subject to analysis of results obtained.

3.4 Six-Layer Architecture

Over the years, IoT's fundamental architecture has undergone a lot of modifications to keep up with the latest technology and to meet dedicated application's requirements. A new six-layer architecture of IoT has been proposed in the literature. This architecture has focus layer, cognizance layer, and competence business layer in

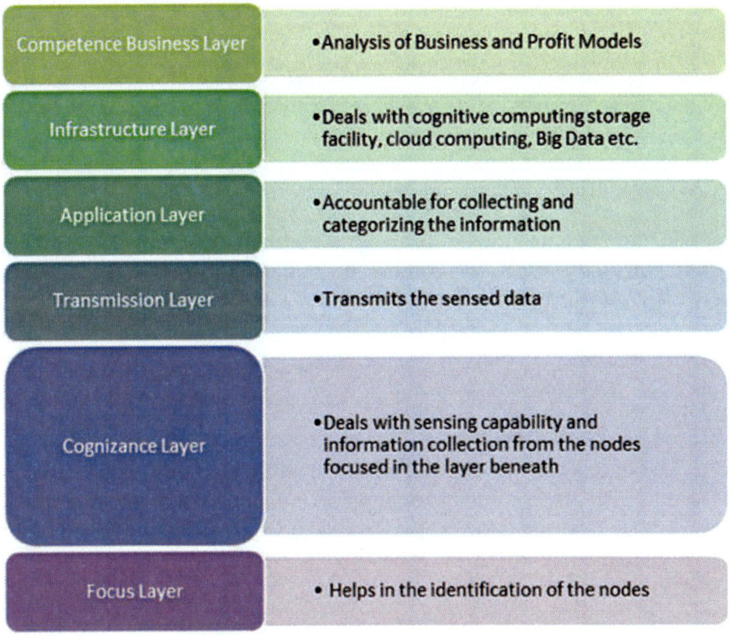

Fig. 7 Six-layer architecture of IoT

addition to transmission, application, and infrastructure layer [6]. Figure 7 depicts a six-layer architecture of IoT.

3.5 *Seven-Layer Architecture*

Following continuous changes in the IT world, seven-layer architecture of IoT was proposed in the Internet of Things World Forum (IoTWF) [12]. It consists of the following layers:

- **Physical Devices and Controllers**

Here, a varied range of physical devices, sensors, actuators, and device controllers are present, which are referred to as "Things" in IoT. These things are capable of sensing and gathering information about the surrounding environment.

- **Connectivity**

This layer specifies the communication protocols as it is responsible for making connections and data transfers in the IoT system.

- **Edge Computing**

Edge computing, or more specifically "Cloud Edge" or "Cloud Gateway" computing layer is responsible for protocol conversion and routing. It focuses on data analysis and transformation.

- **Data Accumulation**

This layer deals with data storage. It could use simple SQL for implementation or may be implemented through some more sophisticated solutions or some NoSQL solutions.

- **Data Abstraction**

At this layer, we "make sense" from the data using the data mining techniques or machine learning. Here, data from different sources is collected to make it accessible regardless of location through data virtualization, and multiple data formats are reconciled. It is also responsible for organizing incoming data into suitable schema and movement for upstream/downstream processing.

- **Application Layer**

At this layer, users can use the information about the environment that is taken by the things.

- **Collaboration and Processes**

Finally, at this layer, users can use the input processed at lower layers to make business decisions.

Figure 8 illustrates IoT World Forum Reference Model.

In this section, we presented multi-layered architectures of IoT ranging from three layers to seven layers. There is no standard architecture of IoT which is adopted globally. Depending upon the domain at hand and to meet application's specific requirements multiple architectures of IoT have been developed so far.

4 IoT Enabling Technologies

IoT can be realized with a number of enabling technologies/protocols [13]. In this segment, we discuss about state-of-the-art of essential communication technologies in IoT. IoT could use a range of communication protocols which are available as both short and long-range standards. These technologies/protocols [14, 15] vary in terms of transmission range, transmission rate, frequency and are defined by several standards, as shown in Table 1. Figure 9 displays few enabling protocols at various layers of IoT architecture.

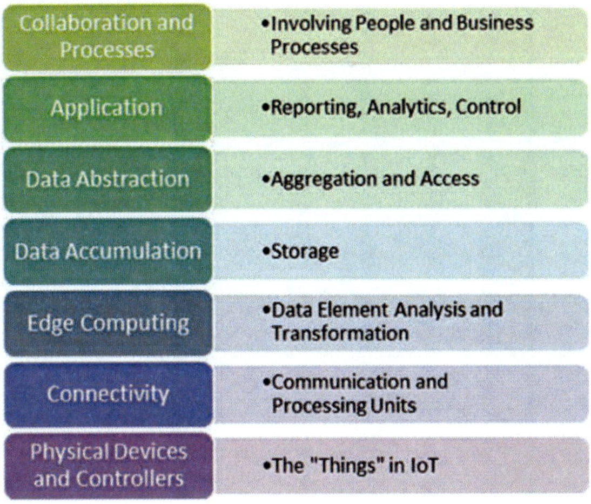

Fig. 8 Seven-layer architecture of IoT (IoT world forum reference model)

In Table 2, various routing protocols used in IoT domain have been compared. In Table 3, various application layer protocols have been compared.

5 Conclusion

The IoT will enable development of a smart world where everything will be connected to one network in the future. From this survey, we understand that a uniform IoT architecture and its related standards need to be designed with a holistic approach. Also, the intricacy involved designing the IoT system, handling communication between various layers and its unification with the real contexture becomes challenging as the number of layers in IoT architecture rises. In this paper, we have presented analysis of existing IoT architectures. Due to inherent openness, we found it to be vulnerable to attacks. In addition to innumerous advantages offered by IoT, it also endures various challenges and problems which need to be managed in forthcoming years. Notable development in the field of IoT has been realized because of numerous enabling and emerging technologies [16] for a wide range of applications, but have yet not been able to be exploited to its full aptitude. All the above-discussed issues laid the foundation for research opportunities in the field of IoT and addressing these issues can assist achieving improved future for IoT.

Table 1 IoT enabling technologies

Technology	Standard	Transmission range	Transmission rate	Frequency
RFID	ISO/IEC, EPCglobal	< 10 cm up to 100 m	2 kbps to > 100 kbps	125–134.3 kHz/13.56 MHz/860–960 MHz/2.45 GHz
NFC	ISO/IEC 18092 ECMA-340	< 0.2 m	424 kbps	13.56 MHz
Bluetooth LE	IEEE 802.15.1	~ 50 m	~ 200 kbps	2.4–2.5 GHz
Wi-Fi	IEEE 802.11 a/b/g/n	Max 300 m	Max 300 Mbps	2.4–5 GHz
WiMAX	IEEE 802.16 d/e	Max 50 km	Max 70 Mbps	2–11 GHz
ZigBee	IEEE 802.15.4	100 m	20 kbps 40 kbps 250 kbps	915 MHz/2.4 GHz
Z-wave	ITU-T G9959	50 m	9.6 kbps, Max 100 kbps	918 MHz/916 MHz
Cellular	EDGE/GPRS/GSM—2G, HSPA/UMTS—3G, LTE—4G	GSM: Maximum 35 km HSPA: Maximum 200 km	GPRS: 35–170 kbps EDGE: 120–384 kbps HSPA: 600 kbps–10 Mbps UMTS: 384 kbps–2 Mbps LTE: 3–10 Mbps	900/1800/1900/2100 MHz
Sigfox	Proprietary Sigfox	Urban range: 3–10 km Rural range: 30–50 km	10–1000 bps	868–868.6 MHz

(continued)

Table 1 (continued)

Technology	Standard	Transmission range	Transmission rate	Frequency
LoRaWAN	LoRaWAN	Urban range: 2–5 km Rural range: 15–20 km	0.3–50 kbps	Various
NB-IoT	Open	~ 22 km	Max 250 kbps	180 kHz
Neul	Neul	10 km	Few bps to 100 kbps	ISM: 900 MHz White space: 470–790 MHz
Thread	Thread: IEEE802.15.4, 6LowPAN	30–100 m	250 kbps	2.4 GHz

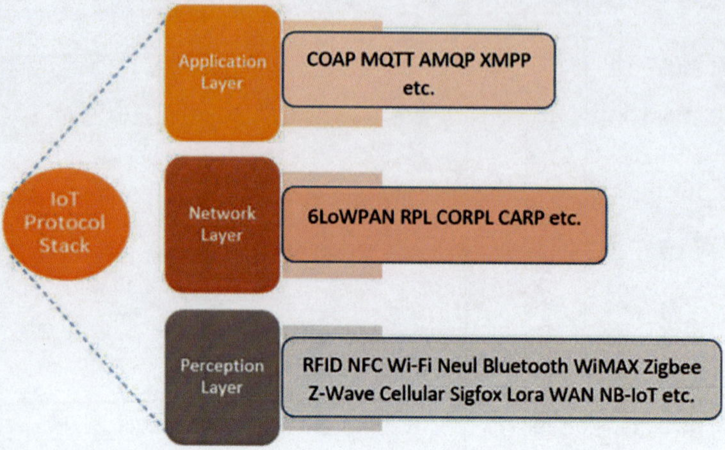

Fig. 9 IoT protocol stack

Table 2 Comparison of various routing protocols used in IoT

Protocol	Channel-aware routing protocol (CARP)	Routing protocol for low-power and lossy networks (RPL)	Cognitive RPL (CORPL)
Designed for	Underwater communication, IoT sensor network applications	Constrained network, supports a range of data link protocols	Cognitive networks
Strategy adopted	Reactive	Proactive	Proactive
Routing	Distributed routing protocol	It is based on destination-oriented directed-acyclic graphs (DODAG), distance vector routing protocol	Cognitive RPL makes use of DODAG but with variations to RPL: in that it coordinates between nodes to choose the best next hop for the packet to be sent to
Advantages	Data management: supported	1. Data management: supported 2. Storage management: supported 3. Server technologies: supported	1. Data management: supported 2. Server technologies: supported
Disadvantages	Security: Unsupported Server technologies: Unsupported Does not support previously collected data	Security: unsupported	Security: unsupported Storage management: unsupported

Table 3 Comparison of application layer protocol

Protocol	Constrained application protocol (CoAP)	Message queue telemetry transport (MQTT)	Advanced message queuing protocol (AMQP)	Extensible messaging and presence protocol (XMPP)
Communication model	Request—Response, 1:1	Publish—Subscribe, M:N	Publish—Subscribe	Request—Response Publish—Subscribe
Transport protocol	UDP	TCP	TCP—UDP	TCP
QoS level	Moderate	High	High	Moderate
Advantages	1. Easily integrated with the web 2. Stateless HTTP 3. Light weight 4. Multi-cast support 5. Low latency	1. Compact message size 2. High performance and energy efficiency 3. Support several levels of quality of services 4. Asymmetric client–server relationship 5. Easy to implement and to integrate new device	1. End-to-end encryption 2. Symmetric client–server relationship 3. Messages over TCP—UDP	1. Supports numerous extensions 2. Decentralized networks with high fault tolerance 3. Easy understandable and extensible 4. Client server model
Disadvantages	1. Lack of topic publication/subscription approach 2. Less existing libraries and solution support 3. Unencrypted protocol	1. Lack of encryption 2. No error-handling 3. Hard to add extensions	1. High resource utilization, in terms of power and memory usage 2. Does not support last value queue	1. Text-based messaging 2. No quality of service 3. No end-to-end encryption 4. Not suitable for embedded IoT applications

References

1. Al-Fuqaha A, Guizani M, Mohammadi M, Aledhari M, Ayyash M (2015) Internet of things: a survey on enabling technologies, protocols, and applications. IEEE Commun Surv Tutorials 17(4):2347–2376
2. Gartner, Inc. (2021) Gartner forecasts global devices installed base to reach 6.2 billion units in 2021. Accessed on: 1 Nov 2021 [Online]. Available: https://www.gartner.com/en/newsroom/press-releases/2021-04-01-gartner-forecasts-global-devices-installed-base-to-reach-6-2-billion-units-in-2021
3. Gartner, Inc. (2020) Gartner survey reveals 47% of organizations will increase investments in IoT despite the impact of COVID-19. Accessed on: 31 Sept 2021 [Online]. Available: https://www.gartner.com/en/newsroom/press-releases/2020-10-29-gartner-survey-reveals-47-percent-of-organizations-will-increase-investments-in-iot-despite-the-impact-of-covid-19
4. Markets and Markets Research Private Ltd. (2020) Covid-19 impact on internet of things (IoT) market by components (software solutions, platforms, services), vertical (BFSI, healthcare, manufacturing, retail, transportation, utilities, government and defense) and region—global forecast 2021. Accessed on: 20 Oct 2021 [Online]. Available: https://www.marketsandmarkets.com/Market-Reports/covid-19-impact-on-iot-market-212332561.html
5. Jurcut A, Niculcea T, Ranaweera P, Le Khac NA (2020) Security considerations for internet of things: a survey, pp 1–19
6. Kumar NM, Mallick PK (2018) The internet of things: insights into the building blocks, component interactions, and architecture layers. Procedia Comput Sci 132:109–117
7. Kakkar L, Gupta D, Saxena S, Tanwar S (2021) IoT architectures and its security: a review. In: 2nd international conference on information management and machine intelligence. Lecture notes in networks and systems, vol 166
8. Giri A, Dutta S, Neogy S, Dahal K, Pervez Z (2017) Internet of things (IoT): a survey on architecture, enabling technologies, applications and challenges. In: ACM international conference proceeding series
9. International Telecommunication Union—Telecommunication Sector (2012) Series Y: global information infrastructure, internet protocol aspects and next generation networks—frameworks and functional architecture models—overview of the internet of things, Y.2060
10. Said O, Masud M (2013) Towards internet of things: survey and future vision. Int J Comput Sci 5(1):1–17
11. Khan R, Khan SU, Zaheer R, Khan S (2012) Future internet: the internet of things architecture, possible applications and key challenges. In: Proceedings of 10th international conference on frontiers of information technology, FIT, pp 257–260
12. http://cdn.iotwf.com/resources/71/IoT_Reference_Model_White_Paper_June_4_2014.pdf
13. Tournier J, Lesueur F, Le Mouël F, Guyon L, Ben-Hassine H (2020) A survey of IoT protocols and their security issues through the lens of a generic IoT stack. In: Internet of things, p 100264
14. Ray PP (2018) A survey on internet of things architectures. J King Saud Univ Comput Inf Sci 30(3):291–319
15. Maple C (2017) Security and privacy in the internet of things. J Cyber Policy 2(2):155–184
16. Kakkar L, Gupta D, Saxena S, Tanwar S (2019) An analysis of integration of internet of things and cloud computing. J Ambient Intell Humaniz Comput 16(10):4345–4349

A Solar-Powered IoT System to Monitor and Control Greenhouses-SPISMCG

Sunilkumar Hattaraki and N. C. Jayashree

Abstract Agriculture benefits from technological advancements. Adopting technology in agriculture makes life easier and more secure for farmers. Growing vegetables in a greenhouse is a unique farm method. Greenhouses' major goal is to create good growing conditions for crops while also safeguarding them from of the adverse weather and pests. The idea of this research is to integrate the Internet of Things (IoT) and solar power to monitor and regulate a greenhouse, allowing the farmer to maintain optimum environmental conditions within the greenhouse at any time. In this scenario, the photovoltaic module provides power to Arduino UNO, which is equipped with a LDR, a fire sensor, a temperature sensor, a soil moisture sensor, and a humidity sensor. These sensor readings are used to check day and night circumstances to turn on the light, detect smoke in the greenhouse, turn on the fan to reduce the excessive warmth, enable the water pump to maintain soil moisture level, and spray water to maintain optimal humidity, respectively. These data are sent to the farmer via the Node MCU Wi-Fi module, and these procedures can be automated or controlled using the Blynk app. The green house entrance system uses an RFID-enabled tag as an identifying badge to protect from unauthorized access. The use of photovoltaic modules in the implementation of this system can boost self-sufficiency and lower production costs.

Keywords Greenhouse · Sensors · NodeMCU · Photovoltaic · IoT · Blynk app · SPISMCG

1 Introduction

A greenhouse system absorbs solar radiation because it contains clear glass that allows the sun's rays to penetrate. When planning plants in greenhouse environments, a proper environment is created that encourages the development of strong

S. Hattaraki · N. C. Jayashree (✉)
Department of Electronics and Communication Engineering, BLDEA's V. P. Dr. P. G. Halakatti College of Engineering and Technology, Vijayapura, India
e-mail: ncjayashree@gmail.com

P. Chatterjee et al. (eds.), *Computational Intelligence for Engineering and Management Applications*, Lecture Notes in Electrical Engineering 984,
https://doi.org/10.1007/978-981-19-8493-8_35

plants, thereby gradually improving the standard of the plants. And also the performance of farmers. In countries with adverse climatic conditions, barren soils, or an external threat from pests, a complicated, inexpensive greenhouse system with optimal environmental conditions could significantly increase agricultural productivity. Humidity in greenhouses makes it possible to increase productivity, reduce production risks, and provide food all year round [1, 2].

The farmers benefit from the green house because it boosts agricultural yield and production rate, as illustrated in Fig. 1. It is inconvenient for farmers since they must visit the greenhouse on a regular basis to maintain ideal climatic conditions, and failing to do so would result in crop destruction and a decrease in production rate [3, 4].The green house's structure is implanted with sensors (Arduino UNO, RFID tag, and Node MCU) in the paper to enable an automated monitoring and control system in the solar-powered green house.

Electricity is the main concern everywhere without electricity, we cannot turn on any electrical device. The energy requirement is indispensable and expensive. In the

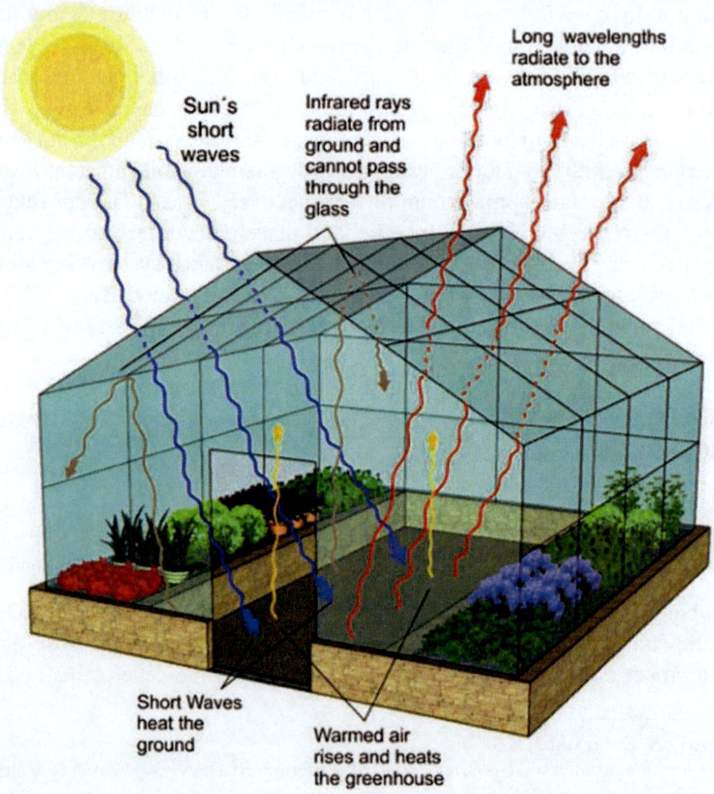

Fig. 1 Greenhouse environment

proposed system, the energy source is replaced by solar energy using a photovoltaic cell. Solar energy is converted into electrical energy, which can then be stored in a rechargeable battery. These batteries power the automated greenhouse monitoring and control system, which allows the user to remotely control their greenhouse with the Blynk app. The fundamental component of the system, NodeMCU, is the Arduino UNO, and all sensors are integrated into it. This device can record essential plant growth parameters such as temperature, humidity, soil moisture, fire sensor, and light intensity, among others [5]. The sensors' output is transferred to the Arduino-based control system for additional data processing and greenhouse environment management. For the protection and security of the greenhouse, adequate measures must be taken by installing external RFID access. This can help farmers grow fruits, vegetables, and crops in all environmental conditions and get the best yield [1].

2 Related Work

Control climatic conditions in the greenhouse automatically, allowing any variety of plant to be cultivated all year. Atmospheric sensors are sensing the changes in greenhouse areas. For these sensors, threshold values are to be set. These values will fluctuate according to the climate, based on this, the actuators will take some actions which are connected to outside the world. The parameters can be tracked using a PC/laptop from any location through GPRS. If there is a variation in the network data is transmitted through SMS by the GSM module to farmers [5]. The disadvantage is not user-friendly; the user has to carry the PC/laptop for tracking, and they can't control the actuator functions if any malfunction occurs in it. They have to solve it manually. In this paper, the first problem is to overcome by using Blynk mobile app which can show the information and provides associate choice for farmer to regulate actuators. The solar panel is utilized to provide electricity to the system, which makes it cost-effective and self-sufficient. Although the RFID tag is used to control the entry of unauthorized persons to the greenhouse. This information can send to the farmer through the Wi-Fi module to Blynk app to maintain the safety of crops in the greenhouse.

3 Hardware Requirements

The components which are used are given below:

Solar cells, solar panels, light dependent resistor, moisture sensor, humidity sensor, fire sensor, relay module, water pump, RFID MFRC 522, Arduino UNO, and NodeMCU ESP8266.

Fig. 2 Solar cell

3.1 Solar Cells

A solar collector is another name for a solar cell [6]. As shown in Fig. 2, it is electric equipment that converts light energy into electricity via the photovoltaic effect.

3.2 Solar Array

Solar modules, or groups of photovoltaic cells, are installed in a system for installation in Fig. 3. Solar cells generate electricity by using the sun's luminance as an energy source. A panel is a group of photovoltaic modules, and a panel system is a matrix. A photovoltaic system's arrays transmit solar energy to electrical equipment.

3.3 Solar Panels

The solar panel [4, 7] consists of multiple layers including solar cells, glass, frame, junction box, encapsulant, and backsheet, as shown in Figs. 4 and 5. A photovoltaic module is commonly referred to as a solar panel.

A photovoltaic module [8, 9] is a frame that holds a collection of solar cells. Photovoltaic cells use daylight as a supply of energy to come up with DC. A photovoltaic module is a group of solar modules, while a panel system is a matrix. A photovoltaic system's matrix provides solar electricity to electrical equipment.

Fig. 3 Pictorial representation from solar cell to solar array (panel)

Fig. 4 Solar panels

Fig. 5 Layers of solar PV module

Fig. 6 Light-dependent resistor

3.4 Light-Dependent Resistor

A light-dependent resistor (also referred to as a photograph resistor, LDR, or electrical conduction cell) could be a passive part that lowers the resistance to receiving luminance (light) at the component's sensitive surface, as shown in Fig. 6. As the incidence of light increases, the photo resist's resistance falls. In other words, photoconductivity is controlled by intensity. In light-sensitive detector circuits, a photo resist can be used. Circuits that are actuated by light and dark and behave as resistive semiconductors. A photo resistor in the dark can have a resistance of several million ohms (M), whereas in the light, it can only have a resistance a few hundred ohms.

3.5 Moisture Sensor

Soil moisture sensors, as shown in Fig. 7 [10–12] are sensors that monitor volumetric water content.

Another type of sensor monitors the water potential of soils, which is a feature of wetness. Tensiometers and gypsum blocks are examples of these sensors, which are also known as soil water potential sensors.

Fig. 7 Moisture sensor

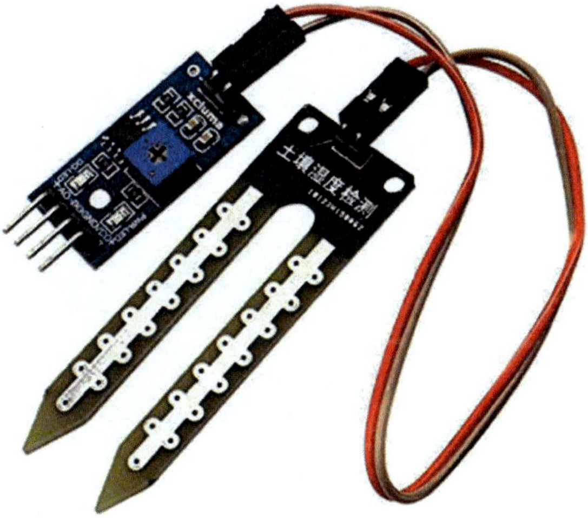

3.6 *Humidity Sensor*

Humidity sensors, as shown in Fig. 8 [10–12], measure the relative humidity of the environment in which they are placed. They measure the air's humidity and temperature, with relative humidity defined as the ratio of the air's humidity to the maximum quantity that the air can hold at its current temperature. As the air becomes hotter, it holds more moisture, resulting in a change in humidity level as a function of temperature.

3.7 *Fire Sensor*

Fire detectors, as shown in Fig. 9, are a type of sensor which can detect and respond to the existence of fire and flames which can be able to detect smoke, heat, infrared and/or UV radiation, gas, etc.

3.8 *Relay Module*

A relay is an electrically operated switch, as seen in Fig. 10. It comprises of a set of control signal input terminals and a set of operating contact terminals [11].

Fig. 8 Humidity sensor

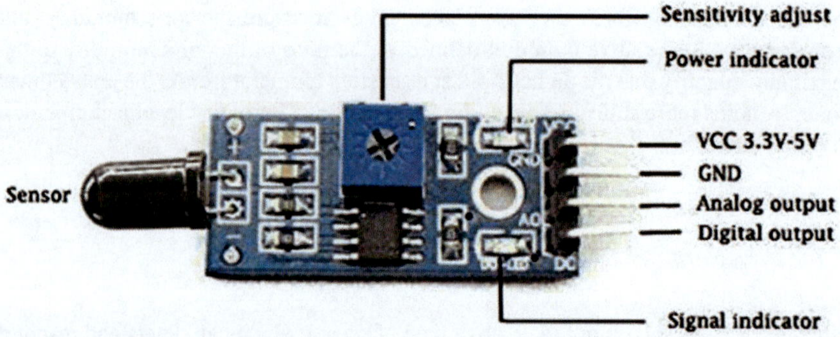

Fig. 9 Fire sensor

3.9 RFID RC 522

Radio Frequency Identification and Tracking (RFID) [6] uses electromagnetic fields to detect and track tags that may be affixed to items, as shown in Fig. 11. An RFID system consists of a radio transponder, a radio receiver, and a radio transmitter. When triggered by an electromagnetic interrogation pulse from a nearby RFID reader, which is typically recognized by an inventory number, the tag transmits digital data to the reader. This number can be used to track inventory.

Fig. 10 Relay module

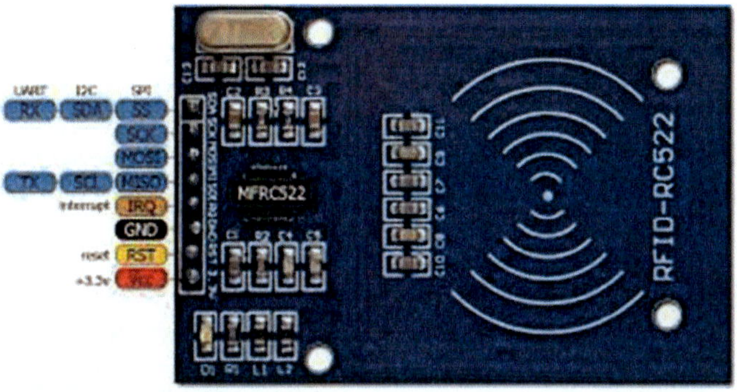

Fig. 11 RFID RC522

3.10 Arduino UNO

Arduino UNO [11–14] boards as shown in Fig. 12, use different types of controllers and microcontrollers. It contains input and output contacts for both digital and analog signals. It contains 14 digital I/O, 6 analog input pins, a 16 MHz crystal oscillator, a reset button, a USB connection and connectors. Because it uses the ATmega328, the Arduino UNO differs from all previous boards.

Fig. 12 Arduino UNO

3.11 NodeMCU

NodeMCU [11, 12] is a Lua-based free software and development board for IoT-based applications, as seen in Fig. 13. If the software and hardware are based on the ESP12 module, it includes firmware that runs on Espress's ESP8266 Wi-Fi SoC.

Fig. 13 NodeMCU ESP8266

4 Proposed Methodology

The proposed methodology demonstrates how a solar-powered, controlled, and monitored IOT-based greenhouse system works in detail. Solar energy is converted into electrical energy using a photovoltaic system. The Aurdino Uno, which is equipped with LDR, temperature, humidity, soil moisture, and fire sensors, is powered by this energy stored in a rechargeable battery. The sensors are set to a threshold voltage. If the sensor values exceed the threshold value, the corresponding actuators, which are connected to the outside world, are switched on, but in order for the fire sensors to detect some fires in the greenhouse, this information is sent to the farmer via the Node MCU Wi-Fi module, which is located in the Blynk-App and can be displayed. Using this data, the farmer can manually take action. The actuators can control automatically by using sensor data or remotely by the farmers using the Blynk app by analyzing the sensor values in it. To avoid unauthorized entry, RFID tags and readers are used to give entry access to the outside door of the greenhouse. This can assist farmers in maintaining overall control and monitoring of the greenhouse system.

Flowchart of SPISMCG is shown in Fig. 14, and block diagram of SPISMCG is shown in Fig. 15.

5 Results and Discussion

The working module of this project successfully fulfills the greenhouse needs by properly controlling and monitoring the greenhouse with a solar-powered IoT system and protecting the greenhouse with an RFID reader as shown in Fig. 16.

The system is integrated with soil moisture, DHT11 temperature, LDR, and fire sensors. These sensor values are displayed in the Blynk app shown in Figs. 17 and 18.

Setting up a greenhouse using the IoT monitoring and control system will minimize human error as well as labor costs and maximize production. India has agriculture as its main occupation and also the most important aspect for its economy; this paper will strengthen the farmers and the country financially. It's so affordable and convenient that anyone can use this system to grow the plants they want in any environmental condition. It is also environmentally friendly and does not harm the environment.

In comparison to current work by other researchers on this system, Table 1 demonstrates how the suggested SPISMCG system delivers automation as well as all potential services.

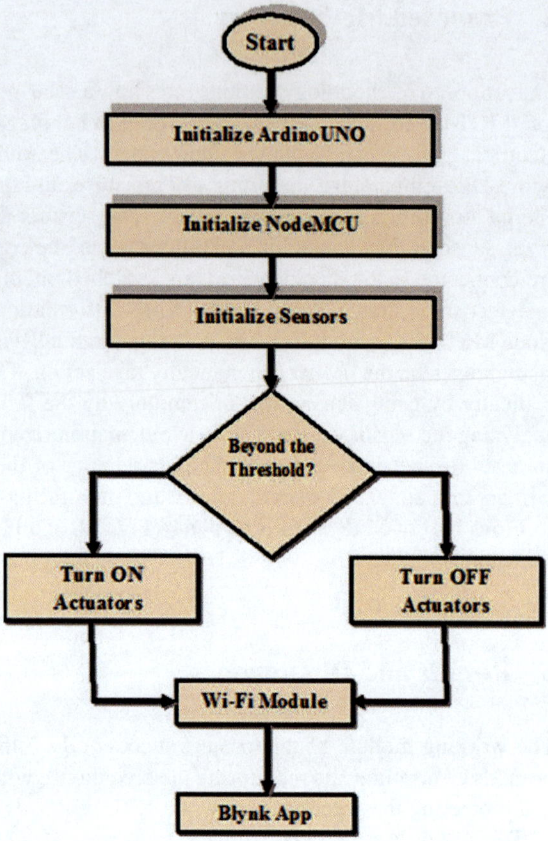

Fig. 14 Flowchart of SPISMCG

6 Conclusion

In order to boost agricultural productivity, the low-cost greenhouse model developed can be used to monitor and adjust the temp, light intensity, humidity, and soil moisture of a greenhouse.

Setting up a greenhouse using the IoT monitoring and control system will minimize human error as well as labor costs and maximize production. India has agriculture as its main occupation and also the most important aspect for its economy; this paper will strengthen the farmers and the country financially. It's so affordable and convenient that anyone can use this system to grow the plants they want in any environmental condition. It is also environmentally friendly and does not harm the environment.

However, some additional work will be required in the future by growing a variety of plants in the greenhouse. The large data sets can be used for computer vision and machine learning to check the soil texture and soil fertility.

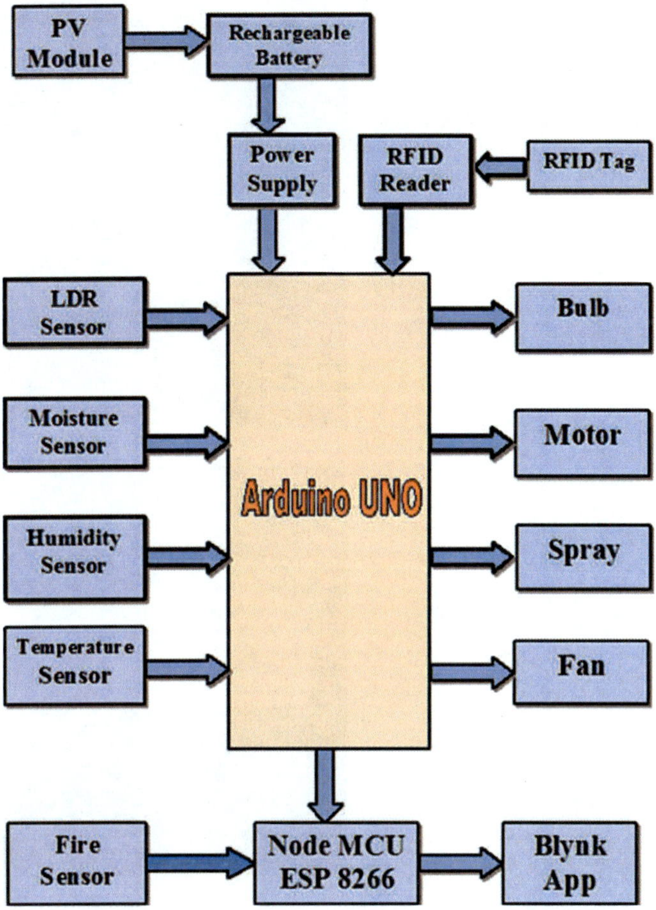

Fig. 15 Block diagram of SPISMCG

Fig. 16 Working module of proposed system

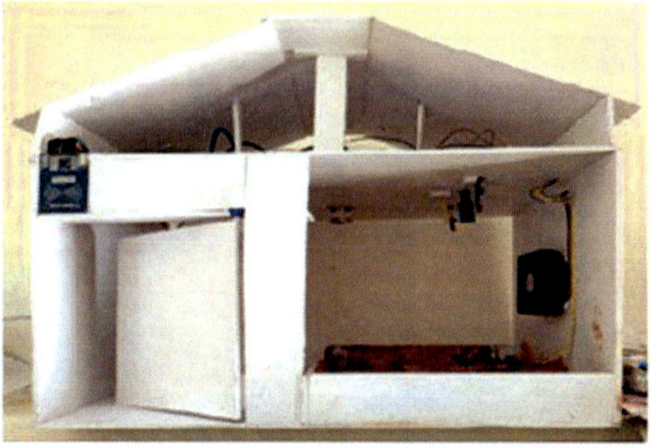

Fig. 17 Sensors value displayed on the Blynk app

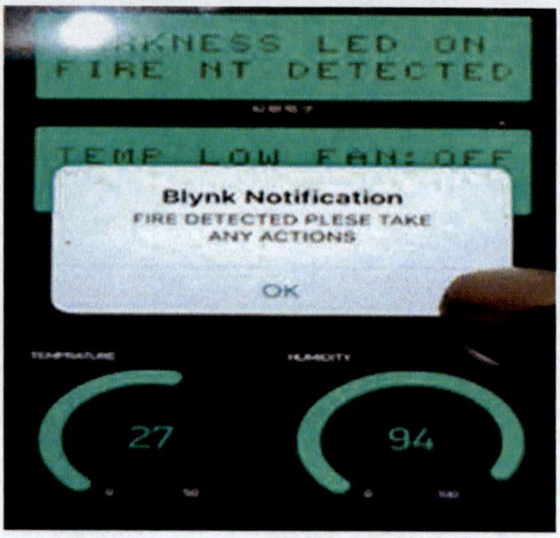

Fig. 18 Fire detection alert message on the Blynk app

Table 1 Comparison of the SPISMCG with recent works

Parameters	Ref No.			Proposed system
	[1]	[3]	[5]	
Solar power	✓			✓
Temperature	✓	✓	✓	✓
Humidity	✓	✓	✓	✓
Moisture	✓	✓	✓	✓
LDR	✓	✓	✓	✓
Fire				✓
RFID tag				✓

References

1. Hoque MDJ, Ahmed MR, Hannan S (2020) An automated greenhouse monitoring and controlling system using sensors and solar power. Eur J Eng Res Sci 5(4):510–515. http://doi.org/10.24018/ejers.2020.5.4.1887
2. Selmani MO, El khayat M, Guerbaoui M, Ed-Dahhak A, Lachhab A, Bouchikhi B (2019) Towards autonomous greenhouses solar-powered. Procedia Comput Sci 148:495–501. ISSN 1877-0509
3. Raj JS, Vijitha Ananthi J (2019) Automation using IoT in greenhouse environment. J Inf Technol Digit World 01(01). http://doi.org/10.36548/jitdw.2019.1.005
4. Khedkar A, Bhise B, Shinde D, Hingane S (2021) Solar powered greenhouse monitoring using IoT. J Sci Technol 06(01). http://doi.org/10.46243/jst.2021.v6.i04.pp254-261. ISSN: 2456-5660
5. Shirsath DO, Kamble P, Mane R, Kolap A, More RS (2017) IOT based smart greenhouse automation using arduino. Int J Innov Res Comput Sci Technol (IJIRCST) 5(2). http://doi.org/10.21276/ijircst.2017.5.2.4. ISSN: 2347-5552
6. Sahana B, Sravani DK, Prasad DR (2020) Smart green house monitoring based on IOT. Int J Eng Res Technol (IJERT) 8(14)
7. Aakanksha A, Anand C, Shameem Akhter S, Nikitha S, Pavan Kumar N (2021) Smart greenhouse automation using solar power. IJCRT 9(6). ISSN: 2320-2882
8. Gao L, Cheng M, Tang J (2013) A wireless greenhouse monitoring system based on solar energy. TELKOMNIKA Indonesian J Electr Eng 11. http://doi.org/10.11591/telkomnika.v11i9.3305
9. Ding J, Zhao J, Ma B (2009) Remote monitoring system of temperature and humidity based on GSM. In: 2nd international congress on image and signal processing
10. Al-Ali AR, Al Nabulsi A, Mukhopadhyay S, Awal MS, Fernandes S, Ailabouni K. IoT-solar energy powered smart farm irrigation system. J Electron Sci Technol. http://doi.org/10.1016/j.jnlest.2020.100017
11. Hattaraki S, Patil A, Kulkarni S (2020) Integrated water monitoring and control system-IWMCS. In: 2020 IEEE Bangalore humanitarian technology conference (B-HTC), pp 1–5. http://doi.org/10.1109/B-HTC50970.2020.9297890
12. Devanath S, Hemanth Kumar AR, Shettar R (2019) Design and implementation of IOT based greenhouse environment monitoring and controlling system using arduino platform. Int Res J Eng Technol (IRJET) 06(09)
13. Yoon C, Huh M, Kang S, Park J, Lee C (2018) Implement smart farm with IoT technology. In: 2018 20th international conference on advanced communication technology (ICACT), pp 749–752. http://doi.org/10.23919/ICACT.2018.8323908
14. Asolkar PS, Bhadade US (2015) An effective method of controlling the greenhouse and crop monitoring using GSM. In: International conference on computing communication control and automation, pp 214–219

An IoT-Based Smart Band to Fight Against COVID-19 Pandemic

Moumita Goswami and Mahua Nandy Pal

Abstract Today, IoT has drawn convincing research ground as a new research topic in the educational, industrial, and especially in healthcare disciplines. In the healthcare domain, it can deliver better quality of services and advanced user experiences. In the present pandemic situation, all the countries, including India, are fighting with COVID-19, and are still looking for a practical and cost-effective solution to face the problems from several dimensions. Apart from inventing a vaccine, different measures like lock-down, mask-wearing, and social distancing implementation were adopted to slow down the spread of the pandemic. According to the World Health Organization (WHO), social distancing has been proven to be the only solution to avoid the contact of a contagious patient. In this work, an IoT-based wearable smart band device is implemented, that detects the body temperature and oxygen level of the person who wears it and also measures safe distancing between persons. So, the band will work as a safety device. The device continuously monitors the body temperature and oxygen level, and if the results cross a standard threshold, then it sends a warning message to the nearest person of the infected one. It also generates an alarm if someone is detected within the range of 1 m around the person.

Keywords Internet of things · COVID-19 · Arduino nano · Oximeter sensor · PIR sensor

1 Introduction

The current situation caused by coronavirus manifests the greatest global public health crisis. According to the World Health Organization (WHO) report, the number of confirmed COVID-19 cases crossed 31 million people with an alarming death toll of over 960,000 people [1]. This disease has common symptoms such as fever, cough, and fatigue, which are essential for recognizing the initial diagnosis [2]. Surprisingly, a patient without any symptoms may also be responsible for spreading the virus. This

M. Goswami (✉) · M. N. Pal
MCKV Institute of Engineering, Howrah, India
e-mail: moumita02314@gmail.com

P. Chatterjee et al. (eds.), *Computational Intelligence for Engineering and Management Applications*, Lecture Notes in Electrical Engineering 984,
https://doi.org/10.1007/978-981-19-8493-8_36

disease has a high potential to be spread easily in comparison with other diseases. Many ongoing efforts and research are being carried out presently to alleviate the spread of the virus. In this situation, IoT technology has been used as a safe and efficient way to deal with the COVID-19 pandemic.

Our goal is to implement an IoT-based small wearable device that can detect all the common symptoms of COVID-19 in the early phase of diagnosis. Early detection and diagnosis may lead to reduced infection and better health services for infected patients. Since early 2020, the world has been struggling [3] to find a treatment or control the spread of COVID-19 but no acceptable result has been obtained. There is a high demand for devices monitoring patients with symptomatic and asymptomatic COVID-19 infection. In recent years, IoT has proven to have an important role in different phases of various infectious diseases [4]. So, in the current pandemic situation, [5], there is an essential need for faster diagnosis due to the high rate of contagiousness of COVID-19. The sooner the patient is diagnosed, the spread of the virus can be better controlled and the patient can get proper treatment. IoT devices can increase the detection process by acquiring information from patients. This can be implemented by capturing body temperatures, oxygen levels using different sensors, and so on. To stop the spread of the virus, social distancing may be handled by using IoT devices to monitor and track people and to make sure that the appropriate distance is measured and maintained.

2 Literature Review

Since early 2020, the world has been suffering from the pandemic situation caused by the coronavirus. Now the causes, signs, and remedial actions of this virus attack are acquainted with most people. IoT technology is a safe and efficient way to deal with the COVID-19 pandemic. This literature review evolved around the discussions related to different IoT-based applications like health management systems, drone-based systems, smartphone-based applications, and ultimately the literature of wearable devices, to fulfill the requirements of this work implementation. The following subsections are analyzed as mentioned.

Implementation of health management systems using IoT technology is a broad application area. Paper [6] investigates the role of IoT technology in three different phases of COVID-19—early diagnosis phase, quarantine phase, and post-recovery phase. Paper [7] also discussed the technology which uses a large number of interconnected devices for exchanging data and to create a smart network for a health managing system. It also enables the social workers and health workers to be connected with patients, civilians, etc. IoT is used to capture health data from various locations and manage all the data using a virtual management system [8, 9]. This technology helps to control the follow-up process based on the reports obtained. The internet-connected hospital is able to inform the concerned medical staff during any emergency. Automated treatment process, wireless healthcare network to detect

COVID-19 patients, smart tracing of infected patients, etc. help to track the patients and mitigate the requirements of them.

Another type of IoT application to fight against COVID-19 with technological aid is the use of a drone. Drone, also known as an unmanned aerial vehicle (UAV), works with the help of sensors, GPS, and communication services and makes it possible to do a variety of tasks such as searching, monitoring, delivering, etc [10, 11]. Different smart drones can be controlled by a smartphone and a controller with a minimum of time and energy. These include Thermal Imaging Drone [12], Disinfectant Drone [13], Medical Drone [14], Surveillance Drone [15], Announcement Drone [16], etc. But the purpose of using a drone is different from the purpose of using a wearable device. The wearable device mainly helps to detect different symptoms of COVID 19 in individuals. So, to make the people more aware of this pandemic situation and to control the disease spreading, different smartphone-based applications, wearable IoT devices are tried to be developed.

Different IoT-based smartphone applications are designed to do efficient tasks in various domains such as healthcare, retail, agriculture, etc. [17–19]. Different IoT-based smartphone applications have been developed in the area of healthcare, and some of them are meant to use in this pandemic situation like Social Monitoring, StayHomeSafe, TraceTogether, Hamagen, Coalition, and AarogyaSetu, etc. [20–25].

Now different wearable devices, which are either worn or stuck to the body, maybe designed for different purposes in different domains like healthcare, fitness, lifestyle, and so on [26–28], such as bands, glasses, and watches. These wearable devices are used in [29] for remote treatment and monitoring of patients. Other different IoT wearable devices have been observed in literature like smart thermometers [30, 31], smart helmets [32], smart glasses [33], IoT-Q-Band [34], EasyBand [35], and Proximity Trace [36]. Developing these devices is having a notable impact on the initial detection of diseases. For example, a wearable IoT device can confirm whether the respirational signs of a patient are normal or not. With this information, the patient can notice any changes in his/her health condition and then decide to make a medical appointment before any other symptoms appear [37]. Thus, the COVID-19 pandemic situation might be easier to fight with an appropriate wearable device. A smart thermometer [30, 31] is used to record constant measurements of body temperatures. Smart helmets with a thermal camera [12] are much safer compared to an infrared thermometer gun due to lesser human interactions. When the temperature is detected by the smart helmet, then the image of the person's face and the location are captured by an optical camera and sent to the assigned mobile number with an alarm. Another wearable device is IoT-based smart glasses [33], which have lesser human interactions compared with thermometer guns [30, 31]. In smart glasses, optical and thermal cameras have been used to scan crowds and the inbuilt face detection technology makes the tracking process easier after detecting doubtful cases. So, different wearable devices are developed to detect the different parameters of COVID-19, but it is not always possible to use different devices for measuring different parameters.

Looking into the above-mentioned aspects, a wearable IoT application is developed which associates the benefits of different methods in an analytical way. This IoT-based smart band device is relatively small in size compared to other wearable devices. It is portable and capable of capturing three parameters at a time. This device is able to detect human body temperature and blood oxygen percentage as well as maintain the physical distance between individuals to prevent the spread of infection. In the case of wearable devices, size is one of the important factors as otherwise, it is very difficult to wear and carry continuously. In the case of smart helmets [32], smart glasses [33], it's very difficult to wear continuously. The easy-to-wear IoT-based Smart Band device remains continuously connected with our body and is used for continuous monitoring of the captured values. If these values are not within a predefined safety threshold, then an SMS alert will be generated automatically and sent to an acquainted person. The device also detects human presence within a certain range and displays an alert for maintaining social distancing. So, this is a single device that can detect all the known signs of COVID-19 as well as is useful for maintaining social distancing.

3 Process Flow of the System

Temperature and oxygen level are two major parameters of coronavirus infection that need to be measured for detecting the health condition of a patient. Here, an IoT enables smart band has been introduced that will help us to fight against the COVID-19 pandemic by capturing patient's data remotely. It helps to predict the health condition of infected people and is also used for preventing and controlling the spreading of this contagious virus by detecting the distance between the wearer and other individuals. The system generates a mobile alert to the near one of the detected persons. This wearable IoT device can detect symptoms of a patient and thus helps people in the early arrangement of medical appointments before any significant advancement of the disease takes place.

The device is implemented based on the working principles of a max30100 pulse oximeter sensor, DHT11 a temperature, and a PIR sensor which measures the oxygen level, body temperature, and social distancing. Arduino nano is used as a microcontroller that will control every sensor module. So, the main operation is executed by the Arduino Nano according to the input sensor data. The oximeter sensor senses the oxygen saturation in blood from the fingertip and sends the data to Arduino. The body temperature sensor will measure the body temperature of the user from the wrist. The PIR sensor will ensure the social distancing of the user by receiving infrared radiation from the human body. From the given data, the microcontroller will operate in the following way.

It will check if the oximeter data (Oxygen saturation data) is lower than a given threshold saturation (95%).

It will check if the body temperature data is greater than a certain value (98.4 °F).

It will also check if the PIR sensor gives any value (more than 1 m or not).

A Bluetooth module is also used to transfer sensor data from device to smartphone. If any of these conditions satisfy, the Arduino will send SMS through GSM and SIM card module to a saved contact. Figure 1 describes the process flow diagram of our

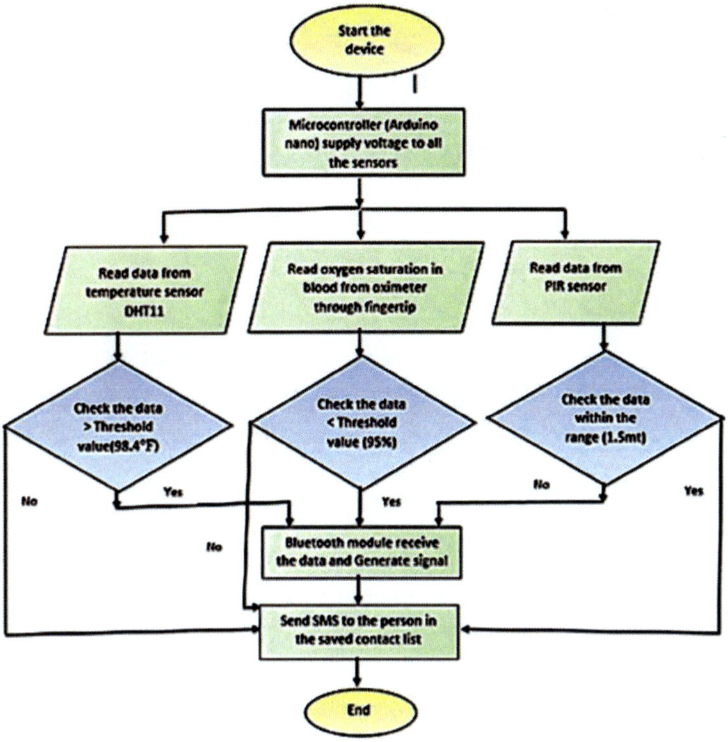

Fig. 1 Process flow diagram

project. In this way, the device will not only detect the health condition but also measure the social distancing within a given range. Figure 2 represents the block diagram of the system.

4 Working Principles with Technical Details

The Arduino lord works on 3.3 V DC to 5 V logic. All the sensors also work on either 3.3 V DC or 5 V DC. In this implementation, the temperature sensor works on 5 V logic, the PIR sensor also works on 5 V logic, oximeter works on 3.3 V logic, and the bluetooth module works on 5 V logic.

Arduino Nano gives supply voltage to all other sensors and devices, 5 V or 3.3 V according to their requirement, receives data from all other sensors, and operates Bluetooth data transmission process. Temperature Sensor (DHT11) sends data to Arduino via Digital pin 3. PIR Sensor sends data to Arduino via Digital pin 2. Oximeter Sensor (MAX30100) directs data to Arduino via Analog pin 5 (SCL) Analog pin 4 (SDA). And Bluetooth module receives data from Arduino via Digital

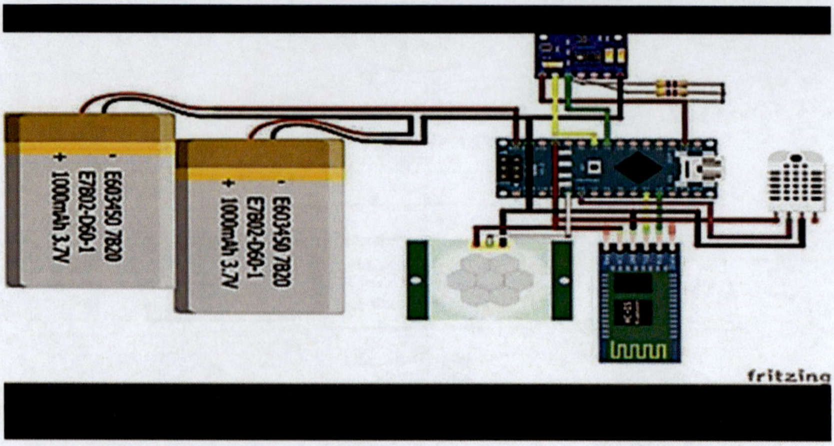

Fig. 2 System block diagram using fritzing

pin 9 (Rx), and Digital pin 8 (Tx). After processing the raw data according to the coded logic written inside the Microcontroller Arduino sends the data to the app installed in the smartphone using Bluetooth. Connections are given in Fig. 2 and each part is described below.

Temperature Sensor (DHT11)—Temperature sensor output is connected to Digital Pin 3, V_{cc} was connected to 5 V, and Ground was connected to the common ground with the other sensors and Arduino. DHT11 gives a digital signal according to the body temperature as output and this data was given to Arduino via Digital pin 3.

A Temperature and Humidity Sensor DHT11 is used here which produces a digital output. These sensors have two parts; a capacitive humidity sensor and a thermistor. It can be interfaced with a microcontroller to get instantaneous results. This sensor uses a negative temperature coefficient thermistor, which decreases its resistance value with an increase in temperature. Temperature range of DHT11 is carrying from 0 to 50 °C with 2 degrees of accuracy and the range of Humidity of this sensor is from 20 to 80% with 5% of accuracy. The rate of sampling of this sensor is 1 Hz, i.e., it gives one reading for every second. DHT11 is a small size sensor with an operating voltage from 3 to 5 V. The utmost current used while measuring is 2.5 mA.

PIR Sensor—Passive Infrared Sensor is termed as PIR sensor output is connected to Digital Pin 2, V_{cc} was connected to 5 V, and Ground was connected to the common ground with the other sensors and Arduino. PIR has a variable range to operate, but here it will operate within 1 m. PIR gives digital signal 1 if any human body is present in front of the sensor within the range, and 0 if the human body is not present, as output. And this data was given to Arduino via Digital pin 2.

These sensors are used to detect human movement when it comes within a particular range. The coverage range depends upon the type and design of the Sensor. Passive Infrared Sensors are consisting of pyroelectric sensors and have a round metal with a rectangular crystal in the center that is used to detect levels of infrared radiation generated from the human body to create an alert into a mobile phone. An object is constantly radiating infrared rays to the outside world. The temperature of the human body is between 36 and 27 °C and most of its radiant energy is concentrated in the wavelength range of 8–12 μm. If the human infrared radiation is directly irradiated on the radar, it will reason of a temperature change to output a signal and activates the detection with an alarm signal.

Once the sensor detects any motion, the microcontroller (Arduino) will send a message via the serial port. PIR delays for a certain time to check if there is a new motion. If there is no motion detected, Arduino sends a new message that the motion has ended.

Oximeter Sensor (MAX30100)—This sensor sends data using I2C PROTOCOL (Inter-Integrated Circuit Protocol). This kind of device uses two other connections other than V_{cc} and Ground line. Those two connections are Serial Clock Line (SCL) and Serial Data Line (SDA) lines. This sensor operates on 3.3 V logic. So, V_{cc} was connected to 3.3 V and Ground was connected to the common ground with the other sensors and Arduino. SCL and SDA were connected to Arduino built-in SCL (Arduino A5, i.e., Analog pin 5), SDA (Arduino A4, i.e., Analog pin 4). MAX30100 sends measured blood oxygen saturation data to Arduino via SCL, SDA Line.

When the heart pumps blood by contraction and expansion, there is an increase and decrease in oxygenated blood. Ultimately, by observing the time between the increase and decrease of oxygen-rich blood, the device calculates the pulse rate. Oxygenated blood consumes more infrared light and passes more red lights whereas deoxygenated blood consumes red light and passes more infrared light. The important feature of MAX30100 is that it reads the levels of absorption for both light sources and stores them in a buffer that can be read via I2C.

Microcontroller—The system can be implemented with Arduino or raspberry pie. Arduino is less powerful than Raspberry Pi, but it doesn't need any OS and software applications to run. Arduino has a 'real-time' and 'analog' ability that Raspberry Pi does not have and this flexibility allows it to work with almost any kind of sensor or chips. The simulation of the device was accomplished using the Arduino nano because of its small size. The architecture of Arduino nano is given in Fig. 3 with all its pin descriptions.

Bluetooth module—Bluetooth module connects with the phone Bluetooth and sends the acquired sensor data to an application. The mobile application is implemented by us. The data transmission process is controlled by the microprocessor Arduino according to the coded logic. Bluetooth works on 5 V logic. So V_{cc} was connected to 5 V and Ground was connected to the common ground with the other sensors and Arduino. Bluetooth module receives data from Arduino via Rx, Tx pin, to send data to smartphone. Arduino Built-in Tx pin is Digital pin 8, and Rx pin is Digital pin 9.

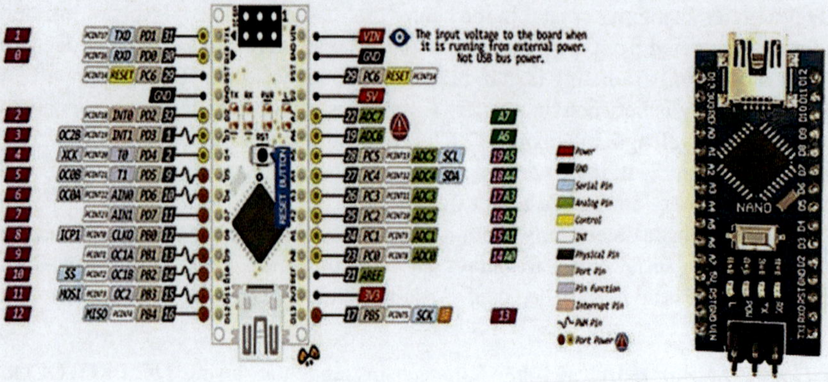

Fig. 3 Arduino nano

Bluetooth Module (HC-05) is a Serial Port Protocol (SPP) module that is used to establish a transparent wireless serial connection. It communicates via serial communication which interfaces with the controller or PC. HC-05 Bluetooth module gives switching mode between master and slave mode.

Smartphone application—The name of the application is OxyTemp. This application is solely implemented by us with no rights are violated. This application is supposed to send an SMS alert to any selected contact (User Selected) if temperature rises above 98.4 °F or blood oxygen level fall below 95% stating "Name: Patient Name and phone no: 0123456789 at a high risk from the perspective of COVID-19 infection. Body temperature 99 °F and blood O2 level 94%" (all the given data within this message are just for example purpose).

5 Experimental Results

The sample result generated by the DHT11 sensor is given below

Humidity: 82.00% Temperature: 28.70 °C 83.66 °F Heat index: 34.32 °C 93.78 °F
Humidity: 82.00% Temperature: 28.70 °C 83.66 °F Heat index: 34.32 °C 93.78 °F
Humidity: 82.00% Temperature: 28.70 °C 83.66 °F Heat index: 34.32 °C 93.78 °F
Humidity: 82.00% Temperature: 28.70 °C 83.66 °F Heat index: 34.32 °C 93.78 °F
Humidity: 82.00% Temperature: 28.70 °C 83.66 °F Heat index: 34.32 °C 93.78 °F
Humidity: 82.00% Temperature: 28.70 °C 83.66 °F Heat index: 34.32 °C 93.78 °F
Humidity: 82.00% Temperature: 28.70 °C 83.66 °F Heat index: 34.32 °C 93.78 °F
Humidity: 82.00% Temperature: 28.70 °C 83.66 °F Heat index: 34.32 °C 93.78 °F
Humidity: 82.00% Temperature: 28.70 °C 83.66 °F Heat index: 34.32 °C 93.78 °F
Humidity: 82.00% Temperature: 28.60 °C 83.48 °F Heat index: 34.04 °C 93.27 °F
Humidity: 82.00% Temperature: 28.60 °C 83.48 °F Heat index: 34.04 °C 93.27 °F

Humidity: 81.00% Temperature: 28.60 °C 83.48 °F Heat index: 33.84 °C 92.92 °F
Humidity: 81.00% Temperature: 28.60 °C 83.48 °F Heat index: 33.84 °C 92.92 °F
Humidity: 81.00% Temperature: 28.60 °C 83.48 °F Heat index: 33.84 °C 92.92 °F
Humidity: 81.00% Temperature: 28.60 °C 83.48 °F Heat index: 33.84 °C 92.92 °F
Humidity: 81.00% Temperature: 28.60 °C 83.48 °F Heat index: 33.84 °C 92.92 °F
Humidity: 81.00% Temperature: 28.60 °C 83.48 °F Heat index: 33.84 °C 92.92 °F
Humidity: 81.00% Temperature: 28.60 °C 83.48 °F Heat index: 33.84 °C 92.92 °F
Humidity: 79.00% Temperature: 28.60 °C 83.48 °F Heat index: 33.46 °C 92.23 °F
Humidity: 79.00% Temperature: 28.60 °C 83.48 °F Heat index: 33.46 °C 92.23 °F
Humidity: 79.00% Temperature: 28.60 °C 83.48 °F Heat index: 33.46 °C 92.23 °F
Humidity: 79.00% Temperature: 28.60 °C 83.48 °F Heat index: 33.46 °C 92.23 °F
Humidity: 79.00% Temperature: 28.60 °C 83.48 °F Heat index: 33.46 °C 92.23 °F
Humidity: 79.00% Temperature: 28.60 °C 83.48 °F Heat index: 33.46 °C 92.23 °F
Humidity: 79.00% Temperature: 28.60 °C 83.48 °F Heat index: 33.46 °C 92.23 °F
Humidity: 79.00% Temperature: 28.60 °C 83.48 °F Heat index: 33.46 °C 92.23 °F

Result recorded by MAX30100 PULSE Oximeter is given below

16:29:43.513 → Heart rate: 45.77 bpm/SpO2: 99%
16:29:44.535 → Heart rate: 45.77 bpm/SpO2: 99%
16:29:44.581 → Beat!
16:29:45.279 → Beat!
16:29:45.514 → Heart rate: 63.33 bpm/SpO2: 99%
16:29:46.028 → Beat!
16:29:46.530 → Heart rate: 73.01 bpm/SpO2: 98%
16:29:46.812 → Beat!
16:29:47.554 → Heart rate: 76.05 bpm/SpO2: 98%
16:29:47.554 → Beat!
16:29:48.342 → Beat!
16:29:48.528 → Heart rate: 77.31 bpm/SpO2: 97%
16:29:49.552 → Heart rate: 77.31 bpm/SpO2: 97%
16:29:49.879 → Beat!
16:29:50.530 → Heart rate: 48.96 bpm/SpO2: 97%
16:29:50.576 → Beat!
16:29:51.274 → Beat!
16:29:51.553 → Heart rate: 75.67 bpm/SpO2: 98%
16:29:51.880 → Beat!
16:29:52.531 → Heart rate: 88.76 bpm/SpO2: 98%
16:29:52.577 → Beat!
16:29:53.320 → Beat!
16:29:53.552 → Heart rate: 83.30 bpm/SpO2: 97%
16:29:54.109 → Beat!
16:29:54.575 → Heart rate: 80.76 bpm/SpO2: 97%

The output of the PIR sensor is either 0 (low) or 1(high). If it detects any object then it returns 1 otherwise returns 0.

The real-time simulation of the device using a micro-controller and sensors is shown in Fig. 4. When someone comes inside the critical field of radius 1 m, the alert will be given on the mobile with the indication that a person is in the range. Mobile app screenshots are given in Fig. 5.

Fig. 4 Real-time circuit implementation

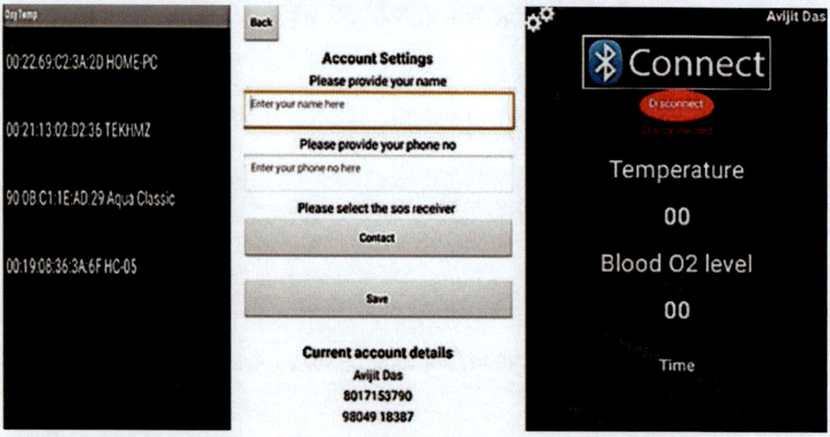

Fig. 5 Mobile app interfaces

6 Conclusion

IoT is an innovative technology. It enables devices to be connected over a network to fight against the COVID-19 pandemic. The smart band will help general people to identify symptoms of the infectious disease and manage an infected case of COVID-19. The smart band is a small offline and portable device and in this pandemic situation, this offline smart band will help to mitigate the spread of the coronavirus as well as reduce the complexity and turnaround time for effective management of the pandemic.

COVID-19 is a global health crisis and is having a devastating effect on businesses, marketplaces, economy, society, and our lives. The social, economical, and health effects of this pandemic and its restrictions will take time to be fully accepted and quantified; however, there are lots of ongoing efforts in research that are being carried out to detect, treat, and trace the virus and its symptoms to reduce its impacts. Internet of Things (IoT) has given positive results in three phases of COVID-19 which are the early detection phase, quarantine phase, and after recovery phase. However, by learning more about the virus and its behavior, the technology is required to be adjusted and improved. Artificial intelligence (AI) can be integrated with IoT technology to use AI power to reap the research results more efficiently. This may be considered as the future scope of this work.

References

1. WHO (2020) Coronavirus disease (COVID-19). https://bit.ly/2ZU5x08. Accessed 09 July 2020
2. Symptoms of coronavirus (2020). https://www.cdc.gov/coronavirus/2019-ncov/symptoms-testing/symptoms.html. Accessed 26 June 2020
3. Zhang SX, Wang Y, Rauch A, Wei F (2020) Unprecedented disruption of lives and work: health, distress and life satisfaction of working adults in China one month into the COVID-19 outbreak. Psychiatry Res 112958
4. Christaki E (2015) New technologies in predicting, preventing and controlling emerging infectious diseases. Virulence 6(6):558–565
5. Phelan AL, Katz R, Gostin LO (2020) The novel coronavirus originating in Wuhan, China: challenges for global health governance. JAMA 323(8):709–710
6. Nasajpour M, Pouriyeh S, Pariziy RM, Dorodchiz M, Valero M, Arabniax HR (2020) Internet of things for current COVID-19 and future pandemics: an exploratory study. J Healthc Informatics Res 4:325–364
7. Singh RP, Javaid M, Haleem A, Suman R (2020) Internet of things (IoT) applications to fight against COVID-19 pandemic. Diabetes Metab Syndr Clin Res Rev 14:521–524
8. Stoessl AJ, Bhatia KP, Merello M (2020) Movement disorders in the world of COVID-19. Mov Disord Clin Pract (in press)
9. Gupta M, Abdelsalam M, Mittal S (2020) Enabling and enforcing social distancing measures using smart city and its infrastructures: a COVID-19 use case. arXiv preprint arXiv:2004.09246
10. Rouse M (2019) Drone (UAV). https://bit.ly/2ZHuonE. Accessed 04 July 2020
11. Nayyar A, Nguyen B-L, Nguyen NG (2020) The internet of drone things (IoDT): future envision of smart drones. In: First International conference on sustainable technologies for computational intelligence. Springer, Berlin, pp 563–580

12. Mohammed M, Hazairin NA, Al-Zubaidi S, AK S, Mustapha S, Yusuf E (2020) Toward a novel design for coronavirus detection and diagnosis system using IoT based drone technology. Int J Psychosoc Rehabil 24(7):2287–2295
13. Shaw KK, Vimalkumar R (2020) Design and development of a drone for spraying pesticides, fertilizers and disinfectants. Eng Res Technol (IJERT)
14. Zema NR, Natalizio E, Ruggeri G, Poss M, Molinaro A (2016) Medrone: on the use of a medical drone to heal a sensor network infected by a malicious epidemic. Ad Hoc Netw 50:115127
15. Ding G, Wu Q, Zhang L, Lin Y, Tsiftsis TA, Yao Y-D (2018) An amateur drone surveillance system based on the cognitive internet of things. IEEE Commun Mag 56(1):29–35
16. Marr B (2020). Robots and drones are now used to fight COVID-19. Pozyskano z: https://www.forbes.com/sites/bernardmarr/2020/03/18/how-robots-and-drones-are-helping-to-fight-coronavirus
17. El Khaddar MA, Boulmalf M (2017) Smartphone: the ultimate IoT and IoE device. In: Smartphones from an applied research perspective, p 137
18. Sinha D (2019) IoT-based mobile applications and their impact on user experience. https://www.iotforall.com/mobile-iot/. Accessed 04 July 2020
19. Parizi RM, Guo L, Bian Y, Azmoodeh A, Dehghantanha A, Choo K-KR (2018) CyberPDF: smart and secure coordinate-based automated health PDF data batch extraction. In: 2018 IEEE/ACM international conference on connected health: applications, systems and engineering technologies (CHASE). IEEE, pp 106–111
20. Kelion L (2020) Coronavirus: Moscow rolls out patient-tracking app. https://www.bbc.com/news/technology-52121264. Accessed 19 June 2020
21. Hui M (2020) Hong Kong is using tracker wristbands to geofence people under coronavirus quarantine. https://qz.com/1822215/hong-kong-usestracking-wristbands-for-coronavirus-quarantine/. Accessed 19 June 2020
22. TraceTogether, safer together (2020). https://www.tracetogether.gov.sg/. Accessed 06 June 2020
23. Stub ST (2020) Israeli phone apps aim to track coronavirus, guard privacy. https://www.usnews.com/news/best-countries/articles/2020-04-20/new-tech-apps-in-israel-aim-to-track-coronavirus-guard-privacy. Accessed 27 June 2020
24. Tokenpost (2020) IoT blockchain platform launches a COVID-19 contact tracing app. https://bit.ly/3eS2VGt. Accessed 27 June 2020
25. Aarogyasetu. https://www.mygov.in/aarogya-setu-app/. Accessed 06 June 2020
26. Juniper Research. Smart wearables market to generate $53bn hardware revenues by 2019. https://www.juniperresearch.com/press/press-releases/smart-wearables-market-to-generate-53bn-hardware. Accessed 04 July 2020
27. Wright R, Keith L (2014) Wearable technology: if the tech fits, wear it. J Electron Resour Med Libr 11(4):204–216
28. Berglund ME, Duvall J, Dunne LE (2016) A survey of the historical scope and current trends of wearable technology applications. In: Proceedings of the 2016 ACM international symposium on wearable computers, pp 40–43
29. Rahman MS, Peeri NC, Shrestha N, Zaki R, Haque U, Ab Hamid SH (2020) Defending against the novel coronavirus (COVID-19) outbreak: how can the internet of things (IoT) help to save the world? Health Policy Technol 9(2):136
30. Chamberlain SD, Singh I, Ariza C, Daitch A, Philips P, Dalziel BD (2020) Real-time detection of COVID-19 epicenters within the United States using a network of smart thermometers. MedRxiv, 2020-04
31. Tamura T, Huang M, Togawa T (2018) Current developments in wearable thermometers. Adv Biomed Eng 7:88–99
32. Mohammed M, Syamsudin H, Al-Zubaidi S, AKS RR, Yusuf E (2020) Novel COVID-19 detection and diagnosis system using IoT based smart helmet. Int J Psychosoc Rehabil 24(7)
33. Mohammed M, Hazairin NA, Syamsudin H, Al-Zubaidi S, Sairah A, Mustapha S, Yusuf E (2019) Novel coronavirus disease (COVID-19): detection and diagnosis system using IoT based smart glasses. Int J Adv Sci Technol 29(7) (Special Issue, 2020)

34. Singh VK, Chandna H, Kumar A, Kumar S, Upadhyay N, Utkarsh K (2020) IoT-Q-band: a low cost internet of things based wearable band to detect and track absconding COVID-19 quarantine subjects. EAI Endorsed Trans Internet Things 6(21)
35. Tripathy AK, Mohapatra AG, Mohanty SP, Kougianos E, Joshi AM, Das G (2020) EasyBand: a wearable for safety-aware mobility during pandemic outbreak. IEEE Consum Electron Mag 9(5):57–61
36. Contact tracing IoT solution (2020). https://bit.ly/2B60rF2. Accessed 06 June 2020
37. Haghi M, Thurow K, Stoll R (2017) Wearable devices in medical internet of things: scientific research and commercially available devices. Healthc Inf Res 23(1):4–15

Anomaly Detection in Blockchain Using Machine Learning

Gulab Sanjay Rai, S. B. Goyal, and Prasenjit Chatterjee

Abstract With the increasing use of Blockchain, community has become increasingly worried about its security, that has led to substantial research by academics, with anomaly detection being a major issue. Regardless of the fact that they can provide availability and integrity, the bulk of public Blockchain systems are decentralized and have minimal confidentiality. The Blockchain network is vulnerable to transaction privacy breaches since all of the network's keys are exposed to everyone. Various security flaws in Ethereum and smart contracts have recently been discovered. As a result, it's critical to improve Blockchain's security features. This research will mainly use the literature survey method and the inductive analysis method to analyze the relevant research works for anomaly detection in Blockchain technology and to find out the trends and characteristics of the development and application of the anomaly detection models and explore its feasibility in ensuring security in Blockchain technology. This paper also proposes a framework for anomaly detection in Blockchain technology.

Keywords Blockchain · Security issues · Anomaly detection · Machine learning

1 Introduction

As a consequence of the enormous business potential of virtual currencies, blockchain technology has been one of the most popular topics in the advanced method of ledger controls. Scholars, professionals, or even investment funds are all on the lookout for new Blockchain systems such times. Even though Blockchain began as a decentralization network for means of exchange [1], since then it has grown into somewhat more. It is a cryptocurrency method that enables funds to be forwarded even without

G. Sanjay Rai · S. B. Goyal (✉)
City University, 46100 Petaling Jaya, Malaysia
e-mail: drsbgoyal@gmail.com

P. Chatterjee
Department of Mechanical Engineering, MCKV Institute of Engineering, Howrah, West Bengal, India

P. Chatterjee et al. (eds.), *Computational Intelligence for Engineering and Management Applications*, Lecture Notes in Electrical Engineering 984,
https://doi.org/10.1007/978-981-19-8493-8_37

involvement of a trusted intermediary. The most typical applications of blockchain technology are personal, social and multinational exchanges, that also start taking advantage of the low cost, avoiding the expenses of a trusted intermediary, including a banks or lender. Low latency, tamper resistance, confidentiality as well as anomaly are also features of Blockchain. Blockchain is probable to be used for non-currency asset transactions [2, 3], distributing uses [4], file storage technologies, and land registrations due to such character traits. A Blockchain is composed of a main chain with each block's hash values linking it together. No one (except the transactions originator) can edit or delete a Blockchain transaction with for this framework. This suggests that structure is extremely opposed to change. This suggests, nevertheless, that it would be unable to overturn suspicious charges engendered by theft with private keys. As an outcome, appropriate preventative measures, including the rapid detection and correction of unauthorized exchanges even before to permission, are required to minimize the damage [5].

Once functioning with Blockchain, the first issue is figuring out how to verify financial transactions. Unlike currency transactions, in which verifying a transaction requires only verifying the giver and receiver's names and financial information, establishing a strategy to verify a general corporate reasoning is much more challenging. This is especially true if a part of the company or organization is conducted beyond the Blockchain. One illustration is food supply chain networks [6]. Using blockchain's hashing keys, one can validate that a logged entry was never modified, but "what might prevent bad operators in the supply chain from manipulating with during first place" [7]? We need to have a sophisticated algorithm that automatically de-weights items that aren't as reliable. The second problem is determining how to come to an agreement on transactional data. The bitcoin core system is built on the concept of a probabilistic public Blockchain: It is envisioned that system would consist of independent random verifiers (or mining) who will fight for a reward. As a result of this hypothesis, hackers may estimate their chances of being successful faking most recent transactional information. Probability is modest as soon as node predominate the system. In permissionless systems, however, the assumption of stochastic competitiveness is manifestly false. The Byzantium fault tolerant mechanism, or a variant of it, is frequently used to secure agreement between endorsers [8].

The paper concentrates on anomaly detection, which is the process of detecting odd behaviors of assets using sensor data before they break or experience unanticipated service outages in Blockchain-based devices [9]. Embedded device attacks may be detected using a variety of security technologies. An intrusion detection system is one such instrument (IDS). Anomaly-based IDSs learn a network's or host's regular behavior and identify deviations from that behavior. As a result, these systems may be able to identify new dangers without having been explicitly trained to do something like that. This method is intriguing since it needs no vertical management [10], in addition to the capacity to discover new "nil" attacks. In constructing an anomaly-based IDS, the network should collect and adapt upon ordinary conclusions drawn during a timed "training period." Based to one fundamental concept, the data collected during learning phase would both be harmless and record all the device's likely activities. That statement might be valid in some situations. This assumption, however, is problematic in terms of the Blockchain for the reasons listed:

The Model's Generalization A lab setup could be used to train the IDS securely. Replicating all possible placements and engagements with the device, on the other hand, is a difficult task. That's the case since some functionality is dependent solely on a single or even more sensing devices, social interaction, or event-based triggers. This procedure is much costlier and time consuming. The simulation could also be modified on during implementation. The model, however, will not be available for implementation till the training process is done (threat detection). It's also uncertain if the trained algorithm will be able to recognize benign but unusual activities. Consider how a smart camera's motion detection system behaves or how a smoke detector reacts when a fire is detected. These rare but lawful behaviors will result in misleading alarms during ordinary execution.

The next kind of attack is adversarial. Although on-site training is a more natural method to learn about the typical behavior of a Blockchain device, the model must assume that all observations collected during the training phase are benign. This technique opens the system to fraudulent data, making it possible for an intruder to use the hardware to evade suspicion or cause additional issues.

The bitcoin cryptocurrency's Blockchain idea has shown to be a promising technology for decentralized peer-to-peer transactions, data integrity, and transparent storage. Some of the specific hazards and threats associated with Blockchain technology in diverse applications include: double spending attack [11], Sybil attack [12], 51% attacks [13], Phishing attack [14], Byzantine fault detection [15], and Routing attack [16]. As transaction might be changed prior authorization, quick transaction detection system is essential to avoid the harm caused by fraudulent transactions and different attacks. Current anomalous detection systems, on the other hand, need all transaction to be processed in Blockchain, which takes much longer than that of the approved period.

2 Literature Review

In [17], author presented a subgraph-based anomaly detection approach that uses a portion of the Blockchain data to accomplish the detection. The suggested subgraph structure is suited for graphical processing unit (GPUs) to use parallel processing to speed up detection. When the amount of targeting transactions was 100, suggested technique was 11.10 times quicker than a previous works without reducing detecting accuracies in a test utilizing actual bitcoin transaction data. In [18], author proposed anomaly detection algorithms, with accuracy, precision, and recall as the study's criteria. Researchers identified the greatest deal with almost 95% assurance at a reputation rating of 5. The accuracy was determined to be 77.56%, with a 2.76% false positive rate. Proposed solution decreases the amount of transactions that cannot be added to the Blockchain by 1.60%, resulting in a significant speed boost. In [19], author used LSTM to detect abnormalities. We find four patterns that may be utilized to discriminate between various kinds of contracts and aid our knowledge of contracts transactional behavior. The 14 fundamental components of a contract then are assembled. The exploratory database is then created using an information slice approach. The smart contract datasets are also trained and tested using an LSTM network. In [20], researchers created a novel ADS based on One-Class SVM to detect network intrusions. Proposed SVM yields higher detection rates for different kinds of attacks and has an average better performance in terms of accuracy, recall, and F-value when compared to other approaches. The problem is that for low-frequency attacks like R2L and U2R, both the single-class SVM and the other have inadequate specificity and sensitivity. Inadequate data is somewhat to blame for impacting the outcome's dependability. The detection model, on the other hand, may be enhanced. The F-value is 0.9518, the precision is 0.9903, the recall is 0.9161, and the precision is 0.9903. In [21], researchers observed a variety of ML algorithms to detect suspicious transactions in various virtual currency exchanges. Researchers observed that supervised learning tactics provide favorable results; however, uncontrolled methods are more difficult to categorize. Furthermore, to better the study of a directed graph, modern and powerful machine learning methods may be applied. These technologies' autonomy might be exploited to give important input network observations. The recommended work has a detection rate of 48.3, a sensitivity of 84.2, an accuracy of 72.6, a specificity of 62.6, and a sensitivity of 84.2. In this research, used network theory to examine several networks of varied sizes and complexity. Researchers start by assessing graph networking representations learning methodologies with two key objectives in mind: user information and transactions. In [22], author identified abnormalities in the Ethereum Blockchain network using a One-Class Graph Neural Network-based anomaly detection framework. The suggested technique achieves greater anomaly detection accuracy than standard non-graph-based machine learning methods, according to empirical assessment. The suggested model's greatest accuracy and $F1$-score are 86 and 83.46%. These findings show that standard algorithms fail to identify abnormalities with smaller training samples. As the percentage of data increases, the accuracy decreases. chosen for training the model becomes small.

3 Blockchain Architecture

Architecture of Blockchain can be categorized as Blockchain 1.0 (cryptocurrencies), Blockchain 2.0 (smart contracts), and Blockchain 3.0 (Blockchain application) as shown in Fig. 1.

3.1 Blockchain 1.0

Blockchain 1.0 is a distributing ledger that is used to effectively maintain digitalized monetary transaction among users. The transactions are saved as "Blocks," which are

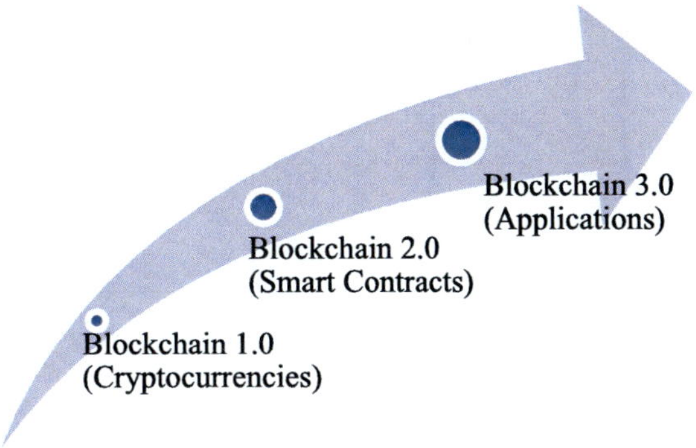

Fig. 1 Architecture of blockchain

a growing collection of entries. These blocks are impervious to any change and may be verified indefinitely. The validation of ledger entries is usually handled by a variety of consumers connected via a peer-to-peer network. A majority of the network's users must agree on any modifications inside blocks before they can be implemented. This section delves into the specifics of the design, operation, and ongoing study of certain Blockchain components, such as blocks, networks, and consensus.

3.2 Blockchain 2.0

The idea of Blockchain 2.0 (Smart Contracts) isn't really novel, as it has been discussed in this study since 1994. "A computerized transaction procedure that implements the contract terms" is how it is described. The goal was to convert contract agreements (collateral, bonding, etc.) into code and incorporate those as hardware and software to self-enforce people with least amount of help from trusted third parties. Smart contracts, in aspects of Blockchain, instantaneously regulate contracts among two or more entities. In Blockchain technologies like Ethereum and Hyper ledger, smart contracts are embedded as computer programmers. It has blocks, nodes, and consensus, same like Blockchain 1.0.

3.3 Blockchain 3.0

Blockchain 3.0 refers to a set of innovative Blockchain applications. Aside from the financial sector, Blockchain technology is being used in a variety of industries to construct decentralized applications, such as gaming, users-generating contents networking, IoT, intelligent hardware, supply chain management, sources tracking, and economic share credit. Blockchain 3.0 would be the age in which this technology is fully integrated into our everyday lives. Such distributed systems benefit

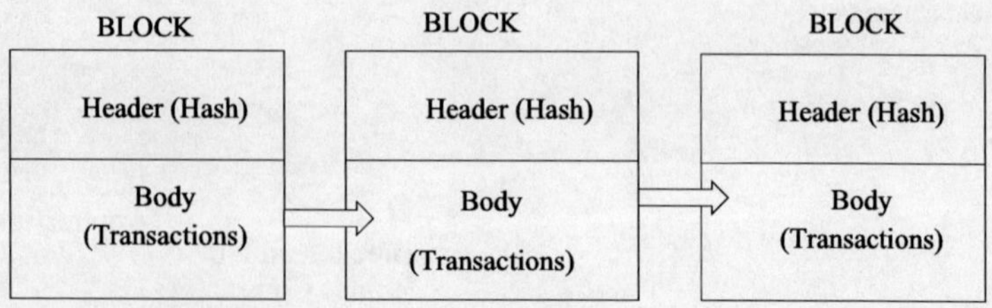

Fig. 2 Blocks in a blockchain

from Blockchain technology's numerous characteristics, like improved results in terms of reduced latencies, easier identity authentication, the capacity to perform offline payments, and customizable manageability for system upgrades including bug recovery.

Following are the key components of a Blockchain:

Block: In Blockchain, a block is the data structure used to store transaction information. It is divided into two components, as indicated in Fig. 2, Block Header and Block Body. The transaction as well as transaction counter are contained in the block body. The blocks size and the amount of every transaction inside it define the block's maximum ability to hold transactions.

Network: In a Blockchain network, there are two kinds of nodes in general: (1) A full node and (2) Fig. 3 shows a lightweight node. A full node is a fully functioning node that serves as a server. When contrasted to the archive node, pruned nodes have a lower functionality. When PN approach the defined limit, they only preserve the header of blocks, although still retain the blocks data from the beginning. The capacity to add blocks to the Blockchain is available to archival nodes. All transactions are verified by miner nodes, which are specialized nodes. Authority nodes are the ones that will execute a consensus mechanism called Proofs-of-Authorities. In a Blockchain network, the master node retains a complete record of all transactions and validates them. A small payment verification node is often referred to as a lightweight node.

Consensus: To verify a transaction as well as upgrade the ledger, the consensus is necessary. Proof of Work was the very first consensus protocol utilized in Blockchain 1.0. PoW is regarded as bitcoin's most significant breakthrough in terms of achieving consensus on a distributed decentralized Blockchain network with up to 1000 node. The consensus algorithm produces how Ethereum nodes decide to attach a new block to the Blockchain as well as how the validator works. In smart contracts, the consensus process is being used to settle any disputes between members and to log transaction for a specific contract. Various smart contract systems or frameworks are accessible.

As depicted in Fig. 4, a generic Blockchain design is offered in the research as a hierarchical architecture for developing distributed applications. There are just a few levels to it. Business apps built on the Blockchain serve as the application layer. The contact layer displays the many programmable techniques for Blockchain. The

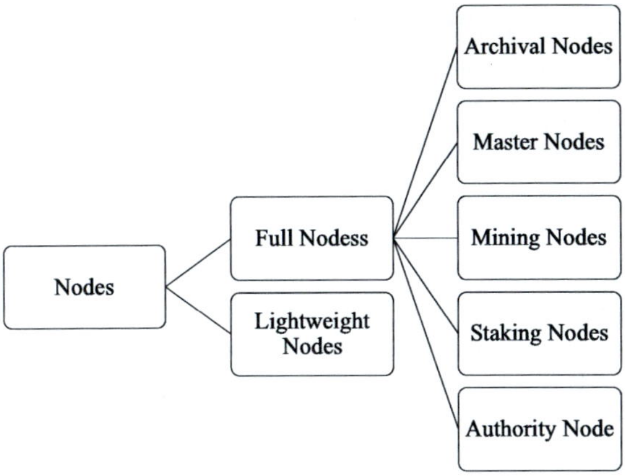

Fig. 3 Types of blockchain nodes

nodes involved in application management get incentives based on the mechanisms specified in the incentive layer. For Blockchain systems, the consensus layer makes several consensus methods accessible. Information propagating and data validation techniques, as well as distributed networking technologies, make up the network layer. The data layer includes time stamped data chunks. The security of these blocks is managed via a chained mode, a Merkle tree, encryption, and hashing function.

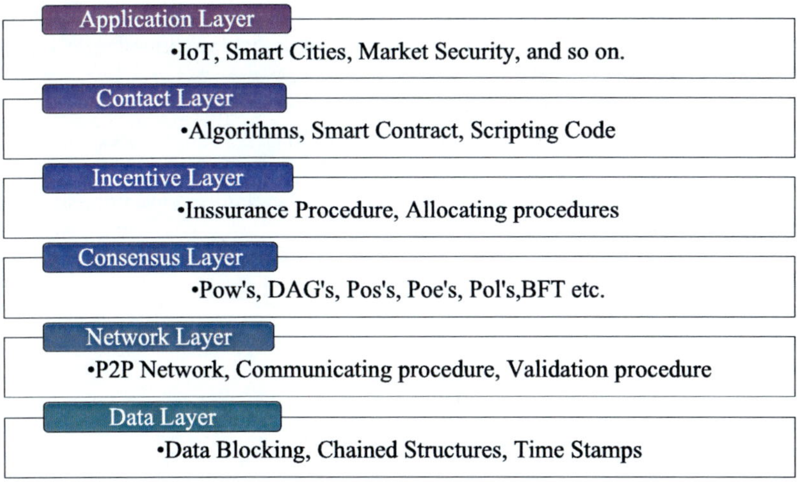

Fig. 4 Different layers in blockchain architecture

4 Vulnerabilities in Blockchain

In this section, six key vulnerabilities are discussed; they may be exploited to cause anomalous control flow paths [23]:

- Reentrancy: Whenever an innocuous contract contacts or delivers ether to an external contract, hackers may hijack the external calls, causing the contract to run extra code, including answering calls to the innocuous contract itself. If the innocuous contract does not anticipate reentrancy but does not perform the necessary checks, it will be vulnerable to unlawful contract state changes. This type of attack was utilized in the notorious DAO breach.
- Default Visibilities: The functional visibility specifiers in solidity contracts are set to open by default, enabling additional contracts or users to invoke the function externally. As a result, whenever programmers fail to specify visibility selectors for functions and shouldn't be accessible to external calls, fatal vulnerabilities might occur.
- Superficial randomness: Contracts that use Ethereum state variables as random seeds, such as block time stamps, are vulnerable to attacks since these variables may all be changed by miners.
- Unchecked send: If the contract makes an unsuccessful external call, the transaction will ordinarily revert.
- Tx. Origin Authentication: Contracts that employ the TX. Origin variable to authenticate users are prone to phishing attacks, which may lead to users completing authenticated activities on the susceptible contract.
- Denial of Service: This is a wide category, but it essentially refers to assaults in which users may render a contract unworkable for a short amount of time, or even permanently.
- As can be seen, aberrant control flow routes are connected with the bulk of the vulnerabilities. This is preliminary proof that IDS based on control flow anomalies may be useful in protecting smart contracts. The immediate attack target for the remaining vulnerabilities is either the Ethereum execution environment or the contracts themselves.

5 Research Problem

In the process of research, it was discovered that Blockchain is also vulnerable to security issues as discussed in above sections. In practice, it is found that very few contributions of researchers to study these aspects. So, following flow should be used to identify proper solution for research, as illustrated in Fig. 5.

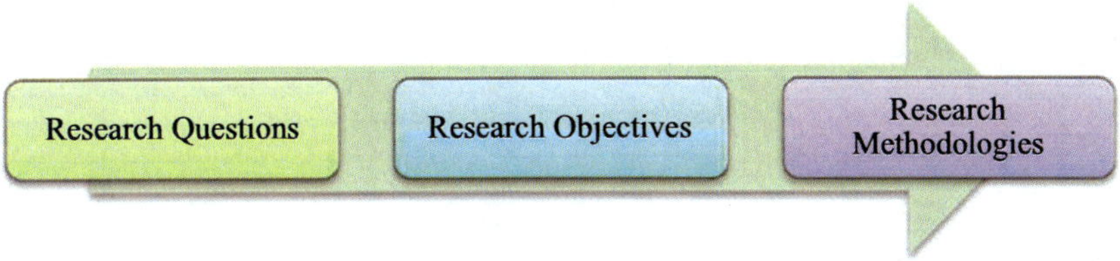

Fig. 5 Pathway for research problems

5.1 Research Questions

To solve the problems proposed in the background, this research proposes the following four questions:

RQ1:	The majority of public Blockchain systems are decentralized systems with limited secrecy, despite the fact that they offer integrity and availability, What and how?
RQ2:	What are the available anomaly detection models for Blockchain?
RQ3:	What are the issues faced by existing models?
RQ4:	How will machine learning affect future anomaly detection models for Blockchain?

5.2 Research Objectives

To protect the anomaly issues in Blockchain, machine learning model will be used to solve this problem.

RO1:	To identify existing security issues in Blockchain and anomaly detection tools and models
RO2:	To analyze the features of existing models and their performance
RO3:	The limitations of existing models can be solved by using machine learning
RO4:	To proposed more accurate machine learning model for anomaly detection in Blockchain. By evaluating a trust score of transactions to detect fraudulent transactions and to improve the false positive rate (FPR) of benign unknown traffic

6 Research Methodology

As per RQ's and RO's presented in above section, this research proposes four broad steps methodology as illustrated in Fig. 6.

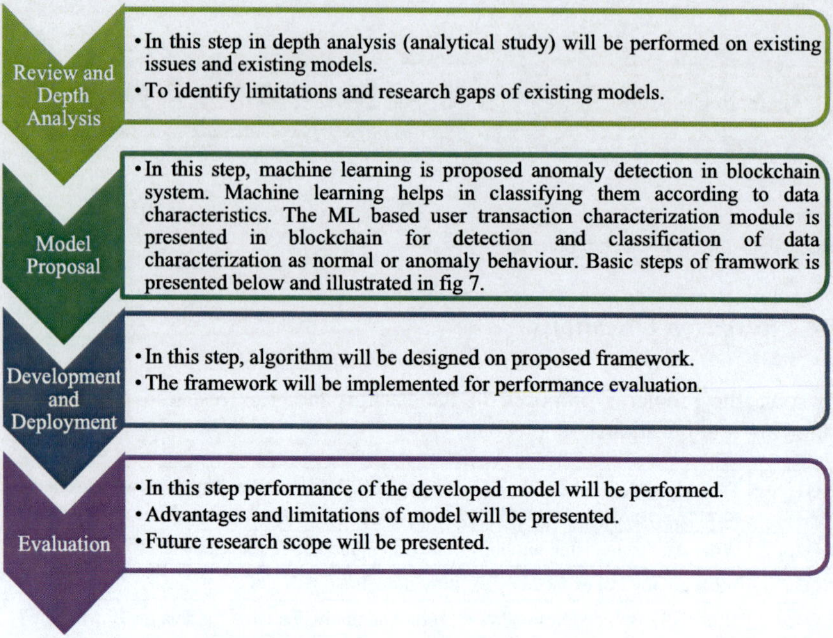

Fig. 6 Research proposal steps

The steps of the proposed model is presented as below:

- In first step, all the data characteristics or transaction characteristics are collected.
- Send to anomaly detection model.
- In this model, first of all the data attributes are extracted for behavior analysis by clustering them according to their trust levels.
- In second step, the clustered data are fed into ML-based data models for anomaly characterization.
- Finally, using this collaborative approach, the final result is evaluated (Fig. 7).

7 Expected Impact

From past few years, anomaly detection has been considered to be well-studied area for researchers. To resolve anomaly detection issues and problems, several tools and techniques were developed. Some techniques exhibit good performance for network anomaly detection but very few research works have proposed explicit superior solution for anomaly in Blockchain. This research methodology is designed

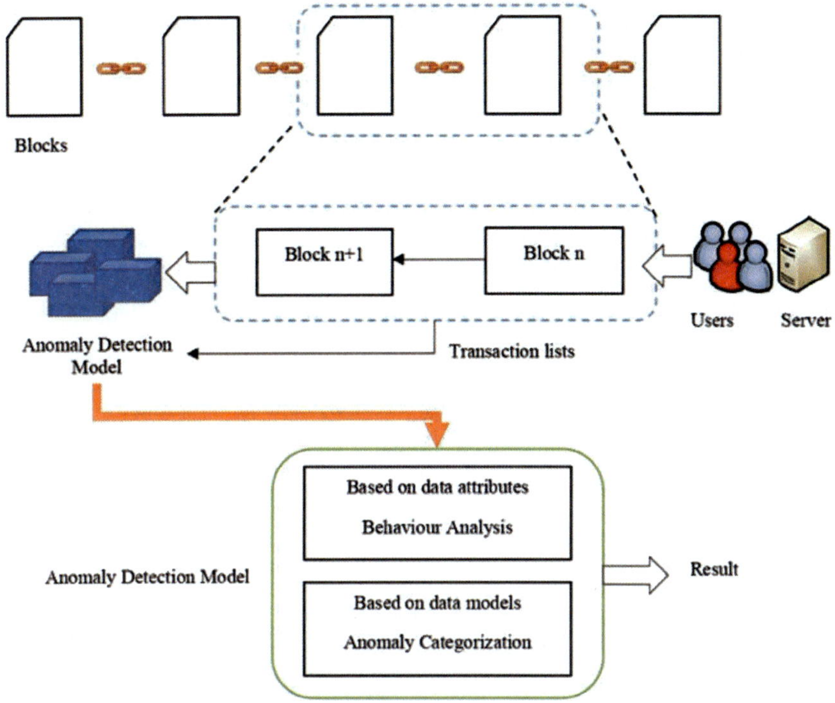

Fig. 7 Model for anomaly detection in blockchain

to bridge machine learning to Blockchain for more accurate malicious activity identifications on transactional network data. The work intends for detection and identification of malicious transaction versus non-malicious transactions. Anomaly detection within Blockchain can improve in the future if certain elements progress, such as the availability of relevant and rich data.

8 Conclusion and Recommendations

Blockchain is a game-changing technology that paves the way for the development of distributed and secure applications in every fields. Blockchain is expected to do for trustable transactions what online world appears to have done for communication systems due to its huge and fast app development. Unlike conventional bank transactions, authorized transactions on the Blockchain, even illegitimate ones, cannot be amended. Because transactions might be amended before approval, quick anomaly detection of transactions is essential to avoid the harm caused by illicit transactions. Existing anomaly detection systems, on the other hand, require process all transactions in Blockchain, which takes longer than the approval period. In this paper, we propose a machine learning based anomaly detection method to perform the detection using a part of the Blockchain data.

References

1. Bodkhe U et al (2020) Blockchain for Industry 4.0: a comprehensive review. IEEE Access 8:79764–79800. https://doi.org/10.1109/ACCESS.2020.2988579
2. Salah K, Rehman MHU, Nizamuddin N, Al-Fuqaha A (2019) Blockchain for AI: review and open research challenges. IEEE Access 7:10127–10149. https://doi.org/10.1109/ACCESS.2018.2890507
3. Franciscon EA, Nascimento MP, Granatyr J, Weffort MR, Lessing OR, Scalabrin EE (2019) A systematic literature review of blockchain architectures applied to public services. In: 2019 IEEE 23rd international conference on computer supported cooperative work in design (CSCWD), pp 33–38. https://doi.org/10.1109/CSCWD.2019.8791888
4. Tahir M, Habaebi MH, Dabbagh M, Mughees A, Ahad A, Ahmed KI (2020) A review on application of blockchain in 5G and beyond networks: taxonomy, field-trials, challenges and opportunities. IEEE Access 8:115876–115904. https://doi.org/10.1109/ACCESS.2020.3003020
5. Ahmed M, Mahmood AN, Hu J (2016) A survey of network anomaly detection techniques. J Netw Comput Appl 60:19–31
6. Yu Y, Li Y, Tian J, Liu J (2018) Blockchain-based solutions to security and privacy issues in the internet of things. IEEE Wirel Commun 25(6):12–18. https://doi.org/10.1109/MWC.2017.1800116
7. Dai F, Shi Y, Meng N, Wei L, Ye Z (2017) From bitcoin to cybersecurity: a comparative study of blockchain application and security issues. In: 2017 4th international conference on systems and informatics (ICSAI), pp 975–979. https://doi.org/10.1109/ICSAI.2017.8248427
8. Huynh TT, Nguyen TD, Tan H (2019) A survey on security and privacy issues of blockchain technology. In: 2019 international conference on system science and engineering (ICSSE), pp 362–367. https://doi.org/10.1109/ICSSE.2019.8823094
9. Meng W, Tischhauser EW, Wang Q, Wang Y, Han J (2018) When intrusion detection meets blockchain technology: a review. IEEE Access 6:10179–10188. https://doi.org/10.1109/ACCESS.2018.2799854
10. Alexopoulos N, Vasilomanolakis E, Ivanko NR, Muhlhauser M (2017) Towards blockchain-based collaborative intrusion detection systems. In: Proceedings of international conference on critical information infrastructures security, pp 1–12
11. Xing Z, Chen Z (2021) A protecting mechanism against double spending attack in blockchain systems. In: 2021 IEEE world AI IoT congress (AIIoT), pp 0391–0396. https://doi.org/10.1109/AIIoT52608.2021.9454224
12. John R, Cherian JP, Kizhakkethottam JJ (2015) A survey of techniques to prevent sybil attacks. In: 2015 international conference on soft-computing and networks security (ICSNS), pp 1–6. https://doi.org/10.1109/ICSNS.2015.7292385
13. Gupta KD, Rahman A, Poudyal S, Huda MN, Mahmud MAP (2019) A hybrid POW-POS implementation against 51 percent attack in cryptocurrency system. In: 2019 IEEE international conference on cloud computing technology and science (CloudCom), pp 396–403. https://doi.org/10.1109/CloudCom.2019.00068
14. Andryukhin AA (2019) Phishing attacks and preventions in blockchain based projects. In: 2019 international conference on engineering technologies and computer science (EnT), pp 15–19. https://doi.org/10.1109/EnT.2019.00008
15. Wang X, WeiLi J, Chai J (2018) The research on the incentive method of consortium blockchain based on practical byzantine fault tolerant. In: 2018 11th international symposium on computational intelligence and design (ISCID), pp 154–156. https://doi.org/10.1109/ISCID.2018.10136

16. Perazzo P, Arena A, Dini G (2020) An analysis of routing attacks against IOTA cryptocurrency. In: 2020 IEEE international conference on blockchain (Blockchain), pp 517–524. https://doi.org/10.1109/Blockchain50366.2020.00075
17. Morishima S (2021) Scalable anomaly detection in blockchain using graphics processing unit. Comput Electr Eng 92:107087. https://doi.org/10.1016/j.compeleceng.2021.107087. ISSN 0045-7906
18. Maskey SR, Badsha S, Sengupta SS, Khalil I (2021) ALICIA: applied Intelligence in blockchain based VANET: accident validation as a case study. Inf Process Manag 58(3):102508. https://doi.org/10.1016/j.ipm.2021.102508. ISSN 0306-4573
19. Hu T, Liu X, Chen T, Zhang X, Huang X, Niu W, Lu J, Zhou K, Liu Y (2021) Transaction-based classification and detection approach for ethereum smart contract. Inf Process Manag 58(2):102462. https://doi.org/10.1016/j.ipm.2020.102462. ISSN 0306-4573
20. Zhang M, Xu B, Gong J (2015) An anomaly detection model based on one-class SVM to detect network intrusions. In: 2015 11th international conference on mobile ad-hoc and sensor networks (MSN), pp 102–107. https://doi.org/10.1109/MSN.2015.40
21. Martin K, Rahouti M, Ayyash M, Alsmadi I (2021) Anomaly detection in blockchain using network representation and machine learning. Secur Priv e192
22. Patel V, Pan L, Rajasegarar S (2020) Graph deep learning based anomaly detection in ethereum blockchain network. In: Kutyłowski M, Zhang J, Chen C (eds) Network and system security. NSS 2020. Lecture notes in computer science, vol 12570. Springer, Cham
23. Wang X, He J, Xie Z, Zhao G, Cheung S-C (2020) ContractGuard: defend ethereum smart contracts with embedded intrusion detection. IEEE Trans Serv Comput 13(2):314–328

QoS-Aware Resource Allocation with Enhanced Uplink Transfer for U-LTE–Wi-Fi/IoT Using Cognitive Network

A. Muniyappan and P. B. Pankajavalli

Abstract Unlicensed Long Term Evolution (U-LTE) is one of the unique scientific areas that enable 5 GHz unauthorized accessibility, thus strengthening throughput of the data transfer. But, the limitation is the requirement to cooperate with Wi-Fi/IoT systems with effectively scheduled uplink transfer (UT) in unauthorized spectrums. To solve this challenge, a Desirable U-LTE–Wi-Fi (D-U-LTE–Wi-Fi) has been developed to enhance the uplink efficacy in unauthorized bands. However, a total client-perceived Quality of Service (QoS) was not enhanced which affects the scheduling task which allocates channels or subframes to the clients depending on their QoS demands. Therefore, this article further enhances the D-U-LTE–Wi-Fi system with QoS-aware-Based Resource Allocation (QBRA) by solving the appropriate distribution of resources for LTE-Cognitive Radio Network (CRN). The LTE-CRN is constructed by primary (authenticated) eNodeBs (eNBs) which distribute their spectrum resources with the secondary (unauthenticated) eNBs. The major aim is to ensure channel access or allocation to the unauthenticated eNBs without compromising the QoS for the authenticated eNB. This is achieved by using a 2-stage process. In the primary stage, the spectrum resources are distributed to the authenticated eNBs for increasing the QoS for the authenticated clients. In the secondary stage, the additional service ability of the main spectrum is allocated amid the unauthenticated eNBs for increasing the QoS of the unauthenticated clients. Further, the simulation results exhibit the D-U-LTE–Wi-Fi–QBRA system accomplishes better resource efficiency and system reliability than the conventional systems.

Keywords CRN · D-U-LTE–Wi-Fi · QoS · Queue size · Resource allocation · U-LTE · Uplink transfer

---

A. Muniyappan (✉) · P. B. Pankajavalli
Department of Computer Science, Bharathiar University, Coimbatore, Tamilnadu, India
e-mail: muniyappanma@gmail.com

P. Chatterjee et al. (eds.), *Computational Intelligence for Engineering and Management Applications*, Lecture Notes in Electrical Engineering 984,
https://doi.org/10.1007/978-981-19-8493-8_38

1 Introduction

Wireless technologies of the fourth generation (4G), also known as LTE networks, have primarily progressed with the purpose of broadcasting huge data via mobile channels. Cell size reduction is not only a solution to service requests; also it necessarily involves extra computing resources. The distribution of unlicensed band alongside licensed spectrum to satisfy the channel requests of their clients continues to expand a significant future for LTE services [1]. The Third Generation Partnership Project (3GPP) has aided in the deployment of LTE to meet extremely higher data transmission constraints in emerging streaming technologies [2]. In essence, 3GPP LTE-A has enforced the coordination of authenticated and unauthenticated band in femtocells in order to support authentication knowledge [3–5]. But, Internet suppliers are constrained by the distribution of licensed bandwidths. As a result of such constraints, LTE is allowed to operate on the unapproved bandwidth used by Wi-Fi/IoT systems [6–8]. Because 5 GHz seems to have hundreds of MHz of bandwidth availability, U-LTE is often marketed to provide 5 GHz Industrial, Scientific, and Medical (ISM) attention [9, 10]. The goal of U-LTE is to take properties from one of the most recent LTE 3GPP guidelines and develop them to uncertified deployment in developed countries which do not enable Listen-Before-Talk (LBT) [11]. It always promotes unauthenticated data transmission in the bandwidths 5150–5250 MHz and 5725–5850 MHz, while the bandwidths 5250–5725 MHz preferred for long term usages [12, 13]. The major problem of applying these spectrums is that certain Wi-Fi/IoT systems encourage the adoption of LBT [14]. To address this condition, U-LTE strengthened the CRN-LBT guidelines for appropriate use of the 5 GHz band [15]. This principle has following objectives: (a) fulfill U-LTE-LBT guidelines [16] and (b) develop their integration with Wi-Fi/IoT users by lessening the Wi-Fi back-off limit in a non-interference circumstance. In addition, by leveraging the general principle of channel estimation and assignment, U-key LTE's functionalities for new channel accesses and the cooperation of supplementary clients with authenticated users were redesigned. Despite this, the illegitimate band was simply utilized for downlink traffic. The band quality and robustness have also not been strengthened. As a result, an enhancement to the opportunistic cooperation of U-LTE–Wi-Fi/IoT with Enhanced Conflict Tolerance (ECT)-LBT by CRN has been implemented (called E-U-LTE–Wi-Fi) [17]. The License Assisted Access (LAA) strategy was initially designed to promote the LBT mechanism. The channel control process can be simplified in sophisticated deployment scenarios by using the LTE duty cycle. As well, a Low Amplitude Stream Injection (LASI) technique was introduced for supporting parallel Wi-Fi–LTE access control on a shared spectrum, as well as data healing from collisions. The Conflict-Tolerant Channel Assignment (CTCA) methodologies have been improved for modifying channel estimation and improving spectrum usage at 5 GHz. Indeed, it has been proposed that an enhanced Cell ON/OFF strategy be executed, which might significantly boost spectrum usage and strengthen internet connectivity by developing LTE to illegal spectral range. As a result, the Spectrum Efficiency (SE) and efficiency have been boosted while the

Transmission Delay (TD) was limited. Nonetheless, the estimated UT efficiency in the unpermitted bands was adversely influenced by eNB and User Equipment (UE) dual LBT constraints. Additionally, performance failure was happened as a result of degradation in UE's channel allocation possibility. So, D-U-LTE–Wi-Fi [18] has been designed as an advancement of UT in E-U-LTE–Wi-Fi/IoT systems. A new UT mechanism was supposed which does not engage eNB assignment method for improving uplink efficiency for LTE in illegal spectrum depending on Grant-less Multiple Subframe Scheduled Uplink (GMSSUL) transfer. Primarily, the Cat4.5 LBT-based channel identification was applied by the UE with the aid of uplink information to ensure satisfactory cooperation. After that, a preamble signal was needed prior to the transfer of information to identify the Uplink Burst (UB) at an anchored eNB, and a reservation signal was needed to maintain the preset limit.

Following that, an overall Highest Channel Possession Time (HCPT) was considered by the UE to exchange information, and a feedback data was requested by the eNB for Amalgam Automatic Repeat Request (AARQ). Also, the Multi-Subframe Scheduling (MSS) was used to improve the uplink subframe through enabling the UE for entailing several channel identification opportunities and continuously exchange information for several subframes. Though GMSSUL boosts the UT on both authorized and unauthorized spectrums, an overall client-perceived QoS was not enhanced which affects the scheduling task since the scheduling can be used to allocate channels or subframes to the clients depending on their QoS demands. Also, the solution of uplink and downlink transmission does not effective for resource allocation in considered CRN structure wherein the primary eNBs have absolute prioritized access to their bandwidth resources, while the secondary eNBs can utilize only the remaining capacity.

Hence in this paper, D-U-LTE–Wi-Fi–QBRA is developed to distribute the appropriate resources for 3GPP–LTE–CRN. The major objective of this system is to guarantee QoS demands of the unauthorized eNBs without compromising the QoS demands of authorized eNBs. So, a 2-stage procedure is applied: (1) The band resources are shared to the authorized eNBs in the first stage to enhance the QoS for the authorized clients, and (2) the additional service ability of the primary spectrum is shared amid the unauthorized eNBs in the second stage to enhance the QoS of the unauthorized clients. As a result, it can be used to enhance the resource distribution and total QoS of the system.

The remaining sections of this manuscript are the following: Sect. 2 presents the works related to the resource distribution in LTE-based CRN. Section 3 explains the methodology of D-U-LTE–Wi-Fi–QBRA system and Sect. 4 displays its simulation efficiency. Section 5 summarizes the entire work and suggests future scope.

2 Literature Review

A QoS-aware collaborative energy management and RA mechanism [19] has been developed for preserving the QoS of client restraints and minimizing the interference in femtocell networks. But, the efficiency of resource use and average throughput were not improved. Two resource distribution methods [20] have been discussed for femtocell systems which effectively distribute resources like spectrum and transfer energy among hostile mobile customers in the LTE–Wi-Fi frequencies. But, the energy use was high if the UE density was increased and also the computational burden was high. A new 2-step method [21] has been developed for dynamically allocating the resources in LTE downlink cloud-radio access network. But, its computational complexity and CPU duration were high. A statistical framework [22] has been recommended, which has nonlinear restraints on binary parameters for devising the stochastic optimization issue. Also, a submodular-based greedy method was presented for solving the high-dimensional NP-hard assignment challenge. However, its computational time complexity was high. A scheme [23] has been developed for distributing Resource Blocks (RBs) to clients. But, it's QoS performance was not effective while cooperating LTE with CRN. Robust energy efficiency-based maximization resource sharing issue [24] has been focused by mutually optimizing the energy distribution, subcarrier assignment, and transfer time by accounting band observation faults and medium ambiguities concurrently. But, it has high computational time complexity and less energy efficiency. A model of clients association and energy distribution [25] has been developed in the 2-tier heterogeneous networks with non-orthogonal multiple access. But, it does not consider the system fairness and also has high computational difficulty.

3 Proposed Methodology

In this section, the D-U-LTE–Wi-Fi–QBRA system is described in detail. Originally, the authenticated and unauthenticated clients are started through interacting UEs with Wi-Fi access points and eNB, respectively, using unapproved bands. The uplink transfer is enhanced by Cat.4 LBT which preserves better cooperation. Additionally, the resources and service rates are properly distributed by the 2-stage process for enhancing the QoS of both authenticated and unauthenticated clients. This process is explained in below subsection.

3.1 System Model

A standard CRN structure deployed in a typical U-LTE–Wi-Fi/IoT network is illustrated in Fig. 1.

Consider an U-LTE–Wi-Fi/IoT network with CRN having n authorized (primary) eNBs (PBs) denoted as $PB_1, \ldots, PB_n$ and m unauthorized (secondary) eNBs (SBs) denoted as $SB_1, \ldots, SB_m$. The eNBs are interacted to the CRN through a centralized system controller (CSC). The transfer among the eNBs, CRN, and the Wi-Fi Access Point (AP) is established by the direct IP connections for enabling quick data transfer.

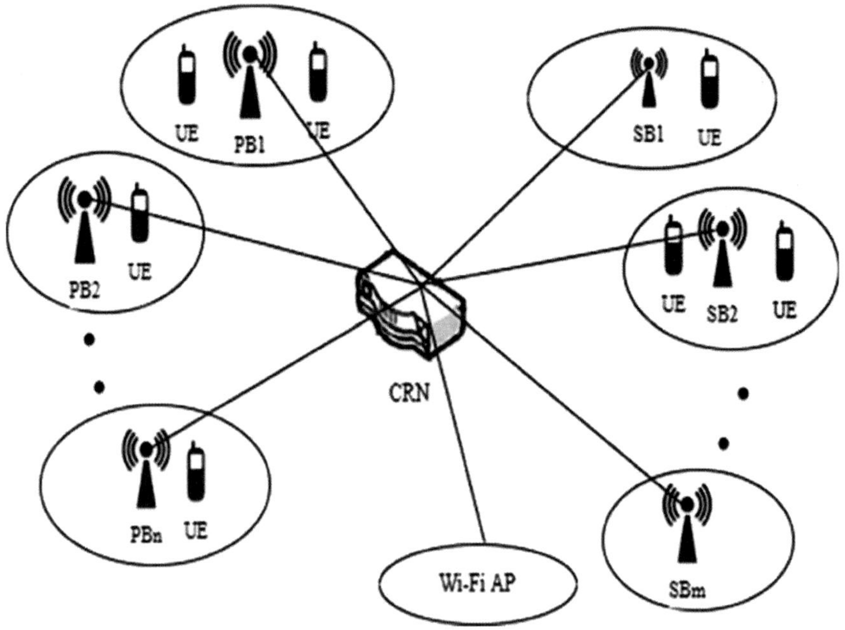

Fig. 1 Standard CRN with U-LTE–Wi-Fi/IoT network

Every PB runs on its constant authorized spectrum band with few specified distribution ability. It will distribute its main spectrum with single or multiple SBs, which do not contain a constant authorized frequency. The PB has prioritized connectivity to their main spectrum. The ability that a PB distributes with specific SB based on the spectrum assignment strategy utilized by CSC.

The system runs on a slotted-interval manner, i.e., the interval slot is split into jointly disjoint interval slots $\{T_S, (t+1)T_S\}, t = 0, 1, \ldots$ where T_S is the slot length and t is the slot index. Every eNB acts a wireless client situated in its service region called cell. The client-produced data is enqueued in the UEs and sent to corresponding eNBs by the data allocation process. During this process, the detail regarding the number of information enqueued in the UEs buffers are regularly sent to the eNB; thus, the eNB recognizes the appropriate number of information produced by means of clients at t. This data is utilized by the eNB for distributing the uplink transfer resources to UEs by specified distribution mechanism. During the downlink, the transfer resources are distributed depending on the number of information sent from CSC through IP connection. In summary, the considered network structure comprises the collection of n primary channels belonging to $PB_1, \ldots, PB_n$ and m secondary channels belonging to $SB_1, \ldots, SB_m$. All eNBs (i.e., both PB and SB) in this framework are defined by the uplink (primary) and downlink (secondary) queues. Table 1 presents the notations used in this study.

Table 1 Notations used for D-U-LTE–Wi-Fi–QBRA system

Parameters	Description
n	Primary channels
m	Secondary channels
$\mathrm{PB}_1, \ldots, \mathrm{PB}_n$	Primary (authorized) eNBs
$\mathrm{SB}_1, \ldots, \mathrm{SB}_m$	Secondary (unauthorized) eNBs
$\mathcal{T}_\mathrm{S}$	Time slot length
t	Time slot index
$\mathcal{Q}_a^{\mathrm{PB}}(t)$	Size of the queue at the starting of t in PB_a
$\mathcal{Q}_b^{\mathrm{SB}}(t)$	Size of the queue at the starting of t in SB_b
$\mathcal{U}_a^{\mathrm{PB}}(t)$	Service rate/amount of RBs distributed to the authenticated clients during t in PB_a
$\mathcal{U}_{ab}^{\mathrm{SB}}(t)$	Capacity/amount of RBs that PB_a distributes with SB_b during t
$\mathcal{D}_a^{\mathrm{PB}}(t)$	Overall number of data produced during t by the wireless clients of PB_a
$\mathcal{D}_b^{\mathrm{SB}}(t)$	Overall number of data produced during t by the wireless clients of SB_b
ψ_a	Service ability of the authenticated spectrum belonging to PB_a
$\mathcal{U}_b^{\mathrm{SB}}(t)$	Anonymous service rate distribution vector
$f\{\mathcal{U}^{\mathrm{SB}}(t)\}$	Function of anonymous service rate distribution vector
k_a	Target queue size
$\mathcal{S}(t)$	Group of each authenticated eNB indices
$\mathcal{K}_a$	Variable
$\mathcal{R}_a$	Amount of RBs for a bandwidth of the unauthorized channel of PB_a
ψ_{RB}	Data rate
$\mathcal{F}_{\mathrm{RB}}$	Frequency of single RB
c_{BW}	Adjustment coefficient for band quality of the LTE
c_{SNR}	Adjustment coefficient for SNR of the LTE
ζ_a	Mean data rate of single RB of PB_a

3.2 Problem Formation for Service Rate Capacity Distribution Process

Consider $\mathcal{Q}_a^{\mathrm{PB}}(t)$ and $\mathcal{Q}_b^{\mathrm{SB}}(t)$ are the size of the queue at the starting of t in PB_a and SB_b, accordingly, $\mathcal{U}_a^{\mathrm{PB}}(t)$ is the service rate during t in PB_a, $\mathcal{U}_{ab}^{\mathrm{SB}}(t)$ is the capacity that PB_a distributes with SB_b during t, $\mathcal{D}_a^{\mathrm{PB}}(t)$ and $\mathcal{D}_b^{\mathrm{SB}}(t)$ are the overall number of data produced during t by the wireless clients of PB_a and SB_b, accordingly and ψ_a is the service ability of the authenticated spectrum belonging to PB_a.

The overall amount of bits distributed by an authenticated spectrum is not greater than its highest distribution ability:

$$\mathcal{U}_a^{\text{PB}}(t) + \sum_{b=1}^{m} \mathcal{U}_{ab}^{\text{SB}}(t) \leq \psi_a, \quad \forall a \in A \tag{1}$$

In Eq. (1), $A = \{1, \ldots, n\}$. Observe that for any $a \in A$, the appropriate ranges of $\mathcal{Q}_a^{\text{PB}}(t)$ and $\mathcal{D}_a^{\text{PB}}(t)$ are calculated at PB_a. Likewise, for any $b \in B$, the ranges $\mathcal{Q}_b^{\text{SB}}(t)$ and $\mathcal{D}_b^{\text{SB}}(t)$ are calculated at SB_b at t. Principally, the CSC must distribute the service rates $\mathcal{U}_a^{\text{PB}}(t)$ and $\mathcal{U}_{ab}^{\text{SB}}(t)$ for increasing the QoS for the clients of PBs and SBs. Also, it must retain the service priorities of PBs in their channels.

To satisfy this criterion, the service rate distribution is executed independently for PBs and SBs using 2-stage process. In the initial stage, the service rate for each PB within CRN is assigned to increase the QoS for primary clients (Wi-Fi/IoT). After that, the additional service ability of the primary channels is allocated among SBs.

The service rates $\mathcal{U}_a^{\text{PB}}(t)$ and $\mathcal{U}_{ab}^{\text{SB}}(t)$ are distributed depending on the queue sizes at the succeeding time intervals $\mathcal{Q}_a^{\text{PB}}(t+1)$ and $\mathcal{Q}_b^{\text{SB}}(t+1)$. In this scenario, an appropriate distribution of service rate can reduce the future queue size which prevents the network congestion or failure. The queue length is decided as an optimization objective as it is linked to the major QoS parameters like round-trip delay and loss. Here, $\mathcal{Q}_a^{\text{PB}}(t+1)$ and $\mathcal{Q}_b^{\text{SB}}(t+1)$ are determined from $\mathcal{Q}_a^{\text{PB}}(t)$, $\mathcal{Q}_b^{\text{SB}}(t)$, $\mathcal{D}_a^{\text{PB}}(t)$ and $\mathcal{D}_b^{\text{SB}}(t)$ as:

$$\mathcal{Q}_a^{\text{PB}}(t+1) = \left[\mathcal{Q}_a^{\text{PB}}(t) + \mathcal{D}_a^{\text{PB}}(t) - \mathcal{U}_a^{\text{PB}}(t)\right]^+, \quad \forall a \in A \tag{2a}$$

$$\mathcal{Q}_b^{\text{SB}}(t+1) = \left[\mathcal{Q}_b^{\text{SB}}(t) + \mathcal{D}_b^{\text{SB}}(t) - \sum_{a=1}^{n} \mathcal{U}_{ab}^{\text{SB}}(t)\right]^+, \quad \forall b \in B \tag{2b}$$

where

$$[u]^+ = \max[0, u] \tag{2c}$$

Observe that in Eqs. (2a–2c), $\mathcal{U}_a^{\text{PB}}(t)$ and $\mathcal{U}_{ab}^{\text{SB}}(t)$ have to be distributed; $\mathcal{Q}_a^{\text{PB}}(t)$, $\mathcal{D}_a^{\text{PB}}(t)$ are determined in PB_a; $\mathcal{Q}_b^{\text{SB}}(t)$, $\mathcal{D}_b^{\text{SB}}(t)$ are determined in SB_b at t. Initially, $\mathcal{Q}_a^{\text{PB}}(t+1)$ is reduced for each PB for increasing the QoS for the clients of PBs and SBs as well as preserving the service priority of PBs. After, the additional service capacity is used such that $\mathcal{Q}_b^{\text{SB}}(t+1)$ is reduced. This 2-stage process is proposed for distributing $\mathcal{U}_a^{\text{PB}}(t)$ such that $\mathcal{Q}_a^{\text{PB}}(t+1)$ is reduced, i.e.,

$$\min \sum_{a=1}^{n} \left[\mathcal{Q}_a^{\text{PB}}(t) + \mathcal{D}_a^{\text{PB}}(t) - \mathcal{U}_a^{\text{PB}}(t)\right]^+ \tag{3a}$$

$$\text{s.t.} \quad \mathcal{U}_a^{\text{PB}}(t) \geq 0, \quad \forall a \in A \tag{3b}$$

$$\mathcal{U}_a^{\text{PB}}(t) \leq \psi_a, \quad \forall a \in A \tag{3c}$$

After, it is simple to certify that,

$$\mathcal{U}_a^{\mathrm{PB}}(t) = \begin{cases} \psi_a, & \text{if } \mathcal{Q}_a^{\mathrm{PB}}(t) + \mathcal{D}_a^{\mathrm{PB}}(t) \geq \psi_a \\ \mathcal{Q}_a^{\mathrm{PB}}(t) + \mathcal{D}_a^{\mathrm{PB}}(t), & \text{or else} \end{cases} \tag{4}$$

Normally, the network is not completely loaded; therefore, it will have $\psi_a > \mathcal{U}_a^{\mathrm{PB}}(t)$ for few ranges of a. It facilitates the respective PBs for serving few unauthenticated clients. The service rates for the unauthenticated clients are computed by resolving the min-max issue. The anonymous service rate distribution vectors and the function are defined as

$$\mathcal{U}_b^{\mathrm{SB}}(t) = \begin{bmatrix} \mathcal{U}_{1b}^{\mathrm{SB}}(t) \\ \vdots \\ \mathcal{U}_{nb}^{\mathrm{SB}}(t) \end{bmatrix}, \; \mathcal{U}^{\mathrm{SB}}(t) = \begin{bmatrix} \mathcal{U}_1^{\mathrm{SB}}(t) \\ \vdots \\ \mathcal{U}_m^{\mathrm{SB}}(t) \end{bmatrix} \tag{5}$$

$$f\{\mathcal{U}^{\mathrm{SB}}(t)\} := \max_{b \in B} \left[\mathcal{Q}_b^{\mathrm{SB}}(t) + \mathcal{D}_b^{\mathrm{SB}}(t) - \sum_{a=1}^{n} \mathcal{U}_{ab}^{\mathrm{SB}}(t) \right]^+ \tag{6}$$

The value of $\mathcal{U}^{\mathrm{SB}}$ is determined by resolving the optimization issue as

$$\min f\{\mathcal{U}^{\mathrm{SB}}(t)\} \tag{7a}$$

$$\text{s.t. } \mathcal{U}_{ab}^{\mathrm{SB}}(t) \geq 0, \quad \forall a \in A, b \in B \tag{7b}$$

$$\mathcal{U}_a^{\mathrm{PB}}(t) + \sum_{b=1}^{m} \mathcal{U}_{ab}^{\mathrm{SB}}(t) \leq \psi_a, \quad \forall a \in A \tag{7c}$$

This issue (7a–7c) is illustrated to be a convex problem. So, it is probable to resolve (7a–7c) in a polynomial interval. This is accurate while $\mathcal{U}_a^{\mathrm{PB}}(t)$ and $\mathcal{U}_{ab}^{\mathrm{SB}}(t)$ are facilitated to be random real numbers. But, for D-U-LTE–Wi-Fi, it is not possible. So, for every PB_a, a target queue size $k_a > 0$ is assigned such that

$$\mathcal{Q}_a^{\mathrm{PB}}(t+1) = \mathcal{Q}_a^{\mathrm{PB}}(t) + \mathcal{D}_a^{\mathrm{PB}}(t) - \mathcal{U}_a^{\mathrm{PB}}(t) \leq k_a, \quad \forall t \tag{8}$$

But, it may not possible for every a, i.e., if

$$\mathcal{Q}_a^{\mathrm{PB}}(t) + \mathcal{D}_a^{\mathrm{PB}}(t) > \mathcal{K}_a + \psi_a \tag{9}$$

On the other hand, it cannot fulfill (1). In this scenario, ability of PB_a can be utilized for serving its individual data, and it does not distribute its ability with the unauthenticated clients. Consider $\mathcal{S}(t)$ is the group of each index a such that (3a–3c) keeps:

$$\mathcal{S}(t) := \{a \in A : \mathcal{Q}_a^{\text{PB}}(t) + \mathcal{D}_a^{\text{PB}}(t) \leq \mathcal{K}_a + \psi_a\} \tag{10}$$

Hence, $\mathcal{S}(t)$ denotes the group of each authenticated eNB indices to distribute their ability with unauthenticated eNBs. Moreover, this process is generalized for distributing the service rates for PBs and SBs by mutual optimization as

$$\min f\{\mathcal{U}^{\text{SB}}(t)\} \tag{11a}$$

$$\text{s.t. } \mathcal{U}_{ab}^{\text{SB}}(t) \geq 0, \quad \forall a \in A, b \in B \tag{11b}$$

$$\mathcal{U}_a^{\text{PB}}(t) \geq 0, \quad \forall a \in A \tag{11c}$$

$$\mathcal{U}_a^{\text{PB}}(t) \geq \mathcal{Q}_a^{\text{PB}}(t) + \mathcal{D}_a^{\text{PB}}(t) - \mathcal{K}_a, \quad a \in \mathcal{S}(t) \tag{11d}$$

$$\mathcal{U}_a^{\text{PB}}(t) = \psi_a, \quad a \notin \mathcal{S}(t) \tag{11e}$$

$$\mathcal{U}_a^{\text{PB}}(t) + \sum_{b=1}^{m} \mathcal{U}_{ab}^{\text{SB}}(t) \leq \psi_a, \quad \forall a \in A \tag{11f}$$

In Eq. (11d), the variable $\mathcal{K}_a$ is constant or adaptive, whereas its range is selected depending on certain requirements of authenticated eNBs, spectrum efficiency, or few additional exterior qualities.

3.3 Resource Distribution Process

For a bandwidth of the unauthorized channel of PB_a, the corresponding amount of RBs $\mathcal{R}_a$ are obtained. In LTE, the RA is executed simply in the aspects of the RBs. So, at any interval, an eNB can distribute just an integer amount of RBs to every client. As per the principle, LTE utilizes link adjustment for transfer among the UEs and eNBs. It indicates that based on immediate spectrum circumstances, every spectrum between the UEs and eNB is allocated various Modulation and Coding Scheme (MCS) index. So, the ability of RB is varied within the system. But, it assumes that in an eNB, the spectrum ability is fixed because it is highly complex for estimating the appropriate amount of spectrum and their efficiency in QoS limited RA latency and throughput estimation. During this scenario, the total network ability is approximated by the mean data rate of an eNB. Consider a complex network having an unauthorized and authorized eNBs serving several clients. In this system, every eNB is assigned specified amount of RBs which are after allocated to the clients. Prior to the frequency assignment, it is not promising for estimating where and how every RB is utilized. So, consider that the RB ability is fixed and similar to few mean data rate. During this scenario, the predicted service ability of single RB is computed by modified Shannon capacity as

$$\psi_{\mathrm{RB}} = \mathcal{F}_{\mathrm{RB}} \cdot c_{\mathrm{BW}} \log_2(1 + c_{\mathrm{SNR}} \cdot \mathrm{SNR}) \tag{12}$$

In Eq. (12), ψ_{RB} denotes the data rate, and $\mathcal{F}_{\mathrm{RB}} = 180\,\mathrm{kHz}$ denotes the frequency of single RB. The adjustment coefficients c_{BW}, c_{SNR} are the band quality and Signal-to-Noise Ratio (SNR) of the LTE, accordingly. In LTE, the RB's band quality is minimized due to many overheads on the connection and network stage. The SNR is degraded through the constrained code block size, MCS, aerial settings, and few efficiency problems associated with the source and destination. The determination of SNR is highly difficult compared to the band quality computation; so, it is applied to utilize arc fitting to connection adjustment arc denoted by the step factor where every step relates to every of the utilized MCS.

Consider ζ_a is the mean data rate of single RB of PB_a computed by (12). Consider $\mathcal{U}_a^{\mathrm{PB}}(t)$ is the amount of RBs distributed to the authenticated clients in PB_a. After,

$$\mathcal{U}_a^{\mathrm{PB}}(t) = \zeta_a u_a^{\mathrm{PB}}(t), \quad \psi_a = \zeta_a \mathcal{R}_a \tag{13}$$

Likewise, if $\mathcal{U}_{ab}^{\mathrm{SB}}(t)$ denotes the amount of RBs which PB_a distributes with SB_b at t, then

$$\mathcal{U}_{ab}^{\mathrm{SB}}(t) = \zeta_a u_{ab}^{\mathrm{SB}}(t) \tag{14}$$

Based on this policy, $u_a^{\mathrm{PB}}(t)$ is distributed for reducing $\mathcal{Q}_a^{\mathrm{PB}}(t+1)$. So,

$$\overline{\mathcal{R}}_a = \frac{\mathcal{Q}_a^{\mathrm{PB}}(t) + \mathcal{D}_a^{\mathrm{PB}}(t+1)}{\zeta_a} \tag{15}$$

It is simple to certify that

$$u_a^{\mathrm{PB}}(t) = \begin{cases} \mathcal{R}_a, & \text{If } \overline{\mathcal{R}}_a \geq \mathcal{R}_a \\ \lceil \overline{\mathcal{R}}_a \rceil, & \text{or else} \end{cases} \tag{16}$$

Normally, there are few additional capacity, i.e., $\overline{\mathcal{R}}_a < \mathcal{R}_a$ maintained for several ranges of a. After, this additional capacity is used for serving the authorized clients. To obtain the separate distributions, the following Eqs. (7a–7c) is solved subject to the restraint that $u_{ab}^{\mathrm{SB}}(t)$ is an integer. Also, the anonymous service rate distribution vectors and the function are defined as

$$u_b^{\mathrm{SB}}(t) = \begin{bmatrix} u_{1b}^{\mathrm{SB}}(t) \\ \vdots \\ u_{nb}^{\mathrm{SB}}(t) \end{bmatrix}, \quad u^{\mathrm{SB}}(t) = \begin{bmatrix} u_1^{\mathrm{SB}}(t) \\ \vdots \\ u_m^{\mathrm{SB}}(t) \end{bmatrix} \tag{17}$$

$$\overline{f}\{u^{SB}(t)\} := \max_{b\in B}\left[\mathcal{Q}_b^{SB}(t) + \mathcal{D}_b^{SB}(t) - \sum_{a=1}^{n} \zeta_a u_{ab}^{SB}(t)\right]^+ \quad (18)$$

After that, u^{SB} is determined by solving the optimization problem given below:

$$\min \overline{f}\{u^{SB}(t)\} \quad (19a)$$

$$\text{s.t.} \quad u_{ab}^{SB}(t) \in Z^+, \quad \forall a \in A, b \in B \quad (19b)$$

$$u_a^{PB}(t) + \sum_{b=1}^{m} u_{ab}^{SB}(t) \le \mathcal{R}_a, \quad \forall a \in A \quad (19c)$$

Observe that Z^+ is the group of each optimistic integers in (19b) and (19c) is found by splitting either terms of (7c) through ζ_a. The corresponding resource distribution method operates as: For every t,

1. Each PBs/SBs gather and transmit $\mathcal{Q}_a^{PB}(t)/\mathcal{Q}_b^{SB}(t)$, $\mathcal{D}_a^{PB}(t)/\mathcal{D}_b^{SB}(t)$ to CSC.
2. CSC computes and transmits the optimal $u_a^{PB}(t)/u_{ab}^{SB}(t)$ to each SBs/PBs.
3. The resources of the authenticated spectrum belonging to PB_a are distributed by PB_a that engages $u_a^{PB}(t)$ RBs and SBs which occupy $u_{ab}^{SB}(t)$ RBs.

Figure 2 illustrates the overall flow diagram of D-U-LTE–Wi-Fi–QBRA system.

4 Simulation Results

This part evaluates the effectiveness of D-U-LTE–Wi-Fi–QBRA system with D-U-LTE–Wi-Fi using MATLAB 2016a. The comparison is conducted based on Spectral Efficiency (SE), throughput, Average Transmission Number (ATN), Transmission Delay (TD), interference, and Signal-to-Interference Noise Ratio (SINR). This system is modeled by considering the simulation parameters in [17].

4.1 Spectral Efficiency

The percentage between transfer number and spectrum bandwidth is called SE.

$$\text{SE} = \frac{\text{Transmission number}}{\text{Spectrum bandwidth}} \quad (20)$$

The outcomes of SE for D-U-LTE–Wi-Fi and D-U-LTE–Wi-Fi–QBRA systems are given in Table 2.

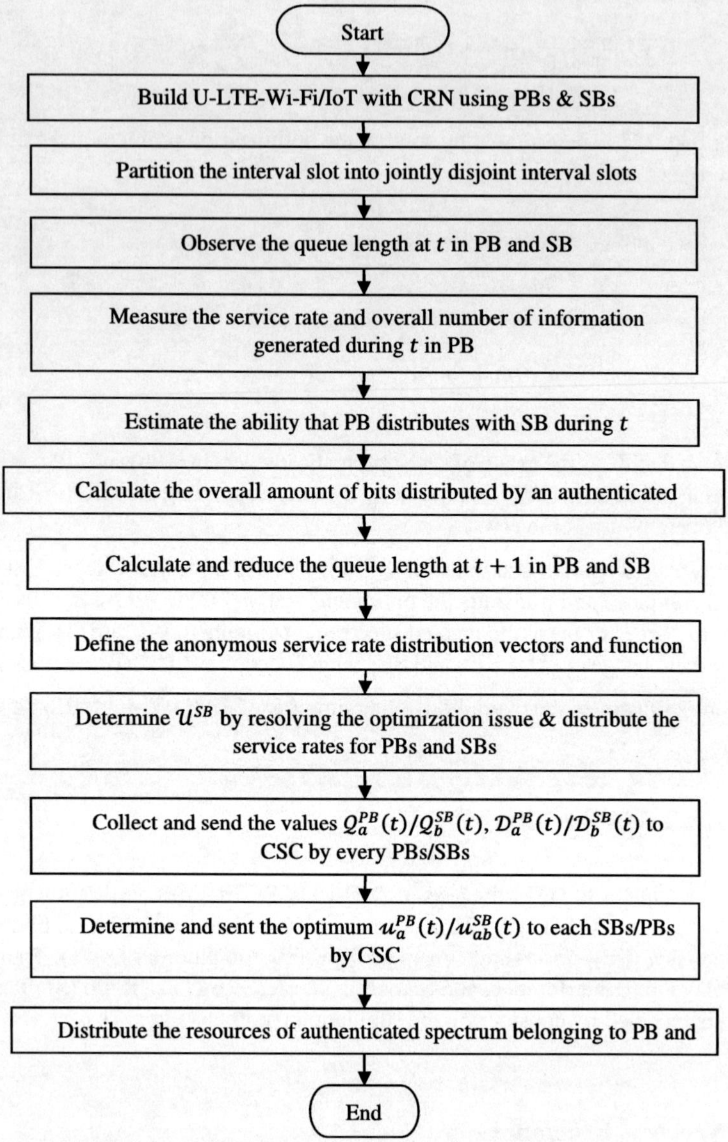

Fig. 2 Overall flow diagram of D-U-LTE–Wi-Fi/IoT–QBRA system

Figure 3 depicts the SE (in MHz) of D-U-LTE–Wi-Fi and D-U-LTE–Wi-Fi–QBRA. If 500 UEs are considered, then the SE of D-U-LTE–Wi-Fi–QBRA is 3.46% greater than the D-U-LTE–Wi-Fi. It observes that the D-U-LTE–Wi-Fi–QBRA system can attain high SE compared to the D-U-LTE–Wi-Fi.

Table 2 Analysis of SE

No. of UEs	SE (MHz)	
	D-U-LTE–Wi-Fi	D-U-LTE–Wi-Fi–QBRA
50	173.061	187.72
100	186.95	200.65
150	195.549	207.79
200	215.442	228.16
250	241.524	254.99
300	260.730	270.44
350	291.235	303.31
400	331.57	343.05
450	356.45	368.31
500	399.425	410.84

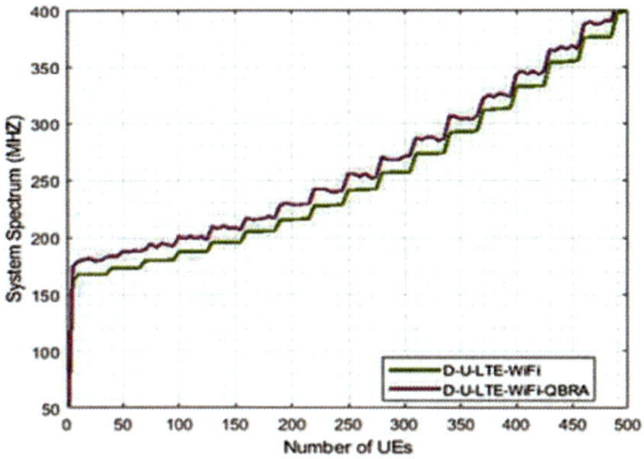

Fig. 3 SE versus number of UEs

4.2 *Throughput*

The number of data sent through the wireless channels in a given time is called throughput. The outcomes of throughput for D-U-LTE–Wi-Fi and D-U-LTE–Wi-Fi–QBRA are given in Table 3.

Figure 4 displays the throughput (in Mbps) of D-U-LTE–Wi-Fi and D-U-LTE–Wi-Fi–QBRA. For 500 UEs, the throughput of D-U-LTE–Wi-Fi–QBRA is 10.07% greater than the D-U-LTE–Wi-Fi system. So, it observes that the D-U-LTE–Wi-Fi–QBRA system has the highest throughput compared to the D-U-LTE–Wi-Fi.

Table 3 Analysis of throughput

No. of UEs	Throughput (Mbps)	
	D-U-LTE–Wi-Fi	D-U-LTE–Wi-Fi–QBRA
50	62	75.58
100	75	88.17
150	93	103.25
200	105	117.59
250	120	128.92
300	129.5	143.43
350	135	150.80
400	130	143.09
450	126	140.09
500	122.5	133.87

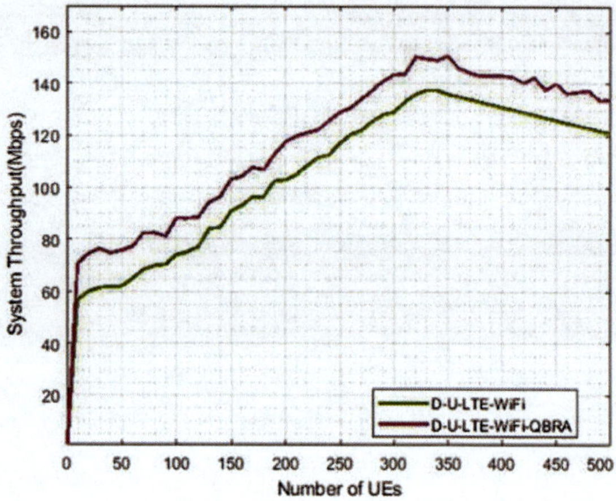

Fig. 4 Throughput versus number of UEs

4.3 *Transmission Delay*

The interval taken for sending a data among UEs is called system TD. The outcomes of system TD between various number of UEs during data transmission based on the D-U-LTE–Wi-Fi and D-U-LTE–Wi-Fi–QBRA are presented in Table 4.

Figure 5 shows the system TD (in s) for D-U-LTE–Wi-Fi and D-U-LTE–Wi-Fi–QBRA systems. When 500 UEs are considered, the TD of D-U-LTE–Wi-Fi–QBRA is 13.38% lower than D-U-LTE–Wi-Fi. Thus, it notices that the D-U-LTE–Wi-Fi–QBRA system has the minimum TD compared to the D-U-LTE–Wi-Fi comparably.

Table 4 Analysis of system TD

No. of UEs	System TD (s)	
	D-U-LTE–Wi-Fi	D-U-LTE–Wi-Fi–QBRA
50	23.31	18.54
100	23.25	18.61
150	23.22	18.68
200	23.18	18.84
250	23.16	20.09
300	23.15	20.11
350	23.13	19.15
400	23.10	20.01
450	23.08	18.14
500	23.05	18.39

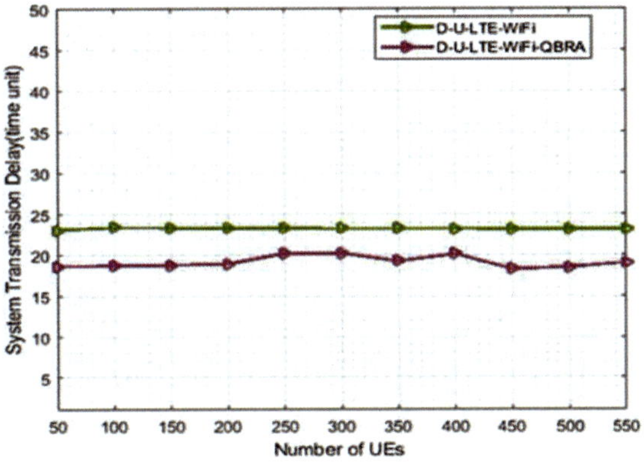

Fig. 5 System TD versus number of UEs

4.4 Average Transmission Number

The number of data will be transmitted/second/connection is called ATN.

$$\text{ATN} = \sum_{i=1}^{n} \text{TN}_i + \sum_{j=1}^{m} \text{TN}_j \tag{21}$$

In Eq. (21), n is the number of Wi-Fi connections, m is the amount of LTE connections, and TN_i and TN_j are the transfer integer of Wi-Fi connection i and LTE connection j, accordingly. Table 5 presents the outcomes of ATN for D-U-LTE–Wi-Fi and D-U-LTE–Wi-Fi–QBRA.

Figure 6 portrays the ATN for D-U-LTE–Wi-Fi and D-U-LTE–Wi-Fi–QBRA systems. If the number of UEs is 450, the ATN of D-U-LTE–Wi-Fi–QBRA is 6.93%

Table 5 Analysis of ATN

No. of UEs	ATN	
	D-U-LTE–Wi-Fi	D-U-LTE–Wi-Fi–QBRA
50	254.24	267.72
100	246.14	259.84
150	228.62	239.04
200	224.19	234.84
250	223.29	233.54
300	215.28	229.20
350	198.79	212.19
400	198.26	212.00
450	192.56	203.72

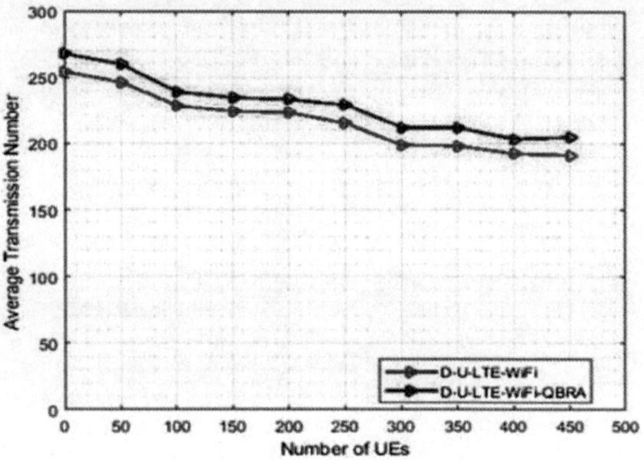

Fig. 6 ATN versus number of UEs

greater than D-U-LTE–Wi-Fi. From this scrutiny, it observes that the D-U-LTE–Wi-Fi–QBRA can attain the maximum mean amount of transfer compared to the D-U-LTE–Wi-Fi.

4.5 SINR

The SINR for user i in the D-U-LTE–Wi-Fi–QBRA configuration is determined by

$$\text{SINR}_i = \frac{p_{\text{D-U-LTE-WiFi-RA,i}} \times G_{\text{D-U-LTE-WiFi-RA}}}{I_{\text{D-U-LTE-WiFi-RA network}}} \quad (22)$$

In Eq. (22), $p_{\text{D-U-LTE-Wi-Fi-RA,i}}$ is the overall energy at eNB from UE i, $G_{\text{D-U-LTE-Wi-Fi-RA,i}}$ is the channel gain between U-LTE and Wi-Fi and $I_{\text{D-U-LTE-Wi-Fi-RA network}}$ is the overall interferences in the D-U-LTE–Wi-Fi–QBRA.

Table 6 Analysis of SINR

UE distance from eNB (m)	SINR (dB)	
	D-U-LTE–Wi-Fi	D-U-LTE–Wi-Fi–QBRA
5	9.9	7.5
20	8.2	6.3
40	6.0	5.1
60	4.9	3.2
80	3.7	2.1
100	2.0	1.5
120	1.1	0.82

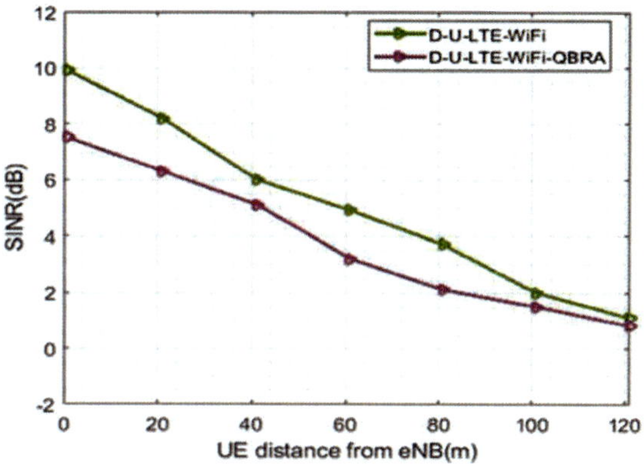

Fig. 7 SINR versus UE distance from eNB

The outcomes of SINR for D-U-LTE–Wi-Fi and D-U-LTE–Wi-Fi–QBRA are listed in Table 6.

Figure 7 demonstrates the SINR (in dB) for D-U-LTE–Wi-Fi and D-U-LTE–Wi-Fi–QBRA systems. For 100 m distance of UEs from eNB, the SINR of D-U-LTE–Wi-Fi–QBRA is 25% lower than the D-U-LTE–Wi-Fi. It observes that the D-U-LTE–Wi-Fi–QBRA system can minimize the SINR compared to the D-U-LTE–Wi-Fi significantly.

4.6 *Interference*

It is the overall interferences caused by the D-U-LTE–Wi-Fi–QBRA. The outcomes of interferences for D-U-LTE–Wi-Fi and D-U-LTE–Wi-Fi–QBRA are provided in Table 7.

Figure 8 depicts the interference (in dB) for D-U-LTE–Wi-Fi and D-U-LTE–Wi-Fi–QBRA systems. If the distance between UEs and eNB is 100 m, then the

Table 7 Analysis of interference

UE distance from eNB (m)	Interference (dB)	
	D-U-LTE–Wi-Fi	D-U-LTE–Wi-Fi–QBRA
0	−200	−231
20	−185	−201
40	−168	−173
60	−152	−168
80	−132	−153
100	−111	−124
120	−89	−98

Fig. 8 Interference versus UE distance from eNB

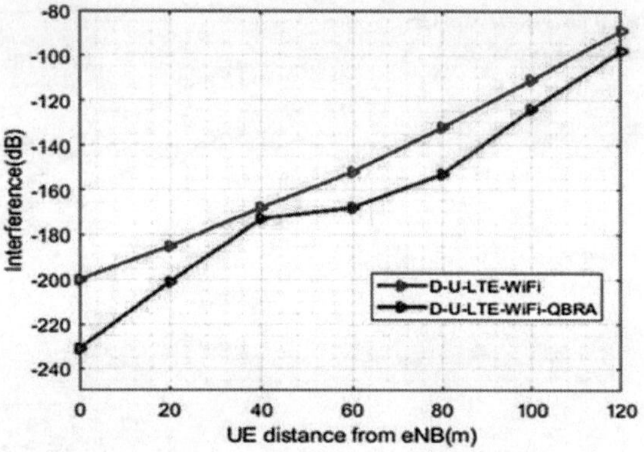

interference of D-U-LTE–Wi-Fi–QBRA is −124 dB which is less than the interference of D-U-LTE–Wi-Fi. It observes that the D-U-LTE–Wi-Fi–QBRA system can significantly decrease the interference compared to the D-U-LTE–Wi-Fi due to the distribution of appropriate resources and service rate capacities of both PBs and SBs.

5 Conclusion

In this paper, D-U-LTE–Wi-Fi–QBRA system is presented with the aim of guaranteeing the wireless connectivity to the unauthenticated eNBs with no cooperation of the QoS for the authenticated eNBs by using a 2-stage process. First, the band resources are distributed to the authenticated eNBs for increasing the QoS for the authenticated clients. Then, the additional service ability of the main spectrum is allocated amid the unauthenticated eNBs for increasing the QoS of the unauthenticated clients. Thus, the RA in D-U-LTE–Wi-Fi system can useful to increase the

QoS of the network effectively. To conclude, the findings revealed that the D-U-LTE–Wi-Fi–QBRA system has higher SE and throughput when maintaining less TD and SINR than the other classical network systems.

Acknowledgements The authors thank the Department of Science and Technology Interdisciplinary Cyber Physical System (DST-ICPS), New Delhi, (DST/ICPS/IoTR/2019/8) for the financial support to this research work.

References

1. Grigoriou E, Chatzimisios P (2015) An overview of 3GPP long term evolution (lte). IGI Global Publications
2. Vaigandla KK, Venu N (2021) A survey on future generation wireless communications-5g: multiple access techniques, physical layer security, beam forming approach. J Inf Comput Sci 11(9):449–474
3. Mustafa S, Alam KA, Khan B, Ullah MH, Touseef P (2019) Fair coexistence of LTE and WiFi-802.11 in unlicensed spectrum: a systematic literature review. In: Proceedings of the 3rd international conference on future networks and distributed systems
4. Bojović B, Giupponi L, Ali Z, Miozzo M (2019) Evaluating unlicensed LTE technologies: LAA vs LTE-U. IEEE Access 7:89714–89751
5. Mekonnen Y, Haque M, Parvez I, Moghadasi A, Sarwat A (2018) LTE and Wi-Fi coexistence in unlicensed spectrum with application to smart grid: a review. In: IEEE/PES transmission and distribution conference and exposition
6. Sun H, Fang Z, Liu Q, Lu Z, Zhu T (2017) Enabling LTE and Wi-Fi coexisting in 5 GHz for efficient spectrum utilization. J Comput Netw Commun 2017
7. Alhulayil M, Lopez-Benitez M (2018) Coexistence mechanisms for LTE and Wi-Fi networks over unlicensed frequency bands. In: IEEE 11th international symposium on communication systems networks & digital signal processing
8. Hafaiedh HB, El Korbi I, Saidane LA, Kobbane A (2017) LTE-U and Wi-Fi coexistence in the 5 GHz unlicensed spectrum: a survey. In: IEEE international conference on performance evaluation and modeling in wired and wireless networks
9. Pang Y, Babaei A, Andreoli-Fang J, Hamzeh B (2017) Wi-Fi coexistence with duty cycled LTE-U. Wirel Commun Mobile Comput 2017
10. Mehrnoush M, Sathya V, Roy S, Ghosh M (2018) Analytical modeling of Wi-Fi and LTE-LAA coexistence: throughput and impact of energy detection threshold. IEEE/ACM Trans Netw 26(4):1990–2003
11. Sathya V, Mehrnoush M, Ghosh M, Roy S (2018) Association fairness in Wi-Fi and LTE-U coexistence. In: IEEE wireless communications and networking conference
12. Reddy SRV, Roy SD (2021) SBT (Sense before transmit) based LTE licenced assisted access for 5 GHz unlicensed spectrum. Wirel Pers Commun 119:2069–2081
13. Zhang R, Wang M, Cai LX, Zheng Z, (Sherman) Shen X, Xie L-L (2015) LTE-unlicensed: the future of spectrum aggregation for cellular networks. IEEE Wirel Commun 22(3):150–159
14. Maglogiannis V, Naudts D, Shahid A, Moerman I (2018) An adaptive LTE listen-before-talk scheme towards a fair coexistence with Wi-Fi in unlicensed spectrum. Telecommun Syst 68:701–721
15. Sumathi AC, Vidhyapriya R, Vivekanandan C, Sangaiah AK (2019) Enhancing 4G co-existence with Wi-Fi/IoT using cognitive radio. Cluster Comput 22(Suppl 5):11295–11305
16. ETSI ET (2012) Electromagnetic compatibility and radio spectrum matters (ERM); short range devices (SRD), radio equipment to be used in the 25 MHz to 1000 MHz frequency range with power levels ranging up to 500 mW. European Harmonized Standard EN 300

17. Muniyappan A, Pankajavalli PB (2019) Enhancement of opportunistic coexistence of U-LTE and Wi-Fi/IoT in 5GHz using cognitive radio. Int J Innov Technol Exploring Eng 2278–3075
18. Muniyappan A, Pankajavalli PB (2021) Enhanced grant-less multiple sub frame scheduled uplink transmission for enhancement on U-LTE-Wi-Fi/IoT through cognitive radio. Natural Volatiles Essent Oils 1333–1347
19. Wang C, Kuo W-H, Chu C-Y (2017) QoS-aware cooperative power control and resource allocation scheme in LTE femtocell networks. Comput Commun 110:164–174
20. Thakur R, Kotagi VJ, Murthy CSR (2017) Resource allocation and cell selection framework for LTE-unlicensed femtocell networks. Comput Netw 129:273–283
21. Lyazidi MY, Aitsaadi N, Langar R (2018) A dynamic resource allocation framework in LTE downlink for cloud-radio access network. Comput Netw 140:101–111
22. Liu J-S (2018) Joint downlink resource allocation in LTE-advanced heterogeneous networks. Comput Netw 146:85–103
23. Kaur M, Randhawa NS, Bansal R (2019) Enhancement of proportional scheduling in lte using resource allocation based proposed technique. Procedia Comput Sci 155:797–802
24. Xu Y, Yang Y, Liu Q, Li Z (2020) Joint energy-efficient resource allocation and transmission duration for cognitive HetNets under imperfect CSI. Signal Process 167:107309
25. Liu Z, Hou G, Yuan Y, Chan KY, Ma K, Guan X (2020) Robust resource allocation in two-tier NOMA heterogeneous networks toward 5G. Comput Netw 176:107299

A Secure Key Management on ODMRP in Mesh-Based Multicast Network

Bhawna Sharma and Rohit Vaid

Abstract The essential agreement for any multicast group communication has proven to be problematic due to the dynamic nature of multicast. Distributed, centralized, or mixed key management architectures are available. Although various strategies to handle key changes have been presented, this paper focuses on the features of rekeying that those systems execute. The primary security method for group communication under the present scheme is performed via traditional encryption algorithms, the Group Controller is in charge of key distribution and rekeying for the group key. As MANET is infrastructure-less network, so it requires security for communication among nodes. For cellular ad hoc networks, this study provides a mesh-based totally multicast key control mechanism in ODMRP. These are steady and subjectively greater trustworthy individuals of the multicast group form a mesh and provide important services with excessive safety and availability. The system gives perfect overall performance-based totally on numerous factors average packet delay, control overhead, average end-to-end latency, and Normalized Routing Load (NRL) demonstrates that the key management approach is both secure and efficient.

Keywords Mesh-based · Tree-based · Multicast · Key distribution · ODMRP

1 Introduction

A group of wireless communication nodes is known as mobile ad hoc network that self-configure in a dynamic manner to set up a network without using centralized or fixed infrastructure supervision [1]. The practice of transmitting data from alternate sources to several targets is known as multicast routing. There are two type of routing in MANET-unicast and multicast [2]. To attain high efficiency, path delay because of choking and the value of the path tree to reach destinations should be minimized [3].

B. Sharma (✉) · R. Vaid
Department of Computer Science and Engineering, MMEC, MM (Deemed to be University), Mullana, Ambala, India
e-mail: bsbhawna90@gmail.com

P. Chatterjee et al. (eds.), *Computational Intelligence for Engineering and Management Applications*, Lecture Notes in Electrical Engineering 984,
https://doi.org/10.1007/978-981-19-8493-8_39

MANET multicast strategies may be classified as tree-based or mesh-based totally on how they disseminate information [4, 5].

In mesh-based protocols, a subset of hubs (mesh) is responsible for bring data to all multicast receiver nodes, at the same time as tree-based protocols unfold facts over a tree pass-over all multicast group individuals [6–8]. Multicast routing protocol based on mesh. A mesh network is formed using a mesh-based multicast routing protocol that continues many pathways. ODMRP and PUMA are mesh-based multicast routing technologies [9].

Data transfer via an insecure network necessitates security. There are a number of strategies for dealing with unicast security concerns, but none of them can be immediately applied to a multicast system. Multicasting is inherently more vulnerable than unicast since it uses many network channels to transmit data. Only authorized members of multicast group share a session key, that's dynamically changed to make certain forward and backward secrecy, referred to as "organization rekeying."

2 On-Demand Multicast Routing Protocol (ODMRP)

It is a mesh-based protocol and use the idea of forwarding groups (simplest a group of nodes those forward the multicast packets through scoped flooding). Use an on-demand strategy to dynamically generate routes and maintain membership in multicast groups. ODMRP is ideal for wireless ad hoc networks with mobile hosts that have limited bandwidth, constantly changing topologies, and limited power consumption. Use simulation to evaluate the scalability and performance of ODMRP.

– Finding a route "On spot"

  - Each node broadcasts to its neighbors a packet indicating that the following hop to a set of sources is unavailable.
  - If a node has a route to the multicast supply when it receives this packet, it unicasts a Join respond to its next hop neighbors.
  - If there is no route to known, the packet is simply broadcast, indicating that the next hop is unavailable.
  - It sets its FG FLAG in both circumstances.
  - It aids in the establishment of an alternate path until the next refresh phase, when a more efficient route is developed.

2.1 Algorithm

The source-node on-demand establishes group membership and multicast the routes in ODMRP. A multicast source transmits a Join-Query manipulate data to the whole network when it has data/packet to send, however no direction toward the multicast

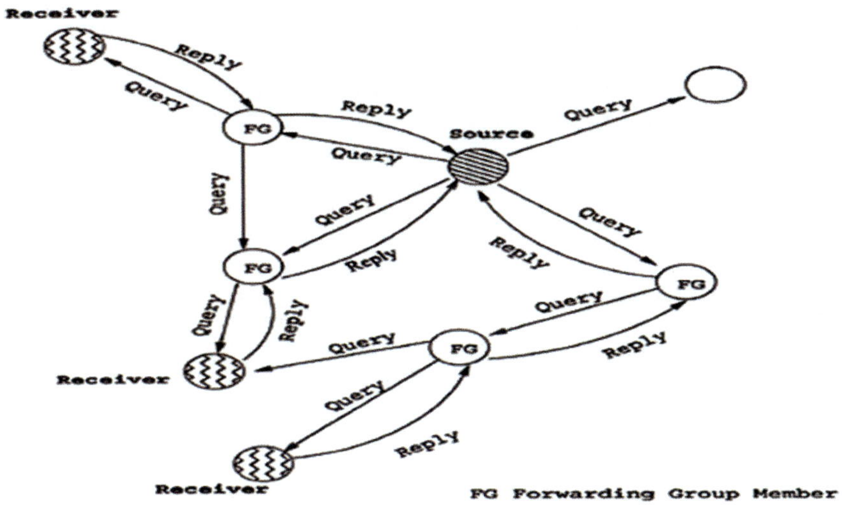

Fig. 1 On-demand process for route establishment and maintenance

group. This manipulate packet is transmitted on a regular basis to replace membership information and routes. When a multicast receiver gets a Join-Query packet, it produces and broadcasts Join-Reply to its neighbors. When node receives it, it checks to peer if the subsequent hop id is similar to its own. If they match, the node recognizes that it is at the source's path and joins the forwarding group by setting the Forwarding Group Flag (FG FLAG). When a node gets a multicast data packet, it only forwards it if it isn't a duplicate, reducing traffic overhead. Finding the correct flooding interval is important to ODMRP performance due to the nodes maintain a smooth state. ODMRP forecasts the period of time that routes can be valid based on location and movement facts [10]. When direction breaks of ongoing data sessions are approaching, a "join facts" packet is flooded with the predicted time of direction disconnection. In phrases of bandwidth usage, it indicates that ODMRP is extra suitable for ad hoc networks. Figure 1 indicates the approach for developing and maintaining a membership on demand.

2.2 *Merits of ODMRP*

- Simplicity.
- Low storage and channel overhead.
- Use of up-to-date shortest routes.
- Reliable route and forwarding group design.
- Robustness in terms of hosting mobility.
- The upkeep and use of a large number of channels.

- Taking use of the broadcast nature of the wireless surroundings router with unicast abilities.
- The multicast tree construction may be accomplished fastly and efficaciously by the use of unicast path information [11].

2.3 *ODMRP's Drawbacks*

- High overhead it incurs as a result of the reply packets being broadcast to a large number of nodes.
- Topology is complicated since it is mesh-based.
- Single point failure problem [11].

3 Multicast Security in MANET

The following are the basic security characteristics of MANET.

The network information will not be provided to the unlawful unit due to confidentiality. Integrity is required to ensure that the data being transported between nodes does not change or degrade. The term "availability" refers to the availability of requested services in a timely manner, with no systemic difficulties. Due to a loss of authentication, an attacker can impersonate any node and benefit manage of the complete network. Nonrepudiation ensures that the message initiator can't refuse to deliver the message [12].

3.1 *Key Management*

Key management refers to the processes of create, distribute, and update the keys to secure group communication application [13]. Traffic Encryption Keys (TEKs) and the Key Encryption Keys (KEKs) are used for encryption and de-encryption process (KEKs). Each member in a secure multicast connection has a key for encrypt and decrypt the data. The rekeying operation corresponds to the procedure of update and distribute keys to group members. The rekey process is carried out when each membership changes. Key management, on the other hand, requires multiple transfer per unit of time to maintain the forward and backward privacy during continuous membership modification [14]. There are two types of secure multicasting schemes: centralized and distributed schemes. In the event of a centralized system, the Group Controller (GC) manages group keys and only applies minimal loads on the group's users. To reduce the load on the user, key management is conducted by each user in a distributed architecture [15, 16].

Distributed key management protocols: In distributed key management protocols, no specific manager and key creation are done by the members themselves. Access control can be performed by all members, and key generation can be support, that is all members provide some information to create the group key, or it can be done by the one of members. In the case of key updates, distributed protocols have a scalability challenge because they necessitate massive calculations and have high communication overheads. They also require all members of the network to have access to formidable resources. Octopus Protocol, Distributed Logical Key Hierarchy, and Diffie–Hellman Logical Key Hierarchy are some examples of distributed key management protocols [10, 17].

4 Related Work

The Mesh Certification Authority (MeCA), described in [18], tackles the challenge of wireless mesh network authentication and key management in the absence of a trusted third party. It uses WMN's self-configure and self-organize capabilities to distribute certification authority functions among multiple mesh routers. Threshold cryptography is used for key functions like secret sharing and key distribution, which reduces the risk of data disclosure even if some nodes are compromised by attackers. To reduce operation overhead, MeCA uses the multicast tree building approach described in [5].

A certificateless design for multicast wireless mesh networks is presented in reference [19]. It uses a certificateless proxy de-encryption system. It makes use of a change technique to minimize the scale of cypher texts which represents the duration of the optimal path from source to the target multicast users and indicates the variety of legitimate receivers in the group. To decide session keys, all communication entities are participating in key settlement protocols. The Diffie–Hellman (DH) key settlement protocol is the maximum typical key settlement protocol utilized in maximum distributed group key control systems. Here are some examples: Bresson et al. [9] advanced a generic authenticated group DH Key change with a provably secure method. Katz and Yung [20] proposed the primary provably stable group DH method; this is constant-round and absolutely scalable in the popular model. The essential gain of the group DH key change is that it lets in all institution individuals to set up a secret group key without relying on a mutually trusted KGC [21].

The present research has primarily focused on standard routing procedures. However, the realities of group communication present numerous problems. In this part, we'll go over various multicast routing protocols. In comparison with protocols multicast steering protocols, Ravindra Vaishampayan et al. developed a PUMA directing protocol that promotes excessive information transfer percentage with little control overhead and additionally will increase better package transfer proportion.

Menaka Pushpa and K. Kathiravan offered answers for two kinds of assaults: guard dog-based data package drop attack identifiable evidence and MA parcel manufacture attack. Multicast is an idea proposed via way of means of Elizabeth M. Royer et al.

impromptu MAODV [22] is an on-demand multicast protocol that creates mesh tree to support multiple sender nodes and collectors in a multicast session. A guiding convention for this sort of dynamic self-configure system should be organized for unicast, multicast and to be able to offer optimal data transmission.

Using (*t*, *n*) threshold cryptography, Zhou and Hass [23] offer a secure key management approach. T 1 hacked servers can be tolerated by the system. However, this system does not specify how a node can communicate with t servers in a secure and efficient manner if the servers are dispersed throughout the area. To combat mobile opponents, a share refreshing method is offered. However, there is no mention of how to distribute secret shares in an effective and secure manner.

URSA is a localized key management technique proposed by Luo, Kong, and Zerfos [24]. All nodes are servers in this system. The benefits of this method include local communication efficiency and secrecy, as well as system availability; however, it weakens system security, particularly when nodes are not sufficiently protected. One issue is that nodes will have to continuously relocating to obtain certificates updated if the threshold k is considerably bigger than the network degree d. Convergence in the share updating phase is the second significant issue. The third major issue is that there is far too much offline setting required before connecting to the networks.

MOCA key management is a scheme proposed by Yi, Naldurg, and Kravets [25]. Certificate service is provided through Mobile Certificate Authority (MOCA) nodes, which are physically more secure and strong than other nodes, according to their strategy. According to their approach, a node could find k + MOCA nodes at random, through the shortest way, or via the most recent path in its route cache. However, as most secure routing methods rely on the provision of a key service, the essential question is how nodes can securely identify those paths.

5 Results and Discussion

Table 1 represents result evaluation metrics using key management in ODMRP. Different parameters such as average packet delay, end-to-end delay, control overhead, and network routing load are used to measure the efficiency with or without key management.

5.1 *Average Packet Delay*

The average packet delay is calculated by multiplying the time it takes for successful data packets to travel from their origins to their destinations by the total number of successful packets. In miliseconds, it is measured.

Figure 2 illustrates that average packet delay with or without key management in ODMRP is varying between 4.41 and 5.24 ms.

Table 1 ODMRP metrics using key management

Protocol used	No. of receiver nodes	Parameters	Without key management	With key management
ODMRP	5	Average packet delay (ms)	4.28	5.12
		End-to-end delay (ms)	3.81	4.35
		Control overhead	3.39	4.27
		Network routing load	5.24	4.41

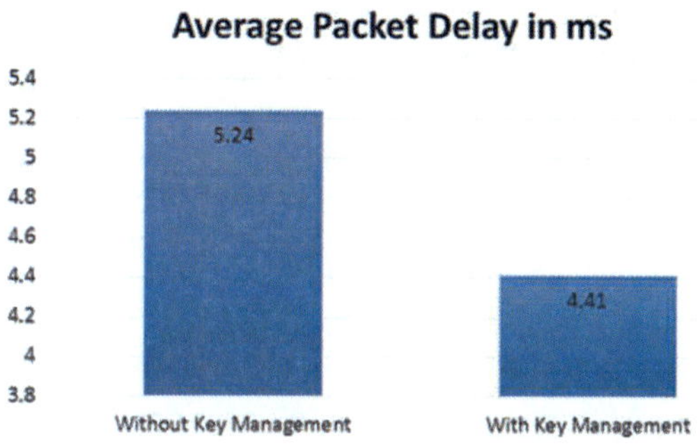

Fig. 2 Average packet delay with or without key management in ODMRP

5.2 *Control Overhead*

The control overhead is decided on this simulation as the overall range of control packets required to construct a strong path between sources to the destination (multicast receiver).

Figure 3 describes that control overhead with key management in ODMRP is less as compared to without key management.

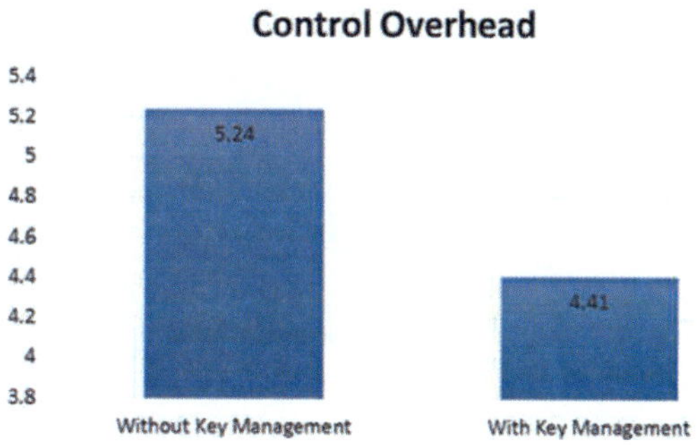

Fig. 3 Control overhead with or without key management in ODMRP

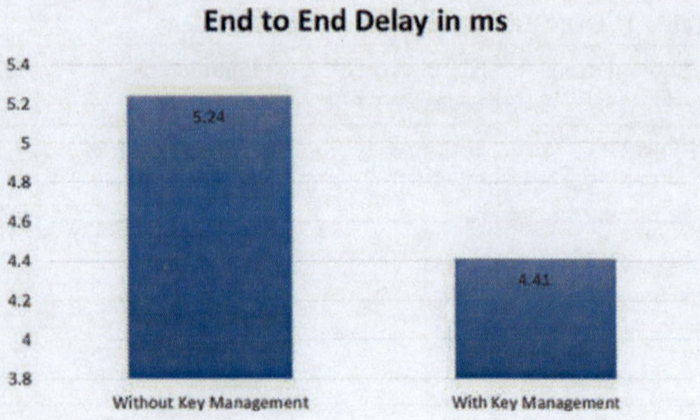

Fig. 4 End-to-end delay with or without key management in ODMRP

5.3 *Average End to End Delay*

It's the time it takes for a data packet to travel from its origin to its destination on average.

$$\text{EED average} = \text{total EED}/\text{number of the packets sent}$$

Figure 4 describes that end-to-end delay without key management in ODMRP is high as compared to using key management.

5.4 *Normalized Routing Load (NRL)*

The number of data packets received divided by the number of routing packets received is known as NRL.

$$\text{NRL} = \text{number of data packets received}/\text{number of routing packets received}$$

In Fig. 5, it is clear by using key management in Normalized Routing Load, the performance is improved.

6 Discussion

From the analysis of ODMRP, a multicast routing protocol, it can be seen that by applying key management for security purpose the Average Packet Delay(ms), end-to-end delay (ms), and control overhead slightly increased, whereas Network Routing Load has been decreased. Using key management on this protocol, the rate of successfully reached packet at destination has been increased as compared to the without using key management.

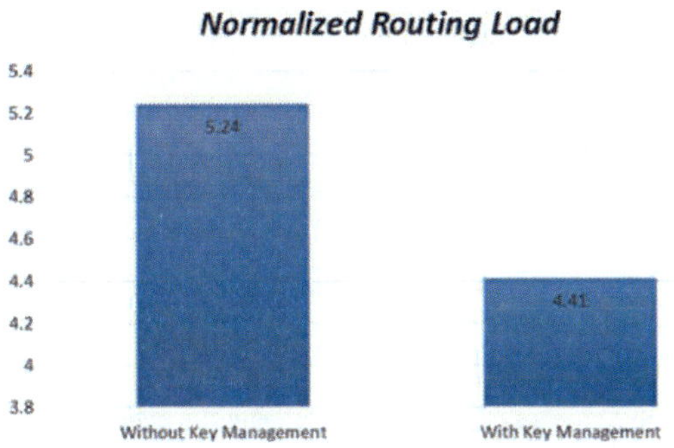

Fig. 5 NRL with or without key management in ODMRP

7 Conclusion

MANET is a network without infrastructure and no central authority. Group communication is one of the most essential forms of communication in these types of networks for a variety of applications. To achieve this group communication, the Multicasting Mechanism might be used. To select out keys and privately supply keys to all communication nodes, key switch protocols rely upon a mutually depended on key generation center (KGC). In this research, we used the ODMRP routing protocol, which yields advanced outcomes, while security keys are applied and create a secure communication network between two nodes.

In the future, we will consider more receiver nodes to check the efficiency of the proposed approach. In addition to this, we can implement secure key management on various routing protocols with performance comparison among them.

References

1. Junhai L, Liu X, Danxia Y (2008) Research on multicast routing protocols for mobile ad-hoc networks. Comput Netw 52(5):988–997
2. Shaveta Jain and Kushagra Agrawal, "Prevention Against Rushing Attack on MZRP in Mobile AdHoc Networks" in IJCST Vol. 5, Issue 3, July - Sept 2014.
3. S Sarkar, T.G. Basavaraju, C. Puttamadappa, Ad Hoc Mobile Wireless Networks: Principles, protocols and applications, Auerbach Publications, 2008
4. Zifen Yang, Deqian Fu, Lihua Han, Seong Tae Jhang, Improved route discovery based on constructing connected dominating set in MANET, Int. J. Distrib. Sens. Netw. (2015)
5. P. M. Ruiz and A. F. Gomez-Skarmeta, "Heuristic algorithms for minimum bandwith consumption multicast routing in wireless mesh networks," in Ad-Hoc, Mobile, and Wireless Networks, V. R. Syrotiuk and E. Chávez, Eds., vol. 3738 of Lecture Notes in Computer Science, pp. 258–270, 2005.
6. Danladi Ali, Michael Yohanna, W.N. Silikwa, Routing protocols source of self-similarity on a wireless network, Alexandria Eng. J. 57 (4) (2018) 2279–2287.
7. Yaser M. Khamayseh, Shadi A. Aljawarneh, Ensuring survivability against Black Hole Attacks in MANETS for preserving energy efficiency Sustainable Computing: Informatics and Systems, Alaa Ebrahim Asaad 18 (2018) 90–100

8. Ansuman Bhattacharya, Koushik Sinha, in: An Efficient Protocol for Load-Balanced Multipath Routing in Mobile ad hoc Networks, 63, Ad Hoc Networks, 2017, pp. 104–114
9. Bresson E, Chevassut O, Pointcheval D (2007) ProvablySecure Authenticated Group Diffie-Hellman Key Exchange. ACM Trans. Information and System Security 10(3):255–264
10. S Sumathy, Beegala Yuvaraj, E Sri Harsha "Analysis of Multicast Routing Protocols: Puma and Odmrp" in "International Journal of Modern Engineering Research (IJMER)" Vol.2, Issue.6, Nov-Dec. 2012 pp-4613–4621
11. Shaveta Jain and Kushagra Agrawal, "A Survey on Multicast Routing Protocols for Mobile Ad Hoc Networks" in International Journal of Computer Applications (0975 – 8887) Volume 96– No.14, June 2014.
12. Rajan C, Shanthi NS (2013) Misbehaving attack mitigation technique for multicast security in mobile ad hoc networks (MANET). J Theor Appl Inf Technol 48(3):1349–1357
13. Devi DS, Padmavathi G (2009) A reliable secure multicast key distribution scheme for mobile Adhoc networks. World Academy of Science, Engineering and Technology 56:321–326
14. Devaraju S, Padmavathi G (2010) Dynamic clustering for QoS based secure multicast key distribution in mobile Ad hoc networks. International Journal of Computer Science Issues 7(5):30–37
15. R. Srinivasan, V. Vaidehi, R. Rajaraman, S. Kanagaraj, R. Chidambaram Kalimuthu, and R. Dharmaraj, "Secure group key management scheme for multicast networks," International Journal of Network Security, vol. 10, no. 3, pp. 205–209, 2010
16. B. Madhusudhanan, S. Chitra, and C. Rajan, "Mobility Based Key Management Technique for Multicast Security in Mobile Ad Hoc Networks" in Hindawi Publishing Corporation Scientific World Journal Volume 2015
17. Kim Y, Perrig A, Tsudik G (2004) Tree-Based Group Key Agreement. ACM Trans Inf Syst Secur 7(1):60–96
18. Kim J, Bahk S (2009) Design of certification authority using secret redistribution and multicast routing in wireless mesh networks. Comput Netw 53(1):98–109
19. Wang H, Cao Z, Wei L (2014) A scalable certificateless architecture for multicast wireless mesh network using proxy re-encryption. Security and Communication Networks 7(1):14–32
20. Katz J, Yung M (2007) Scalable protocols for authenticated group key exchange. J. Cryptology 20:85–113
21. S.Sasikala Dev and Dr. Antony Selvadoss Danamani, "Dr. Antony Selvadoss Danamani" in , Int. J. Comp. Tech. Appl., Vol 2 (3), 385–391 ISSN:2229-6093
22. Liu J, Li J (2010) A better improvement on the integrated Diffie-Hellman-DSA key agreement protocol. Int J Netw Secur 11(2):114–117
23. Zhou, L. and Z. Haas. Securing Ad Hoc Networks. IEEE Network Magazine, Vol. 13, 1999
24. H. Luo and S. Lu. URSA: Ubiquitous and Robust Access Control for Mobile Ad-Hoc Networks. UCLA, 2004
25. S. Yi, P. Naldurg and R. Kravets Security-Aware Ad-hoc Routing for Wireless Networks. Report No.UIUCDCS-R-2002-2290, UIUC, 2002

Detecting Cyber-Attacks on Internet of Things Devices: An Effective Preprocessing Method

Ngo-Quoc Dung

Abstract Cyber-attacks targeting Internet of Things (IoT) devices are still attracting the interest of network security researchers as hackers always improve their cyber-attack methods to bypass intrusion detection systems. Machine learning plays an important role in IoT security from cyber-attacks. In other words, machine learning algorithms that are widely applied to detect anomalous network flows from IoT devices are still limited. This is because the actual data is often complex, noisy, and has many redundancies that are not readily available for machine learning classifiers. Meanwhile, IoT devices are diverse and heterogeneous, so the actual network flow data of IoT devices is more complex. Therefore, two feature sets common to network traffic data (Tshark and CICFlowMeter) have been evaluated through two real datasets, i.e., ToN-IoT and Agent-IoT. This research result contributes to providing a reference to the research community in choosing the suitable method to preprocess actual network traffic data from IoT devices when using machine learning to cyber-attack detection. Experimental results show that there are many differences between training and testing on the same dataset and different datasets. Also, experimental results obtained by our algorithms are promising and have achieved more than 99% F1-score with CICFlowMeter.

Keywords IoT device · Preprocessing · Cyber-attack detection · CICFlowMeter · Tshark

1 Introduction

Internet of Things (IoT) is understood as a network of physical and virtual objects or 'things' embedded with electronic equipment, software [1]; each of these objects is provided with its identifier, and all can transmit and exchange information and data with each other over the Internet. According to this approach, Cisco has classified IoT devices into three main groups constrained devices, gateways or border routers,

N.-Q. Dung (✉)
Posts and Telecommunications Institute of Technology, Hanoi, Vietnam
e-mail: dungnq@ptit.edu.vn

P. Chatterjee et al. (eds.), *Computational Intelligence for Engineering and Management Applications*, Lecture Notes in Electrical Engineering 984,
https://doi.org/10.1007/978-981-19-8493-8_40

and the cloud platform [2]. Thus, IoT devices are behind many popular solutions and technologies such as smart home, smart cities, autonomous cars, and Internet of Medical Things. So, IoT devices are very diverse and are forecasted to explode in the number of IoT devices. According to a report by IoT Analytics [3], from 2019 to 2025, the number of non-IoT devices increased very little, about 3%. Meanwhile, IoT devices increased by more than 209% from 10 billion to 30.9 billion devices connected to the Internet. The connection is implemented via a Wi-Fi network, broadband network (3G, 4G), infrared or Bluetooth, and so on. Today, IoT devices are the 'promising land' for cyber-criminals to attack, infiltrate, and exploit due to the enormous number of devices increasing the probability, frequency, and severity of attacks. In the event of a successful attack, the hacker will quickly take control of the entire network and deaden many IoT devices at the same time. Therefore, in the context of the explosion of IoT devices, the detection of cyber-attacks is one of the tasks that organizations and individuals must be addressed. The cyber-attacks on an IoT device are diversity such as distributed denial of service, data type probing, scanning, malicious operation, and spying. However, hackers often use network port scanning attacks and denial of service attacks. Also, the existing security protocols and mechanisms are staggering to avoid cyber-attacks efficiently. One of the commonly used techniques to detect cyber-attacks is the use of intrusion detection systems (IDSs). IDSs [4] are defined as software systems built to monitor and analyze the behavior of networks and systems with the objective of detecting anomalies and intrusions. IDS is categorized under two classes, namely signature-based (or misuse) IDS and anomaly-based IDS [5].

Signature-based IDSs apply different rules/signatures of known attacks' rules to detect a cyber-attack over a network or a host. While, anomaly-based IDSs, monitor network traffic and can differentiate between normal and malicious flow based on previously learned patterns to spot anomalous activities. However, signature-based methods have shown to be unable to recognize new or unknown cyber-attacks. Anomaly-based IDSs overcome the limitation of signature-based ones, but they often have high false alarms (false-positives) rates [6]; therefore, expert knowledge is required to evaluate and validate these false positives. However, this process is considered time consuming and requires highly qualified specialists. In fact, the technology and techniques used to build IDS are mostly publicly available, so hackers are always looking for improved forms of cyber-attacks to bypass IDS without being detected. It creates the motivation to design an anomaly-based IDS solution using artificial intelligence (AI), such as machine learning (ML) and deep learning (DL) in order to improve cyber-attacks detection [7]. Deep learning is a type of machine learning. Machine learning methods are very useful for identifying or classifying malicious, intrusion, and cyber-attacks on IoT devices. Input datasets play an important role in machine learning techniques. Therefore, in order to accurately detect cyber-attacks against IoT devices using machine learning, it is essential to select an efficient input feature set, which selects relevant features and eliminates redundant and irrelevant features which perform degradation in the operation of the classifier [8]. For this purpose, feature selection is the preprocessing method that helps select and remove unwanted features from the input feature set [9]. Data preprocessing is an important task in data mining applications because it helps to create suitable datasets which mining and algorithms can be applied. For example, Lu et al. [10] proposed a data preprocessing method for network security situational awareness based on conditional random fields. Larriva-Novo et al. [11] proposed the study and evaluation of several preprocessing techniques based on traffic categorization for a machine learning neural network algorithm. This research uses for its evaluation two benchmark datasets, namely UGR16 and the UNSW-NB15, and one of the most used datasets, KDD99.

However, in classifying network flows of IoT devices, there is still no effective machine learning model proposed yet. Therefore, it is important to study the issue of effective data preprocessing for IoT device cyber-attacks, especially with actual data input from IoT devices. In addition, actual datasets often present many challenges for researchers because it is often incomplete due to omissions in the data collection process, constraints in the data acquisition process, or cost limitations, which restrict the storage of some values that should be present in the dataset. This paper is focused on the preprocessing step in an intrusion detection system. This paper presents comparative analysis using two different data preprocessing methods, CICFlowMeter [12] and Tshark (or Wireshark) [13] for cyber-attacks in IoT network traffic and to improve the performance of machine learning techniques. This research result contributes to providing a reference to the research community in choosing the suitable method to preprocess actual network traffic data from IoT devices when using machine learning to cyber-attack detection.

The rest of the proposed work is organized as follows. In Sect. 2, we discuss the related works on detecting cyber-attacks on IoT devices, especially the problem of data preprocessing for IoT device cyber-attack detection. The proposed method is detailed in Sect. 3. Next, the results are discussed in Sect. 4. Finally, the conclusion and future directions are highlighted in Sect. 5.

2 Related Works

Data preprocessing for cyber-attack detection problems has been studied quite a lot [14–17]. Davis and Clark [14] presented a review of the state of the art on data preprocessing techniques with 33 attributes of each packet which were collected from header packets and used for anomaly-based network intrusion detection systems. Revathi and Malathi [15] proposed an efficient data preprocessing method for network traffic data based on a hybrid Simplified Swarm Optimization (SSO) and Random Forest algorithm. SSO is a simplified Particle Swarm Optimization (PSO), which can filter data and reduce incomplete, noisy, and dimensionality problems for both discrete and continuous variables in a dataset. The proposed method is evaluated by testing with the KDD99cup dataset. The experimental result shows that the proposed method reduces feature selection attribute that reduces the false-positive rate and efficiency for an intrusion detection system. Ahmad and Aziz [16] proposed a method based on a combined correlation-based features selection (CFS) and Particle Swarm Optimization (PSO) to implement preprocessing to the data and selecting relevant features. Firstly, the model was designed to get a new range of data by data normalization with the min–max method. The next step is selecting features by using CFS–PSO for optimizing the selection process. Finally, three machine learning classifiers utilized for classification include Naive Bayes, K-NN, and SVM. The experiment was performed on three different datasets, including KDDCUP 99, Kyoto 2006, and UNSW-NB15. The best results with SVM achieved 99.9291% accuracy on the KDD99cup dataset. These studies show us the importance of data preprocessing in cyber-attack detection. Ham et al. [17] proposed an efficient preprocessing method for big-data mobile Web log information for detection cyber-attacks on mobile Web servers. Specifically, the authors build an algorithm using the divide-and-conquer mechanism on the Weblog dataset using the multi-threading method. The algorithm works by dealing with overlapping strings in Web log data. Considering the overlapped strings, the characteristics of strings are an indexing method overlapped strings in Weblog data files. In addition to improving the performance of the existing preprocessing, the paper configures the index information based on the B-tree structure. In order to evaluate the effectiveness of the proposed method, the paper is experimental on the IIS web log dataset with eight different log sizes from 500,000 log lines to 4,000,000 log lines. However, these are all studies of cyber-attacks on computer networks in general, not networks of IoT devices.

Recently, many studies have paid attention to the problem of detecting network attacks for IoT such as Sriram et al. [18], Su at el. [19]. Sriram et al. [18] proposed a deep learning approach that works on network traffic to detect IoT cyber-attack. Specifically, network flow snapshots from connected IoT devices in the packet capture (PCAP) format. Then convert network flow information into connection records. During training, these connection records are labeled manually and fed to various traditional machine learning algorithms and deep learning models. Experimental method on two different datasets include N-BaIoT and DS1. The results were positive and the deep learning-based method outperformed the classical machine learning classifiers. Su at el. [19] introduced a correlation change-based feature selection method. In their feature selection method, clustering first correlated sensors to recognize the duplicated deployed sensors according to sensor data correlations. Then, monitor the data correlation changes in real time to select the sensors with correlation changes as the representative features for anomaly detection. In their experimental analysis, they proved their proposed technique is able to capture the effective features for anomaly detection in IoT networks. Most previously proposed solutions used deep learning algorithms to detect IoT cyber-attacks. However, these solutions do not use CICFlowMeter and Tshark that are common in network traffic data preprocessing [20–24]. Alsamiri and Alsubhi [21] introduced an approach by utilizing machine learning algorithms for detecting cyber-attacks in IoT devices. It describes a method of performing CICFlowMeter which helps in extracting flow-dependent features from IoT network traffic. The authors performed an experimental study on the Bot-IoT dataset in which they performed various machine learning models (i.e., Naïve Bayes and Decision Tree) to evaluate their performance. Sarhan et al. [21] presented an automated tool, called SHAP, to show which features are most influential in machine learning models to detect cyber-attacks on IoT device networks. In this paper, NetFlow and CICFlowMeter are two feature sets that have been evaluated over various datasets, such as CSE-CIC-IDS2018, ToN-IoT, and BoT-IoT. During the experiments, Deep Feed Forward (DFF) and Random Forest (RF) classifiers are used to distinguishing between benign and abnormal network traffic present in the datasets. DFF and RF classifiers achieved a better cyber-attack detection accuracy in a less predictable time. In addition, the essential features impacting the predictions of the machine learning models have been identified for each dataset.

Iqbal and Naaz [22] introduced the importance of Wireshark as a sniffing application which can detect which device is sending a request to which service, their IP address, and port in a computer network. Wireshark is capable of capturing network packets and displays it as humanreadable. In the paper, the ability of Wireshark has been demonstrated by analyzing many cyber-attacks such as DOS attacks, ARP poisoning, MAC flooding, and DNS. Hussain et al. [23] proposed a framework for the security of healthcare IoT devices. IoT-Flock allows scripting to use IoT devices to generate both benign and malicious network traffic. The framework then uses the Wireshark tool to intercept network traffic and storage in packet capture (PCAP) format and utilize to create an IoT dataset for training the machine learning models. Finally, machine learning models are trained and tested by implementing six machine learning classifiers to classify the network traffic in the underlying IoT use case. Experimental results showed the Random Forest classifier outperformed all other classifiers with precision, recall, accuracy, and F1-score at 99.7068%, 99.7952%, 99.5123%, and 99.6535, respectively. Rizal et al. [24] performed network forensics investigation with analysis method using Wireshark to detect flooding attacks on the Internet of Things device. To demonstrate the proposed method, they experiment flooding attack to an infected IoT Bluetooth Arduino device. Then, the log data files obtained with the pcap extension are analyzed by the Wireshark application.

To the best of our knowledge, CICFlowMeter and Tshark are popular tools in network flow data preprocessing but from the literature survey, it has been observed that most of the works have not compared the effectiveness of CICFlowMeter and Tshark in detecting cyber-attacks on IoT devices based on network traffic utilizing deep learning models with many large and real datasets. This will be described in detail in the next section of this paper.

3 Proposed Method

3.1 Overview

The overall model of the proposed method is presented as shown in Fig. 1. To collect network traffic data from IoT devices network, we use two methods: (1) install agent directly on IoT device; (2) using some network traffic dataset available on public

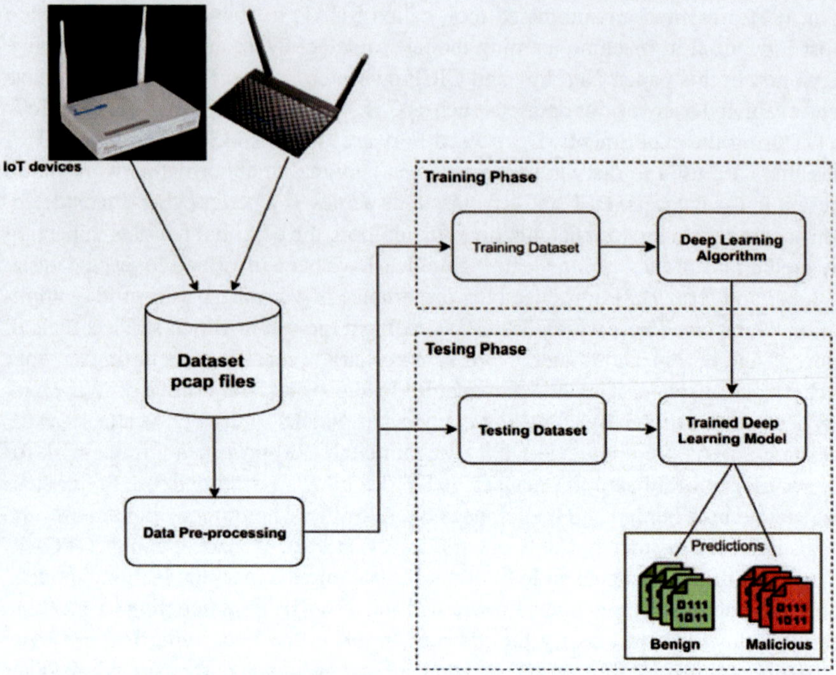

Fig. 1 Overview of the proposed model

repositories. The dataset generated after capturing the traffic was in the form of PCAP files. To preprocess and analyze the generated dataset, we converted the PCAP files into CSV files using a python script. The next step is to train the deep learning model. Then, we test the trained models over the test set to evaluate the model's performance to detect cyber-attack traffic and normal traffic in IoT environments.

3.2 Dataset

To reliably assess deep learning-based IDS performance, the model needs to be evaluated on various datasets using the same feature set.

- ToN-IoT dataset [25]: The heterogeneous data sources from datasets of Internet of Thing and Industrial IoT, operating systems datasets of Windows 7 and 10 as well as Ubuntu 14 and 18 TLS and Network traffic datasets which designed at the Cyber Range and IoT Labs, the School of Engineering and Information technology (SEIT), UNSW Canberra, Australia. Various cyber-attack techniques are implemented against web applications, IoT/IIoT networks, IoT gateways, known as Backdoor, DoS, DDoS, Injection, MITM, Password, Ransomware, Scanning,

and XSS. The packet capture data includes Injection 4 Pcap files (2.6 GB); Man in the Middle 4 Pcap files (2.7 GB); Backdoor 1 Pcap file (546.8 MB); DDoS 12 Pcap files (6.5 GB); DoS 5 Pcap files (4.1 GB); Ransomware 2 Pcap files (1.2 GB); Scanning 6 Pcap files (4.5 GB).; XSS 10 Pcap files (6.7 GB); Password 5 file Pcap (3.8 GB); Normal 15 Pcap files (15.6 GB).
- Agent-IoT dataset: This data set is generated by the agent installed directly on the IoT device, including IP Camera and Router (TP-Link, VNPT GPON). This agent is developed and installed directly on IoT devices to collect network-level and system-level data from IoT devices by us. The agent is less than 1 MB in size and is capable of running on multiple architectures. This dataset contains the types of attack as DDoS and Scanning. There are 4058 Pcap files normal (1 GB), 193 Pcap files Scanning (9.4 MB), and 421 Pcap files Slowloris (43.5 MB).

3.3 Dataset Preprocessing

The data preprocessing part consists of two main parts: feature extraction and data normalization.

- **Normalization**: Since all initial features are numeric data, the data after feature extraction is always numeric data. Therefore, encoding methods are not needed. Some popular data normalization methods like min–max scaling, Z-score normalization, and batch normalization. In this paper, we use batch normalization. The training process is divided into many iterations (epoch). Each epoch consists of many steps. In each step, a small data packet extracted from the original full dataset (batch) will be fed to the learning network using gradient descent. The mini-batch gradient descent method is used when training deep learning models with large datasets that cannot be fully loaded into memory. Batch normalization, as the name implies, will normalize data in batches according to the same formula as z-score and update moving_mean and moving_var indexes. These indices are used to normalize the data when predicting. Batch data when training will be normalized as follows:

$$z = \frac{x_{batc} - \mu_{batc}}{(\sigma_{batc} + \epsilon) \times \gamma} + \beta \tag{1}$$

where

- μ_{batc}: Average value of the batch.
- σ_{batc}: Variance of batch
- x_{batc}:
- ϵ: As a small constant that can be changed, the purpose is to avoid division by 0.
- β: is a learnable bias coefficient.
- γ: As a scale factor that can be learned.

Then the moving_mean and moving_var indexes will be updated as follows:

$$\text{moving}_{\text{mean}} = \text{moving}_{\text{mean}} * \text{momentum} + \text{mean(batch)} * \quad (1 - \text{momentum})$$

$$\text{moving\_var} = \text{moving\_var} * \text{momentum} + \text{var(batch)} * \quad (1-\text{momentum})$$

Batch data when predict will be normalized as follows (can also predict batch by batch)

$$x = \frac{x_{\text{batch}} - \text{moving\_mean}_{\text{batch}}}{(\text{moving}_{\text{var batch}} + \epsilon) \times \gamma} + \beta \tag{2}$$

- **Feature extraction**: The input to this step will be CSV files containing network flow features—each column of the CSV file corresponds to a feature. If the CSV files are training data, they will remove duplicate lines or contain NULL values, as well as handle infinity values, but if used for predicting, skip this step. In this paper, we use CICFlowMeter and Tshark.
  - **CICFlowMeter** is a network traffic flow generator and analyzer. It can be used to create a bi-directional flow where the first packet defines the inbound (source to destination) and outgoing (destination to source) directions. Therefore, there are more than 80 network traffic analysis features that can be calculated separately in the incoming and outgoing directions. The output of CICFlowMeter is a CSV format file with six labeled columns for each flow (FlowID, SourceIP, DestinationIP, SourcePort, DestinationPort, and Protocol) with more than 80 network traffic analysis features. However, TCP streams are usually terminated on disconnection (by a FIN packet), while UDP streams are terminated by a timeout. The working mechanism of the CICFlowMeter applied data preprocessing method is as follows: (1) extract the characteristics of the packets in the Pcap file; (2) combine packets with the same [source_ip, dst_ip, source_port, dst_port] into one stream, then compute further features of that flow; and (3) using batch normalization to normalize data. Then, the preprocessed data is fed into feedforward models for training and testing.
  - **Tshark** is a network protocol analyzer. It allows to capture packet data from an active network or read packets from a previously saved capture file, a print decoded form of those packets to standard output or write packets to a file. Tshark is Wireshark's command-line interface (CLI), which can be used directly at the terminal. The working mechanism of the data preprocessing method applied Tshark is as follows: (1) Extract the characteristics of the packets in the Pcap file; (2) Combine *n* consecutive packets into a flow, adding other features to each flow. Then, preprocessed data is fed into regression models (LSTM, GRU, …) for training and testing.

As part of the experiment, the CICFlowMeter and Tshark tool have been utilized to extract the features from the Agent-IoT and ToN-IoT datasets. The generated data flow from the packet capture files has been labeled in multilayer, using the ground truth events.

4 Experiments and Evaluation

4.1 Evaluation Criteria

To measure the classification performance of classifiers results. Confusion metrics are the primary and important basis for measuring performance, including a confusion matrix; the numbers of true positive (TP), true negative (TN), false positive (FP), and false negative (FP) instances.

- TP: The number of data instances correctly detected the attack flow as an attack
- FN: The number of data instances incorrectly classified the attack flow as a normal
- FP: The number of data instances wrongly detected the normal flow as an attack
- TN: The number of data instances correctly detected the normal flow as a normal

To evaluate the effectiveness of the proposed method, we use

$$\text{Precision} = \frac{\text{TP}}{\text{TP} + \text{FP}}$$

$$\text{Recall} = \frac{\text{TP}}{\text{TP} + \text{FN}}$$

$$F1\text{-score} = \frac{2 * \text{Precision} * \text{Recall}}{\text{Precision} + \text{Recall}}$$

4.2 Experiment and Discussion

The experiment method proceeds as follows:

- Training and testing on the same datasets (Sect. 3.2) are preprocessed by CICFlowMeter and Tshark
- Training and testing on different datasets (Sect. 3.2) preprocessed by CICFlowMeter and Tshark.

The experimental environment is conducted on a computer with the following configuration, Linux operating system, CPU: Intel(R) Core(TM) i5-8500 CPU @ 3.00 GHz, GPU: GeForce GTX 1080 Ti and Ram: 32 GB (available ~ 20 GB).

Table 1 Number of network flows extracted from Agent-IoT dataset

	Normal	Scanning	Slowloris
Number of Pcap file	4058	193	421
Feature extracted with CICFlowMeter	51,460	12,589	50,071
Feature extracted with Tshark	1,828,658	98,447	213,197

Table 2 Configuring the training model

	CICFlowMeter	Tshark
Training model	– Input (input_dim) – Batch normalization – Dense (input_dim / 1.5, 'relu') – Dense (input_dim/2.25, 'relu') – Dense (3, 'softmax')	– Input (input_dim) – Batch normalization – LSTM (64) – Dense (32, 'relu') – Dense (3, 'softmax')

a. *Training and testing on the same dataset*

The process is performed on the Agent-IoT dataset, and the processed data set includes (Table 1).

It can be seen that this flow data is quite unbalanced, with normal data flow accounting for the majority, scanning data flow accounting for a minority, and the difference between flow data types is quite large (except for CICFlowMeter). With CICFlowMeter tools, we use more decision-tree-based (Recursive Feature Elimination or RFE) to reduce 69 features to 60 and 40 features. The initial data is divided into train/test sets with a ratio of 4/1 (Table 2).

Train in 20 epochs, each epoch consists of 1000 steps, each step trains a data packet of 768 threads equally divided per class (batch size 256) to minimize data imbalance. Batch size has been tested before and shows that a larger batch size will help the network converge faster, but the batch size of 256 and above does not show much difference. Perform the same for the CICFlowMeter and Tshark use cases.

From the obtained results as given in Table 3, it can be seen that the Recall, Precision, and F1-score of the experiment with the CICFlowMeter tool in data preprocessing and feature extraction for the Feedforward Neural Network model are possible, and the results decrease proportionally to the number of features. The number of features at 69 gives the best results, and the higher number of features helps the model to converge faster. Continue testing with Tshark on the Agent-IoT dataset. Test with flow lengths of 100, 60, and 20. Flow length is the length of a sequence of consecutive packets. Initial data is divided into train/test sets with a ratio of 4/1. The configuration of the training model is presented in Table 4.

To have a multi-dimensional view, when preprocessing and feature extraction using the Tshark tool, we use Long short-term memory (LSTM) network. The experimental results obtained show that the processing with CICFlowMeter gives better results than Tshark, and the convergent LSTM model is not as good as the Feedforward neural network model.

Table 3 Experimental results of classification using CICFlowMeter with F1-score

Epoch	69 features			60 features			40 features		
	Scanning	Slowloris	Normal	Scanning	Slowloris	Normal	Scanning	Slowloris	Normal
1	94.06	97.52	97.84	93.21	95.94	96.21	89.79	96.24	96.26
2	94.94	97.98	98.25	95	97.79	97.79	92.88	96.7	96.88
3	94.95	98	98.41	95.2	98.03	98.4	93.68	95.96	96.03
4	94.98	98.11	98.42	95.78	98.11	98.17	94.52	97.45	97.54
5	93.94	97.93	98.38	95.54	98.2	98.26	94.68	97.56	97.84
6	95.43	98.41	98.6	94.51	98.16	98.53	94.45	97.62	97.93
7	95.5	98.38	98.64	95.65	98.38	98.65	93.09	97.48	97.54
8	95.37	98.35	98.79	95.52	98.47	98.7	94.86	97.77	98.03
9	94.72	98.23	98.69	96.18	98.32	98.55	94.46	97.83	98.03
10	96.04	98.47	98.84	95.9	98.6	98.81	94.59	97.79	98.13
11	96.06	98.56	98.79	96.56	98.58	98.88	94.43	97.96	98.39
12	95.84	98.63	98.92	96.36	98.47	98.78	94.86	97.91	98.14
13	96.6	98.7	98.92	96.32	98.53	98.93	94.71	97.91	98.26
14	96.39	98.57	98.75	96.14	98.55	98.75	94.13	97.75	98.13
15	96.15	98.57	98.9	96.27	98.72	99.06	95.32	98.06	98.33
16	96.38	98.69	99	95.58	98.57	99.03	94.94	97.84	98.23
17	96.51	98.68	98.95	96.69	98.62	98.89	94.4	97.9	98.35
18	95.73	98.58	99.02	96.88	98.53	98.76	94.75	97.92	98.2
19	96.87	98.79	99.02	96.43	98.72	98.99	94.67	97.89	98.32
20	96.87	98.81	99.06	96.55	98.74	99.03	94.82	97.9	98.38

Table 4 Experimental results of classification using Tshark with F1-score

Epoch	100			60			20		
	Scanning	Slowloris	Normal	Scanning	Slowloris	Normal	Scanning	Slowloris	Normal
1	55.05	50.27	94.83	88.25	93.64	99.64	91.62	96.83	99.8
2	72.99	84.35	99.01	85.78	92.62	99.56	91.99	96.23	99.83
3	68.58	68.87	98.23	85.02	92.86	99.61	90.95	95.56	99.77
4	71.06	68.99	96.89	85.98	93.8	99.65	92.62	96.94	99.86
5	70.21	80.23	98.99	84.28	92.63	99.62	93.54	97.53	99.79
6	71.13	81.33	98.98	86.05	93.67	99.68	93.47	96.74	99.89
7	72.85	80.14	99.04	87.5	94.76	99.64	92.89	97.16	99.88
8	71.5	81.04	98.76	85.16	93.57	99.65	92.62	97.06	99.85
9	69.69	80.97	99.06	87.44	93.06	99.62	91.71	96.37	99.89
10	70.51	77.55	98.67	87.1	94.4	99.7	93.3	96.71	99.88
11	72.92	83.03	98.79	87.5	94.75	99.7	90.96	95.71	99.89
12	76.15	84.88	98.9	88	94.38	99.71	92.58	96.83	99.85
13	74.61	86.23	99.18	86.19	94.37	99.68	92.75	96.93	99.85
14	77.72	83.98	98.86	87.8	95.01	99.73	91.95	96.24	99.86
15	78.48	85.15	99.35	87.98	94.66	99.73	92.89	96.67	99.88
16	80.18	87.35	99.06	88.56	94.5	99.77	92.03	96.14	99.85
17	81.57	86.66	99.15	86.85	94.38	99.65	91.28	95.84	99.83
18	80.09	85.75	99.04	88.4	94.85	99.76	93.02	96.83	99.94
19	82.96	88.51	99.32	88.18	94.72	99.82	92.66	96.57	99.91
20	80.73	88.26	99.26	88.24	95.31	99.8	92.67	96.62	99.92

Table 5 Number of network flows extracted from ToN-IoT dataset using CICFlowMeter

Type of data	Normal	Scanning	Slowloris
Number of flows	81,278	16,222,608	8,225,012

b. *Training and testing on the different dataset*

Cyber-attacks are complex and increasingly diverse. Therefore, the actual dataset often encounters many limitations, such as incomplete due to omissions in the data collection process, constraints in the data acquisition process, or cost constraints. In this section, we experiment on two different data sets presented in Sect. 3.2. The training data is combined from the Agent-IoT and ToN-IoT datasets, but the testing process uses the Agent-IoT dataset. However, the ToN-IoT dataset is very diverse in attacks, so to match the Agent-IoT dataset, we only use DDoS, Scanning, and Normal data in the ToN-IoT dataset. The process is performed on the ToN-IoT dataset, and the processed data set includes (Table 5).

In this process, we will only use CICFlowMeter to conduct experiments. After removing unnecessary features and typeface features, 69 features remain. The training method here uses mini-batch gradient descent. The training process is divided into several iterations (epochs). Each epoch consists of many steps. In each step, a small data packet extracted from the original full dataset (batch) will be fed to the learning network using gradient descent. This method is suitable when large amounts of data cannot be loaded into memory. A data batch consists of three layers (Normal, DDoS, and Scanning), each layer consists of 1024 flows, combining 512 flows of ToN-IoT and 512 Agent-IoT flows. Each sample batch is a random sample. We use the empirical model with the regression network as in the previous section. In addition, we conduct experiments with the neural network model. Since the original data is also very large, a smaller portion of the data will be sampled (sampling method). The distribution of data after sampling is as follows (Table 6).

After training, the model will be tested against the test dataset. We use Keras, which is a high-level API of TensorFlow to build a neural network model for the classification of network flows. Train in 5 epochs, 1000 steps each, batch size 1024×3.

As can be seen, the proposed method of using the CICFlowMeter tool in data preprocessing and feature extraction with real datasets of IoT devices is appropriate and gives positive results. Because the actual dataset has an imbalance between normal and cyber-attack data, we use the Recall to evaluate the effectiveness in this experimental section. As given in Table 7, after testing, the ability to classify the

Table 6 Distribution of data after sampling

	ToN-IoT dataset	Agent-IoT dataset
Normal	48,766/32,512	7200/4800
DDoS (Slowloris)	720,000/480,000	7200/4800
Scanning	348,064/232,044	7200/4800

Table 7 Experimental results with training and testing on different datasets

	Recall (ToN-IoT dataset)					Recall (agent-IoT dataset)				
Class\epoch	1	2	3	4	5	1	2	3	4	5
DDoS (Slowloris)	86.41	97.68	97.59	97.65	97.98	96.79	97.58	97.67	97.71	98.77
Scanning	96.09	95.66	96.14	94.44	95.89	95.69	96.71	96.92	97.12	96.9
Normal	97.39	97.54	97.59	98.33	97.63	87.54	90.85	89.02	96	93.5

network traffic in the ToN-IoT dataset with DDoS, Scanning, and Normal classes achieved Recall at 97.98%, 95.89%, and 97.63%, respectively. Meanwhile, the results achieved in the Agent-IoT dataset are 98.77%, 96.9%, and 93.5%, respectively.

5 Conclusion

In the digital transformation era, IoT is one of the core technology platforms to help users have many experiences in digital life. However, hackers can also take advantage of the potential of IoT to carry out cyber-attacks that threaten users' privacy and security. Therefore, security for IoT devices is necessary. Like in the traditional computer network, the anomaly detection system is one of the important security solutions for IoT. However, to improve the efficiency of IDS, it is necessary to apply machine learning, and machine learning-based network IDS has obtained notable cyber-attack detection performance in the security research community. In this paper, we present research focusing on the preprocessing of IDS. CICFlowMeter and Tshark tools are popular in network traffic data preprocessing. In this paper, the efficiency of the proposed Tshark-based network traffic data processing is evaluated and compared with the feature set designed by CICFlowMeter. The evaluation has been conducted over two datasets (Agent-IoT, ToN-IoT) using a deep learning model. Experimental results show that training and testing on the same actual data set collected from IoT devices by the proposed method yielded positive results with F1-score reaching over 96% with CICFlowMeter. With training and testing on different real datasets, the proposed method also gives positive results, Recall reaching over 93%. However, in this study, we have not provided a list of meaningful features in network attack detection of IoT devices handled by CICFlowMeter, which is the direction of our future research.

Acknowledgements This work has been supported by the Cybersecurity Lab, Posts and Telecommunications Institute of Technology, Hanoi, Vietnam, and funded by the Ministry of Science and Technology, Vietnam grant number KC-4.0-05/19-25.

References

1. ITU (2012) T, Recommendation ITU-T Y.2060: Overview of the Internet of things
2. Nagasai (2017) Classification of IoT devices. https://www.cisoplatform.com/profiles/blogs/classification-of-iot-devices
3. Lueth KL (2020) State of the IoT 2020: 12 billion IoT connections, surpassing non-IoT for the first time. https://iot-analytics.com/state-of-the-iot-2020-12-billion-iot-connections-surpassing-non-iot-for-the-first-time/
4. Hindy H et al (2018) A taxonomy and survey of intrusion detection system design techniques, network threats and datasets
5. Gupta D, Garg S, Singh A, Batra S, Kumar N, Obaidat MS (2017) ProIDS: probabilistic data structures based intrusion detection system for network traffic monitoring. In: GLOBECOM IEEE global communications conference, pp 1–6
6. Aljawarneh S, Aldwairi M, Yassein MB (2018) Anomaly-based intrusion detection system through feature selection analysis and building hybrid efficient model. J Comput Sci 25:152–160
7. Salih A et al (2021) A survey on the role of artificial intelligence, machine learning and deep learning for cybersecurity attack detection. In: 7th International engineering conference "research & innovation amid global pandemic" (IEC)
8. Dash M, Liu H (1997) Feature selection for classification. Intell Data Anal 1:131–156
9. Gusti Bagus Dharma Putraa I, Gusti Agung Gede Arya Kadyanana I (2021) Implementation of feature selection using information gain algorithm and discretization with NSL-KDD intrusion detection system. Jurnal Elektronik Ilmu Komputer Udayana 9(3):359–364
10. Lu A, Li J, Yang L (2010) A new method of data preprocessing for network security situational awareness. In: 2nd International workshop on database technology and applications, pp 1–4
11. Larriva-Novo X et al (2021) An IoT-focused intrusion detection system approach based on preprocessing characterization for cybersecurity datasets. Sensors 21(2):656
12. Asad M, Asim M, Javed T, Beg MO, Mujtaba H, Abbas S (2020) Deepdetect: detection of distributed denial of service attacks using deep learning. Comput J 63(7):983–994
13. Mohamed T, Otsuka T, Ito T (2018) Towards machine learning based IoT intrusion detection service. In: International conference on industrial, engineering and other applications of applied intelligent systems, pp 580–585
14. Davis JJ, Clark AJ (2011) Data preprocessing for anomaly based network intrusion detection: a review. Comput Secur 30(6–7):353–375
15. Revathi S, Malathi A (2013) Data preprocessing for intrusion detection system using swarm intelligence techniques. Int J Comput Appl 75(6)
16. Ahmad T, Aziz MN (2019) Data preprocessing and feature selection for machine learning intrusion detection systems. ICIC Express Lett 13(2):93–101
17. Ham YJ, Lee HW (2014) Big Data preprocessing mechanism for analytics of mobile web log. Int J Adv Soft Comput Its Appl 6(1)
18. Sriram S, Vinayakumar R, Alazab M, Soman KP (2020) Network flow based IoT botnet attack detection using deep learning. In: IEEE INFOCOM conference on computer communications workshops, pp 189–194
19. Su S, Sun Y, Gao X, Qiu J, Tian Z (2019) A correlation-change based feature selection method for IoT equipment anomaly detection. Appl Sci 9(3):437
20. Alsamiri J, Alsubhi K (2019) Internet of Things cyber attacks detection using machine learning. Int J Adv Comput Sci Appl 10(12):627–634
21. Sarhan M, Layeghy S, Portmann M (2021) An explainable machine learning-based network intrusion detection system for enabling generalisability in securing IoT networks
22. Iqbal H, Naaz S (2019) Wireshark as a tool for detection of various LAN attacks. Int J Comput Sci Eng 7(5):833–837
23. Hussain F, Abbas SG, Shah GA, Pires IM, Fayyaz UU, Shahzad F, Garcia NM, Zdravevski E (2021) A framework for malicious traffic detection in IoT healthcare environment. Sensors 21(9):3025

24. Rizal R, Riadi I, Prayudi Y (2018) Network forensics for detecting flooding attack on internet of things (IoT) device. Int J Cyber-Secur Digit Forensics 7(4):382–390
25. Moustafa N (2019) Ton-iot datasets. https://doi.org/10.21227/fesz-dm97

A Survey on Web 3.0 Security Issues and Financial Supply Chain Risk Based on Neural Networks and Blockchain

Praveen Singh, Rishika Garg, and Preeti Nagrath

Abstract With the development of Web 3.0, issues of privacy and ownership have arisen. Supply chain finance (SCF) has given rise to a risk of credit for mid-cap and small-cap businesses. Regulating the risk involved is a major assignment, and it should be minimized as it is applicable in all the fields of finance in most situations. Web 3.0 privacy and ownership concerns are to be addressed for its advancement. This study focuses on Web 3.0 security issues and SCF and uses a fuzzy neural network (FNN) and blockchain to study the risk of credit for small-cap and mid-cap businesses. The FNN algorithm is devised to appraise the risk involved in SCF and develop the supply chain's efficiency.

Keywords Blockchain · Supply chain · Fuzzy neural network (FNN) · Supply chain finance (SCF) · Web 3.0

1 Introduction

The financial institutions are trailblazing the way forward. A country's economy majorly depends upon this sector [1]. However, several unreliable situations exist in the supply chain (SC) [2]. SCF is a form of financial transaction, and it has several phases, such as commercial finance and bill discounting. Against these phenomena, SCF is burgeoning and expanding.

Many researchers found SC financial risk interesting [3]. Researchers believe that SCF can be beneficial in the field of finance [4]. During the starting of the research, researchers analyzed the several risks that were with SCF models. As the research advanced, an exhaustive examination of the risk was conducted by the researchers of

P. Singh (✉)
Department of Electronics and Communication Engineering, Bharati Vidyapeeth's College of Engineering, New Delhi, Delhi, India
e-mail: praveensingh3129@gmail.com

R. Garg · P. Nagrath
Department of Computer Science and Engineering, Bharati Vidyapeeth's College of Engineering, New Delhi, Delhi, India

P. Chatterjee et al. (eds.), *Computational Intelligence for Engineering and Management Applications*, Lecture Notes in Electrical Engineering 984,
https://doi.org/10.1007/978-981-19-8493-8_41

the unified SCF from a broader angle, and in the past few years, researchers analyzed the risk from an industrial viewpoint [5]. To save the cost of the SC, its complexity should be reduced [6]. Bern concluded the halt for SCF from the viewpoint of small-cap and mid-cap enterprises loans [7]. The major elements that affect the SCF are simulated [8]. According to them, it is arduous to receive the credit due to bad credit support for small-cap and mid-cap businesses, so they planned a new financing model [9]. The main highlight of their model was that large-cap businesses or financial institutions would serve the demands of small-cap and mid-cap businesses that are difficult to finance [10].

Abrin describes SCF from the viewpoint of financial organization [11]. According to him, financial institutions, information sources, major businesses form the backbone of SCF [12]. In SC, the expenses of the businesses can be reduced efficiently by concentrating on financing in SC [13]. Data privacy and data sharing are the two major advantages of SCF [14]. Against these phenomena, SCF is burgeoning and expanding. Some institutions describe it as a financial institution that caters to the needs of financial services of businesses in the SC to help them maintain their business.

In contrast to the conventional financing system, SCF has effective credit and low expenses [15]. It is one of the key factors for the growth of SCF [16]. SCF has enhanced the efficiency of the banks that had difficulties in functioning and has also augmented the problems of financing that were faced by Small and Medium-sized Enterprises (SMEs) [17]. Despite this, there is a risk of credit with the devising of SCF [18]. Assessing the risk of credit of the SC should be devised upon the commercial banks to minimize financial risk. [19] The advancement of SCF has given birth to the risk of credit of small-cap and mid-cap businesses [20], and the large-cap businesses can cater to the demands of small-cap and mid-cap businesses [21].

The objective of this study is to highlight the model that will allow large-cap businesses or financial institutions to cater to the demands of small-cap and mid-cap businesses that are difficult to finance and are not able to be managed by the huge financial institutions directly. The security issues with Web 3.0 technology, which uses blockchain technology, are also addressed. This paper, using Fuzzy Neural Network algorithms (FNN), studies Web 3.0 security issues and the risk of credit for small-cap and mid-cap businesses.

2 Blockchain Technology and Web 3.0

Currently, research and advancements in the field of blockchain have developed the interest of people [22]. Blockchain is also becoming the building block and key behind the future of the Internet in the form of Web 3.0 [23]. As one of the fastest developing nations, India should grasp this opportunity, increase its investment in blockchain, harvest talents in universities, and develop the infrastructure for blockchain [24]. Its growth has been revolutionary; it has gained popularity and is currently being widely used throughout the world [25]. Blockchain uses multiple

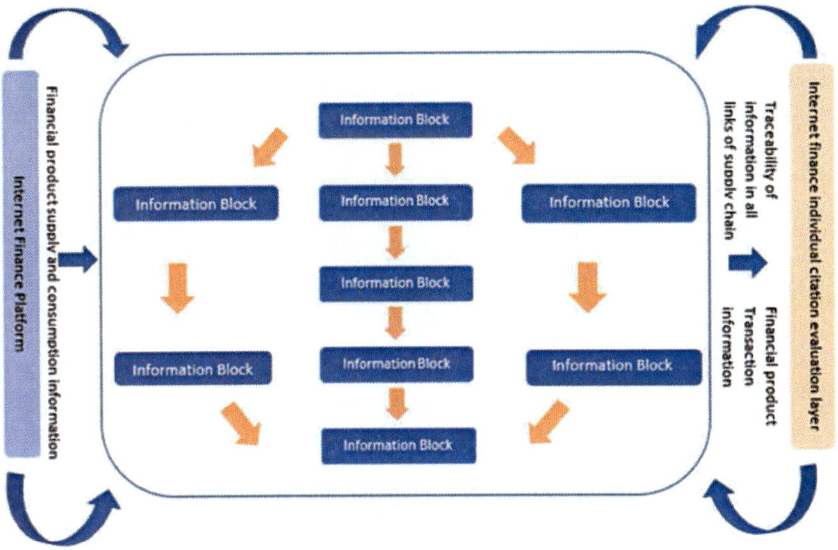

Fig. 1 Research on the SCF risk appraisal and information flow based on blockchain

data points or nodes and thereby reduces the dependence on a central system, and it makes blockchain durable preventing data loss [26]. Research on the appraisal of financial risk of the supply chain is shown in Fig. 1.

Presently, the blockchain version is between 2.0 and 3.0. Blockchain is currently transitioning toward 3.0. If we want to reach it, it demands the global adoption of blockchain and solving various constraints [27]. There are various issues that develop with extensive industrial application. There are two major building blocks of blockchain: intra-block and inter-block [28].

The body contains information such as identity digits, cryptic code for the previous block, random float value, time stamp, and block hash cipher [29]. The technology that is used was first introduced in 2008. Blockchain is said to be a uniformly dispersed database [30]. It is decentralized, cannot be easily meddled with, detectable, and can be effectively managed [31].

3 Literature Review

This study was inspired by and attempts to extend the study by Wang in 2021, wherein they studied SME financing from the SCF viewpoint, focusing on the credit risks involved, and how blockchain and FNNs could be employed for the same. Their study established a credit risk assessment model which can be applied to Internet SCF. Drawing from the above, in our study, we study Web 3.0 security issues and the risk of credit for small-cap and mid-cap businesses [32].

Xie et al. proposed using neural networks and deep learning to process financial data and using stock indices and future prices to evaluate these in-depth learning models. They found that their research could potentially contribute toward building investment strategies and an automated investment model. However, their research could not be backed by real-life applications and remains a theoretical proposition due to a large number of unpredictable factors involved [33].

Sahoo and Pattanaik studied the application of the Internet of Things (IoT) and cloud computing in blockchain. They discussed the ubiquity of IoT and blockchain and concluded that the application of Artificial Intelligence (AI) in blockchain can help it achieve its full potential. This study was limited by its lack of attention to implementation methods [34].

Zhou et al. proposed a self-organizing FNN with an adaptive learning algorithm, which they called SOFNN-ALA, for nonlinear modeling and recognition with the intent of enhancing convergence speed, modeling operation, and comprehensive performance [1].

Kim and Henderson studied (mainly relational and structural) embeddedness to understand how the performance outcomes of focal firms are affected by the resource dependency of suppliers and customers along with the effects of quality and the network architecture of exchange relationships on economic behavior and outcomes. The limitations to their study were their surface study of relational embeddedness, sample selection (reflecting power imbalances), and data that was aggregated at the firm level (rather than the division level) [5].

As discussed in this section, previously done research paid attention to combining Artificial Intelligence with business-oriented strategizing and decision-making for automation and practicality. However, most of them did not account for the risks involved, especially since there are several volatile factors involved, thereby rendering their research highly theoretical. In this study, we study these risks and highlight a model to allow large-cap businesses or financial institutions to cater to the demands of small- and mid-cap businesses that are difficult to finance and be managed directly by large financial bodies, taking into account the concerns that may entail.

4 Assessment of Web 3.0 Security Issues and SCF Risk

The concept of Web 3.0 was originated in 2006 when John Markoff coined the term Web 3.0, and SCF originated in the 1980s when the motion of "SCF" received scholastic recognition [35]. Subsequently, the theoretical definition of SCF was put forward by other scholars [36]. Hofmann suggests that SCF is a new integrative constituting supply chain development, commerce, and engineering. Its intent is to create a valuable environment for the representatives and members of the supply chain. A supply chain is weak if there is even a single vulnerable element in the SC. Therefore, the failure risk is more in longer SC [37].

According to Lamoureux's definition, in today's corporate ecosystem, where the power to take the decision is held by a few people, and they dominate the entire industry, SCF is a revolution that provides systematic optimization of the availability of funds. Prior to the adoption of this technology, the enterprises used to sell the bill to an intermediatory before the bill expires [38]. By the 1990s, financial institutions and banks developed methods to grant credit to enterprises by trade financing from bill discounting. Based on deposits, advance payments, and accounts receivable in the transaction of commodities, banks devised structured short-term financing tools [39]. Figure 2 shows the construction framework of the system.

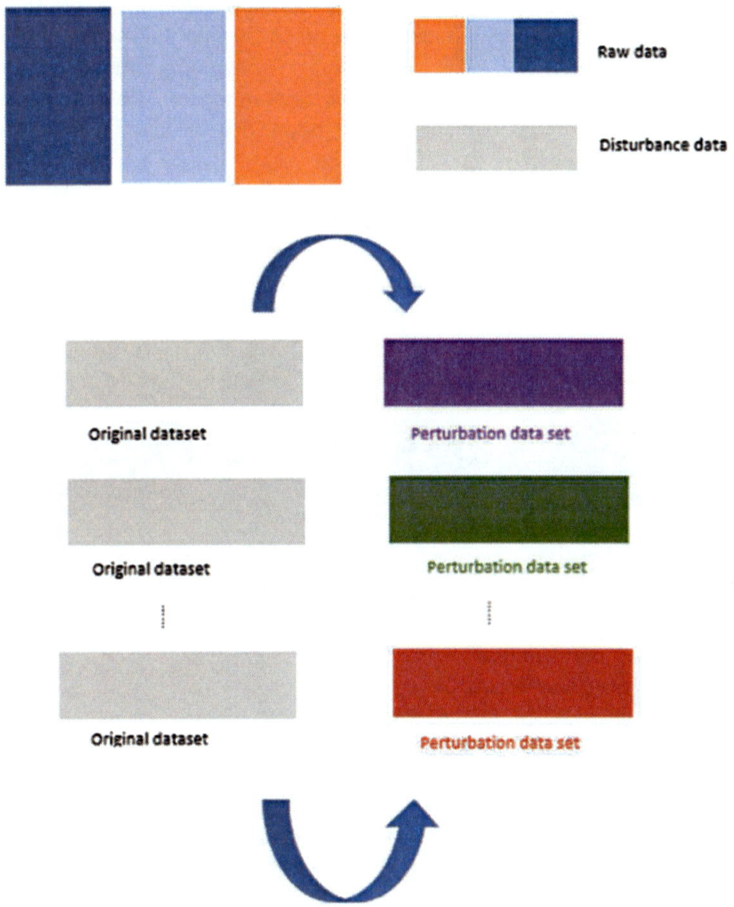

Fig. 2 Development framework of SCF risk indicator system

The financial appraisal of the risk of commercial businesses is based on the benchmark of the economy of businesses, and the management of small-cap and mid-cap financial threats is inspected thoroughly [40]. It can be introduced with chain economics but with some shortcomings. There are several advantages of SC.

Web 3.0 is to be launched using blockchain, which will lead to more privacy and ownership to the consumers and take power from top executives running the Internet. However, there are problems. Too much privacy gives rise to crimes. When things become completely private and anonymous, cybercrime can increase. If any scam occurs in Web 3.0, then who would be held liable and who would take the responsibility?

Billions of neurons make up the human brain, and based on this, the fuzzy set theory was devised and laid the fundamentals for fuzzy neural network (FNN). The leverage of a neural network is that it can improve by attaining and acquiring information through experience and training. In the training process, when the input data is provided, it is converted into important data followed by the learning process using a neural network (Fig. 3).

In control systems, biologic neurons are implemented using artificial neurons. Taking $\alpha 1$, $\alpha 2$, $\alpha 3$, and so on (n inputs) with an associated weight of $\omega 1$, $\omega 2$, $\omega 3$, and so on, respectively. Based on the data, we apply the training mechanism which uses a function as shown in Fig. 4. The neural function implemented in this paper is $\Sigma \alpha_i \, \omega_i$.

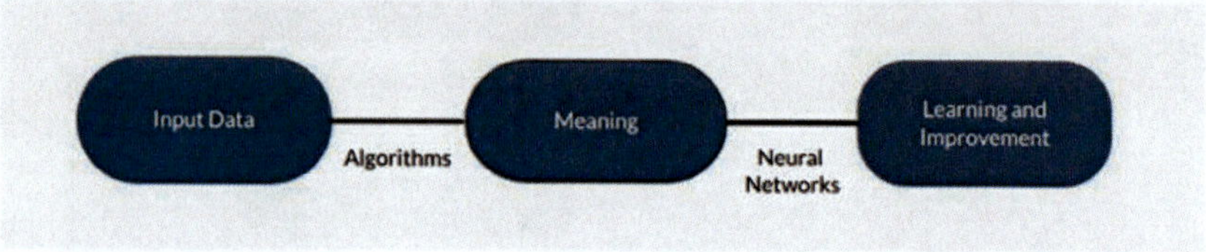

Fig. 3 Research on the appraisal of financial risk of supply chain and training process of the machine

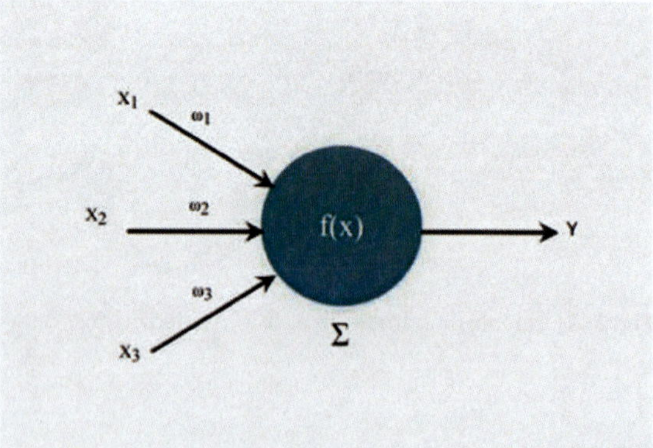

Fig. 4 Construction framework of the system and the function used as the input for the machine

The output generated is compared with the threshold to evaluate it with the most accurate value and train the machines.

Human understanding and intentions are measured in uncertainty and discrete values such as 0 and 1 seldom come into the picture. For representing the intensity and degree of human intentions, discrete values flounder. This is why the fuzzy set theory, which handles uncertainty and represents the intensity of a situation using a degree, was introduced.

Mathematics has been rediscovered through the advancement of the fuzzy theory [41].

Fuzzy neurons are the fundamental block of FNN. The calculated value is converted into a fuzzy output value by the fuzzy neuron.

$$yt = Ft\theta t + vt, vt \sim N[0, Vt] \tag{1}$$

Fuzzy neurons have a very interesting property and give opposite results, though they are officially akin to former neurons. They convert fuzzy input into certain signals.

$$\theta t = \theta t - 1 + \omega r, \omega r \sim N[0, Wt] \tag{2}$$

Because the scope of input varies, the NN learning algorithm needs to evaluate the domain based on the input. The efficiency cannot be determined precisely due to the difference in these results, which make the domain of these NN complicated and difficult to evaluate.

$$(\theta 0|D0) \sim N[m0, C0] \tag{3}$$

To overcome this complication, it standardizes all the data in the range. It is depicted by

$$Dt = \{yt, Ft, Dt - 1\} \tag{4}$$

$$(\theta t|Dt - 1) - N[at, Rt] \tag{5}$$

The flexible alteration of the NN has great upside, and still, people find it complicated.

$$(YT, Dt - 1) \sim N[ft, Qt] \tag{6}$$

In this research paper, we have discussed dynamic FNN. It can be depicted by

$$X = 0, 1 \text{ and } X * x[2U(x) - c - v] - U(x)x = 0+ \tag{7}$$

$$G(x) = [2U(x) - c - v] - U(x) \tag{8}$$

Each node of this layer has the main job to serve as the anterior of respective fuzzy rules. Each of the nodes depicts

$$\left\{ \begin{array}{l} \text{Max } G(x) = 2U(0) - c - V \\ \text{Min } G(x) = U(1) - c - v \end{array} \right\} \quad (9)$$

$$[2U(x-) - c - v] - U(x-)x = 0; \quad (10)$$

As one of the biggest economies in the world, India should increase its investment in blockchain and, instill knowledge in students, develop infrastructures, and expertise heights in the universal blockchain. With the FRY method, the speed of convergence of FNN has drastically increased.

5 Case Study of Assessment of SCF

Since the understanding of the input data is provided for training the system is the only source of information for enhancing the control system, the quality of the sample data has a crucial role in the end performance of the system and the test results.

There are many checks, and it is crucial to discard the checks with higher interrelationships. SPSS 25 is used in the article to handle the interrelationship between indicators in the system. In the 96% certainty test cases, the checks with the factual interrelationship coefficient more than or equivalent to 0.7 are eliminated. Furthermore, for enterprises with comparatively high credit levels, the grade is 5 points, so eliminating it is not mentioned in this research paper. Figure 5 depicts the training analysis.

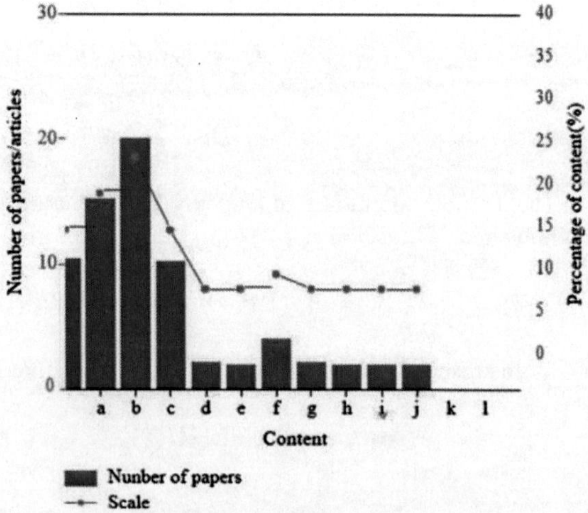

Fig. 5 Training analysis of FNN and the plot of the number of papers versus content

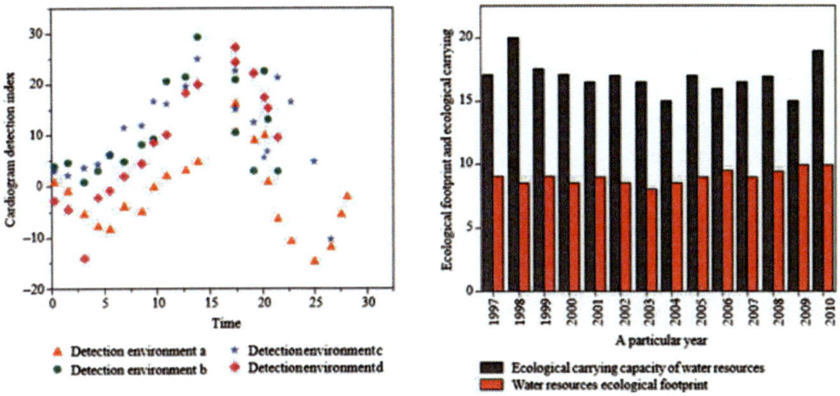

Fig. 6 Dissemination of FNN test results and efficiency of distinct index systems

First, select the generated five thousand samples in the model and separate them into two groups of three thousand three hundred training samples and one thousand seven hundred test samples. The dissemination of the FNN test result is shown in Fig. 6.

Ultimately, this article formulates a financial risk appraisal model pertinent to web SCF. The FNN model is trained by the training set consisting of three thousand three hundred samples followed by the learning of the FNN through the categorization of the training set. The analogy of the efficiency of different index systems is also depicted in Fig. 6. After that, it is evaluated against the experimental results of the conventional SCF risk of credit appraisal index system.

In the conventional system, there are no web-based SCF credit risk abstraction systems for the buying and selling circumstances. The first twenty-four 3 level indicators comprise the conventional system. Still, 0 is used to indicate overdue loans and 1 to indicate that the amount of loan which is overdue is zero. Differentiation of the consequences of FNN in the reset of data is depicted in Fig. 7. Differentiation of the consequence of various approaches in data reset is depicted in Fig. 8.

In order to defeat the sturdiness of the design, ten abstractions are performed in this paper. For training, the abstraction is an aggregation of ten difference three thousand three hundred training sets and one thousand seven hundred test sets. There is no web-based SCF risk abstraction in the conventional SCF credit risk system. By the 1990s, to provide financial loans to enterprises, banks developed trade crediting for unpaid invoices which are due to be paid at a future date are sold to a financier. Based on debits and account credits in the transaction of items, financial institutions implemented financing instruments. There is a single indicator for total profit and buying selling quantity of the SC in the analysis.

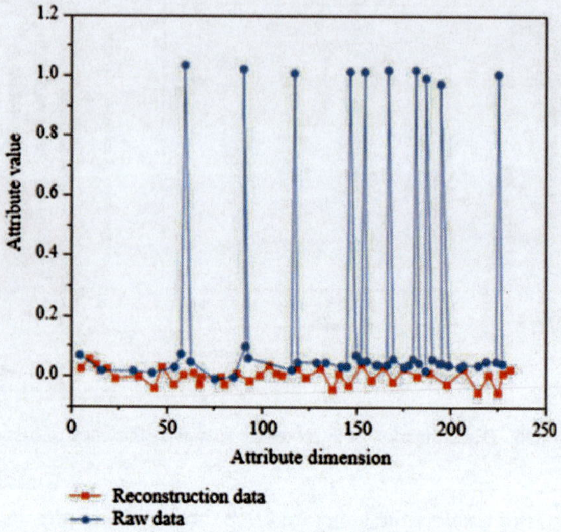

Fig. 7 Differentiation of the consequences of FNN in the reset of data and differentiation of consequence of the different approaches in data reset

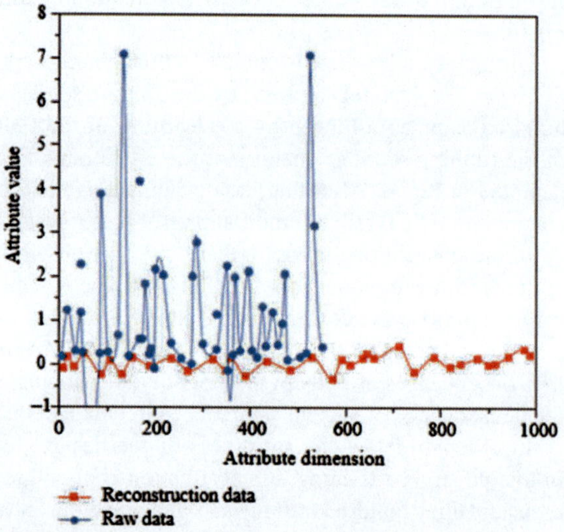

Fig. 8 Differentiation of the consequences of FNN in the reset of data and differentiation of consequence of the different approaches in data reset

6 Conclusion

This article includes concepts of Web 3.0 security issues and theoretical analysis of the potential threats and opportunities of Web 3.0. SCF risk of credit is discussed, along with regulating the risk involved and minimizing it using FNN. In all the aspects of businesses and financial institutions, credit risk has a very significant role; thus, appraisal of the credit risk of SCF on these financial institutions and banks is crucial. With the advancements of SCF, SMEs' credit risk has arisen. Maintaining the risk is crucial for SCF's efficiency, and it effectively applies to the aspects of all the financial institutions. This article addresses Web 3.0 security issues and SCF and uses a fuzzy neural network (FNN) and blockchain to study the risk of credit for small-cap and mid-cap businesses. The FNN algorithm is devised to appraise the risk involved in SCF and develop the supply chain's efficiency.

Future work includes exploring more methods of studying credit risks, studying more applications of SCF as well as extending the research to cover objectives like risk attenuation and possibly, risk elimination.

References

1. Zhou H, Zhao H, Zhang Y (2020) Nonlinear system modeling using self-organizing fuzzy neural networks for industrial applications. Appl Intell 50(5):1657–1672. https://doi.org/10.1007/S10489-020-01645-Z
2. Zhao S, Zhu Q (2015) Remanufacturing supply chain coordination under the stochastic remanufacturability rate and the random demand. Ann Oper Res 257(1):661–695. https://doi.org/10.1007/S10479-015-2021-3
3. Jiang YG, Dai Q, Mei T, Rui Y, Chang SF (2015) Super fast event recognition in Internet videos. IEEE Trans Multimedia 17(8):1174–1186. https://doi.org/10.1109/TMM.2015.2436813
4. Zhang M, Tse YK, Dai J, Chan HK (2019) Examining green supply chain management and financial performance: roles of social control and environmental dynamism. IEEE Trans Eng Manage 66(1):20–34. https://doi.org/10.1109/TEM.2017.2752006
5. Kim YH, Henderson D (2015) Financial benefits and risks of dependency in triadic supply chain relationships. J Oper Manag 36:115–129. https://doi.org/10.1016/J.JOM.2015.04.001
6. Choi TY, Krause DR (2006) The supply base and its complexity: Implications for transaction costs, risks, responsiveness, and innovation. J Oper Manag 24(5):637–652. https://doi.org/10.1016/J.JOM.2005.07.002
7. Sheng Y, Lewis FL, Zeng Z, Huang T (2020) Lagrange stability and finite-time stabilization of fuzzy memristive neural networks with hybrid time-varying delays. IEEE Trans Cybern 50(7):2959–2970. https://doi.org/10.1109/TCYB.2019.2912890
8. Yangyong G, Juan W (2020) Modeling of false information on microblog with block matching and fuzzy neural network. 32(2). https://doi.org/10.1142/S0129183121500194
9. Xie W, Zhu Q (2018) Input-to-state stability of stochastic nonlinear fuzzy Cohen–Grossberg neural networks with the event-triggered control 93(9):2043–2052. https://doi.org/10.1080/00207179.2018.1540887
10. Wu X, Han H, Liu Z, Qiao J (2020) Data-knowledge-based fuzzy neural network for nonlinear system identification. IEEE Trans Fuzzy Syst 28(9):2209–2221. https://doi.org/10.1109/TFUZZ.2019.2931870
11. Wang RM, Zhang YN, Chen YQ, Chen X, Xi L (2020) Fuzzy neural network-based chaos synchronization for a class of fractional-order chaotic systems: an adaptive sliding mode control approach. Nonlinear Dyn 100(2):1275–1287. https://doi.org/10.1007/S11071-020-05574-X
12. Thompson KM, Kalkowska DA (2021) Potential future use, costs, and value of poliovirus vaccines. Risk Anal 41(2):349–363. https://doi.org/10.1111/RISA.13557
13. Stachowski M, Fiebig A, Rauber T (2021) Autotuning based on frequency scaling toward energy efficiency of blockchain algorithms on graphics processing units. J Supercomputing 77(1):263–291. https://doi.org/10.1007/S11227-020-03263-5/FIGURES/12
14. Shabani M (2019) Blockchain-based platforms for genomic data sharing: a de-centralized approach in response to the governance problems? J Am Med Inform Assoc 26(1):76–80. https://doi.org/10.1093/JAMIA/OCY149

15. Rubin J, Ottosen A, Ghazieh A, Fournier-Caruana J, Ntow AK, Gonzalez AR (2017) Managing the planned cessation of a global supply market: lessons learned from the global cessation of the trivalent oral poliovirus vaccine market. J Infectious Dis 216(suppl_1):S40–S45. https://doi.org/10.1093/INFDIS/JIW571
16. Pavão LV, Pozo C, Costa CBB, Ravagnani MASS, Jiménez L (2017) Financial risks management of heat exchanger networks under uncertain utility costs via multi-objective optimization. Energy 139:98–117. https://doi.org/10.1016/J.ENERGY.2017.07.153
17. Nasir V, Nourian S, Avramidis S, Cool J (2019) Stress wave evaluation for predicting the properties of thermally modified wood using neuro-fuzzy and neural network modeling. Holzforschung 73(9):827–838. https://doi.org/10.1515/HF-2018-0289/MACHINEREADABLECITATION/RIS
18. Monasterolo I, Battiston S, Janetos AC, Zheng Z (2017) Vulnerable yet relevant: the two dimensions of climate-related financial disclosure. Clim Change 145(3–4):495–507. https://doi.org/10.1007/S10584-017-2095-9
19. Liu L, Du M, Ma X (2020) Blockchain-based fair and secure electronic double auction protocol. IEEE Intell Syst 35(3):31–40. https://doi.org/10.1109/MIS.2020.2977896
20. Li H, Li C, Ouyang D, Nguang SK, He Z (2021) Observer-based dissipativity control for T-S fuzzy neural networks with distributed time-varying delays. IEEE Trans Cybern 51(11):5248–5258. https://doi.org/10.1109/TCYB.2020.2977682
21. Kusi-Sarpong S, Gupta H, Sarkis J (2018) A supply chain sustainability innovation framework and evaluation methodology 57(7):1990–2008. https://doi.org/10.1080/00207543.2018.1518607
22. Kees MC, Bandoni JA, Moreno MS (2019) An optimization model for managing the drug logistics process in a public hospital supply chain integrating physical and economic flows. Ind Eng Chem Res 58(9):3767–3781. https://doi.org/10.1021/ACS.IECR.8B03968/SUPPL_FILE/IE8B03968_SI_001.XLSX
23. Huh JH, Seo K (2018) Blockchain-based mobile fingerprint verification and automatic log-in platform for future computing. J Supercomputing 75(6):3123–3139. https://doi.org/10.1007/S11227-018-2496-1
24. Hanson-Heine MWD, Ashmore AP (2020) Computational chemistry experiments performed directly on a blockchain virtual computer. Chem Sci 11(18):4644–4647. https://doi.org/10.1039/D0SC01523G
25. Hamdaoui B, Alkalbani M, Znati T, Rayes A (2020) Unleashing the power of participatory IoT with blockchains for increased safety and situation awareness of smart cities. IEEE Netw 34(2):202–209. https://doi.org/10.1109/MNET.001.1900253
26. Wang R, Wu Y (2021) Application of blockchain technology in supply chain finance of Beibu Gulf Region. Math Prob Eng 2021. https://doi.org/10.1155/2021/5556424
27. Ghadge A, Jena SK, Kamble S, Misra D, Tiwari MK (2020) Impact of financial risk on supply chains: a manufacturer-supplier relational perspective. 59(23):7090–7105. https://doi.org/10.1080/00207543.2020.1834638
28. Higham LE et al (2018) Effects of financial incentives and cessation of thinning on prevalence of Campylobacter: a longitudinal monitoring study on commercial broiler farms in the UK. Vet Rec 183(19):595–595. https://doi.org/10.1136/VR.104823
29. Das BS, Khatua KK (2022) Prediction of flow in non-prismatic compound channels using adaptive neuro-fuzzy inference system. 2017, Accessed 11 Feb 2022 [Online]. Available: http://dspace.nitrkl.ac.in:8080/dspace/handle/2080/2848
30. d'Amore F, Sunny N, Iruretagoyena D, Bezzo F, Shah N (2019) European supply chains for carbon capture, transport and sequestration, with uncertainties in geological storage capacity: insights from economic optimisation. Comput Chem Eng 129:106521. https://doi.org/10.1016/J.COMPCHEMENG.2019.106521
31. Han JH, Lee IB (2013) A comprehensive infrastructure assessment model for carbon capture and storage responding to climate change under uncertainty. Ind Eng Chem Res 52(10):3805–3815. https://doi.org/10.1021/IE301451E/SUPPL_FILE/IE301451E_SI_001.PDF

32. Wang Y (2021) Research on supply chain financial risk assessment based on blockchain and fuzzy neural networks. Wirel Commun Mob Comput 2021. https://doi.org/10.1155/2021/5565980
33. Xie M, Li H, Zhao Y (2020) Blockchain financial investment based on deep learning network algorithm. J Comput Appl Math 372. https://doi.org/10.1016/J.CAM.2020.112723
34. (PDF) Converging block chain and nextgeneration artificial intelligence technologies | IAEME Publication - Academia.edu. https://www.academia.edu/45632205/CONVERGING_BLOCK_CHAIN_AND_NEXTGENERATION_ARTIFICIAL_INTELLIGENCE_TECHNOLOGIES. Accessed 17 Feb 2022
35. Chen SG, Lin FJ, Liang CH, Liao CH (2021) Intelligent maximum power factor searching control using recurrent Chebyshev fuzzy neural network current angle controller for SynRM drive system. IEEE Trans Power Electron 36(3):3496–3511. https://doi.org/10.1109/TPEL.2020.3016709
36. Lin FJ, Huang MS, Chen SG, Hsu CW (2019) Intelligent maximum torque per ampere tracking control of synchronous reluctance motor using recurrent Legendre fuzzy neural network. IEEE Trans Power Electron 34(12):12080–12094. https://doi.org/10.1109/TPEL.2019.2906664
37. Gurtu A, Johny J (2021) Supply chain risk management: literature review. Risks 9(1):1–16. https://doi.org/10.3390/RISKS9010016
38. Chen CH (2020) A cell probe-based method for vehicle speed estimation. IEICE Trans Fundam Electron Commun Comput Sci E103A(1):265–267. https://doi.org/10.1587/TRANSFUN.2019TSL0001
39. Barrette J, Thiffault E, Achim A, Junginger M, Pothier D, de Grandpré L (2017) A financial analysis of the potential of dead trees from the boreal forest of eastern Canada to serve as feedstock for wood pellet export. Appl Energy 198:410–425. https://doi.org/10.1016/J.APENERGY.2017.03.013
40. Azimi S, Azhdary Moghaddam M, Hashemi Monfared SA (2019) Prediction of annual drinking water quality reduction based on groundwater resource index using the artificial neural network and fuzzy clustering. J Contam Hydrol 220:6–17. https://doi.org/10.1016/J.JCONHYD.2018.10.010
41. McCook A (2016) Duke fraud case highlights financial risks for universities. Science 353(6303):977–978. https://doi.org/10.1126/SCIENCE.353.6303.977/ASSET/F5818234-834A-48C7-BD0E-ACD35743E834/ASSETS/GRAPHIC/353_977_F1.JPEG

Computational Intelligence in Special Applications

Productive Inference of Convolutional Neural Networks Using Filter Pruning Framework

Shirin Bhanu Koduri and Loshma Gunisetti

Abstract Deep neural networks have shown phenomenal performance in many domains including computer vision, speech recognition, and self-driving cars in recent years. Deep learning model's high performance normally comes at the cost of computation time and significant size of the model. These factors ultimately become a bottleneck for the deployment of deep learning models on battery and memory constrained devices, for example embedded systems or mobile phones. Over the past few years network acceleration is a burning topic. To address this task of compressing deep learning models, in the past few years many researchers have come up with compression techniques like pruning. The authors, in this paper, have tried to implement the filter pruning technique on VGG 16 architecture by using a clustering methodology to compress and deploy the model on resource-constrained devices like smartphones. The authors have also aimed at improving the inference time of the model for fast and considerably accurate predictions.

Keywords Deep learning · Deep neural networks · VGG16 · Filter pruning · Convolutional neural network (CNN) · Inference

1 Introduction

Today's world is governed by Machine Learning and Artificial Intelligence (AI). The living standards of the human race have shown exponential growth recently due to the influence of Machine Learning and AI. Artificial Intelligence finds its scope in every field such as medicine, defense, and economy and has also spread its roots in our daily life applications like Netflix suggestions, virtual Chatbots, Google assistant, Alexa, and whatnot. With the help of reinforcement learning, machines

S. B. Koduri (✉) · L. Gunisetti
Sri Vasavi Engineering College, Tadepalligudem, India
e-mail: shirin.koduri@srivasaviengg.ac.in

L. Gunisetti
e-mail: loshma@srivasaviengg.ac.in

P. Chatterjee et al. (eds.), *Computational Intelligence for Engineering and Management Applications*, Lecture Notes in Electrical Engineering 984,
https://doi.org/10.1007/978-981-19-8493-8_42

have become capable enough to perform the tasks at a superhuman level and also unlock new possibilities of solving a certain problem that even humans could not figure out.

In few years ago, Machine Learning and deep learning have exhibited outstanding development in Stock Market Prediction, Weather Forecasting, Natural Language Processing, Audio Processing and Computer Vision. Nowadays, research in the computer vision field is given huge importance. Many research projects are being developed in computer vision to solve complex tasks like surgical operations in medicine, self-driven vehicles, industrial automation, security, and surveillance where timing and precision matters a lot. Using computer vision in such complex real-time applications helps reduce human efforts and also improve the performance significantly at the same time.

The foundation of machine vision actually began in the late 1960s in academia which were groundbreaking research in Artificial Intelligence. In the early 1970s, research formed robust foundations for various machine vision algorithms that are present now such as polyhedral modeling, non-polyhedral modeling, and feature extraction from images, etc. [1]. But due to the lack of suitable hardware, graphic support, and enough computing power, it didn't show much significance. Current technological advancement has reversed that path. From 2010 (beyond), an improvement in deep learning techniques and technology [2] was observed. With the development in deep learning, it is possible to program supercomputers to train themselves, self-improvement over a time period, and providing such capabilities as online applications, viz. cloud-based applications.

Moreover, due to improved technological progress in recent years, Artificial Intelligence has also proved to outperform human-level performance in some of the applications. For, instance on January 14, 2011, IBM Watson, outplayed two all-time best Jeopardy champions, at the popular television quiz show Jeopardy [3]. IBM Watson beats them both in real time in live competition with no tricks. Even Google has started working on its AI called Google DeepMind back in 2010 which now has become a driving force of Google to solve complex real-life problems. DeepMind using Genetic Algorithms has figured out how to defeat humans in Quake III Arena's Capture the Flag matches. One of the most notable triumphs of DeepMind's AI is its capability to traverse through a city without a map [4]. The market in AI is predicted to explode in the near future. It is possible to grow from \$643.7 million at the current time, to \$36 billion by 2025, says Tractica a market research firm [5]. From all the above discussion, it can be said that ML and AI are getting immense popularity nowadays. Imagine having such a remarkable superpower that can transform the human lives and bring the impossible into reality right within your hands. If such a thing ever happens AI would not only be a revolution, but it will be more of an evolution. Though the thought of AI influencing every walk of human life seems very interesting but in reality even today with much technological advancement, this seems to be impractical. Deploying computer vision models and other powerful AI algorithms everywhere irrespective of the device is not promising yet today.

With the progression in deep learning, the CNN models state is attaining more accuracy but this advancement certainly comes with a price. Nowadays, deep learning models require a substantial amount of power, computation, and memory which restricts its applications only for high-end devices like supercomputers and computers with good GPU support. Deploying bulky and heavy models like VGG16, VGG19 on resource-constrained devices is practically impossible. For instance, VGG19 is a popular CNN model. It consumes the memory size of 549 MB and has 143,667,240 parameters [6]. This ultimately becomes a bottleneck in the conditions where it is required to deploy computer vision models on edge devices like smartphones, embedded systems, microprocessors, and also browsers with restricted computational resources. For current deep learning models, efficiency of energy has also become a key concern. Thus, such problems are now switching the attention of researchers to compress these bulky models and make them compatible with recourse constrained devices. The challenge we are to face in such a scenario is to compress these CNN models without compromising the model's accuracy. To deal with this challenge, in the past few years, researchers have come up with compression techniques like pruning. Pruning is needed to produce efficient models, viz. smaller in dimensions, faster at inference, enhanced memory efficiency, and high accuracy. Pruning is inspired by a biological activity of brain called synaptic pruning [7]. Synaptic pruning is the biological process where dendrite and axon decay completely and die off causing synapse rejection. Pruning will start close the time of birth and lasts up to mid-twenties. It is believed that synaptic pruning in the mammalian brain occurs to eliminate unnecessary neuronal structures from the brain. During the human brain development process, there is an important need to understand more complex structures, and simpler associations formed at childhood are supposed to be replaced by complex structures. The similar concepts of pruning are applied for the compression and acceleration of deep learning models by eliminating redundant weights or filters without compromising the model's accuracy.

There are different pruning techniques existing like weight pruning, neuron pruning, layer pruning, and filter pruning. When pruning weights are compared throughout the network, an obvious way of pruning without initiating sparsity not involving the usage of any specialized hardware and sparse libraries is filter pruning [8]. More focus is put on reduction of the computation cost of well-trained CNNs by filter pruning techniques.

We have tried to implement the filter pruning technique on VGG16 architecture by using the clustering methodology. Clustering is a technique to group similar data objects into a same group and dissimilar data objects into different groups. For each individual layer of VGG16 different clusters of filters were formed by using agglomerative clustering with cosine similarity check. The filters of maximum similarity are clustered together. Like this from each layer k number of clusters are formed. From each cluster, one filter is kept, and rest all are considered to be redundant. By doing so, similar filters are pruned from the architecture thereby reducing the size and number of trainable parameters. The resulting model is compressed from 63 to 4 MB, and also the trainable parameters are reduced from 14,987,722 to 732,898.

2 Related Work

Deep neural networks have shown phenomenal performance in many domains including computer vision, speech and audio recognition, and self-driving cars. Using neural networks in such precision-oriented applications has caused increase in many applications by its increasing power to model, subsequent analysis, and finally solve the problems in many diverse areas. However, as discussed they are very computationally expensive and memory-intensive which brings major challenges when it comes to deploying them on resource-limited environments. Network acceleration is now a burning topic for bringing the significant task for deploying such networks in real-time applications or in resource constraint devices. In this section, different approaches on how to compress DL models are discussed.

Li et al. [9] proposed a method of Hessian approximation-based incremental pruning which can be used to decrease the size of a deep neural network. The proposed method measures the "importance" of each weight in a deep neural network by using Hessian. Firstly, to measure "importance" of every weight and to evade computing the Hessian matrix, it was proposed to utilize the second moment in Adam optimizer. Next, an incremental model is proposed to prune a neural network phase after phase. An incremental method can adjust the remaining non-zero weights of the whole network after each pruning and helps boost the performance of the pruned network. Lastly, an automatically generated global threshold is applied to all the weights among all the layers, which achieves the inter-layer bit allocation automatically. In [10], Abdullah Salama et al. proposed an innovative pruning method that focuses on eliminating whole neurons and filters based on their relative L1-norm values as compared to the remaining network. This yields added compression and reduced parameters' redundancy. The resultant network is non-sparse, more compact, and needs no special infrastructure for deployment after pruning. Han et al. [11] recommended an iterative pruning technique to eliminate the redundancy in deep neural network models. The key intuition is that small-weight connectivity under a threshold ought to be rejected. In practice, *l1* or *l2* regularizations can be applied to reduce connectivity values. The main weakness of this approach is the loss of flexibility and universality which seems impractical in real world applications [12]. In this paper, Zhang Chiliang et al. proposed Channel Threshold-Weighting (T-Weighting) modules to prune insignificant feature channels at the inference level. As dynamic pruning is used, it is known as Dynamic Channel Pruning (DCP). Dynamic Channel Pruning contains the actual Convolutional Neural Network and an amount of "Channel T-Weighting" modules in some layers. The "Channel T-Weighting" module allocates weights to analogous unique channels and prunes zero weighted channels. Thus, pruning the redundant channels makes the Convolutional Neural Network accelerated, and that remaining channels when multiplied with weights enhance feature expression. In [13], Dong J. et al. worked on deeper Convolutional Neural Network trained to perform face recognition, then explored sparse neuron connections for compressing the density of network. In [16], Holistic Filter Pruning (HFP) uses a pruning loss that takes accurate pruning rates for the number of both parameters and multiplications into account. Ding et al. [17] proposed a novel Centripetal SGD (C-SGD) to make some filters identical, resulting in ideal redundancy patterns, as such filters become purely redundant due to their duplicates; hence, removing them does not harm the network.

It is proposed to use an activation-based weight significance criterion. This criterion approximates the impact that every weight causes in the activations of next layer neurons. It eliminates the weights that make the least contribution first. A precise and better technique formulated for pruning tightly connected neural network parameters.

From all the compression techniques discussed above, it can be said that individual weight pruning and neuron pruning seems to be time consuming, whereas layer pruning might have a chance of over pruning the model. Moreover, while working with CNN models filter pruning approach based on their similarity might yield significant results. The authors in this paper have discussed compressing VGG16 architecture by using a filter pruning technique aided by agglomerative clustering.

3 Methodology

A Filter Pruning Framework is proposed for efficient inference of Deep Neural Networks. The biggest limitation of CNN is that convolutions take a lot of time to process and are also RAM intensive. Another drawback observed in CNN architectures is that in the initial layers most of the filters are showing significant similarity with each other. This is one of the challenges that we have tried to resolve using our approach in reduction of the storage size and the amount of parameters of a bulky architecture.

Before working on the filter pruning approach, a detailed study on each layer of the VGG16 model was done. VGG16 model is trained on the > 500 MB ImageNet dataset. Table 1 shows the comparison of different CNN models in terms of features, parameter, FLOP, and accuracy. It can be observed that VGG16 shows an accuracy of 90.1% with about 138 million trainable parameters. Thus, it can be concluded that VGG16 is a huge model and cannot be deployed in resource-constrained devices.

In our project, we have trained the VGG16 model on CIFAR10 dataset, and the model size was 63 MB. By using agglomerative clustering methodology for filter pruning, we compressed this VGG16 model of 63–4 MB by maintaining a constant accuracy of 93.6%. By reducing the memory requirement of the model to such a

Table 1 Comparison of CNN models [6]

Model	Parameters	Size	Depth	Top-1 accuracy	Top-5 accuracy
Xception	22,910,480	88 MB	126	0.790	0.945
VGG16	138,357,544	528 MB	23	0.713	0.901
VGG19	143,667,240	549 MB	26	0.713	0.900

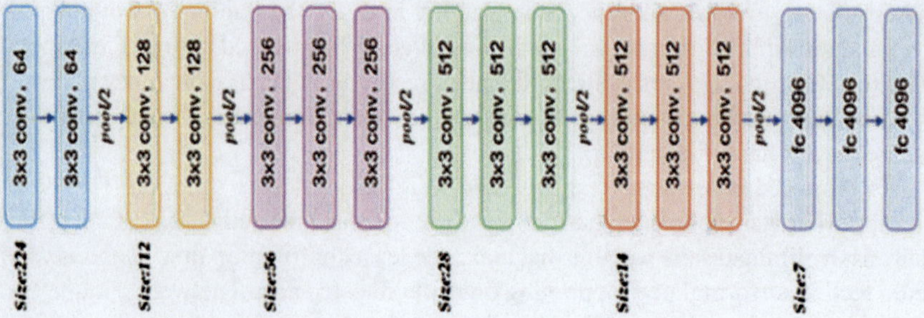

Fig. 1 VGG16 model architecture [14]

lower extent without compromising its accuracy, we can easily deploy it on devices having less RAM or processing capacity.

VGG16 model consists of 13 convolutional layers, five pooling layers, and three fully connected or dense layers. Figure 1 shows VGG16 architecture.

An algorithm that focuses on agglomerating and pruning convolutional filters is proposed. The objective is to determine the number of filters that are representative of the original filter using the agglomerative hierarchical clustering approach in each convolutional layer of VGG16. Agglomerative clustering is one of the most common categories of hierarchical-based clustering for grouping of objects in clusters based on the similar patterns. Achieving effective clustering of filters requires choosing a suitable similarity metric that expresses the inter-feature distances between filters. To evaluate the similarity of filters, we used cosine similarity metric. Number of appropriate agglomerative similarity testing algorithms can be applied like Euclidean distance in addition to Manhattan distance for localizing inessential filters. Clustering methodology of similar filters was developed on the basis of comparative evaluation.

Cosine similarity between two vectors [15] is given below:

$$\vec{a}.\vec{b} = \|\vec{a}\| \left\|\vec{b}\right\| \cos\theta \tag{1}$$

$$\cos\theta = \frac{\vec{a}.\vec{b}}{\|\vec{a}\| \left\|\vec{b}\right\|} \tag{2}$$

By using the above formula, similarity between two filters was determined. By means of grouping filters that are approximately identical in a given layer, algorithm 1 explains the redundant filter based pruning in detail. In general, pruning an enormous fraction of filters normally results in the performance decline of the model. In fact, it is witnessed that some convolutional layers are immensely delicate to pruning than others, and this must be taken into contemplation when pruning such layers in addition to models.

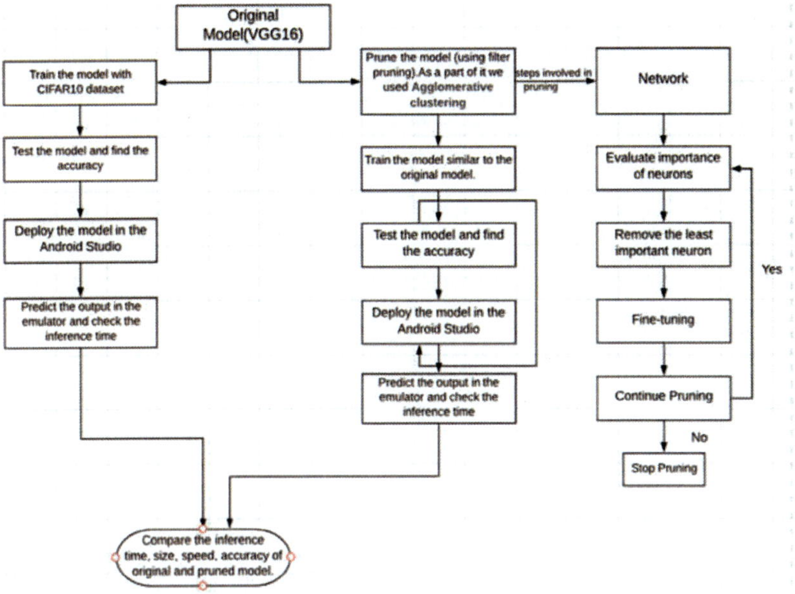

Fig. 2 Block diagram of the proposed compression technique

Figure 2 shows the block diagram of the proposed compression technique. The original VGG16 model is pruned by performing agglomerative clustering and thereby applying the configuration mask which extracts only required filters and discards redundant filters. The pruned model is fine-tuned/re-trained until maximum accuracy is achieved.

Algorithm 1

1. for ith convolutional layer of VGG16 (i ranges from 1 to 13) do
2. Reshape the filters to 1D vectors
3. Compare the reshaped filters based on cosine similarity
4. Form clusters of similar filters using agglomerative clustering.
5. Extract required filters from each cluster discard redundant filters using configuration mask.
6. Increment i and goto step 1
7. end

4 Experimental Results

By clustering, methodology for pruning the resulting model turned out to be of size 4 MB which is just 6.3% of the actual size of VGG16. Thus, the compression was

Table 2 Comparison of the original model and pruned VGG16 model

Layer number	Clusters	In shape	Out shape	Parameters	Parameters after pruning
conv_1	9	3	9	1728	243
conv_2	28	9	28	36,864	2,268
conv_3	41	28	41	73,728	10,332
conv_4	25	41	25	147,456	9,225
conv_5	47	25	47	294,912	10,575
conv_6	78	47	78	589,824	32,994
conv_7	51	78	51	589,294	35,802
conv_8	131	51	131	1,179,648	60,129
conv_9	132	131	132	2,359,296	155,628
conv_10	135	132	135	2,359,296	160,380
conv_11	131	135	131	2,359,296	159,165
conv_12	137	131	137	2,359,296	161,523
conv_13	117	137	117	2,359,296	144,261

done significantly without over pruning the model. Table 2 represents the comparison of parameters of the original model to that of the pruned model for all 13 convolutional layers of VGG16. The clusters column shows the number of clusters formed using agglomerative clustering for each layer. By comparing the parameters and parameters after pruning, it can be concluded that removing similar filters from each layer redundancy in VGG16 model can be reduced to a great extent thereby accelerating the model and reducing its inference time on edge devices.

The total number of trainable parameters in the VGG16 model trained on CIFAR10 dataset is around 14.9 million. By using our pruning approach, we scaled the parameters to 732,898 which makes the model run easily on smartphones without consuming much RAM.

In Fig. 3, the horizontal axis shows the layers of the VGG16 model, whereas the vertical axis represents the number of trainable parameters for the original and pruned model (Table 3).

Model Deployment: In order to check the inference time required for the efficient performance of the model, a smartphone emulator was developed on an android studio. A comparison between the inference of both original and pruned models was done. From Fig. 4, it can be observed that the inference time required for the original model is 18 s 140 ms, whereas the pruned model has inference time of just 3 s 842 ms. Thus, according to the objective of the project, the model is compressed and accelerated as required.

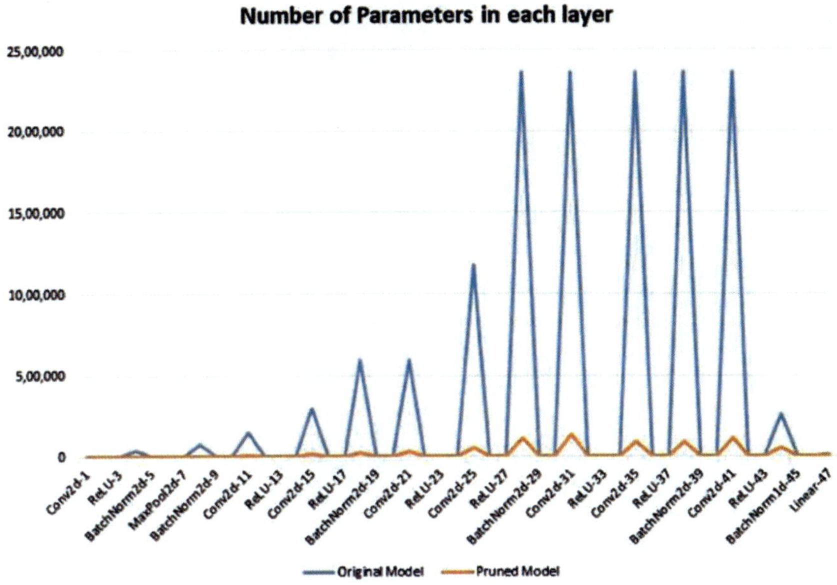

Fig. 3 Graphical representation of trainable parameters for original and compressed model

Table 3 Comparison of total trainable parameters and memory size

Parameters	Original model	Pruned model
Total params	14,987,772	732,898
Trainable params	14,987,772	732,898
Non-trainable params	0	0
Input size	0.01	0.01
Forward/backward pass size (MB)	6.57	1.20
Params size (MB)	57.17	2.80
Estimated total size (MB)	63.76	4.01

5 Conclusion

Filter pruning based on sampling approach for CNNs exhibits provable guarantees on the performance and size of the pruned neural network. The inference time of the pruned and original models are compared, and proposed pruned model has relatively low inference time, i.e., it gives away the result faster when compared to the original model (In our case inference times of original model and pruned model are around 18 secs and 2 secs, respectively). And also filter pruning method results in the reduction of the size of the model. (In our case it got reduced from 64 to 4 MB).

Fig. 4 Comparison of inference time for original and pruned VGG16 model

6 Future Scope

The proposed filter pruning algorithm can be used to compress and accelerate any CNN model. Therefore, it can be installed in handheld and wearable gadgets like smartphones, watches, and smart goggles. By doing so we can make deep learning models to work efficiently on any device irrespective of its computation power, RAM, and battery life thereby enabling AI to work in every aspect of human life.

References

1. Computer vision. Available at https://en.wikipedia.org/wiki/Computer_vision. Viewed 17 June 2020
2. Computer Vision History. Available at https://www.pulsarplatform.com/blog/2018/brief-history-computer-vision-vertical-ai-image-recognition/. Viewed 12 June 2020
3. Jeopardy! As a Modern Turing Test: Did Watson Really Win? Available at https://thebestschools.org/magazine/watson-computer-plays-jeopardy/. Viewed 18 June 2020

4. 5 Amazing Things Google's DeepMind AI Can Already Do. Available at https://www.makeuseof.com/tag/google-deepmind-ai/#:~:text=What%20Is%20DeepMind%3F,and%20deep%20reinforcement%20machine%20learning.&text=The%20deep%20reinforcement%20learning%20of,both%20research%20and%20applied%20contexts. Viewed 16 June 2020
5. The Age of Artificial Intelligence (3): The Future. Available at https://www.bbvaopenmind.com/en/science/research/the-age-of-artificial-intelligence-3-the-future/. Viewed 19 June 2020
6. Keras Applications. Available at https://keras.io/api/applications/. Viewed 18 June 2020
7. Pruning Deep Neural Networks. Available at https://towardsdatascience.com/pruning-deep-neural-network-56cae1ec5505. Viewed 18 June 2020
8. Li H, Kadav A, Durdanovic I, Samet H, Graf HP, Pruning Filters for Efficient ConvNets. Available at https://arxiv.org/abs/1608.08710.s
9. Li L, Li Z, Li Y, Kathariya B, Bhattacharyya S (2019) Incremental Deep Neural Network Pruning Based on Hessian Approximation. In: Proceedings of 2019 data compression conference (DCC), 26–29 March 2019 Snowbird, UT, USA, USA. Viewed 17 June 2020
10. Salama A, Ostapenko O, Klein T, Nabi M (2019) Prune your neurons blindly: neural network compression through structured class-blind pruning. In: Proceedings of ICASSP 2019—2019 IEEE international conference on acoustics, speech and signal processing (ICASSP), 12–17 May 2019, Brighton, United Kingdom, United Kingdom. Viewed 18 June 2020
11. Han S, Pool J, Tran J, Dally W (2015) Learning both weights and connections for efficient neural network. Proc Adv Neural Inf Process Syst 2015:1135–1143
12. Chiliang Z, Tao H, Yingda G, Zuochang Y (2019) Accelerating convolutional neural networks with dynamic channel pruning. In: Proceedings of 2019 data compression conference (DCC), 26–29 March 2019, Snowbird, UT, USA, USA
13. 5 Dong J, Zheng H, Lian L (2017) Activation-based weight significance criterion for pruning deep neural networks. In: Zhao Y, Kong X, Taubman D (eds) Image and graphics. ICIG 2017. Lecture Notes in Computer Science, vol 10667
14. VGG16 architecture. Available at https://www.google.com/url?sa=i&url=https%3A%2F%2Fwww.quora.com%2FWhat-is-the-VGG-neuralnetwork&psig=AOvVaw25jUr0AdbZlvYj2X7jblSG&ust=1592634700689000&source=images&cd=vfe&ved=0CAIQjRxqFwoTCIC-k-GgjeoCFQAAAAAdAAAAABAD
15. Machine Learning: Cosine similarity for vector space models (part III). Available at http://blog.christianperone.com/2013/09/machine-learning-cosine-similarity-for-vector-space-models-part-iii/
16. Enderich L, Timm F, Burgard W (2021) Holistic filter pruning for efficient deep neural networks. In: Proceedings of the IEEE/CVF winter conference on applications of computer vision
17. Ding X et al (2021) Manipulating identical filter redundancy for efficient pruning on deep and complicated CNN. arXiv preprint arXiv:2107.14444

Gate-Enhanced Multi-domain Dialog State Tracking for Task-Oriented Dialogue Systems

Changhong Yu, Chunhong Zhang, Zheng Hu, and Zhiqiang Zhan

Abstract Dialogue state tracking (DST) task is a core component of task-oriented dialogue systems. Recently, several open-vocabulary-based models were proposed in multi-domain setting, which relies on copy mechanism and slot gate classification modules. However, as the ontology gets more and more complex, it becomes challenging to fill values to slots with type-specific features fully utilized, apply appropriate operation, and tackle the over-trivial carryover problem which used to be neglected. To address the above issues, a hierarchical gate-enhanced DST framework called DSA-Gate DST is proposed in this paper. Domain activity prediction and semantic confirming recognition modules are introduced to track slots from different domains discriminately. Experiment results on multi-domain task-oriented dialog corpora are conducted to show that our model outperforms various baseline algorithms in widely various language settings. Meanwhile, we conduct a comprehensive analysis on the noisy annotation in the MultiWoZ dataset from multiple aspects to explore the potential reasons which limiting DST's performance.

Keywords Dialogue state tracking · Dialogue systems · Multi-domain · Scalability · Deep learning

1 Introduction

Task-oriented dialogue systems aim to help a human user accomplishes diverse tasks such as restaurant booking or travel planning through multi-turn conversations with a virtual assistant agent. *Dialogue State Tracker* (DST), one of the key components in dialogue systems, tracks the user's intents from utterances by filling values into

C. Yu · C. Zhang (✉) · Z. Zhan
Key Laboratory of Universal Wireless Communications, Ministry of Education, Beijing Univerity of Post and Telecommunications, Beijing 100876, China
e-mail: zhangch@bupt.edu.cn

Z. Hu
State Key Laboratory of Networking and Switching Technology, Beijing University of Post and Telecommunications, Beijing 100876, China

P. Chatterjee et al. (eds.), *Computational Intelligence for Engineering and Management Applications*, Lecture Notes in Electrical Engineering 984,
https://doi.org/10.1007/978-981-19-8493-8_43

slots, recorded as dialogue belief state in form of slot-value pairs (such as {*price*, *expensive*}), which would help the agent to decide next actions which match the user goal. More recently, the *multi-domain* dialogue systems, which span across mixed domains and tasks, have gained increasing research interests to scale traditional single-domain dialogue systems [1, 2] to a more complex ontology.

For mixed-domain conversations, traditional approaches generally follow the *fixed-vocabulary* (i.e., *predefined ontology*)-based approach in single-domain setting to track dialogue states. This approach formulates DST as a classification process by selecting a suitable value from all possible candidates in a fixed list for each predefined slot type. However, this approach faces difficulty in multi-domain settings, where the size of ontology can rapidly grow. For instance, the answers for certain domain-slot pairs like Taxi-departure would have innumerable potential choices in realistic scenarios. Besides, this approach is incapable to do inter-turn reason and transfer shared knowledge across domains. To mitigate the above issues, the *open-vocabulary-based* approach [3–5] is proposed, which use a copy-based Seq2Seq [6, 7] network to generate or copy answers of slots from context instead of computing the scores over the list of predefined candidates. As illustrated in Fig. 1, five domains are predefined for dialogue systems. Each domain has its nested slots (e.g., *name* to Attraction) to track. In this case, a *user* would first reserve a hotel and then book a taxi from an attraction to the hotel. Since *user* prefers hotels with Internet service, slot *Internet* in domain Hotel if filled with Boolean value *yes*. By contrast, slot *departure* in domain Taxi is filled with value *Express by holiday Inn Cambridge*, which can be directly copied from dialogue context or previous dialogue state.

It is also notable that, at each turn, not all slots are actively involved by the user, and different slots should not be operated by one single way. For instance, in Fig. 1, only the four colored slots need to be focused. Intuitively, it is imperative to determine specific operations for distinct slot types to build a scalable multi-domain dialogue system. Some models [8, 9] use a slot-level classifier to predict the type of slot values before tracking. Unfortunately, despite success on single-domain dialogue systems, almost all existing multi-domain DST algorithms remain imperfect. First, in multi-domain setting, the conversation topics users concern might change back and forth with dialogue ongoing. It means that, subject to the length of daily utterances, only a subset of predefined domains (Attraction, Restaurant and Train in Fig. 1 for example) would be activated at a turn, and all slot values belonging to the deactivated domains need not update. However, the previous works did not explicitly observe domain-level activity and treat slots from different domains indiscriminately. Second, slot-value pairs show significant differences in answering ways. For instance, the value of Hotel-name should be answered by a named entity extracted from dialogue context, while the candidate choices of Hotel-parking are limited in a few values like *yes*, *no*, *dontcare*. Some researches [8, 10, 11] artificially divided all predefined slots into two categories and performed independent DST algorithms separately. However, they fail to propose a unified methodology on this division strategy.

To deal with the above problems, we propose an open-vocabulary-based DST model with hierarchical Domain-Slot-Answer Gate (i.e., **DSA-Gate DST**). Grounded in the hierarchy nature of intention recognition process in multi-domain

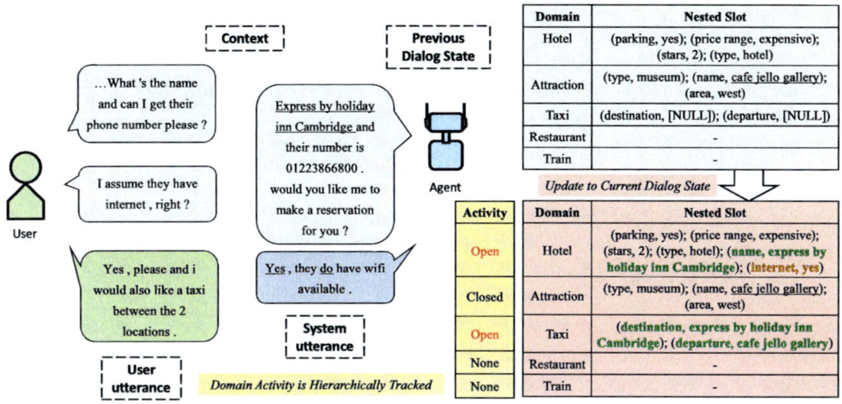

Fig. 1 Example fragment in MultiWoZ with dialog states updating across two turns in DSA-Gate DST. "Agent" means utterances from the system agent while "User" denotes utterances from a user. Dashed boxes make up the inputs of our model. Three kinds of slot-operations (black for CARRYOVER, yellow for UPDATE-CONFIRM, green for UPDATE-TEXT) at slot-level and domain-operations (NONE, CLOSED, OPEN) at domain-level are hierarchically performed to predict values. Names of unmentioned slot-value pairs are omitted here. Although there are 5 domains and 30 domain-specific slots predefined in MultiWoZ, only the operations of nested slots in OPEN domains (Hotel and Taxi in this scenario) are to be predicted

scenarios, DSA-Gate DST consists of three-level operation prediction classifiers: *Domain Gate, Slot Gate, and Answer Gate*. The *Domain Gate* predictor learns the activity of each domain at each turn to fully exploit the topic transition information across dialogue context. For domain predicted as activated, the slots in the domain are passed on to *Slot Gate* for operation prediction. If *Slot Gate* thinks a slot would update its value, it further sends the slot to *Answer Gate* so that each slot can be tracked distinctively with answering ways identified in real time. With the *Domain Gate* explicitly introduced, the trivial noisy information is filtered out from beginning in domain level. In this way, the efficiency of DST is improved since fewer slots would be processed by subsequent gates. Moreover, the unbalanced distribution over slot operations (to be depicted in Table 4) in natural conversation is alleviated, which would ameliorate the accuracy of slot gate.

We evaluate the effectiveness of our model on MultiWoZ dataset and achieve a joint accuracy rate of 51.71% on MultiWoZ2.0 and 53.14% on MultiWoZ2.1, which outperforms previous open-vocabulary-based DST baselines. Besides, we conduct an analysis on the *Domain and Slot-updating* annotation in MultiWoZ2.1 to explore the factors that limit DST's performance. We believe these works would be helpful for research communities. Contributions of this paper are summarized as follows:

1. Proposing a hierarchical dialogue state tracking framework, DSA-Gate DST, for multi-domain setting by explicitly exploiting the domain-level properties of conversation, and designing a semantic confirming predictor to supplement traditional copy mechanism for generating slot values.

2. Comprehensively analyzing the statistic properties of the MultiWOZ 2.1 dataset to provide insight on how they inspire the design consideration of our proposed model. Besides, discussing the potential effects of the annotation errors in MultiWOZ to DST algorithms.
3. Outperforming previous open-vocabulary-based DST baselines on MultiWOZ 2.0 and 2.1 corpus.

2 Related Work

Traditional DST research is restricted in single-domain setting, where the size of ontology (i.e., the number of domains, slots, and possible values predefined) is limited. Recently, dialogues spanning multiple domains attract the interest of task multi-domain DST. Challenges around *multi-domain DST* can be summarized as follows: (1) The predefined slot types are numerous and domain-specific; (2) traditional models fail to transfer knowledge between cross-domain slots; (3) it is challenging to generalize to new domains and unseen slot types. In a multi-domain setting, a dialog state is composed of several triplets as {domain, slot, value} which representing the goals a user informs. Earlier, researchers perform *predefined ontology* (i.e., *fixed vocabulary*)*-based DST*, where several slot-specific classifiers are trained to estimate the probability distributions of all predefined candidate values. It works on predefined slot values with a fully accessible ontology, but fails in real life where a slot could have an infinite number of candidate values. To mitigate this issue, some works [12, 13] proposed a slot-utterance match method to predict all slot-value labels in a non-parametric way.

Currently, the *open-vocabulary-based DST* approach is actively utilized. It treats DST as a machine reading comprehension (MRC) problem by generating answers to the question "what is the correct value of the slot." The copy-based Seq2Seq [6, 14] network is utilized to answer slots through generating or directly extracting values from dialogue context without a predefined ontology required. It makes models not rely on an accessible ontology and scalable to unseen slot-value pairs. The slot gate [4] (i.e., slot-operation classifier) is generally applied based on the idea of answering different slots with different tragedies. TRADE [5] designs a three-way slot-classifier to determine at each turn which method to use to fill each slot. Some other models [7, 8] add a CARRYOVER operation category to the slot gate to monitor slots whose values keep not updated (i.e., *carryover slot*). In this way, a portion of slot values from previous dialogue states could be "carry over" rather than repetitively generated. However, all existing models neglect or underutilize the information in domain-level activity. The intuition is that in real life, subject to the length of utterances, users usually only focus on a limited number of domains (i.e., conversation topics) per turn. Naturally, the nested slots that belong to uninterested domains could be directly tagged as *carryover* in domain level, while previous models treat all slots indiscriminately.

Meanwhile, since the *open-vocabulary-based DST* approach works without a given ontology, it fails to exploit slot-specific semantic characteristics (diverse answering manners) contained within external databases. For instance, values of Hotel-name are always named entities stored in databases, while Hotel-parking can only be answered by Boolean "*yes*" or "*no,*" which could be expressed diversely by users. Train-arrive, moreover, should resolve to values restricted in a specific format like "*08:45.*" To exploit the above category-specific features, several works [8, 10, 11] proposed a dual-strategy DST paradigm, which artificially divides all predefined slots into two categories based on hand-crafted features so *fixed-vocabulary-* and *open-vocabulary-based DST* approaches could be combined by a hybrid framework. However, research communities have not reached an agreement on how to divide the slot types. Some recent researchers [15, 16] also develop user simulators to optimize dialogue agents.

3 DSA-Gate DST Architecture

The architecture and working procedure of our network is shown in Fig. 2. The encoder takes in the previous dialogue state and dialogue context and encodes them into the embedding space. The generator consists of a text generation (i.e., a copy-augmented decoder) and a confirming prediction module for slot answering. The hierarchical gate predictor, peculiarly, is proposed to answer slots on conditions, which hierarchically leverages information in domain, slot, and answer levels.

3.1 Dialogue History Encoder

The dialogue state, which records all mentioned domain-slot-value (QUOTE) triplets per turn, is represented in nested approach [17, 18]. Assuming there are QUOTE domains and QUOTE nested slots for the QUOTE th domain, the total number of domain-slot pairs is $N = \sum_{i=1}^{M} N_i$. In MultiWoZ, QUOTE = 5, QUOTE = 30 in Table 1. The QUOTE th dialogue state QUOTE is arranged as $b_t^1 \oplus \cdots \oplus b_t^M$. Each subsequence QUOTE is represented as follows:

$$b_t^i = [\text{DOM}]_t^i \oplus d_i \oplus S_{i,t}^1 \oplus \cdots \oplus S_{i,t}^{N_i} \Rightarrow S_{i,t}^j = [\text{SLOT}]_{i,t}^j \oplus s_i^j \oplus - \oplus v_{i,t}^j \quad (1)$$

where QUOTE and QUOTE are special tokens followed by domain name QUOTE and slot name QUOTE at QUOTE th turn, respectively.—is a boundary token. At QUOTE -turn, the dialogue utterance QUOTE is represented by concatenating the agent utterance QUOTE and the user utterance QUOTE as $D_t = A_t \oplus U_t$. The input sequence for the encoder consists of the current dialogue utterances QUOTE , the

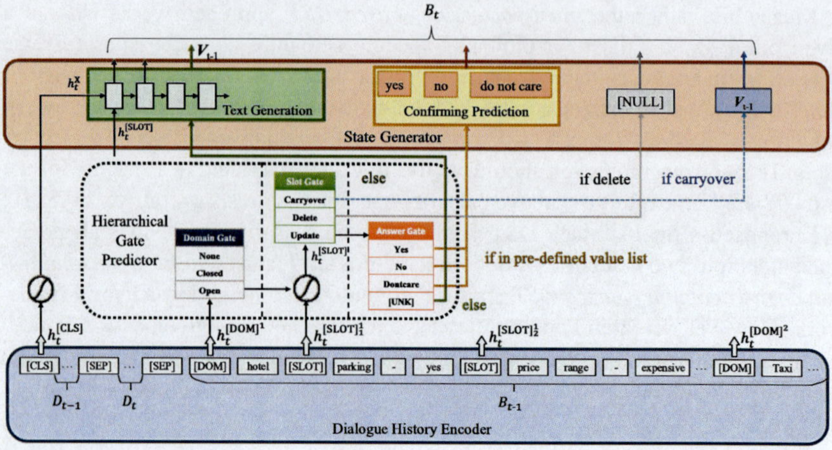

Fig. 2 DSA-Gate DST architecture consists of three major components: Dialogue History Encoder, Hierarchical Gate Predictor, and State Generator. The encoder gets hidden states of the input dialogue history sequence. Then, the gate predictor hierarchically performs classification in domain, slot, and answer level. The first domain-level gate distributes each domain into {NONE, CLOSED, OPEN}. For OPEN domains, operations for nested slots are further predicted among {DELETE, CARRYOVER, UPDATE} by the second slot gate. For other domains, all nested slots are directly tagged with CARRYOVER. The third answer gate is used to answer UPDATE-slots with special values related to confirmation expression, including {*yes*, *no*, *dontcare*}. The generator consists of two separated parts called Confirming Prediction Module and Text Generation Module. The model outputs the current belief state with referenced values for all domain-slot pairs turn-by-turn

Table 1 Statistics for all 5 domains and 30 slots in MultiWoZ in experiments

Domains	Slots	Count
Attraction	Area, name, type	3
Hotel	Book day, book people, book stay, area, internet, name, parking, price range, stars, type	10
Restaurant	Book day, book people, book time, area, food, name, price range	7
Taxi	Arrive by, departure, destination, leave at	4
Train	Book people, arrive by, day, departure, destination, leave at	6

previous utterances QUOTE and the previous dialogue state QUOTE The encoder exploits BERT [19] pre-trained model to encode the input sequence QUOTE and output the hidden state $H_t \in \mathbb{R}^{|X_t| \times d_{\text{emb}}}$ as:

$$X_t = [\text{CLS}] \oplus D_{t-1} \oplus [\text{SEP}] \oplus D_t \oplus [\text{SEP}] \oplus B_{t-1};\ H_t = \text{BERT}(X_t) \qquad (2)$$

where [CLS] and [SEP] are auxiliary tokens defined in BERT. The vectors $h_t^{[\text{DOM}]^i}$ and $h_t^{[\text{SLOT}]^i}$ in QUOTE that correspond to tokens QUOTE and QUOTE are obtained as domain activity semantics and slot-operation semantics for the hierarchical gate, respectively. The vector $h_t^{[CLS]}$ that corresponds to [CLS] is further processed as $h_t^X = \text{Tanh}\left(W h_t^{[CLS]}\right)$ to represent the entire semantics of the input sequence.

3.2 *Hierarchical Gate Predictor*

The gate predictor is aimed to hierarchically determine the domain activity, slot operations, and answering way for slot values by capturing features from the dialogue context. This module operates in this order: (1) A domain-level classifier predicts each domain's activity to capture the conversation topics focused by a user. (2) A slot-level classifier further determines the operations of all slots that are nested in the active (OPEN) domains. (3) An answer-level classifier answers all update slots differently.

Domain Activity Prediction The domain-level gate is designed to predict the focused conversation topic (i.e., active domains) at each turn. We identify three classes of domain activities with the development of the dialogue. NONE implies the domain is kept not mentioned up to the current turn, OPEN denotes the user is concerned with the domain currently, and CLOSED means the domain has been answered by *Agent* and is no longer the focused conversation topic. This paradigm matches the expectations of task-completion process in real life. OPEN domains are viewed as active, where all nested slots require further prediction of operations, while NONE and CLOSED domains are deactivated as they provide no information to the current dialogue state. The probability distribution over three activities for the QUOTE th domain is calculated as:

$$P_{\text{dom},t}^{i} = \text{softmax}\left(W_{\text{dom}}\left(h_t^{[\text{DOM}]^i}\right)\right) \tag{3}$$

For a domain along with its nested slots, we construct the bridge between domain activity and slot-operation in the following rules: (1) If answers for all nested slots are not-mentioned values QUOTE , the domain is tagged with NONE; (2) if at least one nested slot's operation is not CARRYOVER (i.e., slot's value is unchanged in between turns), the domain is tagged with OPEN; (3) other domains, whose all nested slots are unchanged and at least one slot's value is not NONE, are tagged with CLOSED. In light of the above connections, only the slots specific to OPEN domains participate in the slot-operation prediction, while all other slots are directly tagged with CARRYOVER.

Slot-Operation Prediction The slot-level gate assigns each slot in OPEN domains to distribution {CARRYOVER, DELETE, UPDATE}. CARRYOVER implies that the slot-value keeps unchanged in this turn. The non-carryover slots are divided into two groups according to whether a new acceptable answer is discovered (UPDATE) or not (DELETE) in context in this turn. To fully exploit domain-level information, the *fusion gate mechanism* [20] is utilized by merging corresponding hidden vectors $h_t^{[DOM]}$ and $h_t^{[SLOT]}$ for each domain-slot pair to predict its operation as follows:

$$v_t^{i,j} = \sigma\left(W_v \odot \left[h_t^{[DOM]^i}, h_t^{[SLOT]_i^j}\right]\right)$$

$$g_t^{i,j} = v_t^{i,j} \otimes h_t^{[DOM]^i} + \left(1 - v_t^{i,j}\right) \otimes h_t^{[SLOT]_i^j} \tag{4}$$

$$P_{slot,t}^{i,j} = \text{softmax}\left(W_{slot} g_t^{i,j}\right)$$

where $\odot$ and $\otimes$ mean pointwise and element-wise multiplications, respectively. If the operation is predicted as DELETE, the slot would be directly filled by the not-mentioned value [NULL]. If the operation is CARRYOVER, the slot keeps its previous value unchanged. If the operation is UPDATE, the slot has a meaningful value, which would be sent to the answer-based gate to predict its answering manner.

Answer Gate The answer-based gate is utilized to allocate a suitable answering manner for *update-slots*. Two methods, *Confirming Prediction* and *Text Generation*, are provided along with a predefined list as **{yes, no, do not care}**. *Confirming Prediction* method is used for slots related to confirmative expressions including affirmation "*yes,*" negation "*no,*" and unconcern "*do not care,*" while *Text Generation* method is used for slots whose values are named entities (recorded as [UNK]) appearing in dialogue context, which is to be generated by the copy-augmented generator. Another BERT model, whose weight is frozen during training, is used to encode above four terms into hidden states as follows:

$$Y_i = \text{BERT}_{\text{fixed}}([\text{CLS}] \oplus v_i \oplus [\text{SEP}]),\ v_i \in \left[\text{yes, no, dontcare, [UNK]}\right] \tag{5}$$

Then, $y^{[CLS]_i}$, the first vector in Y_i is used to represent the semantics of word v_i. The process of Answer Gate would be introduced in more detail in the next subsection.

3.3 State Generator

If a slot's operation is predicted as UPDATE, the State Generator is utilized to predict its value based on the classification output of the gates. A predefined list {*yes, no, dontcare*}, as mentioned above, contains several special values that can be diversely expressed by users. For a slot to update, one value is selected from the list by *Confirming* module if it is predicted in it. Otherwise, the value is recorded as [UNK], corresponding to an unknown named entity to be identified by *Text Generation* Module.

Confirming Prediction Module To answer a slot whose value belong to {*yes, no, dontcare*} separately, a semantic Confirming Prediction Module is employed. Precisely, the probability distribution on the four terms is obtained like this:

$$q = \mathrm{Tan}h\left(W_h\left(\left[h_t^X; h_t^{[\mathrm{SLOT}]}\right]\right)\right)$$

$$P^i_{\mathrm{answer},t} = \frac{\exp\left(y_i^{[\mathrm{CLS}]} W_a q\right)}{\sum_i \exp\left(y_i^{[\mathrm{CLS}]} W_a q\right)} \tag{6}$$

Text Generation Module The Gated Recurrent Unit (GRU) [14] and soft-gated copy mechanism [5] are used to generate each slot values with the context and vocabulary information utilized. At the k th time step of decoding, the hidden state of GRU is updated recursively by taking a word embedding $e_t^{j,k}$ as the input until [EOS] token is generated in the following ways:

$$g_t^{j,0} = \mathrm{h}_t^{\mathrm{X}}, e_t^{j,0} = h_t^{[\mathrm{SLOT}]^j}$$

$$g_t^{j,k} = \mathrm{GRU}\left(g_t^{j,k-1}, e_t^{j,k}\right) \tag{7}$$

$$P^{j,k}_{vcb,t} = \mathrm{softmax}\left(E g_t^{j,k}\right)$$

$$P^{j,k}_{ctx,t} = \mathrm{softmax}\left(H_t g_t^{j,k}\right)$$

3.4 Loss Function

The joint loss is optimized as a combination of four parts:

$$L = \lambda_d L_{\mathrm{dom}} + \lambda_s L_{\mathrm{slot}} + \lambda_a L_{\mathrm{answer}} + L_{\mathrm{generate}} \tag{8}$$

the loss unction for *Text Generation* L_{generate} is cross-entropy loss during the whole training period, while for the three components of *Hierarchical Gates*, cross-entropy loss is used in the first several epochs and focal loss [21] is used in the following epochs to meet sample-imbalance issues in classification.

4 MultiWoZ Dataset Analysis

In general, the DSA-Gate DST optimizes its performance heavily relying on the assumption that the training data has correct annotations on all three levels of hierarchy. Unfortunately, we have found substantial noises in the *domain-level* annotations in the original MultiWoZ [22] dataset, which will negatively affect the performances of *Domain Gate* and then propagate prediction errors to the subsequent modules. In this section, we try to analyze the statistics on *Domain* and *Slot Updating* annotations of the original MultiWoZ2.1 dataset, which is not explicitly mentioned by previous works.

4.1 Domain Label Noise

The original MultiWOZ 2.1 dataset provides Domain Label for user utterances per turn. However, the dataset keep assigns one exact domain label for each turn, which neglects the case of "non-cooperative" user behavior in real-life dialogues and leads to incorrect annotation. In this subsection, we specify four kinds of common domain label errors in the dataset as follows and show corresponding examples in Table 2:

(1) Mixed-domain Intent: The user's intents might span over mixed-domains. In Row 1 of Table 2, after finding the wanted train, the user made a booking for 2 people and then looked for a hotel to stay, while both Hotel and Train domains are mentioned in fact, the original MultiWoZ 2.1 only annotates Hotel domain but ignores Train domain.
(2) General Intent: The user might make chit-chat utterances unrelated to any domains such as "*bye*" or "*thanks*" as in Row 2 of Table 2. Actually, the domain label for this case is simply inherited from previous turns, which causes noise.
(3) Ambiguous Intent: The user utterance might be brief or ambiguous for annotators to identify which domain is active. In Row 3 of Table 2, the term "*cityroomz*" could refer to a Hotel-name or a restaurant name, but the annotator only annotated the former.
(4) Inconsistent Annotation: The annotation rule itself might be inconsistently, controversial, and even erroneous. In Row 4 of Table 2, since the user called a taxi to dine out before informing the destination restaurant, the agent had to request the user to make a restaurant reservation. In this case, the ground-truth domain is switched to Restaurant, but the annotator mistook it as Taxi.

Compared with significant domain annotation errors, Table 2 illustrates that the slots annotations could accurately indicate the *non-carryover* slots. In view of this, we discard the original domain labels and predict each domain's activity among {NONE, CLOSED, OPEN}. With the inherent constraints between slot and domain predefined by the ontology, we take the *Slot* annotations provided by MultiWoZ 2.1 as a ground-truth benchmark to align the corresponding *Domain* annotations. The

Table 2 Some examples of four types of noise in terms of domain labels in MultiWoZ2.1

Noise type	Example			Domain label
	Utterances	Non-carryover slots	True domains	
Mixed-domain intent	*Agent*: The fare is 10.10 gbp per ticket *User*: I need to make a booking for 2 people and can you find me a place to stay in the north?	Train: bookpeople-2 Hotel: area-north	Train hotel	Hotel
General intent	*Agent*: Glad i could be of help *User*: Yes, thank you	–	–	Hotel
Ambiguous intent	*Agent*: *User*: I need info about cityroomz	Restaurant: name-cityr Hotel: name-cityroomz	Hotel restaurant	Hotel
Inconsistent annotation	*Agent*: I am sorry we did not reserve a table for you at any restaurant, would you like to do that now? *User*: Yes, expensive French food for 3 people at 12:30 on Monday	Restaurant: food-French, pricerange-expensive, bookpeople-3, bookday-Monday, booktime-12:30	Restaurant	Taxi

Agent corresponds to the agent utterance while *User* corresponds to the user utterance. According to the utterances and ground-truth *non-carryover slots*, we provide *True Domains* that is active in these turns by ourselves, which differs from the original domain label (right-most column of the table) in dataset construction

percentage of domain annotations changed by our correction for MultiWoZ 2.1 is about 14.28%. After correction, the re-annotated domain labels are well consistent with the slot labels and would not introduce noise in the training stage.

Table 3 The number of active domains involved per turn in MultiWoZ2.1 with percentages reported as well

	0	1	2	>2	Avg.
Train	18,410 (33.48%)	36,105 (65.66%)	461 (0.84%)	8 (0.01%)	0.67
Dev	2500 (33.92%)	4805 (65.19%)	66 (0.90%)	0 (0.00%)	0.68
Test	2456 (33.33%)	4827 (65.51%)	85 (1.15%)	0 (0.00%)	0.68

The average number of active domains per turn is about 0.67, which is far less than 5, the number of domains predefined in the ontology

4.2 *Over-Trivial Carryover Problem*

This subsection discusses how utilizing *Domain Gate Predictor* as a filter for *Slot Gate* could mitigate the *over-trivial carryover* problem. This issue refers that in the previous baselines, the slots nested in uninterested domains, whose values are trivially unchanged, get unnecessarily analyzed by *Slot Gate* [3, 4]. As is shown in Table 4, the CARRYOVER operation accounts for an unbalanced vast majority in the dataset. It can be interpreted as in real life, subject to the length of utterances, the domains that user interest are usually only a limited part of the whole 5 domains in a single turn. To illustrate more concretely, we identified three types of domain activities, {NONE, CLOSED, and OPEN}, with the dialogue progresses. Table 3 reports the number of OPEN (i.e., active) domains per turn in MultiWoZ2.1. Nearly, 65% turns have only 1 OPEN domain, while other 33% are some kinds of chit-chats. This ratio conforms to real-life scenarios and supports our interpretation. Besides, the average number of active domains per turn is computed as 0.67, which is far less than 5 and leads to the unbalanced ratio of 3.96% of carryover slots. With the number of domains included in ontology increased, this gap would widen and harm models' scalability. To tackle this issue, we employ the *Hierarchical Gate* methodology in our model to predict domain activities as a filter for follow-up *Slot Gate*. For MultiWoZ2.1, with the uninterested domains detected by *Domain Gate*, only 11.54% out of 96.04% samples are passed on to the *Slot Gate*, where the unbalanced ratio of data distribution is dramatically reduced.

5 Experiments and Discussions

We evaluate our model on MultiWOZ2.0 and 2.1 datasets. Following previous baselines, 5 domains and 30 domain-specific slots are used in our experiments. We pre-process data similar to TRADE [5] and replace the not-mentioned value "*none*" by a special token [NULL] following SOM-DST [4]. Our model is based on *open-vocabulary-based* approach.

Table 4 Turn-level statistics distribution of three types of domain activities and two types of slot operations under each domain type in MultiWoZ2.1 test set

Domain activity	Count (%)	Slot operation		
		Carryover	Non-carryover	Ratio
None	25,345 (68.80%)	147,415	0	–
Open	4997 (13.56%)	24,490	8743	73.69/26.31
Closed	6498 (17.64%)	40,392	0	–
Sum	36,840	212,297	8743	96.04/3.96

With domain activity taken into account, the ratio of *carryover slots* is decreased from 96.04 to 73.69% and ratio of *non-carryover slots* is increased from 3.96 to 26.31%

5.1 Implementation Details

Our model employs the pre-trained BERT-base-uncased model [19], where the hidden size is 768, to encode dialog history. The max sequence length for all inputs is fixed to 256. During training, the model is optimized with Adam [23] optimizer for 50 epochs with a batch size of 32. We use a learning rate of 1e−4 for text generation and 2e−5 for hierarchical gates. The dropout [24] probability is 0.1, and the warmup proportion is 0.1. The teacher forcing rate for text generation is 0.5. As the Domain Gate results directly determine subsequent components' performance, the hyper-parameters in loss objective are set as $\lambda_s = \lambda_a = 1$ and $\lambda_d = 2$ after a grid search on the validation set.

5.2 Evaluation Metrics

Joint Accuracy Joint accuracy (JA) is a common metric to evaluate DST performance, which measures the percentage of exactly matched dialogue states. For a turn, if all the ⟨domain, slot, value⟩ triplets are correctly predicted, the dialogue state is considered as matching to the ground truth.

F1 Score for Gate Classification F1 score is also used to measure the classification performance of the Hierarchical Gates in DSA-Gate DST, which would influence the final Joint Accuracy of DST. Unfortunately, few of previous baselines had released their classification performances.

5.3 Experimental Results

Table 5 reports the comparison Joint Accuracy results of DSA-Gate DST with other baseline models. Our proposed model outperforms the prior best *open-vocabulary-based* baseline by + 0.14% and + 0.36% in MultiWOZ 2.0 and 2.1 [25], respectively. We attribute the performance gain to the *Hierarchical Gate* method. Table 6 shows the results of ablation studies of our model on MultiWOZ 2.1. The slot-level prediction is always required since the dialog state at the previous turn serves as part of the input of the *Dialogue History Encoder*. Therefore, we remove *Domain Gate* and *Answer Gate* optionally to explore the effect of our Hierarchical Gate method. In comparison, the Joint Accuracy is decreased by 0.78% and 0.55%, respectively, without domain or answer gate participated. The results suggest the effect of our method to apply hierarchical domain-, slot-, and answer-level gates before generating values.

Error Analysis The source of DST errors in the evaluation stage is analyzed as well in Table 6. It is observed that, with the domain-, slot-, answer-level gates replaced by oracle counterparts, the final performance rises, respectively, by 2.01%, 35.93%, and

Table 5 Comparison results of DSA-Gate DST and baselines on MultiWoZ2.0 and 2.1 measured by JA

Type	Models	BERT used	2.0	2.1
Predefined ontology	HyST [11]		42.33	38.10
	SUMBT [12]	√	48.81	52.75
	DS-DST [10]	√	–	51.21
	DST-picklist [10]	√	–	53.30
	CHAN-DST [13]	√	**52.68**	**58.55**
Open vocabulary	TRADE [5]		48.60	45.60
	HDCN [9]		–	46.76
	COMER [18]	√	48.79	48.79
	SOM-DST [4]	√	51.38	52.57
	DSA-Gate-DST (ours)	√	**51.71**	**53.24**

All models are categorized in two types according to whether the predefined ontology is utilized or not. √ indicates a pre-trained BERT language model [19] is exploited for word embeddings. Note that DST-picklist and CHAN-DST (scores in bold) employ extra supervision by human-crafted rules in a predefined ontology-based setting, which hampers DST's accessibility and scalability

Table 6 Ablation studies and error analysis for different settings on MultiWoZ 2.1.

Models	JA
DSA-Gate DST	**53.24**
(−) domain gate	52.46 (– 0.78)
(−) answer gate	52.51 (– 0.55)
(−) domain & answer gate	52.03 (– 1.21)
(+) oracle domain activity	55.25 (+ 2.01)
(+) oracle slot operation	89.17 (+ 35.93)
(+) oracle answer confirming	54.04 (+ 0.80)
(+) oracle text generation	56.31 (+ 3.07)
(+) oracle previous state	81.11 (+ 27.87)
(+) training set revised	53.40 (+ 0.16)

The bolded score is the basic JA result of our DSA-Gate DST model. (−) Means removing the component in DSA-Gate DST. (+) Denotes replacing the predicted results of the component by oracle counterparts in evaluation

0.80%, while the oracle text generation, which means all value strings of *update-slots* are precisely generated by the decoder, leads to an improvement of 3.07%. Overall, it shows that utilizing oracle *Slot Gate* results in the most significant improvement to the model's performance. We attribute this to the use of the previous dialog state as part of the input of Dialogue History Encoder. To prove it, we further conduct an experience that the predicted previous dialogue state is replaced by oracle one. Results show that Joint Accuracy is raised to 81.11%, which is nearly to 89.17%.

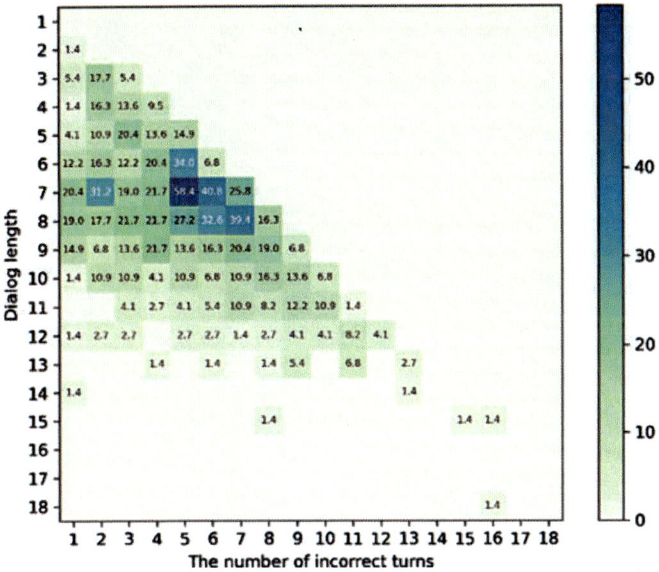

Fig. 3 Heatmap of prediction errors of our model in MultiWoZ2.1 test set. Each row of the matrix represents the number of turns of the dialogue sample (i.e., dialogue length), which range from 2 to 18. Each column represents the number of turns in which our model makes erroneous predictions. The value in each cell is the proportion (%) of the error case. It is noticeable that most erroneous cases are close to the diagonal of the heatmap, which demonstrates that the error propagation across turns is severe

It indicates that the error propagation in slot-operation level occurring across turns is to blame for performance limitation. Figure 3 reports the visualization of the overall error distribution. As expected, the majority of erroneous cases are close to the diagonal of the heatmap, which verifies the significant error propagation trouble in multi-turn dialogue scenarios.

F1 Scores for Hierarchical Gate Table 7 reports the F1 scores in domain-, slot-, and answer-level gates in MultiWOZ2.1 with the statistical distribution of labels, respectively. It shows that *open* in domain-level and *update* in slot-level have relatively high F1 scores, which proves the effect of our *Hierarchical Gate* algorithm. Moreover, it can be seen that the less portion of a type accounts for, the lower the F1 score it has. For instance, *delete* accounts for only 0.05% of *Slot Gate*, and no accounts for 0.13% of *Answer Gate*. We attribute this to that the corpus collectors neglect those unusual but critical user cases, such as withdrawing values mentioned before or informing negation intent when dialogue progressing.

Table 7 F1 scores of three-level hierarchical classification gates on MultiWOZ 2.1 test set

Gate	Type	Count	(%)	F1
Domain	None	25,345	68.80	99.65
	Closed	6498	17.64	73.36
	Open	4997	13.57	75.34
Slot	Carryover	212,297	96.15	98.59
	Delete	109	0.05	3.10
	Update	8399	3.80	80.82
Answer	Yes	322	3.73	83.63
	No	11	0.13	28.95
	Dontcare	235	2.72	48.55
	[UNK]	8066	93.42	97.44

In each gate, statistics of all types in dataset are displayed. The underlined type means the target class to be further sorted by subsequent gate

5.4 *Open Discussion*

This subsection analyzes several DST results predicted incorrectly by our model in MultiWOZ 2.1 to answer 2 research questions: $Q1$: : Why all existing DST models fail to reach a performance ceiling for JA of 60%? $Q2$: Is it proper to pursue a high performance with the annotation schema irrational??

As already explained, with *carryover* operation [3] employed, slot values could be borrowed from previous dialogue states but the accuracy might decrease because of the **error propagation** occurs across turns. As shown in Table 6, using oracle value-string generation or replacing predicted slot gate operation results by ground truth can rise JA by + 3.07% or + 35.93%, respectively. It reveals that the noisy annotation mainly comes from incorrect slot operations, instead of imprecise value strings.

We further research some incorrect prediction cases in the slot-updating procedure. As the instances displayed in Table 8, the noisy slot-operation annotation in corpus causes disputable *value-updating* prediction. The erroneous cases are intuitively categorized into three types: *missed updating*, *incorrect updating*, and *over updating*. It is noticeable that *missed updating* and *over updating* in algorithm are equivalent in *over annotating* and *missed annotating* in corpus. Since these noisy annotations usually occur at dialogues beginning, they cause a more severe error propagation in dialogue state tracking, as Fig. 3 shows. With algorithm performance unchanged, the larger the average number of turns is, the lower Joint Accuracy would be, especially in multi-domain dialogue scenes. Note that no existing models could reach a performance ceiling of 60%, while the above issue keeps neglected by the research community.

Table 8 Some cases of slot-operation prediction errors of DSA-Gate DST in MultiWoZ2.1 test set

Case	No	Example	GT update	Pred update
Missed update	$C1$	*Agent*: Is there something else i can help you with then? *User*: Nope. that should cover it	Restaurant: {area, centre}	–
	$C2$	*Agent:* Yes, the Cambridge belfry is a cheap 4 star hotel that includes free parking *User*: Do they have internet there?	Hotel: {internet: yes, name: Cambridge belfry}	Hotel: {name: Cambridge belfry}
Incorrect update	$C3$	*Agent:* Many of the guesthouses in the north offer internet, will you require parking? *User*: Parking is not important. could you recommend 1 for me?	Hotel: {parking: no}	Hotel: {parking: dontcare}
	$C4$	*Agent*: Ok, let s start with which day you would like to have this reservation? *User*: Book a table for 8 people at 17:15 on Sunday	Restaurant: {book day: Sunday, book time: 1715}	Restaurant: {book day: Sunday, book time: 17:15}
Over update	$C5$	*Agent:* Before i book your restaurant would you like to book your lodging? i think you will like Hamilton lodge. it meets your needs *User*: Does it have internet?	–	Hotel: {name: Hamilton lodge}
	$C6$	*Agent*: Any type of food you would like? *User*: I am open to suggestions. i just would prefer it to be in the moderate range	Restaurant: {pricerange: moderate}	Restaurant: {pricerange: moderate, food: dontcare}

(continued)

Table 8 (continued)

Case	No	Example	GT update	Pred update
	$C7$	*Agent:* *User:* What can you tell me about the good luck Chinese food takeaway?	Restaurant: {name: good luck}	Restaurant: {name: good luck, food: Chinese}

The symbol *Agent* denotes an agent utterance while *User* denotes a user utterance. GT means ground-truth slots to be updated in the current turn, which sometimes contains controversial or noisy annotation compared with dialogue utterances

In addition, we dedicatedly study the erroneous prediction cases as enumerated in Table 8 to explore how the noisy annotation is produced. We identify the following three categories of causes of the issue:

(1) **Confusion between Attributes and Intents**: It refers to cases that the semantic information in slot value is inconsistent in terms of subjective users' intent or objective entity attributes. As examples $C2$ and $C5$, the annotation *hotel-internet-yes* sometimes refers to the given hotel surely equipped with Internet, but at other times refers to that the user wants a hotel to have Internet.
(2) **Confirming Ambiguity**: This ambiguity comes from the diverse interpretation in user expression. It particularly occurs in the ambiguity among "*not mentioned,*" "*no*" and "*do not care*" as $C6$ demonstrates, where the user intent can be described in various kinds of expression.
(3) **Annotation Error**: Crowd-workers might carelessly produce faulty annotation, which contribute to the data noise naturally, as the example $C1$, $C4$, and $C7$.

The above discussions concludes that the noisy annotation inside datasets is the key factor limiting Slot Gate performance. These issues cannot get resolved by crowd-sourcing correction, but require a novel, consistent, and unified annotation strategy. Simply pursuing a high joint accuracy would enforce models to over-fit these noise.

6 Conclusion

In this paper, we propose a novel hierarchical-enhanced dialogue state tracking model called DSA-Gate DST to predict dialogue states with extra domain activity information and answer slot values in a specific manner. Some domain- and slot-level annotation noise are identified and analyzed in MultiWoZ 2.1. Experiment results show that our model can effectively track dialogue states in mixed-domain scenarios and outperform previous open-vocabulary-based baselines. We further propose some suggestions to researchers to reach one novel and proper annotation strategy. We hope all above contributions would be beneficial for research communities in the future.

Acknowledgements This work was supported by the National Key R&D Program of China under grant 2019YFF0302601.

References

1. Nouri E, Hosseiniasl E (2018) Toward scalable neural dialogue state tracking model. In: 32nd Conference on neural information processing systems (NeurIPS 2018)
2. Ramadan O (2018) Large-scale multi-domain belief tracking with knowledge sharing. arXiv preprint arXiv:1807.06517v1
3. Gao S, Sethi A, Aggarwal S, Chung T, Hakkani-Tur D (2019) Dialog state tracking: a neural reading comprehension approach. arXiv preprint arXiv:1908.01946.
4. Kim S, Yang S, Kim G, Lee S-W (2019) Efficient dialogue state tracking by selectively overwriting memory. arXiv preprint arXiv:1911.03906
5. Wu C-S, Madotto A, Hosseini-Asl E, Xiong C, Socher R, Fung P (2019a) Transferable multi-domain state generator for task-oriented dialogue systems. arXiv preprint arXiv:1905.08743
6. Wu C-S, Socher R, Xiong C (2019b) Global-to-local memory pointer networks for task-oriented dialogue. arXiv preprint arXiv:1901.04713
7. Lei W, Jin X, Kan M-Y, Ren Z, He X, Yin D (2018) Sequicity: simplifying task-oriented dialogue systems with single sequence-to-sequence architectures. In: Proceedings of the 56th annual meeting of the association for computational linguistics, vol 1, Long Papers, pp 1437–1447
8. Ma Y, Zeng Z, Zhu D, Li X, Yang Y, Yao X, Zhou K, Shen J (2019) An end-to-end dialogue state tracking system with machine reading comprehension and wide & deep classification. arXiv preprint arXiv:1912.09297v1
9. Zhang L, Wang H (2021) Learn to focus: hierarchical dynamic copy network for dialogue state tracking. arXiv preprint arXiv:2107.1177825
10. Zhang J, Hashimoto K, Wu C, Wan Y, Yu PS, Socher R, Xiong C (2019) Find or classify? Dual strategy for slot-value predictions on multi-domain dialog state tracking. arXiv preprint arXiv:1910.03544
11. Goel R, Paul S, Hakkani-Tür D (2019) Hyst: a hybrid approach for flexible and accurate dialogue state tracking. arXiv preprint arXiv:1907.00883
12. Lee H, Lee J, Kim T-Y (2019) SUMBT: slot-utterance matching for universal and scalable belief tracking. arXiv preprint arXiv:1907.07421
13. Shan Y, Li Z, Zhang J, Meng F, Feng Y, Niu C, Zhou J (2020) A contextual hierarchical attention network with adaptive objective for dialogue state tracking. arXiv preprint arXiv:2006.01554
14. Cho K, van Merrienboer B, Bahdanau D, Bengio Y (2014) On the properties of neural machine translation: encoder-decoder approaches. arXiv preprint arXiv:1409.1259
15. Tsengy B-H, Daiz Y, Kreyssigy F, Byrne B (2021) Transferable dialogue systems and user simulators. arXiv preprint arXiv:2107.11904
16. Kim S, Chang M, Lee S-W (2021) NeuralWOZ: learning to collect task-oriented dialogue via model-based simulation. arXiv preprint arXiv:2105.14454
17. Le HT, Sahoo D, Liu C, Chen NF, Hoi SCH (2020b) Uniconv: a unified conversational neural architecture for multi-domain task-oriented dialogues. arXiv preprint arXiv:2004.14307
18. Ren L, Ni J, McAuley J (2019) Scalable and accurate dialogue state tracking via hieorarchical sequence generation, pp 1876–1885
19. Devlin J, Chang M, Lee K, Toutanova K (2018) Bert: pre-training of deep bidirectional transformers for language understanding. arXiv preprint arXiv:1810.04805
20. Hochreiter S, Schmidhuber J (1997) Long short-term memory. Neural Comput 9(8):1735–1780
21. Lin T-Y, Goyal P, Girshick R, He K, Dollár P (2017) Focal loss for dense object detection. In: Proceedings of the IEEE international conference on computer vision, pp 2980–2988

22. Eric M, Goel R, Paul S, Kumar A, Sethi A, Ku P, Goyal AK, Agarwal S, Gao S, Hakkani-Tur D (2019) MultiWOZ 2.1: a consolidated multi-domain dialogue dataset with state corrections and state tracking baselines. arXiv preprint arXiv:1907.01669v1.
23. Kingma DP, Ba J (2014). Adam: a method for stochastic optimization. arXiv preprint arXiv:1412.6980
24. Srivastava N, Hinton G, Krizhevsky A, Sutskever I, Salakhutdinov R (2014) Dropout: a simple way to prevent neural networks from overfitting. J Mach Learn Res 15:1929–1958
25. Budzianowski P, Wen T-H, Tseng B-H, Casanueva I, Ultes S, Ramadan O, Gašić M (2018) Multiwoz—a large-scale multi-domain wizard-of-oz dataset for task-oriented dialogue modelling. arXiv preprint arXiv:1810.00278.

Anomaly Based Intrusion Detection Systems in Computer Networks: Feedforward Neural Networks and Nearest Neighbor Models as Binary Classifiers

Danijela Protic, Miomir Stankovic, and Vladimir Antic

Abstract Anomaly based intrusion detection systems monitor the computer network traffic and compare the unknown network behavior with the statistical model of the normal network behavior. The anomaly detection is mainly based on binary classification. Machine learning models are common tools for determining the normality of the network behavior. Binary classifiers like feedforward neural network and the nearest neighbor models have proven to be the best classification option in terms of both processing time and the accuracy when the instances were normalized and the features selected to reduce the data. The results of the experiments carried on the six daily records from the Kyoto 2006+ dataset show the apparent decrease in accuracy of ~ 1% for a number of instances greater than ~ 100,000 per day.

Keywords Anomaly detection · Binary classification · Feedforward neural network · Nearest neighbors · Machine learning

1 Introduction

Intrusion detection systems (IDSs) monitor computer networks for malicious attacks or abnormal network behavior. IDSs are often classified according to how they detect attacks. The signature or misuse-based IDSs compare the network behavior with rules. They maintain a database with known attack signatures, compare incoming data traffic with these signatures and exploit weakness of the system to detect intrusions. Once the attack is detected, the associated network traffic pattern is added to the

D. Protic (✉) · V. Antic
Center for Applied Mathematics and Electronics, Belgrade, Serbia
e-mail: danijelaprotic318@gmail.com

V. Antic
e-mail: avladimir@ptt.rs

M. Stankovic
Mathematical Institute of SASA, Belgrade, Serbia

P. Chatterjee et al. (eds.), *Computational Intelligence for Engineering and Management Applications*, Lecture Notes in Electrical Engineering 984,
https://doi.org/10.1007/978-981-19-8493-8_44

database of known attack signatures [1, 2]. One of the main drawbacks of signature-based detection is that if the pattern is not known as the signature the attack may not be visible at all. Anomaly based IDSs compare the network behavior with profiles that detect when the network behavior is outside of an acceptable range. They measure the current state of the network traffic and compare it to the statistical model of normal behavior. The main advantage of anomaly based IDSs is that it can detect both unknown attacks and unusual network behavior. One of the main disadvantages of anomaly based detection is the difficulty in defining the rule set. For correct detection, it is necessary to develop a detailed model of the accepted behavior, which can be computationally expensive [3]. Anomaly detection is often based on machine learning models that perform binary classification, analyze the inputs and predict which of the two possible classes the input instance belongs to. Machine learning (ML)-based algorithms are common tools for deciding what should be considered 'normal' [4, 5]. Nonlinear optimization algorithms have mainly been used in the detection of anomalies because they minimize the objective functions when fitting a model to a set of the data points [6–10]. Supervised ML is widely used to train the classifiers that decide whether or not the network traffic is normal or not. Some ML models are known as lazy learning models, which store all of the data from the training set and wait for the test set to appear. For this reason, the evaluation of the classifier can be time-consuming and the storage space consumption can be considerable. On the other hand, eager learners like neural networks do not wait for the test data to appear, but instead initiate the evaluation process immediately after the training data are connected to the inputs. The main advantage of a neural network is that it can be easily adapted to nonlinear data without trying different learning algorithms, as is the case with the traditional machine learning models. In this article, we present the evaluation of three types of binary classifiers based on (1) feedforward neural network (FNN), (2) k-nearest neighbor (k-NN) algorithm and (3) weighted k-nearest neighbor (wk-NN) algorithm. The FNN contains an input layer and a hidden with nine nodes each and output layer with one output node. The Levenberg–Marquardt (LM) algorithm is used to train the FNN. The algorithm combines two algorithms: the gradient descent algorithm (GDA), which gives the model stability and the Gauss–Newton algorithm (GNA), which accelerates training close to the global minimum of the objective function [11, 12]. The k-NN algorithm is one of the simplest ML algorithms that assigns objects to the class to which most their nearest neighbors in the feature space belong [13]. The wk-NN algorithm extends the k-NN algorithm in such a way that the instances of the training set that are closest to a new one have higher decision weights. In this study, the nearest neighbors are calculated based on Euclidean distances, while the weights are calculated based on inverse distances [14].

One of the main problems with supervised learning is the large number of instances that are required to train classifiers. To reduce the size of the data set used to train the models, preprocessing techniques such as feature selection and instance normalization can be used. The techniques proposed depend on the data set used to evaluate the models and the purpose of the model. The feature selection algorithm presented here is based on performing tangent hyperbolic function (tan h) after the instances have been normalized in the range [− 1, 1] as suggested in [4] and [5]. In this way, the dataset was reduced, which shortened processing time and increased accuracy of the models. In addition, instances normalization rescaled features so that the effects of one feature could not affect the other. Another problem in supervised ML is how to avoid overfitting and overtraining. To do this, an initial data set must be separated into training and test set. If the model classifies the data in the training set much better than in the test set, an overfitting is more likely. To avoid overfitting, some common methods can be used as a solution: cross-validation, training with more data, removing features, early stopping, regularizing, or using ensemble methods. In contrast to overfitting, which shows the complexity of the model, overtraining points to the problem of an oversized test set. Overtrained models predict the training set with high precision, but cannot generalize to new data. To avoid overtraining, the penalty can be added to the training error depending on the weight size. A higher weight can be a sign of overtraining. The regularization brings the models to smaller weights. In addition, the methods such as bagging can be used. Bagging is a sampling ensemble method that creates different training sets for each model. As the accuracy calculated of the validation data peaks after training for the number of steps and then stagnates or begins to decline, this could be a good indicator of the overtraining.

The aim of this study is to show the accuracy changes of binary classifiers related to the size of the data set. For this purpose, this study uses the Kyoto 2006+ dataset built on actual packet-based network traffic recorded on five different computer networks inside and outside of Kyoto University from 2006 to 2009. The dataset is intended for anomaly based IDS research and development. Fourteen statistical features from the Kyoto 2006+ dataset were derived from the KDD-Cup'99 dataset. The authors ignored all features that contained redundant data and added ten additional features that were used exclusively for the detection of anomalies [15]. The IDS Bro software is used to transform traffic data into 24-feature sessions containing ~ 90 million unbalanced instances classified as normal traffic (~ 50 million), known attacks (~ 43 million) and unknown attacks (~ 425,000) [16–18]. The experiments are carried out on six daily records, and the results are presented in terms of accuracy of the models and the processing time. Before starting the training, the features are selected in such a way that all categorical features and all features used for further analyses are discarded, except for the 'Label' that was used to indicate the normal network traffic or the anomaly. Data that could not be scaled in the range [− 1, 1] were cut off. Finally, nine features were used to train the classifiers.

The results show that the processing time of the FNN is approximately ten times shorter than the processing time of the nearest neighbor models, with a high level of accuracy being demonstrated for all models. The accuracy varies with the data set and indicates the need to reduce the number of instances if it exceeds ~ 100,000 per day. The results also show that a large number of instances results with lower accuracy, suggesting overtraining that can be avoided with algorithms that reduce the size of the data sets or with hybrid models that aim to increase the accuracy based on FNN and NN classifiers and principle of nonparametric correlation analysis described in [19].

2 Literature Review

Anomaly based IDSs, binary classification and machine learning models have been explored in a number of studies over several decades. In [20], the authors explain that the binary classification for the detection of anomalies in the computer network comprises four phases: preparation of the dataset, training and tests, construction of the classification model and determination of the key performances of measurement metrics. First, in order to maintain the reliability of the model, a selection and modification of the features must be made. The label is used to determine if instance is as either normal or abnormal. The aim of the feature selection methods is to select the most useful features to build the compact classification models with a better generalization ability [21]. The feature selection methods can be broadly divided into supervised, unsupervised and semi-supervised [22–24]. Furthermore, the supervised feature selection can be divided into the filter, wrapper and embedded [25]. There are two basic techniques for scaling the data in different dimensions to eliminate the influence of one feature to others: normalization and standardization [26]. The normalization is a scaling method in which the values are shifted and rescaled to end up ranging into the ranges [0, 1] or [− 1, 1]. Standardization is a scaling technique in which the values are centered around the mean with a unit standard deviation. Normalization is proper to use when the distribution of the data set does not follow the Gaussian distribution, which can be useful in nearest neighbor algorithms or in neural networks. Standardization is useful when data are following the Gaussian distribution. Second, the machine learning task which has three schemes (training, validation and testing) is performed. ML algorithms are used to determine performance of classifiers in terms of classification rate and processing time. Finally, in the context of machine learning, a metric measure like accuracy is used to represent the performance of binary classification. In [27], Dudani introduced and showed the advantages and disadvantages of widely used the distance-weighted-k-nearest neighbor model. In [28], the authors compared the k-NN and weighted k-NN classifiers and demonstrated combining schemes for accuracy improvement when the dataset size is large. In [29], the authors refer to the high-dimensional data source in determining the performances of network intrusion detection. They present the classification results by taking into account a feature selection algorithm and propose a feature selection combination strategy based on an integrated optimization algorithm and wk-NN, in order to improve the performance of the detection of network intrusions. The results show that wk-NN increases the efficiency at the expense of poor accuracy. In [30], the authors presented various distance measures that are used as weights in various nearest neighbor algorithms. They showed that the Euclidean offers a better recognition performance compared to the cosine distance measurement. The chi-square test method is used for anomaly detection. In [31], the authors have shown the experimental comparison of the machine learning techniques in intrusion detection based on MATLAB, the selection of features for FNN and ML models and confusion matrices. They have demonstrated the high accuracy of the classification for the FNN and NN models. The FNN is widely recognized as an accurate classifier in anomaly detection because of its learning capability, parallelism, adaptability, simplicity and fault tolerance [32]. In [33], the authors analyzed the performance of the k-NN and FNN for the binary classification of anomalies. The results on accuracy showed that, for 42 features, the accuracy of both models did not exceed 95% for the training set and 87% for the test set. In addition, the accuracy of the training and test sets did not exceed 96% for 19 features.

There are many data sets that the authors use for intrusion detection studies. In [34–39], the authors examined, described and compared ADFA, AWID, CAIDA, CIC-IDS-2017, CSE-CIC-2018, DARPA, ISCX2012, KDD CUP'99, NSL-KDD and UNSW-NB15 datasets. With the exception of CAIDA, however, all data sets represent simulated network traffic. In addition, all data sets are intended for specific attacks detection rather than anomaly detection. The Kyoto 2006+ dataset is the only one that was collected from actual network data and is intended solely anomaly based IDSs research and development [15].

3 Binary Classification

Binary classification is the process of dividing a data set into two referred classes so that one class represents the normal condition and the other represents the opposite condition (true/false, one/zero). The binary classification problems consider the assignment of a new data instance to one of the classes, by measuring a number of features [40]. The aim of binary classification is to learn a function that minimizes the probability of a misclassification [41]. Classification is a task that requires the use of ML algorithms that assign a class label to an example of the problem domain [42]. Many popular binary classification methods are used in learning, such as logistic regression (LR), naïve Bayes (NB), nearest neighbors, decision trees (DT), support vector machines (SVM) and neural networks [43]. The LR algorithm uses a logistic function to model the probability that an outcome will occur. The algorithm is most useful for determining whether a new instance is the best fit for a class, but fails if there is no linear separation of values [44]. The NB algorithm classifies data based on historical results; it is quick to build and does not require an extensive training set. Since the performance of the Bayesian algorithm depends on the accuracy of its assumptions, the results can be very poor [45]. In [13], the authors explain that the k-NN algorithm assigns objects to the class to which most of their closest neighbors in the feature space belong to. The algorithm requires the k points that are closest to the sampling point and averages their feature value as a prediction value. The DT algorithm predicts a class label from the input data, by following the decisions in the tree starting from the root of the tree down to the leaf nodes. The possibility space is divided into subsets according to the branching conditions assigned to each node [46, 47]. The SVM algorithm uses a hyperplane in n-dimensional space to classify the instances [48]. The instances that fall on the different sides of the hyperplane are assigned to different classes. The selected hyperplane directly affects the accuracy of the results. The neural network outcomes are based on the prediction probability and the classification threshold. The threshold defines how strict the model is in assigning an input to the positive class. The computation is done by forward propagation for computing outputs and backward pass to adjust the weights [49].

For the purposes of this research, two types of binary classifiers are presented. The feedforward neural network is an eager learner who creates a classification model based on the training data before receiving the data to predict. The nearest neighbor models are lazy learners who do not focus on building a general model, but instead store instances of the training data and waits for the test data to appear. The FNN and the nearest neighbor models are often used in the binary classification because they are very precise and have a high predictive power in the detection of anomalies [4, 33].

3.1 Classification Based on the Feedforward Neural Network

The feedforward neural network is a human brain-inspired structure made up of a series of layers with highly connected neurons in each layer. The first layer has a connection to the inputs of the network, while all the subsequent layers have connections to the previous layer. The final layer produces the outputs that match the inputs with the desired result. The FNN transfer function is given by the following expression:

$$y_i = F_i\left(\sum_{j=1}^{q} W_{ij} f_j\left(\sum_{l=1}^{m} w_{ji} x_l + w_{j0}\right) + W_{f0}\right) \tag{1}$$

where f_j and F_i denote transfer functions of hidden and output layers, m represents the number of inputs x_l, q is number of outputs y_i, w_{ji} and W_{ij} are weights, while w_{j0} and W_{f0} represent biases [50]. The FNN is trained in epochs by iteratively modifying the weights so that the given input pattern reflects the correct output in order to classify the inputs according to the target classes. The size of the FNN must be large enough to be able to generalize and provide a suitable fit of the data for all unknown patterns [51]. The FNN can have a large number of parameters which, due to the convergence on a suitable set of parameter values, can lead to the estimation problems. The overall performance of the FNN decreases as the number of parameters increases [52]. In this study, the weights are updated with the LM nonlinear iterative algorithm, which uses the concept of 'neighborhood' to improve the model's performance in both memory and in terms of processing time. The LM algorithm solves nonlinear least squares problems and provides a numerical solution in a series of calculations [11, 12]. The algorithm is a hybrid technique used to balance the properties of the GDA, which makes it possible to find an optimal solution when the initial guess is far from the optimal value and the GNA, which converges faster near the minimum [53, 54]. The change between the GN and GDA is referred to as a damping strategy, which is controlled by a damping factor: If the damping factor is large, the LM adjusts the weights as with the GDA; if it is small, the adjustment is carried out as according to the GNA. In [10], the authors presented the basic idea of the LM algorithm: Choose the kth iterate $x^{(k)}$ to minimize the expression:

$$\left\| \hat{f}\left(x; x^{(k)}\right) \right\|^2 + \lambda^{(k)} \left\| x - x^{(k)} \right\|^2, \ \lambda^{(k)} > 0 \quad (2)$$

The goal is to minimize the first and the second objectives in (Eq. 2) [10, 55, 56]. The LM algorithm describes the exchange between these two objectives with the damping parameter $\lambda^{(k)}$ so that $(k + 1)$th iterate $x^{(k+1)}$ can be derived from the formula:

$$x^{(k+1)} = x^{(k)} - \left(\mathbf{H} + \lambda^{(k)}\mathbf{I}\right)^{-1}\mathbf{J}^{\mathrm{T}} f\left(x^{(k)}\right) = x^{(k)} - \left(\mathbf{J}^{\mathrm{T}}\mathbf{J} + \lambda^{(k)}\mathbf{I}\right)^{-1}\mathbf{J}^{\mathrm{T}} f\left(x^{(k)}\right) \quad (3)$$

where **I** represent the identity matrix, **H** is the Hessian and **J** is the Jacobi matrix. For $\lambda^{(k)} \rightarrow \infty$, the LM algorithm tends toward the GDA, while for $\lambda^{(k)} \rightarrow 0$ the LM algorithm behaves like the GNA.

3.2 *The Nearest Neighbor Classifiers*

The nearest neighbor algorithms are supervised learning algorithms which are often used for both classification and regression. The nearest neighbor model stores instances of the training set and ranks the new data based on similarity. The NN models are widely known as lazy learners because they do not learn from the training set right away, but instead store the data and take actions at the time of classification. The ranking is calculated by averaging the results of the nearest neighbor majority vote to predict the class of a new point. The principle behind the NN methods is to find a number of training instances that are closer in distance to the new instance and to predict the label from this. Distance can be any metric measurement in general, but Euclidean distance is the most common choice. Consider the instance of a feature vector of known instances $x = \{x_1, \ldots, x_k\}$ x_i, which is denoted by the label y_i of a feature vector of known observations $y = \{y_1, \ldots, y_k\}$ then $\{(x_i, y_i), i = 1, \ldots, k\}$ can be assigned as the training data. Predictions of the k-NN for a new instance are made by searching the entire training set for the k-nearest neighbors and summarizing the output variable for those cases. Unfortunately, there is no specific method for finding the best value for k. A very low value of k can be noisy and cause outlier effects in the model. Large values of k are good, but encounter some difficulties and can be complex. The computational cost is high due to the calculation of the distance between the data points for all instances in the training set, which is the main disadvantage of the k-NN algorithm. The wk-NN algorithm extends the k-NN algorithm in such a way that instances of the training set that are closer to a new instance have greater weight in the decision than a more distant [14]. For this reason, the distances are converted into the weights that represent the degree of similarity. Consider the distance $d_w(\mathbf{x}, \mathbf{y})$

$$d_w(\boldsymbol{x}, \boldsymbol{y}) = \sqrt{\sum_{i=1}^{p}(x_i - y_i)^2} \tag{4}$$

where w is the weight. One of the simplest wk-NN functions is the inverse distance given with

$$w = \frac{1}{d_w(\boldsymbol{x}, \boldsymbol{y})^2} \tag{5}$$

The wk-NN training phase consists only in storing the features and class labels of the training set. The classifier adapts immediately as the new training data is collected so that the algorithm reacts quickly to input changes in real-time use. In contrast to the fast training phase, the algorithm requires very expensive tests. The overall cost of the algorithm is in the prediction because for each test sample to calculate distances and find the closest neighbors: If the instance x to be classified is changed, the neighborhood changes too, which changes the weights [27].

4 Results and Discussion

The experiments are carried on six daily records form the Kyoto 2006+ dataset consisting of 158,572, 128,740, 90,129, 81,807, 58,317 and 57,278 instances, respectively. After preprocessing the Kyoto 2006+ dataset as in [4], the features 5–13 were left, with normalized instances, for training the models. The instances are normalized (1) to change the values of the numerical features on the common scale when they have a different range and can affect other features (2) to avoid the saturation of the FNN at the beginning of the training and (3) to ensure that the distance measurement does not lose accuracy due to a small difference between the most distant and closest neighbors. Feature 18 (Label) is used to indicate whether the network traffic was malicious or not. The description of the selected features can be found in Table 1 [16].

All models are simulated in MATLAB, installed on an Intel Core i7 processor with 2.7 GHz CPU and 16 GB RAM. The classification learner is used to train the NN classifiers. The distance metrics is set to 'Euclidean,' and the number of nearest neighbors is set to 10. Weights are calculated based on Eq. 5 (inverse distance). The Neural Network Toolbox is used to evaluate the FNN-based classifiers. The FNN with one hidden layer has nine inputs, nine hidden nodes and one output node, and a hyperbolic tangent activation function is applied to the nodes. The hyperbolic tangent activation function is monotonic, differentiable, centered at zero, its values are in the range of − 1 to 1 and its derivatives are not monotonous (limited to movement in

Table 1 Selected features from the Kyoto 2006+ dataset

No.	Feature	Description
5	Count	The numbers of connections whose source IP address and destination IP address are the same to those of the current connection in the past two seconds
6	Same_srv_rate	% of connections to the same service in the Count feature
7	Serror_rate	% of connections that have 'SYN' errors in Count feature
8	Srv_error_rate	% of connections that have 'SYN' errors in Srv_count (% of connections whose service type is the same to that of the current connections in the past two seconds) features
9	Dst_host_count	Among the past 100 connections whose destination IP address is the same to that of the current connection, the number of connections whose source IP address is also the same to that of the current connection
10	Dst_host_srv_count	Among the past 100 connections whose destination IP address is the same to that of the current connection, the number of connections whose service type is also the same to that of the current connection
11	Dst_host_same_src_port_rate	% of connections whose source port is the same to that of the current connection in Dst_host_count feature
12	Dst_host_serror_rate	% of connections that have 'SYN' errors in Dst_host_count feature
13	Dst_host_srv_serror_rate	% of connections that have 'SYN' errors in Dst_host_srv_count feature
18	Label	Indicates whether the session was attack or not; '1' means normal. '− 1' means known attack was observed in the session, and '− 2' means unknown attack was observed in the session

a certain direction). The last characteristic is useful for LM optimization since the algorithm starts as the GDA [19].

The instances of each daily record are divided into the training, validation and test set containing 70%, 15% and 15% instances of the data set, respectively. The training set is used to adjust the parameters of the models, the validation set is used to compare estimated output of the model to its predicted value, and the test set is used to access the performance of the finally assessed classifier [57]. The results are presented in terms of the accuracy and processing time. The accuracy of the model is analyzed using the confusion matrix and represents the ratio between the number of correctly predicted instances and the total number of instances, given by the formula

$$\text{Acuracy} = \frac{\text{TP} + \text{TN}}{\text{TP} + \text{TN} + \text{FP} + \text{FN}} \tag{6}$$

where TP, TN, FP and FN denote true positives, true negatives, false positives and false negatives. TP and TN denote correctly recognized anomalies and normal

Table 2 Accuracy of the FNN, k-NN and wk-NN classifiers

Number of instances		Accuracy [%]		
		FNN_ACC	k-NN_ACC	wk-NN_ACC
Day_1	158,572	98.8	98.3	98.4
Day_2	128,740	98.3	98.2	98.1
Day_3	90,129	98.9	99.0	99.1
Day_4	81,807	98.3	98.8	98.8
Day_5	58,317	99.0	99.1	99.2
Day_6	57,278	99.2	99.4	99.5

data, while FP and FN denote incorrectly classified normal network behavior as abnormal and vice versa. The classification accuracies FNN_ACC, kNN_ACC and wkNN_ACC of the FNN, k-NN and wk-NN models, respectively, are presented in Table 2.

The results show the higher accuracy of the wk-NN model compared to the other classifiers. In addition, the accuracy values of the NN models are similar. Figure 1 shows that the accuracy of the classifiers, expressed as percentage. It is shown that the they do not depend directly on the number of instances.

It is interesting that the highest accuracy of all models is achieved in very few instances in the daily set. Although it was expected that more instances in the training set would increase the accuracy, this was not the cases. The results show that a large number of instances results in lower accuracy, suggesting the negative impact of model overtraining as machine learning models try to find an equation that better fits the data. The algorithmic simplicity of the learning process or the optimization process is a prerequisite for a high-precision solution. Since the k-NN and wk-NN models are relatively simple structures, the loss of accuracy cannot emerge due the overfitting, as overfitting indicates the complexity of the model structure. Hence, the accuracy indicates overtraining, which indicate problems of large and oversized test sets. In [58], several architectures are presented that improve the general properties of the nearest neighbor models.

Fig. 1 Accuracy of the evaluated binary classifiers given in percent

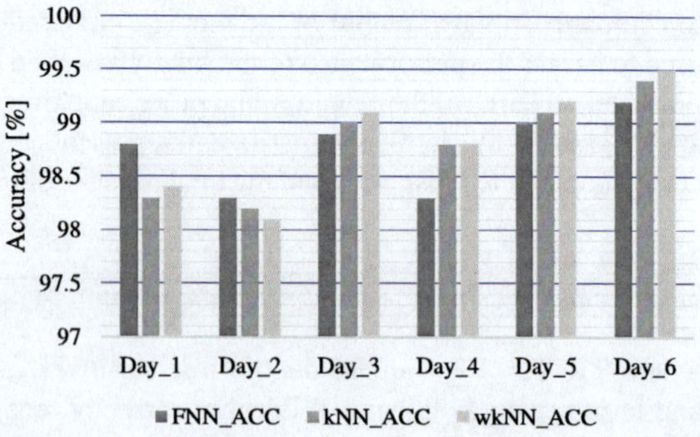

Table 3 Processing time of the proposed models

Number of instances	PT [s]			
	FNN_PT	kNN_PT	wkNN_PT	
Day_1	158,572	26	275.7	277.3
Day_2	128,740	18	193.8	194.8
Day_3	90,129	11	101.3	101.8
Day_4	81,807	10	91.2	91.3
Day_5	58,317	3	31.7	31.7
Day_6	57,278	2	43.7	43.3

The salient properties of the FNN can lead to a poor generalization of the trained FNN to fit the new data. This problem arises from (1) the size of the FNN, which can cause overfitting and (2) the processing time (PT), which leads to poor predictability and overtraining. Table 3 shows the processing time required for the model evaluation, which depends not only on the model, but also on the size of the selected data sets and the size of the training, test and validation sets.

We use the sum of the time spent for training, validating and testing the models as processing time. FNN_PT, kNN_PT and wkNN_PT represent the PTs of each of the models. Figure 2 shows the decrease in processing time with decreasing size of the data set. In addition, the significant difference between the PT of the FNN and the other two models can be seen. The result shows the difference between the eager and lazy learners, which is described in the text above.

The results do not indicate overtraining or overfitting and do not provide meaningful information about whether or not the classifier generalizes well, but the processing time may be an important factor in order to decide on the structure of the models. The learning dynamics can indicate the stability of the training in terms of the gradient, problems with decision-making, the need for a training break (early stop) or the like [59]. Based on the presented results, when the number of instances in the daily set exceeds ~ 100,000, model training must be ended.

Fig. 2 Processing time of the FNN and NN models

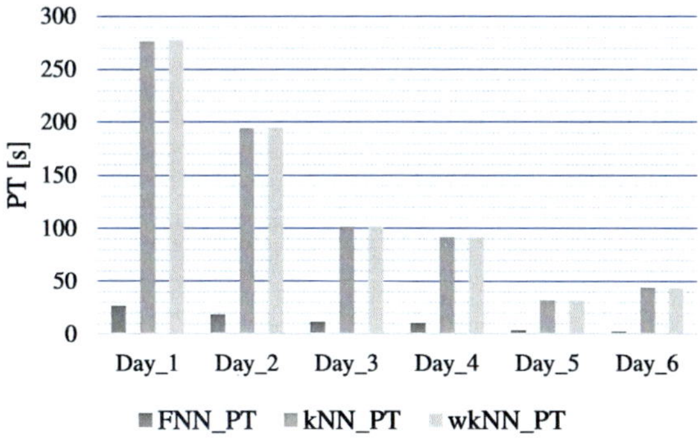

5 Conclusion

The constant need to improve IIDSs has led to the development of many new algorithms and models that classify network traffic as normal or abnormal. The binary classifiers based on the FNN and NN models have been shown to be very accurate in classifying network traffic, divided into two basic classes, one defined as 'normal' and the other as 'anomaly.' The preprocessing algorithm resulted in a high accuracy of the models, which is over 98.2%. However, there is apparent ~ 1% decrease in accuracy when the number of instances reaches around a thousand instances per day, indicating overtraining. The processing time results show less time as the number of instances per day decreases. Based on these results, our future work looks at both online optimization algorithms that reduce a number of instances and the hybrid models to further improve accuracy. The hybrid model represents a combination of FNN and one of the NN classifiers working parallel. The accuracy increases if the principle of nonparametric correlation analysis is taken into account. The principle of the correlation between two Boolean variables is based on a logical XOR operation. When executing on the outputs of the classifiers, the logical one indicates the opposite decision. In this way, the accuracy of the models can be improved.

References

1. Hanumantha Rao K, Srinivas G, Damodhar A, Krishna VM (2011) Implementation of anomaly design technique using machine learning algorithms. Int J Comput Sci Telecommun 2(3):25–31
2. Ciric V, Cvetkovic D, Gavrilovic N, Stojanovic N, Milentijevic I (2001) Input splits design techniques for network intrusion detection on Hadoop cluster. Facta Univ, Ser: Electron Energ 34(2):239–257
3. Jyothsha V, Prasad R (2011) A review of anomaly based intrusion detection systems. Int J Comput Appl 28(7):26–35
4. Protic D, Stankovic M (2018) Anomaly-based intrusion detection: feature selection and normalization influence to the machine learning models accuracy. In: Proceedings of 4th international conference on engineering and formal science, Amsterdam 14–15 Dec 2018, pp 46–51
5. Protic D, Stankovic M (2020) Detection of anomalies in the computer network behavior. In: Proceedings of 5th international conference on engineering and formal science, Brussels, 24–25 Jan 2020, pp 40–46
6. Lampton M (1997) Damping-undamping strategies for the Levenberg-Marquardt least-squares method. Comput Phys 11(1):110–115
7. Gavin H (2020) The Levenberg-Marquardt algorithm for nonlinear least squares curve fitting problems. Duke University: Department of Civil and Environmental Engineering. 18 Sept 2020
8. Sotteroni AC, Galski RL, Ramos FM (2013) The q-gradient method for continuous global optimization. AIP Conf Proc 1558:2389–2393
9. Croeze A, Pittman L, Reynolds W (2021) Solving nonlinear least squares problems with Gauss-Newton and Levenberg-Marquardt methods (2021)
10. Protic D, Stankovic M (2021) The q-Levenberg-Marquardt method for unconstrained nonlinear optimization, 1–5. http://arxiv.org/abs2017.03304
11. Levenberg K (1944) A method for the solution of certain problems in least squares. Q Appl Math 5:164–168

12. Marquardt D (1963) An algorithm for least-squares estimation of nonlinear parameters. SIAM J Appl Math 11(2):431–441
13. Hechenbichler K, Schliep K (2004) Weighted k-nearest-neighbor techniques and ordinal classification. Sonderforschungsbereich 386:339
14. Subasi A (2020) Machine learning techniques. In: Practical machine learning for data analysis using Python
15. SIGKDD—KDD Cup, KDD Cup 1999: Computer network intrusion detection (2018)
16. Song J, Takakura H, Okabe Y, Eto M, Inoue D, Nakao K (2011) Statistical analysis of honeypot data and building Kyoto 2006+ dataset for NIDS evaluation. In: Proceedings of the 1st workshop on building analysis dataset and gathering experience returns for security, Salzburg, 10–13 April 2011, pp 29–36
17. Protic D (2020) Intrusion detection based on the artificial immune system. Vojnotehnički glasnik/Mil Tech Courier 68(4):790–803
18. Demertzis K (2018) The Bro intrusion detection system, Project: Machine Learning to Cyber Security, November 2018
19. Protic D, Stankovic M (2020) A hybrid model for anomaly-based intrusion detection in complex computer networks. In: 21st International Arab conference on information technology, 6 Oct 2020, Giza, Egypt 2020, pp 1–8
20. Maeder M, McCann N, Norman S (2009) Model-based data fitting. Compr Chemometr: Chem Biochem Data Anal 3:413–436
21. Solorio-Fernandez S, Corrasco-Ochoa A, Fco J (2020) Martinez-Trinidad, Ateviev of unsupervised feature selection methods. Artif Intell Rev 53:907–948
22. Song L, Smola AJ, Gretton A, Borgwardt KM, Bedo J (2007) Supervised feature selection via dependence estimation. In: International conference on machine learning
23. Dy JG, Brodley CE (2005) Feature selection for unsupervised learning. J Mach Learn Res 5:845–889
24. Mitra P, Murthy CA, Pal S (2002) Unsupervised feature selection using feature similarity. IEEE Trans Pattern Anal Mach Intell 24:301–312
25. Porkodi R (2014) Comparison on filter based feature selection algorithms: an overview. Int J Innov Res Technol Sci 2(2):108–113
26. Liu X, Li T, Zhang R, Wu D, Lu Y, Yang Z (2021) A GAN feature selection-based oversampling technique for intrusion detection. Secur Commun Netw, Article ID 9947059, 15p
27. Dudani SA (1976) The distance-weighted k-nearest-neighbor rule. IEEE Trans Syst, Man, Cybern, SMC-6(4):325–327
28. Bicego M, Loog M (n.d.) Weighted k-nearest neighbor revisited
29. Xu H, Ptzystupa K, Fang C, Marciniak A, Kochan O, Beshley M (2020) A combination strategy of feature selection based on integrated optimization algorithm and weighted k-nearest neighbor to improve the performance of network intrusion detection. MDPI Electron 9(8):1206
30. Wang W, Gombault S (2007) Distances measures for anomaly intrusion detection
31. Tait K-A, Khan JS, Alqahtani F, Shah AA, Khan FA, Ur Rehman M, Bouila W, Ahmad J (2021) Intrusion detection using machine learning techniques: an experimental comparison. arXiv:2015.13435v1 [cs.CR] 27 May 2021
32. Haddadi F, Khanchi S, Shetabi M, Derhami V (2010) Intrusion detection and attack classification using feed-forward neural network. In: 2nd International conference on computer network and technology, pp 262–266
33. Kasongo SM, Sun Y (2020) Performance analysis of intrusion detection systems using a feature selection method on the UNSW-NB15 dataset. J Big Data 7:105
34. Protic D (2018) Review of KDD CUP '99, NSL-KDD and Kyoto 2006+ datasets. Mil Tech Courier/Vojnotehnički glasnik 66(3):580–595
35. Bohara B, Bhuyan J, Wu F, Ding J (2020) A survey on the use of data clustering for intrusion detection system in cybersecurity. Int J Netw Secur Appl 12(1):1–18
36. Ankit Thakkar, Ritika Lohiya (2020) A review of the advancement in the intrusion detection datasets. In: International conference on computational intelligence and data science (ICCIDS 2019), Procedia computer science, vol 167, pp 636–645

37. Khraisat A, Gondal I, Vamplew P, Kamruzzaman J (2019) Survey of intrusion detection systems: techniques, datasets and challenges. Cybersecurity 2–20
38. Ferryian A, Thamrin AH, Takeda K, Murai J (2021) Generating network intrusion detection dataset based on real and encrypted synthetic attack traffic. MDPI Appl Sci 11:2–17
39. Serkani E, Gharaee H, Mohammadzadeh N (2019) Anomaly detection using SVM as classifier and DT for optimizing feature vectors. ISeCure 11(2):159–171
40. Parmigiani G (2001) International Encyclopedia of the social & behavioral sciences
41. Zhou SK (2016) Medical image recognition, segmentation and parsing
42. Brownlee J (2020) 4 types of classification tasks in machine learning. In: Phyton machine learning 8 April 2020
43. Karabiber F (2021) Binary classification. What is binary classification?
44. Nawir M, Amir A, Lynn OB, Yaakob N, Badlishah Amad R (2018) Performances of machine learning algorithms for binary classification of network anomaly detection system. J Phys: Conf Ser 1018
45. Rice DM (2013) Causal reasoning. In: Calculus of thought: neuromorphic logistic regression in cognitive machines. Accademic Press Inc.
46. Sebastiani F (2002) Machine learning in automated text categorization. ACM Comput Surv 34(1):1–47
47. Breiman L (2001) Random forest. Mach Learn 24(2):123–140
48. Burges M (1998) Computer immunology. In: 12th USENIX conference on system administration, Boston, MA, USA, 6–11 Dec 1998, pp 283–298
49. Hardesty L (2017) Explained: neural networks. MIT News on campus and around the world, April 17
50. Schmidt W, Kraaijveld M, Duin R (1992) Feed forward neural networks with random weights. Delft University of Technology, Faculty of Applied Phisics, The Nederlands, pp 1–4
51. Singh S, Khan Y, Saxena AK (2014) Intrusion detection based on artificial intelligence technique. Int J Comput Sci Trends Technol 2(4):31–35
52. Protic D (2015) Feedforward neural networks: the Levenberg-Marquardt optimization and the optimal brain surgeon pruning. Vojnotehnički glasnik/Mil Tech Courier 3(63):11–28
53. Osborne MR (1992) Fisher's method of scoring. Int Stat Rev 86:271–286
54. Young-tae K, Ji-won H, Cheol-jung Y (2011) A new damping strategy of Levenberg-Marquardt algorithm for multilayer perceptrons. Neural Netw World 4(11):327–340
55. Stanford ENGR 108, Intro to Applied Linear Algebra. Lecture 51. Levenberg Marquardt, 18 (2021)
56. Lai KK, Mishra SK, Panda SK, Ansary MAT, Ram B (2020) On q-steepest descent method for unconstrained multiobjective optimization problems. AIMS Math 5(6):5521–5540
57. Bobic A (n.d.) Model selection. In: CS7616 Pattern Recognition
58. 9 Adaptive soft K-nearest-neighbor classifiers with large margin (n.d.) 1–21
59. Pamucar D, Marinkovic D, Kar S (2021) Dynamisc under uncertainity: modeling simulation and complexity. Mathematics 9(12):1416

Flexible Reverse Engineering of Desktop and Web Applications

Shilpi Sharma, Shubham Vashisth, and Ishika Dhall

Abstract The day-to-day increase in the number of cyber-criminal activities raises the demand for strengthening and refining our computer security systems. Reverse engineering plays a decisive role in upholding essential security standards. The method of reverse engineering was formerly applied to hardware, but currently, it is also being applied on software applications, databases and, even in the domain of natural sciences. In cybersecurity, reverse engineering enables the finding of the breach details attempted by the attacker. This additionally helps in the detection of bugs, vulnerabilities and loopholes present in the software application and thereby solidifying the security aspects of the application. This paper discusses the various application of reverse engineering in the field of cybersecurity, appropriate tools for its implementation and practical demonstration of reverse engineering a software application.

Keywords Reverse engineering · Reverse engineering tools · Web application · Desktop application · Cybersecurity

1 Introduction

In today's modern era, there is a need for an enriched interface and improved functionality of web applications. To meet those demands, discoveries in the field of scripting languages like Php, Node.js, Ajax, etc., are made constantly. These discoveries simultaneously create more opportunities for the attackers to discover vulnerabilities and bugs to exploit the Web sites by several attacks making them less secure and unsafe.

S. Sharma (✉) · S. Vashisth · I. Dhall
ASET, Amity University, Noida, Uttar Pradesh, India
e-mail: ssharma22@amity.edu

P. Chatterjee et al. (eds.), *Computational Intelligence for Engineering and Management Applications*, Lecture Notes in Electrical Engineering 984,
https://doi.org/10.1007/978-981-19-8493-8_45

To overcome various impediments, many preventive measures have been reported. Reverse engineering is one of those promising measures which is used to re-scan, debug and abstract the Web sites to reveal the architecture, model, data structure and all over the behavior of the web application. The process of reverse engineering can be defined as the analysis of operations, functions and structure to rediscover the technological principles of an entity or an object. Unique methods and techniques along with the reverse engineering testing tools as presented by [1] are instigated which enable tracing and detection of various vulnerabilities that can be used by the attacker to exploit the Web site. Determination of vulnerabilities in a system can be done using reverse engineering. It can also provide an adequate abstraction to Web sites and proper management of a system. It can help in the generation of UML diagrams. Also, it helps in web pattern recognition as presented by [2] and improving the overall management of the system. Hence, reverse engineering can be used to extract imperative information in order to avoid the parlous conditions and establishing a trusted and secure system.

This paper acknowledges the reader with the versatile applications of reverse engineering in the field of cybersecurity and discusses various methodologies, techniques and advancement made in the science of reverse engineering over the last decade.

2 Literature Review

The reverse engineering software engineers follow a typical style of working in order to get the best out of their work:

1. Time sensitivity: they have to be quick and attentive while dealing with any sort of malware and have to find measures to block and remove the malware as soon as possible.
2. Co-workers being off the network: this is done to prevent the risk of infecting the confidential data and other computer systems.
3. Designing own artifacts: the reverse engineering process being a great challenge itself, many a time requires creating artifacts for their own internal usage. [1] explains that the ideal process of reverse engineering consisting of the following phases:
   (a) Analyzing is the root of the reverse engineering followed by
   (b) Documenting: provides support for pre- and post-analyses
   (c) Transferring knowledge
   (d) Articular work which implies to coordination and scheduling tasks and subtasks and finally
   (e) Reporting, (the processes are being divided among different groups) parallel to the process the reverse engineering tools consist of disassemblers, visualization and communication tools, along with the process and tools, artifacts are also an important part behind the science of reverse engineering which provides annotations and cognitive support to the whole idea.

The paper concludes by providing the basic structure behind the complete process of reverse engineering and demands new reverse engineering tools to safeguard data and compete with the anonymous and complex world of hackers over the web.

Usage of reverse engineering to extract design patterns used in representing web applications in a diagrammatic manner is done in order to upsurge the readability and understanding of code for a secondary developer for maintenance purposes.. Reverse engineering thereby provides a source for developing pattern detection tools which further allows product analysis, digital updates, documenting system codes, competitive technical intelligence and academic learning. [2] Discusses the techniques used for detecting design patterns like:

- E. Gamma proposed a technique commonly known as Gang of Four (GoF) which provided a fruitful solution for detecting patterns; divides design patterns into creational, structural and behavioral design patterns.
- Dirk Heuzroth proposed a technique which was able to structure the system's behavior and analyze the code statically and dynamically with respect to patterns which could further prove the existence of a certain pattern in the code.

Further, the paper practically performs a static behavioral analysis with the aim to detect the basic structure and the behavior driven pattern taking the example of the singleton pattern. Firstly, a CFG was built where the control flow was represented via directed edges followed by backward data flow analysis, tracking variable activities. Based on the extracted design pattern finally, a UML class diagram is developed. The paper presents a conceptual diagram of IDIPDetect system application which is implemented in Java, and it provides detection and visualization approach for design pattern recognition.

PHP2XMI is an automated tool to test dynamic web applications using source information technology to recover a dynamic behavior model by reverse engineering UML 2.1 sequence diagrams for PHP-based web applications. Reverse engineering a sequence diagram generally faces five major problems:

(1) Identification of application entities since web applications doesn't follow Object Oriented approach.
(2) Identification of conditions and loops.
(3) Identification of similar execution trace patterns from run time information.
(4) Single page execution can affect several components such as session ID, cookies, etc.
(5) Analysis of multilingual documents: PHP2XMI as presented in [3] is designed keeping these challenges in mind.

PHP2XMI recovers role permissions at the level of page access but currently not at the entity level; it has shown confirming results in detection of security vulnerabilities such as XSS attack and SQL injection by the application of behavioral model recovery technique. The paper discusses the approaches used in the designing of PHP2XMI, that are:

(1) Parsing and dynamic instrumentation
(2) Filtering and sorting
(3) Database analysis and model generation.

The paper does contain the practical demonstration of generating a UML sequence model by analyzing the internet bulletin board system PhpBB2.0 successfully using PHP2XMI but still requires more working to upgrade the sequence diagram from the page level to the entity level.

The implementation of the concept of reverse engineering via penetration testing is possible, i.e., a simulated attack on a computer system done with the objective to evaluate its security or threat level, [4] states that it is being performed in order to do the vulnerability analysis through automated or normal system. Then, the paper contains a practical demonstration of using reverse engineering tools like "3d trace out," "Angry IP Scanner," "Spy-net," etc., on "ebc.com," an e-book Web site involved in selling online books and presented the vulnerabilities the Web site consisted of like XSS attacks, cookie disclosure and usage of older versions of the Apache server along with the recommendations such as filtering input to prevent injection or XSS attacks, upgrading the version of server, etc. Finally, the paper concludes by practically demonstrating the application and significance of reverse engineering in the field of cybersecurity by backtracking the footprints of the attacker using reverse engineering tools.

Reverse engineering can provide the abstraction of web applications in a more managed, structured and precise manner. Reference [5] proposes the solutions to various problems encountered while applying the process of revere engineering (Fig. 1).

The paper concludes by mentioning the need for developing reverse engineering tools required for instrumentation of the source code along with the need for migration of web applications to web services as the future work.

WebAppViewer (WAVI) is a reverse engineering tool for web applications. The implementation of WAVI is done on Node.js to take advantage of efficient modules as well as to easily share the tool for enabling precious feedbacks. WAVI uses filter-based mechanism and static analysis to recover and document the structure of the web applications. Reference [6] mentions that currently most of the reverse engineering tools are outdated and does not consider the technology used in modern web development like Ajax, which makes WAVI unique and advanced as compared to the other tools. Further, the paper discusses the functioning of WAVI: 1) taking input (HTML, CSS, Js) then, 2) extraction of the source code, which further consists of searching elements, recovering file dependency, finding element interaction, and finally 3) presenting the structure of the web application with two diagrams, i.e., force directed diagram and class diagram. On practically experimenting, WAVI resulted better in resolving JavaScript calls. The paper concludes by mentioning the extension of WAVI by integrating the analysis of more languages such as PHP, Ruby, etc., and including the feature of dynamic analysis which was missing in the current version.

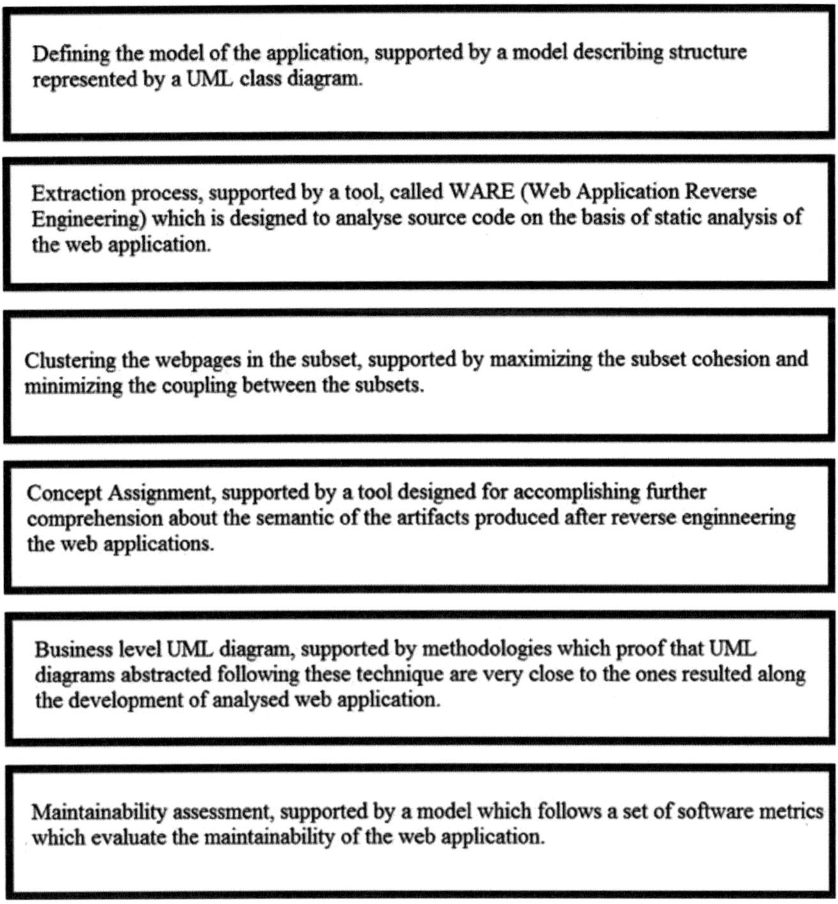

Fig. 1 Cloud computing system

The evolution of web applications from static Web sites to the modern Rich Internet Applications (RIA) in the past years has highly effected the advancement in the domain of reverse engineering. Early approaches to promote the concept of reverse engineering were made at WSE 2000 supported by RMM methodology. The basic framework for reverse engineering tools was presented by Kienle and Muller in 2001. WARE, as presented by [7] is a reverse engineering tool which used the concept of static analysis was one of the main contributors in the field of reverse engineering. The model proposed by Ricca and Tonella at WSE 2002 gathered information based on static as well as dynamic analysis. Then, Jim Connellan at WSE 2006 proposed a model which included the extended feature of UML diagrams making it more of a general model. M. L. Bernardi, at WSE 2008, proposed a model compliant with WARE and clone detection tool aiming toward the abstraction of a UWA hyperbase model. Further, M. H. Alalfi at WSE 2009 proposed a model which aimed at the analysis of dynamic interactions on the server side of web applications developed on PHP and other scripting languages. Afterward, in 2011 edition of WSE, D. Amalfitano presented a tool with the feature of automatic generation for RIA. The paper concludes by mentioning that the large usage of libraries and framework on the client and server side practically makes the complete analysis impossible. Moreover, the techniques that are applied to the RIA still suffer from the problem of state explosion because of the application of clustering techniques which remains unresolved.

In order to prevent the tampering attacks on many cyber systems including Android which is done by the attacker to interfere with the system. The analysis after reverse engineering is finally used to propose a few approaches to obfuscation by introducing some redundancies into the program. Furthermore, the paper discusses anti-debugging and watermarking to strengthen tamper-proof design. To apply the Automated N-version Generation, a candidate solution consisting of three algorithms is proposed in the paper. Reference [8] proposes the concept of "N-version Obfuscation," being inspired by the idea of the classical "N-version Programming." The resulted NVO comes out to be more tamper resistant (O(N) security), less costly (O(1)) and automated.

It is possible to reverse engineer the web applications by abstracting and using UML diagrams. Reverse engineering is used to enhance the web application which is characterized by the help of goals, tools and models; here goals specify the revere engineering motivations; i.e., setting the abstraction level, models define the information to be extracted, and the reverse engineering tool is used for the maintenance of the recovering process. Reference [9] proposes a tool for the same and uses a real web application for a case study to confirm their approach. This tool widely accepts various client- and server-side scripting languages like JS and VBScript, etc. Further static and dynamic analyses followed by behavioral analysis are performed in order to present the conceptual model for the web application. The paper discusses a case study of a research project named as LINK, where reverse engineering was successfully implemented in order to detect various components, and then, UML diagrams and class diagrams were analyzed to predict the behavior of the web application. The paper concludes by mentioning the future work as up gradation of the tools for accepting other scripting languages like PHP and implantation of an abstractor in order to provide a UML diagram automatically and also to enhance the annotation and visualization of the diagrams.

Reverse engineering can also be defined as the process of understanding programs logic and behavior deprived of the source code of a program. DynStruct is an open source reverse engineering automatic tool which recovers the data structure used by an executable binary. Unlike most of the reverse engineering tools, it uses dynamic analysis rather than the static analysis which frees it from the need to understand the control flow of a program. Secondly, DynStruct is also able to analyze an obfuscated program which is one of the main obstacles in the analysis performed by static analysis-based tools. Reference [10] discusses the data gathering phase of DynStruct which further comprises of allocation monitoring, access monitoring, function call, context and finally data recording and output. The paper concludes by presenting promising results obtained by reverse engineering small as well as big applications by DynStruct in order to detect the vulnerabilities by understanding their data structure.

SRCYUML is a reverse engineering tool developed in C++. This tool is able to produce yUML (text format for UML class diagram) as output by taking srcML (XML representation of the abstract syntactic information of source code) as input. The present version of SCRYUML supports many features like:

(1) Separation of classes and interfaces
(2) Attribute details including the type, visibility and multiplicity
(3) Operation details
(4) Parameter details and
(5) Associations comprising of aggregation, composition, generalization and realization.

The tool works run as a command line interface (CLI) tool; it is an open source tool and is made available at srcML.org under the download page. Reference [11] mentions SRCYUML as a highly efficient tool for reverse engineering the class diagrams because of its mapping technique. In the end paper mentions adding new features like providing user-friendly layout and adding support of more languages like C# and JAVA as the major of the future work.

Reverse engineering is one of the efficient methods for adding the security to the system still creates many difficulties when we talk about large applications; to tackle this situation, [12] proposes a system security performance model for dynamic web applications which adds security and makes the maintenance of the application easy. The proposed model uses a UML-based secure software maintenance procedure, and it uses reverse engineering and security engineering for mining the web application patterns and finding out the vulnerabilities from the application. The paper further proposes an SPF-based framework of TOS web application which can improve the speed effectiveness and server's efficiency. Further, the paper presents the code developed in C++ in support of the model and offers secured forward and reversed engineering. In the end, the paper concludes by testing their model successfully for enabling easy maintenance of large-scale applications and adding security to old web applications which are based on object-oriented software development technique.

WebUml is an automatic reverse engineering tool which constructs UML diagrams for existing web applications. WebUml can generate state and class diagrams by source code analysis and web server communication. This tool uses static analysis and dynamic analysis by exploitation of the server-side execution engine. Reference [13] discusses the review of the current works in modeling, testing, reverse engineering and testing in the field web application. The paper further discusses several web application components and divides them into three major parts (Fig. 2).

Further, the meta model of the class diagram is presented. The class diagram generated using WebUml is used to describe the structure of web application, e.g., Java applets, cookies, sessions, frames and objects, whereas the state diagrams generated are used to describe the navigational structure (client/server pages, scripting code flow, form inputs, navigational links and other static and dynamic components) and

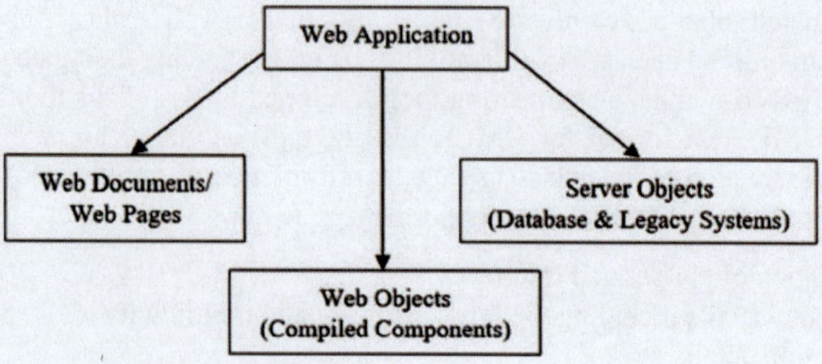

Fig. 2 Components of web application

behaviors of a web application model. Paper further discusses a running example to reverse engineer of a web application using WebUml; it was able to generate a class diagram, state diagram to derive the data structure of the web application. The paper concludes by successfully explaining the architecture of WebUml, which is able to describe the UML models with insignificant user interaction.

3 Methodology

The above-reviewed papers are able to present several concepts of reverse engineering. There are different approaches styled by the authors to implement the concept of reverse engineering to analyze and synthesize software applications. In broader terms, reverse engineering differentiates an entire application into segregated components, which makes analysis and synthesis of these components possible. This is done by construction of distinctive and resourceful reverse engineering tools or reverse engineering kits, as presented by [6, 10] and [11]; these tools ensure tranquil and orderly enactment of the theory of reverse engineering in the real-world complications. Furthermore, the concept of reverse engineering is also helpful in abstracting an application by generating various types of UML diagrams which enable elucidation of the entire application and helps in understanding the behavior of the software applications. Overall reverse engineering has proved to be one of the most systematic and efficient approaches that can be used to nonconcrete and comprehend the existing desktop and web application we have today. Applications of the concept of reverse engineering primarily in the field of cybersecurity are enormous, consisting of pattern recognition, structure recovery, debugging and software analysis to apprehend the inclusive edifice of the web application which empowers one to resolute numerous vulnerabilities and helps in solidification of the overall system.

4 Result

The reviewed papers were useful to use various concepts of reverse engineering. The existing, insecure web and desktop applications can be modified by using reverse engineering. Flexible reverse engineering can be performed on desktop and web platforms in order to execute vulnerability analysis, fix bugs and improve the performance of a program even in those cases where the source code is unavailable. With the assistance of reverse engineering tools, the life cycle of creating software could be reversed as shown in Fig. 3.

The concept of reverse engineering is being used to develop distinctive reverse engineering tools, with the help of which we are able to unravel various imperfections and errors that exist in the applications. An existing application could be decomposed using a reverse engineering tool allowing us to generate UML diagrams which can be used to interpret the sense and working of the application. The same existing application can then be analyzed and re-engineered with the motive to make it robust and more secured.

Table 1 shows of the applications of reverse engineering in the domain of cybersecurity along with the reverse engineering tool for implementing the same practically.

Below is a practical demonstration of reverse engineering a "trail" piece of the application, using a reverse engineering tool named as OllyDbg with the intention to reveal the vulnerabilities and differentiate the components contained within the application. OllyDbg is one of the most popular reverse engineering tools which is used for binary code analysis. It is a 32-bit debugger/dissembler. The most important feature of Olly is that it is able to patch native binaries, i.e., in those situations where source code is unavailable. Olly provides a user-friendly interface and an improved functionality by allowing the extension of third-party plugins. It also enables one to patch the corresponding binaries to bypass various restrictions imposed on the software. Olly also makes it convenient for one to fix the inherent bugs present in

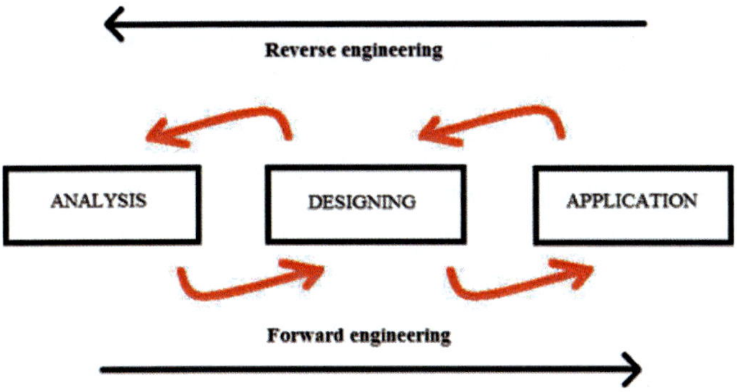

Fig. 3 Block diagram of reverse engineering a software

Table 1 Reverse engineering tools and their applications

Application area	Suggested tool
Determination of bugs and loopholes	OllyDbg, WinDbg
Web pattern recognition	Fujaba
Web applications reverse engineering	WAVI, WARE
Dynamic web applications	PHP2XMI, WebUml
Generation of accurate UML class diagrams	SRCYUML
Structure recovery and memory in use analysis	DynStruct
Android and Java (.class) files	Dex2jar
Efficient java Decompilation	Jad Debugger
Deobfuscation	Jackstab

the software, making it the most suitable reverse engineering tool for accomplishing our goal [14, 15].

The downloaded sample used has some copy protections. When executed, this sample application pops up an error message stating that the software has been expired and denies further access to it, as shown in Fig. 4.

The labeled components of Olly are shown in Fig. 5.

The first step for reverse engineering the Unfixed_sample.exe application is by opening it in OllyDbg by clicking the menu bar and navigating to the location of Unfixed_sample.exe. Then, Olly automatically disassembles the binary files, as shown in Fig. 6.

After the above step, this will take us to the entry point. In the next step, our goal is to search errors by scrolling the opcodes. Rather than going through each and every line of code for searching the errors instead, we will use Olly for tracing the error message. For this, we need to run Olly by hitting F9 or by pressing the run button to encounter the error message as shown in Fig. 7.

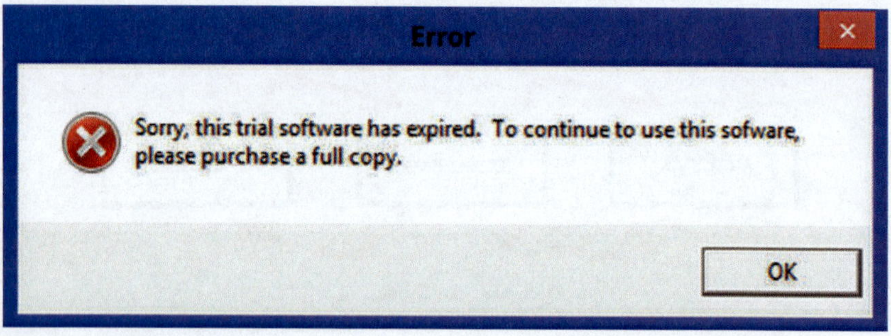

Fig. 4 Executing the sample file

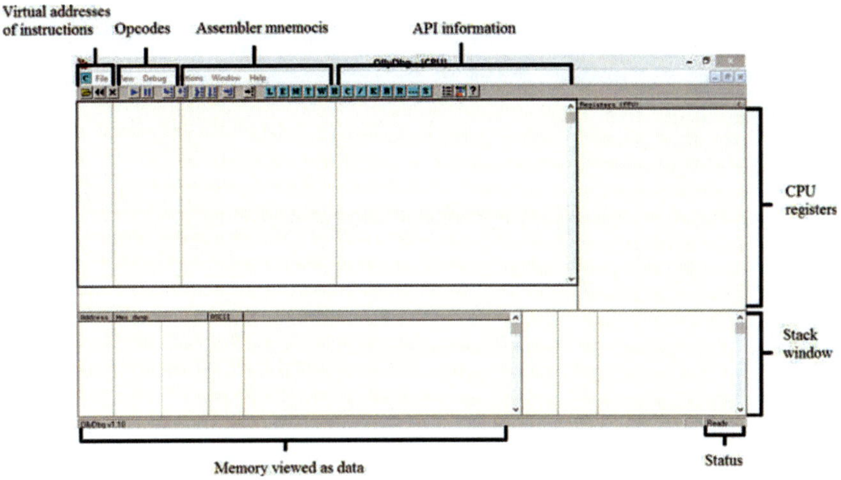

Fig. 5 Layout of OllyDbg

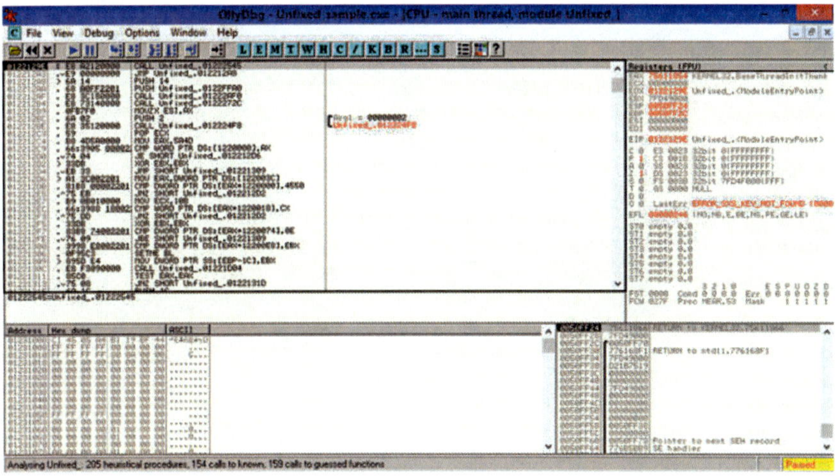

Fig. 6 Disassembling the binary file

Next, we need to pause the execution by pressing the "pause" button or by hitting the F12 button in order to search for the code that causes the error. We need to examine the current call stack for resolving the cause of the error. This can be done by pressing Alt + K; by doing this, we can note that the error message string is a parameter of the MessageBoxA function call and then right-click and select "Show call" at USER32.MessageBoxA, see Fig. 8.

After selecting the Show call, note the highlighted line with the ">" symbol as shown in Fig. 9.

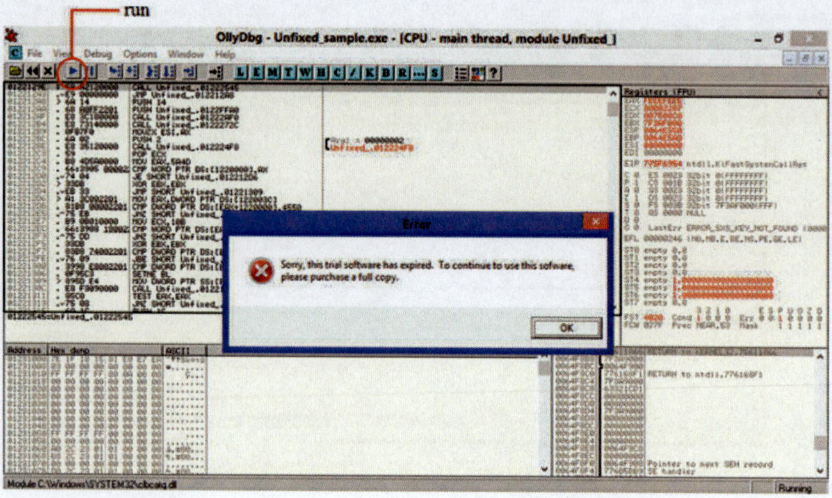

Fig. 7 Running Olly

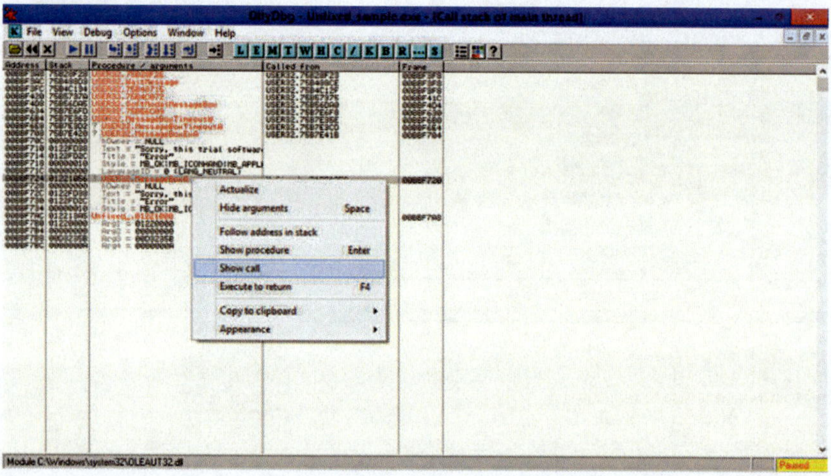

Fig. 8 Call stack window and select Show call

The ">" symbol specifies the other fragment of code that jumps to this position. Because we are at PUSH 10 instruction (indicated by the gray line), see Fig. 9. Next, we need to inspect the Hints pane to discover the fragment of code that references this call. By right clicking on the Hints pane, a context menu which permits us to reach out the code where the jumps are made appears. In Fig. 9, there are two jumps:

(1) Go to JNZ from 01221060
(2) Go to JE from 01221064.

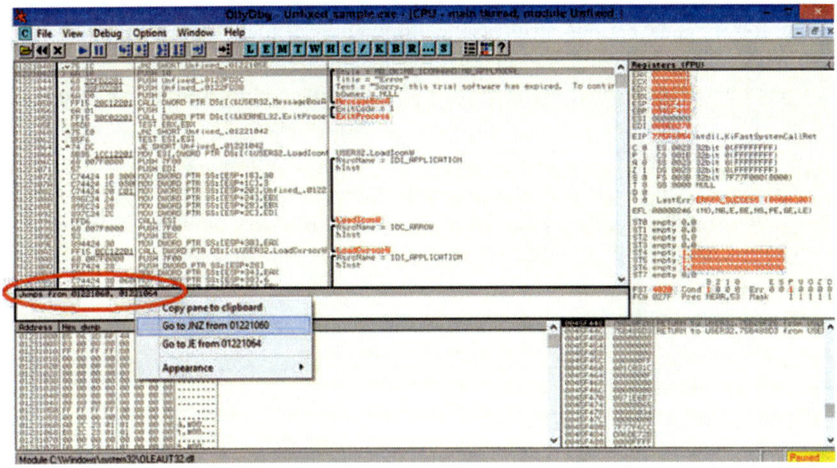

Fig. 9 MessageBoxA call and Hints pane

Therefore, there is need to amend these parts of the code. By selecting "Go to JNZ from 01221060" from the context menu, we will reach to command at 01221060, see Fig. 10.

Next, replace the jump command to no operation (NOP), i.e., right-click the command at 01221060, then select binary and then further select "Fill with NOP's (see Fig. 10). Hit '–', i.e., the minus key to return to the "PUSH 10" command. Repeat the above steps in a similar order for "Go to JE from 01221064" jump statement [16].

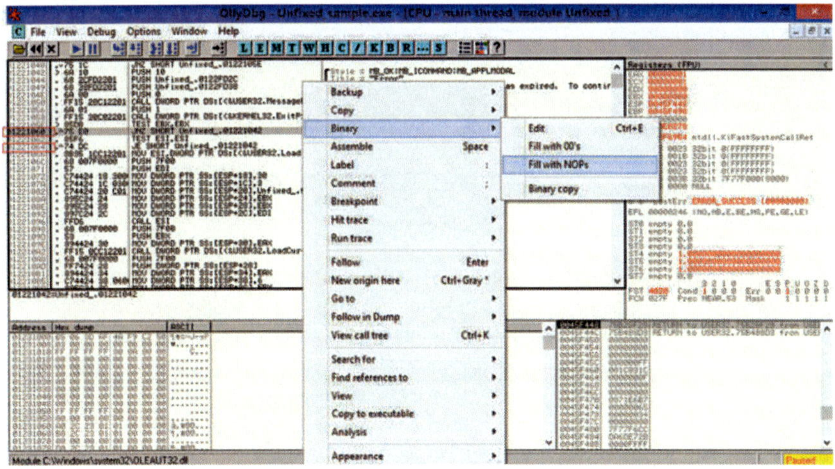

Fig. 10 Patching jumps with NOP's

The above modifications result in the elimination of the code path that directed us to the error message. After successfully patching jumps with NOP's, save all the amendments by right-clicking in the CPU window and then selecting "Copy to executable" followed by "All modifications" as done in Fig. 11.

Then select the operation "Copy all" in the dialog box (see Fig. 12).

By carrying out the above steps, a fresh window will pop up. Right-click and choose the option to "Save file" and then finally rename and save the file at the desired location as shown in Fig. 13.

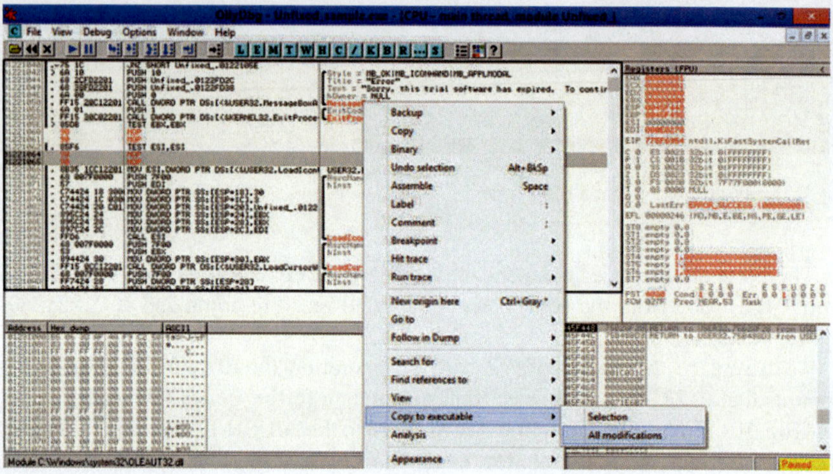

Fig. 11 Saving the modifications

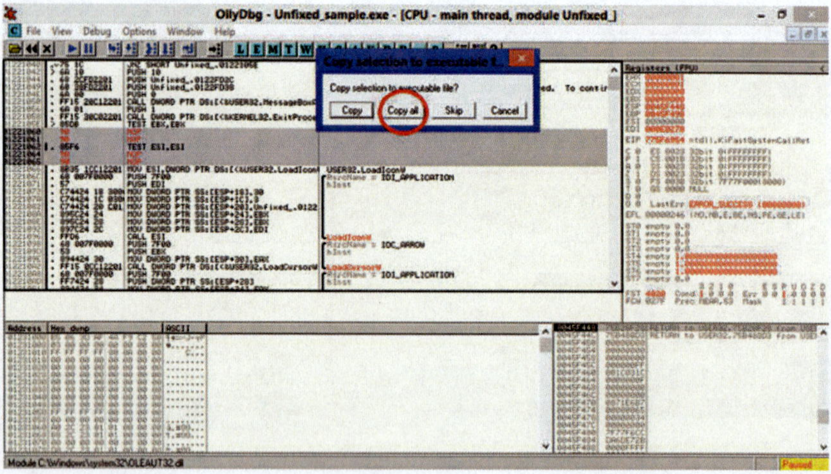

Fig. 12 Copying selection to executable file

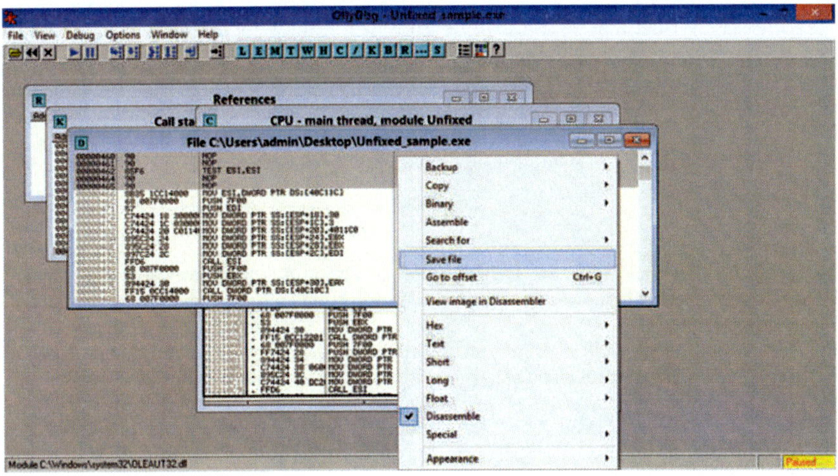

Fig. 13 Saving

Hence, with the help of OllyDbg, we were able to successfully reverse engineer the desktop application (source code unavailable) in order to reveal the vulnerabilities and destabilized components present in it.

5 Conclusion

For recent years, reverse engineering has proved to be a definitive approach for strengthening the web and desktop applications by enabling detection of vulnerabilities, bugs, loopholes, etc. This process is implemented using various reverse engineering tools that work on static analysis and dynamic analysis. These tools can generate various illustrations (UML diagrams) allowing one to understand the structure and behavior of the application in a defined manner. This further allows one to achieve more abstractness in the representation of the application and helps in understanding the data structure and the overall model of the same. Since new scripting languages and technologies like Ajax are being developed for facilitating an advanced interface and service for a web application.

6 Future Perspectives

Today, most of the existing reverse engineering tools are restricted to a single platform and tools for reverse engineering web application lack the ability to recognize the modern scripting technology used in web application development. Thus, there is a need for such tools that support the feature of platform independence and have a ubiquitous design in order to produce more efficient and an unabridged outcome for a better understanding of an application.

References

1. Treude C, Figueira Filho F, Storey MA, Salois M (2011) An exploratory study of software reverse engineering in a security context. In: Working conference on reverse engineering, pp 184–188
2. Thankappan J, Patil V (2015) Detection of web design patterns using reverse engineering. In: Second international conference on advances in computing and communication engineering, pp 697–701
3. Alalfi MH, Cordy JR, Dean TR (2009) Automated reverse engineering of UML sequence diagrams for dynamic web applications. In: International conference on software testing, verification, and validation workshops, pp 287–294
4. Kumar M (2017) Reverse engineering and vulnerability analysis in cyber security. Int J Adv Res Comput Sci
5. Tramontana P (2005) Reverse engineering web applications. In: International conference on software maintenance, pp 705–708
6. Cloutier J, Kpodjedo S, El Boussaidi G (2016) WAVI: a reverse engineering tool for web applications. In: International conference on program comprehension
7. Tramontana P, Amalfitano D, Fasolino AR (2013) Reverse engineering techniques: from web applications to rich Internet applications. In: IEEE International symposium on web systems evolution (WSE) pp 83–86
8. Xu H, Zhou Y, Lyu M (2016) N-version obfuscation. In: International workshop on cyber-physical system security, pp 22–33
9. Di Lucca GA, Di Penta M, Antoniol G, Casazza G (2001) An approach for reverse engineering of web-based applications. In: Working conference on reverse engineering, pp 231–240
10. Mercier D, Chawdhary A, Jones R (2017) dynStruct: an automatic reverse engineering tool for structure recovery and memory use analysis. In: International conference on software analysis, evolution and reengineering, pp 497–501
11. Decker MJ, Swartz K, Collard ML, Maletic JI. A tool for efficiently reverse engineering accurate UML class diagrams, International Conference on Software Maintenance and Evolution (2016) 607–609
12. Pathak N, Sharma G, Singh BM (2017) Towards designing of SPF based secure web application using UML 2.0. Int J Syst Assur Eng Manag, pp 208–218
13. Bellettini C, Marchetto A, Trentini A (2004) WebUml: reverse engineering of web applications. In: Proceedings of the 2004 ACM symposium on applied computing, pp 1662–1669
14. Jung YK, Chang K, Park SH, Ho VT, Shim HJ, Kim MW (2021) Reverse engineering and database of off-the-shelf propellers for middle-size multirotors. Unmanned Syst, pp 321–332
15. Oh SJ, Schiele B, Fritz M (2019) Towards reverse-engineering black-box neural networks. In: Explainable AI: interpreting, explaining and visualizing deep learning, pp 21–144
16. Sabharwal S, Sharma S (2020) Ransomware attack: India issues red alert. In: Emerging technology in modelling and graphics, pp 471–484

Sentence Pair Augmentation Approach for Grammatical Error Correction

Ryoga Nagai and Akira Maeda

Abstract The deep learning model requires a large amount of data when learning a task. The sentence proofreading task requires the text before and after proofreading as the training data. Usually, most of the available publications are after proofreading and are easily accessible. However, the text before proofreading is rarely seen in the general publications and is highly difficult to obtain. In this study, we assume a case where we cannot prepare sufficient amounts of data for training a sentence proofreading task, such as work procedure manuals. We propose a method that automatically generates both pre- and post-proofreading sentences. We generate pseudo post-proofread sentences by Markov chains or GPT-3. The sentences generated by Markov chains are often semantically incorrect. We identify and remove these incorrect sentences by gated recurrent unit (GRU). Then, we generate pseudo pre-proofread sentences by adding noises to the pseudo post-proofread sentences using three different methods. In the experiments, we have compared the case where seq2seq-based grammatical error correction method is trained with a small corpus only, and the case where it is trained with the pseudo-sentence pairs generated in this study in addition to the small corpus. As a result, one of our methods improved the accuracy by 12.1% in the metric BLEU.

Keywords Deep learning · Natural language processing · Proofreading · seq2seq

1 Introduction

Many people, both beginners and professionals, use word processors like Microsoft Word. The word processors often have the ability not only to write sentences but

R. Nagai
Formerly Graduate School of Information Science and Engineering, Ritsumeikan University, 1-1-1 Noji-Higashi, Kusatsu 525-8577, Shiga, Japan

A. Maeda (✉)
College of Information Science and Engineering, Ritsumeikan University, 1-1-1 Noji-Higashi, Kusatsu 525-8577, Shiga, Japan
e-mail: amaeda@is.ritsumei.ac.jp

P. Chatterjee et al. (eds.), *Computational Intelligence for Engineering and Management Applications*, Lecture Notes in Electrical Engineering 984,
https://doi.org/10.1007/978-981-19-8493-8_46

also to automatically proofread the written sentences. Proofreading is the process of correcting incorrect sentences into correct sentences. Automated proofreading may improve the readability and semantic correctness of sentences. Most traditional automated proofreading systems are implemented by rule-based systems. They record the history of a wide variety of revisions as rules and revise the sentences that match those rules. In recent years, an increasing number of them have been implemented based on deep learning. Inui et al. [1] have shown the effectiveness of seq2seq for sentence proofreading.

Deep learning models predict future events from past cases. Highly accurate deep learning models require a large amount of data. The sentence proofreading task requires the sentences before proofreading as the input and the sentences after proofreading as the target. The sentences after proofreading can easily be obtained as a publication, even through the Internet. However, it is generally difficult to obtain the sentences before proofreading, as these sentences are usually not publicly available, but kept in publishing companies. Therefore, when training a deep learning model optimized for automatic proofreading of a domain, the data may not be sufficiently available. Feng et al. [2] have reported that there is a lot of research being done on data augmentation in natural language processing (NLP), but there are still many areas that have not been explored.

In this study, we propose a method to automatically augment the sentence data before and after proofreading. First, a Markov chain or GPT-3 will automatically generate the post-proofreading sentences. Sentences generated from Markov chains are often grammatically correct but not semantically correct because they are generated based on the order probability of words in given sentences. Gated recurrent unit identifies and deletes sentences that appear to be generated from a Markov chain. The remaining sentences are used as the pseudo post-proofread sentences. We generate pseudo pre-proofread sentences from pseudo post-proofread sentences using three different methods: Wikipedia, Word2vec word substitution combined with rule-based editing, and back translation. We have compared the case where seq2seq-based grammatical error correction method is trained with a small corpus only, and the case where it is trained with the pseudo-sentence pairs generated in this study in addition to the small corpus.

We have presented the previous version of this study in [3]. The major improvements from it include the corpus size, the use of GPT-3, pre-proofreading sentence augmentation method, deep learning model optimization, and data formatting. We used only 10,000 sentences in SNOW T15: Japanese Simplified Corpus with Core Vocabulary [4] corpus instead of 50,000. It is a parallel corpus that includes easy and difficult Japanese and their English translations. In our previous work, only Markov chains were used to generate pseudo post-proofread sentences, but in this study, we also use GPT-3 to generate them. We modified the learning rate of our deep learning model and reshaped the data to make it suitable for deep learning training. The method proposed in this study has a significant improvement over the accuracy of the method we proposed in our previous work.

2 Related Work

2.1 Deep Learning in Natural Language Processing

In the natural language processing field, time-series processing models are generally used. A recurrent neural network (RNN) is the most popular one of them. It is a model that refers to past states when predicting the output from the current state. Such a model can be trained on time-series data such as a sentence. Seq2seq, which is used in machine translation and text proofreading, is a combination of two time-series processing models that are connected as encoder and decoder. This is appropriate for translation and proofreading tasks. In this study, we have used seq2seq as a baseline model.

2.2 Markov Chain

The Markov chain model generates sentences based on the ordering probabilities of words in a given corpus. It follows the Markov property. The Markov property means that the future state is determined by the present state only and is independent of past state transitions. In this study, we use the Markov chain model to generate pseudo post-proofread sentences.

2.3 GPT-3

GPT-3 [5] is the deep learning model that has been released by OpenAI. GPT-3 has 175 billion parameters and is trained with 45 TB of text data. It can be used for a variety of tasks and is very accurate. In this study, we use GPT-3's ability to "predict the future sentence from a given sentence" to generate pseudo post-proofread sentences. We have compared the accuracy of using pseudo post-proofread sentences generated from GPT-3 with that of using sentences generated from the Markov chains.

2.4 Data Augmentation for NLP

When there is not enough training data to train a deep learning model, often the amount of data is augmented by adding noise to the data. For example, in the field of image processing, images are rotated, cropped, or changed in color for data augmentation. In the field of natural language processing, the amount of data is also increased by converting words into similar words, randomly deleting words, and so on. Coulombe

[6] reported that the rule-based augmentation of sentence data can improve the accuracy of deep learning models. In this study, we employ three different methods for generating pseudo pre-proofread sentences: word substitution using the Wikipedia revision history corpus [7] combined with a rule-based method, word thesaurus substitution using Word2vec combined with a rule-based method, and back translation using the CopyNet model. We generate pseudo pre-proofread sentences by applying one of these three methods to the pseudo post-proofread sentences generated by the Markov chain or GPT-3.

2.5 CopyNet

Edunov et al. [8] reported that the dataset augmentation by back translation is effective in improving the accuracy of deep learning models for translation tasks. It is a method of increasing the amount of data by using a translator to generate input sentences from the target sentences when there are no input sentences for the target sentences. Gu et al. [9] also reported a model that learns to copy the words from the input sentences, focusing on the fact that there is no significant difference between input and output sentences in the proofreading task. It is called CopyNet. In this study, we use CopyNet for back translation.

3 Proposed Method

3.1 Overview

In this study, we augment the sentence pair corpus for the proofreading task. The overall flow of the proposed method is shown in Fig. 1. First, the Markov chain model or GPT-3 is used to generate pseudo post-proofread sentences. Since sentences generated from the Markov chain model may have semantic errors, we classify them according to whether they appear to be generated from Markov chains or not by GRU. The sentences that are classified as not appearing to be generated from Markov chains are used as pseudo post-proofread sentences. Insertion of noise is done on the pseudo post-proofread sentences using one of the three augmentation methods. The sentences generated by these methods are used as pseudo pre-proofread sentences. We aim to improve the accuracy of the sentence proofreading task by adding the sentence pairs of pseudo pre- and post-proofread generated by the above processes to the 10,000 sentences of SNOW T15 as the augmented data.

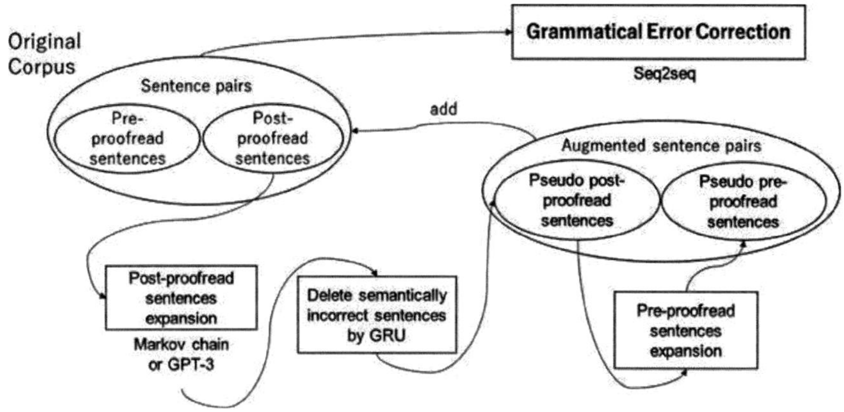

Fig. 1 Overview of the proposed method

3.2 Generating Pseudo Post-proofread Sentences

The first thing to do is to augment the target sentences, which are the post-proofread sentences. The post-proofread sentences are sentences in which errors have been corrected. Therefore, the pseudo post-proofread sentence generation must always generate grammatically correct sentences. We use Markov chain or GPT-3 to generate pseudo post-proofread sentences. Since it is important in this study that the output of the Markov chain is grammatical, we set n to 3 for the nth-order Markov chain.

The sentences generated from GPT-3 are assumed to be correct and are not classified by GRU. The sentences generated by one of the methods described above are treated as pseudo post-proofread sentences, and their effectiveness is compared.

3.3 Deletion of Sentences by Gated Recurrent Unit (GRU)

The sentences generated from the Markov chain model may contain semantically incorrect sentences. Therefore, such sentences must be removed. In this study, GRU classifies the sentences that appear to be generated from the Markov chain model and those that do not, removes those that appear to be so, and retains those that do not as pseudo post-proofread sentences. The output is set to 0–1 by sigmoid. The GRU identifies that 0 is generated by Markov chain, and 1 is not.

3.4 Generating Pseudo Pre-proofread Sentences

We generate pseudo pre-proofread sentences from the generated pseudo post-proofread sentences. We use three different methods to generate erroneous sentences by inserting noise into the pseudo post-proofread sentences.

The first method is a combination of word substitution using the Wikipedia revision history and a rule-based method. If a word matching post-revised word in the Wikipedia revision history corpus is found in a pseudo post-proofread sentence, it is converted into the post-revised word with a probability of 3%.

The second method is a combination of word substitution using word2vec and a rule-based method. Word2vec is a method of representing the features of a word by vectorizing it. We convert nouns in pseudo post-proofread sentences into synonyms with a probability of 3% using word2vec. Our word2vec model is trained on the Wikipedia Japanese corpus [10]. The rule-based method is a probabilistic method of substitution, insertion, and deletion of words in sentences. The probability of each occurrence is shown in Table 1. We have significantly reduced the probability of replacement compared to our previous study. In our previous study, there was a 66% chance of replacement occurring more than once. Realistically, replacement errors do not occur with such frequency, and therefore, we reduced this probability.

The third method is back translation. In back translation, the data that is supposed to be the output is used as the input and the data that is supposed to be the input is used as the output. We adopt CopyNet as our model for back translation. The 10,000 sentence pairs from SNOW T15 are used as the training data. The trained model takes the pseudo post-proofread sentences as the input and outputs the pseudo pre-proofread sentences.

Table 1 Editing rules

Rule	Explanation
Replacement	3% probability of swapping the positions of two words determined by uniform random numbers, 1% probability of doing this operation twice, and 96% of doing nothing
Insertion	For each word, there is a 5% probability that the same word will be inserted one word after it
Deletion	For each word, there is a 5% probability of deletion

4 Experiments

4.1 Dataset

Overview. In this study, we assume that we do not have enough pre- and post-proofed sentence pairs, which is necessary when training for the sentence proofreading task. For this reason, we use 10,000 randomly selected sentence pairs of easy and difficult Japanese out of the 50,000 sentence pairs in the SNOW T15 corpus. We treat easy Japanese as the post-proofread sentences and difficult Japanese as the pre-proofread sentences. In this section, "SNOW T15 corpus" refers to the 10,000 sentence pairs described above. The SNOW T15 corpus is used as the training data for the baseline model. For training the Markov chain model, and for the input to GPT-3, we use easy Japanese sentences from the SNOW T15 corpus. We generate 40,000 sentences from 10,000 sentences of the SNOW T15 using the Markov chain model and retained 32,645 sentences after removing duplicates. We use SentencePiece [11] as the Japanese morphological analyzer. SentencePiece can morphologically analyze the sentences by building a dictionary to keep the corpus to a specified number of words using unsupervised learning. It has the advantage that it eliminates the out-of-vocabulary problem and keeps all the words in the vocabulary.

Deletion of Sentences by GRU. The training data for the GRU model to identify sentences generated from the Markov chain model is the Wikipedia Japanese corpus. The Wikipedia Japanese corpus is separated into sentences by periods. We pick one million randomly selected sentences that are between 20 and 30 characters in length (1). The Markov chain model is trained with these sentences as the input to generate one million pseudo-sentences (2). GRU is trained on (1) as semantically correct sentences and (2) as semantically incorrect sentences.

4.2 Experimental Settings

GPT-3. We use Curie as the engine for GPT-3. The parameter settings for GPT-3 are as follows: temperature is 0.7, top P is 1, frequency penalty is 1.5, presence penalty is 1.5, and the stop sequence is "。". "。" means a period in Japanese. It uses only the first five characters of the 10,000 sentences as the input to generate 10,000 pseudo post-proofread sentences. We also compare the results with the case where five sentences are generated per sentence under the same conditions.

Deletion of Sentences by GRU. The GRU that identifies semantically incorrect sentences takes a sentence as the input and a number from 0 to 1 as the output. It is bidirectional and has two layers, the hidden layer dimension is 256, and dropout rate is 0.2. The sentences with an output of less than 0.5 are considered to be human-like sentences, while the sentences with an output of 0.5 or more are considered to be sentences generated by a Markov chain.

Seq2seq Model for Experiments. The proofreading model used in the experiment is seq2seq, in which both encoder and decoder are GRUs. This model is bidirectional, with a hidden layer dimension of 256 and a dropout rate of 0.2. The output layer is softmax. The batch size is 256, the optimization function is Adam, and the loss function is Pytorch's NLLoss, which is common to all models in this study.

4.3 Results

The sentence pairs of SNOW T15 are divided into 7:2:1 to be used as training, validation, and test data. We compare the accuracy of seq2seq with and without the addition of pseudo pre- and post-proofread sentence pairs generated by each method. For the evaluation index, we adopt accuracy, which measures simple word matching, and BLEU, which measures the degree of matching by N-grams. BLEU outputs a number between 0 and 1. In this paper, it is multiplied by 100 and expressed as a percentage of 100. Table 2 shows the evaluation results of our proposed method.

The pseudo post-proofread sentences generated by the Markov chain model and the pseudo pre-proofread sentences generated by back translation are the most effective of all the methods. This combination is 12.1% higher than the baseline method on BLEU. All the other methods also achieved better results than the baseline method. Relatively poor results of GPT-3 may be due to a problem with the augmentation method in this study. In the experiments, the first five characters of the SNOW T15 corpus are given to GPT-3. It is to predict what comes after that to generate pseudo post-proofread sentences, but the sentences generated by this method do not have originality because the first five characters always match the original sentences. Since GPT-3 is a sophisticated model, we need to figure out how to use it effectively.

Table 2 Experimental Results

Pseudo post-proofread augmentation method	Pseudo pre-proofread augmentation method	Accuracy	BLEU
Baseline		70.5	59.8
Markov chain	Wikipedia	73.9	68.9
	Word2vec	74.6	70.2
	Back translation	75.4	71.9
GPT-3 (one-to-one)	Wikipedia	72.2	67.5
	Word2vec	72.8	68.8
	Back translation	72.6	68.3
GPT-3 (one-to-five)	Wikipedia	72.0	68.3
	Word2vec	72.9	71.0
	Back translation	72.5	71.2

5 Conclusion

This study shows that pseudo pre- and post-proofread sentence pairs generated by a combination of several methods can contribute to the accuracy of the automatic proofreading task. The results in Table 2 show that the most effective sentence pairs are the pseudo post-proofread sentences generated by the Markov chain model and the pseudo pre-proofread sentences generated by back translation. Back translation is the most effective in all cases of pseudo pre-proofread sentences augmentation. We assume that this is because the other two methods add errors into a pseudo post-proofread sentence, while back translation successfully simulates the actual errors. In future work, we will consider effective augmentation methods using GPT-3.

References

1. Hitomi Y, Tamori H, Okazaki N, Inui K (2017) Proofread sentence generation as multi-task learning with editing operation prediction. In: Proceedings of the Eighth international joint conference on natural language processing, pp 436–441
2. Feng YS, Gangal GV, Wei J, Chander S, Vosoughi S, Mitamura T, Hovy E (2021) A survey of data augmentation approaches for NLP, arXiv preprint arXiv:2105.03075v5
3. Nagai R, Maeda A (2021) Dataset augmentation for grammatical error correction using Markov chain. In: Proceedings of the world congress on engineering 2021, pp 97–100
4. Maruyama T, Yamamoto K (2018) Simplified corpus with core vocabulary. In: Proceedings of the 11th international conference on language resources and evaluation, LREC 2018, pp 461–466
5. Brown BT, Mann B, Ryder N, Subbiah M, Kapalan J, Dhariwal P, Neelakantan A, Shyam P, Sastry G, Askell A, Agarwal S, Ariel H-V, Krueger G, Henighan T, Child R, Ramesh A, Ziegler MD, Wu J, Winter C, Hesse, Chen M, Sigler E, Litwin M, Gray S, Chess B, Clark J, Berner C, MacCandlish S, Radford A, Sutskever I, Amodei Dario (2020) Language models are few-shot learners, arXiv preprint arXiv:2005.14165
6. Coulombe C (2018) Text data augmentation made simple by leveraging NLP cloud APIs, arXiv preprint arXiv:1812.04718
7. Tanaka Y, Murawaki Y, Kawahara D, Kurohashi S (2020) Building a Japanese Typo dataset from Wikipedia's revision history. In: Proceedings of the ACL 2020 student research workshop, pp 230–236
8. Edunov S, Ott M, Auli M, Grangier D (2018) Understanding back-translation at scale. In: Proceedings of the 2018 conference on empirical methods in natural language processing, pp 489–500
9. Gu J, Lu Z, Li H, Li VOK (2016) Incorporating copying mechanism in sequence-to-sequence learning. In: Proceedings of the 54th annual meeting of the association for computational linguistics, pp 1631–1640
10. Database backup dumps of Japanese Wikipedia. https://dumps.wikimedia.org/jawiki/. Last accessed 2021/11/24
11. SentencePiece. https://github.com/google/sentencepiece. Last accessed 2021/11/24

Hardware in the Loop of a Level Plant Embedded in Raspberry

Luigi O. Freire, Brayan A. Bonilla, Byron P. Corrales, and Jorge L. Villarroel

Abstract This chapter presents a plant prototype for level control of a physical module converted into a virtual module and developed within the MyOpenLab software for a didactic interaction with the user. It is complemented with a hardware in the loop that allows the implementation of mathematical model of the plant while integrating electrical signals (read and write) that are used within the process using Arduino nanoboard as data acquisition card and programmable logic controller S7-1200 for process control. The system presents a two-dimensional (2D) virtual environment where the behaviour of the primary signals (process variable CV, control variable CV and set point SP) is shown, along with a graphical representation of the plant including tank level, pump drive and variable behaviour graphs. The user will thus be able to interact in a realistic way and complete their training.

Keywords Hardware in the loop · MyOpenLab · Level control · Virtual module

1 Introduction

Complex real-time embedded systems are developed and tested using a simulation method known as hardware in the loop (HIL). HIL provides a productive testing environment by bringing the complexity of the process-actuator system, sometimes referred to as a plant, to the test platform [1]. By lowering and eliminating the cost barriers of implementing laboratory equipment and hardware incompatibility, HIL

L. O. Freire (✉) · B. A. Bonilla · B. P. Corrales · J. L. Villarroel
Universidad Técnica de Cotopaxi, Latacunga, Ecuador
e-mail: luigi.freire@utc.edu.ec

B. A. Bonilla
e-mail: brayan.bonilla4225@utc.edu.ec

B. P. Corrales
e-mail: byron.corrales@utc.edu.ec

J. L. Villarroel
e-mail: jorge.villarroel3491@utc.edu.ec

P. Chatterjee et al. (eds.), *Computational Intelligence for Engineering and Management Applications*, Lecture Notes in Electrical Engineering 984,
https://doi.org/10.1007/978-981-19-8493-8_47

has grown to be a crucial step in the system development process that enables real-time simulation [2]. It also permits the use of sensors and actuators that are virtually represented. HIL technology is used for testing level, temperature, pressure, etc. Because it allows to establish mathematical models of processes, the obtained results are very close to real life [3]. Since large industrial plants handle various liquids or gases that directly and indirectly affect health, HIL helps in training technical personnel [4]. Most embedded systems require the use of controls such as PI/PD/PD. Therefore, it is necessary to study the behaviour of these electrical and electronic systems under possible circumstances that may occur and it is possible to reduce disadvantages or detect errors before the manufacturing process, optimization of resources, greater reliability and compliance with external regulatory requirements [5, 6]. The design of models (MBD) is one of the most popular since there are all kinds of information about them, which results in their frequent use. It is used with physical laws and mathematical equations or is performed from experimental data which allows fast and accurate designs in what are dynamic systems, control systems and signal processing systems [7].

The HIL test techniques are great alternatives to conventional testing since they enable the prediction of events like the proper controller adjustment, plant startup, or emergency shutdowns [8]. When a simulation is executed, a technically accurate mathematical model that operates in real time in a simulator can replace the function of the plant or physical module [9]. The HIL simulator can enable extensive testing of the closed loop system without requiring real systems by properly simulating the plant or physical module, its dynamics, coupled with a collection of sensors or devices that are in the process [10]. The HIL technique is used in this article to create a virtual environment for a level process that enables monitoring and control. Bernoulli's equation is used for mathematical modelling, and the user can control the variable that sets the level tank as well as the opening of a valve that will act as a disturbance. The most important process variables in a level plant are visible in the interface made in the MyOpenLab programme, running on the operating system on Linux.

2 System Structure

2.1 Problematic

The project is based on generating a HIL module for level control; it starts from a mathematical modelling that is performed using the Bernoulli's principle; the operation is intended to resemble a real physical plant for which through Arduino acts as data acquisition card with values of 0–5 V; in the PWM, output signals are conditioned to obtain an analog signal through an I2C BUS. 2D interface created by the MyOpenLab software installed on the Raspbian operating system in its Linux version for Raspberry as shown in Fig. 1 [11].

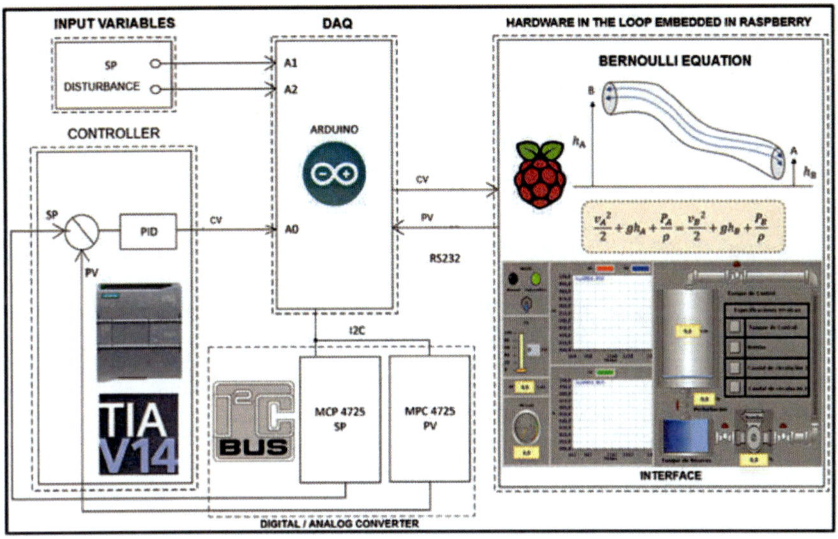

Fig. 1 System structure

2.2 *Development*

The design of the 2D virtual environment is based on the P&ID diagram of the piping and instrumentation used from a physical tank level control module shown in Fig. 2.

The design of the plant is done in three main parts: reading, data acquisition and the development of the graphical environment that will present the module making references to the physical module to be replicated. Figure 3 shows a flowchart showing the programming used.

2.3 *Mathematical Modelling of Level Plant*

The behaviour of the plant is simulated using Bernoulli's principle:

$$\frac{dh}{dt} = \frac{q_{\text{in}} - q_{\text{out}}}{A} \tag{1}$$

$$q_{\text{in}} = k_1 a_1 \tag{2}$$

$$q_{\text{out}} = k_2 a_2 \sqrt{2gh} \tag{3}$$

where

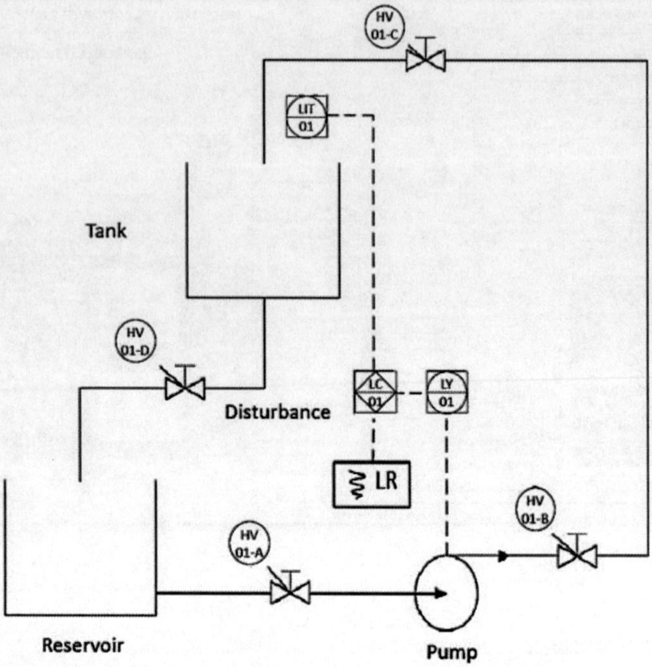

Fig. 2 P&ID diagram of a training module for the level control of a tank

h: tank level (0–1 m)
q_{in}: tank inlet flow (0–50 gpm)
q_{out}: tank outlet flow rate
A: tank area (0.09 m^2)
k_1: valve constant at tank inlet (0.05)
k_2: : valve constant at tank outlet (0.015)
a_1: valve opening at tank inlet (0–100%)
a_2: opening of tank outlet valve (0–100%)
g : gravity (9.8 m/s^2)

2.4 Controller Design

To test the response of mathematical model applied to the simulator, a PID Lambda control will be implemented by means of a PLC S7 1200. [12]

$$G(s) = \frac{\text{km} * \text{e}^{s*tm}}{s+1}$$

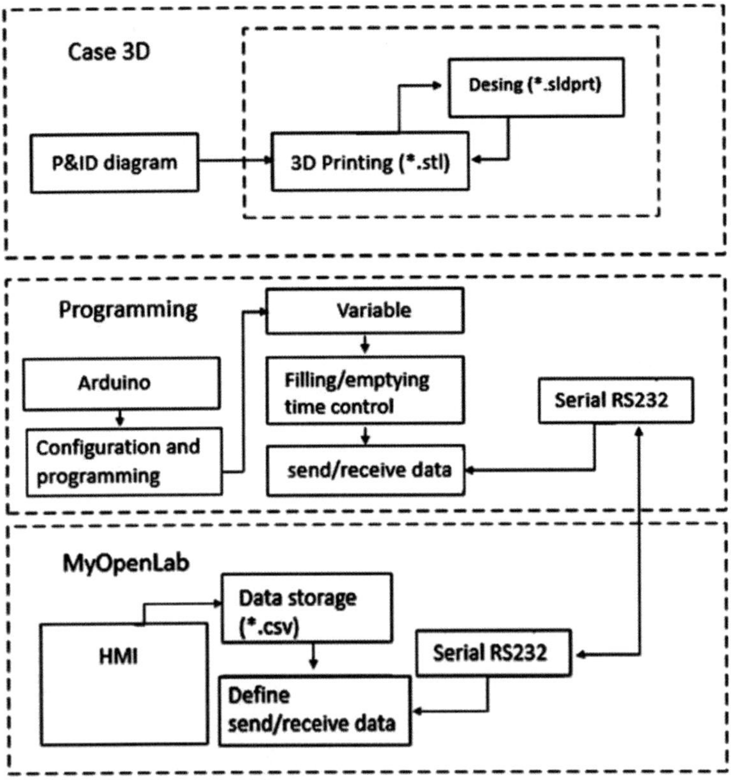

Fig. 3 Serial plant communication

$$Kp = \frac{T}{\text{Km} * \left(\frac{L}{2} + \lambda\right)}$$

$$Ti = t \quad Td = \frac{l}{2}$$

$$Ti = 0.2Td = 0.3$$

$$Kp = \frac{0.2}{0.04 * \left(\frac{0.6}{2} + 0.2\right)}$$

2.5 Results

The graphical interface shown on the Raspberry display allows intuitive navigation through the environment allowing users easy adaptation and manipulation of the prototype as shown in Fig. 4.

(a) Main panel

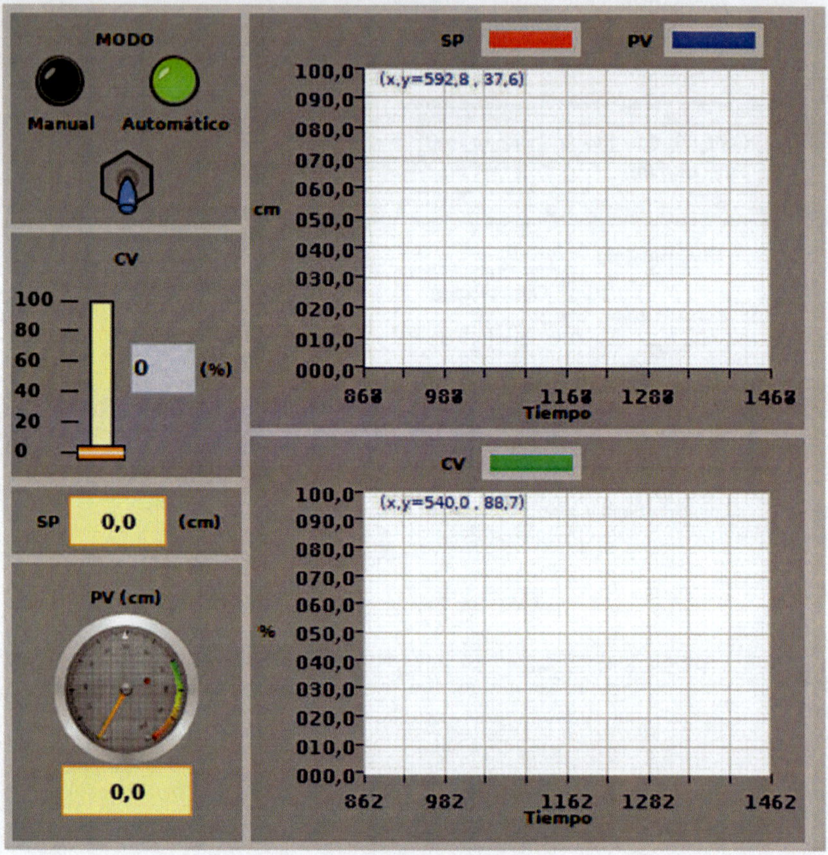

(b) Control panel

Fig. 4 Software environment HIL

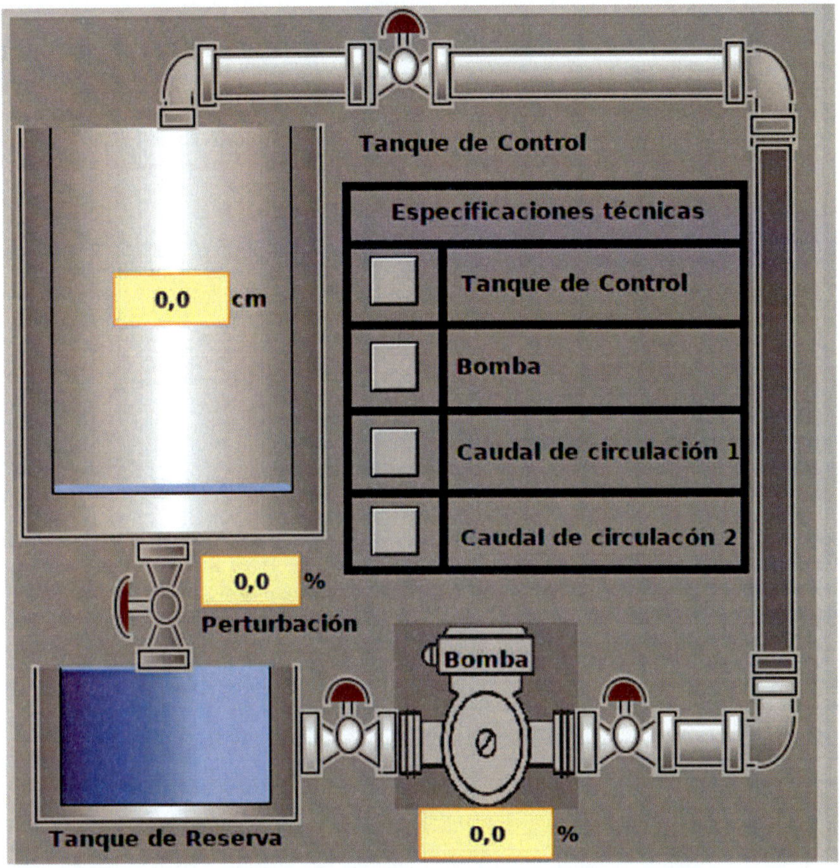

(c) Front panel

Fig. 4 (continued)

The results of the implemented PID Lambda control system are shown in Fig. 5, and this control system does not allow ideal overshoot to exist for the level process.

3 Conclusion

The implemented HIL approach offers a real-time response to the traditional PID Lambda control system and is comparable to that of an actual level plant. The virtual environment displays components with which the user can interact, such as sensors, actuators and environmental factors. The system has been developed using open-source hardware and software and has the benefit of working with any commercial programmable logic controller and responds like a physical module. It is also very

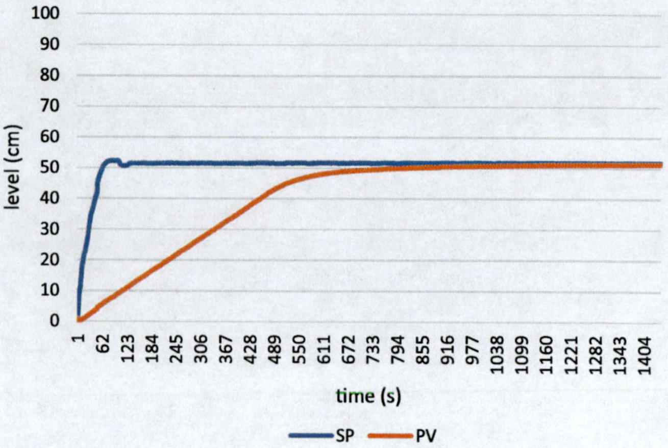

Fig. 5 Control implemented in S7 1200 PLC, for "HIL" level plant control

simple to deploy. Without consuming time and resources on physical testing, HIL testing can simulate hundreds or thousands of scenarios. Situations that would be too risky or impractical to test physically can be accommodated through HIL testing. HIL tests are also reproducible and offer a regular software release cycle with predictable system behaviour.

References

1. Casellas F, Esteban JA, Guinjoan FS, Pique R, Martinez H, Velasco G, Universitària E (2014) Simulación mediante "hardware in the loop: de un convertidor buck. In: Proceedings of the XXI annual seminar on automation, industrial electronics and instrumentation. Universitat Rovira i Virgili, pp 1–5
2. Oscar C, Sebastian J, Edilbetro M, Oscar A, Dario H (2013) Control system of a plant embedded in fpga using hardware-in-the-loop. Scielo 80(179)
3. Paula L, Angel C, Alberto S, Gustavo R, Francisco A, Alberto P (2021) Hardware-in-the-Loop and digital control techniques applied. Electronics 10(1563)
4. Juan A, Nicolás M (2018) Hardware-in-the-loop for the control of pressure, flow and level processes through the embedded system my river to be carried out in the laboratory of industrial networks and process control. Latacunga
5. Hipólito G, Miguel A (2015) Hardware-In-the-Loop simulation environment for fault tolerance studies in complex electronic systems. Universidad de Sevilla, Sevilla
6. Rosaura S, Noemi H, Cuauhtemoc M, Maria C, Gabriel C, Edgar A (2018) Hardware-In-the-Loop implementation of a complete control scheme for a simple pendulum system using a PID controller. AMCA 1(176):210–212
7. Arkadusz M, Kierdelewivz A (2018) Fractional-Order water level control based on PLC: hardware-in-the-loop simulation and experimental validation. MDPI 11(2928)
8. Mundo Electrónico (2016) Mundo Electrónico. (En línea). Available: http://www.mundo-electronico.com/?p=376296. Último acceso: 18 Nov 2021
9. Rozzana A, Gabriela H (2021) Mathworks. (En línea). Available: https://la.mathworks.com/solutions/power-electronics-control/hardware-in-the-loop.html. Último acceso: 16 Nov 2021

10. OPAL (2021) Opla-rt technologies. (En línea). Available: https://www.opal-rt.com/hardware-in-the-loop/. Último acceso: 15 Nov 2021
11. Vacacela SG, Freire LO (2021) Implementation of a network of wireless weather stations using a protocol stack. Intell Manufact Energy Sustain
12. Guambo W, Corrales BP, Freire LO, AMD (2021) Performance evaluation of a predictive and PID control applied to a didactic flow regulation process. Intell Manufact Energy Sustain 265:559–570

Multi-label Classification Using RetinaNet Model

Sandeep Reddy Gaddam, K. R. Kruthika, and Jesudas Victor Fernandes

Abstract Object detection and classification have several versions which include image classification, object localization within an image, to multiple object detection and localization. Object recognition or classification has moved further to multi-label classification of an image and of an object. In present state of art models, multi-label image classification or a single label for object localization in images is available. Multiple labels for object classification and localization are not available. We present a framework accomplished by the study and customization of the RetinaNet architecture, to do object classification and localization with multiple labels for multiple classes of objects.

Keywords Object detection · Object classification · Multi-label classification multi-object classification · Multi-task learning RetinaNet

1 Introduction

The vast range of approaches including recent technical advancements in object detection has drawn growing interests in recent years. In academics as well as the real world, this problem is under comprehensive study, such as defense surveillance, autonomous vehicles, transport monitoring, drone scenes analytic including computer vision [1]. Object recognition is the classification of instances and objects belonging to a class within an image. Two major types of SOTA methods are commonly recognized: one-stage methods as well as two-stage approaches. Infusion speeds are prioritized for one-stage processes, single shot detector (SSD), You

S. R. Gaddam (✉) · K. R. Kruthika · J. V. Fernandes
Sandlogic Technologies Pvt. Ltd., Bengaluru, Karnataka, India
e-mail: sandeep@sandlogic.com

K. R. Kruthika
e-mail: kruthika.kr@sandlogic.com

J. V. Fernandes
e-mail: jesu@sandlogic.com

P. Chatterjee et al. (eds.), *Computational Intelligence for Engineering and Management Applications*, Lecture Notes in Electrical Engineering 984,
https://doi.org/10.1007/978-981-19-8493-8_48

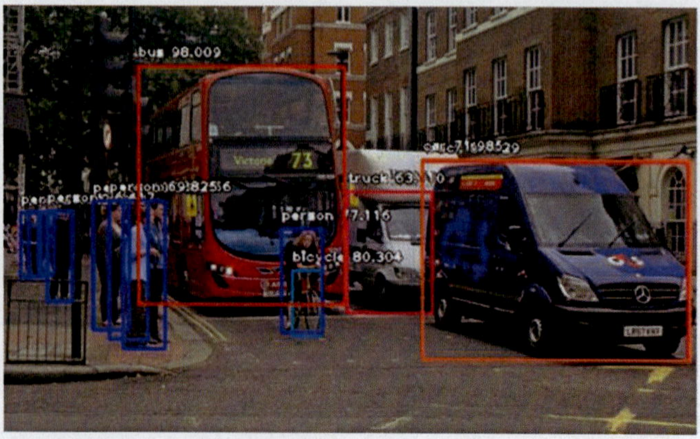

Fig. 1 Multi-class object detection and classification with bounding boxes

Only Look Once (YOLO), as well as RetinaNet are excellent examples [2]. Two-step detection emphasizes precision approaches, and examples include Mask R-CNN, Fast R-CNN, etc.

A significant number of labeled datasets have recently significantly increased. Figure 1 shows an example of a multi-class classification with bounding boxes featuring a variety of transports and humans in various positions of multi-class classification processes. Special success for visual classification/recognition functions is obtained by deep learning convolutional neural networks (CNNs). Thus, the features derived from CNNs will have a strong global representation of the problem of individual object detection [3]. The advent of deep convolutional neural networks and GPUs processing resources are many of the reasons contributing to the rapid growth of object detection methods.

2 Form of Object Detection and Classification

2.1 Image Classification

With artificial intelligence (AI) becoming omnipresent advances, we presently have tremendous volumes of information being produced. Contrasting in structure, information could be discourse, text, picture, or a blend of any of these. As photographs or recordings, pictures compensate for a critical portion of worldwide information creation [4–7]. Image classification and allotting marks to gatherings of pixels or vectors inside a picture are subject to specific principles. The arrangement law can be applied through one or numerous ghostly or textural portrayals. Picture characterization strategies are predominantly categorized into two classifications: supervised and unsupervised techniques [8–10]. Table 1 discusses the classifications.

Table 1 Image classification

Technique	Features
Supervised	Utilize recently grouped reference tests (the ground truth) to prepare the classifier and hence arrange new, obscure information Picking tests of preparing information inside the picture and allotting them to pre-picked classifications, including vegetation, streets, water assets, and structures
Unsupervised	Fully automated method that does not leverage training data Machine learning algorithms are used to analyze and cluster unlabeled datasets by discovering hidden patterns or data groups without the need for human intervention

2.2 *Object Detection*

Object detection seems to be a fast progressive change in the field of computer vision. Its contribution in the mix of item grouping just as article confinement makes it one of the most difficult themes in the space of PC vision. In basic words, the objective of this discovery procedure is to figure out where articles are situated in a given picture called object confinement and which class each item has a place with, which is called object classification [11–13]. Table 2 discusses the emerging type of object detection algorithm.

Let us look at the different types of object detection, localization, and classification in Fig. 2.

Single-class object detection is a simpler multi-class object detection—because it is enough to locate where the object is in a source image since we know what it really is. As the name suggests, single and multi-class, multi-label detection of objects means Fig. 2a to detect a specific physical entity in an input picture and Fig. 2b to determine and label the class of detected physical entity with the predicted attributes(object). For example, in Fig. 2f, the object of class 'cat' is labeled with the name of the breed and color. Here the object class is 'cat,' and it has two attributes: name and color. In Fig. 2g, objects of two classes ('dog' and 'cat') are available, and they are labeled with the name of the breed and their color.

3 Current Development

Deep learning models in the larger field of computer vision, involving general recognition of objects as well as domain-based object detection, are currently frequently implemented. Most modern object recognition systems are used to retrieve features from data (photographs or videos), classification, and localization using deep learning networks as their core as well as detection network [13, 14]. Object recognition is a functional vision- and image-processing system that deals with the detection in

Table 2 Object detection (OD) algorithms

Algorithms	Key features
Fast R-CNN	Detection quality is high High speed and accurate Training in single-stage Update all network layer with the help of training
Faster R-CNN	Cost effective Full conventional network can predict object bound and position Trained end-to-end High-quality detection
HOG- Histogram of Oriented Gradients	Gradient orientation in the localized portion of the image Easier to understand the key information present in the image
R-CNN	Localizing object with deep network Training high capacity model Accuracy is high using deep ConvNet
RF-CNN	Region-based detection Full convolutional architecture Learnable weight layers are Convolutional
SSD	Use single DNN Discretize the bounding boxes into different aspect ratios of boxes Easy to train and integrate to the system
SPP-net	Generate fixed length of image Features can be detected only once
YOLO	High speed and accurate Model can process 45–115 frames per second

digital videos and photographs of semantic objects of a certain type (e.g., individuals, houses, animals, or cars) [14, 15]. Multi-category identification, edge detection, outstanding object detection, pose detection, graphic scene detection, facial detection, as well as vehicle detection, etc., are all well-studied areas of object detection [15, 16].

4 Methodology

Most common object detectors are based on a two-stage process. The very first stage creates a limited collection of candidate object positions as it is pioneered in the R-CNN framework [16], as well as stage two, which characterizes each candidate as an advanced class or a context via a CNN model.

Fig. 2 Different forms of detection, localization, and classification: **a** Image classification, **b** single-class detection and localization, **c** single-class multi-instance detection, localization, **d** multi-class classification (single label), **e** multi-label image classification, **f** single-class, multi-label classification, **g** multi-class and multi-label classification

5 RetinaNet

RetinaNet is one among various efficient object detection models which are built using a unified backbone network and two sub-networks that are specific to the task [17, 18]. RetinaNet addresses the problem of class imbalance with the concept of focal loss function during the training phase. The focal loss is a comparative method for cross-entropy which actually looks for the training of difficult negative examples.

5.1 Focal Loss

The classic object detection like the single-stage detection SSD method suffers from a class imbalance problem which is nothing but models map up to 104–105 locations per image among which only a few locations evaluate object locations (foreground) and the remaining are backgrounds. This issue is solved by using focal loss [18–20]. The class imbalance overwhelms the cross-entropy loss and computed gradients which leads to inefficient training of dense detectors with more easy negatives. Thus,

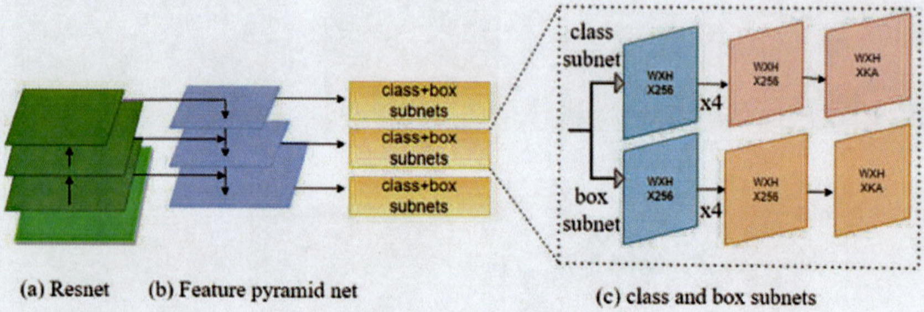

Fig. 3 Single-stage RetinaNet architecture: the FPN is modified as follows. Construction of a pyramid from P3 to P7 (see Fig. 2). Major differences are: the P2 is removed to decrease computations. In P6 instead of down-sampling stridden convolution is applied. An added layer of P7 to increase the accuracy for large objects detection

loss function is modified as a focal loss to decrease the weight of easy samples and give more weightage to hard negatives during training [21].

5.2 *Architecture*

The structure of RetinaNet consists of three main components: a backbone, a feature pyramid network (FPN), and a backend detection. Figure 3 depicts that FPN is used above ResNet to create a rich multi-scale input image characteristic pyramid. FPN is multi-level, and easy to measure at all scales. The backbone with the following FPN forms a network similar to encoder–decoder. The value of the FPN is that it merges the characteristics of successive stages from the grossest to the finest, which essentially spread the characteristics on various levels and to the corresponding sheet. The multi-scale pyramid (P3-P7) characteristics will then be fed into the backend of the bounding box correlation as well as the description of the items by two detection groups [22–24].

5.3 *Anchors*

Anchor boxes are fixed-size boxes used by the model to detect the object's bounding box. This may be achieved by reversing the offset between the position of the middle of the object and that of the anchor box, and then by estimating the relative size of the object by the width and height of the anchor box [25, 26]. For RetinaNet, there are nine anchor boxes in each position on the specified map (at three scales and three ratios). The pyramid-level anchors are from 322 to 5122 from P3 to P7. Three {1:2, 1:1, 2:1} aspect ratios are used. The size anchors {20, 21/3, 22/3)} at each pyramid level are applied for denser scale coverage. Nine anchors per floor in all. The size from 32 to 813 pixels is distributed through the stages. As each object box will be assigned by each anchor the corresponding entry in its length label vector K to 1 and other values to 0; if any of these anchors are not assigned which is possible due to overlap in [0.4, 0.5], then it is ignored during the training phase. In a similar way, box regression values are calculated by the offset between the anchor and its assigned object box or it is ignored if not assigned [27, 28]

5.4 Classification of Sub-net

The classification subnet is a small FCN that is simple in its design. This block actually gives the probability of the object detection at each spatial position for every A anchor and K object class. This classification subnet FCN is connected to each FPN level, and its parameters are shared across all pyramid levels. For a given pyramid level, the subnet with input feature map with a C channel applies four 3 × 3 Conv layers each with C filters along with RELU activations and followed by a 3 × 3 layer with K and A filters. Then at the end, sigmoid activation is added to the K and A for 2 class predictions per spatial location [26].

5.5 Box Regression Sub-net

In addition to the classification subnet, there is also one fully connected network at each pyramid level which is basically for regression offset for each anchor box and to the ground truth object, if it is present. The network architecture of the box regression subnet is very similar to the classification subnet in the major part, and the only change is at the end, there are four linear outputs per spatial location. The four outputs predict the relative offset of the anchor box with the ground truth box at each anchor per spatial location. Even though the design of classification and regression subnet are very similar to each other, they use different parameters [18, 26, 27].

6 Our Need and Approach

6.1 Multi-label Classification for Multiple Object Needs

RetinaNet architecture has been efficient in object detection and classification. However, we needed a multi-label classification using RetinaNet. Our search and queries on the same did not find any response, which led us to make the changes to the RetinaNet architecture to meet our needs for multi-label classification. The customer needed object detection and classification, which would detect the object

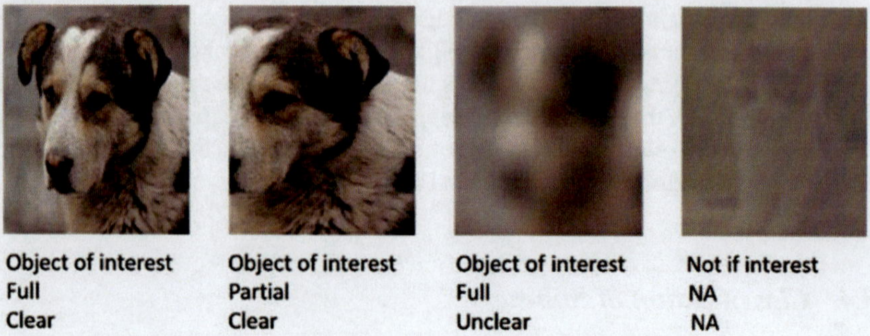

Fig. 4 Sample of multi-labels required

of interest, and label the object with the attributes of object type, completeness, and visual clarity. Figure 4 depicts a sample of the output labels required.

In our search for possible solutions, we have referred to multiple websites and GitHub, posted queries [10]. However, there was no positive outcome from our queries and search, and hence, we went ahead to modify the RetinaNet for the required multi-label classification for one or more objects. Such a model is also called multi-task learning model, where separate subnets are created for specific tasks as per the need [28, 29].

6.2 *Customized RetinaNet Architecture for Multi-label Classification for Multi-objects*

The implementation of the RetinaNet architecture was taken from [9] which was used as the base code for our customization. The implementation is PyTorch. The existing RetinaNet architecture was modified to enable it for multi-label classification for multiple object classes. The RetinaNet architecture was configured for the following:

a. Backbone—ResNet50 with FPN
b. CSV file used as annotations input file format.

Steps for making the modified RetinaNet work for multi-label and multi-object classification.

1. Update the annotations to add the additional labels as required, to ground truth bounding boxes.
2. Update the class config file with the additional class labels as required.
3. Modify the data loader to handle the additional labels provided in the annotations which are required for training for multi-labels.
4. Add classification and regression subnets as per the number of classes of labels required.
5. Modify the loss function to handle the losses for the additional classes.

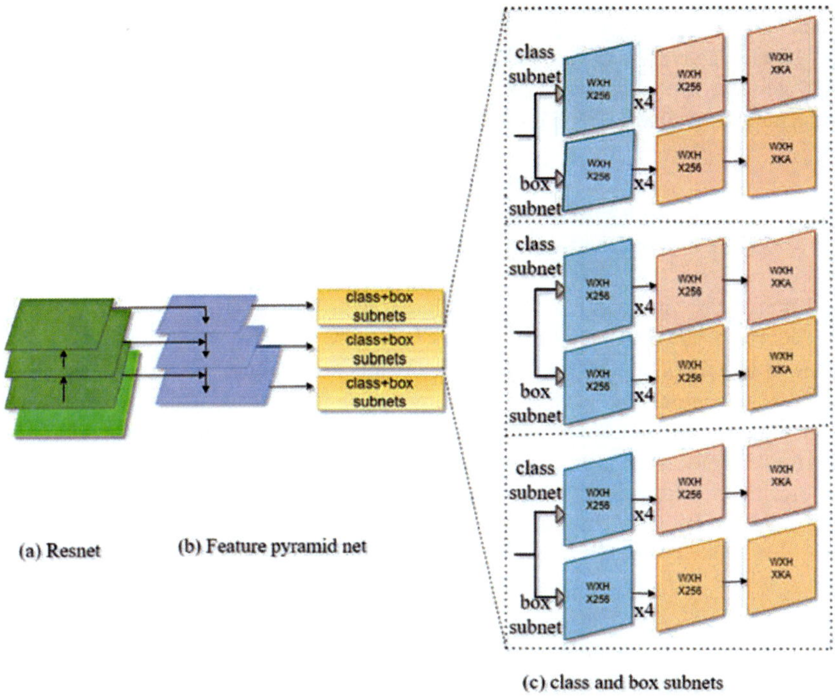

Fig. 5 Customized RetinaNet architecture based on ResNet50-FPN-800 × 800

6. Change the training and prediction methods to take additional training inputs for additional classes and return the predictions for the same.
7. Add additional logic in the NMS method to handle the top candidates for all classes and return a single bounding box, values of each class label, and their corresponding scores (Fig. 5).

We also automated the code changes to be made as per the annotations file. Once the annotations file is read, the code is updated based on the number of labels required.

6.3 Datasets

6.3.1 Datasets for a Single Object Type with Multiple Attributes. UTKFace Datasets

Here the object is the human face. For each face detection, we will predict the race and the gender of the person. The custom model was trained with the UTKFace dataset [11]. This dataset is a collection of very large-scale face dataset which consists of faces that ranges from 0 to 116 years of age span. The size of the face dataset is around 20,000 images which are labeled as age, gender, and ethnicity. There are up to 10,000 images for training and 5000 images each for test and validation. In this work gender, race labels were taken and the age label was truncated, and also face was added as another label [11, 29–33].

6.3.2 Datasets for a Single Object Type with Multiple Attributes. UTKFace Datasets

A custom dataset was created which has three labels. First is the pet type, i.e., cat and dog, its color, and the breed type. For each pet type, we have three colors and three types of breeds. The dataset consists of 670 images. We took 600 training images and 70 test images. The datasets we used for building the custom datasets were taken from [12]. The annotations were modified as per our requirements. AuVi.io tool [13] was used for dataset creation, import of data and annotations, annotating the data, and export of annotations in the required format. This tool helped speed up the creation of the dataset as required for testing the modified RetinaNet model [34, 35].

6.4 Training

The modified RetinaNet model was trained on the above two datasets separately.

a. The UTKFace dataset was trained for 33 epochs.
b. The modified model with the UTKFace dataset was trained on Azure Linux VM with Ubuntu 18.04 and K80 GPU.
c. The dataset of labels with Dogs and Cats was trained for 40 epochs.
d. The modified model was trained on a Windows 10 laptop with RTX2060 for dogs and cats dataset (Fig. 6; Table 3).

6.5 Limitations and Issues

a. The modified RetinaNet model enforces that all objects must have the same number of labels.
b. The modified RetinaNet model does not support more than three labels per object.
c. The modified RetinaNet model has not been tested for multi-labels for more than two objects.

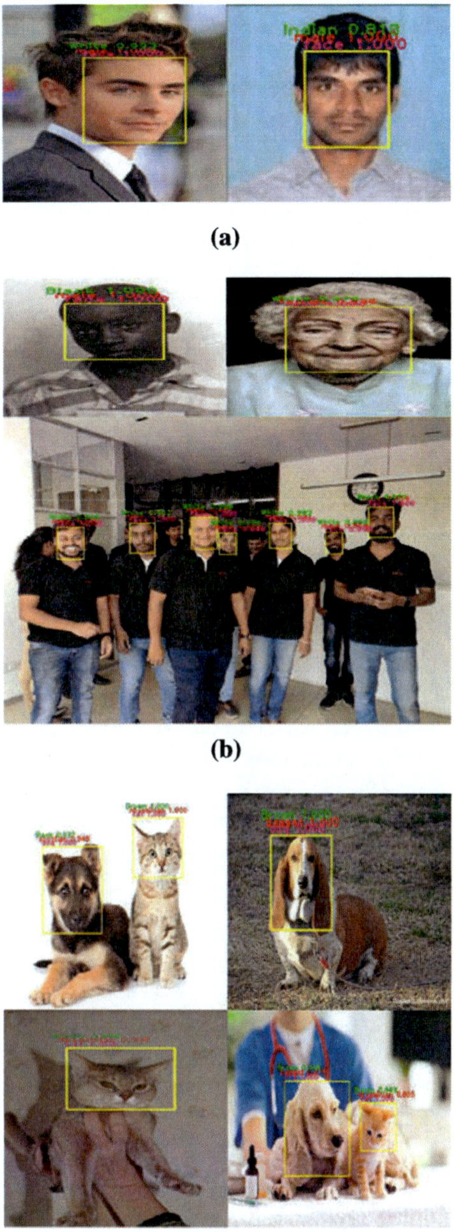

Fig. 6 **a** Prediction on sample test image using UTKFace datasets, **b** prediction on real image using UTKFace datasets, **c** prediction on sample test image: custom cats and dogs datasets

Table 3 Results of the training and testing on various datasets

Dataset name	Backbone	mAP on train data	mAP on test data
UTKFace	Resnet50-FPN	82	64.6
Custom dataset (cats and dogs with breed)	Resnet50-FPN	88	77

6.6 Future Scope

The modified RetinaNet model can be improved for the following and made more generic for multi-label and multi-object classification.

a. Extend the model to support more than two object classes.
b. Extend the model to support more than three labels per object class.
c. Automate the customization of the model for the above-mentioned steps.

7 Conclusion

The work's main focus is to implement multi-label classification for multiple object classes by modifying the model architecture of RetinaNet. The results are demonstrated by training the customized UTKFace dataset by adding the extra-label with modified RetinaNet architecture for 33 epochs with a mAP value of 82%. The same model was also trained and tested with custom datasets with labels of dogs' and cats' images that achieve a mAP of 88%. These two experiments demonstrate RetinaNet as an efficient object detection model that can be used for multi-label classification for multiple object classes.

References

1. Zhao ZQ, Zheng P, Xu ST, Wu X (2019) Object detection with deep learning: a review. IEEE Trans Neural Netw Learn Syst 30(11):3212–3232
2. Wu X, Sahoo D, Hoi SC (2020) Recent advances in deep learning for object detection. Neurocomputing
3. Yan Z, Liu W, Wen S, Yang Y (2019) Multi-label image classification by feature attention network. IEEE Access 18(7):98005–98013
4. Ge W, Yang S, Yu Y (2018) Multi-evidence filtering and fusion for multi-label classification, object detection and semantic segmentation based on weakly supervised learning. In: Proceedings of the IEEE conference on computer vision and pattern recognition 2018, pp 1277–1286
5. Gong T, Liu B, Chu Q, Yu N (2019) Using multi-label classification to improve object detection. Neurocomputing 370:174–185
6. Girshick R, Donahue J, Darrell T, Malik J UC Berkeley (2014) Rich feature hierarchies for accurate object detection and semantic segmentation. arXiv:1311.2524v5 [cs.CV] 22 Oct 2014

7. Lin TY, Maire M, Belongie S, Hays J, Perona P, Ramanan D, Dollár P, Zitnick CL (2014) Microsoft coco: common objects in context. In: European conference on computer vision, pp 740–755
8. Lin T-Y, Goyal P, Girshick R, He K, Dollar P, Focal loss for dense object detection
9. GitHub - yhenon/pytorch-retinanet: Pytorch implementation of RetinaNet object detection.Khan, B. H., A framework for web-based learning. Englewood Cliffs, NJ: Educational Technology Publications (2000)
10. 11.UTKFace—Large scale face dataset: UTKFace I Large Scale Face Dataset (susanqq.github.io)
11. Cats and Dogs Breeds Classification Oxford Dataset: Cats and Dogs Breeds Classification Oxford Dataset I Kaggle
12. AUVI.io—SandLogic Artificial Intelligence Foundation—SandLogic
13. Henderson P, Ferrari V (2016) End-to-end training of object class detectors for mean average precision. In: Asian conference on computer vision. Springer, Cham, pp 198–213
14. Hönes F., Lichter J (1994) Layout extraction of mixed mode documents. Mach Vis Appl 7(4):237–246
15. Hu H et al (2018) Relation networks for object detection. In: Proceedings of the IEEE conference on computer vision and pattern recognition, pp 3588–3597
16. International conference on neural information processing. Springer, Cham, pp 713–722
17. Jain AK, Zhong Y (1996) Page segmentation using texture analysis. Pattern Recogn 29(5):743–770
18. Kisantal M et al (2019) Augmentation for small object detection, arXiv preprint arXiv:1902.07296
19. Laganière R (2014) OpenCV computer vision application programming Cookbook, 2nd edn. Packt Publishing Ltd.
20. Li K et al (2014) A maximal figure-of-merit learning approach to maximizing mean average precision with deep neural network based classifiers. In: 2014 IEEE International conference on acoustics, speech and signal processing (ICASSP). IEEE, – pp 4503–4507
21. Lin TY et al (2017) Focal loss for dense object detection. In: Proceedings of the IEEE international conference on computer vision, pp 2980–2988
22. Nagy G (2000) Twenty years of document image analysis in PAMI. IEEE Trans Pattern Anal Mach Intell 22(1):38–62
23. Perner P, Imiya A (ed) Machine learning and data mining in pattern recognition: 4th international conference, MLDM 2005, Leipzig, Germany, July 9–11, 2005, Proceedings. Springer Science & Business Media, vol 3587
24. Randriamasy S, Vincent L (1994) A region-based system for the automatic evaluation of page segmentation algorithms. In: Proceedings of the international association for pattern recognition workshop on document analysis systems DAS94, pp 29–41
25. Review: RetinaNet—Focal loss (object detection). Towards Data Science. https://towardsdatascience.com/review-retinanet-focal-loss-object-detection-38fba6afabe4 (режим доступа: 17.01.2021)
26. Setitra I et al (2017) Text line segmentation in handwritten documents based on connected components trajectory generation. In: International conference on pattern recognition applications and methods. Springer, Cham, pp 222–234
27. Szegedy C, Toshev A, Erhan D (2013) Deep neural networks for object detection. Adv Neural Inf Process Syst, pp 2553–2561
28. Taheri S et al (2018) OpenCV. js: Computer Vision processing for the open Web platform. In: Proceedings of the 9th ACM multimedia systems conference, pp 478–483
29. Wang Y et al (2019) Automatic ship detection based on RetinaNet using multi-resolution Gaofen-3 imagery. Remote Sens 11(5):531
30. Wood SL, Marks JP, Pearlman J (1980) A segmentation algorithm for ocr application to low resolution images. In: Conference Record of the Fourteenth Asilomar Conference on Circuits, Systems and Computers, pp 411–415

31. Wei H et al (2013) Evaluation of SVM, MLP and GMM classifiers for layout analysis of historical documents. In: 2013 12th International conference on document analysis and recognition. IEEE, pp 1220–1224
32. Yi X et al (2017) CNN based page object detection in document images. In: 2017 14th IAPR international conference on document analysis and recognition (ICDAR). IEEE, vol 1, pp 230–235
33. Zeng N (2018) RetinaNet explained and demystified [Электронный ресурс]. blog.zenggyu.com/en/post/2018-12–05/retinanet-explained-and-demystified
34. Zhang H et al (2019) Cascade retinanet: maintaining consistency for single-stage object detection, arXiv preprint arXiv:1907.06881
35. Zou Z et al (2019) Object detection in 20 years: a survey. arXiv preprint arXiv:1905.05055

Unsupervised Process Anomaly Detection Under Industry Constraints in Cyber-Physical Systems Using Convolutional Autoencoder

Christian Goetz and Bernhard G. Humm

Abstract To realize complex industrial processes, the application of cyber-physical systems (CPS) in modern manufacturing is rising. Thereby, CPS perform repetitive and recurrent process steps, e.g., in production, packaging, and transportation. With the growing complexity of these processes, the relevance of each process step to realize the desired outcome is increasing. A failure in one step can result in an entire faulty process that can interrupt the current operation of the CPS or even the whole production line. Therefore, it is essential for modern and safe CPS to monitor each single process step and detect potential failures as soon as possible. With the help of a detected anomaly, the emergence of a failure could be prevented, or an occurred failure could be solved as quickly as possible. The following paper introduces a concept for unsupervised process anomaly detection in CPS under industry constraints. We focus on repetitive process tasks with a fixed duration and industry constraints like near-time requirements, prediction quality, configurable design, data-driven limitations, processing limitations, and communication interface constraints. The concept is evaluated using an industrial application, which simulates a transportation process. The test result confirms that the presented concept can be successfully applied to detect anomalies in the investigated process. Therefore, the concept is a promising approach for anomaly detection in repetitive process tasks for CPS.

Keywords Anomaly detection · Cyber-physical systems · Convolutional autoencoder · Machine learning · Unsupervised learning

C. Goetz (✉) · B. G. Humm
Hochschule Darmstadt—University of Applied Sciences, Haardtring 100, 64295 Darmstadt, Germany
e-mail: christian.goetz@yaskawa.eu

B. G. Humm
e-mail: bernhard.humm@h-da.de

C. Goetz
Yaskawa Europe GmbH, Hauptstraße 185, 65760 Eschborn, Germany

P. Chatterjee et al. (eds.), *Computational Intelligence for Engineering and Management Applications*, Lecture Notes in Electrical Engineering 984,
https://doi.org/10.1007/978-981-19-8493-8_49

1 Introduction

Cyber-physical systems (CPS) are playing an increasingly important role in modern manufacturing due to the possibility of realizing complex industrial processes [1]. CPS consist of mechanical and electrical systems combined with software and electronic components [2, 3]. In industrial manufacturing, CPS perform recurring processes like production, transportation, packing, and quality insurance [4].

Due to the rising complexity of these processes, the relevance of each process step to realize the desired outcome is increasing. A single fault in one step could influence the complete production line. This could result in faulty products, a breakdown of the overarching process, or a carryover of the failure throughout the entire production.

Anomaly detection (AD) in CPS refers to identifying unnormal system behavior, i.e., behavior that is not shown under the regular operation of the system. Thereby normal operation of a CPS in manufacturing can be defined as a regular process, resulting in the desired outcome while each process step behaves like expected. Anomalies can be taken as an essential indication of process failures which can be used for early failure detection. Therefore, by detecting anomalies in a process step, there is the possibility to recognize, react early, and in the best case to fix the anomaly to prevent the rise or the carryover of the failure.

Techniques for anomaly detection can be differentiated into model-based [5] and data-driven approaches [6]. Considering model-based approaches, manually creating precise models, is a challenging task. Engineering these models is time-consuming and requires deep expert knowledge. In contrast, data-driven approaches can detect anomalies only based on the collected data. Therefore, no proper expert knowledge is needed. Data-driven techniques can further be divided into supervised and unsupervised methods [7]. However, it is challenging to collect anomalous data and label them by considering CPS, a significant drawback for supervised methods.

The contribution of this paper is an unsupervised anomaly detection concept for process tasks in CPS under industry constraints. We focus on repetitive process tasks with a fixed duration and industry constraints, including near-time requirements, prediction quality, configurable design, data-driven limitations, processing limitations, and communication interface constraints.

We employ a 1D convolutional Autoencoder (1D-ConvAE) for CPS to reach an adequate prediction quality and fulfill near-time requirements. Current approaches do not consider the limitations and constraints of industrial setups. They mainly follow a centralized approach, where the installation and execution of the anomaly detection are done on a central unit. We are focusing on a decentralized approach by splitting the installation and the execution over two units, namely the onboard processing unit and the backend processing unit, to consider the processing limitations and the communication interface constraints of CPS in industry. The installation is fully automated to tackle data-driven restrictions. Thereby, no expert knowledge and explicit known anomalies are needed.

The paper is structured as follows. Section 2 summarizes related work about anomaly detection in CPS. The problem statement is specified in Sect. 3. In Sect. 4, a concept for unsupervised process anomaly detection in CPS is presented. Information about a prototypical implementation is provided in Sect. 5. In Sect. 6, the evaluation of the approach is presented based on an industrial setup. Finally, a conclusion and an outlook of future work are given in Sect. 7.

2 Related Work

Surveys about anomaly detection in CPS can be found in [8–10]. Approaches can be differentiated into model-based anomaly detection (e.g., [11, 12]) and data-driven approaches[1] [13]. However, as mentioned before, model-based approaches require a deep prior knowledge of the CPS to manually create a sufficient model. The field of data-driven approaches can be divided into supervised and unsupervised techniques [6].

Supervised techniques require labeled data, in the case of anomaly detection in CPS, data about unnormal system behavior. Collecting such data is a difficult task because the generation can be hazardous for the CPS itself. Additionally, defining all possible anomalies in advance is nearly impossible. In this paper, we focus on unsupervised methods.

In [14], a generative adversarial network (GAN) for anomaly detection in multivariate time series data is proposed. The GAN structure is realized using long short-term memory (LSTM) and recurrent neural networks (RNN) as the base model, resulting in high model complexity. While this approach works well, it cannot be applied to CPS with limited computational resources. In the study of [15], two different methods are compared, namely deep neural networks (DNN) and one-class support vector machines (OC-SVM), to detect cyber-attacks on CPS. While the authors pointed out that the computational costs of the DNN are much higher, it performs slightly better than the OC-SVM. At the same time, both methods show limitations in detecting gradual changes in sensor data and detecting anomalous actuator behavior. A dual isolation forest-based (DIF) approach to detect cyber-attacks in industrial control systems is introduced in [16]. The authors implemented two isolation forests trained independently, using the normalized and a preprocessed version of the data. For preprocessing the data, a principal component analysis (CPA) has been applied.

While traditional CNNs are commonly used on images, they can also be successfully utilized in time series processing. By extracting the temporal dependencies within the time series data, CNNs can be used for anomaly detection [17, 18]. In [19], a fault diagnosis by detecting anomalies in wheelset bearings of high-speed

---

[1] It shall be noted that data-driven model approaches are also based on models; however, those models are generated automatically based on data and not manually by domain experts.

trains was realized by applying a 1D-CNN. The authors used the vibration acceleration data of 11 wheelset bearings and a model structure consisting of three parts, including a feature extraction layer, auxiliary tasks boosted layer, and a multi-loss function. In [20], the authors detect motor faults in real time by implementing a 1D-CNN. They use the motor signal only and prove that 1D-CNN can achieve highly accurate fault detection. A 1D-CNN is also successfully applied to detect cyber-attacks in industrial control systems. In [21], a 1D-CNN predicts cyber-attacks with adequate precision by learning the system features. The survey highlighted that 1D-CNN outperforms the more complex RNN while being much smaller and faster to train and execute.

Other approaches are autoencoders [22, 23] and variants thereof [24, 25]. By learning the latent features of the input data, they can reconstruct their input as output. Therefore, they are regarded as reconstruction-based techniques. In [26], a variational auto-encoder for unsupervised anomaly detection for seasonal KPIs in web applications is introduced. While the methods mentioned above can be applied to detect the spatial characteristics of the input data, they are missing to consider the temporal dependencies, which are necessary indicators for anomalies in process data.

One approach to overcome this drawback is to use a recurrent neural network (RNN) in combination with a CNN. While the first can extract the temporal information, the latter can detect local and spatial features of the input data. In [27], a multi-scale convolutional recurrent encoder–decoder architecture for anomaly detection in industrial power plants is proposed. The authors of [28] use nearly a similar architecture, applying an LSTM structure for anomaly detection in time series. A variational recurrent autoencoder with attention is used in [29] to detect anomalies in energy time-series data produced by photovoltaic systems. While those approaches show promising results, due to the high complexity caused by the combination of CNN and an RNN, they are not usable for process anomaly detection for CPS with limited processor performance.

Another concept is to use convolutional autoencoders (ConvAE) due to their ability to detect temporal anomalies with the help of the convolutions and spatial anomalies by the autoencoder structure. In [30], a convolutional autoencoder is used to detect outliers in time series data. While evaluating the method on a different dataset, no real data of a CPS is used. The authors in [31] show that a ConvAE can effectively detect features and local structures from data. They point out that only convolutional layers are needed to realize the proposed goal for the autoencoder structure. In [32], the authors introduce a convolutional variational autoencoder for detecting anomalies in industrial robots. They use a sliding-window approach and 2D convolutional layers to detect anomalies from an unseen pattern of data. An abnormal sensor signal detection is shown in [33]. The authors utilize the ability of a ConvAE to reconstruct the input signal channel-wise and detect anomalies in automobile sensor data by applying the channel-wise reconstruction error to additional machine learning techniques like the local outlier factor.

In summary, there are several approaches for data-driven unsupervised anomaly detection. Only a few are evaluated in CPS, and even fewer are applied for anomaly detection on process data of applications used in industrial production lines. Overall, there is no work that deals with anomaly detection on repetitive process tasks while considering all the different industrial limitations of a CPS.

In this work, we propose a concept that addresses all the requirements that must be considered to realize a usable anomaly detection in CPS under industrial constraints. Our contribution in this paper is summarized as follows. We employ a 1D-ConvAE for unsupervised anomaly detection in CPS for repetitive process tasks. We introduce a novel concept that splits the installation and the execution of the anomaly detection over two units to meet industrial requirements. The concept is fully automated, and no expert knowledge or explicit known anomalies are needed.

3 Problem Statement

The problem statement of this paper is as follows. The aim is to implement an unsupervised anomaly detection concept for repetitive process tasks with a fixed duration under industry constraints. Therefore, the problem statement can be described by the different industrial requirements that must be considered to implement such a concept.

1. **Anomaly detection**: An anomaly detection shall be performed for multi-variate time series data of repetitive process tasks with fixed time length in CPSs; for example, in a transportation system, like a conveyer belt in an industrial production line.
2. **Near-time**: To react early on anomalies, the result of the anomaly detection should be available in near-time. Therefore, one requirement is the execution of the anomaly detection during production, e.g., immediately after one repetitive process, like the transportation of a component from the start position to the destination position.
3. **Prediction quality**: For an application in an industrial environment, adequate prediction performance is required. This depends on the different use cases where the anomaly detection is applied, e.g., an F1 Score of 0.95 or better for detecting anomalies in a transportation system.
4. **Configurable**: To apply anomaly detection in different use cases, it should be configurable for various applications. The possibility of adapting the anomaly detection to different process variables and different process lengths with diverse sample rates should be given. For instance, a transportation system provides features like torque, speed, position at a sample rate of 2 ms for a complete process of 14 s.
5. **Data-driven**: As mentioned before, manually creating models is time-consuming and requires deep expert knowledge. Simultaneously recording anomalous data from CPS can be dangerous for the system itself. Therefore, anomaly detection should only be trained with regular production data and without the need of expert knowledge.

6. **Feasible**: To allow a generalist integration in different scenarios, the anomaly detection should be compatible with current production settings. This includes constraints and limitations of commonly used CPS in industrial environments, including:
   a. Processing limitations, due to the design of CPS in industry, that are not able to execute process-intensive tasks in parallel to the motion program, e.g., 256 MB RAM as limitation of a Yaskawa two-axis Servopack Sigma-7C with built-in motion controller MP3300.
   b. Communication interface constraints of commonly available CPS in industry, that are not designed to transfer high amounts of collected data in a short time to other devices, e.g.:
      i. FTP Server/Client: FTP access to process data, for example, with a transfer up to 8 MB each 20 ms for the motion controller.
      ii. Modbus TCP/IP: For instance, transfer of 50 float values each 20 ms.

4 A Concept for Unsupervised Process Anomaly Detection in Cyber-Physical Systems

4.1 Overview

This section describes a concept for unsupervised process anomaly detection based on a 1D-ConvAE which fulfills the requirements specified in the problem statement. The concept consists of two different cycles, which are split between two processing units in order to comply with industry constraints of CPS. This is shown as a BPMN diagram in Fig. 1. The first cycle, named *AD Installation*, realizes the data sampling, preprocessing, model generation, and the generated anomaly detection pipeline export. The second cycle, named *AD Production*, implements and executes the pipeline as a part of the CPS. Both cycles are developed to be executed automatically, and only a configuration file needs to be changed by an operator. This enables implementing anomaly detection without deep expert knowledge. While the processing unit backend performs all computationally intensive tasks in the first step, it can be removed when only the AD Production cycle has to be executed.

4.2 AD Installation

The AD Installation cycle can be split into four different parts, *data recording, preprocessing, model generation*, and *pipeline export*.

Data recording: Data recording is triggered by the operator on the processing unit onboard; additional information from a configuration file is needed. Due to the

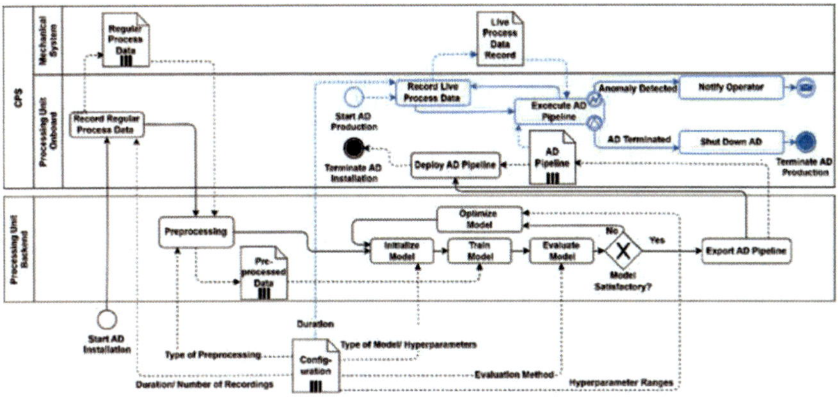

Fig. 1 Overview of AD Installation and AD Production

handling of repetitive process tasks with fixed durations, the starting point and the total runtime of the overwatched process must be specified. Regular process data is saved inside the processing unit onboard until the specified number of records is reached. Then all data is transmitted from the processing unit onboard to the processing unit backend. Regular process data consists of features like torque, speed, and position, sampled over a fixed period. This data can be defined as a process window containing the sampled features from the mechanical system over the defined task duration.

Preprocessing: The data provided by the mechanical system (e.g., speed, torque, position) consists of features with different ranges and units. To bring these features to an equal range, preprocessing of the transferred regular process data is needed. The type of the desired preprocessor is defined in the configuration file. This enhances a configurable setup, which can handle various process variables with different units and ranges.

Model generation: The model generation is based on four different steps, *initialization, training, evaluation,* and *optimization.* In the initialization step, the desired model type defined in the configuration file is instantiated. Additional hyperparameters are the number of layers, filters per layer, loss function, and optimizer. After initialization, the model is trained on the preprocessed data. Then, the model is evaluated using the method specified in the configuration file. In the optimization step, the hyperparameters are changed, influenced by the hyperparameter ranges and tuning parameters defined in the configuration file. This is done by a Bayesian optimization to search for the best possible parameters. These steps are executed iteratively until the specified reconstruction performance (e.g., the desired MAE Value) is reached.

Deployment: After optimization is finished, the preprocessor and the model are combined into an anomaly detection pipeline (AD pipeline). The AD pipeline is exported and deployed to the processing unit onboard. This terminates the AD Installation.

4.3 AD Production

The second cycle starts after deployment is finished. Live process data is sampled regularly and executed by the AD pipeline. All process data will be preprocessed and evaluated by the AD pipeline. The anomaly detection will be executed cyclically, and detected anomalies will be shown to the operator by a notification. Additionally, the operator can shut down the anomaly detection with an external command.

4.4 Convolutional Autoencoder

In order to meet the industry constraints, we choose a 1D-ConvAE as model type (see Fig. 2). Autoencoder is reconstruction-based neural networks, which reconstructs their input as output. The main idea is that by focusing on learning the reconstruction of the regular pattern only, every data consisting of unseen, abnormal patterns cannot be correctly reconstructed and will result in a higher reconstruction error. To gain adequate prediction performance and meet the processing limitations, 1D convolutional layers are used. Adding these layers to the autoencoder allows the model to learn spatially invariant features and capture spatially local correlations from the data. This means that it can recognize patterns of high-dimensional data without the need of feature engineering in advance. At the same time, the computational complexity of a 1D convolutional layer is significantly lower than the comparable 2D convolutional layers. The 1D-ConvAE can be trained without expert knowledge or explicitly known anomalies, only with regular process data. This fulfills the data-driven requirement.

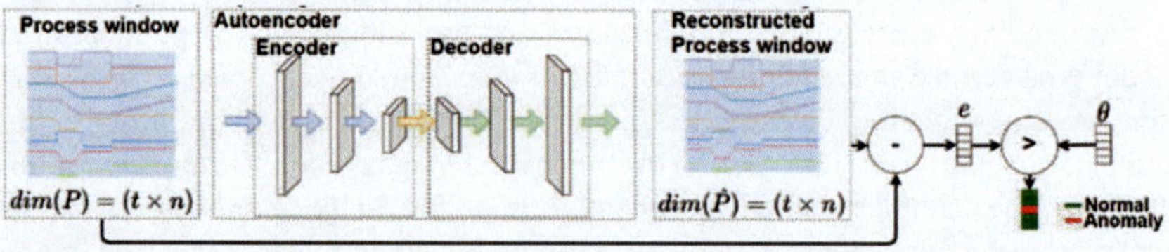

Fig. 2 Convolutional autoencoder

4.5 Anomaly Detection

As described in Sect. 4.2, regular process data can be defined as a process window containing the sampled features $F = (f_1, f_2, \ldots, f_n)$ from the mechanical system, like torque, speed, and position, over the defined task duration T. An example can be seen in Fig. 2. The process window P can be specified as a matrix with $\dim(P) = (t \times n)$, where n defines the number of different features and t the count of measured time steps. The output of the model is the reconstructed process window $\widehat{P}$ with the same dimensions as the input signal $\dim\left(\hat{P}\right) = (t \times n)$. To calculate the reconstruction error vector $\boldsymbol{e}$, the respective reconstruction error of each feature e_{f_i} can be calculated as the mean absolute error (MAE) $e_{f_i} = \frac{1}{t}\sum_{j=1}^{T}\left|f_{i,j} - \hat{f}_{i,j}\right|$ $(1 \leq i \leq n)$ between the input and the reconstruction of feature f_i. This results in a vector $\boldsymbol{e_F}$ representing each feature as a value of the differentiation between the input and reconstructed process window. We also tested other measures to evaluate the reconstruction error (e.g., mean-squared error and root-mean-squared Error) but achieved the best results with MAE.

To evaluate at which value a reconstruction error indicates if an anomaly is detected, threshold values must be defined. We employ the following method for automatically computing threshold values. After training the model, all training data is evaluated. This results in a number of reconstruction vectors, depending on the amount of training samples. Taken from all vectors, the maximum reconstruction error of each feature is used to construct a threshold vector θ. After each evaluation in the AD Production cycle, the reconstruction error vector $\boldsymbol{e}$ will be compared to the threshold vector θ. If one value of e_{f_i} exceeds the related threshold value θ_{f_i} $(1 \leq i \leq n)$, the repetitive process task will be declared anomalous.

5 Prototype Implementation

The concept has been implemented prototypically. As programming language, Python 3 is used. A MongoDB is established on the processing unit backend to save and export the regular process data. The preprocessing is realized with the library of Scikit-learn [34]. The model is implemented using the Keras library, running on top of TensorFlow [35]. For hyperparameter tuning, the Python library Mango [36] is used. Finally, to track the training results, the library MLflow [37] is applied.

The configuration file is written in YAML and can be changed by the operator. We have implemented an experimental setup which is described in the next section. The configuration file for the experimental setup consists of 125 lines of YAML code containing, for example, the data structure of the process windows, sample count, model, and preprocessor type. It is shared between the processing units and includes all necessary information to realize the entire concept. To satisfy Requirement 4 (configurable), the preprocessor and the anomaly detection model are implemented as abstract classes. All different considered models and preprocessors can be implemented, as long as they follow the structure of the abstract classes.

6 Evaluation

6.1 Experimental Setup

To evaluate the described concept, an experimental setup was chosen. The experimental setup consists of the processing unit backend and the CPS, which is a mechanical system combined with the processing unit onboard (see Fig. 3). The mechanical system simulates a transportation process of different components from a start to an end position and vice versa. This is realized by a motor connected to a ball screw that moves a carriage mounted on a guide system. A Yaskawa two-axis Servopack Sigma-7C with built-in motion controller MP3300 controls the motor. Combined with an absolute encoder, the Servopack measures different features like speed, torque, and position of the motor and sends this continuously in a multivariate data stream to the motion controller. The mechanical system repeatedly performs a transportation process. After waiting in an initial start position, the carriage moves with high speed to an end position. Here the carrier stops again and simulates a loading process (e.g., a component is placed on the carriage). After a constant time, the carriage moves back to the start position, and the next cycle begins.

During the process, an internal data logger in the motion controller records the multivariate data stream. After the process is finished, the data logger saves the recorded data to an internal FTP server. A microcontroller, an Nvidia Jetson Tx2, constantly imports the data from the FTP server and saves it in a database. Later, the stored data can be transferred to the processing unit backend, an Intel i5-6600 Windows 10 PC in the experimental setup. Eight features were recorded for training and testing, including the feedback and target position, reference and feedback speed, position deviation, calculated and coordinated position, and feedback torque. The internal sample period of the data logger is 2 ms. Therefore, with a process execution time of 14 s, the whole process results in 7000 measurements for each feature and 56,000 data points per process window.

6.2 Data Recording

To evaluate the proposed model, it was tested on regular data and realistic fault data. To simulate faults, different error cases were produced and differentiated into two error types. The first error type, namely *faulty product*, simulates the transportation

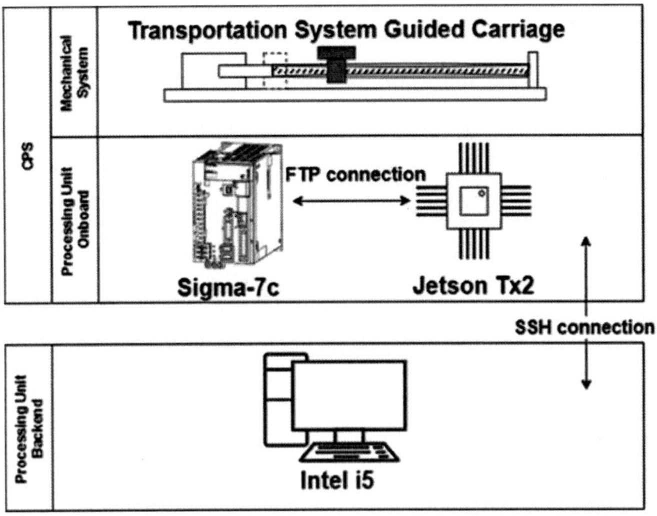

Fig. 3 Experimental setup

of a faulty or incorrect product through the industrial production line. To simulate this error, diverse objects with different weights are placed on the carriage during the transportation process. The second error type, defined as *friction*, simulates an increasing motion resistance due to the abrasion of the carrier, which can be caused by constant friction between the carriage and the guide system. To create such an error, a resistor band that acts as a brake to the motion of the carriage was used. Five different error cases have been generated, and 50 samples for each case were recorded (see Table 1). Finally, 2000 regular samples as training data and 449 test data containing 250 anomalous samples and 199 regular samples were recorded to evaluate the anomaly detection.

Table 1 Simulated error cases

Error	Error type	Description	Number of samples
Error case 1	Faulty product	1 kg object	50
Error case 2	Faulty product	2 kg object	50
Error case 3	Faulty product	4 kg object	50
Error case 4	Friction	1 kg resistor band	50
Error case 5	Friction	2 kg resistor band	50

6.3 Model Configuration

Only regular process data is used to train the model. A MinMaxScaler was chosen to preprocess the data by scaling the features between zero and one. An Adam [38] optimizer was used, and the loss function was set to MAE.

The hyperparameter tuning results in a decoder consisting of two convolutional layers, with strides one and a kernel size of 9, and an encoder with two transposed convolutional layers, with strides two and kernel size of 9. As activation function the rectified linear unit (ReLU) [39] was chosen. Two dropout layers as regularization and one max pooling layer to reduce the dimensionality are applied.

The best result was reached with 64 filters in the first layer and 32 filters in the second layer of the decoder, respectively; the first layer of the encoder has 32 filters, and the last layer has 64 filters.

6.4 Experimental Results

This section validates the described concept applied in the experimental setup against the requirements defined in the problem statement.

Anomaly detection: Fig. 4 shows the reconstruction errors of test data, separated by the different features. The anomalous samples (red) have a much higher reconstruction error than the regular samples (blue). For most features, there is a gap between the reconstruction error of regular and anomalous samples. Combined with the results shown in Fig. 5, this confirms that the model can successfully be applied to detect anomalies in the defined experimental setup by evaluating the reconstruction error.

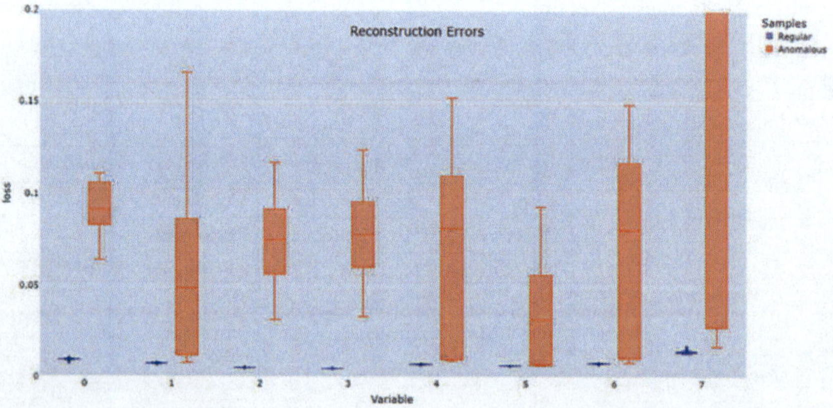

Fig. 4 Reconstruction error

Fig. 5 Confusion matrix

		Predicted: positive	Predicted: negative	
True (Anomaly detected)	positive	248	2	Precision: 0.988
	negative	3	196	
		Recall: 0.992		F1 Score: 0.989

Near-time: To detect an anomaly in a given process window, the AD pipeline needs on average 44 ms, with a maximum value of 120 ms and a minimal value of 25 ms. The time was measured during the evaluation of the test data. This confirms that the concept fulfills the near-time requirement in the experimental setup by checking the regular process data immediately after the process is finished.

Prediction quality: To evaluate the performance of the model, the F1 Score is used. The detailed performance can be seen from the confusion matrix in Fig. 5. By reaching an F1 Score of 0.989, an adequate prediction performance for the application is realized. This confirms that the model can reliably detect anomalies in the given setup.

Configurable: The anomaly detection can be configured for various applications. Only the YAML configuration file has to be changed to adapt the concept to different processes.

Data-driven: The model is trained with regular process data only. Therefore, no anomalous data or feature engineering is needed. No values are added or changed. Only the data from the motor/encoder/amplifier is used. The model is created in an automated way by the pre-defined configuration file, without the need of expert knowledge.

Feasible: The method utilized a standard communication interface of the motion controller, which is also available in other setups. By splitting the tasks between both processing units, the concept enables applying anomaly detection for the CPS, even with the processing power limitations and the constraints given by the communication interface.

7 Conclusion and Future Work

This paper presents an unsupervised anomaly detection concept for repetitive process tasks in CPS under industry constraints. The concept is configurable and feasible to apply anomaly detection in different use cases under the limitations of commonly used CPS in industrial environments. The anomaly detection is executed during production, and the evaluation is carried out immediately after the process is finished.

The model is generated and tuned in an automated way. No expert knowledge and anomalous data are needed. Overall, the experiment shows that the model achieves stable and accurate results. Thus, it presents a promising approach for process anomaly detection in CPS under industry constraints.

However, despite the apparent success of the concept, there are several directions for future research. In this work, the concept was only tested by one model in a single scenario. Therefore, more studies with different models and under different scenarios will be performed in future work. Secondly, up to now, the concept is only evaluated on the data provided by one motor. Additional tests and improvements to include several axes simultaneously are another research direction in the future. Finally, the output of the anomaly detection is up to now a notification to inform the operator that an anomaly is detected. Adding more information may be helpful for the operator. Ways to gather and provide this additional context information will be evaluated and investigated.

References

1. Monostori L, Kádár B, Bauernhansl T, Kondoh S, Kumara S, Reinhart G, Sauer O, Schuh G, Sihn W, Ueda K (2016) Cyber-physical systems in manufacturing. CIRP Ann 65:621–641
2. Rajkumar, Lee I, Sha L, Stankovic J (2010) Cyber-physical systems. In: Sapatnekar S (ed) Proceedings of the 47th design automation conference, ACM, New York, NY, p 731
3. Shi J, Wan J, Yan H, Suo H (2011) A survey of cyber-physical systems. In: 2011 International conference on wireless communications and signal processing (WCSP 2011): Nanjing, China, 9–11 Nov 2011, IEEE, Piscataway, NJ, 2011, pp 1–6
4. Liu Y, Peng Y, Wang B, Yao S, Liu Z (2017) Review on cyber-physical systems. IEEE/CAA J Autom Sinica 4:27–40
5. Adepu S, Mathur A (2021) Distributed attack detection in a water treatment plant: method and case study. IEEE Trans Dependable Secure Comput 18:86–99
6. Chandola V, Banerjee A, Kumar V (2009) Anomaly detection. ACM Comput Surv 41:1–58
7. Pimentel MA, Clifton DA, Clifton L, Tarassenko L (2014) A review of novelty detection. Signal Process 99:215–249
8. Cook AA, Misirli G, Fan Z (2020) Anomaly detection for IoT time-series data: a survey. IEEE Internet Things J 7:6481–6494
9. Goldstein M, Uchida S (2016) A comparative evaluation of unsupervised anomaly detection algorithms for multivariate data. PLoS ONE 11:e0152173
10. Ruff L, Kauffmann JR, Vandermeulen RA, Montavon G, Samek W, Kloft M, Dietterich TG, Muller K-R (2021) A unifying review of deep and shallow anomaly detection. Proc IEEE 109:756–795
11. Ekanayake T, Dewasurendra D, Abeyratne S, Ma L, Yarlagadda P (2019) Model-based fault diagnosis and prognosis of dynamic systems: a review. Procedia Manuf 30:435–442
12. Marzat J, Piet-Lahanier H, Damongeot F, Walter E (2012) Model-based fault diagnosis for aerospace systems: a survey. Proc Inst Mech Eng, Part G: J Aerosp Eng 226:1329–1360
13. Bulusu S, Kailkhura B, Li B, Varshney PK, Song D (2020) Anomalous example detection in deep learning: a survey. IEEE Access 8:132330–132347
14. Li D, Chen D, Jin B, Shi L, Goh J, Ng S-K (2019) MAD-GAN: multivariate anomaly detection for time series data with generative adversarial networks. In: Tetko IV, Kůrková V, Karpov P, Theis F (eds) Text and time series. Springer, Cham, pp 703–716

15. Inoue J, Yamagata Y, Chen Y, Poskitt CM, Sun J (2017) Anomaly detection for a water treatment system using unsupervised machine learning. In: Gottumukkala R (ed) 17th IEEE international conference on data mining workshops: 18–21 November 2017. Louisiana proceedings, IEEE, Piscataway, NJ, New Orleans, pp 1058–1065
16. Elnour M, Meskin N, Khan K, Jain R (2020) A dual-isolation-forests-based attack detection framework for industrial control systems. IEEE Access 8:36639–36651
17. Kiranyaz S, Avci O, Abdeljaber O, Ince T, Gabbouj M, Inman DJ (2021) 1D convolutional neural networks and applications: a survey. Mech Syst Signal Process 151:107398
18. Xie X, Wang B, Wan T, Tang W (2020) Multivariate abnormal detection for industrial control systems using 1D CNN and GRU. IEEE Access 8:88348–88359
19. Liu J, Wang H, Liu Z, Wang Z (2020) Fault diagnosis based on 1D-CNN with associated auxiliary tasks boosted for wheelset bearings of high-speed trains. In: Qin Y (ed) Proceedings of 2020 international conference on sensing, diagnostics, prognostics, and control (SDPC): August 5–7, 2020 Beijing, China, IEEE, Piscataway, NJ, pp 98–103
20. Ince T, Kiranyaz S, Eren L, Askar M, Gabbouj M (2016) Real-time motor fault detection by 1-D convolutional neural networks. IEEE Trans Ind Electron 63:7067–7075
21. Kravchik M, Shabtai A, Detecting cyber attacks in industrial control systems using convolutional neural networks. In: Lie D, Mannan M, Rashid A, Tippenhaeur NO (eds) Proceedings of the 2018 workshop on cyber-physical systems security and PrivaCy, ACM, New York, NY, USA, 10152018, pp 72–83
22. Chen Y, Zhang H, Wang Y, Yang Y, Zhou X, Wu QMJ, Net MAMA (2021) Multi-scale attention memory autoencoder network for anomaly detection. IEEE Trans Med Imaging 40:1032–1041
23. Sadaf K, Sultana J (2020) Intrusion detection based on autoencoder and isolation forest in fog computing. IEEE Access 8:167059–167068
24. Park S, Adosoglou G, Pardalos PM (2021) Interpreting rate-distortion of variational autoencoder and using model uncertainty for anomaly detection. Ann Math Artif Intell
25. Jinwon A, Sungzoon C, Variational autoencoder based anomaly detection using reconstruction probability. In: Special Lecture on IE, pp 1–18
26. Xu H, Feng Y, Chen J, Wang Z, Qiao H, Chen W, Zhao N, Li Z, Bu J, Li Z, Liu Y, Zhao Y, Pei D (2018) Unsupervised anomaly detection via variational auto-encoder for seasonal KPIs in web applications. In: Champin P-A, Gandon F, Lalmas M, Ipeirotis PG (eds) Proceedings of the 2018 World Wide web conference on World Wide Web—WWW'18, ACM Press, New York, New York, USA, 2018, pp 187–196
27. Zhang C, Song D, Chen Y, Feng X, Lumezanu C, Cheng W, Ni J, Zong B, Chen H, Chawla NV (2019) A deep neural network for unsupervised anomaly detection and diagnosis in multivariate time series data. AAAI 33:1409–1416
28. Yin C, Zhang S, Wang J, Xiong NN (2020) Anomaly detection based on convolutional recurrent autoencoder for IoT time series, IEEE Trans Syst Man Cybern Syst, pp 1–11
29. Pereira J, Silveira M (2018) Unsupervised anomaly detection in energy time series data using variational recurrent autoencoders with attention. In: Wani MA, Kantardzic M, Sayed-Mouchaweh M, Gama J, Lughofer E (eds) 17th IEEE International conference on machine learning and applications: ICMLA 2018 17–20 December 2018, Orlando, Florida, USA proceedings, IEEE, Piscataway, NJ, 2018, pp 1275–1282
30. Amarbayasgalan T, Lee HG, van Huy P, Ryu KH (2020) Deep reconstruction error based unsupervised outlier detection in time-series. In: Nguyen NT, Jearanaitanakij K, Selamat A, Trawiński B, Chittayasothorn S (eds) Intelligent information and database systems: 12th Asian conference, ACIIDS 2020, Phuket, Thailand, March 23–26, 2020, Proceedings, Part II, 1st edn. Springer International Publishing; Imprint Springer, Cham, pp 312–321
31. Guo X, Liu X, Zhu E, Yin J (2017) Deep clustering with convolutional autoencoders. In: Liu D, Xie S, Li Y, Zhao D, El-Alfy E-SM (eds) Neural information processing: 24th international conference, ICONIP 2017, Guangzhou, China, November 14–18, 2017 proceedings. Springer, Cham, pp 373–382
32. Chen T, Liu X, Xia B, Wang W, Lai Y (2020) Unsupervised anomaly detection of industrial robots using sliding-window convolutional variational autoencoder. IEEE Access 8:47072–47081

33. Kwak M, Kim SB (2021) Unsupervised abnormal sensor signal detection with channelwise reconstruction errors. IEEE Access 9:39995–40007
34. Scikit-learn: Machine learning in Python, 2011
35. Abadi M, Barham P, Chen J, Chen Z, Davis A, Dean J, Devin M, Ghemawat S, Irving G, Isard M, Kudlur M, Levenberg J, Monga R, Moore S, Murray DG, Steiner B, Tucker P, Vasudevan V, Warden P, Wicke M, Yu Y, Zheng X, TensorFlow: a system for large-scale machine learning. In: Papadopouli M (Ed) Papers presented at the 2005 workshop on Wireless traffic measurements and modeling, USENIX Association, Berkeley, CA, 2005, pp 265–283
36. Sandha SS, Aggarwal M, Fedorov I, Srivastava M, Mango: a Python library for parallel hyper-parameter tuning. In: 2020 IEEE International conference on acoustics, speech, and signal processing: proceedings May 4–8, 2020, Centre de Convencions Internacional de Barcelona (CCIB), Barcelona, Spain, IEEE, Piscataway, NJ, USA, 2020, pp 3987–3991
37. Chen A, Chow A, Davidson A, DCunha A, Ghodsi A, Hong SA, Konwinski A, Mewald C, Murching S, Nykodym T, Ogilvie P, Parkhe M, Singh A, Xie F, Zaharia M, Zang R, Zheng J, Zumar C (2020) Developments in MLflow. In: Proceedings of the fourth international workshop on data management for end-to-end machine learning, Association for Computing Machinery, New York, NY, United States, pp 1–4
38. Kingma DP, Ba J (2014) Adam: a method for stochastic optimization
39. Ying Y, Su J, Shan P, Miao L, Wang X, Peng S (2019) Rectified exponential units for convolutional neural networks. IEEE Access 7:101633–101640

3D Virtual System of an Apple Sorting Process Using Hardware-in-the-Loop Technique

Bryan Rocha, Carlos Tipan, and Luigi O. Freire

Abstract This paper deploys an industrial automation process in a virtual environment for the classification of three-dimensional objects through the systematization of programmable automatons in a graphic engine (Unity 3D), integrating hardware-in-the-loop simulation methods to couple the electrical signals of sensors and actuators through a microprocessor and with the use of a Logo 8.1 programmable logic controller. The virtual environment provides a field of 3D modeled objects in order to create an industrial environment, thus linking the external signals of the process control, visualizing them in the indicators of the environment. The industrial simulation project is intended for users to adopt new techniques in the fields of industrial process knowledge, automation, programming languages, communication protocols and thus optimize resources and improve results in the industry.

Keywords Automation · Industrial processes · Signals · Virtual environment · Microcontroller · Hardware in the loop

1 Introduction

Virtual reality is currently occupying a position of high hierarchy among software developers, especially scientific, military and medical research [1]. Virtual reality in the field of industrial process automation encompasses a set of electromechanical systems and elements whose purpose is to control, monitor and execute industrial processes more efficiently and autonomously [2]. An automated system is defined as a set of machines or processes capable of reacting without the intervention of

B. Rocha (✉) · C. Tipan · L. O. Freire
Universidad Técnica de Cotopaxi, Latacunga, Ecuador
e-mail: bryan.rocha8742@utc.edu.ec

C. Tipan
e-mail: carlos.tipan5380@utc.edu.ec

L. O. Freire
e-mail: luigi.freire@utc.edu.ec

P. Chatterjee et al. (eds.), *Computational Intelligence for Engineering and Management Applications*, Lecture Notes in Electrical Engineering 984,
https://doi.org/10.1007/978-981-19-8493-8_50

an operator (automatically) to changes in the different variables involved in the processes, performing the most appropriate operations to perform the function or work for which they have been designed [3].

The simulation of the automation of an industrial process involving sensors, actuators, microcontrollers and PLCs (programmable logic controllers) which using communication protocols and electrical signals interacts with each other to census variables (analog or digital) and control actuators to generate a virtual environment through a graphic engine which shows the process of hardware in the loop. Virtual reality (VR) represents the simulation of a scenario (virtual environment) created through the use of computers, which allows us to interact between the different devices attached to the virtual environment generating the sensation of interaction with it [4]. 3D modeling is the process of creating a mathematical representation of surfaces using geometry. This can be represented in two ways: on screen as a two-dimensional image through a process known as 3D rendering or as a physical object [5]. For the creation of 3D models is used SolidWorks software which creates designs quickly and accurately, including 3D models and 2D drawings of assemblies and complex parts also optimizing the design and manufacturing costs using cost estimation tools and performing manufacturing feasibility checks [6]. These objects when created in an exact mechanical parts modeling program have a large file size; for this reason, these files are compressed by exporting them to a different format using Blender program which is the free and open-source 3D creation suite. It supports the entire 3D pipeline: modeling, assembly, animation, simulation, rendering, compositing and motion tracking, even video editing and game creation [7]. A graphical interface (virtual reality environment) is a manipulable database capable of generating a simulation or environment (3D modeling), generated by a computer, explorable, viewable and manageable in real time in the form of digital images and sounds, which creates in the user the feeling of being inside that world, however, depends on the level of immersion that this can interact with objects and the world. Virtual reality can be classified into immersive and non-immersive; the first achieves a complete link with the environment through peripherals (virtual reality helmets, glasses, positioners, etc.), to the point of disappearing the real world, and in the second (non-immersive), the user interacts with the virtual world, but without being completely immersed in it, for example, through a computer monitor [8].

In the exploration to achieve a more efficient automation that involves the optimization of processes in which PLCs have been one of the most essential mechanisms, currently several companies are dedicated to the development, manufacture and distribution of these devices. According to National Electrical Manufacturers Association (NEMA), a PLC is defined as: a digitally operated electronic device, which uses a programmable memory for internal storage of instructions to implement specific functions, such as logic, sequencing, recording and timing control, counting and arithmetic operations to control, through digital (ON/OFF) or analog (0–5 VDC, 4–20 mA) input/output modules [9]. The use of sensors is present in all areas of society, such as automation, home automation, medicine, industrial control or agriculture [10], in the output of the controllers are coupled actuators which transform electrical signals into another type of energy (mechanical, light, heat, etc.) as required by the operator and the programmer of the system. The sensors and actuators of an industrial process are focused on achieving greater comfort, safety, efficiency and total control by the operator in the industrial process that directs [11].

Microcontrollers are integrated circuits used to transform sensed information and knowledge useful for decision making. By connecting sensors, actuators and communication between devices, it is possible to automate tasks based on input data [12], these devices operate in a loop. Once the process in the microcontroller concludes its loop, it can send the sensing and control data through the different communication protocols (MODBUS, serial, etc.) to transmit them to the different coupled devices (PLCs, virtual reality environments, computers with a graphical interface) that allow the interpretation of data received by the communication ports and in the same way send a response to take the appropriate action to the process. Since there are a variety of useful protocols for data reading and control of industrial processes, it is necessary to be aware of the advantages and disadvantages offered by each one when choosing the protocol for different applications [13]; for this type of application, serial communication was selected, which consists of sending one bit at a time, sequentially through a communication channel or bus.

2 System Structure

This paper presents the development of a non-immersive virtual reality application for the simulation of an apple sorting process according to its size using the following tools:

- Environment construction (3D modeling)
- Graphics engine (Unity 3D software)
- Code compiler
- Programming code
- Microcontroller (Atmega328)
- Programmable logic controller (LOGO 8.1)
- Communication protocols (serial)
- Microcontroller power circuit (indicators, pushbuttons, etc.).

Using 3D modeling software (SolidWorks, Inventor, Blender), the necessary physical instruments will be recreated (conveyor belt, sensors, buckets, indicators, controllers, infrastructure, etc.) to be exported to a 3D graphics engine (Unity 3D) to create a non-immersive virtual environment with motion animations and environmental sound) to export them to a 3D graphics engine (Unity 3D) to create a non-immersive virtual environment with motion animations and environmental sound, implementing a serial communication port to establish a serial communication protocol for reading and writing data with the Atmega328 microcontroller, which will have a control circuit (push buttons) and indicators (LEDs), in turn detect the sensor signals sending electrical pulses to the programmable logic controller (LOGO 8. 1) so that it activates or deactivates the actuators according to the type of variable to be classified.

3 Virtual Environment

For the development of the non-immersive virtual environment of an industrial process of classification of apples according to their variable (size) 3D objects were designed using SolidWorks CAD Software (computer-aided design) according to the dimensions required for proper operation, once designed and assembled the different parts of the required objects we proceed to export them to a compatible format (. stl) to be able to work in a more intuitive way their textures in the 3D design program Blender and thus save the objects in an even lighter format (.fbx) and compatible with the graphics engine (Unity 3D) (Fig. 1).

3.1 Stage 1. Process Initialization

For the start-up of the simulation of the industrial process, there are two ways to start it; the first will be with a physical button (push button) connected to the microcontroller, which will have two indicators (LED lights), when the simulation is started, the green indicator will light up, and while the process does not have a start pulse, the red LED will remain on. The second way to start the process will be through the interaction between the user and the environment; within the simulation, a control board has been created with different drives and indicators; and in this way, it will be possible to assign an action to the push button within the environment replacing the physical drive and in the same way simulating the lighting of an indicator within the environment.

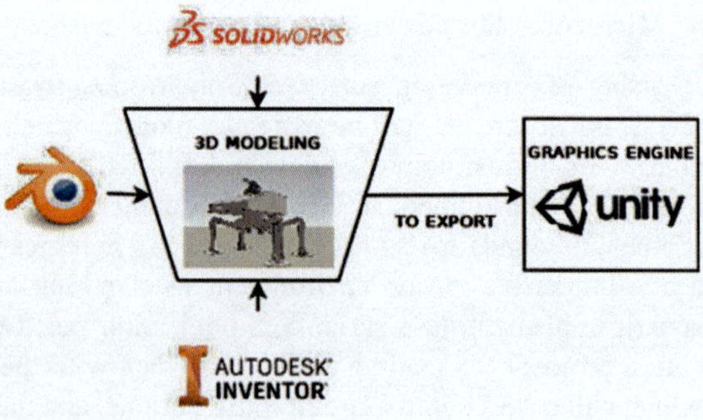

Fig. 1 Virtual environment creation process

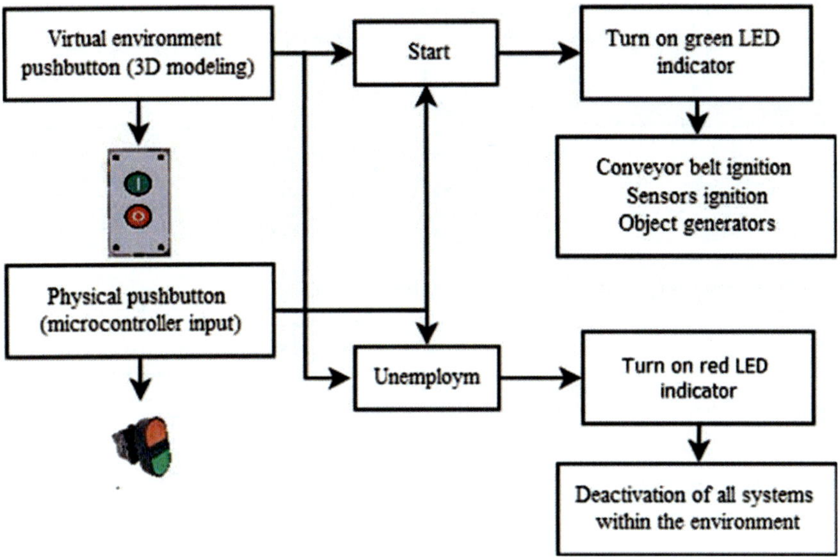

Fig. 2 On/OFF control diagram of the process

Once the process is started, the actuators within the environment will be activated, causing the belt to start turning, the sensors to turn on, the object generators to start their action, and in this way, the process will start its loop (Fig. 2).

3.2 *Stage 2. Random Creation of the Variables to Be Classified*

For this stage, it is necessary the 3D modeling of a container that by means of the programming of the 3D graphic engine, this can produce objects (apples) of different sizes in a random way, coupling it to the beginning of the conveyor belt, and thus fulfills the sequence of classification of objects so that the subsequent stages activate the sensors according to their variable.

The object generation process will be carried out every certain time interval as long as the sensors are deactivated and the objects have reached their destination, working together with the conveyor belt. (If one of the sensors detects the presence of an object, both the variable generator and the conveyor belt will stop running) (Fig. 3).

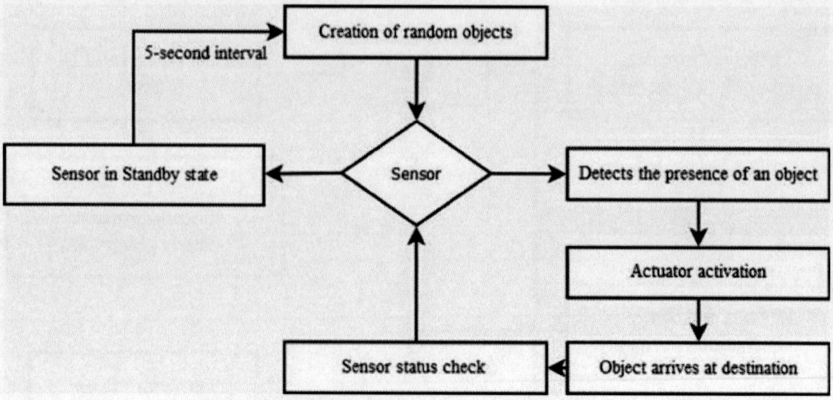

Fig. 3 Diagram for creating objects randomly

3.3 Stage 3. Define the Type of Variables

For this stage, the types of variables will be defined by the object generation process at the beginning of the sequence. At this point, two proximity sensors will work, each time an object (apple) is generated, and it goes horizontally through the moving conveyor belt and arrives at a point where it is classified by its size. Three sets of objects are generated (large, medium and small), and once it passes through the proximity sensors, this will trigger a pneumatic piston pushing the object to another conveyor belt, thus complying with the respective classification of variables by size. The conveyor belt stops completely when one of the sensors is activated or one of the pneumatic pistons is in motion.

The order of classification of the objects is descending (the sensor of the largest objects is located at the beginning of the conveyor belt, the second sensor of medium-sized objects is located at a prudent distance from the first sensor to avoid interference and finally, the objects that do not activate any of the sensors will fall at the end of the conveyor belt allowing to classify them in three different sizes) (Fig. 4).

3.4 Stage 4. ON/OFF Control of Actuators for Classification

In the next point to be discussed, the data stored in stage four by the proximity sensors that are located on one side of the conveyor belt. For this stage we proceed to use two pneumatic pistons that will be activated when the process requires it and these pneumatic pistons have the operation of classifying the objects (apples) that are coming out of the generator and change direction to comply with the established in the sequence of sorting process.

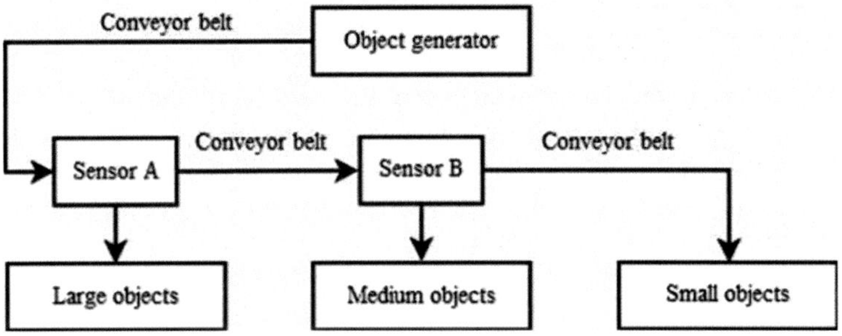

Fig. 4 Diagram operation of the sensors

The pneumatic pistons work together with the proximity sensors, once the objects reach a dead point, and with the animation of the pneumatic piston, the apple will change conveyor belt. The conveyor belt comes to a complete stop when one of the sensors is activated or one of the pneumatic pistons is in motion (Fig. 5).

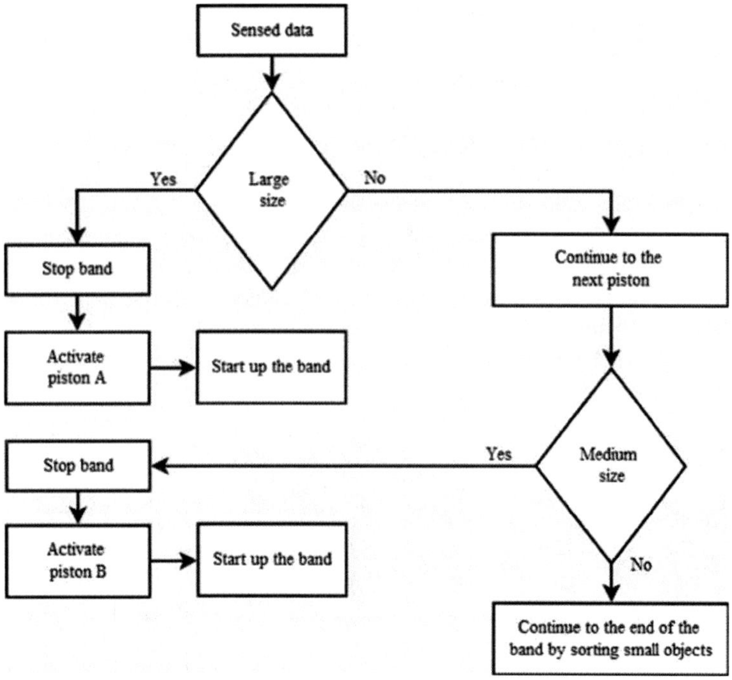

Fig. 5 Diagram operation of actuators

3.5 Stage 5. Result Indicator

For the final stage of the industrial process of sorting objects (apples), the results of each of the previous stages are obtained in a control panel created in the same virtual environment; when the industrial process starts, this will be reflected in a green indicator (LED). Each time an object (apples) is created, the data will be sent, saved and visualized on a display, which serves as an object count. When the conveyor belt is running, a sensor detects an object or one of the pistons starts moving; each of these will have its own LED indicator both in the virtual environment and in the outputs connected to the microcontroller. At the end of each band, there will be a counter showing more detailed information on how many objects of each set have been sorted and displaying the quantities in the virtual environment.

Once the cuvettes are filled with the different types of variables classified by size, these will disappear automatically showing us in a new indicator the quantity of total cuvettes classified in a work cycle.

4 Simulation Communication

In this simulation project of an industrial process, the communication protocol used is RS-232 (serial) and transmission of high and low electrical signals.

In virtual environment, it will generate a type of variable (sizes) assigning to each one of them a code in the same way with the indicators and actuators of the process, generating a line of encrypted code which will be transmitted through the serial port to the microcontroller which will decrypt the message and process the data sending electrical signals to the programmable logic controller to control the actuators responding with electrical signals to the microcontroller with the actions to be taken transmitting it back to the virtual environment through the serial port displaying the results on a screen (bidirectional communication) (Figs. 6 and 7).

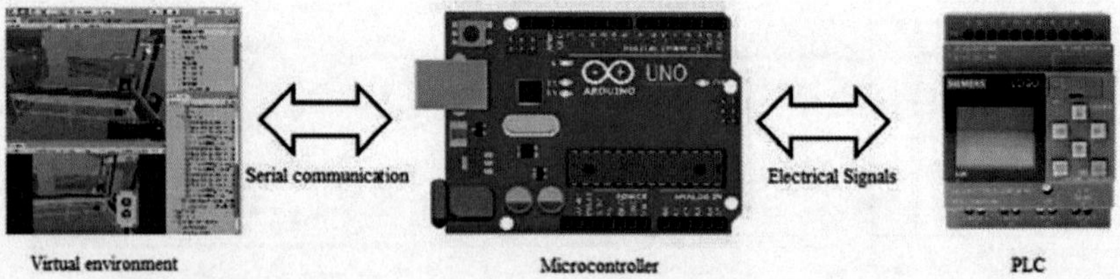

Fig. 6 Communication diagram between the environment, the microcontroller and the PLC

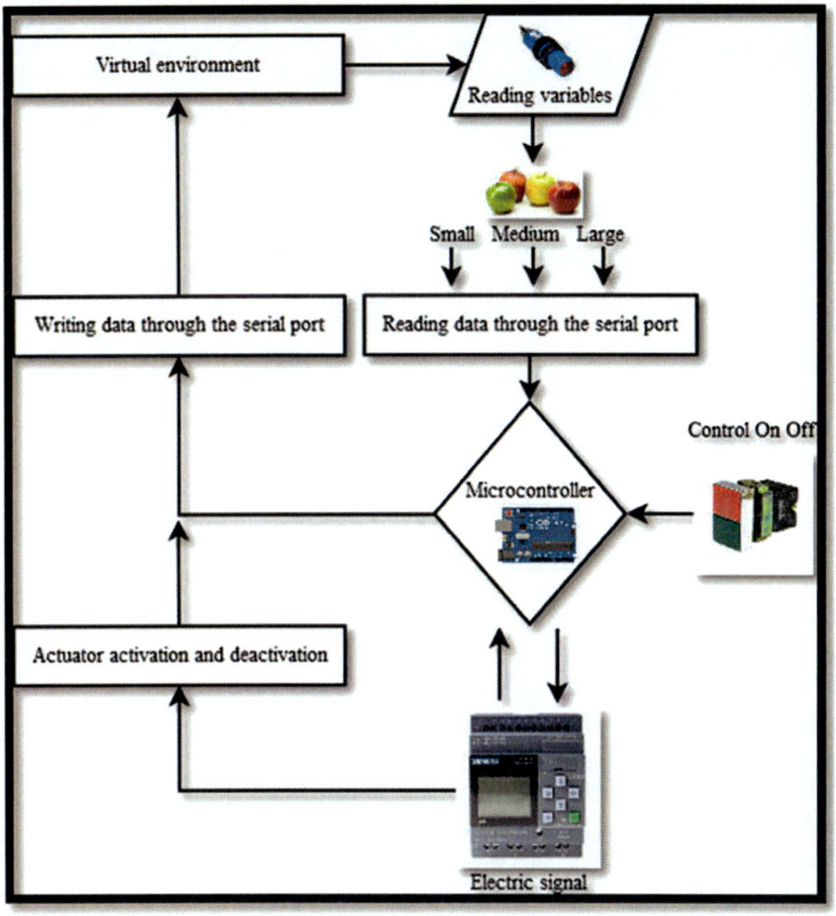

Fig. 7 Diagram of operation of the environment and physical devices

5 Results

The tests that are deployed below are mainly focused on the results obtained in each stage of the project which indicates an industrial process of classification of apples according to their caliber. The automation of the process is performed using a programmable logic controller LOGO 8.1 (Siemens) to proceed to verify what is the main function of each of the stable stages working together with the communication protocols obtaining electrical signals (sensors, actuators) for their different actions to be performed (Fig. 8).

Test 1. Once the process has started, the actuators within the environment are started, causing the belt to start turning, the sensors to turn on, the object generators to start

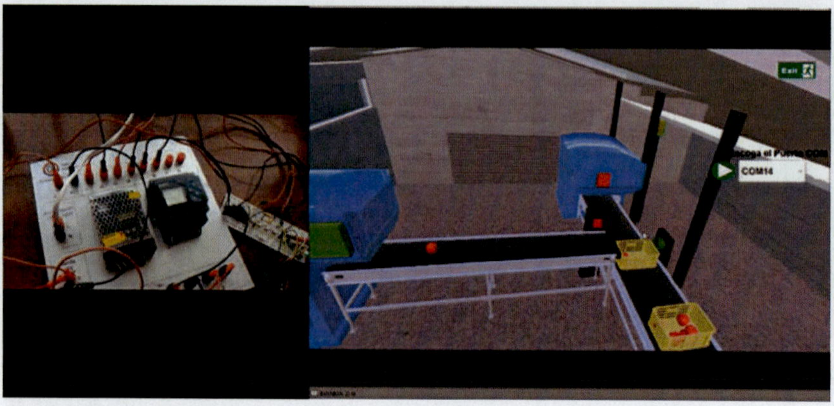

Fig. 8 Communication between the virtual environment and controllers

their action, and in this way, the process will start its loop, visualizing all the work through internal (modeled) and external (physical) led indicators. See Fig. 9.

Test 2. The following Figs. 10 and 11 show the process of object generation, which is carried out every certain time interval as long as the sensors are deactivated and the objects have reached their destination, working together with the conveyor belt. (If one of the sensors detects the presence of an object, both the variable generator and the conveyor belt will stop working).

Test 3. The order of classification of the objects is descending (the sensor of the largest objects is located at the beginning of the conveyor belt, the second sensor of medium-sized objects is located at a prudent distance from the first sensor to avoid interference and finally, the objects that do not activate any of the sensors will fall at

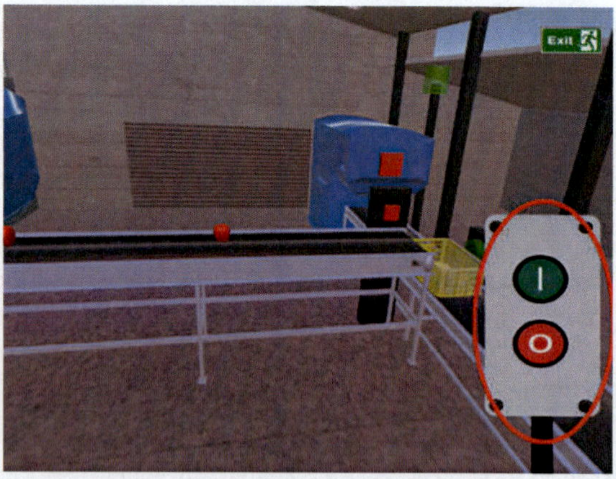

Fig. 9 Initializing the virtual process

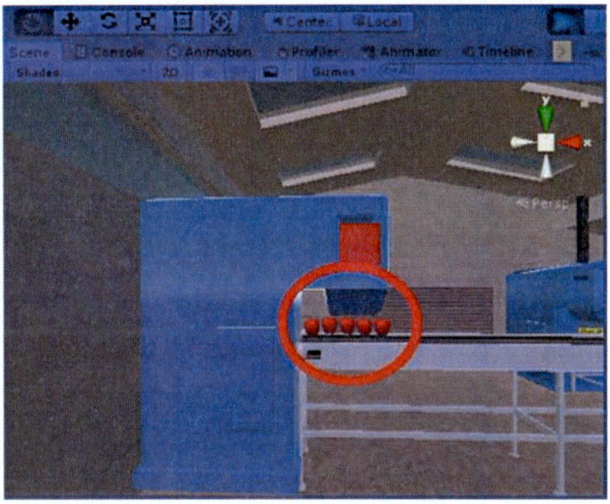

Fig. 10 Variable generator

Fig. 11 Vessel generator

the end of the conveyor belt allowing them to be classified into three different sizes), showing the action of the sensors in the following Figs. 12 and 13.

Test 4. In the following Fig. 14, the pneumatic pistons work together with the proximity sensors, once the objects reach a dead stop, and with the animation of the pneumatic piston, the apple will change conveyor belt. The conveyor belt stops completely when one of the sensors is activated or one of the pneumatic pistons is moving.

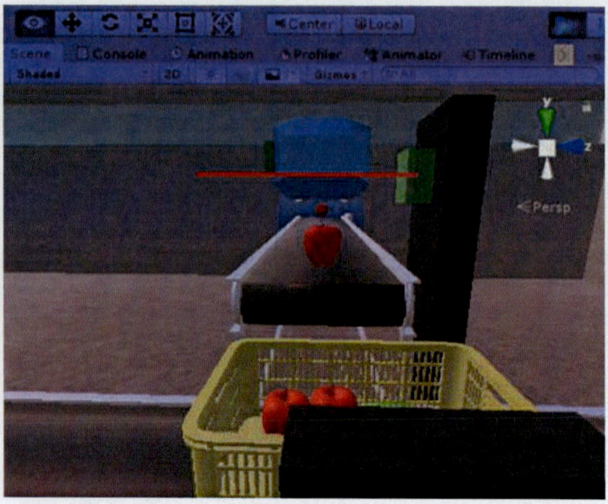

Fig. 12 Proximity sensor 1 (high)

Fig. 13 Proximity sensor 2 (medium)

Test 5. Once the cuvettes are filled with the different types of variables classified by size, they will disappear automatically, showing us in a new indicator the total amount of cuvettes classified in a work cycle, as shown in the following Fig. 15.

6 Conclusions

It was successfully developed the automation of an industrial process of classification of variables according to their size within a virtual environment thus achieving visualize through indicators the correct operation of each of the stages of the process, showing the creation of objects, controls, activation and deactivation of sensors and

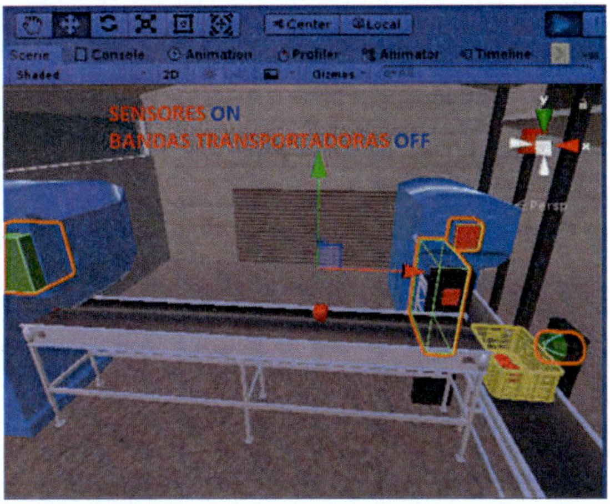

Fig. 14 Sensors on standby, paralyzed bands

Fig. 15 Classification of apples according to size

actuators, counting how many objects have been classified, thus checking the correct communication between the environment and peripherals connected to it, obtaining a double control, within the virtual environment (simulation) and externally by means of electrical signals (pushbuttons).

The implemented system allows the user to adopt new techniques in different fields such as automation, instrumentation, 3D modeling, programming logics and communication protocols to optimize and improve resources in industrial processes; the system is low cost and easy to use compared to real industrial models for related practices.

References

1. García Vélez RS (2021) Diseño y desarrollo de ambientes de realidad virtual para reducir el estrés en el ámbito ácademico. Universidad Politécnica Salesiana vol I, no 1, pp 1–17
2. Aladrén VS (2021) Proyecto de diseño y automatización de una línea logística de clasificación, paletizado y almacenamiento de productos empleando FACTORY I/O para su evaluación mediante gemelo digital. Universitat Politécnica de Valéncia vol I, no 1, pp 1–137
3. Llopis RS, Romero Pérez JA, Ariño Latorre CV (2010) Automatización industrial. España: Publicacions de la Universitat Jaume I
4. Calvopiña Estrella FE, Paredes Almach JE, Vaca Cárdenas LA (2021) Entorno Virtual de Aprendizaje para el Ensamblaje de Computadoras. Revista de tecnologías de la informática y las telecomunicacones vol I, no 1, pp 25–43
5. Jorquera Ortega SA (2010) Fabricacion Digital: Introduccion al modelado e impresion 3D. España: Secretaria Editorial Tecnica
6. Corporation DSS (2021) SOLIDWORKS. Dassault Systèmes SolidWorks Corporation, 2020–2021. [Online]. Available https://www.solidworks.com/es/domain/design-engineering. Accessed 16 Dec 2021
7. Blender (2021) Blender, Artistic freedom starts with Blender, 2020–2021. [Online]. Available https://www.blender.org/about/. Accessed 16 Dec 2021
8. Arroyave TAF, Espinosa GDA, Duque MND (2015) DESARROLLO DE OBJETOS DE APRENDIZAJE USANDO TÉCNICAS DE REALIDAD VIRTUAL. CORE vol I, no 1, pp 74–81
9. Solorio Alvarado AI (2021) Reconversión de máquina ponedora a modelo hidráulico e interfaz de control hombre-máquina. Universidad de Guanajuato vol 1, no 1, pp 1–100
10. Salguero Barceló FJ (2021) La sectorización basada en criterios energéticos como herramienta para la gestión hídrica de redes de distribución de agua. Universitat Politécnica de Válencia vol I, no 1, pp 1–189
11. Gordillo Rojas OD (2021) Desarrollo de una aplicación móvil para el rastreo de ubicación y comando remoto vehicular. Universidad Politécnica Salesiana vol I, no 1, pp 1–114
12. Rodríguez Sorní M (2021) Diseño e implementación de un robot móvil basado en microcontrolador. Universitat Politècnica de València vol 1, no 1, pp 1–72
13. Arias Martijena A, Castillo Ruíz AA, Roa Arias ÁI, Bidó Cuello E, García Maimó J, Mariano-Hernández D, Aybar Mejía ME (2021) Protocolos y topologías utilizadas en los sistemas de comunicación de las microrredes eléctricas vol IV, no 1, pp 81–95

Application of Augmented Reality for the Monitoring of Parameters of Industrial Instruments

Jairo L. Navas, Jorge L. Toapanta, and Luigi O. Freire

Abstract The development of this augmented reality application on mobile devices with Android operating system is oriented for the identification and visualization of industrial equipment and instruments, the interaction through the characteristics and labels of its parts, exploded views and assembly that make up a process of a level control panel. The generation of this mobile application was done with the use of recognition of each element through the generation of a QR code, to be subsequently focused on 3D models using CAD software linked through a multiplatform. Additionally, the simulation and visualization of data from industrial sensors in real time of the process are based on the operation of the instruments, and these generated data contribute to the development of skills for the manipulation and control of industrial processes.

Keywords Augmented reality · Mobile application · Virtual environment · Digitalization · Industrial process · Control module

1 Introduction

The advance of technology is a fact, and the reflection of this advance is currently reflected in mobile devices and their daily evolution during the last few years [1]. The technological advances in smartphones are on the rise; as a result, it has established itself as the platform with the largest number of users. The use and advancement of applications in the devices have made people achieve great advances in both their professional and personal profile in the family since the new technologies allow

J. L. Navas (✉) · J. L. Toapanta · L. O. Freire
Universidad Técnica de Cotopaxi, Latacunga, Ecuador
e-mail: jairo.navas5896@utc.edu.ec

J. L. Toapanta
e-mail: jorge.toapanta2138@utc.edu.ec

L. O. Freire
e-mail: luigi.freire@utc.edu.ec

P. Chatterjee et al. (eds.), *Computational Intelligence for Engineering and Management Applications*, Lecture Notes in Electrical Engineering 984,
https://doi.org/10.1007/978-981-19-8493-8_51

to develop a quieter life [1, 2]. Two of the technologies that are gradually gaining more and more presence in society are virtual reality and augmented reality. These software proposals make use of the technological advances achieved by hardware in processing units [3, 4].

The most innovative developments are augmented reality (AR) and virtual reality (VR) technologies, each of which brings together information in a different way, but both fields focus on providing the user with an immersive 3D environment [5].

Focusing on augmented reality, it can be defined as the additional information obtained by observing an environment through the camera of a mobile or fixed device to which a specific program or application has been pre-installed [6, 7]. A set of sophisticated AR applications that could be called immersive augmented reality applications and not applicable to consumption, which add quite advanced and expensive technologies, is military or academic application utilities that recreate virtual universes that allow, for example, to execute certain capabilities and that are able to reproduce 3D images [8]. For this reason, we seek to develop an immersive AR application that allows the development of people's skills in a learning environment, based on an industrial process that simulates the work environment of a person in a company. [9]. The AR technology in the field of education in a training format is explicit in part by the characteristics presented by this, since being a hybrid reality that integrates several factors, you can have an integration of different sources of information in various formats such as text, images, videos, websites among others favoring the person who needs it [10]. The use of AR in the context of Industry 4.0 allows industrialists to control systems to expand their capabilities. The objective of the AR applied to real-time line research systems in the framework of Industry 4.0 is to develop a prototype that can be integrated into a Supervisory Control and Data Acquisition (SCADA) system [11, 12]. It is intended that the mobile application has an interface that allows to see the operation of equipment and instruments in the industry, in order to identify conditions and changes in the operation and thus be able to know what is happening through digital support [13]. A design and digitalization software is Unity, which is a digital tool that helps the design and creation of video games, due to this capacity and ease of the digital environment it has, industrial processes can be generated which can integrate the work of different production areas, infrastructure, assemblies and elementary machinery, allowing the digital interface of the subject with the digital area before it is projected to the physical area, helping the savings and availability of capital [14].

In this article, the development of an augmented reality application of an industrial process is carried out, which is digitally conformed of each of the elements involved in this process, such as a flow sensor, a water pump, frequency converter among others, these elements are presented with a mobile device virtually through a QR code, where you can perform its manipulation while their different characteristics will be exposed, having a communication interface where you can see the operating values in real time.

2 Methodology

2.1 Augmented Reality

The development of this augmented reality system allows us to superimpose the elements of an industrial process, the same that serves as a reference of an application process in an industry or company, in which its characteristics and fundamental parts are appreciated with the generation of a virtual environment of transposition by means of a mobile device and a QR code.

As shown in Fig. 1, the interaction between the AR environment and the person is done through graphic activators, in which, by characterizing their operation, QR codes can be generated that will allow us to connect with the physical environment through a mobile device, thus providing virtual information of the element in the augmented reality immersion.

The recognition of the elements has been elaborated in a 3D virtual environment by means of a design program to obtain the characterization and animation in the augmented reality environment of the elements, together with digital tools. The different steps that were carried out for the generation of this application in AR facilitate the knowledge about the operation of this process with the manipulation of the elements giving to know their characteristics and manipulation, as well as favoring the intuitive development of the person or operator since it is not necessary to have the element in a physical way, but it is provided through visual information to the real one, resulting in an interesting tool for obtaining practical physical knowledge about an industrial process as shown in Fig. 2.

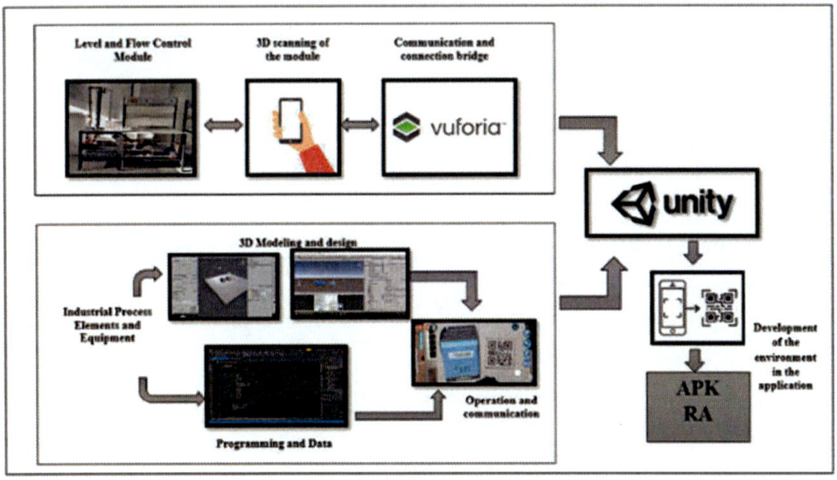

Fig. 1 Application development flowchart

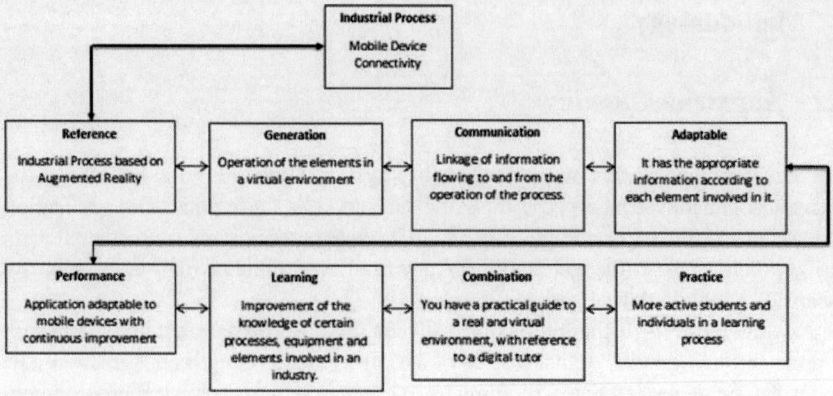

Fig. 2 Application process

2.2 *System Development*

The design and modeling of the elements that make up the multivariable process module were carried out in a 3D CAD design program. These elements were designed based on the characteristics of the PID diagram of the level control module, in addition to the characteristics of the elements that were extracted from books and manufacturer's manuals as shown in Fig. 3.

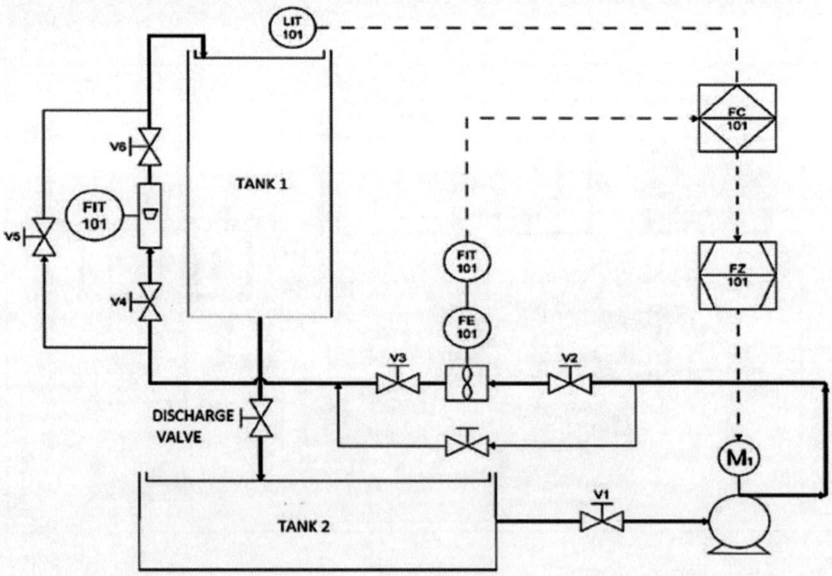

Fig. 3 Multivariable process diagram

For the virtual environment generated, it is necessary to know the operation of the process as well as that of each of the elements that make it up, since the animation has a scale of compatibility of the elements, since the models that will intervene belong to an industrial process. Once the animation to be performed is clear, the model of each element is presented in a programming language to obtain the linkage link, which will have the same coded data package of each of the elements. The elements involved in this process are the following:

- Frequency inverter
- Three-phase water pump
- Level sensor
- Ultrasonic flow sensor
- Rotameter
- PLC
- Tank
- Manual valves

2.3 *Functionalities*

Scene. According to the model obtained, an individual scene is generated for each of the elements; this will have several labels and functions that will determine the specifications of the element, and this scene will be in charge of managing the database in the device where the connection interface is established in order to have a stable transition of the scene in operation.

Animation. The animation of the components is based on the operation of each one of them; these animations manage the labels of the names of the elements, the exploded view and union of the same, the option to start, pause and stop the element that has movement as well as its restart at any time you want. Information boxes. In these, information bridges are added or designed, which can be expressed in dialog tabs, images or videos.

The animation model allows the visualization of technical data and characteristics, in addition to obtaining real-time operating data. These sections are important because they seek to obtain the maximum interest of the person while improving their concentration on the functions presented by the animated model in augmented reality.

Figure 4 shows how the application has been generated following four development stages where the most relevant tasks contained in each marker are framed, both in its generation and execution.

First layer. Signals the recognition of a code which determines the 3D object; it contains an information link defined for each object and the application proposes the use of a trigger, i.e., features and marks.

Fig. 4 Application layer diagram

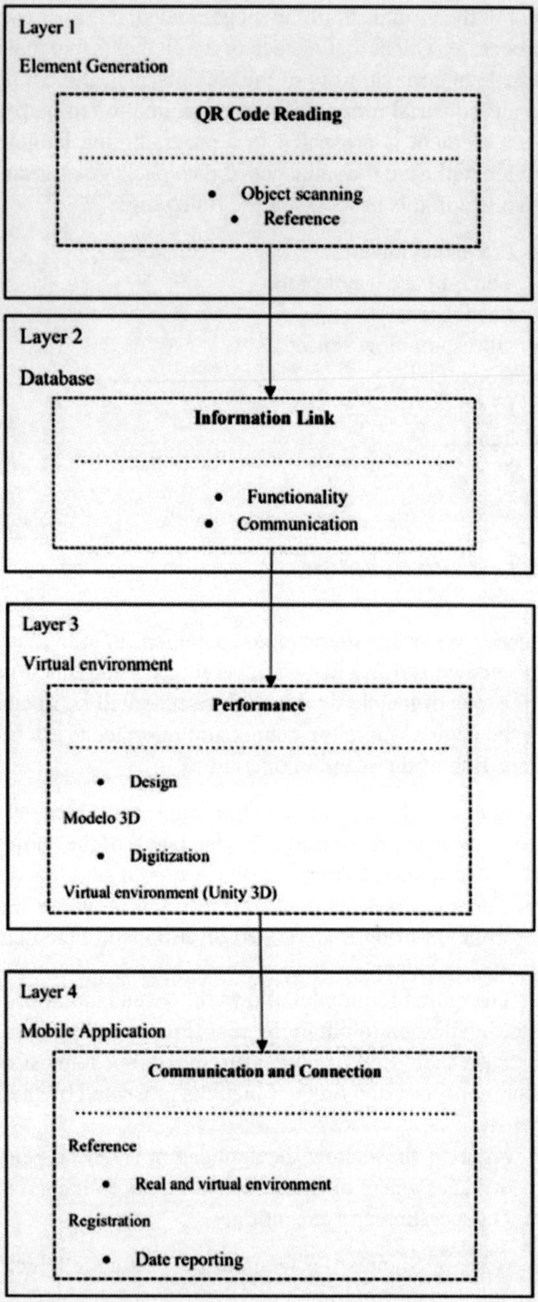

Second layer. Contains the database of the elements, tags and characteristics; these are made known through an online developer, Vuforia through its digital tools package allows us to generate compatible records for the application developer.

Third layer. Is the virtual environment where all the components of the industrial process were developed, it has animations and control buttons in Unity, and the elements of this environment are generated by means of CAD software.

Fourth layer. Is where all the required parameters are configured—elements, name labels, nameplate, command controls such as rotate, disarm, reset. All these are configured in a programming line which helps us to generate the communication interface. This configuration will have a line of compatibility of the Android system, since this system will generate the package of files for installation and execution on the mobile device.

2.4 Animation and Operation

The virtual environment that manages the application in its execution has three scenarios; in the first instance, you will find the main scene for reading the code, the second scene contains the sections for manipulation and finally the simulation.

Execution buttons: Table 1 shows the characteristics of the commands that will be displayed at the moment the user starts the application.

(1) Main scene of the application where the QR code recognition will be performed Fig. 5
(2) Second scenario where the augmented reality element is shown and can be interacted with by means of control buttons arranged in the scenario through the screen of the mobile device as shown in Fig. 6
(3) The last scenario is the simulated environment of the industrial process, where you can interact with it.

2.5 Analysis of Results

The augmented reality environment and its use in industrial processes is presented as a technological tool for handling and obtaining measurement results between the interaction of both instruments, the physical equipment and the mobile application. The facility that these digital tools show us in the identification of the control elements is the ability to read the graphic triggers with the mobile device, since these will give us the data capture generated through communication between the web server database and the data generated on the board; these data capture will allow us to have a control of the variables measured within a value equal or close to that generated by the process. The result stages are presented (Fig. 7).

Table 1 Control commands

Virtual environment buttons		
Name	Operation	Icon
Labels—names	Displays the names of the parts of the selected element	
Characteristics	Displays the specifics of the selected element	
Information	Displays the technical characteristics of the selected element	
Exploded view	Disassembles the element into different parts	
Armed	Joins all the parts of the element after quartering	
Play	Displays an informative video about the selected item	
Control	Allows the movement of the element	
Turn	Rotate the element in the virtual environment	
Reset	Resets the element to the initial position	
Off—exit	Exit the application	

2.6 Preparation and Operation of the Level Module

First of all, the application must be installed on the device to be used, which can be a cell phone or a tablet. Once the installation is complete, the QR code corresponding to the element that you want to know must be read, as shown in Fig. 8.

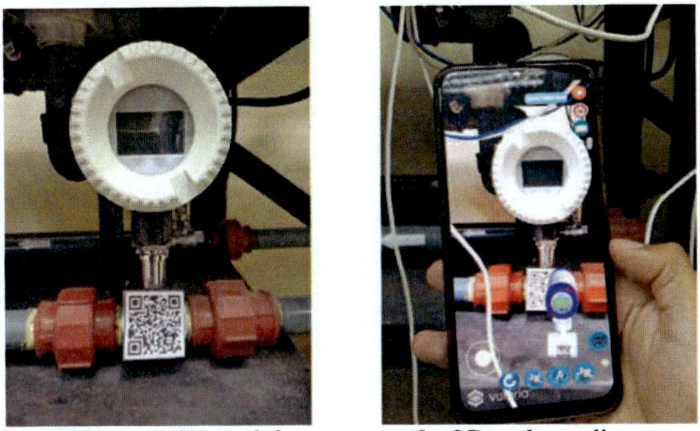

a. Element of the module **b.** QR code reading

Fig. 5 Main application scenario

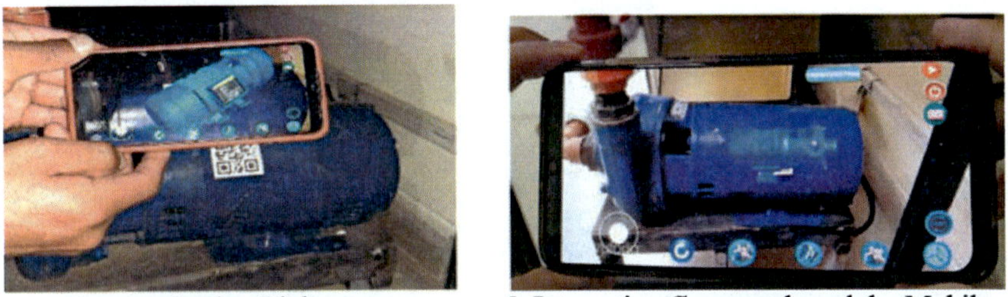

a. Communication Link between the QR Code **b**.Interaction Commands and the Mobile Device

Fig. 6 Mobile application manipulation commands

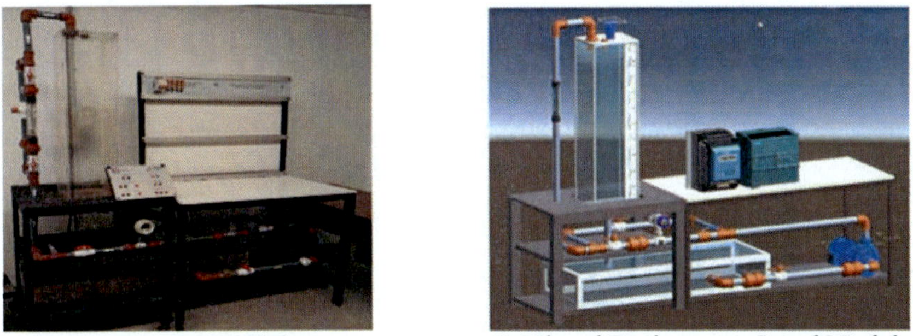

a. Physical Level Module **b.** Virtual Environment Level Module

Fig. 7 Interaction in the virtual environment

Fig. 8 Use of the mobile application

2.7 Data Conditioning and Acquisition

For the acquisition of the data previously, the elements that will intervene in the operation of the control panel must be calibrated, as it would be done in a normal work environment within an industry or company; for this, the parameters of the level sensors, ultrasonic and frequency variator have been configured, the same that will give us in first instance the values of operation of the process (Fig. 9).

2.8 Parameter Processing and Visualization

The data obtained in the physical element, in operation, are shown in contrast to the data obtained in the augmented reality application (Fig. 10).

2.9 Module Operation

The results of the validity of the application for the use of people as an aid in learning and manipulation of the elements and equipment involved in an industrial process based on augmented reality are presented for consideration. However, the assistance of this in the handling of the operation of the level and flow module by the person has all the characteristics and functionalities of the equipment and elements in a simulated environment (Table 2 and Fig. 12).

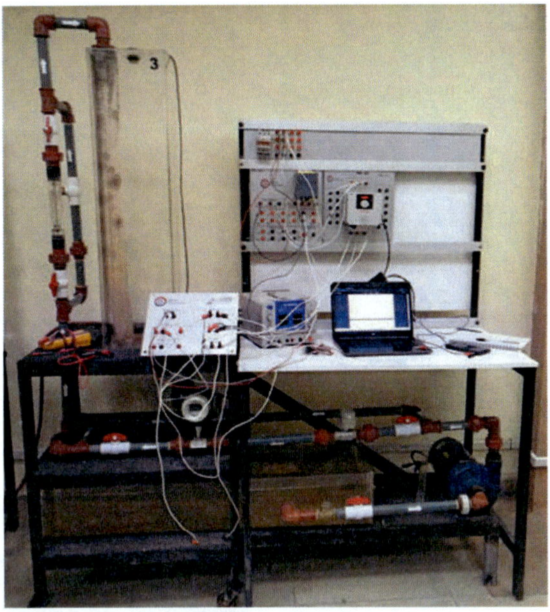

Fig. 9 Connection of the level control board

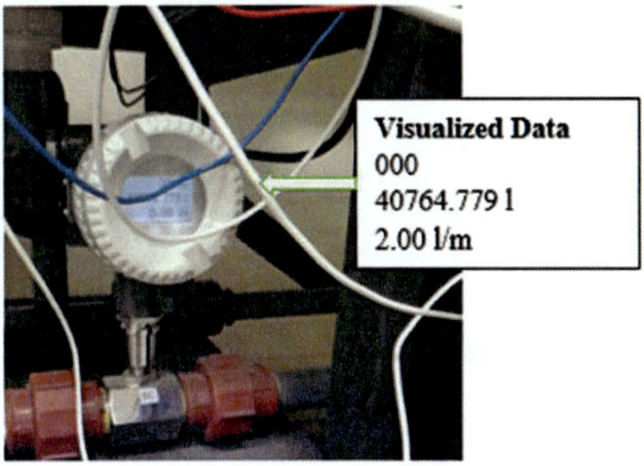

Fig. 10 Actual operating data

Figure 11 shows the comparison graph of the water levels obtained, where it can be seen that the referred or operating values are in blue, while the values obtained through the mobile application are in gray, taking into account that the measurement of the main tank is 62.00 L represented in a measuring tape on the main face of 100 cm^3, measuring 90 cm^3 due to the range of the sensor.

Table 2 Measured values of the control module

Time (S)	Actual measurements (cm^3)	Application measurements AR
1	1	1
3	10	9.97
6	20	19.98
9	30	30.5
12	40	40
15	50	49.97
18	60	60.6
21	70	70
24	80	80.98
27	90	89.98

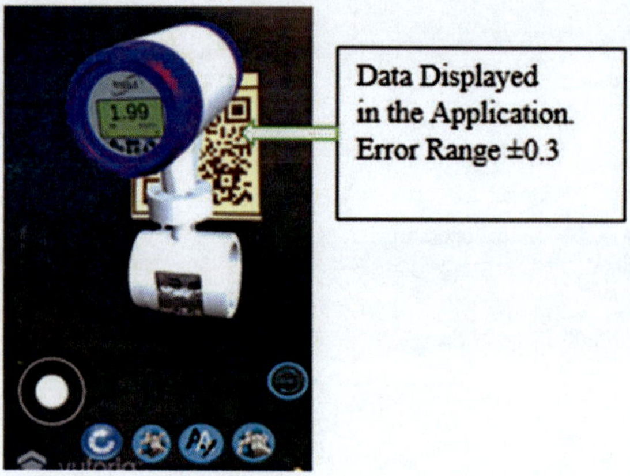

Fig. 11 Mobile application data

3 Conclusions

The augmented reality that is applied to an object or element is subject to some components, physical, visual and components established in its function; based on these, the construction of the required technological resources is created, where with the different programming parameters, the necessary requirements for the creation and observation of objects in augmented reality are obtained. The development of the application for a mobile device based on augmented reality is a technological tool that allows people to learn about the operation of an industrial process, where they can access the manipulation of the elements and instruments of the equipment, as well as learn more about the characteristics and data sheets that present each of these. In turn, the most practical way to learn about the operation of industrial processes in an augmented reality environment is the visualization in real time, allowing the understanding of the operation of the process.

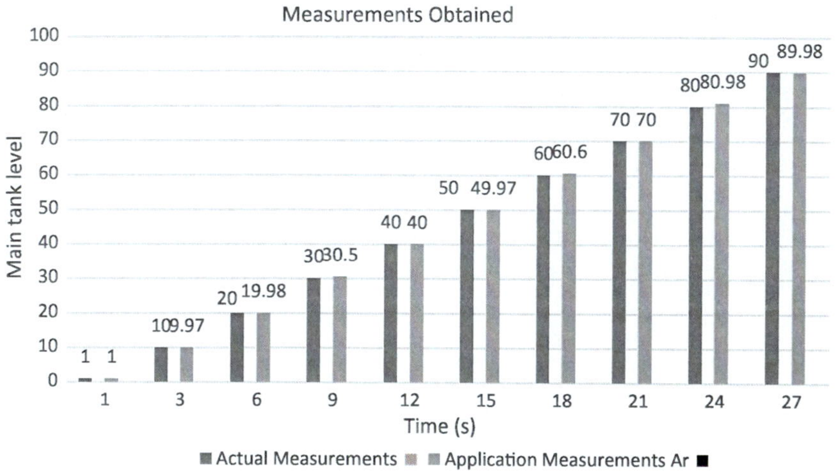

Fig. 12 Water level comparison chart

References

1. López B (2019) Advances in mobile technology and its impact on society. vol 2, no 4, pp 5–19
2. Llorente JS, Giraldo IB, Toro SM (2016) University of Zulia, 2016 Jul 04. [Online]. Available https://www.redalyc.org/journal/737/73749821005/html/
3. Sebastián Rovira WPNS (2021) eLAC.2022. [Online]. Available https://repositorio.cepal.org
4. Castillo JO (2017) Virtual reality and augmented reality in the marketing process. Management and business administration magazine, pp 155–229
5. Góngora B (2017) Universidad Autónoma de Madrid. 8 Sep 2017. [Online]. Available http://hdl.handle.net/10486/681646
6. Sevilla AB (2017) Augmented reality in education. Tele-Education Office of the Vice-Rectorate of Technological Services of the Universidad Politécnica de Madrid
7. Erbes A (2019) Industry 4.0: opportunities and challenges for the productive development of the province of Santa Fe. Economic Commission for Latin America and the Caribbean, Santiago, 2019
8. García GV (2013) Media convergence and new forms of communication. Polytech Mag 9(16):117–130
9. Hormechea Jimenez KD, Lopez Pulido CA, Gonzalez Rodriguez LA, Camelo Quintero YA (2019) Cooperative University of Colombia. 2019. [Online]. Available https://repository.ucc.edu.co/handle/20.500.12494/14569
10. Román V (2016) Industry 4.0: the digital transformation of industry. Valencia: conference of directors and deans of computer engineering, CODDII reports
11. Sánchez-Pinilla MD Information and communication technologies: their options and limitations Y Sus Efectos En La Enseñanza. Nomads. Critical J Soc Juridical Sci
12. Sanchez DO (2020) Industry 4.0: the challenge on the road to digital organizations. Manage Stud Int J Manage 8
13. Carneiro R, Toscano JC, Díaz T (2021) The challenges of ICT for educational change. OEI
14. Unity (2021) Unity 3D [Online]. Available https://unity.com/es

Analysis of Spectrum Sensing Techniques in Cognitive Radio

Chandra Mohan Dharmapuri, Navneet Sharma, Mohit Singh Mahur, and Adarsh Jha

Abstract Cognitive radio is an intelligent radio that is leap ahead of the conventional wireless communication mechanism. In cognitive radio, underutilized licensed frequency bands are efficiently utilized by means of dynamic spectrum allocation (DSA). This paper reviews the three major spectrum sensing techniques, namely (1) energy detection, (2) matched filter detection and (3) covariance-based detection in detail along with their software implementation. Analysis of these techniques is formulated by using their respective probability detection (Pd) vs. signal-to-noise ratio (SNR), and using these Pd vs. SNR curve, comparison is carried out between the three techniques on the basis of (a) performance with respect to SNR, (b) sensing time, (c) complexity and (d) practicality. The motivation for this paper is to choose the optimum spectrum sensing technique out of all the included techniques.

Keywords Cognitive radio · Spectrum sensing technique · Energy detection · Matched filter detection · Covariance-based detection

1 Introduction

In cognitive communication, there are two types of users, primary users (licensed users) and secondary users (unlicensed users). The primary users are provided with certain frequency bands for their operations to take place, whereas studies prove that most of this allotted spectrum is underutilized and the idle spectrum band called white space in spectrum can be provided for the operations of secondary users [1–3], for example, the 2.4 and the 5 GHz bands which are used for Bluetooth, IoT and

C. M. Dharmapuri · N. Sharma (✉) · M. S. Mahur · A. Jha
Department of Electronics and Communications Engineering, G. B. Pant Government Engineering College New Delhi, New Delhi 110020, India
e-mail: navneet.sharma.312000@gmail.com

C. M. Dharmapuri
e-mail: chandu.dharmapuri@gbpec.edu.in

P. Chatterjee et al. (eds.), *Computational Intelligence for Engineering and Management Applications*, Lecture Notes in Electrical Engineering 984,
https://doi.org/10.1007/978-981-19-8493-8_52

Wi-Fi. But, due to the rising number of devices in the present era, managing all these secondary users in the free bands is hectic and troublesome. It is also important to keep the rights of primary user intact without any interference to its transmission. Cognitive radio [4] is the solution to the aforementioned problem. It senses the whole RF spectrum at each instant and detects the licensed user bands which are not under use and transfer the secondary users to those bands, and the secondary users continue their operations until the original primary user comes back to its band; if it detects the primary users, it immediately transfers the secondary user to some other band and this cycle continues. With time as the number of secondary users is increasing exponentially and scarcity of spectrum is bottleneck for these wireless applications, various researches are progressing in this field to make this system more reliable and secure. One of the crucial steps in cognitive radio cycle is spectrum sensing. The accurate and fast spectrum sensing is the key in efficient utilization of spectrum. In literature, several spectrum sensing techniques are proposed in majorly two categories, i.e., non-cooperative and cooperative spectrum sensings [4]. In non-cooperative sensing, each node decides spectrum availability on its after sensing, whereas cooperative sensing, group of nodes takes collective decision based on their individual sensing reports.

The contribution of this paper is to review three important major non-cooperative spectrum sensing techniques, namely, (1) energy detection, (2) matched filter detection and (3) covariance-based detection in detail and their software implementation. This work has been motivated from the selection of the optimum spectrum sensing technique out of these included techniques.

2 Literature Survey

In literature, Amir Zaimbashi offered a composite hypothesis to formulate based on the received signal covariance matrix. They use the likelihood-ratio test (LRT), idea to create a set of new LRT test [5]. Juha Kalliovara devised the LSA system, which was created to provide predictable quality of service (QoS) and exclusive access to shared spectrum resources to LSA licenses. Long-term static licenses will be used to solve the challenge of using unoccupied spectrum resources in the 2.3–2.4 GHz band which is for mobile broadband (MBB) [6]. For cognitive radio networks, Amandeep Kaur offered a full classification and survey of several machine learning algorithms for intelligent spectrum management, as well as their optimization paradigms [7]. Xue Wuag presented a multiband signal reconstruction blind method, to solve the problem in reconstruction multiband signal [8]. Sureka proposed that a detection and defense against PUEA are realized using yardstick-based threshold allocation (YTA) technique, by assigning threshold level to the base station, thereby efficiently enhancing the spectrum sensing ability in a dynamic CR network [9]. Zhu suggested a lightweight privacy-preserving location verification protocol to secure each secondary user's (SU) identity and location privacy, as well as to verify SU location. Furthermore, the SU does not need to supply location information in order to request an available channel from the DB using this protocol [10]. Zheng devised short preamble cognitive MAC (SPC-MAC) protocol for CRSN to promote reliable and rapid spectrum access while lowering energy consumption. Cognitive radio sensors' networks (CRSNs) have a limited amount of energy [11].

Fang offers a new model called semi-tensor product compressed spectrum sensing (STP-CSS), which is a generalization of classical spectrum sensing, for energy compression and reconstruction of spectral signals [12]. Rathee suggested a secure and trustworthy routing and handoff technique tailored to the CRN environment, in which rogue devices are recognized at the lower layers and excluded from the communication network [13]. To secure the secondary transmission, Bai presented a PLS approach based on random beamforming for MISOME CRNs, so that secondary transmissions in cognitive radio network are more secure [14]. Marwanto made a proposal. A cognitive radio co-channel monitoring prototype (CCMCR) makes use of a node MCU Arduino, which analyzes not only co-channel but also adjacent-channel interference [15]. A model for combining CSS and PUEA detection was proposed by Banerjee. As a multiclass hypothesis, it employs fuzzy conditional entropy maximization, in order to reduce primary user emulation attack (PUEA) in cognitive radio [16]. Darwhekar made a suggestion. A wideband triangular patch antenna for the use of cognitive radio in the T. V. white space (470–806 MHz), which can aid in long-distance transmission. Because the unlicensed 2.4 and 5 GHz bands are becoming overcrowded, they are being exploited for short-range communication [17]. Ahmad proposed a paper which gives an in-depth analysis of all the major 5G-related spectrum sharing techniques as well as 5G enabling techniques, such as HetNets, SS and flexibility, massive MIMO, ultra-lean design URLLC, convergence of access and backhaul, mMTC, mmWave and NM [18]. Katta proposed a teaching learning-based optimization (TLBO) to optimize the sensing error probability throughput and blocking probability which gives better optimum solutions compared to genetic algorithm and differential evolution [19].

Xie proposed a deep learning-based CNN-LSTM spectrum sensing detector. Different from the traditional detectors, the CNN-LSTM structure is free of the signal noise model assumption; and moreover, it is able to simultaneously learn the signal energy correlation features and the temporal primary user activity pattern features to promote the detection performance [20]. To identify MSUs, Kuldeep Yadav presented a modified delivery-based scheme, in which only a small number of samples is examined for sensing. The suggested security system is proven to effectively reduce the effect of MSU on global-decision making, hence increasing the throughput achievable by a trustworthy secondary user [21]. Xu Haitao He created a differential game model to describe it logically and symbolically. Solving the game model yielded open-loop Nash equilibrium and feedback Nash equilibrium solutions, suggesting that efficient resource allocation approaches for SUs indeed exist. Secondary users with the energy harvest function can use their saved energy and the spectrum resource leased from the PU to complete the information transmission task [22]. Grigorios Kakkavas proposed a resource allocation approach for cognitive radio networks based on Markov random field framework, realizing a distributed cross-layer computation for the secondary nodes of the cognitive radio network. The framework implementation consists of self-contained modules developed in G.N.U. radio realizing cognitive functionalities, such as spectrum sensing, collision detection [23]. Alexandru Martian came up with a brilliant idea. For spectrum sensing, an ED with three events' (3EED) method is used. Newton's method is used to provide a precise estimate of the ideal decision threshold that minimizes the chance of decision mistake in a single iteration via forced convergence (DEP) [24]. The cognitive radio sensing and the interference mechanism, as well as how and why the cognitive configuration is better to conventional radio, were proposed by Divya Lakshmi. Adaptive radio and software-defined radio, two technologies employed in the cognitive radio configuration, were also researched [25].

3 Motivation

The motivation behind choosing this topic for our research paper is to first of all come to deduce which of the spectrum sensing techniques is more practical to use in real-world scenarios and to carry forward the conclusions from this paper and deep dive into the problems a real-world cognitive radio faces and to come up with solutions with the help of the models implemented in this paper.

4 Overview of the Working of Cognitive Radio

4.1 Spectrum Sensing

It assists in detecting idle spectrum and sharing it with users without causing detrimental interference [25].

4.2 Spectrum Decision

It aids in the acquisition of the best available spectrum to suit the communication requirements of users.

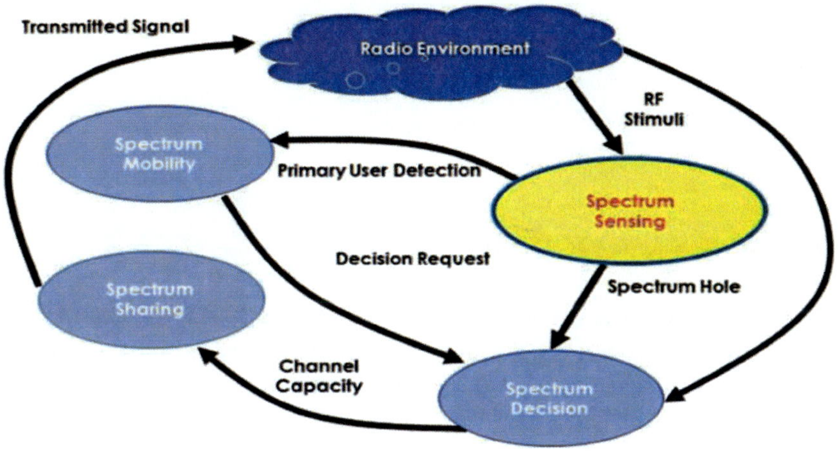

Fig. 1 Cognitive radio cycle

4.3 Spectrum Sharing

It permits the cognitive user (unlicensed user) to transmit along the selected spectrum band provided that the licensed user is not harmed [24].

4.4 Spectrum Mobility

It helps in seamless transmission of cognitive user from one channel to another whenever a primary user is detected in the same channel or better communication channel is detected (Fig. 1).

5 Spectrum Sensing Techniques in Cognitive Radio

5.1 Energy Detection

This is a simple sensing technique that does not require any knowledge of the PU signal. As a result, this decision method offers number of benefits in terms of application and computing complexity 3]. The received energy is a measurement of a certain band of the spectral region. To determine whether a channel is available, the detector compares the observed energy to a threshold value.

$$\lambda = \left(Q - 1(Pf)\sqrt{2N} + N\right)\delta w^2 \tag{1}$$

λ is sensing threshold
Q is the Q-function
Pf is probability of false alarm
N is the sample number
δw^2 is variance of white noise.

5.2 Matched Filter Detection

In matched filter-based detection, a coherent sensor that detects known primary user signals is required. It increases the SNR at the detector's output, but it necessitates prior knowledge of the principal user signal [7]. As a result, it is anticipated that the primary user transmitter provides a pilot stream along with the data.

$$\lambda = \left(Q - 1(Pf)\sqrt{E\delta w^2}\right) \quad (2)$$

λ is sensing threshold
Q is the Q-function
Pf is probability of false detection
E is the PU signal energy
δw^2 is variance of white noise.

5.3 Covariance-Based Detection

To detect the presence of the principal user signal, a covariance-based detection technique uses the sample covariance matrix of the received signal and singular value decomposition. This is determined by analyzing the structure of the incoming signals' covariance matrix. The principal user's signals are correlated and can be distinguished from the background noise [13].

$$\gamma = \left(\sqrt{Ns} + \sqrt{L}\right)^2 \Big/ NS(1 + \left\{\left(\sqrt{Ns} + \sqrt{L}\right)^{-2/3} \Big/ (NsL)^{1/6}\right\}$$
$$F1^{-1}(1 - Pfa) \quad (3)$$

γ is the threshold value.
Ns is the available sample size.
L is the smoothing factor.
F1 is the cumulative distribution function.
Pfa is the probability of false alarm.

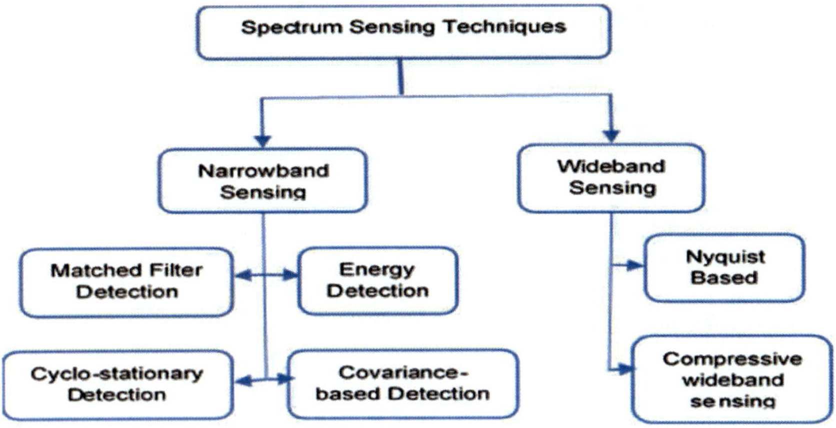

Fig. 2 Classification of spectrum sensing techniques

5.4 Cyclostationary Feature Detection

The most often utilized signals have sinusoid carrier resulting in statistical features that are periodic. Primary user signals have this type of periodicity and can be easily detected by using cyclostationary detection technique at a very low signal-to-noise ratio due to its noise resistance properties [15].

5.5 Cooperative Spectrum Sensing

A method for improving detection performance in which secondary users work together to perceive the spectrum and locate spectrum holes. The procedure through which a secondary user confirms the state of a channel is by working with other secondary users or a fusion center. A method in which cognitive radio share their unique sensing data in order to improve the total sensing data for the principal user [2] (Fig. 2).

6 Problem Statement

1. Studying the following three spectrum sensing techniques, namely,
   i. Energy detection
   ii. Matched filter detection
   iii. Covariance detection.

2. Plot Pd versus SNR curve for all the three techniques.
3. Based on the curve, derive comparisons between the aforementioned spectrum sensing techniques on the basis of:
   i. Performance with respect to SNR
   ii. Complexity
   iii. Sensing time
   iv. Practicality.

7 Methodology

The study of this paper is divided into seven parts as follows:

1. Selecting a suitable range of SNR values for evaluation of probability of detection
2. Adding random noise to the input signal
3. Calculating threshold for the current SNR value
4. For energy detection, calculating the signal energy; for covariance-based detection, calculating the max eigenvalue from the covariance matrix generated from the signal; and for the matched filter detection, generating the test statistic with the help of input signal and the template signal
5. Comparing the technique specific parameters with the threshold values. If parameter is greater than threshold, then signal is present else absent
6. For each SNR value, simulating the results 1000 times for better accuracy
7. Deducing conclusions based on probability of detection vs. SNR plots for the three used techniques.

8 Algorithms

1. **Energy Detection**
   i. Start
   ii. snr_db = [−16: 0.5: −4]
   iii. l = length (snr_db)
   iv. Sample = 100;
   v. pf = 0.1;
   vi. For i = 1: l d = 0;
   vii. For j = 1: 1000
   viii. Generate input signal, generate received signal by adding noise to the input signal.
   ix. snr = 10snr_db(i)/20;
   x. var = 1/snr;
   xi. Calculate threshold, calculate energy of the received signal.
   xii. If energy > threshold d = d + d;

xiii. End the internal loop.
xiv. pd (i) = d/1000;
xv. End the outer loop.

2. **Matched Filter Detection**

i. Start.
ii. snr_db = [-16: 0.5: -4];
iii. l = length (snr_db);
iv. a = numel (snr_db);
v. pf = 0.1;
vi. n = 10;
vii. pd = zeros (1, a);
viii. For i = 1: a d = 0;
ix. For j = 1: l
x. Generate input signal, generate received signal by adding noise to the input signal.
xi. snr = 10snr_db(i)/20;
xii. var = 1/snr;
xiii. If energy > threshold d = d + 1;
xiv. End the internal loop.
xv. pd(i) = d/(l *n);
xvi. End the outer loop.

3. **Covariance-Based Detection**

i. Start.
ii. snr_db = [-16: 0.5: -4];
iii. l = length (snr_db);
iv. snr = 10snr/10;
v. l = 8;
vi. pd = zeros(1, l);
vii. For i = 1: a d = 0
viii. For j = 1: l
ix. Generate input signal, generate received signal by adding noise to the input signal.
x. var = 1/snr;
xi. Calculate threshold, calculate max eigenvalue of the received signal.
xii. If max eigenvalue > threshold
xiii. End the internal loop.
xiv. pd (i) = count/2000;
xv. End the outer loop.

MATLAB and its signal processing tools are used for the implementation.

9 Flow Sequence

Figure 3 shows the flowchart for energy detection. The energy of the received signal is computed and then compared to the threshold level; if the energy of the received signal is greater than the threshold level, the signal is present; otherwise, it is absent.

This technique employs the use of coherent detector which provides the signal characteristics to the filter. When the signal is received the matched filter, filter-out the noise and the original signal is received, which is then compared to a certain threshold level. If the signal parameters are greater than threshold, then the signal is considered present; else, it is absent (Fig. 4).

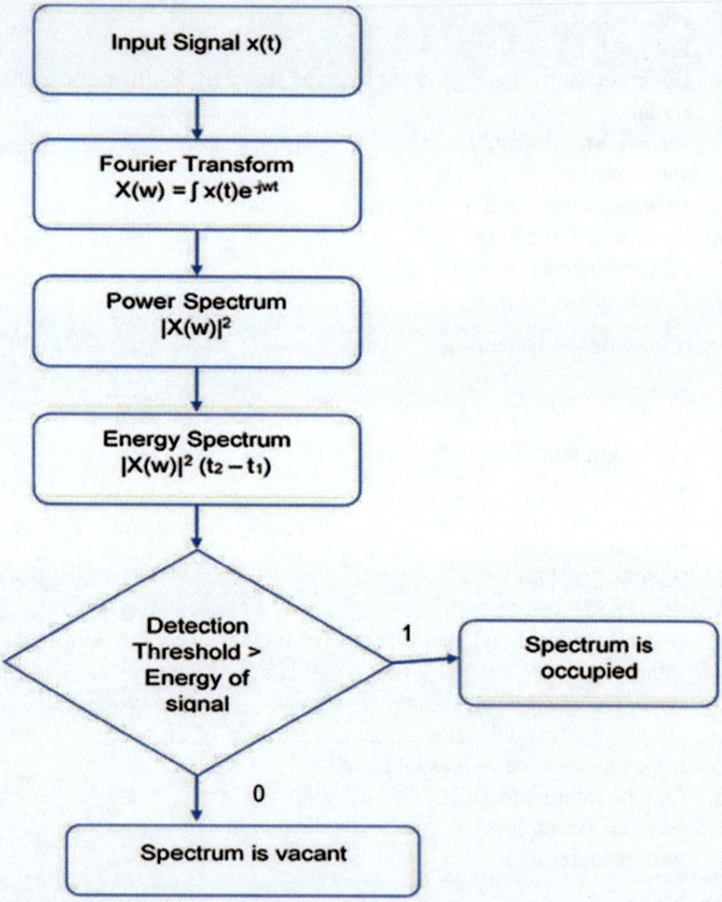

Fig. 3 Flowchart for energy detection

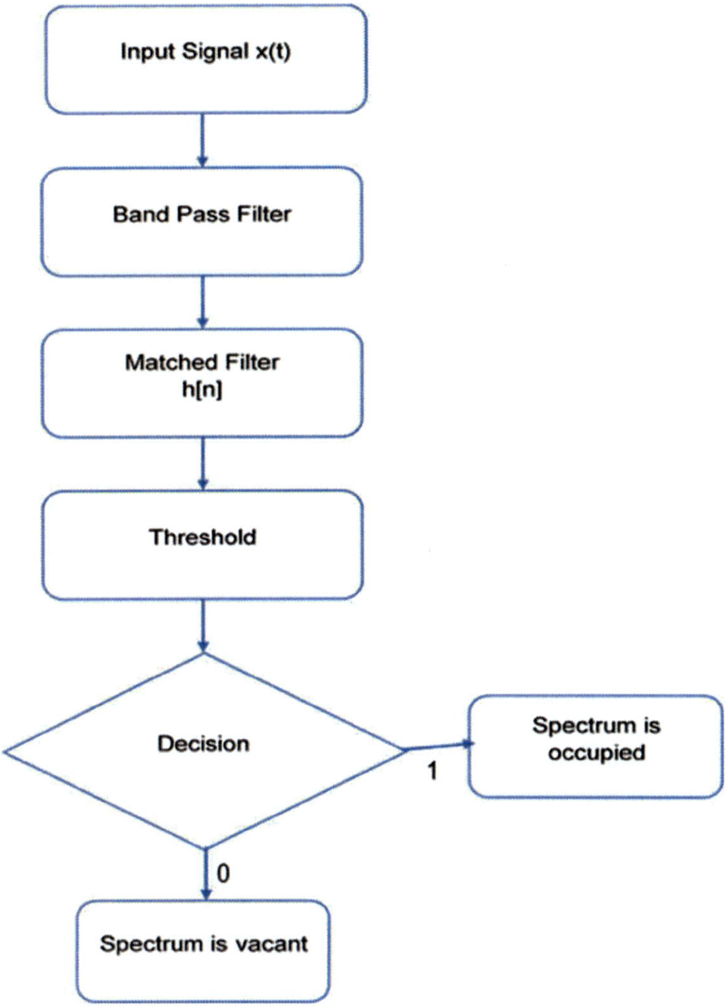

Fig. 4 Flowchart for matched filter detection

Figure 5 shows the flowchart for covariance-based detection. In this technique, with the help of received signal, a covariance matrix is constructed, and then, all the eigenvalues of the obtained matrix are abstracted. Then, the maximum of all the eigenvalues is compared with the threshold, and if the max eigenvalue is greater than the threshold, then signal is present; else, it is absent.

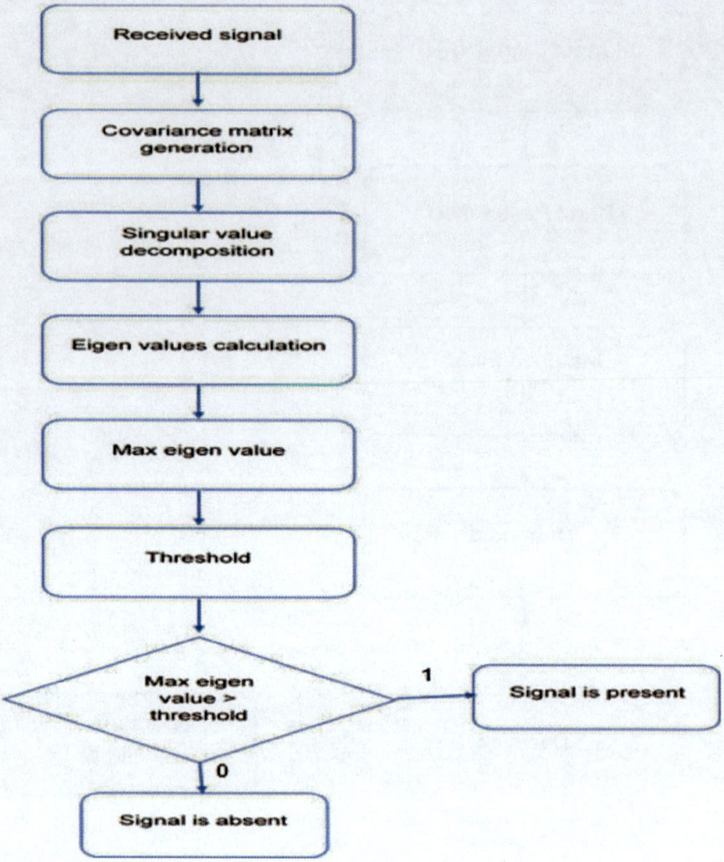

Fig. 5 Flowchart for covariance-based detection

10 Results and Analysis

The output shown in Fig. 6 shows that in energy detection, at lower level of signal-to-noise ratio, the probability of detection (Pd) is almost close to 0, and it increases steadily when the signal-to-noise ratio increases.

The output (shown in Fig. 7) shows that in matched filter detection, at lower level of signal-to-noise ratio, the probability of detection is still relatively good, and with further improvement in signal-to-noise ratio, the probability of detection rises rapidly.

In covariance-based detection (Fig. 8), at lower levels of signal-to-noise ratio, the probability of detection is comparable to matched filter detection, whereas as the signal-to-noise ratio increases, the matched filter detection takes the lead in this regard.

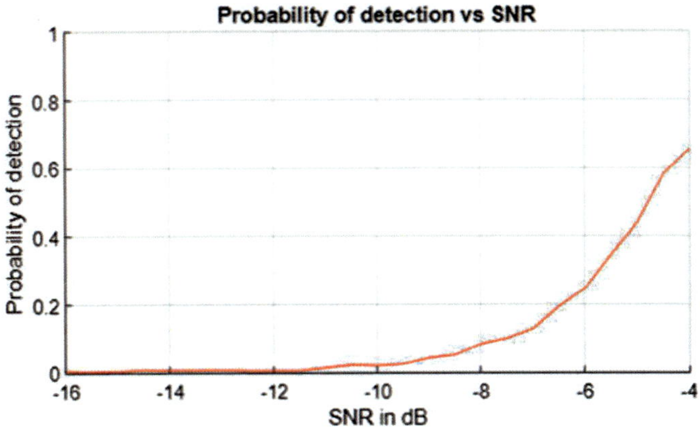

Fig. 6 Pd versus SNR in energy detection

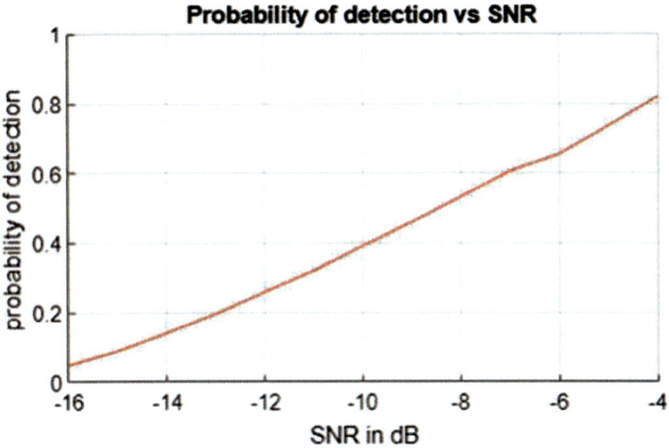

Fig. 7 Pd versus SNR in matched filter detection

Table 1 shows the comparison between all the three techniques.

11 Conclusion

In this paper, at first, we discussed about cognitive radio and how it proved to be one of the best developments in wireless communication as it helps in improving the efficient utilization of scarce RF spectrum. The process of cognitive radio was then addressed, which included spectrum sensing, decision, sharing and mobility. We also talked about energy detection, matched filter detection, covariance-based detection,

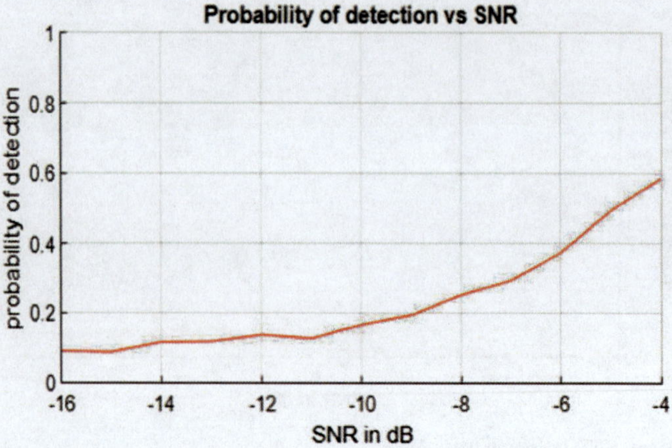

Fig. 8 Pd versus SNR in covariance-based detection

Table 1 Comparison between the three techniques based on above results

Parameter	Energy detection	Matched filter detection	Covariance-based detection
Performance with respect to SNR	Average	Good	Good
Complexity	Low	High	Low
Sensing time	Moderate	Least	Moderate
Practicality	High	Low	High

cyclostationary feature detection and cooperative spectrum sensing, which are some of the most prominent and used spectrum sensing approaches. The main area of focus of this research paper includes analyzing and comparing energy detection, matched filter detection and covariance-based detection on their performance, complexity, sensing time and their practicality. Furthermore, this paper leaves a scope for further research by implementing some new ways to improve the efficiency of the techniques and furthermore studying the loopholes in these techniques and finding solution for that.

References

1. Meelu R, Anand R (2010) Energy efficiency of cluster-based routing protocols used in wireless sensor networks. In: AIP conference proceedings (vol 1324, no 1, pp 109–113). American Institute of Physics
2. Backens J, Xin C, Song M, Chen C (2014) DSCA: Dynamic spectrum co-access between the primary users and the secondary users. IEEE Trans Veh Technol 64(2):668–676

3. Paliwal KK, Israna PRA, Garg P (2011) Energy efficient data collection in wireless sensor network-a survey. In: International conference on advanced computing, communication and networks' II, pp 824–827
4. Wang B, Liu KR (2010) Advances in cognitive radio networks: a survey. IEEE J Sel Top Sign Process 5(1):5–23
5. Zaimbashi A (2019) Spectrum sensing in a calibrated multi-antenna cognitive radio: exact LRT approaches. J Electron Commun, 152968
6. Kalliovaara J, Jokela T, Kokkinen H, Paavola J (2018) Licensed shared access evolution to provide exclusive and dynamic shared spectrum access for novel 5G use cases
7. Kaur A, Kumar K (2022) A comprehensive survey on machine learning approaches for dynamic spectrum access in cognitive radio networks. https://doi.org/10.1080/0952813X.2020.1818291
8. Wang X, Jia M, Gu X, Guo Q (2018) Sub-Nyquist spectrum sensing based on modulated wideband converter in cognitive radio sensor networks. Digital Object Identifier. https://doi.org/10.1109/ACCESS.2018.2859229
9. Sureka N, Gunaseelan K (2019) Detection and defence against primary user emulation attack in dynamic cognitive radio networks
10. Zhu R, Xu L, Zeng Y, Yi X (2019) Lightweight privacy preservation for securing largescale database-driven cognitive radio networks with location verification. Hindawi Secur Commun Netw vol 2019, Article ID 9126376
11. Zheng M, Wang C, Du M, Chen L, Liang W, Yu H (2019) A short preamble cognitive mac protocol in cognitive radio sensor networks. IEEE Sens J. https://doi.org/10.1109/JSEN.2019.2908583
12. Fang Y, Li L, Li Y, Peng H, Yang Y (2020) Low energy consumption compressed spectrum sensing based on channel energy reconstruction in cognitive radio network. Sensors 20:1264
13. Rathee G, Ahmad F, Kerrache CA, Azad MA (2019) A trust framework to detect malicious nodes in cognitive radio networks. Electronics 8:1299
14. Bai S, Gao Z, Hu H, Liao X (2018) Securing secondary transmission in cognitive radio networks using random beamforming. IEEE
15. Marwanto A, Nuha MU, Hapsary JP, Triswahyudi D (2018) Cochannel interference monitoring based on cognitive radio node station. In: Proceeding of EECSI 2018, Malang—Indonesia, 16–18 Oct 2018
16. Banerjee A, Maity SP (2019) Joint cooperative spectrum sensing and primary user emulation attack detection in cognitive radio networks using fuzzy conditional entropy maximization. Trans Emerg Tel Tech, pp e3567. https://doi.org/10.1002/ett.3567
17. Darwhekar IS, Peshwe PD, Surender K, Kothari AG (2019) Wideband triangular patch antenna for cognitive radio in TV white space
18. Ahmad WSHMW, Radzi NAM, Samidi FS, Ismail A, Abdullah F, Jamaludin MZ, Zakaria MN (2020) 5G technology: towards dynamic spectrum sharing using cognitive radio networks. IEEE Access https://doi.org/10.1109/ACCESS.2020.2966271
19. Hoque S, Talukdar B, Arif W (2021) Impact of buffer size on proactive spectrum handoff delay in cognitive radio networks from © Springer Nature Singapore Pte Ltd. 2021 Mandloi M et al (eds) 5G and beyond wireless systems, Springer Series in Wireless Technology
20. Lakshmi JD, Rangaiah L (2019) Cognitive radio principles and spectrum sensing. Int J Eng Adv Technol (IJEAT) ISSN: 2249–8958, vol 8 Issue 6
21. Chakraborty D, Sanyal SK (2021) Time-series data optimized AR/ARMA model for frugal spectrum estimation in Cognitive Radio. Phys Commun 44:10
22. Darabkh KA, Amro OM, Al-Zubi RT, Salameh HB (2021) Yet efficient routing protocols for half- and full-duplex cognitive radio Ad-Hoc Networks over IoT environment. J Netw Comput Appl 173:102836
23. Ata SÖ, Erdogan E (2019) Secrecy outage probability of intervehicular cognitive radio networks from © 2019 John Wiley & Sons, Ltd
24. Khan AA, Rehmani MH, Rachedi A (2016) When cognitive radio meets the internet of things?" from 978-1-5090-0304-4/16/$31.00 ©2016 IEEE
25. Garg P, Anand R (2011) Energy efficient data collection in wireless sensor network. Dronacharya Res J 3(1):41–45

Computational Intelligence in Management Applications

Monitoring of Physiological and Atmospheric Parameters of People Working in Mining Sites Using a Smart Shirt: A Review of Latest Technologies and Limitations

Sakthivel Sankaran, Preethika Immaculate Britto, Priya Petchimuthu, M. Sushmitha, Sagarika Rathinakumar, Vijay Mallaiya Mallaiyan, and Selva Ganesh Ayyavu

Abstract Smart shirt is widely used in measuring the vital physiological parameters of the user. This idea of a smart shirt is especially for miners. It is a well-known fact that a lot of people working in mines are losing their lives due to mine collapse. The ultimate goal of this idea is to help miners to stay alert from all the risks in mines. According to statistics given by ENVIS Center on Environmental Problems of Mining, 155 died and got severely injured due to mine collapse between 2017 and 2021. The official number of deaths in the mining community per year is more than 15,000. The idea is to continuously monitor the barometric pressure, temperature, humidity, and the presence of carbon monoxide using sensors, to alert the user if over-experiencing any of these factors with the help of an alert system, and to enable the users and the supervisor to continuously monitor these parameters as it can possibly turn out to be a threat to the workers. After a thorough survey on the related works, the idea of developing a smart shirt for the mining community will be the better replacement for the previous work. This idea of developing a smart shirt is

S. Sankaran (✉)
Faculty of Biomedical Engineering, Kalasalingam Academy of Research and Education, Virudhunagar, India
e-mail: sakthivelsankaran1992@gmail.com

P. I. Britto
Faculty of Biomedical Engineering, College of Engineering (Women), King Faisal University, Al-Hofuf, Kingdom of Saudi Arabia

P. Petchimuthu
Faculty of Biotechnology, Kalasalingam Academy of Research and Education, Virudhunagar, India

M. Sushmitha
Department of Biomedical Engineering, Saveetha Engineering College, Thandalam, Chennai, India

S. Rathinakumar · V. M. Mallaiyan · S. G. Ayyavu
Student of Biomedical Engineering, Kalasalingam Academy of Research and Education, Virudhunagar, India

P. Chatterjee et al. (eds.), *Computational Intelligence for Engineering and Management Applications*, Lecture Notes in Electrical Engineering 984,
https://doi.org/10.1007/978-981-19-8493-8_53

very useful as it would be greatly helping them to ensure safety during their work underground.

Keywords IoT · Smart shirt · Detection of body's vital parameters · Combustible gases

1 Introduction

Typically, smart shirts help in monitoring the vital parameters of the person's vital physiological parameters for constantly keeping track of their health condition. There are various fields where these smart shirts can be used.

Nowadays, people started using smart devices for various purposes. Heart patients to keep track of their heart rate, athletes for monitor their calories, etc. Almost everyone has smart wearable systems in their hand to monitor or measure their own desired parameters. This project is specially made for miners as there no serious concern about them as of now and so many deaths are occurring in the mining sites also. This project is designed with the ultimate aim of ensuring a safer environment for the workers working in the mining site.

2 Literature Survey

Sardini et al. [1] have discussed the problems and inconvenience in monitoring the health parameters of elderly patients. Biomedical signals in non-clinical applications have been discussed and acquisition requires different monitoring devices between the other characters with low power and compatibility to the environment. To overcome the problems, a new wearable device has been developed to monitor the physiological parameters of a patient. This is a non-invasive process. The sensors have contactless characteristics that enable monitoring without direct contact with the patient's body. The sensors they have used are an accelerometer and respiratory sensor to monitor electrocardiogram (ECG) and heart rate (HR) which are derived from the ECG signal, respiratory rate, and three-axis motions. They concluded their project as a home-based monitoring system that is highly sensitive and non-invasive. By this product, the health parameters of elderly people or patients can be monitored at home. The major drawback is that only ECG and heart rate are monitored and no other vital parameters can be measured with this device.

Fang-Yie et al. [2] have discussed the use of wireless body sensor networks (WBSNs) which are used to measure physiological parameters and for disease monitoring, treatment, and prevention. This project proposed a mobile physiological sensor system that collects the physical parameters using the sensors that are embedded in a smart shirt. They have used a temperature sensor, motion detector, and ECG sensor. These sensors continuously gather the patient's physical parameters and are delivered to remote health center cloud via Wi-Fi. The health parameters are analyzed and stored in the cloud. By using this, the caretakers can reach the patients as fast as possible and they can provide health assistance to the patients. The drawback is only ECG and temperature can be monitored and no other vital parameters can be measured with this device.

Catarino et al. [3] have discussed the importance of strength training in rehabilitation and sports injury prevention. They have proposed an idea of constructing a t-shirt that has embedded textile electrodes in it to monitor muscle and cardiac activity. They have designed a sleeve knitted t-shirt with EMG and ECG knitted electrodes. The ECG and EMG graphs are obtained with and without pressure while exercising and are analyzed. By this product, they have demonstrated maintaining esthetics of the garments. The drawback is that only ECG graphs can be obtained and no other vital parameters can be measured with this device.

van Helvoort et al. [4] have developed a project to monitor the patients with chronic obstructive pulmonary disease at home, which may increase the quality of life and also decreases the cost of health care. They came up with the idea of using a smart shirt that contains respiratory inductance plethysmography (RIP) which helps in measuring the changes in the thoracic and abdominal cross-sectional areas. By this, the smart shirt will be more feasible to monitor lung hyperinflation (LH). They have concluded that RIP will be a promising method to monitor LH. The drawback is that the patients may falsely be diagnosed with lung hyperinflation due to a temperature change.

Thilagaraj et al. [5] discussed the usage of IoT technology in Sign Language translator. IoT is supported by using TCP protocol in data communication in NI LabVIEW program to enable single-server multi-client access across LAN. This allows all networked systems to receive the same data at the same time, with no delay. IoT is used to transfer the data to the cloud as an additional feature.

Roudjane et al. [6] have discussed designing a smart textile that monitors the human breathes in a real-time manner. The smart textile will be a wearable T-shirt that can be stretchable, which features an array of six breath detection sensors placed in thoracic-abdominal walls of the patient which are contactless and non-invasive. Those sensors will communicate simultaneous data to the base station through Bluetooth technology. With this product, it is possible to detect a pause in breathing and breath patterns. This will be useful in monitoring the sleep apnea and clinical monitoring of the patients. The drawback is that only respiratory rate and breathing patterns are monitored and no other vital parameters can be measured with this device.

Redoutéer et al. [7] have discussed the use of Internet of Things (IoT) with a wearable health monitoring system as an alternative for conventional healthcare systems. They have proposed a small wearable and flexible real-time electrocardiogram monitoring system which is integrated with a T-shirt in this work. They used a biopotential analog front end (AFE) chip which helps to collect the ECG data with satisfactory quality. The collected data are transmitted to an end device through Bluetooth low energy (BLE) for real-time display. The major drawback is that only ECG is monitored and no other vital parameters can be measured with this device.

ButhaynahAlShorman et al. [8] have discussed the problems and challenges faced by the Internet of Medical Things (IOMT) related to the bandwidth, data processing, analytics availability, acquisition of data, cost-effectiveness, power efficiency, privacy, data acquisition. They have proposed a solution to enhance the healthcare living facility using remote health monitoring and IOMT. By this, they have concluded that using RHM and IOMT, the feasible monitoring and processing of important parameters of diabetic patients are possible. The drawback is that it may cause security challenges due to the use of wireless medical devices.

Nagai et al. [9] have discussed the challenges in designing smart clothing which can measure health parameters accurately and also provides comfortable wearing. They have proposed an idea of designing smart clothes that apply knowledge in infant clothing design, machine learning, and sensor technology in this project. They have used temperature sensors for body temperature monitoring in a real-time manner. The use of skin-tight sensors can be more accurate, but it may not be comfortable. By this project, they have concluded that to overcome the trade-off between wear comfort in smart clothing and the accuracy of sensors, multiple skin-loose sensors can be used in health monitoring. Through multiple skin-loose sensors, the accuracy can be limited.

Nowak et al. [10] have discussed the use of biomedical sensors in the wearable smart clothing. They presented the design of wearable smart clothing sensors for health parameters measuring systems. The system was designed especially for telehealth in cardiology. They have designed smart clothing with sewn textile electrodes in it to monitor the ECG of the patient. By this project, they have designed long usage and comfortable smart clothing which is also easy to wash. The major drawback is that only ECG and heart rate are monitored and no other vital parameters can be measured with this device.

Cappon et al. [11] have discussed the increased number of patients affected by diabetes due to sedentary lifestyles and aging. To overcome this, daily monitoring of diabetes in patients using various sensors is important. They have designed a wearable continuous glucose monitoring (CGM) system. They have concluded that the CGM sensors are invasive devices that measure the blood glucose concentration level and provide measurements every 1 to 5 min. The drawback is that it is minimally invasive, but it may not be comfortable and may cause irritation.

Chung-Chih et al. [12] have proposed a smart cloth health monitoring system using various sensors. The system consists of smart clothing, sensing components and control platforms, and mobile devices. The system is a wearable device for heart rate monitoring and electrocardiography signal collection. Empirical mode decomposition algorithm is ensured by Markov model-based algorithm for fall detection and electrocardiography. They have concluded that their project can provide kinds of services like tracking of physiological functions, device wearing detection, monitoring activity field, device low battery warning, anti-lost, an emergency call for help, and also for fall detection. The drawback is that only ECG and heart rate are monitored and no other vital parameters can be measured with this device.

Petz et al. [13] have discussed the development of integrating electronic textiles in the field of wearable smart clothing. They came up with the idea of developing a sensor shirt to record the position and movement using several sensors to monitor the upper body. The collected sensor data will be transmitted through Wi-Fi. They have used an accelerometer, gyroscope, and magnetometer in their smart T-shirt to record the position and movements of the upper body. They have concluded that by measuring the movement and posture detection, it is possible to measure and monitor the biomedical loads and repetitive actions using their smart shirt. The drawback is that the accuracy is nearly 80% and not 100%.

Weihua Chen et al. [14] have discussed the rise in the wearable sensor industry in achieving intervention, prediction functions, and real-time monitoring, in the field of monitoring the health parameters. They have come up with an idea to use biodegradable sensors to monitor health parameters. They have used capacitive sensors, piezoelectric sensors, resistive sensors, and triboelectric sensors. They have used these to sense capacitive, piezoelectric, resistive, and triboelectric effects. They have used biodegradable polymers and capacitive-based biodegradable sensors in this. They have concluded that the use of biodegradable polymers provides flexibility, is enough stretchable, and has an elastic modulus to match with human skin or tissues. The drawback is that the sensitivity and range detection are less and the detection range under high pressure or detection under large strain will be difficult.

Young-Dong et al. [15] have discussed the fast increase in the number of patients due to aging-related diseases and wearable health monitoring systems using sensors. They have come up with the idea of developing a physiological health monitoring system that is wearable to monitor the health parameters of elderly people. They have developed a smart vest for measuring various physiological parameters like electrocardiogram (ECG), heart rate and plethysmograph (PPG), body temperature, galvanic skin response, and blood pressure. The measured ECG data and other parameters are transmitted through an ad-hoc network to the health station for remote monitoring. They have concluded that the designed shirt was convenient to wear and it also transfers the data without trouble in the wireless network environment. The disadvantage is that ad-hoc networks are slower than other traditional networks.

Chen et al. [16] have discussed the struggles in giving proper treatment to newborns and continuous health monitoring. Flexible material-based non-invasive sensors have been organized with wearable sensor systems for neonatal monitoring. The system provides a very comfortable environment in the clinic for neonates, and it measures all their vital parameters. They have used a polydimethylsiloxane-graphene (PDMS-graphene) compound-based stretching sensor to detect the infants' respiratory rate, a textile-based electrode to measure electrocardiogram signal (ECG), and also inertial measurement units to obtain movements. They have concluded that the developed system provides high-quality signals and also continuous long-term monitoring for neonatal infants. The drawback is that only ECG is monitored and no other vital parameters can be measured with this device.

Pandian et al. [17] have discussed developing a wearable smart shirt that can be washed. The smart shirt consists of an array of sensors connected to a CPU for continuously monitoring signals. They have used a temperature sensor to measure body temperature, an ECG belt to monitor the electrocardiogram (ECG), plethysmograph

(PPG) sensor to measure the plethysmograph in their smart vest. The other physiological signals that can be measured are blood pressure and galvanic skin response (GSR). They have concluded that the smart vest is comfortable to wear and the sensor data. The drawback is that the data transmitted to the station may interrupt.

Abro et al. [18] have discussed the problems faced by the coal miners while working inside deep mines. They had come up with the idea of design a wearable smart jacket for protecting the life of coal miners. They have used the temperature, humidity sensor, pulse sensor, gas detector, and a GPS module in the jacket. The smart jacket senses the vital parameters of miners like pulse rate, temperature, humidity, exact depth location, and also the presence of hazardous gases in the mines. They have concluded that the proposed smart jacket can not only be used to monitor the health parameters but can also be used to rescue the miners in cases of emergencies. The drawback is that the usage of various sensors may require a higher power supply.

Krueger-Ziolek et al. [19] have discussed the respiratory measurement during breathing which can be done through spirometry or plethysmography. They came up with an idea to develop a smart shirt for respiratory measurement. This system provides efficient and accurate data from athletes during training to ill patients suffering from the supine. This project uses a motion tracking system and a body plethysmograph to analyze the optimal number of position sensors in a smart shirt. They have concluded that they have used 16 sensors to measure the tidal volume with more accuracy. The drawback is that only the respiratory rate is monitored and no other parameters can be monitored.

Dittmar et al. [20] have discussed the need for improving the quality of health in medicine and home care, ambulatory measurements, and permanent monitoring. They have developed smart cloth and gloves which fit and use for the subject. They have developed a smart t-shirt that measures ribcage, EKG, abdominal respiration, core temperature, and body heat and a smart glove that measures skin temperature and conductance. All sensors are used as non-invasive. They have concluded that smart clothing also provides information on emotional, sensorial, intellectual, and task reactions. The drawback is that the usage of various sensors may require a higher power supply.

Matuska et al. [21] have discussed the usage of wearable sensors and electrically conductive yarns in everyday life. They have proposed a design with conductive yarns in a smart device for health monitoring, and the proposed system collects data from two separate accelerometers. The collected data are transferred to the server via Wi-Fi for processing and storage. The conductive yarns are sewn with an embroidery machine. They have concluded that the proposed system can be used as a monitoring device for drivers while transport, as a fall detector for elderly people. The drawback is that the e-textiles will be stiff and not flexible and may cause discomfort.

Nithya et al. [22] have discussed the risks involved in health and workplace hazards associated with working in coal mining operations. They have come up with an idea to develop a smart helmet that is used to detect hazardous gas in the mining industry. They have used a temperature sensor to measure the miner's body temperature, a heartbeat sensor to measure heartbeat, and a gas analyzer to analyze the hazardous gases like CO, SO2, NO2. They have concluded that a smart helmet provides a better solution for a safety system for coalmines with a ZigBee wireless specification. The mine gas concentration detection which is based on the ZigBee network transmission can greatly improve the intrinsic safety of the miners. The drawback is that the ZigBee network requires additional devices which may increase the cost.

Malhotra et al. [23] have discussed the accidental deaths and dangers involved in coal mining. They have come up with an idea to design a system which is used on the helmets of the miners. The helmet is used to analyze the hazardous health parameters found in mines. The parameters are temperature, humidity, methane, and sulfur dioxide. These parameters give a certain level that can cause roof collapse, poisoning, choking, or explosions. They have concluded that the developed smart helmet system is used to detect the parameters in real time and also alerts the ground control and the mine workers using a buzzer. The demerit is that the usage of multiple sensors in helmets may increase helmet weight and may not be comfortable.

Noorin et al. [24] V have discussed the importance of health and safety in the mining regions. They developed an IoT-based wearable device for safety features. They have a humidity sensor, gas sensor, temperature sensor, and collision detection sensor. The collected parameters are transmitted through the wireless sensor network technology. The IoT platform is used to the real-time data to the cloud, and it can be accessed from anywhere around the restricted area. The miners will be alerted using the LED for temperature and humidity monitoring. They have concluded that the usage of various sensors and communication systems enhances the safety of miners.

Jun Yang and Long Hu et al. [25] have discussed the advantages of smart clothing. They have proposed wearable smart clothing. They have used an ECG monitoring device, heart rate monitor, and fall detection device in the smart clothing. The smart shirt measures the human body signals like oxygen, temperature, heart rate, ECG. The collected data are transmitted through the cloud platform. They have concluded that the smart clothing will collect a variety of human body signals with more accuracy. The drawback is that the usage of various sensors may require more power supply.

AnayMajee [26] have discussed the adverse effects to human health and safety while working in the mines and also about the loss of lives of mine workers due to the inability to communicate about the emergency. The smart helmet uses various sensors to measure temperature, humidity, smoke, and air-quality sensors to monitor the conditions inside the mine. The wireless transmission network is used in this project to transfer the data of the sensor attached to the helmet to the base station using ZigBee technology. He has concluded that this project works based on the IoT-related technology for monitoring and collecting and correlating the data to the server with a graphical representation that can be monitored from anywhere and everywhere. The drawback is that the ZigBee network requires additional devices which may increase the cost.

Yasunoritada [27] suggested that a Holter monitor is used to monitor the ECG signal of a person while he is moving. In this project, six electrodes are attached to the smart shirt which is in the position near the chest to measure the ECG signals with the unipolar precordial leads. A conductive link is used to make the electrodes used in the shirt and is flexible. The portable ECG has four different input channels to measure the ECG signals at 200 Hz and to measure the input impedance. The demerits are it is only used to measure ECG and other vital parameters cannot be measured using this smart shirt.

Tiwari et al. [28] proposed that the main aim of developing wearable devices is to focus on health. It also has gained popularity. The wearable devices help in monitoring heart rate, respiration rate, pressure, and sleep quality. The demerits are the measured values that are unable to be stored for future uses.

Sardini et al. [29] proposed that monitoring any human parameters requires non-invasive sensors during rehabilitation for the patient. A microcontroller is used to acquire the impedance from an ADC and data are to the Bluetooth. The demerits are sending the data through Bluetooth is distance limited.

Kalaoglu et al. [30] have discussed the development of a smart shirt system that can be used by visually challenged. They have developed a smart shirt that helps to avoid obstacles for visually challenged people. In this shirt, they used a neuro-fuzzy algorithm. The device can accurately detect the obstacles in the right, left, and front positions. The lily pad Arduino microcontroller is used in this smart shirt. They have concluded that the smart shirt is a washable device. It gives vibrations to the persons before 2.5 to 3 m above the obstacles. It detects the positions of the obstacles accurately to avoid accidents. The demerits are that the sensors placed in the smart shirt can be only able to detect big obstacles and the smaller obstacles cannot be identified.

Baronetto et al. [31] proposed a gastro digital shirt to capture the sounds that come from the abdomen during the process of digestion. The garment prototype has a low-power wearable computer that consists of eight miniaturized microphones in an array. It enables long-time recording. The extracted instances of BS were structured by hierarchical agglomerative clustering.

Tacchini et al. [32] suggested a smart shirt and wrist band. It was assessed on account of between phase and beat to beat analyses. Domain parameters of time and frequency of shirt have given similar results. The methods used here are data acquisition, phase analysis, beat-to-beat analysis. It extracted from standard ECG measurement which was compared with automatically derived from the shirt and wrist band using ECG electrodes and PPG technology. The demerits are that the beat to beat may vary to the time–frequency.

Mannée et al. [33] measured with QDC-calibrated respiratory inductance plethysmography sensors in a smart shirt with a custom-developed application. Medwear was compared to Oxycon mobile in seven tasks of daily living. All tasks were performed twice in two sessions in between the shirt was removed. Calibration was determined per task in s1 and was applied to the repeated task in s2. The demerits are classical lung function parameters such as lung resistance and compliance or respiratory flow is not available.

Van Leutern et al. [34] proposed that detecting and quantifying the dynamic hyperinflation in COPD patients in real life are an interesting issue, but difficult to perform. This study assessed a smart shirt, capable of measuring respiration parameters. It is a helpful new diagnostic tool compared to the standard equipment which is used to measure lung function. Respiration waveforms produced by the shirt were compared with mobile spirometry. The demerits are the chances of collecting information from the subject and transmit to mobile spirometry is low.

PeterisEizental et al. [35] proposed that musculoskeletal pain and pathology are caused mostly by uncontrolled movement. Movement control improvement can be achieved effectively by retraining. Independent work of a subject leads to an increase in efficiency of movement retaining but is a troublesome task if the user does not understand the movement correctly or does not know to verify the accuracy. The smart shirt contains a compression shirt to which about 11 stretch sensors are attached. It is useful for movement monitoring. The demerits are the subjects who may feel pain while the process is on.

Laufer et al. [36] have discussed the increasing usage of smart shirt and other wearable technologies that provide medical parameters. They have developed a shirt that captures various parameters through a body plethysmograph and optoelectronic plethysmography using a spatial set of sensors. They have fixed sixty-four reflective markers in the compressive shirt and performed different types of respiratory maneuvers. Singular value decomposition has been used to determine the minimum marker sets. Minimum marker sets are required to predict the respiratory mechanics accurately. Using nine markers, the positional data can be used to determine clinical applications accurately and precisely. The drawback is that usage of more markers may not be cost effective.

Sardini et al. [37] have discussed the development of smart clothing in monitoring health parameters. They have developed a smart T-shirt which consists of various sensors to measure the respiratory and heart rate and also measures the acceleration for posture information of the patients. The collected data of the sensors are wirelessly transmitted to a separately located health station and for the telemedicine applications. The system can be used to continuously monitor and analyze health parameters during rehabilitation activities. The drawback is that it is not possible to save all the records during continuous health monitoring.

Karlson et al. [38] have discussed the increasing development of smart T-shirts in monitoring health parameters. They have developed the smart T-shirt which measures muscular and heart activity wirelessly by integrating the textile electrodes. The system might reduce the loss of data which may be caused by the usage of a single channel. They have developed a multichannel heart rate detection device, which prevents the loss of ECG in a single channel. They have used a padded structure device above the trapezius muscle. The position of the electrodes is well defined. The sweat produced in local areas is which increases the recording condition. They have concluded that they have designed a flexible monitoring system. The demerit is that the placement of textile electrodes in many positions is insensitive to the monitoring of ECG.

Sankaran et al. [39] discussed the improvisation of wearable devices for breast cancer detection. This paper suggested techniques to use the microwave sensors with image processing techniques and how effective it can be for detecting the breast cancer on finding whether it is a lump or cancerous tissue.

Sardini et al. [40] suggested a device to monitor the human parameters during rehabilitation exercises. It uses non-invasive sensors for the subject. This paper proposes a wearable garment for monitoring the posture of the user during rehabilitation. An inductive sensor is attached to the fabric to monitor the fabric. The shirt output data are compared with the positions of the subject's back and chest. The constitutive sensor has a copper wire and a separable circuit board. It allows the garment to fulfill the needs of simplicity and non-invasiveness. The demerits are the patients feel uncomfortable to wear continuously.

Kavitha et al. [41] created an IOT-based system to measure the physiological parameters of mine workers. The proposed system makes use of sensors such as a temperature sensor, a gas sensor, a dust sensor, a smoke sensor, and a humidity sensor. The sensors used here are non-invasive and attach to the mine workers' jackets. The acquired data are transmitted via the Xbee module. The Xbee module has a range of up to + 40 miles per hop, making it more efficient than the ZigBee module. This jacket has an alarm system that alerts mine workers when certain parameters are exceeded. The benefit of this system is that it uses a remote monitoring system to monitor the mining environment and transmit data to the central monitoring unit via an Xbee module, which is a very simple low-power consumption portable device with stable performance. Its main disadvantages are high maintenance costs and a lack of a comprehensive solution and slow materialization.

3 Methodology

1. Initially the sensor will sense the data such as temperature, pressure, humidity, and gas level.
2. All the sensor data will be displayed on an OLED display so that the user can keep track of all those parameters.
3. With the help of IoT technology, all the sensor data will be sent to the cloud.
4. A supervisor outside should just install a mobile application called ThingSpeak and can monitor all the employees working environment whether it is safe.
5. An alert system must also be organized with the smart shirt which helps in alerting the user if it is not safe to be in the working environment anymore (Figs. 1).

4 Block Diagram

(See Figs. 2, 3 and 4).

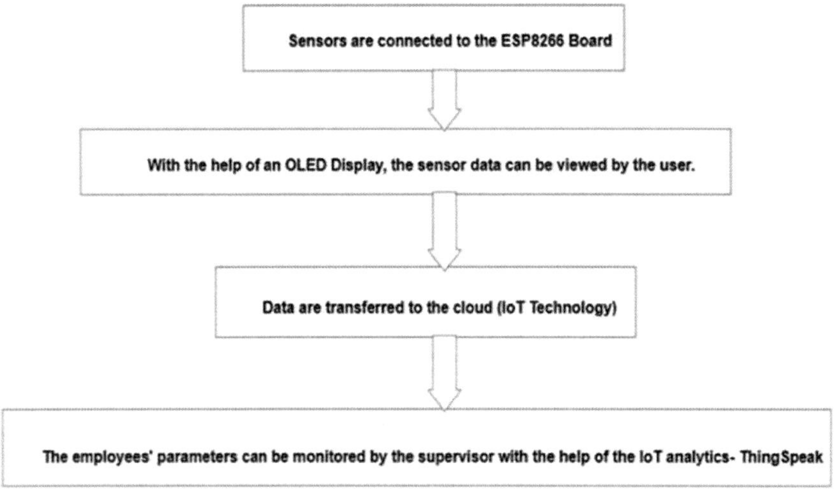

Fig. 1 Methodology of the smart shirt

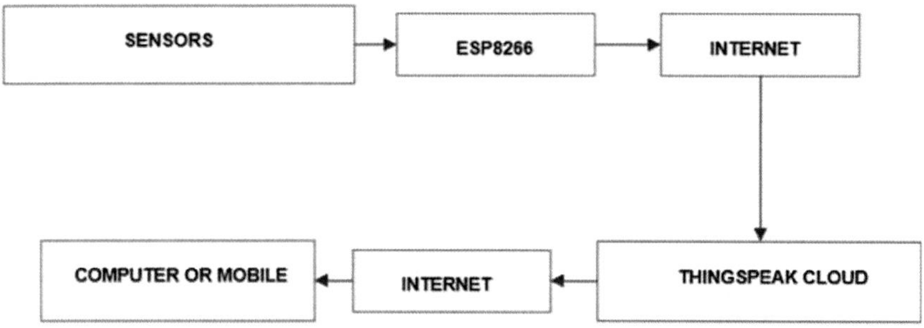

Fig. 2 Block diagram showing the process of displaying the sensor data in ThingSpeak

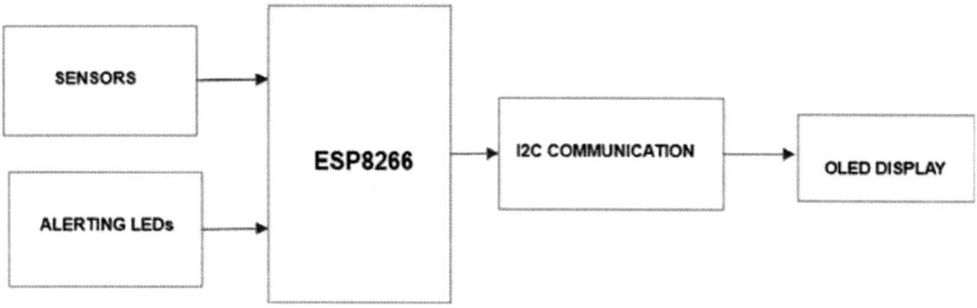

Fig. 3 Block diagram showing the process of displaying the sensor data in the OLED display

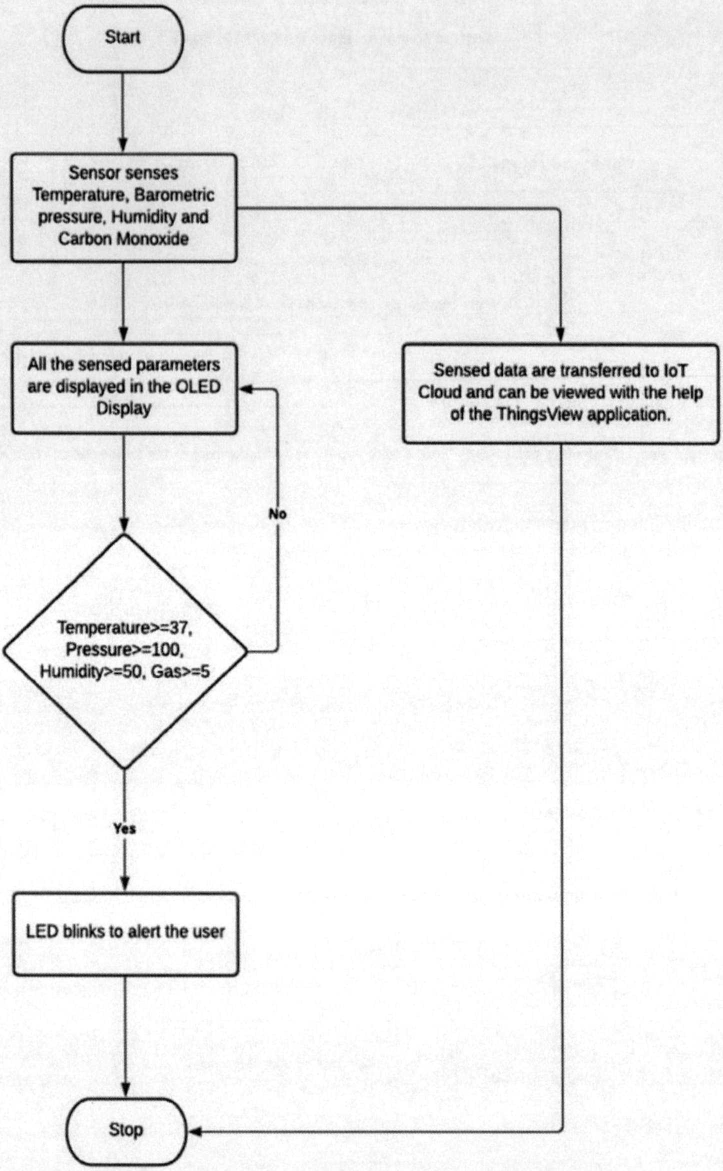

Fig. 4 Algorithm for the working of the different parameters in the smart shirt

5 Conclusion

This paper suggests techniques to overcome problems faced by the mine workers because of the undesirable parameters to facilitate a safer environment deep under the mines and to support the miners for monitoring the parameters rather than using different separate devices and carrying them all the time. After analyzing several literature publications of the past 4–5 years, it has been decided to improve the smart shirt for the workers. The smart safety jacket for laborers working in a mining site is a very useful product. The idea of this jacket is to track all the possible parameters which can be a threat to the workers doing their job underground. It greatly helps the people and gives them immense satisfaction of working in a safer environment.

6 Future Work

It has been planned to implement the idea, add necessary sensors, and develop it into a prototype to monitor the parameters that can possibly become a threat to the workers.

References

1. Sardini E, Serpelloni M, Ometto M (2011) Multi-parameters wireless shirt for physiological monitoring. IEEE Int Symp Med Meas Appl 2011:316–321. https://doi.org/10.1109/MeMeA.2011.5966654
2. Leu FY, Ko CY, You I, Choo K-KR, Ho C-L (2018) A smartphone-based wearable sensors for monitoring real-time physiological data. Comput Electr Eng 65, pp 376–392, ISSN 0045-7906, https://doi.org/10.1016/j.compeleceng.2017.06.031
3. Paiva A, Catarino A, Carvalho H, Postolache O, Postolache G, Ferreira F (2019) Design of a long sleeve t-shirt with ECG and EMG for athletes and rehabilitation patients. In: Machado J, Soares F, Veiga G (eds) Innovation, engineering and entrepreneurship. HELIX 2018. Lecture notes in electrical engineering, vol 505. Springer, Cham. https://doi.org/10.1007/978-3-319-91334-6_34
4. MannéeD, van Helvoort H De Jongh F (2020) The feasibility of measuring lung hyperinflation with a smart shirt: an in vitro study, IEEE Sens J 20(24):15154–15162. https://doi.org/10.1109/JSEN.2020.3010265
5. Kumar MP, Thilagaraj M, Sakthivel S, Maduraiveeran C, Rajasekaran MP, Rama S (2019) Sign language translator using LabVIEW enabled with internet of things. In: Satapathy S, Bhateja V, Das S (eds) Smart intelligent computing and applications. Smart innovation, systems and technologies, vol 104. Springer, Singapore. https://doi.org/10.1007/978-981-13-1921-1_59
6. Roudjane M et al (2020) Smart T-shirt based on wireless communication spiral fiber sensor array for real-time breath monitoring: validation of the technology. IEEE Sensors J 20(18):10841–10850. https://doi.org/10.1109/JSEN.2020.2993286.
7. Wu T, Redouté JM, Yuce M (2019) A wearable, low-power, real-time ECG monitor for smart T-shirt and IoT healthcare applications. In: Fortino G, Wang Z (eds) Advances in body area networks I. internet of things (technology, communications and computing). Springer, Cham. https://doi.org/10.1007/978-3-030-02819-0_13

8. AlShorman O, AlShorman B, Al-khassaweneh M, Alkahtani F (2020) A review of internet of medical things (IOMT): a case study for diabetic patients. Indonesian J Electr Eng Comput Sci 20(1):414–422, ISSN: 2502-4752. https://doi.org/10.11591/ijeecs.v20.i1.pp414-422
9. Wei D, Nagai Y, Jing L, Xiao G (2019) Designing comfortable smart clothing: for infants' health monitoring. Int J Des Creativity Innov 7. https://doi.org/10.1080/21650349.2018.1428690
10. Szczęsna A, Nowak A, Grabiec P, Rozentryt P, Wojciechowska M (2017) Wearable sensor vest design study for vital parameters measurement system. In: Gzik M, Tkacz E, Paszenda Z, Piętka E (eds) Innovations in biomedical engineering. Advances in intelligent systems and computing vol 526. Springer, Cham. https://doi.org/10.1007/978-3-319-47154-9_38
11. Cappon G, Acciaroli G, Vettoretti M, Facchinetti A, Sparacino G (2017) Wearable continuous glucose monitoring sensors: a revolution in diabetes treatment. Electronics 6:65. https://doi.org/10.3390/electronics6030065
12. Lin CC, Yang CY, Zhou Z, Wu S (2018) Intelligent health monitoring system based on smart clothing. Int J Distrib Sensor Netw 14(8) © The Author(s) 2018, Article Reuse Guidelines. https://doi.org/10.1177/1550147718794318
13. Petz P, Eibensteiner F, Langer J (2021) Sensor shirt as universal platform for real-time monitoring of posture and movements for occupational health and ergonomics. Procedia Comput Sci Elsevier B.V 180, ISSN 1877-0509. https://doi.org/10.1016/j.procs.2021.01.157
14. Li Y, Chen W, Lu L (2020) Wearable and biodegradable sensors for human health monitoring. ACS Publications Bio Mater 4(1):122–139. https://doi.org/10.1021/acsabm.0c00859
15. Lee YD, Chung WY (2009) Wireless sensor network based wearable smart shirt for ubiquitous health and activity monitoring, Sens Actuators B Chem Elsevier B.V 140(2), ISSN 0925-4005. https://doi.org/10.1016/j.snb.2009.04.040
16. Chen H et al (2020) Design of an integrated wearable multi-sensor platform based on flexible materials for neonatal monitoring. IEEE Access 8:23732–23747. https://doi.org/10.1109/ACCESS.2020.2970469
17. Pandian PS, Mohanavelu K, Safeer KP, Kotresh TM, Shakunthala DT, Gopal P, Padaki VC (2008) Smart vest: wearable multi-parameter remote physiological monitoring system. Med Eng Phys 30(4), Elsevier 2008, pp 466–477, ISSN 1350-4533. https://doi.org/10.1016/j.medengphy.2007.05.014
18. Abro GEM, Shaikh SA, Soomro S, Abid G, Kumar K, Ahmed F (2018) Prototyping IOT based smart wearable jacket design for securing the life of coal miners. In: 2018 international conference on computing, electronics and communications engineering (iCCECE), pp. 134–137. https://doi.org/10.1109/iCCECOME.2018.8658851
19. Laufer B, Krueger-Ziolek S, Docherty PD, Hoeflinger F, Reindl L, Moeller K (2018) Minimum number of sensors in a smart shirt to measure tidal volumes. IFAC-PapersOnLine 51(27), Elsevier, pp 92–97, ISSN 2405-8963. https://doi.org/10.1016/j.ifacol.2018.11.661
20. Axisa F, Dittmar A, Delhomme G (2003) Smart clothes for the monitoring in real time and conditions of physiological, emotional and sensorial reactions of human. In: Proceedings of the 25th annual international conference of the IEEE engineering in medicine and biology society (IEEE Cat. No.03CH37439) 4:3744–3747. https://doi.org/10.1109/IEMBS.2003.1280974
21. Matuska S, Hudec R, Vestenicky M (2019) Development of a smart wearable device based on electrically conductive yarns, 2019 The Authors. Published by Elsevier B.V., Peer-review under responsibility of the scientific committee of the 13th international scientific conference on sustainable, modern and safe transport (TRANSCOM 2019). https://doi.org/10.1016/j.trpro.2019.07.054
22. Nithya T, Ezak MM, Kumar KR, Vignesh V, Vimala D (2017) Rescue and protection system for underground mine workers based on ZIGBEE. Int J Recent Res Aspects ISSN: 2349-7688 4(4), pp 194–197
23. Mishra A, Malhotra S, Ruchira, Choudekar P, Singh HP (2018) Real time monitoring & analyzation of hazardous parameters in underground coal mines using intelligent helmet system. In:

2018 4th international conference on computational intelligence & communication technology (CICT), pp 1–5. https://doi.org/10.1109/CIACT.2018.8480177
24. Noorin M, Suma K (2018) IoT based wearable device using WSN technology for miners. In: 2018 3rd IEEE international conference on recent trends in electronics, information & communication technology (RTEICT), pp 992–996. https://doi.org/10.1109/RTEICT42901.2018.9012592
25. Hu L, Yang J, Chen M, Qian Y, Rodrigues JJ (2018) SCAI-SVSC: smart clothing for effective interaction with a sustainable vital sign collection. Future Gener Comput Syst 86, Elsevier 2018, ISSN 0167-739X. https://doi.org/10.1016/j.future.2018.03.042
26. Majee A (2016) Iot based automation of safety and monitoring system operations of mines. SSRG Int J Electr Electron Eng 3(9) ISSN: 2348-8379. https://doi.org/10.14445/23488379/IJEEE-V3I9P103
27. Tada Y, Amano Y (2015) A smart shirt made with conductive ink and conductive foam for the measurement of electrocardiogram signals with unipolar precordial leads-2015. https://doi.org/10.3390/fib3040463
28. Tiwari A, Cassani R (2019) A comparative study of stress and anxiety estimation using smart shirt—2019. https://doi.org/10.1109/EMBC.2019.8857890
29. Sardini E, Serpelloni M, Pasqui V (2015) Wireless wearable T shirt for posture monitoring during rehabilitation exercises 2015. https://doi.org/10.1109/TIM.2014.2343411
30. Bahadir SK, Koncar V, kalaoglu F (2019) Smart shirt for obstacles avoidance for visually impaired persons, Elsevier 2019. https://doi.org/10.1016/B978-0-08-100574-3.00003-5ti
31. Baronetto A, Graf LS (2020) Gastro digital shirt: a smart shirt for digestion acoustics monitoring 2020. https://doi.org/10.1145/3410531.3414297
32. Tacchino G, Rocco G (2019) Heart rate variability from wearables: a comparative analysis among standard ECG, a smart shirt and a wristband
33. Mannée D, De Jongh F, Van Helvoort H (2019) Tidal volumes during tasks of daily living measured with a smart shirt—2019. https://doi.org/10.1183/13993003.congress-2019.PA2228
34. van Leuteren R, Fabius T, de Jongh F, van der Valk P, Brusse-Keizer M, van der Palen J (2017) Detecting dynamic hyperinflation in COPD patients using a smart shirt: a pilot study—2017. https://doi.org/10.1183/1393003.congress-2017.PA2214
35. Eizantals P, katashev A, Oks A, Semjonova G (2019) Smart shirt for uncontrolled movement retaining-2019. https://doi.org/10.1007/978-3-030-31635-8_113
36. Laufer B (2020) A minimal set of sensors in a smart-shirt to obtain respiratory parameters—2020 https://doi.org/10.1016/j.ifacol.2020.12.627
37. Sardini E, Serpelloni M (2013) T-shirt for vital parameter monitoring—2013. https://doi.org/10.1007/978-1-4614-3860-1_35
38. Karlsson JS (2008) Urban Wiklund—wireless monitoring of heart rate and electromyographic signals using a smart T-shirt 2008
39. Sankaran S, Sridharan Y, Veerakrishnan B, Murugan PR, Govindaraj V, Britto PI (2020) A literature review: to predict the problem of post analysis in breast cancer determination using the wearable device. Int Conf Commun Signal Process (ICCSP) 2020:0574–0581. https://doi.org/10.1109/ICCSP48568.2020.9182048
40. Sardini E, Serpelloni M, Pasqui V (2014) Wireless wearable t-shirt for posture monitoring during rehabilitation exercises – 2014. https://doi.org/10.1109/TIM.2014.2343411
41. Kavitha D, Chinnasamy A, Devi AS, Shali A (2021) Safety monitoring system in mining environment using IoT. In: Journal of physics: conference series. https://doi.org/10.1088/1742-6596/1724/1/012022

Phishing Site Prediction Using Machine Learning Algorithm

Haritha Rajeev and Midhun Chakkaravarthy

Abstract Theft of sensitive information (phishing) is one of the issues that has plagued the World Wide Web (WWW) and is leading to financial and human and business disasters. It has always been a puzzling issue to identify the crime of theft of sensitive information with high accuracy. Significant developments in the field of classification have been followed by the introduction of the advanced Logistic Regression. This paper is concerned with the precise way of identifying the theft of sensitive web information based on machine learning. Our advanced model has the ability to distinguish criminal websites from stealing sensitive information from official sites. However, due to sample limitations in the database, some machine learning algorithms (SVM, multinomial Nb) are not able to perform well in data analysis. In this regard, our proposed Logistic Regression model has an automated method of predicting sites for the theft of sensitive information in the first instance. The technical results show that a complete accuracy of 98% is obtained in the recommended manner.

Keywords SVM · Multinomial Nb · Logistic regression

1 Introduction

The world is developing fast with the internet, and the web service is becoming more and more popular. Office work, internet business, e-commerce, reading, etc., are almost impossible without the internet. But, all internet users can be distracted by the many web threats that cause economic loss, personal ownership, corporate reputation, and so on. Phishing scams are a form of fraud that obtains personal information such as user id, password, social security number. The developers do this by sending an email to users and asking them to fill in all the information provided by visiting the link. About 65% of all phishing attacks begin with a visit by a mail link. Criminal websites to steal sensitive information almost look like legitimate

H. Rajeev (✉) · M. Chakkaravarthy
Department of Information Technology, Lincoln University College, Petaling Jaya, Malaysia
e-mail: hrajeev@lincoln.edu.my

P. Chatterjee et al. (eds.), *Computational Intelligence for Engineering and Management Applications*, Lecture Notes in Electrical Engineering 984,
https://doi.org/10.1007/978-981-19-8493-8_54

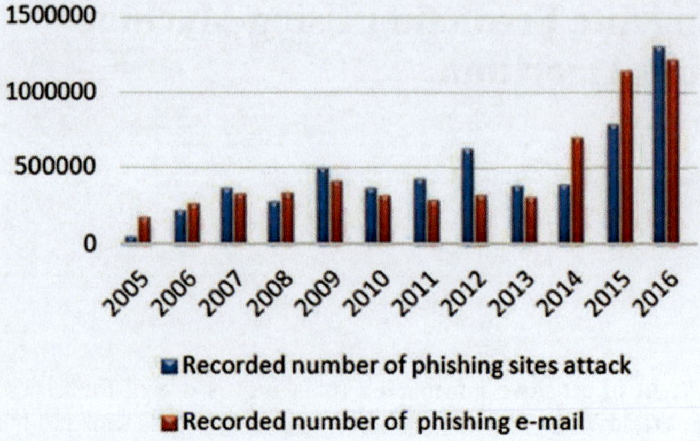

Fig. 1 Phishing growth from 2005 to 2016

websites. By creating a fraudulent site and making people interested in visiting that site, they make the fraudulent attack successful.

The first criminal attack on identity theft took place in America Online (AOL) which can be detected in the early 1995. Figure 1 shows an increase in attacks on identity theft from 2005 to 2016.

The criminal activity of sensitive information was started to be kept under systematic review since 2005, but the highest total number of sensitive identity theft cases was 1,220,523 recorded in 2016. This has been a 65% increase over 2015. Attacks on identity theft were less frequent in 2004 but began an increase in 2012 of 160 percent in 2011. In 2013, the total number of attacks increased to 1% compared to 2012. In Q1 of 2014, the total number of attacks on sensitive identity theft was 125,215, and 10.7, respectively, which was more than Q4 of 2013. About 99.4% of cybercrime sites use 80 holes and 55% of identity theft websites have the same name as real websites to deceive the user. The second highest number of identity theft cases was reported by APWG in Q1 of 2014 between January and March.

Fake e-websites are so similar in comparison to real ones that people with very good internet experience can be fooled by these websites. It creates a million dollar loss. It has a very negative impact on e-commerce. The global criminal justice system for the theft of sensitive information could cost up to $ 5 billion reported by Microsoft Computing Safer Index dated February 3, 2004. Figure 2 shows the financial risks of the crime of theft of sensitive information. The total number of attacks on sensitive identity theft recorded around 450,000 in 2013 resulting in a loss of more than $ 5.9 billion. About 1.2 Billion is the total financial loss in 2011 rising to $ 5.9 billion in 2013 and 4.5 and 4.6 the total financial losses for 2014 and 2015, respectively.

Identity theft is a form of social media attack commonly used to steal user data, including login details and credit card numbers. It happens when an attacker, pretending to be a trusted businessman, tricks the victim into opening an email, instant message, or text message (Fig. 3).

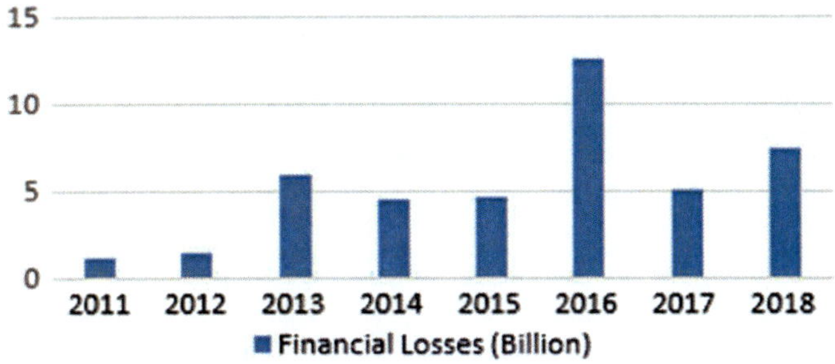

Fig. 2 Financial disasters across world due to phishing attacks

Fig. 3 Classification of cyber attacks

The crime of stealing sensitive information is becoming more complex and often comes across the target site, allowing the attacker to see everything while the victim is roaming the site, and bypassing any additional security barriers with the victim. Since 2020, cybercrime is the most common form of cybercrime, with the FBI Internet Crime Complaint Center recording twice as many cybercrime crimes as any other cybercrime (Fig. 4).

2 Literature Review

A new multifactor verification model has been proposed for Bangladesh to take on cost-effectiveness in the main concern. We have considered two-factor authentication in our previous e-service models that have been proven to be insufficient in relation to the crime of identity theft. Users often fail to identify a site for sensitive identity theft and provide confidential information unintentionally, leading to a successful criminal attempt to steal sensitive information. As a result, the theft of sensitive information can be considered as one of the most sensitive issues and needs to be addressed and minimized. Three things are included to verify multiple items, namely, user ID, secure caption, and one-time password. In a survey, the proposed multifactor model is proved to be 59% better for total users comprising 55% points for technical users and 64% points for non-technical users compared to a two-factor verification

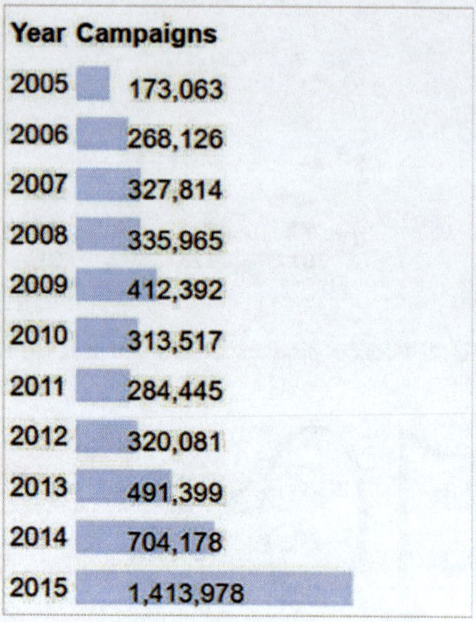

Fig. 4 Unique phishing reports by year

model. As the results and recommendations from the user are shown in the model, user satisfaction was achieved [1].

Cloud computing has become the latest technology computing platform that offers a huge amount of profit to a different organization with its unique business model at low cost. However, there are always concerns about security when uploading sensitive data to a cloud server. Client on the client side is a common and effective solution to assure end-users that an external company user cannot access uploaded data. Cloud service providers maintain a variety of data protection strategies but like google drive, most companies do not use encryption on the client side. In this paper, we suggest how to protect data from looting or loss when storing or uploading data to a cloud server using a combination of advanced encryption level and secure hash algorithm with first vector [2].

The internet has become an integral part of our daily social and financial activities. However, internet users may be at risk of various forms of web intimidation, which may result in financial damage, identity theft, loss of confidential information, damage to product reputation, and loss of customer confidence in online trading and online banking. In this article, we have developed a clever model for predicting the theft of sensitive information based on a neural network of artificial intelligence, especially the sensory networks they construct. Theft of sensitive information is an ongoing problem where the key factors in determining the type of web pages are constantly changing [3].

Identity theft is defined as the credibility of a creditor company's website that aims to capture user personal information such as usernames, passwords, and public security numbers. Crime information theft websites include various references within their content components as well as browser-based security guidelines provided along with the website. Several solutions have been suggested to address the crime of identity theft. However, not a single magic bullet can solve this threat completely. One of the most promising methods of predicting sensitive identity theft is based on data mining, especially "introduction of classification rules" as anti-identity theft solutions aim to accurately predict the website section and exactly the objectives of the data separation strategy [4].

Attack of phishing scams lures web users into visiting fake web pages and providing their personal information. However, exploring criminal websites to steal sensitive information is a challenge. Unlike testing a standard web-based system, we do not know the answer to submitting forms in advance. There is a lack of efforts to assist anti-crime experts who personally verify the identity theft site and take further action. In addition, the current tools do not detect cybercrime attacks on sensitive websites such as site scripts. An attacker may create input forms by entering a text code and stealing information. [5].

Web service is one of the most important Internet communication software services. Phishing scams are one of the many security threats to the web services of the internet. Phishing scams are intended to steal your personal information, such as usernames, passwords, and credit card information, in an effort to legitimize your business. It will lead to the disclosure of information and damage to property. This paper focuses on using an in-depth reading framework to identify criminal websites for stealing sensitive information. This paper begins by designing two types of cybercrime features: real features and interaction features. Discovery model is mainly based on Deep Belief Networks (DBN) [6].

The best results in the segregation area have been achieved with the modernization of deep convolutional neural networks (DCNNs). The paper describe about an accurate way to identify cybercrime attacks based on a convolutional neural network [7].

In this paper, we have introduced an improved version of detection of identity theft attacks with the introduction of the domain name enhancement feature and the inclusion of additional features. Additional features are very useful when the test site does not have a favicon [8].

The number of criminal sites stealing sensitive information is growing and becoming more problematic. Criminal sites for stealing sensitive information often have very short lives. The hackers are thought to have set up crime hotspots using tools such as hacking criminals. Criminal sites for stealing sensitive information are constructed using the same tools that have the same website structures. We suggest a new method based on the similarity of the website design information defined by the types and sizes of web services that make up these websites. Our system can detect criminal sites that steal unregistered sensitive information and blocked lists or that do not have the same URL units that target legitimate sites. [9].

Finding other crime webmasters to steal sensitive information in real time of the day is now a dynamic and flexible topic that combines flexibility with a few requirements. Sensible subtle tactics may be an important factor in locating and evaluating criminal websites to steal sensitive information because of the ambiguity involved in the discovery. Instead of specific principles, fuzzy's mind provides an accurate way to deal with quality fluctuations. The troubleshooting method and model for the open and clever webpage of sensitive identity theft will be proposed in testing the website for phishing scams. This approach is based on the smooth operation of machine learning algorithms that define various aspects of the website for the theft of sensitive information [10].

3 Research Methodology

See (Fig. 5).

3.1 Data Collection

The cybercrime website is publicly accessible at the UCI machine learning facility staffed by our research. This database contains data with 549,346 unique entries. Data are collected from Kaggle. There are two columns: categories A: good—which means the URLs are not containing malicious content and this site is not a phishing site, and B: bad—meaning that the URL contains malicious content and that this site is a site of phishing scams. There is no missing value in the database.

3.2 Respondent Sampling

CountVectorizer is used to transform a corpora of text to a vector of term/token counts. After converting text into token numbers, you sort the data and use an algorithm.

Logistic Regression

- Logistic Regression in the algorithm of the machine learning phase is used to predict the probability of phase-dependent variability. In retrospect, the dependent variance is a binary variance that contains coded data such as 1 (yes, success, etc.) or 0 (no, failure, etc.). In other words, the regression model predicts P (Y = 1) as X function.

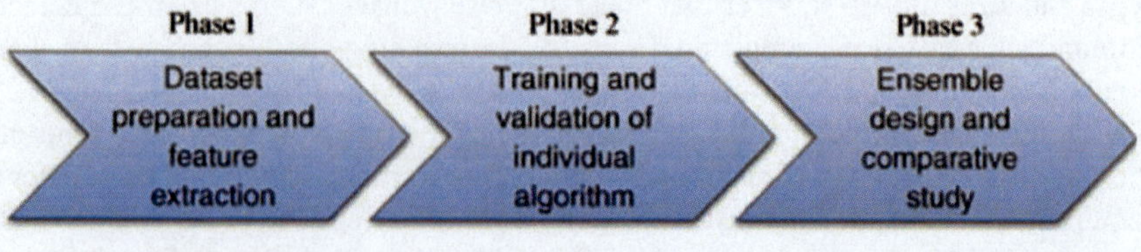

Fig. 5 Conceptual framework of the study

Logistic Regression gives 96% accuracy. Now, we will keep the points in the dict to see which model works best (Fig. 6).

Multinomial NB

Applying multinomial Naive–Bayes to NLP problems. Naive–Bayes classifier algorithm is a family of algorithms that may be based on the application of Bayes theory by the "naive" assumption of conditional independence between both pairs. Multinomial NB gives us 95% accuracy (Fig. 7).

So, Logistic Regression is the best fit model. Now, we make sklearn pipeline using Logistic Regression.

CLASSIFICATION REPORT

	Precision	Recall	F1-score	Support
Bad	0.90	0.97	0.93	36597
Good	0.99	0.96	0.97	100740
Accuracy			0.96	137337
Micro Avg	0.95	0.96	0.95	137337
Weighted Avg	0.97	0.96	0.96	137337

Fig. 6 Classification reports of logistic regression

CLASSIFICATION REPORT

	Precision		Recall	F1-score	Support
Bad	0.91		0.94	0.92	38282
Good	0.98		0.98	0.97	99055
Accuracy				0.96	137337
Micro Avg	0.94		0.95	0.95	137337
Weighted Avg	0.96		0.96	0.96	137337

Fig. 7 Classification reports of multinomial NB

4 Conclusion

This research aims to concern with the precise way of identifying the theft of sensitive web information based on machine learning. Our advanced model has the ability to distinguish criminal websites from stealing sensitive information from official sites In this regard, our proposed Logistic Regression model has an automated method of predicting sites for the theft of sensitive information in the first instance. The technical results show that a complete accuracy of 98% is obtained in the recommended manner.

References

1. Zahid Hasan M, Sattar A, Mahmud A, Talukder KH (2019) A multifactor authentication model to mitigate the phishing attack of e-service systems from bangladesh perspective. In: Emerging research in computing, information, communication and applications, pp 75–86
2. Islam MM, Hasan MZ, Shaon RA (2019) A novel approach for client side encryption in cloud computing. In: 2019 international conference on electrical, computer and communication engineering (ECCE), Cox'sBazar, Bangladesh, pp 1–6
3. Mohammad RM, Thabtah F, Mccluskey L (2013) Predicting phishing websites based on self-structuring neural network. Neural Comput Appl, pp 443–458
4. Mohammad RM, Thabtah F, McCluskey L (2014) Intelligent rule-based phishing websites classification. IET Inf Secur, pp 153–160
5. Shahriar H, Zulkernine M (2012) Trustworthiness testing of phishing websites: a behavior model-based approach. Future Gener Comput Syst, pp 1258–1271
6. Yi P, Guan Y, Zou F, Yao Y, Wang W, Zhu T (2018) Web phishing detection using a deep learning framework. Wirel Commun Mobile Comput, pp 1–9
7. Zubair Hasan KM, Hasan MZ, Zahan N (2019) Automated prediction of phishing websites using deep convolutional neural network. In: 2019 international conference on computer, communication, chemical, materials and electronic engineering (IC4ME2), pp 1–4. https://doi.org/10.1109/IC4ME247184.2019.9036647
8. Chiew KL, Choo JS-F, Sze SN, Yong KSC (2018) Leverage website favicon to detect phishing websites. Secur Commun Netw, pp 1–11
9. Tanaka S, Matsunaka T, Yamada A, Kubota A (2021) Phishing site detection using similarity of website structure. In: 2021 IEEE conference on dependable and secure computing (DSC), pp 1–8. https://doi.org/10.1109/DSC49826.2021.9346256
10. Bhagwat MD, Patil PH, Vishawanath TS (2021) A methodical overview on detection, identification and proactive prevention of phishing websites. In: 2021 third international conference on intelligent communication technologies and virtual mobile networks (ICICV), pp 1505–1508. https://doi.org/10.1109/ICICV50876.2021.9388441

Rating of Movie via Movie Recommendation System Based on Apache Spark Using Big Data and Machine Learning Techniques

Ayasha Malik, Harsha Gupta, Gaurav Kumar, and Ram Kumar Sharma

Abstract Recommendation engines are very useful for businesses to increase their revenue. They are regarded as one of the best types of machine learning models. They are responsible for predicting the choices of people and uncovering all the relationships between items so that discovery of the right choices becomes easy. They help in presenting users with items that they might not even have searched or have known about. Movies are considered as one of the most popular sources of entertainment. It is a very tedious task to search for movies according to the user's taste from a large pool of available movies. The proposed system builds a movie recommendation engine that uses the user's profile to find movies of similar taste as the user. The system recommends the most relevant movies to the user. Apache Spark framework is used for implementing the proposed system via Scala language. The Apache Spark machine learning library (MLlib) is used to ease the implementation. The proposed system provides analysis on various measures. The measures include the total number of ratings by a user, top ten recommended movie ids and names with predicted ratings for a particular user. At last, the performance of the system is evaluated using Root Mean Square Error (RMSE). The value of RMSE gives the accuracy of the model. The results are shown in tabular form. The results show that the model well and after some number of iterations, the value of RMSE is constant.

Keywords Big data · Data extraction · Hadoop · Matrix factorization · Mean square · MLlib Scala

A. Malik (✉)
Delhi Technical Campus (DTC), GGSIPU, Greater Noida, India
e-mail: ayasha07.am@gmail.com

H. Gupta · G. Kumar · R. K. Sharma
Noida Institute of Engineering Technology (NIET) Greater Noida, Greater Noida, India

P. Chatterjee et al. (eds.), *Computational Intelligence for Engineering and Management Applications*, Lecture Notes in Electrical Engineering 984,
https://doi.org/10.1007/978-981-19-8493-8_55

1 Introduction

Big data has become so popular, and it is a wave that everyone wants to ride. The research concludes that many of the big companies have started considering the big data a top priority. It is also predicted that big data market will grow at least five times faster than the overall IT market in the coming years [1]. It is the data that exceeds the storage and processing capability of traditional database systems. The amount of data generated daily is increasing at an alarming rate. For example, approximately 500 million tweets are generated per day. There is an urgent need to handle and process such a huge amount of data. There are many big data processing engines available in the market. For example, Spark, Hadoop, Storm, Oracle, HBase, etc. [2]. There are various Vs that describe big data as shown in Table 1.

Big data is used in various fields such as the Internet of Things (IoT), advertising analysis, predictive analysis, customer churn analysis, Aadhar project, telecom fault detection, natural resource exploration. Research suggests that most of the companies agree that "Big data is important for their day-to-day business operations" [4]. Apache Spark is an effective way to process such a huge amount of data. The frameworks other than Apache Spark are limited in their processing. For instance, MapReduce is limited to batch processing, Apache Storm is limited to real-time stream processing, etc. There is a need for a powerful and general-purpose framework that provides real-time, interactive, graph, in-memory as well as batch processing of data. Apache Spark provides all these qualities and is easy to use. Apache Spark is a lightning-fast cluster computing tool. It is a general-purpose distributed system. It is up to 100 times faster than Hadoop MapReduce (HMR). It is written in Scala, which is a functional

Table 1 Vs of big data

Vi	Attribute name	Action
V1	Volume	It gives an idea about the huge amount of data, generated every second
V2	Velocity	It refers to the fast speed at which data is generated. In order to deal with such a fast speed, we may need real-time data analysis and processing
V3	Variety	It refers to the various forms of data which can be stored and processed using big data. The forms of data include structured data, which stores the data in an organized manner and has fixed format, unstructured data, which does not have any fixed structure or organization, and semi-structured data, which has data, is not in standard database format but still contains some data which has some organizational properties too [3]
V4	Veracity	It refers to noise, anomalies and uncertainty in the data. There are various uncertainties, such as whether the stored data is useful for the problem
V5	Volatility	It refers to the amount of time that the stored data is valid. The stored data may be useful today but may not be of any use tomorrow
V6	Value	It refers to whether the stored data is of any use for analyzing or solving the problems according to our needs. There is no use of storing and processing data which is of no use

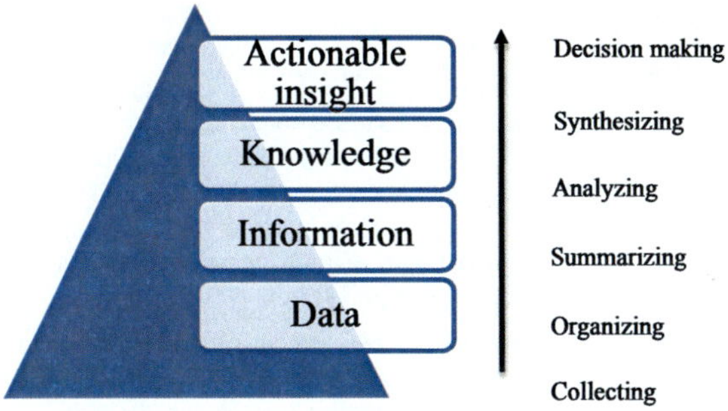

Fig. 1 Transformation of big data

programming language. It has application programming interfaces in Scala, Python and Java. It has the capability to integrate with Hadoop and process existing data [5]. Figure 1 shows the transformation of big data. The main objective of big data is decision-making.

2 Literature Survey

This section discusses some of the research that has been proposed in order to build movie recommendation systems using efficient techniques. Table 2 shows the summary of the studied research in order to accomplish the proposed research work.

3 Prerequisite Knowledge

This section described all the basic knowledge that required previously to implement the model in a better way.

3.1 Matrix Factorization

Matrix factorization models are regarded as one of the best models. These models have good performance in CF. There are two types of matrix factorization models. The first is explicit matrix factorization, and the second is implicit matrix factorization.

3.1.1 Explicit Matrix Factorization

The matrix factorization model deals with sparse data by splitting the two-dimensional matrix into two smaller matrices which are of lower dimension. The

Table 2 Summary of related studies

Authors	Summary	Year
Bokde et al. [6]	Focused on the challenges faced by collaborative filtering (CF) algorithms. Discussion about the different models of matrix factorization (MF). A survey of one of the MF models such as singular value decomposition (SVD) is presented to present a solution to the different challenges faced by CF algorithms	2015
Almohsen and Jobori [7]	Presented one of the recommendation system (RS) techniques referred to as CF and the challenges faced by it. Singular value decomposition (SVD), implemented using big data analytics tools such as Apache Spark and Apache Hadoop, overcomes the problems faced by big data	2015
Panigrahia et al. [8]	A new hybrid algorithm is presented for the CF method which is user-oriented, and Apache Spark implements it. The challenges faced by CF algorithms are overcome by clustering techniques such as K-means and the techniques of dimensionality reduction such as alternating least squares (ALS). The features which correlate the user to products are also presented as one of the techniques to solve the problems faced by CF algorithms	2016
Miryala et al. [9]	Apache Spark machine learning library (MLlib) is used. The performance of the ALS algorithm is compared with various other algorithms. The dataset taken in this setup is from MovieLens. The results show that the performance of the ALS algorithm is the best among the other algorithms	2017
Howal et al. [10]	The goal of this work suggests the presence of the most accurate recommendations to the user and dealing with a large set of information. Apache Spark is used for achieving the desired results	2017
Aljuni et al. [11]	The author introduces many techniques of CF and the solution of their limitations using tools of big data analytics. Various tools of big data such as Apache Flink, Apache Spark, are also introduced	2017
Mali et al. [12]	The author's aim in this paper is to split the computations which are costly in CF into the three phases of MapReduce	2017
Lenka et al. [13]	The authors have used Apache Spark for building the recommendation system. Further, the authors have compared bisecting K-means and K-means clustering algorithms by performing this computation in Apache Spark	2018
Sri et al. [14]	The authors have proposed to use CF techniques to generate recommendations. The authors have used the CF technique which is item based as it performs better than other algorithms	2018

product of the smaller matrices represents the two-dimensional matrix. The dimensionality reduction technique helps in splitting the matrices. If the two-dimensional matrix has dimensions U * I, where U represents the users and I represents the items, then we can split this matrix into two matrices of dimension K. These smaller matrices are called factor matrices. The dimension of the user matrix is U * K, and the dimension of the item matrix is I * K. These factor matrices are generally dense. Factor matrices are referred to as latent feature models as they discover hidden features in the user-item rating matrix. In order to find the rating for a given user and item, a vector dot product is computed between the row of the user's factor matrix and the column of the item's factor matrix [15].

3.1.2 Implicit Matrix Factorization

Most of the collected preference data is implicit feedback, which means that the preferences are not given but are implied on the basis of interaction between a user and an item. Examples of implicit preference data include binary data such as whether a particular movie is viewed by a user and count data such as the number of times a particular movie is watched by a user.

This approach is implemented by MLlib by treating the input rating matrix in the form of two matrices. First is binary preference matrix P, and second is confidence weight matrix C. For instance, if the movie recommendation system preference data is represented in the form of an input rating matrix, then the preference matrix P gives information about whether a movie is being viewed by a particular user and the confidence matrix C represents the count of the number of times a particular movie is being watched by a user. The implicit model also has an item and user factor matrix. The model only approximates the preference matrix P instead of the rating matrix. In this case, recommendation shows the items preferred by a user [16].

3.2 Alternating Least Squares

The cost function of matrix factorization mainly focuses on two variables E and F and their dot product $E * F^T$. The discussion on these two variables is already done in the previous sections of this chapter. Equation (1) depicts the actual cost function.

$$||R - E * F^T||2 = \sum i, j(R_{i,j} - e_i * f_j) \tag{1}$$

In linear regression, the problem is simply to compute β given E and f. The objective is to minimize the squared error $||f - E\beta||^2$. . The least squares formula (LSF) gives the solution. Equation (2) shows the LSF.

$$\beta = (ETE) - 1E^T f \tag{2}$$

The solution of ALS follows a linear regression model. It is an iterative process of optimization and is usually a two-step process. In one iteration, it fixes E for instance and solves for F, while in the next iteration, it fixes F and solves for E. Each ordinary least squares (OLS) solution is unique, and each step guarantees a minimum mean squared error (MSE). The cost function is longer due to the inclusion of the regularization parameter. The two-step process breaks the cost function into two other cost functions.

4 Proposed Approach

Figure 2 shows the implementation steps in a block diagram. All these steps are performed in Spark shell.

4.1 Extracting Features

The dataset which is considered is MovieLens which has 100 k entries. The given inputs are movie IDs, user IDs and ratings for each pair of movie and user.

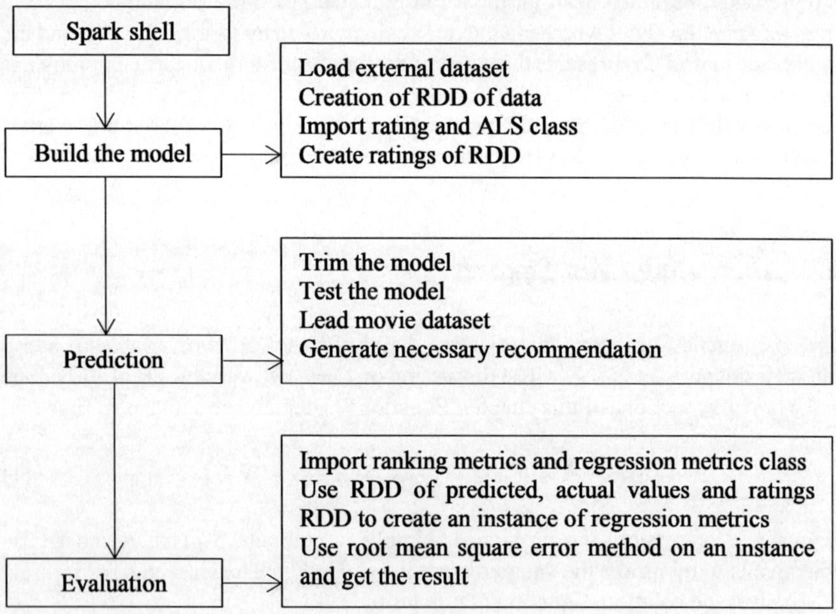

Fig. 2 Implementation steps

Firstly, start the Spark shell. Provide the input path to the MovieLens raw data which consists of user id, movie id, rating, timestamp. The fields have a tab character space between them. As the timestamp field is not needed for training, only the first three fields are extracted. Each record is split on the tab character to give a String array. Also, the take function of Scala is used to retrieve the first three elements of the String array. The first three elements correspond to the user id, movie id and rating. The first record of resilient distributed datasets (RDDs) can be inspected by using the first function which is responsible for returning the first record stored in RDD.

To perform training of the model, Apache Spark's machine learning library (MLlib) is used. For this, the ALS model is imported from MLlib. The methods available in ALS can also be inspected using the appropriate command. Executing the ALS train command gives an error, but the error can give an idea about the method signature. So, it is necessary to provide input arguments such as ratings, rank, iterations and lambda. Also, the rating class needs to be imported. Executing the rating command and inspecting the output suggest that the ALS model needs to be provided with the RDD consisting of rating records. The map method is used for creating the rating dataset and an array of ratings, and IDs are transformed into a rating object. This step creates RDD and can be verified by using the first method [17].

4.2 *Build the Model*

- Loading of the external dataset which is MovieLens 100 k is performed
- In this step, RDD is made consisting of the field's user id, movie id and ratings in array form
- Import necessary classes such as ALS and rating
- Transform all ids and ratings into a rating object in order to create ratings RDD
- Apache Spark provides an MLlib to train the model.

The input should be RDD and the model parameters. The model parameters include rank, iterations and lambda. The rank refers to the factors' count in the ALS model; iterations refer to the number of times the algorithm runs and lambda refers to controlling the regularization in the model.

4.3 *Training*

This section discusses the training of data on the MovieLens dataset which has 100 k entries. After extracting the features and building the model, the model is ready to be trained. MLlib provides facilities for training. In order to train, RDD which was created in the previous section and model parameters is needed. The parameters which are needed for model training include rating, rank, iterations and lambda. Rank gives information about the number of hidden features in the matrices. Iteration tells about the number of times an algorithm runs. Lambda parameter is responsible for controlling the regularization of the model and over-fitting of data [18].

4.4 Training Using Feedback Data

The preceding section discusses training using explicit ratings. This section discusses MLlib's matrix factorization approach dealing with explicit ratings. The train implicit method is used for dealing with implicit data. An additional parameter called alpha is used here. The other parameters such as lambda, regularization parameter should be set in the same way as done in explicit rating data. The alpha parameter is used for controlling the applied level of confidence weighting. If the value of alpha is high, then there is a strong indication that if there is no data, then the user–item pair has no preference.

4.5 Prediction

- Test the model by using the predict method of the matrix factorization model class for user and movie combination pairs
- The recommend products' method of the matrix factorization model class is responsible for generating the recommended movies which are highly demanding. The arguments taken by this method are user id and the count of the number of recommendations to be made. The result is ranked on the basis of the predicted values
- The titles of the top ten recommended movie ids can also be known. The first step then is to load the movie dataset. The data is collected with the help of a map method which takes an integer and a string as a parameter. The map method is used for mapping the movie id to the title
- For a particular user, the total number of ratings done by a user for a different number of users can also be found. This is done by using Spark's keyBy function using ratings RDD in order to create an RDD consisting of key–value pairs. The key consideration here is the user id.

The movies for the user collection are sorted by using the rating's object rating field. The result is a list of top movies with the highest ratings.

4.6 Recommendations

This section discusses generating the appropriate recommendation for a user. The next which comes after training the model is making predictions. These recommendations can take various forms. The forms can be for instance making recommendations for a user or relating to similar items.

4.6.1 Generating Movie Recommendations

The model which generates recommendations for a user predicts the top movies which have the highest probability of liking by the users. In order to do this, a score is predicted for each movie and the list is ranked according to this score. It depends on the model to choose the appropriate method for predicting scores. For instance, if the user-based approach is used, then the score depends on the ratings which are provided by similar users to recommend movies, and if an item-based approach is used, the score depends on the similarity of the items which the user has rated in comparison to other items. Matrix factorization computes the predicted score on the basis of the vector dot product between an item factor vector and a user factor vector. The recommendation model of Apache Spark's MLlib uses factor matrices to predict scores for any given user. Usually, the approach used for explicit and implicit data is the same. To predict the ratings for a user and a movie, predict method of matrix factorization model class is used. This model also has the capability to predict the ratings of an entire RDD consisting of user and movie IDs as the input altogether. The model factorization model provides recommend products' method so that only the top k recommendations are generated where k is an integer. The recommend products' method takes two arguments. The first argument is the ID of the user, and the second argument is the number of items that a user wants to recommend. In this way, every user can compute the predicted rating for each movie. There are methods in this model, using which the titles of the movies recommended for each user can be generated. In order to do this movie, dataset needs to be loaded. The movie data is collected as a map method which has two parameters. The first parameter is an integer, and the second parameter is a string. The map method is responsible for mapping the movie id to the title. This model provides a convenient way to find the movies rated by a particular user and generate the titles of the top movies with the highest ratings. To start performing this computation, keyBy method is used which creates RDD containing key–value pairs using ratings' RDD which has user ID as the key. Then, lookup function is used for returning the ratings for the particular user ID. User ID is also regarded as the key.

Furthermore, the size function can be used to know the number of movies that are rated by the particular user. The next step is to sort the above-created movie collection in order to get the highest movie ratings. This is done by using the rating object's rating field. Movie titles can be extracted for the product ID which is present in the rating class using mapping of the movie titles, and top k titles can be printed with their ratings where k is the user-given number. Top k recommendations can also be generated for a particular user containing movie titles and ratings, where k represents the number of users using the topKRecs method [19].

4.6.2 Item Recommendations

Item recommendations are used to find a set of items matching a particular item. The model decides the computation of similarity. Vector representation of the two items can be compared to find similarities between any two items using similarity measure. Some of the similarity measures are cosine a similarity, Jaccard similarity and Pearson correlation. Item similarity computation is not directly supported by the matrix factorization model. The dot product is computed by using a cosine similarity metric and a dependent of MLlib called the jblas Linear Algebra Library. The only new thing introduced here is the computation of cosine similarity; the rest is similar to the functions performed by recommend products and predict methods. A similarity metric is used for the comparison of a particular item's factor vector with the other entire item's factor vector. The computation of linear algebra requires vector objects to be created from the factor vectors and stored in an array.

The double matrix method of the jblas class takes an array argument. In n-dimensional space, the angle which is between two vectors measures cosine similarity. In order to compute cosine similarity, the dot product is computed between the vectors first, and the result is divided by the multiplication of the length of the two vectors. Cosine similarity is regarded as a normalization of the dot product. The range of cosine similarity can only be from -1 to 1 inclusively. The similarity measure has a value of 1 if the two items are similar and has a value of 0 if the two items are not similar. If the two items are dissimilar completely, the similarity measure has a negative value of -1. In order to just cross-check the value of cosine similarity, it can be tried on any of the item factors of an item. Firstly, an item factor needs to be collected using the lookup function. Then, as the lookup function returns many values, use the head function to get just the first value. Take the value returned from the lookup function as the parameter of the double matrix method so that it is an object can be created. At last compute the cosine similarity, taking two similar parameters, which are objects returned by the double matrix method. If the above steps are correctly done, the value of cosine similarity should be 1 as the two vectors are completely similar. The cosine similarity can also be computed for each item. Firstly, item vector and factor vector objects are created using each of them as the parameter of the double matrix method. Cosine similarity should be computed taking factor vector and item vector as the parameters of the cosine similarity method. Next, each item's similarity score can be sorted using the top function to get the most similar items. The top function uses a special argument that informs Spark to order by value if a key–value pair is given which tells Spark that ordering should be performed on the basis of similarity. Therefore, the computation can be performed using the top function to get the titles and ratings of most similar movies.

4.7 *Evaluating Performance*

There should be some way to know the true validity of a trained model. This section focuses on the ways to evaluate the performance of a trained model. Evaluation metrics are used for predicting the capability of a model. Mean squared error (MSE) is considered as one of the direct ways to predict the capability of a model, and mean average precision at K (MAPK) is used for predicting things that matter in the real world. The performance of a model can be evaluated with different parameter settings using evaluation metrics. These metrics help in selecting the model which performs the best among the other available models. The next sub-section will show mean square error which is used as an evaluation metric in collaborative filtering and recommendation systems.

- In-built evaluation functions are present in Spark's machine learning library.
- Ranking Metrics and Regression Metrics classes evaluate model's performance. Import these classes.
- The root mean square error (RMSE) and MSE are computed by using Regression Metrics. An RDD containing the predicted and actual values for each of the users is passed in order to create an instance of Regression Metrics. Here, another RDD which contains the ratings for each pair of user and movie is also used. Applying MSE and RMSE method on the instance of Regression Metrics gives the result.

4.7.1 Mean Square Error

The MSE measures the error in the predictions which are performed in the rating matrix of user–item. There are several matrix factorization techniques specifically ALS which include it as an objective function to be minimized. This technique is commonly used for explicit rating data. It is usually referred to as the division of the sum of squared errors by the total number of observations. Squaring the differences between the actual and the predicted ratings for a given pair of user and item gives the value of squared error. MSE computation for the entire dataset firstly requires the squared error to be computed for each user. Then, squared error values for each user are summed up. After this, the total sum is divided by the number of ratings. The detailed procedure of performing this computation is as follows. User and product IDs are extracted from the ratings' RDD. Prediction is made for each user and item pair using predict method. The key is the user–item pair, and the predicted rating is the value. The actual rating is extracted and rating RDD is mapped so that key is the user–item pair and value is the actual rating. As the two RDDs are having the same key, they can be joined and a new RDD can be created. Each combination of user and item has both the actual and predicted ratings in the newly created RDD. The MSE is computed by taking the sum of the squared errors by reducing method, and this sum is divided by the number of records which is done using the count method [20].

4.7.2 In-built Evaluation Functions

This section gives an idea about the in-built evaluation functions which are present in Spark's machine learning library. Ranking Metrics and Regression Metrics classes are provided by MLlib to perform an evaluation of the model's performance. An outline of performing the in-built functionalities provided by MLlib is as follows. The RMSE and MSE are computed by using Regression Metrics. An RDD containing the predicted and actual values for each of the users is passed to create an instance of Regression Metrics. Here, another RDD which contains the ratings for each pair of user and movie is also used. Applying the mean square error and root mean square error on the instance of Regression Metrics gives the result.

5 Results and Analysis

The graph analysis is presented on various factors such as the total number of ratings given by a group of users as shown in Fig. 3, top ten movie ids with the predicted rating for user id 186 users as shown in Fig. 4, ten movie names with the highest ratings for user ID 166 users as shown in Fig. 5 and many more. To build the recommendation system, two terminals used are terminal T1 and terminal T2. The dataset used in this experiment is from MovieLens.

The performance of the model is also assessed using RMSE as an evaluation metric. RMSE measures the differences between the actual and the predicted values of ratings. The values of regularization parameter (λ) taken in the model are 0.01, 0.1 and 1, while the values of the number of iterations in the model are 1, 3, 5, 10, 15 and 20. Each iteration takes the given three values of lambda in order to train the model. The discussion of the parameters needed to train the model is already done in the previous section. Table 1 shows the calculated RMSE value for each value of λ and the number of iterations. Table 3 clearly shows that there is almost no change

Fig. 3 Total number of ratings given by ten user ids

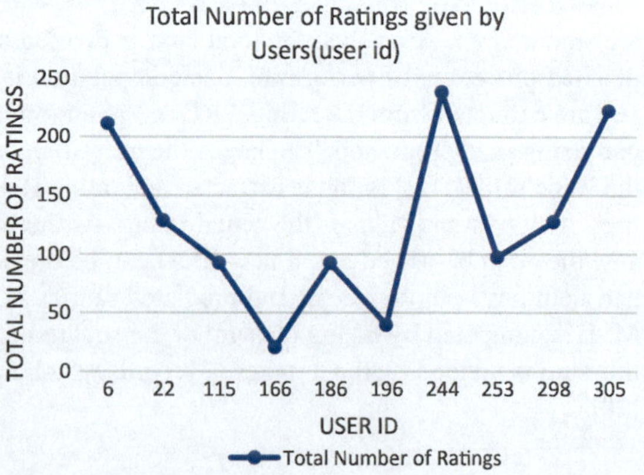

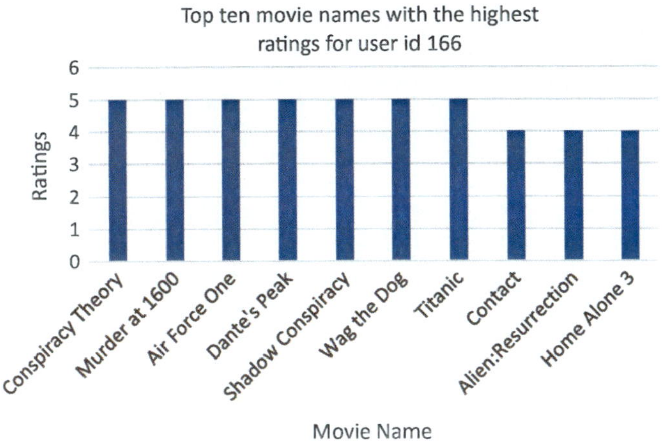

Fig. 4 Top ten movies with the highest ratings for user id 166

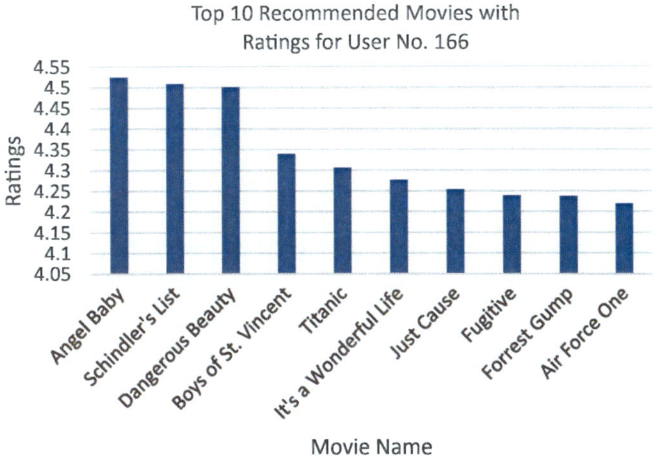

Fig. 5 Top ten recommended movies with ratings for user id 166

in the values of RMSE in the end. This is the reason that further iterations are not taken into consideration.

6 Conclusion and Future Work

Nowadays, many applications like Netflix have started to use recommendation systems in order to increase their business revenue. These recommendations are

Table 3 Computation of RMSE

Number of iterations	λ	RMSE	λ	RMSE	λ	RMSE
1	0.01	0.941672	0.1	1.044468	1	3.629166
3	0.01	0.402474	0.1	0.717591	1	1.369867
5	0.01	0.336146	0.1	0.684007	1	1.359460
10	0.01	0.289859	0.1	0.667755	1	1.358663
15	0.01	0.275645	0.1	0.661485	1	1.358662
20	0.01	0.267996	0.1	0.659192	1	1.358662

very useful for users as they come to know about new items or things, of which they were not having any idea. They are also helping the users ease their search by bringing useful items or things to their notice from a large set of available things or items.

This work has built a simple movie recommendation system. The movie recommendation system is using CF technique. Matrix factorization models are regarded as one of the best models. These models have good performance in CF. Therefore, this work has used the matrix factorization model in CF. The ALS algorithm is used for optimizing and training the model. The analysis is performed on various measures such as the total number of ratings given by a user, top ten recommended movies for a given user id. The model is also evaluated using RMSE. The value of RMSE gives the accuracy of the model. The results are shown in tabular form. The results show that the model well and after some number of iterations the value of RMSE is constant.

The future work is to build a hybrid algorithm that performs better than the proposed model. The hybrid model can be built by combining the strengths of content-based filtering and CF techniques.

References

1. Ahuja S (2018) Big data–introduction, applications and future scope. Int J Emerg Technol Innov Res (IJETIR) vol 5, pp 751–754
2. Acharjya DP, Ahmed KP (2016) A survey on big data analytics: challenges, open research issues and tools. Int J Adv Comput Sci Appl (IJACSA), vol 7, pp 511–518.
3. Elgendy N, Elragal A (2014) Big data analytics: a literature review paper. In: Industrial conference on data mining (ICDM), pp 214–227
4. Vijayarani S, Sharmila S (2016) Research in big data—an overview. Inf Eng Int J (IEIJ) vol 4, pp 1–20
5. Alam JR, Asma Sajid, Ramzan Talib, & Muneeb Niaz, (2014) A review on the role of big data in business. Int J Comput Sci Mobile Comput vol 3, pp 446–453
6. Bokde D, Girase S, Mukhopadhyay D (2015) Matrix factorization model in collaborative filtering algorithms: a survey. In: International conference on advances in computing, communication and control (ICAC3'15), pp 136–146
7. Almohsen KA, Al-Jobori H (2015) Recommender systems in light of big data. Int J Electr Comput Eng (IJECE), pp 1553–1563

8. Panigrahi S, Lenka RK, Stitipragyan A (2016) A hybrid distributed collaborative filtering recommender engine using apache spark. In: International workshop on big data and data mining challenges on iot and pervasive systems (BigD2M 2016), pp 1000–1006
9. Miryala G, Gomes R, Dayananda KR (2017) Comparative analysis of movie recommendation system using collaborative filtering in spark engine. J Global Res in Comput Sci 8:10–14
10. Sadanand H, Vrushali D, Rohan N, Avadhut M, Rushikesh V, Harshada R (2017) Movie recommender engine using collaborative filtering. Int J Adv Res Innov, pp 599–608
11. FadhelAljunid M, Manjaiah DH (2017) A survey on recommendation systems for social media using big data analytics. Int J Latest Trends Eng Technol, pp 048–058
12. Mali KS, Shaikh SI, Shaikh SA, Mohole GP (2017) Generic recommendation engine using spark. Int J Sci Res Dev 5:471–474
13. Lenka RK, Barik RK, Panigrahi S, Panda SS (2018) An improved hybrid distributed collaborative filtering model for recommender engine using apache spark. Int J Int Sys Appl, pp 74–81
14. Sri N, Abhilash P, Avinash K, Rakesh S, Prakash CS (2018) Movie recommender system using item based collaborative filtering technique. Int J Eng Technol Sci Res (IJETSR), vol 5, pp 64–69
15. Shoro AG, Soomro TR (2015) Big data analysis: Ap spark perspective. Global J Comput Sci Technol (GJCST), vol 15, pp 7–14
16. Bhattacharya A, Bhatnagar S (2016) Big data and apache spark: a review. Int J Eng Res Sci (IJOER) vol 2, pp 206–210
17. Jonnalagadda VS, Srikanth P, Thumati K, Nallamala SH (2016) A review study of apache spark in big data processing. Int J Comput Sci Trends Technol (IJCST) vol 4, pp 93–98
18. Zaharia M, Chowdhury M, Das T, Dave A, Ma J, McCauley M, Franklin MJ, Shenker S, Stoica I (2012) Resilient distributed datasets: a fault-tolerant abstraction for in-memory cluster computing. In: Proceedings of the 9th USENIX conference on networked systems design and implementation, pp 15–28
19. Dahiya P, Chaitra B, Kumari U (2017) Survey on big data using apache Hadoop and spark. Int J Comput Eng Res Trends (IJCERT), pp 195–201
20. Pathrikar KK, Dudhgaonkar AA (2017) Review on apache spark technology. Int Res J Eng Technol (IRJET), pp 1386–1388

Application of ISM in Evaluating Inter-relationships Among Software Vulnerabilities

Misbah Anjum, P. K. Kapur, Sunil Kumar Khatri, and Vernika Agarwal

Abstract Complicated digital communication networks generate faults in development, implementation and governance, which are the main cause of software vulnerabilities. The growth of these vulnerabilities urges the software development and application security teams to rely on vulnerability detection tools throughout the development phase of a software. These developers are often overwhelmed with a steady stream of security alerts that must be addressed without slowing down the pace of software development. Vulnerability prioritization often helps the security team to remove the highly severe vulnerabilities first, but to check if the vulnerability is being linked to another vulnerability needs to be taken into the consideration. The purpose of this study is to establish a quantitative evaluation framework to discover and analyze the link between vulnerability categories, leading to an improved elimination of vulnerabilities using the Interpretive Structural Modeling (ISM) technique. The model is validated for an Indian software development.

Keywords Software vulnerabilities · Prioritization · Severity · Multi-criteria decision-making (MCDM) · Interpretive Structural Modeling (ISM)

M. Anjum
Amity Institute of Information Technology, Amity University, Uttar Pradesh, Noida, India

P. K. Kapur
Amity Center for Inter-disciplinary Research, Amity University, Uttar Pradesh, Noida, India

S. K. Khatri
Amity University, Uttar Pradesh, Noida, India

V. Agarwal (✉)
Amity International Business School, Amity University, Uttar Pradesh, Noida, India
e-mail: vernika.agarwal@gmail.com

P. Chatterjee et al. (eds.), *Computational Intelligence for Engineering and Management Applications*, Lecture Notes in Electrical Engineering 984,
https://doi.org/10.1007/978-981-19-8493-8_56

1 Introduction

The prevalence of information security violations has significantly increased [1]. One of the major causes for this increase is security failures in software. Microsoft describes system vulnerabilities as "fault in the product which makes it ineffective even when the product is properly used, to prevent the user from accessing privileges on the system, regulating its operation, compromise data or presume unrequired confidence" [2]. Hackers are making use of software faults to damage organizations severely through restricting system resources for allowed users, changing and altering sensitive data and initiating attacks against all other target organizations [3].

Software vulnerabilities have extensive influence and may cause breakdowns and disruption of organizations worth billions of dollars [4]. One month after Microsoft Chairman "Bill Gates" encouraged employers to work on building more reliable software, its campaign worth $200 m for ".NET" was marked by the revelation of a security flaw in "Visual C++.NET" [5]. The computer emergency response team (CERT) is being informed of many cases of threats and assaults [6]. These instances of vulnerability have led to enormous loss of reputation and money [7]. A total of 1779 incidents of vulnerabilities were reported to the CERT Coordination Center in 2018, which was an increment of the incidents of 1109 in 2017 [8, 9].

Due to the constant revelation of software vulnerabilities, security experts need to decide which issue they should address first [10]. In a recent study, word embedding and convolution neural network (CNN) techniques have been used to prioritize software vulnerabilities using the description of a vulnerability [11]. Another research provides vulnerability risk prioritization, which captures the attacker's desire for exploiting vulnerabilities using a threat modeling technique that employs a neuro-symbolic model, neural network (NN) and probabilistic logic programming (PLP) [12]. The seriousness of the vulnerability might help to select the first vulnerability to be fixed in a software system [13]. The vendor is expected to provide a quick solution and provide the information about the vulnerability and the patch, which will influence program maintenance and its next release [14]. Not all problems, like patches or software upgrades, reported are corrected [15]. Security professionals normally function under budget limitations, and preference should be thus given to discovered vulnerabilities and remedial measures to be implemented [16].

A large number of vulnerability rating techniques have been developed to evaluate the intensity of the software vulnerabilities [17]. Common Vulnerability Scoring System (CVSS) seems to be the de facto standard used to determine the severity of security flaws by the Forum of Incident Response and Security Team (FIRST) [18]. Literature offers high-quality rating systems such as "X-Force", "Microsoft", "Qualys", "Secunia", "RedHat", "VUPEN", "Mozilla" and "Google" [19]. Few more research in this regard include "Weighted Impact Vulnerability Scoring System", which utilizes the same six criteria as CVSS, but takes different weights into consideration for effect measures [20]. By incorporating impact measurements in 2015, they have significantly enhanced it [21]. "Potential value loss" is another quantitative scoring method that employs seven indications to calculate the vulnerability score and makes the algorithm publicly available [22].

The primary aim of this study is to discover software vulnerabilities that impact other vulnerabilities and to examine their relationships with a multi-criterion decision-making method (MCDM). Many MCDM approaches have been used previously in the literature but to support the establishment of a relationship diagram and to understand the degree of influence of one vulnerability over another is yet to be done. Interpretive Structure Modeling (ISM) is therefore utilized in the present study so that the security team can focus on the most influencing vulnerability and remove the same as early as possible.

The following research objectives are to be determined from the previous discussion:

- To discover software vulnerabilities from industry experts as well as through evaluation of literature.
- To understand the contextual relationship among software vulnerabilities.
- To address vulnerabilities affecting additional serious vulnerabilities.

The paper has the following organization: The research technique is explained in Sect. 2. Section 3 provides an analysis of the data followed by results discussion, conclusion of the essay and future work in Sect. 4.

2 Research Methodology

2.1 Interpretive Structure Modeling (ISM)

ISM helps in defining the relationships among specific things, which describes an issue [23]. ISM is a technique where the accompanying factors are arranged into a structured model [24]. The ISM process coverts the ill-defined models into the clearly defined model [25]. It proposes to use expert opinion to identify and form the relationship between different enablers and helps to form a structural relationship between them [26]. In this study, an "ISM" approach is used to detect the inter-relationship among the vulnerabilities. The ISM methodology comprised of the following steps are used in understanding the relationships between the identified vulnerabilities.

Step 1: The first step is to create a structural self-interaction matrix (SSIM). At this stage, the experts were asked to do a pair-wise evaluation of each of the vulnerabilities. Let i and j be the two vulnerability types. The matrix shows the relationship directly between the vulnerabilities using the following symbols as: V—(represents vulnerability i will influence vulnerability j), A—(represents vulnerability j will influence vulnerability i), X—(symbolizes for vulnerability i and j will help achieve each other) and *O*—(symbolizes both vulnerability *i* and *j* are unrelated). Depending on answers, a knowledge base is developed in form of a table in which each row represents the compared pair of pressures and the existing contextual relationship if any are mentioned.

Step 2: The symbols present in the SSIM table are converted into binary numbers 0 and 1 with the help of the following rules:

- If the symbol is *V*, then (*i*, *j*) will be replaced with 1 and (*j*, *i*) entry will be replaced with 0.
- If there is symbol *A*, then (*i*, *j*) will be 0 and (*j*, *i*) will be 1.
- For symbol *X*, both (*i*, *j*) and (*j*, *i*) entries in the reachability matrix will be 1.
- For symbol *O*, both (*i*, *j*) and (*j*, *i*) entries will be 0.

The initial reachability matrix obtained in the previous step has entries for only direct relationships. Based on the rule of transitivity, if *A* influences *B* and *B* influences *C*, then it is considered as *A* influences *C* due to transitivity, though direct relationship does not prevail between *A* and *C*. Each possible transitive link makes entry in the knowledge base, and the logical interpretation is renamed as transitive. In addition, by including every transitivity in the initial accessibility matrix, the final accessibility matrix is built.

Step 3: "The reachability set (RS) and antecedent set (AS) are obtained." In the first iteration, the pressures with same intersection and RS are designated the top most level (Level I), and the level I elements are detached from the entire set. This iterative process is continued until each practice is assigned with levels.

Step 4: "ISM model is drawn" The graph is drawn consistent with the levels obtained by the level partitioning from final reachability matrix. The resulting graph is also called a digraph. A review of the ISM model is done by checking the inconsistency and then making the necessary changes.

3 Numerical Illustration

The current research provides the framework for building the relationship diagram and to understand the degree of influence of one vulnerability over others so that the stakeholders/security managers can focus on the most influencing vulnerability for the removal process. Here, we propose the numerical illustration to validate the given framework.

Table 1 Software vulnerability types

Notations	Vulnerability type
Vt_1	SQL injection (SQLI)
Vt_2	Cross site scripting (XSS)
Vt_3	Buffer overflow (BO)
Vt_4	Cross site request forgery (CSRF)
Vt_5	File inclusion (FI)
Vt_6	Code execution (CE)
Vt_7	Information gain (IG)
Vt_8	Gain of privileges (GP)
Vt_9	Race condition (RC)

3.1 Dataset Description

The dataset utilized in this investigation was collected from the NVD and CVE data base (NVD [9]; CVE Details [8]). Dataset is composed of nine distinct vulnerability kinds. The vulnerabilities and their respective notations are discussed in Table 1.

3.2 Dataset Validation

In our previous study [27], we have already prioritized the software vulnerabilities according to their severity using Fuzzy Best–Worst method (FBWM). In the current study, we are moving a step ahead to determine the degree of influence one vulnerability has over another using Interpretive Structure Modeling (ISM) approach. The Structural Self-Interaction Matrix (SSIM) is initially established on the basis of a common software vulnerability connection. The SSIM matrix is being discussed with stakeholders and professionals. Based on their responses, vulnerability Vt_1 (i.e., SQL injection) leads to the vulnerability Vt_2, Vt_3 and Vt_7 and are denoted by the symbol 'V'. Vulnerability type Vt_6 will influence vulnerability Vt_1 therefore symbol 'A' is allocated. Further, vulnerability Vt_1 and vulnerability Vt_5 do not seem related thus denoted by symbol 'O' and so on as given in Table 2.

As described in step 2 of Sect. 2, the next step is to transform the SSIM matrix into an initial accessibility matrix by substitution of V, A, X, O with the binary numbers (0's and 1's). Table 3 presents the initial accessibility matrix for software vulnerability categories.

The final accessibility is obtained from the initial reachable matrix by adding transitivity in the matrix by considering the rules of transitivity as stated previously in Step 2. Table 4 given below denotes the final reachability matrix where 1* is to incorporate transitivity. The accessible matrix additionally has dependency

Table 2 SSIM matrix

	Vt_1	Vt_2	Vt_3	Vt_4	Vt_5	Vt_6	Vt_7	Vt_8	Vt_9
Vt_1	1	*V*	*V*	*X*	*O*	*A*	*V*	*A*	*A*
Vt_2		1	*O*	*A*	*O*	*V*	*O*	*A*	*O*
Vt_3			1	*O*	*X*	*A*	*O*	*O*	*A*
Vt_4				1	*O*	*X*	*V*	*A*	*O*
Vt_5					1	*O*	*A*	*A*	*O*
Vt_6						1	*O*	*O*	*A*
Vt_7							1	*A*	*O*
Vt_8								1	*O*
Vt_9									1

Table 3 Initial reachability matrix

	Vt_1	Vt_2	Vt_3	Vt_4	Vt_5	Vt_6	Vt_7	Vt_8	Vt_9
Vt_1	1	1	1	1	0	0	1	0	0
Vt_2	0	1	0	0	0	1	0	0	0
Vt_3	0	0	1	0	1	0	0	0	0
Vt_4	1	1	0	1	0	1	1	0	0
Vt_5	0	0	1	0	1	0	0	0	0
Vt_6	1	0	1	1	0	1	0	0	0
Vt_7	0	0	0	0	1	0	1	0	0
Vt_8	1	1	0	1	1	1	1	1	0
Vt_9	1	0	1	0	0	1	0	0	1

power (column) and driving power of each vulnerability (row). The determination of dependency and driving obstacles is based on the final reachability matrix. The final accessibility matrix is also utilized for partitioning the obstacles in the construction of hierarchical model of the ISM.

In the next step, we construct level partitioning as given in Table 5. Reachability and antecedent sets are now calculated for each vulnerability category using the final reachability matrix as described in step 3. The intersections of both of these sets are determined for each vulnerability type. In Table 5, Vt_2 and Vt_3 are at the level 1. Once level I is reached, iteration 1 will be finished and vulnerabilities associated with level I will be eliminated from other vulnerabilities. Similarly, iterations are continued for determining the levels of each vulnerability.

From Table 5, it is observed that vulnerabilities Vt_5 and Vt_7 are placed at second level, Vt_1 and Vt_6 are placed at level third, vulnerabilities Vt_4 is at fourth level and Vt_8 and Vt_9 are placed at level five.

Table 4 Final reachability matrix

	Vt_1	Vt_2	Vt_3	Vt_4	Vt_5	Vt_6	Vt_7	Vt_8	Vt_9	DrP
Vt_1	1	1	1	1	1*	1*	1*	0	0	7
Vt_2	0	1	1*	1*	0	1	0	0	0	4
Vt_3	0	0	1	0	1	0	0	0	0	2
Vt_4	1	1	1*	1	1*	1	1	0	0	7
Vt_5	0	1	1	0	1	0	1	0	0	4
Vt_6	1	0	1	1	1*	1	1*	0	0	6
Vt_7	0	1*	0	0	1	1*	1	0	0	4
Vt_8	1	1	1*	1	1	1	1	1	0	8
Vt_9	1	1*	1	1*	1*	0	0	0	1	6
DeP	5	7	8	6	8	6	6	1	1	

Table 5 Levels of vulnerability iteration

	Reachability set	Antecedent set	Intersection set	Level
Vt_1	1, 2, 3, 4, 5, 6, 7	1, 4, 6, 8, 9	1, 4, 6	III
Vt_2	2, 3, 4	1, 2, 3, 4, 5, 7, 8, 9	2, 3, 4	I
Vt_3	3, 5	1, 2, 3, 4, 5, 6, 8, 9	3, 5	I
Vt_4	1, 2, 3, 4, 5, 6, 7	1, 2, 4, 6, 8, 9	1, 2, 4	IV
Vt_5	2, 3, 5, 7	1, 3, 4, 5, 6, 7, 8, 9	3, 5, 7	II
Vt_6	1, 3, 4, 5, 6, 7	1, 2, 4, 6, 7, 8	1, 4, 6, 7	III
Vt_7	2, 5, 6, 7	1, 4, 5, 6, 7, 8	5, 6, 7	II
Vt_8	1, 2, 3, 4, 5, 6, 7, 8	8	8	V
Vt_9	1, 2, 3, 4, 5, 9	9	9	V

3.3 Building Model

From the level portioning as mentioned in Table 5, the hierarchical structure model is generated and is given in Fig. 1.

The Gain of privileges (Vt_8) and Race condition (Vt_9) are the most independent vulnerabilities. Vt_8 will further lead vulnerability SQLI (Vt1), CSRF (Vt_4) and RC (Vt_9) will lead to CSRF (Vt_4). On the other hand, those on top like XSS (Vt_2) and BO (Vt_3) are the most dependent vulnerabilities and hold the least importance as compared to other vulnerabilities.

With the help of Table 5, MICMAC analysis is formed and depicted in Fig. 2. All the nine-software vulnerabilities are positioned in 4 quarters; linkage, dependent, autonomous and independent. In quadrant I, two vulnerabilities Vt_8 and Vt_9 are classified as independent vulnerabilities. Three vulnerabilities are placed in the linkage quadrant. In quarter III, there is presence of four vulnerabilities. There is no autonomous vulnerability, i.e., none of the vulnerabilities have weak driving and dependence power.

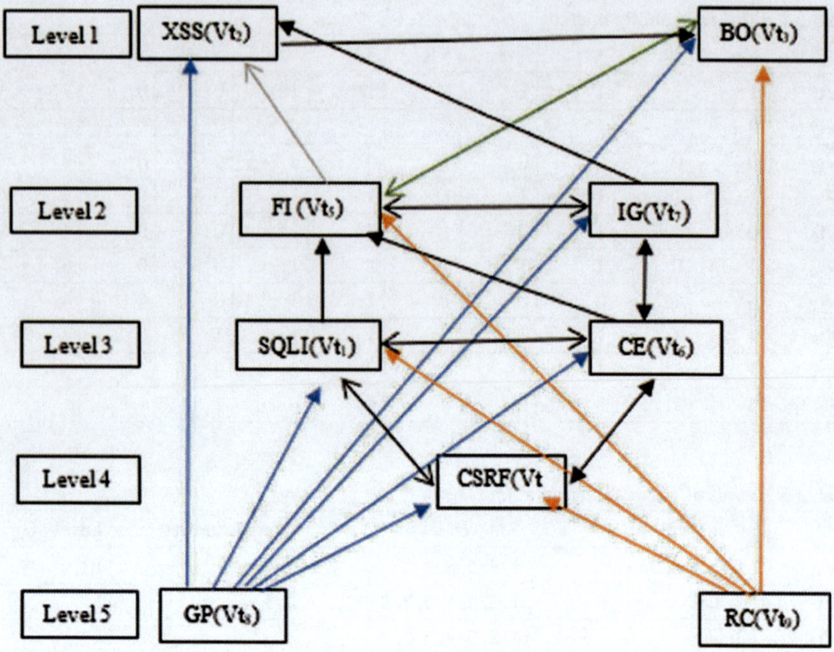

Fig. 1 ISM model

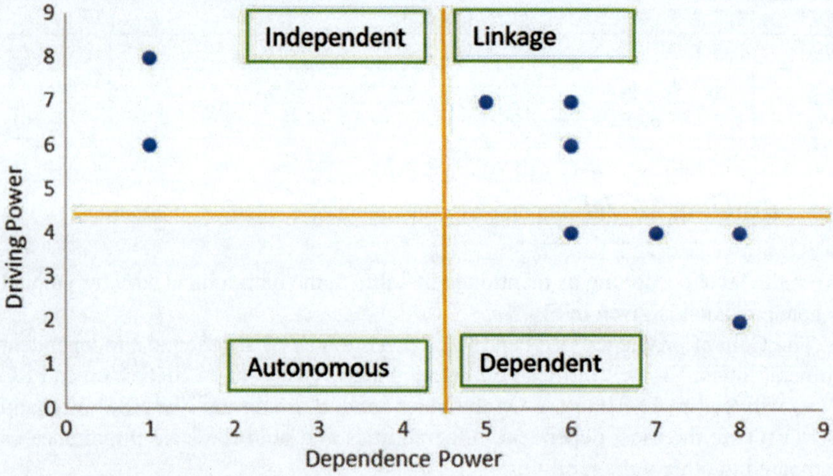

Fig. 2 Representation of MICMAC analysis

4 Conclusion

The rise of software security defects causes serious vulnerability problems. This urges the development and software to install vulnerability detection tools throughout software development. The biggest security team issue is that software developers and testers lack time and resources. In this context, vulnerability prioritization is a key tool that can aid the security team to remove the highly severe vulnerabilities first, but to check if the vulnerability is being linked to another vulnerabilities needs to be taken into the consideration. The present study focuses on establishing a quantitative assessment methodology to quantify the vulnerability correlation across software categories which can lead to an improved vulnerability elimination procedure utilizing the ISM technique. In the study, we have selected 9 major vulnerabilities already discussed in our previous study [25]. The vulnerabilities after prioritization are further evaluated in ISM methodology to understand their inter-relationships. Gain of privileges (Vt_8) and Race condition (Vt_9) are the most independent vulnerabilities. Vt_8 will further lead vulnerability SQLI (Vt_1), CSRF (Vt_4) and RC (Vt_9) will lead to CSRF (Vt_4). On the other hand, those on top like XSS (Vt_2) and BO (Vt_3) are the most dependent vulnerabilities and hold the least importance as compared to other vulnerabilities. This prioritization can help the decision makers to focus on the top vulnerabilities which will indirectly reduce the remaining vulnerabilities.

References

1. Cavusoglu H, Cavusoglu H, Raghunathan S (2004) Economics of IT security management: four improvements to current security practices. Commun Assoc Inf Syst 14:3
2. Scott C (2000) Definition of a security vulnerability. Microsoft TechNet 2000
3. Cavusoglu H, Cavusoglu H, Raghunathan S (2007) Efficiency of vulnerability disclosure mechanisms to disseminate vulnerability knowledge. IEEE Trans Software Eng 33:171–185
4. Kansal Y, Kapur PK, Kumar U, Kumar D (2017) User-dependent vulnerability discovery model and its interdisciplinary nature, life cycle reliability and safety. Engineering 6:23–29
5. Telang R, Wattal S (2005) Impact of software vulnerability announcements on the market value of software vendors-an empirical investigation. Available at SSRN 677427
6. Temizkan O, Kumar RL, Park S, Subramaniam C (2012) Patch release behaviors of software vendors in response to vulnerabilities: an empirical analysis. J Manag Inf Syst 28:305–338
7. Kapur PK, Pham H, Gupta A, Jha PC (2011) Software reliability assessment with OR applications, Springer 364
8. Ozkan S (2019) CVE details: the ultimate security vulnerability data source. Accessed Feb 14 2019
9. NVD: National vulnerability database (2018) Available http://nvd.nist.gov, Accessed Nov 16 2018
10. Huang CC, Lin FY, Lin FYS, Sun YS (2013) A novel approach to evaluate software vulnerability prioritization. J Syst Softw 86:2822–2840
11. Sharma R, Sibal R, Sabharwal S (2021) Software vulnerability prioritization using vulnerability description. Int J Syst Assur Eng Manage 12:58–64
12. Zeng Z, Yang Z, Huang D, Chung CJ (2021) LICALITY–likelihood and criticality: ulnerability risk prioritization through logical reasoning and deep learning. IEEE Transac Netw Serv Manage

13. Ghani H, Luna J, Suri N (2013) Quantitative assessment of software vulnerabilities based on economic-driven security metrics. In: international conference on risks and security of internet and systems (CRISIS) IEEE, pp 1–8
14. Kapur PK, Garg RB (1992) A software reliability growth model for an error-removal phenomenon. Softw Eng J 7:291–294
15. Anjum M, Kapur PK, Agarwal V, Khatri SK (2020) Evaluation and selection of software vulnerabilities. Int J Reliab Qual Saf Eng 27:2040014
16. Sibal R, Sharma R, Sabharwal S (2017) Prioritizing software vulnerability types using multi-criteria decision-making techniques. Life Cycle Reliab Safety Eng 6:57–67
17. Microsoft C (2022) Microsoft security response center security bulletin severity rating system. https://technet.microsoft.com/zhcn/security/gg309177.aspx
18. FIRST, Common vulnerability scoring system (CVSS) version 2.0. https://www.first.org/cvss/v2/guide#i1.2
19. Liu Q, Zhang Y, Kong Y, Wu Q (2012) Improving VRSS-based vulnerability prioritization using analytic hierarchy process. J Syst Softw 85:1699–1708
20. Scarfone K, Mell P (2009) An analysis of CVSS version 2 vulnerability scoring. In: 3rd international symposium on empirical software engineering and measurement IEEE, pp 516–525
21. Spanos G, Angelis L (2015) Impact metrics of security vulnerabilities: analysis and weighing. Inf Security J Global Perspect 24:57–67
22. Wang Y, Yang Y (2012) PVL: a novel metric for single vulnerability rating and its application in IMS. J Comput Inf Syst 8:579–590
23. El-Mokadem AM, Warfield JN, Pollick DM, Kawamura K (1975) Modularization of large econometric models: an application of structural modeling. In: Decision and control including the 13th symposium on adaptive processes IEEE, pp 683–692
24. Thakkar J, Kanda A, Deshmukh SG (2008) Interpretive structural modeling (ISM) of IT-enablers for Indian manufacturing SMEs. Inf Manage Comput Secur
25. Solanki R, Jha PC, Darbari JD, Agarwal V An interpretive structural model for analyzing the impact of sustainability driven supply chain strategies
26. Thakkar J, Deshmukh SG, Gupta AD, Shankar R, Selection of third-party logistics (3PL): a hybrid approach using interpretive structural modeling (ISM) and analytic network process (ANP). In: Supply chain forum: an international journal 6:32–46
27. Anjum M, Kapur PK, Agarwal V, Khatri SK (2020) A framework for prioritizing software vulnerabilities using fuzzy best-worst method. In: International conference on reliability, infocom technologies and optimization (trends and future directions) (ICRITO) IEEE, pp 311–316

A Decision-Making Model for Predicting the Severity of Road Traffic Accidents Based on Ensemble Learning

Salahadin Seid Yassin and Pooja

Abstract A road traffic accident is one of the most common and heinous tragedies that can occur anywhere in the world. Better safety and management of the roadways can only be achieved by investigating the causes of the occurrences. The materials under consideration have been compiled with the objective of addressing a number of themes associated with classifications of road traffic accidents. Nevertheless, the investigators' model and information are not adequate in terms of efficiency or incidence to lessen the catastrophic loss. So this study investigates the possibility of using an ensemble approach to increase accuracy in predicting accident intensity and identifying critical elements. Our work utilizes voting ensemble learning techniques, as well as other underlying base models (Decision Trees, K-Nearest Neighbors, and Naive Bayes) for predicting traffic incidents. Different machine learning metrics were used to compare these models. Comparatively, the Ensemble method surpasses competing base classifiers by 89% accuracy, 89% precision, 89% recall, and 89% F1-scores. Furthermore, the provided informative model excels others on the ROC curve metric, demonstrating that for traffic safety administrators and authorized players, it's a reliable and dependable method to make rational decisions.

Keywords Road traffic accident · Ensemble learning · K-nearest neighbor · Decision tree · Naive Bayes · Voting classifier

1 Introduction

Road traffic crashes are a leading source of injuries, deaths, permanent disability, and property damage in the world. However, this is affecting the healthcare system as hospitals are overwhelmed with patients. Every year, tens of thousands of people around the world are killed or injured in road traffic crashes.

The World Health Organization has recently released some shocking statistics on accident severity and damages that will shock everyone.

S. S. Yassin · Pooja (✉)
Department of Computer Science and Engineering, Sharda University, Greater Noida, India
e-mail: pooja.1@sharda.ac.in

P. Chatterjee et al. (eds.), *Computational Intelligence for Engineering and Management Applications*, Lecture Notes in Electrical Engineering 984,
https://doi.org/10.1007/978-981-19-8493-8_57

Over 1.3 million people have their lives ended in car accidents every year. Most countries lose 3% of their GDP as a result of traffic accidents. Road traffic fatalities affect vulnerable road users such as pedestrians, cyclists, and motorcycle riders. Despite having few percent of the nation's vehicles, low- and middle-income countries are responsible for 93% of road deaths. Road traffic accidents are the greatest risk factor for mortality for teenagers aged 5 to 29 [1]. Especially in under-developing countries, the situation is very terrible. According to the Federal Police Commission, around 1840 Ethiopians were killed in traffic accidents in the first six months of 2020. In addition, 2646 people were critically hurt, with another 2565 suffering from moderate injuries [2]. Unless and until efforts to ameliorate the situation are implemented, the situation will worsen.

In recent decades, scientists have begun to investigate various aspects of the severity of road traffic accidents using real data. In brief studies of road traffic accident prediction and classification, two main model-building methodologies have been identified. The first set of methods consists of model parameters, which are generally simple and straightforward and rely on adjusting input parameters to produce appropriate results. Another category is taking advantage of machine learning (ML) methods to develop appropriate crash prediction models. A recent third method of crash safety involves the use of ensemble learning approaches [3].

The main goal of developing ensemble models is to evaluate the possibility of obtaining more accurate predictions than the individual models used in the ensemble method. In other words, efforts are ongoing to find highly accurate methods for predicting traffic accidents. In this case, ensemble learning will provide an answer by optimizing the model's performance. Road safety teams, drivers, and road users can benefit from pre-operational decision-making with accurate severity prediction models and save lives as well as countries' infrastructure and costs.

Ensemble models have been used to improve the performance of predictive models in a variety of scientific disciplines for a long time. However, ensemble learning is not extensively applied for improved results in accident prediction. In addition, some questions will lead us to use the third astonishing approach to deliver a successful classification.

Regarding the prediction of road traffic accidents, the research needs to address the following questions: Is it possible to employ machine learning models for anticipating traffic accidents? Is there any relevant work area that uses ensemble learners? Road traffic accident prediction area has been begun using machine learning algorithms to make predictions as a response to the first inquiry. The obvious response to the last inquiry is the driving motivation for this research. It's rare to find research related to ensemble tactics to improve the accuracy of ML algorithms used in road traffic accident prediction. Its models can generate more accurate classification models, so the study intends to find an option to create a better accurate classification model using multiple ensemble approaches. Do our machine learning models have the capability to save lives as well? Further, this has also helped the authors of the study in developing ML models for identifying and evaluating traffic accidents. In addition to other analytic and prediction tools, machine learning algorithms play a key role in making reasonable decisions that could help to prevent or reduce unnecessary road accidents. It is also worth mentioning that Ensemble learning models have the ability to deliver an accurate and meaningful outcome.

The current research aims to develop an ensemble-based model to predict the severity of road accidents and compare it to a single classifier. It has been established that an ensemble model can be used as a tool to help to foresee the severity of accidents that result in sudden death and catastrophic injury. As a result, a new approach to aggregating the results of individual models that have been built is presented. In addition, this study helps in developing a more accurate model and identifying factors that influence the severity of road traffic accidents. This could be effective in reducing the frequency and intensity of accidents in the coming decades, thereby saving a lot of money and lives, among other things.

2 Literature Review

Machine learning has lately gained popularity among accident researchers as a way for constructing a predictive road safety model. One of the most significant advantages of machine learning is the capacity to investigate many sources and use complex numerical approaches. This necessitates specialized data pre-processing, fine-tuning, and a thorough understanding of each machine learning technique.

Numerous machine learning methods have been used to determine the extent of traffic accidents. A noteworthy characteristic of this method is the ease with which it can be adapted to the processing of anomalies, messy information, or incomplete data, and its ability to be highly adaptive with little or no prior knowledge of independent factors. These methods are said to make more sense than statistical models when it comes to the problem at hand.

A wide range of machine learning algorithms, including random forest (RF) [4], Artificial Neural Network (ANN) [5], Support Vector Machines (SVM) [6, 7], Decision Tree (DT) [8], logistic regression (LR) [9] were extensively used to make out crucial features of traffic accidents and predict the degree of severity using road traffic accident data sets. Those strong nonlinear techniques were used in road safety research to construct prediction models, bringing in efficient and optimal decision-making. Sirikul et al. used different machine learning models (GBC, RF, KNN, and LR) to improve the effectiveness of driver impairment. The study used driver-related data from the Thai Gov. Road Safety authority [10]. The study showed that the ML system, when combined with pre-crash predictors and BAC, can accurately forecast the drunk driver's road traffic death risk. Their developed models performed well in terms of classification. The study result revealed that GBC performed very well compared to other models using the AUC metric, while logistic regression is the least accurate in mortality risk prediction. Mokoatle et al. used LR and XGBoost. Their research focused on using accident report data from South Africa to predict the severity of road traffic accidents [11]. The results showed that XGBoost outperformed LR in terms of classification criteria. Zhang et al. also used RF and DT comparison to investigate traffic accident prediction at Major road Crossing Points.

A k-fold cross-validation technique was used to test the ML algorithms. The findings demonstrate that RF performed better than DT according to eight classification metrics [12]. Ali et al. developed a probit DT-based method to explore the influence of poor climate on the extent of injuries in signalized intersections accidents. The empirical outcomes were compared with the standard statistical method. It was concluded that the presented techniques excelled the probit approach through the context of estimating correctness, resilience, and reliability [13]. In addition, Hybrid ML Model was established for Predicting Real-Time Secondary Accident Risk Using XGBoost Models with reasonable efficacy [14]. Another study reiterates that to predict crash seriousness on municipal roadways, artificial neural networks and multiple logistic regression were used. When it comes to determining the severity of a road crash, the artificial neural network model outperforms the multiple logistic regression techniques in terms of precision and least faults [15]. To anticipate and examine extremely heavy vehicles, Lin Yi-Hsin used a machine learning paradigm (BPNN, GRNN, WNN). When tested on a large amount of data, GRNN proved to be expensive in terms of execution time. WNN, on the other hand, did well in a short period of time. Shanshan et al. used data mining techniques to analyze risk indicators for Hazmat Road transport [16].

As part of this research, we developed stacking of NB, DT, and KNN to model traffic accidents severity. This study aims to compare the prediction accuracies of base classifiers to those of related stacking ensembles. The results were compared between the base models and their respective ensembles using Addis Ababa data sets. In order to evaluate the model, it will be evaluated in terms of accuracy, precision, recall, F1-score as well as ROC value once the base models have been generated. The assessment of these metrics will demonstrate that ensemble models hold potential as a tool for predicting road traffic accidents.

Hence, it is essential to apply a thorough investigation in order to discover if there is a connection between the influencing factors that affect the number of road deaths and the number of traffic accidents. This investigation was undertaken to acquire a deeper understanding and more accurate results. Due to this, this research is devoted to analyzing the trends of incidents in Addis Ababa, Ethiopia, using ML methods. Although all measures have been taken to prevent car crashes, fatalities, and injuries, the intensity of crashes on Addis Ababa roads remains a major concern. It, therefore, necessitates considerable concern, analysis of both the nature and reality of the situation, as well as the development and implementation of specific and appropriate actions and solutions to lessen the severity of the occurrence.

We will now take a look at how the article is organized: Sect. 3 introduces the data set and validation method for filtering raw data and provides theoretical material on machine learning algorithms such as KNN, NB, DT, and ensemble approaches such as summation. Section 4 contains a discussion of the results and how to build machine learning models, how to test the model's experimental performance, and how to interpret the data. Finally, Sect. 5 concludes the study with new potential directions.

3 Methods and Materials

As part of this section, the RTA data set and ML classifiers, as well as the proposed methodology, are outlined to help to predict the severity of road accidents.

3.1 Data Set Description and Data Pre-treatment

The data sets utilized to evaluate the effectiveness of the presented approach in this research come from Addis Ababa. The features used in the current plan are detailed further below. The relevant factors have been as described in the following: day, driver-age, sex, driver experience, type of vehicle, service year, location, road condition, light condition, weather condition, causality-class, causality-age, causality-sex, and severity (target variable).

The sample's target variable is a two-class worry that can be described as fatal or severe injuries. The data set comprised information from 5000 road traffic accidents, with a total of 14 variables. In order to use samples for the indicated method, data set pre-treatment is required. To achieve a better result, many data pre-processing techniques for information extraction and cleansing are used.

Data must be pre-processed before being fed into the ML models. For instance, data cleansing is essential, ensuring that missing values are filled in, normalization, transformation, and separating training data from test data. The use of these processes can help to significantly improve model performance and can greatly assist in the selection of the most relevant data.

Data sets we use are derived from real-life sources. Data sets from real-life situations are not always well-organized or understandable enough for algorithms to interpret them accurately; particularly, when it comes to "handwritten" data sets, which are extremely vulnerable to damage. It is therefore imperative that the information is in a machine-readable format in order to be able to acquire relevant information and construct an efficient and intelligent system. Our ability to know how severe of a problem we are going to have will be determined by how accurate our data is. During this phase, raw data is transformed or mapped into a format that is able to be easily used by a variety of analytical techniques. In the context of computing, data pre-processing is the act of preparing primary data for a later stage of processing.

Learn more about the data preparation process from our previous work [17].

3.2 Ensemble Learning

It is a generic method for meta-machine learning that aims to improve generalization abilities by aggregating predictions from a number of different models. It is the fundamental concept behind ensemble learning that several weak learners are brought together to generate one powerful learner. Statistical techniques most commonly suffer from variance, noise, and bias. This can be reduced using ensemble techniques, thus increasing the stability of the model as a whole.

This study utilized a Voting Ensemble Classifier to improve road traffic accident severity the prediction performance of models. A Voting Classifier is an ML model that learns over a group of classifiers, then forecasts an outcome (target class) result with the highest likelihood of being the target class. It essentially uses the results of any model that has been submitted to Voting Learner and predicts the final class based on the most votes. This classifier supports two kinds of voting: hard voting and soft voting. The anticipated result class in hard voting is the one with the most votes, in other words, the one with the highest chance of becoming suggested by each of the learners. In contrast, in soft voting, the final prediction class is the weighted mean of the likelihood assigned to each base class.

A variety of metrics, such as classification accuracy and the ROC curve, are used to evaluate classification models.

3.3 Decision Tree

Decision tree algorithms are a well-known machine learning technique that has been applied to a wide range of tasks, especially with respect to classification and regression. A recursive partitioning algorithm using a splitting criterion was used by Breiman et al. [18] to build nodes in a decision tree. In a decision tree building process, the best feature of a data set is placed at the root, and the sample data set is then partitioned into subsets based on its significance. A data set's feature determines its split. Until every leaf node is found at every branch, this method is repeated until all the data have been categorized. Using information gain will help to determine which feature is the most relevant among a group of features. There have been several models developed to assess classification and the value of a target variable in the context of training models that used decision trees.

3.4 Naïve Bayes

The Bayesian method is a supervised machine learning approach for building classifiers. It is based on the Bayesian principle and can be applied to even the most complex classification algorithms. This type of program estimates the probability of an element belonging to a category or group of elements with respect to certain attributes. In a nutshell, it is a probabilistic predictor based on measures of likelihood. It is a method that does not depend on the time that one feature occurs in order for another feature to occur at the same time. This process is very straightforward, and it only requires a small quantity of input samples for categorization, and that all terms can be pre-calculated, which result in a much simpler, faster, and more effective classification.

3.5 K-Nearest Neighbor

It is a technique that is used for both classification and regression. In fact, it is one of the most basic techniques used in supervised machine learning. Each time it receives new data, it stores the instances and compares the distance to the closest k with whom it finds the greatest similarity. Predictions are generated directly from the training sample with K-Nearest Neighbor.

3.6 Performance Evaluation Metrics

In this paper, every learning comparison is made using quantitative measures on numerous parameters such as accuracy, precision, recall, F1-score, and ROC curve. It is a widely used metric for making a distinction between what a model predicts and what actually occurs, as demonstrated in Eqs. 1–4. 75% of the data are used to train a model, meanwhile, 25% are used to test it.

$$\text{Accuracy} = \frac{\text{TN} + \text{TP}}{\text{TP} + \text{TP} + \text{FP} + \text{FN}} \tag{1}$$

$$\text{Recall} = \frac{\text{TP}}{\text{TP} + \text{FN}} \tag{2}$$

$$\text{Precision} = \frac{\text{TP}}{\text{TP} + \text{FP}} \tag{3}$$

$$\text{F1} - \text{score} = 2 * \frac{\text{Precision} * \text{Recall}}{\text{Precision} + \text{Recall}} \tag{4}$$

ROC Curve: ROC curves are also recognized as the AUC-ROC score which uses True positive ratios ($\text{TPR} = \frac{\text{TP}}{\text{TP}} + \text{FN}$) but also false positive ratios ($\text{FPR} = \text{FP}/(\text{FP} + \text{TN})$. TPR/recall is the percentage of positive sample points which are accurately classified as positive when compared to all positive sample points. To put it another way, the greater the TPR, the fewer positive data points that were overlooked or missed. In comparison to all negative sample points, FPR is the percentage of negative sample points which are incorrectly judged positive. To put it another way, the greater the FPR, the much more negative sample data that we misclassify. The ROC value presents the results, and the measurement that uses is the area under the curve, what usually referred to as AUROC.

A "TP" indicates the number of actual positive outcomes, a "TN" indicates the number of true negative outcomes, "FP" indicates the number of false positive outcomes, and "FN" indicates the number of false negative outcomes within the confusion matrix. The graph depicts the relationship between true positive (i.e., sensitivity) and false positive (i.e., 1-specificity). To consider the various learners, the ROC is calculated, and if a classifier's ROC is bigger than all the others, then recognition accuracy will be superior.

4 Result and Discussion

4.1 Experimental System Set up

During the course of this article, we present to you a data set of a real-life traffic accident to evaluate the competence of the classifier. The experiments have been done utilizing a Python script (version 3.7.6). This software works well for computational tasks and provides a large set of tools and libraries. As the first stage of the proposed method, we collect and pre-process data, then plan and execute the experiment utilizing DT, NB, and KNN based learner implementations, as well as ensemble voting models with Python 3.7 on a Jupyter notebook with an Intel Core i7 1.80GHz processor speed, 8GB RAM, and a 1TByte HD system. For each base learner and ensemble model, the default parameters are applied as they have the greatest impact on overall performance.

4.2 Results of the Overall Classification Prediction

In this part, the performance, and accuracy of the model, as well as a comparison of the suggested technique to standard models, are briefly described and evaluated.

Various models are combined to create an ensemble learning model that boosts the performance of supervised learning. The results of this strategy differ from a solo model, in that it produces better prediction performance when compared to it. A key principle underpinning the concept is to learn a set of algorithms and then to allow participants to cast their votes. Three models were developed in this study and they were then combined to achieve superior results. In many instances, ensemble approaches outperform single models in terms of accuracy and other performance metrics. We begin by building separate models using the training data. A voting classifier wraps the models to generate predictions.

In the ensemble model, the K-Nearest Neighbor, Decision Tree, and Naive Bayes models are combined to create a unified model. A voting ensemble model of KNN, DT, and NB is used for predicting road traffic accidents. The performance of each

Table 1 Performance evaluation of base model and ensemble approach

Base learner					
S. No.	Metrics	DT	NB	KNN	Ensemble
1	Accuracy	88	87	88	89
2	Precision	88	87	88	89
3	Recall	88	87	88	89
4	F1-score	88	87	88	89

base learner is compared against its related ensemble model in terms of accuracy, precision, recall, and F1-score. In this section, we provide empirical predictions for base models and related ensembles. As shown in Table 1, ensembles of KNN, DT, and NB significantly improve the prediction accuracy of a base classifier. Ensemble models outperform their underlying ML algorithms in terms of prediction accuracy, demonstrating a clear improvement over the results obtained with single models, demonstrating a qualitative improvement over the results obtained with single models.

4.3 Classification Metrics-Based Predictions

In order to evaluate prediction results, Ensemble Learning, Decision Trees, K-Nearest Neighbors, and Naive Bayes classification techniques have all been employed on the test set. Table 1 gives the results of the base learner and ensemble prediction learner for road traffic accidents in terms of accuracy, precision, recall, and F1-score. As a result of the aforementioned approaches, the majority of the test set samples could be properly categorized. The results of the study showed that virtually all the methods worked brilliantly when it came to predicting the severity of crashes. As far as accuracy is concerned, the Ensemble classifiers have the best performance (up to 89%), followed by Decision Tree and K-Nearest Neighbor (both at 88%), and Naive Bayes (87%). As a result of their total calculating prediction capabilities, Ensemble Learning was ranked as the most effective prediction model for predicting the magnitude of road traffic accidents.

4.4 ROC Curve-Based Prediction Results

In addition, the ROC curve was also plotted to measure the accuracy of each technique's prediction results. For each of the models we looked at, the expected accuracy was clearly an attractive phenomenon.

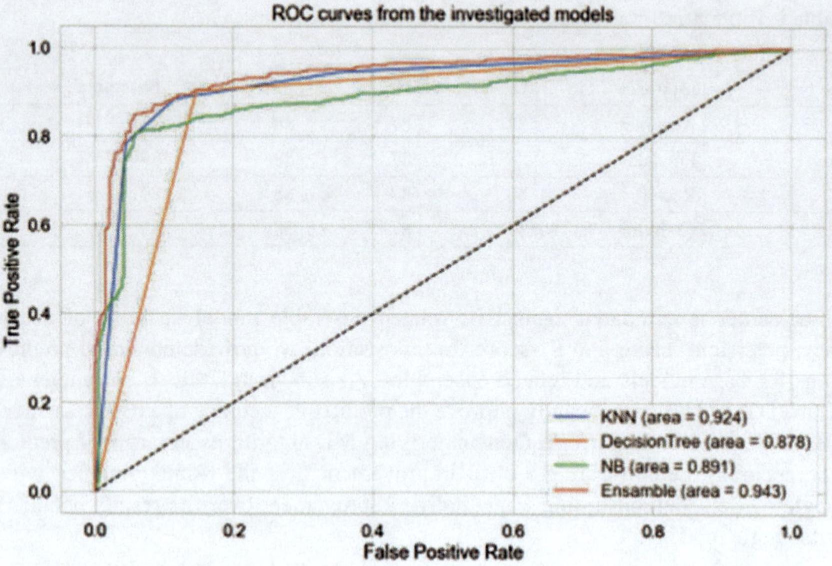

Fig. 1 Overall model ROC curve performance

There is no doubt that Ensemble Learner and K-Nearest Neighbor Classifier are superior to other algorithms in predicting road traffic accidents, as is illustrated by the Fig. 1, with ROC curve values of 94.3 percent and 92.4 percent, respectively. In terms of ROC curves, Naive Bayes produces an 89% ROC curve, which is quite impressive. Compared to comparable approaches, the Decision Tree Classifier's prediction accuracy is less effective when compared to those approaches. It has a value of 87.8% for the ROC curve. Even though it is the least effective, the Decision Tree also performs well. Its performance for severity prediction isn't as bad as it sounds.

5 Conclusion

There is a wide range of scientific disciplines that can benefit from ensemble learning approaches in order to improve the predictive performance of their base learners. This study investigated the ability of ensemble techniques to enhance the effectiveness of individual learners while they are engaged in road traffic accidents. Based on this, we design and evaluate base learners (KNN, DT, and NB) in addition to their associated voting ensemble models based on Ethiopian traffic accident data. We were pleased with the results, the ensemble model looked good and performed better than single learners. Thus, the findings in this study show how ensemble models of KNN, DT, and NB, which are frequently used predictive models in traffic accidents, can be used to enhance the results of prediction classification. Unfortunately, there is no single model that can be used to solve all types of traffic accident problems. Hence, finding different approaches is crucial to obtaining meaningful information and results. It is evident from the findings of this study that ensemble techniques can enhance the predictive accuracy of baseline learners like KNN, DT, and NB. It was hoped that additional combining methods and the development of ensembles from various single models, using this work in the future, would be added to make it even more useful.

References

1. World Health Organization (2018) Global status report on road safety 2018. URL https://www.who.int/publications-detail/global-status-report-onroad-safety-2018.
2. Africa. http://www.xinhuanet.com/english/africa/2021-03/03/c_139781169.htm
3. Xiao J (2019) SVM and KNN ensemble learning for traffic incident detection. Physica A 517:29–35
4. Yassin SS (2020) Road accident prediction and model interpretation using a hybrid K-means and random forest algorithm approach. SN Appl Sci 2(9):1–13
5. Alkheder S, Taamneh M, Taamneh S (2017) Severity prediction of traffic accident using an artificial neural network. J Forecast 36(1):100–108
6. Sharma B, Katiyar VK, Kumar K (2016) Traffic accident prediction model using support vector machines with Gaussian kernel. In: Proceedings of fifth international conference on soft computing for problem solving, pp 1–10. Springer, Singapore
7. Delen D, Tomak L, Topuz K, Eryarsoy E (2017) Investigating injury severity risk factors in automobile crashes with predictive analytics and sensitivity analysis methods. J Transp Health 4:118–131
8. Abellán J, López G, De OñA J (2013) Analysis of traffic accident severity using decision rules via decision trees. Expert Syst Appl 40(15):6047–6054
9. Chen H, Cao L, Logan DB (2012) Analysis of risk factors affecting the severity of intersection crashes by logistic regression. Traffic Inj Prev 13(3):300–307
10. Sirikul W, Buawangpong N, Sapbamrer R, Siviroj P (2021) Mortality-risk prediction model from road-traffic injury in drunk drivers: machine learning approach. Int J Environ Res Public Health 18(19):10540
11. Mokoatle M, Marivate DV, Bukohwo PME (2019) Predicting road traffic accident severity using accident report data in South Africa. In: Proceedings of the 20th annual international conference on digital government research, pp 11–17
12. Zhou X, Lu P, Zheng Z, Tolliver D, Keramati A (2020) Accident prediction accuracy assessment for highway-rail grade crossings using random forest algorithm compared with decision tree. Reliab Eng Syst Saf 200:106931
13. Ghasemzadeh A, Ahmed MM (2017) A probit-decision tree approach to analyze effects of adverse weather conditions on work zone crash severity using second strategic highway research program roadway information dataset (No. 17-06573)
14. Li P, Abdel-Aty M (2022) A hybrid machine learning model for predicting real-time secondary crash likelihood. Accid Anal Prev 165:106504
15. Sarkar A, Sarkar S (2020) Comparative assessment between statistical and soft computing methods for accident severity classification. J Inst Eng (India) Series A 101(1):27–40
16. Wei S, Shen X, Shao M, Sun L (2021) Applying data mining approaches for analyzing hazardous materials transportation accidents on different types of roads. Sustainability 13(22):12773
17. Seid S (2019) Road accident data analysis: data preprocessing for better model building. J Comput Theor Nanosci 16(9):4019–4027
18. Breiman L, Friedman JH, Olshen RA, Stone CJ (2017) Classification and regression trees. Routledge

Factor Analysis Approach to Study Mobile Applications' Characteristics and Consumers' Attitudes

Chand Prakash, Rita Yadav, Amit Dangi, and Amardeep Singh

Abstract This paper aims to identify various characteristics of an online retailer's mobile application and how the identified characteristics affect the consumers' attitude in an online shopping environment. A total of 350 respondents from National Capital Region (NCR), India, who were engaged in online shopping through any mobile application were surveyed with a semi-structured questionnaire. Exploratory Factor Analysis and Regression analysis were used to identify the online retailer's app's key characteristics and to check how the identified characteristics affect the consumers' attitude in an online shopping environment. It was found that respondents of the National Capital Region of India considered Aesthetics, Accessibility, Credibility, Detailed information, Connect with customers, and consistency to be essential characteristics of mobile applications. Among these characteristics, aesthetics was the most crucial factor for creating a positive attitude toward shopping through mobile applications, followed by connection with customers, consistency, and accessibility. Credibility and detailed information characteristics of mobile apps were found to negatively relate to consumers' attitudes. The results may help the mobile application developers and the companies selling on online platforms to understand and prioritize the mobile application characteristics to add upon the most favored characteristics in their applications to combat the tough competition in the online shopping environment.

C. Prakash · A. Singh
SGT University, Gurugram, India

C. Prakash (✉)
School of Management and Commerce, Manav Rachna University, Faridabad, India
e-mail: chandsaini03@gmail.com

R. Yadav
Department of Management, Gurugram University, Gurugram, India

A. Dangi
Amity School of Business, Amity University, Noida, India

A. Singh
Faculty of Commerce, Manav Rachna International Institute of Research and Studies (Deemed University), Faridabad, India

P. Chatterjee et al. (eds.), *Computational Intelligence for Engineering and Management Applications*, Lecture Notes in Electrical Engineering 984,
https://doi.org/10.1007/978-981-19-8493-8_58

Keywords Accessibility · Aesthetics · Credibility · Connect with customers · Consistency · Detailed information · Mobile applications

1 Introduction

Retailing has seen a shift from brick and mortar retail to online retail to mobile app-based retailing. Increased usage of smartphone and internet penetration has enabled customers to shop from the mobile app-based retailer with the advantage of anytime, anywhere [1]. Recently shift of online retailers from websites to mobile apps has enabled the customer to find easily, compare, order, review, access offers, and get loyalty points, which significantly affect the decision-making process [2]. According to Deloitte, influenced by digital mediums in shopping can be understood by the figures that 64 cents of every dollar spent in retail stores are the result of a digital platform [3], and mobile is the obvious catalyst. 76% of prospects first conduct a preliminary search on their smartphone visits an outlet in a day, and 28% of those result in a final purchase [4]. Companies whose apps remember the customers and their past purchase behavior are favored by 58% of smartphone users. 63% favor purchasing from companies whose mobile sites or apps offer them recommendations and the best match of products they have shown interest in earlier. 51% of smartphone users use a company mobile app to get rewards or points [5].

Ever-changing technology always increases the chances of ever-changing consumer experience over time [6, 7]. There have been many studies conducted on the effect of the website on consumers' attitudes [8–10], but since the majority of online retailers have now shifted to the mobile app platform, websites have been converted into apps with minor or significant changes. This study will use the earlier studies to fill the literature gap about the influence of mobile app characteristics on consumers' attitudes to shopping through the mobile app.

2 Review of Literature

Online retailers' mobile application effectiveness can be judged by examining attitudes toward online retailer's apps. Researches show that consumers relate positively to the brand, and purchase them if they find the app effective [11], increasing shopping chances if the app is appealing [1].

Some of the key characteristics that affect consumer's attitude toward an online retailer's app are discussed here:

2.1 *Aesthetics*

In a study on atmospherics, it is stated as a "conscious designing of space to create specific effects on buyers" [12]. Researchers found store aesthetics to play an essential role in buying in addition to the marketing activities [13]. Applying the concept of aesthetics to the context of online retailers' apps, App aesthetics can be understood as the "conscious designing of web environments to create positive effects among users to increase favourable consumer responses" [14, 15] considered the app aesthetics to be similar to the brick and mortar store aesthetics that impact shoppers' attitudes. Hence, it is strongly advised that online retailers build an atmosphere through their app that can favorably influence consumers' perception and improve the chances of buying [16]. Like traditional customers, e-buyers ought to favor encounters that create positive sentiments. Previous researches propose that fun-related components (vividness, aesthetically pleasing design elements, and engaging material) are positively related to attitude to a mobile application [17–19].

Mummalaneni [20] suggested that characteristics like the layout of the app, display fonts (large or small), display quality (good or bad), display colors, and color combinations collectively create aesthetics for the mobile application.

2.2 *Accessibility*

Accessibility means "uncluttered screens, clear organization, logical flow, and simplicity of navigation, in short, a design that encourages one's productive and effective use of the application." Accessibility should upgrade the capacity to handle products and purchase information by minimizing the cost of inquiry, allowing quicker search, increasing the probability of an effective search, and increasing a positive attitude toward the app. The positive relationship between accessibility and attitude toward an app has been reported by many researchers earlier [18, 21, 22]. Accessibility creates confidence in the users for the App [23], to go for online shopping through a mobile application [24], creation of need to buy the product online [25, 26], increased number of orders on an online platform [27], and satisfied customers in the end [28].

2.3 *Detailed Information*

Detailed information incorporates the amount, exactness, and piece of information about the products/items and services offered on an online platform. Since e-buyers cannot look at an item physically, they rely on information available on the platform to select, compare, and finalize items. Online information includes content, tables, diagrams, photographs, sound, and video. Better detailed information enables online customers to settle on better choices, feel more certain about their choices, increase satisfaction with the shopping experience, and enhance their attitude toward an app. A few researchers have identified a positive association between detailed information and the chances of shopping online [17, 18, 29]. Detailed information leads to a positive attitude toward online shopping [24], frequency of online shopping [30], amount spent on purchase [22, 31], and delight with online purchases [28].

2.4 Credibility

Credibility is a significant factor in financial transactions. Many online consumers live in constant anxiety that their personal information will be exploited, that unwanted cookies will be installed, and that they will be bombarded with spam from all corners of the internet. Seventy-one percent of web users in the United States have reservations about online retailers [32]. Chen and Dhillon [33] stated that the key factors that contribute to credibility include likability and trust and situational normalcy and structural reassurance. The qualities of likability and believability are particularly noticeable in sales and advertising writing. Apps achieve situational normalcy by adopting a "professional/expert look," as described above. Structural assurances cover the exchange of merchandise. Other areas covered by structural assurances are security and privacy rules, and third-party confirmations. These preparations and promises imply that the supplier is a trustworthy individual. If retailers neglect to provide them, customers will probably leave without buying. Zeithaml et al. [33] and Chen and Dhillon [34] suggest that credibility is essential for retail mobile applications. Donthu [17] reported that credibility is related to attitude toward an application. Credibility also appears to increase attitude toward online shopping [35], intention to shop online [24, 25], intent to purchase online [26]. Waheed et al. [36] revealed that perceived trust and risk and perceived ease of use and usefulness positively mediate the nexus between the impact of social media apps and consumer online purchase intention.

2.5 Connect with Customer

Connect with customer supplements accessibility [28]. Both elements support the shopping procedure (search, examination, choice, decision, and following). However, while accessibility includes design components that directly support the process, connecting with the customer has to do with unforeseen resources drawn upon only when ordinary shopping processes are not adequate [33]. Similarly, in-store customers search for assistance from salespeople or other customers when something blocks their shopping process. We propose that online connection with customers assumes a comparative part. It allows disrupted e-shoppers to continue shopping. This use of connecting with customers is similar to the definition of "recovery service" proposed by Zeithaml et al. [34]. It is uncertain if generous client support would improve the app's attitude or reduce the probability of dissatisfaction, disappointment, and unfavorable attitudes. That may rely on whether the support provided just meets or surpasses one's expectations.

2.6 Consistency

Consistency implies that everything on a site/mobile application is up-to-date, and it is more than updated information. Consistency implies accuracy, which is a part of the information dimension proposed by Yang et al. [37]. It also contributes to the restoration of normalcy. In other words, if the app appears to be up-to-date, it is assumed to be in proper working order, which is a prerequisite for credibility and trust [33]. Consistency entails more than just having up-to-date information. It includes news, special promotions, and announcements of upcoming events, as well as anything else that helps to keep the app's content and appearance fresh [17]. New page designs, new photographs, and new features would indicate a seller's sense of responsibility to keep their product up-to-date. Traditional retail establishments must constantly update their inventories as well as their shopping environments in order to remain competitive. Web content that confirms that the app has been updated should increase one's confidence in the app and reduce the likelihood of switching. Anything that casts doubt on the app's consistency should be considered a detriment to the vendor's perceived credibility as well as the shopper's perception of the app [17]. According to Fogg et al. [23], consistency increases app credibility; however, no previous research has specifically examined the effect of consistency on attitudes toward mobile applications.

3 Research Questions and Hypothesis Development

1. What are the key characteristics of a mobile application in an online shopping environment?
2. How do the mobile application characteristics affect the consumers' attitude in an online shopping environment?

Ho1: There is no significant impact of mobile application characteristics on consumers' attitudes in an online shopping environment.

4 Research Methodology

4.1 Research Design

The research design is a blueprint for what the researcher intends to do in his or her study. The exploratory and descriptive research designs used in this study are combined. Because it involves an empirical investigation of characteristics of mobile applications and their influence on consumers' attitudes in an online shopping environment, the study is classified as exploratory. It is also classified as descriptive because a large amount of literature has been reviewed in order to make the concept more explicit.

4.2 Sample Design

Time and financial resources are always hurdles in conducting a census. With these limitations, the sample has been drawn so that it may represent the entire population. Therefore, a sample of 350 respondents who were involved in online shopping through the mobile application was taken from the National Capital Region (NCR) of India.

4.3 Data Collection

The study used both primary and secondary data to achieve the research objectives. The questionnaire technique has been used for primary data collection, while books, journals, articles, newspapers, and websites have helped to gather secondary data.

4.4 Questionnaire Design

The questionnaire was divided into two sections by the researcher. Part A was used to collect demographic information, and Part B was used to solicit responses on 27 statements related to the mobile applications' characteristics in an online shopping environment, using a five-point Likert scale with scale anchors ranging from "1" to "5" indicating strong agreement. The researcher used one statement to measure consumer's attitude on a five-point Likert scale with scale anchors ranging from "1"—strongly disagree to "5"—strongly agree.

4.5 *Statistical Tools Used*

For analyzing the data, Exploratory Factor Analysis, Stepwise Regression has been used. Exploratory Factor Analysis was used to explore the mobile applications' characteristics in an online shopping environment; Stepwise regression resulted in the impact of the identified characteristics on consumers' attitudes in an online shopping environment.

5 Results and Discussion

Collected questionnaires were coded and entered into the SPSS 21 spreadsheet for analyzing the data.

5.1 *Factor Analysis Results*

Factor analysis is a tool for reducing data. By applying factor analysis on the recorded responses, the following table was prepared.

The KMO and Bartlett test was used to determine whether or not the sample size was adequate for factor analysis and whether or not factor analysis could be used on the dataset to reduce the amount of data. When factor analysis is used, the acceptable value of KMO should be greater than 0.5 to be considered acceptable [38]. In the current study, the value of KMO was 0.832, which (as shown in Table 1) was deemed meritorious for further investigation. After conducting a sphericity test (approx. chi-square = 5485.467, data set = 351, significance level = 0.000), it was discovered that the original correlation matrix was not unitary. As a result, a sphericity factor analysis has been performed for data reduction based on the KMO and Bartlett tests.

Table 2 showed the factor loading of each statement under each factor. Factor loading of more than 0.5 has been considered for analysis.

Table 2 depicts that S6-S12 came under the first factor (Detailed Information). Factor 2 (Credibility) included S13-S17 variables. Factor 3 (Accessibility) included S18–S21 variables. Factor 4 (Aesthetics) covered the variables S1–S5.

Table 1 KMO and Bartlett's test

Kaiser-Meyer-Olkin measure of sampling adequacy		0.832
Bartlett's test of sphericity	Approx. chi-square	5485.467
	Df	351
	Sig.	0.000

Source Primary data

Table 2 Rotated component matrix

	Component					
	1	2	3	4	5	6
S1				0.620		
S2				0.757		
S3				0.773		
S4				0.850		
S5				0.817		
S6	0.756					
S7	0.792					
S8	0.797					
S9	0.715					
S10	0.778					
S11	0.584					
S12	0.704					
S13		0.762				
S14		0.798				
S15		0.865				
S16		0.806				
S17		0.754				
S18			0.824			
S19			0.823			
S20			0.826			
S21			0.786			
S22						0.856
S23						0.770
S24						0.688
S25					0.633	
S26					0.763	
S27					0.734	

Extraction Method: Principal Component Analysis
Rotation Method: Varimax with Kaiser Normalization
Rotation converged in 6 iterations
Source Primary data

Factor 5(Connect with the customer) comprised S25–S27, variables S22–S24 were included in Factor 6 (Consistency).

Based on the rotated component matrix, a new table (Table 3) has been generated that shows the nomenclature of factors along with their factor loadings, construct reliability, mean value of factors, and variance explained.

Table 3 Features of mobile applications

Constructs	Items	Factor loadings	Cumulative % variance	Mean	Cronbach alpha (α)
F1	Detailed information		15.436	$5.126/7 = 0.732$	0.884
INF1	The app is enabled with visuals of the products or services	0.756			
INF2	The app is updated with useful data on its products/services	0.792			
INF3	The app clearly describes product features	0.797			
INF4	The app mentions the sizes and weights	0.715			
INF5	The app discloses all the terms and conditions of services	0.778			
INF6	The app is enabled with 3D video of the product	0.584			
INF7	The number of images of the product helps to get all the information about the product	0.704			
F2	Credibility		28.653	$3.985/5 = 0.797$	0.833
CR1	I can trust the app with my credit card	0.762			
CR2	I can trust the app in the context of my details	0.798			
CR3	The app ensures the protection of my personal information	0.865			
CR4	I feel free from undesirable cookies while using the app	0.806			

(continued)

Table 3 (continued)

Constructs	Items	Factor loadings	Cumulative % variance	Mean	Cronbach alpha (α)
CR5	Detailed information of the app on the playstore makes it credible	0.754			
F3	Accessibility		41.564	3.259/4 = 0.814	0.886
ACC1	The app is easy to navigate	0.824			
ACC2	The app is designed in a smart manner	0.823			
ACC3	The app assists me in finding information	0.826			
ACC4	The app is free from clutter	0.786			
F4	Aesthetics		53.841	3.817/5 = 0.763	0.882
A1	The app is fun to usage	0.620			
A2	The app is equipped with soothing music	0.757			
A3	The app's design is entertaining	0.773			
A4	The app makes fair use of the video	0.850			
A5	The app has an attractive color combination	0.817			
F5	Connect with customers		61.663	2.13/3 = 0.71	0.759
C SUP 1	The app provides online technical support	0.633			
C SUP 2	The app is updated with customer reviews	0.763			
C SUP 3	The app's format allows for online dialogue	0.734			
F6	Consistency		69.355	2.314/3 = 0.771	0.767

(continued)

Table 3 (continued)

Constructs	Items	Factor loadings	Cumulative % variance	Mean	Cronbach alpha (α)
CON 1	The app has details about upcoming events	0.856			
CON 2	The app updates browsers with a "what's coming" section	0.770			
CON 3	The product availability information is updated on the app	0.688			

Source Primary data

Table 3 categorized various variables under six factors. The cumulative total of the rotated sum of squared loadings was 69.355 that was a good score. It demonstrated that the statements explain the 69.355% variance of the study. The scale's reliability was checked with Cronbach's alpha, the value of Cronbach's alpha was greater than 0.7 for each identified characteristic of a mobile app, and hence the scale was reliable. These six factors are being assigned names based on the nature of variables. These six factors were as follows:

i. **Detailed information**: Detailed information comprised variables like availability of visuals of products, description of the product's size and weights. The mean score of this factor was 0.732. This factor was ranked at no. 5 based on its mean.
ii. **Credibility**: Credibility comprises variables assuring customers about their credit card information, personal information, and no cookies. The mean score of this factor was 0.797, which was the second-highest among the various factors. This factor was ranked at no. 2 based on its mean.
iii. **Accessibility**: It consists of variables like ease of navigation through the app, free from clutter, and ease of finding the product. The mean score of this factor was 0.814, which was on the top among the various factors. This factor was ranked at no. 1 based on its mean.
iv. **Aesthetics**: It consists of variables like fun, interactive features, audio clips, and color schemes. This factor's mean score was 0.763, which was the fourth among the various factors. This factor was ranked at no. 4 based on its mean.
v. **Connect with customers**: Connect with customers consists of variables like customer support, users' review, and query handling process. The mean score of this factor was 0.710, which was the last among the various factors. This factor was ranked at no. 6 (last) based on its mean.

vi. **Consistency:** Consistency consists of variables named a new section, updating new arrivals, and up-to-date content. The mean score of this factor was 0.771. This factor was ranked at no. 3 based on its mean.

5.2 Results of Stepwise Multiple Regression

Table 4 represents that the regression model of apps' characteristics and consumers' attitudes toward mobile applications in an online shopping environment has been found significant ($F = 76.802$; $P = 0.000$). The adjusted R^2 value for the requisite regression model came out 0.566, which was considered substantial for explaining the dependent variable [39, 40]. Adjusted R^2 value (0.566) represents that mobile applications' features measure for 56.6% of the variance in consumers' attitude toward mobile applications in an online shopping environment, and the remaining 44.4% of variance remained unexplained in the desired model.

Stepwise multiple regression analysis was performed to test the mobile app's most critical characteristics affecting consumers' attitudes in an online shopping environment, as shown in Table 5. Retail App was taken as dependent variables while aesthetics, accessibility, credibility, detailed information, connect with customers, and consistency were independent variables on which the retail app was dependent. Based on the *p*-value ($P < 0.05$), all the characteristics were found to contribute significantly to the formation of consumers' attitudes toward mobile applications in an online shopping environment. The negative value of constant represents that the value of the dependent variable is negative when the independent variable is zero. Null hypothesis Ho1, i.e., no significant relationship between mobile app characteristics and consumer attitude, has been rejected here.

Regression equation that showed the impact of app's characteristics on the consumers' attitude has been explained below:

$$\begin{aligned}\text{Consumer attitude} = &-0.934 - 0.128(\text{detailed information})\\ &- 0.145(\text{credibility}) + 0.147(\text{accessibility})\\ &+ 0.833(\text{aesthetics}) + 0.306(\text{connect with customers})\\ &+ 0.311(\text{consistency})\end{aligned}$$

Table 4 ANOVA table

Model		Sum of squares	df	Mean square	*F*	Sig.
1	Regression	206.135	6	34.356	76.802	0.000
	Residual	153.434	343	0.447		
	Total	359.569	349			

($R = 0.757$, $R^2 = 0.573$, Adjusted $R^2 = 0.566$)
Source Primary data

Table 5 Regression analysis

Coefficients						
Model		Unstandardized coefficients		Standardized coefficients	*t*	Sig.
		B	Std. Error	Beta		
	(Constant)	– 0.934	0.300		– 3.115	0.002
	Detailed Information	– 0.128	0.055	– 0.097	– 2.320	0.021
	Credibility	– 0.145	0.059	– 0.102	– 2.463	0.014
	Accessibility	0.147	0.045	0.140	3.247	0.001
	Aesthetics	0.833	0.047	0.643	17.840	0.000
	Connect with customer	0.306	0.049	0.277	6.260	0.000
	Consistency	0.311	0.060	0.203	5.140	0.000

Dependent Variable: Consumers' Attitude
Source Primary data

The higher value of the standardized coefficient beta depicted the most contributing factor on which the consumers' attitude was dependent. Based on standardized beta value (Table 5), the research found that customer's attitude toward a mobile application was primarily because of aesthetics ($\beta = 0.643$) since the standardized beta value was highest followed by connecting with customers ($\beta = 0.277$), consistency ($\beta = 0.203$), accessibility (β = 0.140), credibility ($\beta = -0.102$), and detailed information ($\beta = -0.097$). The negative sign attached with β value of credibility and detailed information depicted their negative relation with consumers' attitudes.

Figure 1 Model for predicting consumers' attitude based on the online retailer's app characteristics

Previous studies supported the results of the study. This research found aesthetics positively related to consumers' attitudes toward online retailers' apps. Previous research of [18, 22] supported the positive relationship between consumers' attitudes toward online retailers' app and aesthetics. Accessibility positive linkage with consumers' attitude toward online retailing app was supported by the researches

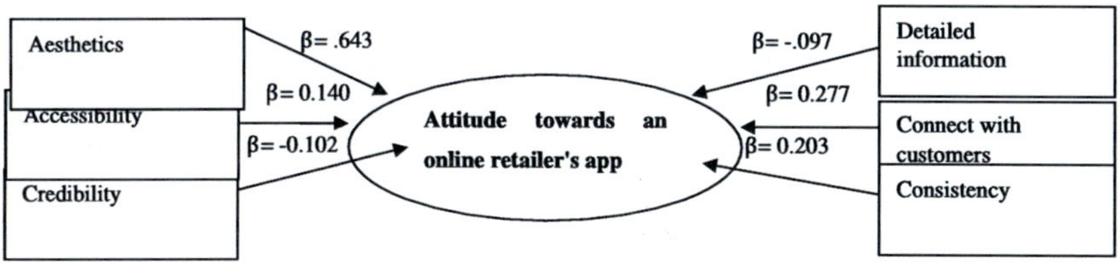

Fig. 1 Model. *Source* Primary data

of [21–23]. Credibility is negatively associated with customers' attitude toward the online retailing app. [41] observed similar behavior of customers. In his study, he found the importance of transaction security, removing the chances of fraud and risk in transactions, and the only probable reason customers would not switch to mobile apps. In the context of detailed information, consumers' attitudes toward online retailing apps were found to be negatively related. This result was supported by Lee and Overby [42], where they highlighted that if the queries are not handled properly or go unattended, the customers will start leaving the mobile app platforms. The present research also showed a positive association between consumers' attitudes and connections with customers in the online retailing app context. Research by Chen and Dhillon [33] supported the result where they highlighted the importance of connecting with customer-aspect in retailing and reason to move the customer toward online platforms. Consistency features were found positively linked with consumers' attitude in online retailing. Research by Donthu [17] and Bellman and Rossiter [23] supported the association between web app consistency and consumers' attitude.

6 Conclusion

Mobile applications have now become an integral part of any business. The existence and expansion of business cannot be done without leveraging the advantage of smartphones and internet penetration. This research explored aesthetics, accessibility, credibility, detailed information, customer support, and consistency as essential characteristics of mobile applications in an online shopping environment. All the characteristics were found to contribute significantly to consumers' attitudes toward mobile applications in an online shopping environment.

Since aesthetics is the characteristic that dominates the consumers' attitude toward mobile applications in an online shopping environment, companies should focus on the availability of fun elements, interactive features, interesting audio clips, and attractive color schemes on their mobile apps. Also, the researchers concluded that customer support negatively affects consumer attitude because customer support plays an essential role in the absence of personal interaction in the online shopping environment. The companies should work upon the query handling process, hassle-free customer support, and the availability of user reviews to make it convenient for potential buyers to continue their visits and shopping on the app.

The result drawn from this research cannot be generalized because of the limitation of the sample, and the geographical area is taken into consideration for the research. The present study has considered online shopping to study consumers' attitude toward mobile applications in other areas like trading and gaming remain unaffected from this research.

References

1. Pantano E (2013) Ubiquitous retailing Innovative Scenario: from the fixed point of sale to the flexible ubiquitous store. J Technol Manag Innov 8(2):84–92
2. Watson C, McCarthy J, Rowley J (2013) Consumer attitudes towards mobile marketing in the smart phone era. Int J Inf Manage 33(5):840–849
3. Deloitte Navigating the Digital Divide, U.S. (2014)
4. Google/Ipsos U.S (2019) Playbook.Omnibus 2019, n = 1610 U.S. online smartphone users, A18+
5. Google/Purchased Digital Diary (2016) How consumers solve their needs in the moment. Smartphone users = 1000, local searchers = 634, purchases = 1140
6. Dennis C, Alamanos E, Papagiannidis S, Bourlakis M (2016) Does social exclusion influence multiple channel use? the interconnections with community, happiness, and well-being. J Bus Res 69(3):1061–1070
7. Verhoef PC, Lemon KN, Parasuraman A, Roggeveen A, Tsiros M, Schlesinger LA (2009) Customer experience creation: determinants, dynamics and management strategies. J Retail 85(1):31–41
8. Demirkan H, Spohrer J (2014) Developing a framework to improve virtual shopping in digital malls with intelligent self-service systems. J Retail Consum Serv 21(5):860–868
9. Hristov L, Reynolds J (2015) Perception and practices of innovation in retailing: challenges of definition and measurement. Int J Retail Distrib Manage 43(2):126–147
10. Pantano E (2014) Innovation drivers in retail industry. J Inf Manag 34(3):344–350
11. Childers TL, Carr CL, Peck J, Carson S (2001) Hedonic and utilitarian motivations for online retail shopping behavior. J Retail 77(4):511–539
12. Kotler P (1973) Atmospherics as a marketing tool. J Retail 49(4):48–64
13. Baker J, Grewal D, Parasuraman A (1994) The influence of store environment on quality inferences and store image. J Acad Mark Sci 22(4):328–339
14. Dailey L (20014) Navigational web atmospherics: explaining the influence of restrictive navigation cues. J Bus Res 57(7):795–803
15. Rayburn SW, Voss KE (2013) A model of consumer's retail atmosphere perceptions. J Retail Consum Serv 20(4):400–407
16. Eroglu SA, Machleit KA, Davis LM (2001) Atmospheric qualities of online retailing: a conceptual model and implications. J Bus Res 54(2):177–184
17. Donthu N (2001) Does your web site measure up? Marketing Management 10(4):29–32
18. Kwon B, Kim C, Lee E (2002) Impact of App information design factors on consumer ratings of web-based auction sites. Behav Inf Technol 21(6):387–40
19. McMillan S, Hwang J, Lee G (2003) Effects of structural and perceptual factors on attitudes toward the app. J Advert Res 43(4):400–409
20. Mummalaneni V (2005) An empirical investigation of web site characteristics, consumer emotional states and online shopping behaviors. J Bus Res 58(4):526–532
21. Stevenson JS, Bruner GC II, Kumar A (2000) App background and viewer attitudes. J Advert Res 40(1):29–34
22. Bellman S, Rossiter J (2004) The app schema. J Interact Advertising 4(2)
23. Fogg BJ, Marshall J, Laraki O, Osipovich A, Varma C, Fang N, Paul J, Rangnekar A, Shon J, Swani P, Treinen M (2001) What makes web sites credible? A report on a large quantitative study. Persuasive Technol Lab 61–68
24. Vijayasarathy LR, Jones JM (2000) Print and internet catalog shopping: assessing attitudes and intentions. Internet Res 10(3):191–202
25. Limayem M, Khalifa M, Frini A (2000) What makes consumers buy from internet? A longitudinal study of online shopping. IEEE Trans Syst Man Cybern-Part A Syst Humans 30(4):421–432
26. Lynch PD, Kent RJ, Srinivasan SS (2001) The global internet shopper: evidence from shopping tasks in twelve countries. J Advert Res 41(3):15–23

27. Winn W, Beck K (2002) The persuasive power of design elements on an e-commerce web site. Techn Commun 49(1):7–35
28. Szymanski DM, Hise R (2000) T e-Satisfaction: an initial examination. J Retail 76(2):309–322
29. Chen Q, Wells WD (1999) Attitude toward the site. J Advert Res 40(5):27
30. Kwak H, Fox RJ, Zinkham GM (2002) What products can be successfully promoted and sold via the internet? J Advert Res 42(1):23–38
31. Korgaonkar PK, Wolin LD (1999) A multivariate analysis of web usage. J Advert Res 39:53–68
32. Pew Foundation America's online pursuits: the changing picture of who's online and what they do. Pew Internet & American Life, Dec 12, found on www.pewinternet.org/report (2003)
33. Chen S, Dhillon G (2003) Interpreting dimensions of consumer Credibility in e-commerce. Inf Technol Manage 4(2–3):303–318
34. Zeithaml V, Parasuraman A, Malhotra A (2002) Service quality delivery through web sites: a critical review of extant knowledge. Acad Mark Sci J 30(4):362–375
35. Kim S, Choi SM (2012) Credibility cues in online shopping: an examination of corporate credibility, retailer reputation, and product review credibility. Int J Internet Mark Advert 7(3):217–236
36. Waheed A, Zhang Q, Farrukh M, Khan SZ (2021) Effect of mobile social apps on consumer's purchase attitude: role of trust and technological factors in developing nations. SAGE Open 11(2):21582440211006710
37. Yang Z, Peterson RT, Huang L (2001) Taking the pulse of internet pharmacies. Mark Health Serv 21(2):4–10
38. Kaiser HF (1974) An index of factorial simplicity. Psychometrika 39(1):31–36
39. Cohen J (1988) Statistical power analysis for the behavioral sciences, 2nd edn. Lawrence Erlbaum Associates, Publishers, Hillsdale, NJ
40. Falk RF, Miller NBA (1992) Primer for soft modeling. University of Akron Press, Akron
41. Shukla PS (2018) E- Shopping using mobile apps and the emerging consumer in the digital age of retail hyper personalization: an insight. Pacific Bus Rev Int 10(10):131–139
42. Lee EJ, Overby JW (2004) Creating value for online 4870(03)00052-7. Shoppers: Implications for satisfaction and loyalty. J Consum Satisfaction Dissatisfaction Complaining Behav 17:54–67

Human Behavior and Emotion Detection Mechanism Using Artificial Intelligence Technology

Zhu Jinnuo, S. B. Goyal, and Prasenjit Chatterjee

Abstract With the rapid development of information technology, the study of human emotion and behavior has become more and more important. The generation of human behavior comes from the response of sensory stimuli from external factors. The current emotion detection process mainly uses artificial intelligence technology to capture the facial features of the target and compares the data through machine learning to judge the emotional state of the target. In the process of face capture, the target emotion is detected and analyzed mainly through the changes in the eyes, mouth, face, nose, and other organs, and the target emotion is determined from it, and finally, the behavior analysis and prediction are carried out in the following ways. Test results. However, the current detection technology is mainly aimed at the data collection and detection of the target face or surface, and cannot detect the psychological and body language emotional changes of the target. This paper starts with the current detection technology, uses traditional detection methods for emotion detection and behavior analysis, and compares the data with the composite emotion detection method combined with wearable technology. Through the implementation of new algorithms and the use of wearable devices, the superiority of wearable technology is reflected, which can improve the accuracy of human emotion detection and more effectively analyze and predict human behavior characteristics.

Keywords Artificial intelligence · Human behavior · Emotion detection · Machine learning · Wearable technology

Z. Jinnuo · S. B. Goyal (✉)
Nanchang Institute of Science and Technology, Nanchang, China
e-mail: drsbgoyal@gmail.com

Z. Jinnuo
City University, Petaling Jaya, Malaysia

P. Chatterjee
Department of Mechanical Engineering, MCKV Institute of Engineering, Howrah, West Bengal, India

P. Chatterjee et al. (eds.), *Computational Intelligence for Engineering and Management Applications*, Lecture Notes in Electrical Engineering 984,
https://doi.org/10.1007/978-981-19-8493-8_59

1 Introduction

In the process of emotion detection, researchers use artificial intelligence to perform machine learning on human facial features, so that artificial intelligence can intelligently match and compare the target face with the learned data, calculate and analyze the target emotion, and then analyze the behavior characteristics of the target. Artificial intelligence emotion detection has now comprehensively covered various fields [1]. Common fields focus on EEG detection, bone detection, prison system detection, driving field, ECG detection, psychological detection, teaching quality detection, face detection, etc. [2]. Through the investigation, we learned that in various common fields, researchers will collect data on the facial features of people or the basic features of things, and perform intelligent detection. However, from the data comparison of the detection results, it can be seen that there is a large deviation between the real emotional characteristics and behavior and the data detected by the artificial intelligence mechanism [3]. The main reason is that at this stage, due to the lack of internal factors in the process of identifying targets through faces or features and the limited range of facial recognition, there is a large detection error in the matching data set of artificial intelligence in the process of machine learning. In addition, some existing detection mechanisms lack the learning of the target's intrinsic emotional and behavioral characteristics [4]. In the traditional model, the judgment of emotions is derived from the face and other aspects, which leads to the inability to detect and analyze the features other than the face, and finally leads to errors and unrecognized situations. To solve the problem of target inner emotion detection, it is necessary to identify and transmit the inner emotion characteristics of the target through the media. By combining the learning of the internal emotional characteristics of the target and the learning of external characteristics in the traditional model, artificial intelligence can make a composite judgment on the real emotion of the target through the internal and external characteristic data. This paper will use wearable technology widely used in the research process to identify and obtain the internal psychological and emotional factors of the target through media such as wristbands. In addition, in the process of face recognition, complex emotion detection is combined with the movement and state of the limbs, which further improves the detection accuracy of the mechanism [5], and analyzes and predicts the behavior characteristics of the samples. In the future, humans and artificial intelligence will realize "human–computer interaction" in the process of detection and recognition, cooperation with humans, and transformation of the world [6]. Figure 1 is a schematic diagram of the complex emotion detection process under artificial intelligence combined with wearable technology.

Through the review of literature information, there is a certain degree of the causal connection between human emotions and production [7]. External factors will have a direct impact on human emotions, and the production of emotions can stimulate the production of behaviors at the same time [8].

In the current research fields, the analysis of human emotion and behavior is in a stage of independent research and development. In the field of emotion research, researchers use artificial or machine learning to determine the capture of target

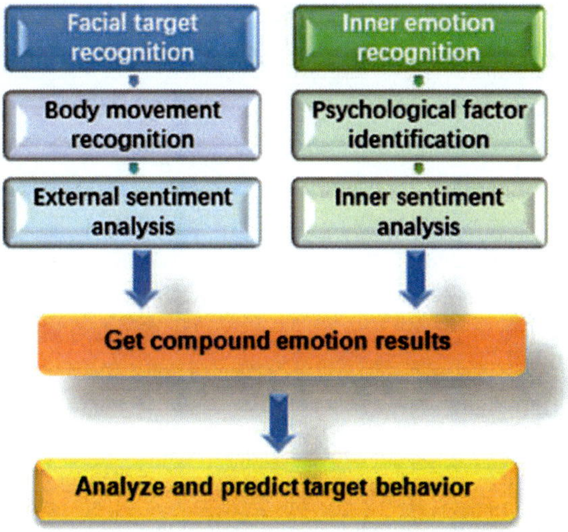

Fig. 1 Block diagram of the emotion detection process

Table 1 Several common behavior generation processes

External factor	Emotional performance	Generated behavior
Happy things	Happiness	Laugh
Troubleshooting	Anger	Boxing
Sad thing	Sad	Weeping
Troublesome	Irritable	Waving hands and feet

facial features or psychological features. In the field of human behavior analysis, researchers conduct research on behavioral characteristics generated by human limbs using 3D skeletons. This paper will improve through experiments, algorithms, and other operations, realize a set of machine learning methods using wearable devices under artificial intelligence technology, and establish a complex emotion detection and behavior analysis mechanism. Table 1 shows the behavioral characteristics of common human emotions [9].

2 Literature Review

2.1 Literature Survey

Related fields' Artificial Intelligence Technology Behavior Analysis and Emotion Detection Case Questionnaire. Table 2 shows the data table of behavioral analysis and emotion detection research in various fields in the past 16 years, including research years, researchers, research areas, and research contributions [10].

Table 2 Behavior analysis and emotion detection research data in various fields

Years	Author	Contribution
2005	Yadav and Vibhute [1]	Related factors of emotion in children's learning process
2005	Da Silva [2]	Design of emotion perception model for virtual human
2006	Khan [3]	Research on the emotion model of an autonomous virtual human
2006	Todd Blick [5]	Research on virtual role module based on Artificial Intelligence
2007	Francese [6]	Numerical calculation of emotion classification description
2007	Kodi and Tene [7]	Realize emotional and humanized intelligent interaction through personalized teaching
2008	Li [11]	Research on 3D face modeling
2008	Imran [12]	Research on virtual human emotional interaction system
2009	McBride [13]	The acquisition, recognition, understanding, and expression of emotion are realized through human–computer interaction
2011	Ma and Huang [14]	Design of emotional interaction model based on BDI cognitive behavior framework
2011	Blom [15]	Research on robot behavior control system
2012	Radmila [16]	Improving agent's affective cognitive decision-making by Q-learning algorithm
2012	Chang [17]	Research on the future theory of computer intelligence technology and interaction technology
2014	Baccarani [19]	Research on the realization of psychological emotion model through differential evolution algorithm and simulation
2014	Annis et al. [20]	The interaction between the new conceptual model and sub-algorithm is constructed to realize simulation reasoning and decision-making
2017	Alo [21]	The autonomy and feasibility of human–computer interaction based on emotional cognition
2018	Allaman [22]	Research on learner behavior method based on face recognition algorithm
2019	Zhang [23]	The intelligence and effectiveness of the rounding algorithm of potential point distribution based on the model concept are verified
2020	Panuwatwanich [24]	Research on human–computer interaction theory under artificial intelligence technology
2021	Barman et al. [25]	Convolutional Neural Network EEG Signal Emotion Classification Model

2.2 *Literature Survey Results and Summary of Researchers' Contributions*

Before 2012, researchers leaned more toward the field of emotion perception and recognition. After 2013, the research focus gradually turned to in-depth research work to realize human behavior and emotion detection through artificial intelligence technology [11]. It can be seen that the study of human behavior and emotion is becoming more and more obvious. It is attracting more and more attention. The research directions from 2015 to 2021 generally range from emotion recognition to emotion detection and the integration of emotion and behavior, which also lays a theoretical foundation for the future research content of this paper [12]. Throughout the collected data, research contributions are reflected in emotion model research, intelligent robot simulation, emotional computing, human–computer interaction, emotion detection, virtual reality, face recognition, behavior research, etc.

2.3 *Analysis of the Deficiencies of Researchers in the Literature Survey*

Because artificial intelligence needs to consider the detection of the target's inner emotions in the process of detecting human emotions, just like the special equipment in the wearable technology to be analyzed below, it can obtain the inner emotions of the target and make a comprehensive judgment with facial emotions [13]. More accurate target emotional characteristics. Whether it is the study of human emotions or through machine learning, artificial intelligence devices need to simulate real human behavior and emotions to accurately identify and judge.

2.4 *Case Analysis of Artificial Intelligence Technology in Facial Emotion Detection*

(1) *Real data analysis of face recognition technology in human emotion detection and behavior analysis*. After the mechanism captures facial features, the recognition degree of the anger index value accounts for 40%, and the other recognition degree values are lower. The mechanism concluded that the detection was biased toward anger [14]. But by detecting the limbs of the target, it can be seen that the arm of the target is raised, showing obvious confidence. Although the AI has already identified the target behavior in the detection process, it only judges the target's facial expressions, ignoring the target's body language, resulting in deviations from the actual results. Figure 2 shows the description of AI Intelligence detection results.

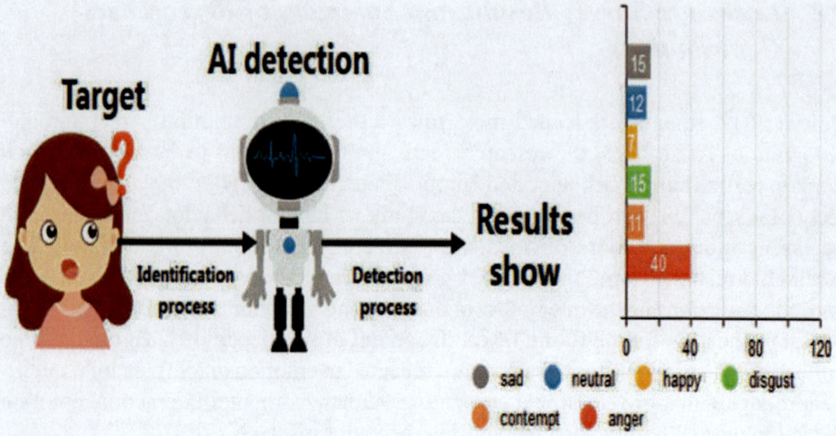

Fig. 2 Results of artificial intelligence detection

After the mechanism captures facial features, the recognition degree of the anger index value accounts for 40%, and the other recognition degree values are lower. The mechanism concluded that the detection was biased toward anger [14]. But by detecting the limbs of the target, it can be seen that the arm of the target is raised, showing obvious confidence. Although the AI has already identified the target behavior in the detection process, it only judges the target's facial expressions, ignoring the target's body language, resulting in deviations from the actual results.

(2) *Analyze the cause of face detection errors*. The above cases show that the recognition and analysis of human facial emotions to determine the behavior of the target cannot be used as the most accurate basis for detection. Because human emotions can appear on the face, and other aspects such as the body and psychology can reflect human emotions, only facial expressions to capture emotions will cause the detection results to be inconsistent with reality. If a corresponding wearable device is used, the internal data of the human body, such as heartbeat, heart rate, blood pressure, and other data, can be obtained through the device. Combining the above data with facial features allows analysis of underlying human emotions [15].

2.5 Analysis of the Advantages of Wearable Technology in Emotion Detection

(1) *Introduce the working principle and performance of wearable technology*. It is stored in human clothing or special parts of the body, and sends data to the detection mechanism through wireless sensing. Human body internal data can be detected during transmission [16]. The main advantages are that the device

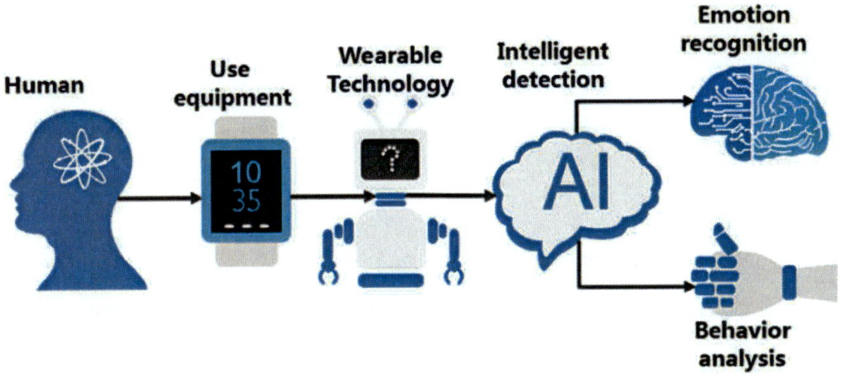

Fig. 3 Process flow of wearable technology

is small in size, easy to carry, high in precision, and data is updated in real time. Wearable technology was originally used in medical institutions to identify patients' physiological signals through wearable media and analyze patients' comprehensive emotions through physiological signals. Figure 3 shows the process flow of wearable technology works.

(2) *Analysis of the advantages of emotion detection in wearable technology and traditional detection modes* [17]. Through the investigation, it was found that the medium under the wearable technology has the function of recognizing the human heartbeat and heart rate. Its working principle is to use its internal contact sensor chip to be close to the human skin, obtain the internal characteristics of the detection target from it, and feedback the data in real time through the display device [18]. At present, the existing detection equipment only supports the capture of facial or surface features, and cannot obtain the intrinsic factors of the target. It lacks detection and analysis of the hidden features and intrinsic emotions of the target, so the detection results also appear to be biased [19].

2.6 Research Questions

To solve a reasonable and effective way for artificial intelligence to detect the emotion of human behavior, we ask the following four questions:

RQ1: Why do we need to develop human behavior and emotion detection mechanisms based on artificial intelligence?

RQ2: What are the features and parameters that capture human behavior and emotion?

RQ3: What are the existing mechanisms for detecting human behavior and emotion?

RQ4: How to develop accurate and efficient mechanisms to capture and explain human behavior and emotions?

2.7 Research Objectives

Based on the above four research questions, through the assumptions and research in this paper, it can be inferred that the research objectives (RO) have the following:

RO1: To query the application development examples and development prospects of artificial intelligence, and develop mechanisms that are more accurate than previous human behavior and emotion detection.

With the development of artificial intelligence technology, the mode of human–computer interaction can allow machines to replace humans to understand human thoughts. Then, develop a human behavior sentiment analysis mechanism based on artificial intelligence technology to judge human behavior and intelligently generate emotions, reducing a lot of human investment in a practical sense.

RO2: To understand the emotional characteristics of people, through literature investigation, through the identification, capture, and detection of human faces, including the face, eyebrows, mouth, eyes, and other organs, as well as the target limbs and inner psychology, from which to analyze emotions and behavior capture methods [20].

The emotional characteristics of different expressions and the emotional changes caused by factors other than the face are analyzed by the changes in human facial expressions.

RO3: To collect research data for Research Objective 2, find out the beneficial parts and areas for improvement of existing AI technologies in detecting human behavior and emotions, and analyze specific factors that lead to bias. Results.

By arranging and analyzing the literature survey data in Research Objective 2, combining them to obtain relevant cases in the existing detection process, and selecting some objects with body language as samples, using existing algorithms to perform emotion detection and analysis in them. Behavior analysis. By analyzing the difference between the detection results and the actual situation, it is verified that the use of wearable technology intervention and machine learning can effectively improve detection accuracy.

RO4: To create a diagram of the human emotion and behavior analysis model under the new algorithm. Combining wearable technology with new algorithms designed to form and integrate. In addition, the detection results under the use of wearable technology are compared with the unused detection data under the existing algorithm, reflecting the advantages of the new algorithm and wearable technology [21].

Summarizes the reasons for the problem in Research Objective 3. Augmenting the wearable part of an existing mechanism for new machine learning based on how wearable technology works. The unique algorithm of wearable technology and the new algorithm designed in this paper are used to carry out experimental tests on the target emotion and behavior, and finally, the difference between the existing algorithm and the new algorithm is compared through SPSS data analysis, from the experimental basis and research results of emotion detection and behavior analysis under the final version of wearable technology.

3 Research Methodology

Based on the research questions and the realization of research objectives, the following three research methods and four steps are proposed:

To achieve Research Objectives 1–4, researchers need to read and summarize extensive literature survey information. The research objectives, it is divided into three steps.

RM1: First, we need to find and analyze relevant literature survey content, and summarize the existing research results of technical human behavior and sentiment analysis under artificial intelligence technology involved in this topic. Through the future direction of human behavior analysis and emotion detection, the problems that need to be solved in the current stage of emotion and behavior development are found, and the initial case and sample data collection work is formed in the literature survey [22].

RM2: Secondly, we need to conduct a preliminary classification of the collected sample data and conduct research on human organ information in emotion recognition and behavior analysis. Through the setting and machine learning of the effective recognition areas in the facial organs and body movements, artificial intelligence can initially identify and analyze the emotional factors of the target through code. It focuses on analyzing the impact of the detection area size and range variation of facial organs such as eyes, nose, and mouth on the detection results, and performs data analysis on the preliminary results.

RM3: Again, we use the existing algorithms in the field of emotion detection to conduct experimental operations on the collected sample data and draw the corresponding human emotion model diagram. Through the error analysis in the test results, the theoretical basis for using wearable technology to improve the test results is demonstrated, and the existing wearable technology is used to carry out preliminary experimental operations on the identification methods of human inner emotions and behaviors. Comparing the emotion and behavior recognition rates of existing algorithms proves the specific contribution to human emotion and behavior research guided by the use of wearable technology, and finds out the research methods for the target's built-in emotion.

RM4: Finally, the recognition algorithm of wearable technology in the existing algorithm is re-improved, and the current human emotional behavior model diagram is re-modified and combined with the corresponding mathematical function formula. It summarizes the shortcomings of existing algorithms and proposes new algorithms, which combine wearable technology with new algorithms to comprehensively evaluate human built-in emotions and facial features, and verify in human behavior analysis to form artificial intelligence under wearable technology, an integrated mechanism for human behavior and emotion detection, and a final SPSS data analysis and evaluation [23].

4 Expected Impact

This research will research theory, conception, analysis, improvement, practice, etc., and finally form an efficient, accurate, and stable detection mechanism. The specific expected impacts are as follows:

(1) Data collection. Select reasonable research samples through random sampling, establish and analyze emotional model graphs according to traditional detection methods, extract facial features, function calculation, SPSS analysis, and compare the corresponding emotional detection results with real data to find out the shortcomings of traditional face recognition detection method in the model proves the technical defects of traditional detection techniques [24].
(2) Preliminary study. By setting the detection range of the characteristics of the target and judging the actual situation of the existing mechanism in the detection, the reasons for the error between the detection results and the actual situation are analyzed. According to the experimental results, the built-in emotional factor modeling and feature extraction are carried out using existing algorithms to perform preliminary operations, and analyze the reasons for data deviation at the current stage and the impact of wearable technology on the results from the actual detection data [25];
(3) Improve operation. Using wearable technology and performing machine learning, use existing algorithms to conduct a feasibility analysis of the target's internal emotions and behavior. Through experiments, the comprehensive emotional evaluation and behavior analysis of the target under the wearable technology is carried out, and the experimental results are compared with the detection data without this technology to illustrate the role and impact of using the wearable technology on improving the detection results. In addition, at the current stage, the capture range of the target's facial and psychological features has been improved, so that the mechanism can further improve the detection accuracy by improving the target recognition area and range.
(4) Final result. Set new algorithms and build new emotion and recognition model graphs based on existing detection processes. Using the calculation method of the new algorithm, the target data is again comprehensively evaluated for emotion and behavioral analysis. Analyze and compare the detection results with the experimental results in the existing algorithms by SPSS data, and summarize

the composite human emotion and behavior detection theory and experimental results after the use of wearable technology through the experimental results.

5 Conclusion

Through the comprehensive evaluation of the existing detection technology and existing algorithms, the current detection methods mainly focus on the feature capture and analysis of the target's facial features, but the analysis of the target's internal emotional factors and physical behavior is lacking. Human beings are a group of sharp thinking. Only by identifying internal factors through external factors can we analyze human groups more effectively. This is also mentioned in this article to use wearable technology to achieve effective detection of intrinsic emotional and behavioral characteristics, summarize the shortcomings of existing algorithms, and propose new algorithms to intelligently unify human emotional and behavioral characteristics with new technologies. Using artificial intelligence technology to realize the integrated mechanism of comprehensive emotion detection and behavior analysis in the use of wearable media, it also lays a solid foundation for future research in the field of human behavior.

In the future, we will continue to conduct in-depth research in this field through wearable technology. It is expected that through the improvement of new algorithms, the mechanism will improve the capture point and recognition accuracy of the target during the movement process, and implement it through time series and LSTM algorithms. Also, it enables emotional and behavioral studies of target dynamics.

References

1. Yadav SS, Vibhute AS (2021) Emotion, detection using deep learning algorithm. Int J Comput Vis Image Process(IJCVIP) 11(4):30–38. https://doi.org/10.4018/IJCVIP.2021100103
2. Da Silva RH, Ferreira Júnior WS, Moura JMB, Albuquerque UP (2020) The link between adaptive memory and cultural attraction: new insights for evolutionary ethnobiology. Evol Biol 47(4). https://doi.org/10.1007/s11692-020-09516-8
3. Khan S (2020) The role of faith-based organizations (FBOs) in the rehabilitation of offenders. Orient Anthropologist A Bi-annual Int J Sci Man 20(2). https://doi.org/10.1177/0972558X20952657
4. Elhai JD, Rozgonjuk D (2020) Editorial overview: cyberpsychology: reviews of research on the intersection between computer technology use and human behavior. Curr Opin Psychol 36. https://doi.org/10.1016/j.copsyc.2020.11.001
5. Todd Blick A (2020) Winners are not keepers: characterizing household engagement, gains, and energy patterns in demand response using machine learning in the United States. Energy Res Soc Sci 70. https://doi.org/10.1016/j.erss.2020.101595
6. Francese R, Risi M, Tortora G (2020) A user-centered approach for detecting emotions with low-cost sensors. Multimedia Tools Appl 79:47–48. https://doi.org/10.1007/s11042-020-09576-0

7. Kodi D, Tene R (2021) Negative emotions detection on online mental-health-related patients texts using the deep learning with MHA-BCNN model. Expert Syst Appl 182:115265. ISSN 0957-4174. https://doi.org/10.1016/j.eswa.2021.115265
8. Thuseethan S, Rajasegarar S, Yearwood J (2020) Complex emotion profiling: an incremental active learning-based approach with sparse annotations. IEEE Access 8. https://doi.org/10.1109/ACCESS.2020.3015917
9. Bu X (2020) Human motion gesture recognition algorithm in video based on convolutional neural features of training images. IEEE Access 8. https://doi.org/10.1109/ACCESS.2020.3020141
10. Guo M (2020) Transportation mode recognition with deep forest based on GPS data. IEEE Access 8. https://doi.org/10.1109/ACCESS.2020.3015242
11. Li M (2020) Multistep deep system for multimodal emotion detection with invalid data in the internet of things. IEEE Access 8. https://doi.org/10.1109/ACCESS.2020.3029288
12. Imran AS (2020) Cross-cultural polarity and emotion detection using sentiment analysis and deep learning on COVID-19 related tweets. IEEE Access 8. https://doi.org/10.1109/ACCESS.2020.3027350
13. McBride SK (2020) Developing post-alert messaging for ShakeAlert, the earthquake early warning system for the West Coast of the United States of America. Int J Disaster Risk Reduction 50. https://doi.org/10.1016/j.ijdrr.2020.101713
14. Ma X, Huang Y (2020) Construction on industrial ecosystem of marine low carbon city based on composite ecosystem. J Coast Res. https://doi.org/10.2112/JCR-SI107-016.1
15. Blom SSAH, Aarts H, Semin GR (2020) Perceiving emotions in visual stimuli: social verbal context facilitates emotion detection of words but not of faces. Experimental brain research. https://doi.org/10.1007/s00221-020-05975-9
16. Radmila J (2020) Machine learning models for ecological footprint prediction based on energy parameters. Neural Comput Appl. https://doi.org/10.1007/s00521-020-05476-4
17. Chang W (2020) Examining the dimensions and mechanisms of tourists' environmental behavior: a theory of planned behavior approach. J Cleaner Prod 273. https://doi.org/10.1016/j.jclepro.2020.123007
18. Grobler M (2020) The importance of social identity on password formulations. Pers Ubiquitous Comput. https://doi.org/10.1007/s00779-020-01477-1
19. Baccarani A (2020) Relaxing and stimulating effects of odors on time perception and their modulation by expectancy. Attention Percept Psychophysics. https://doi.org/10.3758/s13414-020-02182-0
20. Annis J, Gauthier I, Palmeri TJ (2020) Combining convolutional neural networks and cognitive models to predict novel object recognition in humans. J Exp Psychol Learn Mem Cogni. https://doi.org/10.1037/xlm0000968
21. Alo UR(2020) Smartphone motion sensor-based complex human activity identification using deep stacked autoencoder algorithm for enhanced smart healthcare system. Sensors (Basel, Switzerland) 20(21). https://doi.org/10.3390/s20216300
22. Allaman L (2020) Spontaneous network coupling enables efficient task performance without local task-induced activations. J Neurosci Official J Soc Neurosc 40(50). https://doi.org/10.1523/JNEUROSCI.1166-20.2020
23. Zhang Z, Cheng H, Yang T (2020) A recurrent neural network framework for flexible and adaptive decision-making based on sequence learning. PLoS Comput Biol 16(11) https://doi.org/10.1371/journal.pcbi.1008342
24. Panuwatwanich K (2020) Ambient intelligence to improve construction site safety: case of high-rise building in Thailand. Int J Environ Res Public Health 17(21). https://doi.org/10.3390/ijerph17218124
25. Barman P, Thukral T, Chopra S (2020) Communication with physicians: a tool for improving appropriate antibiotic use in the absence of regulatory mechanisms. Curr Treat Options Infect Diseases. https://doi.org/10.1007/s40506-020-00241

An Enhanced Career Prospect Prediction System for Non-computer Stream Students in Software Companies

Biku Abraham and P. S. Ambili

Abstract The stakeholders of the present education system have a key focus on the chances of on-campus employability of their wards. Since software companies play prominent role as campus recruiters, there is a low demand for the traditional Engineering branches. The need to enhance skills for software related employment is a challenge to the private sector professional institutions for the existence of non-computer streams in this scenario. In this study, the authors being academic experts in the similar field, attempt to support institutions, students, curriculum designers, and the faculty by providing insight to better employment possibility in campus. Data set for the study is from the same institution for which the placement prediction was completed. The methodology models the problem as a sequential event prediction problem and employs deep learning techniques. The system extracts data from data set with 19 attributes. The results indicate that this predictive approach provides enhancement methods by which a non-computer stream student can ensure full-time employment in software related jobs through in-campus selection.

Keywords Prediction system · Deep learning techniques · Non-computer stream · Software companies · Campus selection

1 Introduction

The growing pace of digitization has paved way to the boom of Software Industries. Software development and related terms are the buzzwords in the job market now. This has adversely affected the traditional Engineering streams, as majority of the stakeholders in recent years take admission for a stream based on the campus employability demand. The private sector Engineering Colleges with programs in

B. Abraham (✉)
Department of Computer Applications, Saintgits College of Engineering, Kottayam, India
e-mail: biku.abraham@saintgits.org

P. S. Ambili
School of Computer Science and Applications, REVA University, Bangalore, India
e-mail: ambili.ps@reva.edu.in

P. Chatterjee et al. (eds.), *Computational Intelligence for Engineering and Management Applications*, Lecture Notes in Electrical Engineering 984,
https://doi.org/10.1007/978-981-19-8493-8_60

non-computer streams are facing the toughest time for their existence. The US Bureau of Labor Statistics analysis show that 85% of the specialist jobs are in software field and the software developers' job market is expected to increase by 22% by 2029 [1]. This points to the importance of all stream students to be made fit for software related jobs for ensuring increased campus placement possibilities. The authors being educationalist and having vast exposure in the similar field, attempts to help institutions and stakeholders with the proposed prediction system. The study attempts to provide tools/guidance to educators and curriculum developers on how to predict student placement possibilities using performance-based assessments.

We present and calibrate a model that focuses on the job market of undergraduate Engineering students with an exposure of programming. The purpose of this paper is to predict the career of undergraduate non-computer students in South Indian Engineering colleges in software industries. The work carried out the algorithms in Machine Learning and Deep Learning.

1.1 Deep Learning

Deep Learning/Machine Learning technology which includes Artificial Intelligence techniques works well with structured and unstructured types of data. Deep learning practices are very efficient in analyzing and interpreting large amounts of data involving statistical and predictive modeling tasks. Deep learning algorithms provide acceptable level of accuracy as the algorithms work by training data, create a feature set, and then build a predictive model. The initial model may not be perfect and with each iteration, the model will turn more accurate. The deepness of the network is the measure of number of layers it has.

Encouraged by these successful outcomes of deep learning, for a variety of applications, including health prediction systems, image retrieval, and natural language processing, the authors attempt to explore deep learning methods for employability prediction. Despite much research consideration of applying deep learning for many prediction and recommendation systems, still limited amount of attention is noticed focusing on career predictions of non-computer stream students for software jobs, by considering different classes of related attributes.

1.2 Objective of the Research

The non-consideration of attributes contributing to the programming knowledge and design ability of students, will degrade the prediction process in the similar methods employed. Demographic data as well as the mentor index which contribute to the attitude of the person apart from technology skills is necessary for such prediction systems. The authors attempted to employ various ML/DL algorithms and optimization of weights is done for getting the accurate result. The results at various stages are analyzed with actual data available and the true positive cases are proved.

1.2.1 Analysis

Precisely, the authors attempt to address the following open research queries:

(i) Are deep learning techniques effective for learning variety of attributes and making effective predictions?
(ii) How to apply and adapt an existing deep learning model with the data set?

From the experimental studies conducted, the authors were able to attain some encouraging results. The authors were able to introduce the efficient set of attributes to be selected for improving chances of campus placement of non-computer stream students in software related jobs using deep learning techniques which is the key contributions in this work.

2 Literature Review

The aim of the research work of this paper is based on the high potentiality that is diagnosed in machine learning techniques that can be employed for foreseeing the performance of graduates based on several factors including the skills for a programmer. Machine learning skills are properly fit for evaluation and forecast of data collections. The research suggests that some students learn to program with the goal of becoming conversational programmers which means they want to communicate with the developers and improve their marketability [2]. Collaborative filtering methods are similarly effective when compared to machine learning methods [3]. Academic achievements of the pupils primarily depend on the prior actions [4]. The scientists found that data mining classification algorithms also expect the student performance indicator. The neural network model concludes with greater efficiency compared with K-Nearest Neighbor and Decision Tree methods [5]. A unique algorithm called gritnet which uses bidirectional long short-term memory. The investigators used the data set from Udacity students' graduation. Gritnet requires hyper parameter optimization for further accuracy. The primary gain of this research is that it does not use the feature engineering and it can apply it on any student data event. The effort of this work has not covered the activity reports, mentors' value, and other co-curricular values in the data set [6].

The study suggests approaches for student degree plan, faculty interventions, and personalized advising, which aid for retention and better academic performance. Factorization Machines (FM), Random Forest (RF), and the multilinear model give lower error in forecast. A hybrid RM-FM method predicts grades of the graduates.

The work examines the study of recommender system in envisioning the performance. The study has not carried out the statistical evaluation on the rightness of the performance [7]. The authors implemented and related many algorithms of machine learning for interpreting the prediction performance. The task showed that Decision tree, One R, and REPTree give certainty of over 76%. The result provides insight into performance indicator, which stimulate the educationists to get properly trained in the appropriate direction [8].

The researchers carried out a study considering the factors affect the programming. The research included 15 factors that influence the performance of a programming subject. The factors include prior academic and computer experience, self-perception on programming performance, and comfort level of the subject. The work tries to apply correlation and regression on the data. They showed 79% variance in the available data [9]. Bayesian network classifier has a potential to be exploited as a mechanism for forecasting achievement of academic activity. The Bayesian classifier models with various class groups can produce student performance prediction [10]. A new system introduced which uses the non-graded activity of graduates to conclude the behavior in graded systems. The job requires operation of the correlation between auxiliary and target resource type. Better performance will be possible merely by employing other task information [11]. The work viewed the student performance prediction as a short-term sequential behavior. Two-stage classifier uses SVM and HRNN algorithms. The hypothesis in this research shows that the student follows a frequent pattern, and the intention is to improve the grades. The features of student behavior integrated and resulted in a good prediction accuracy [12].

The characteristics of students like behavioral features, academic and demographic performances in the training data set were included for supervised machine learning techniques. The study involves the comparison of many machine learning algorithms and it is observed that logistic regression can give better accuracy. The performances of the algorithms were evaluated by AUC and RUC, which show the real accuracy of the algorithm [13]. Support Vector Machine (SVM) classifier is also featured as one of the best algorithms for prediction analysis [14].

The study suggests a unified predictive model for the student's employability [15] and uses two-level clustering to reduce the data set. For finding the relevant data set, chi-square analysis was used. To predict the employability, four classifier algorithms were integrated and it is proved that the model shows better F1-score than using the simple classification model. Soft max regression and hybrid model deep belief network was used to predict the student's employability and more than 98% accuracy was found [16]. Decision tree classifier shows better accuracy in predicting student employability. The work includes the prediction of employability of students across different disciplines [17]. The performance of top four machine learning algorithms, Decision Tree, Random Forest, Naïve Bayes, and KNN was evaluated. It is found that Random Forest works best for the data set. But the attributes have not included their co-scholastic data [18].

3 Methodology

The experimental data set was the undergraduate Engineering student record of the private Engineering Colleges in South India, for which the prediction system was built. The data set consists of 1655 student data who had passed out from the institution in three consecutive years. The authors performed optimization with Adam Optimizer. After optimization, 19 attributes were selected for prediction of the employability characteristics.

3.1 Data Set Characteristics

The data set is categorized primarily into four: Demographic, Scholastic, Co-Scholastic, and programming skill.

In the outcome-based education scenario, the Co-Scholastic factors contribute a lot to the achievement of Course Outcome (CO) and Program Outcome (PO). Measuring of academic performance is a challenging task, since three factors are connected with the performance and prediction—demographic, academic, and socio-economic. The authors considered all these factors and the data pertaining to programming skill which contribute directly to the possibility of employment was carefully chosen. Technical skills such as index in product development, paper publication, patent, are now a scale for measuring the ability of a student. Mentoring plays a key role in the current education state and the index was concluded based on the overall continuous observation and assessment. The programming skill includes the awards/prizes in the technical fest, training given by the parent institute, certificate programs, online classes, internships, Hackathon participations, and memberships in professional bodies. It is observed that socio-psychological and educational aspects are more effective predictive factors than with the personal and medical [19].

The optimized set of attributes for the data is given in the following table.

Demographic data which points more to the family characteristics can also be measured as a contributor of scholastic levels as well as attitude of a student. The authors were keen in observing the fact that HR level is an inevitable part of the placement process where attitude of a person plays a vital role. The Scholastic factors which are the direct output obtained from evaluation system cannot be replaced in any situation. It is the quantitative measure of eligibility for any selection procedure and has given due importance.

The authors carefully applied the Machine Learning/Deep Learning algorithms as prediction and classification cases. Educational data mining techniques are exploring faster, but need more collaborative studies [20]. Machine learning techniques provide accurate predictions on student's assessment marks and achievements [21]. Dimensionality reduction applied in the data set initially as large number of input features always makes the prediction a challenging task [22].

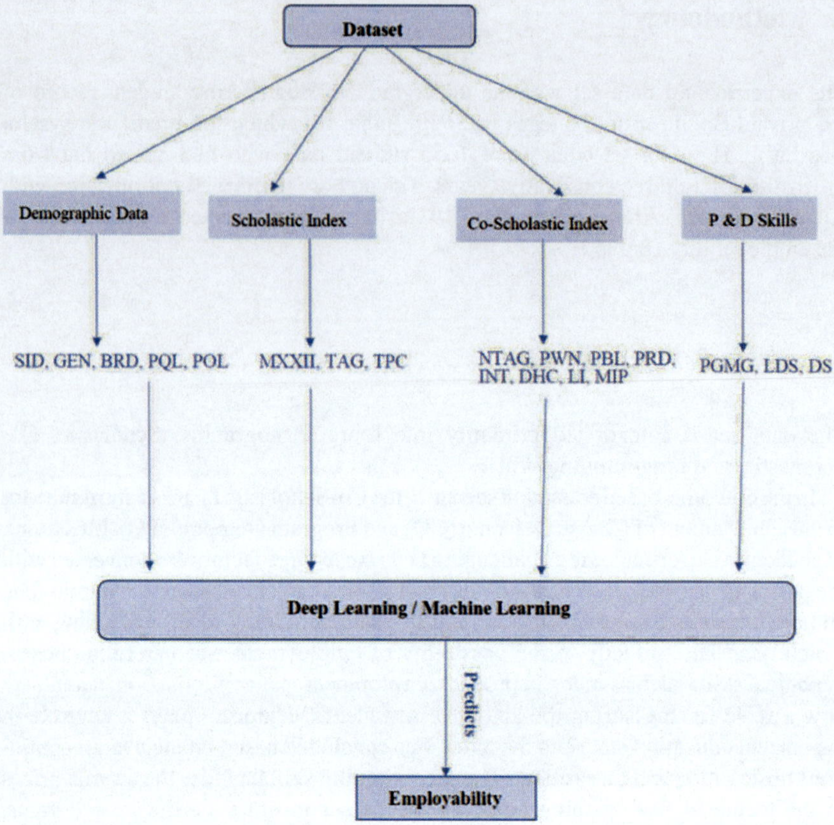

Fig. 1 Proposed system

4 Findings and Discussions

The regression algorithms Linear, Lasso, Ridge, Decision Tree, and Random Forest are used for prediction cases. The five algorithms are applied on the Demographic, Scholastic, Co-Scholastic, and programming index characteristics of the data set. The results of testing with 20 and 30% of the data with Co-Scholastic index and programming index for the R Squared (R^2), Mean Squared Error (MSE), and Root Mean Squared Error (RMSE) are presented in Table 2.

In first set of prediction, with 20% testing data (80% training), best results are obtained with Lasso Regression which gives minimum values for RMSE and maximum values for R^2. In the second set with 30% testing data (70% training), the authors observed best performance with Lasso Regression and Linear Regression falls far behind in producing better prediction performance (Table 3).

The regression algorithms are compared with 30 and 20% testing data. The details are presented in Fig. 2.

Table 1 Data set attributes description

Attributes	Description	Data type
SID	Student ID	object
GEN	Gender	object
BRD	Board	object
PQL	Parent qualification level	object
POL	Parent occupation Level	object
MXXII	10th and 12th marks/percentage/CGPA	float64
TAG	Overall technical achievements (rated on a 5-point scale)	Int64
NTAG	Overall non-technical achievements (rated on a 5-point scale)	int64
PWN	Prizes won in technical events after joining the current course (rated on a 5-point scale)	int64
PBL	Publications/patents filed after joining the current course (rated on a 5-point scale)	int64
PRD	Products developed after joining the current course (rated on a 5-point scale)	int64
INT	Internships (rated on a 5-point scale)	int64
DHC	Domain expertise/hackathons (rated on a 5-point scale)	int64
LI	Leadership initiatives (rated on a 5-point scale)	int64
MIP	Punctuality/Discipline/Mentor Index (values from 0 to 10 include decimal)	float64
TPC	Soft skill ability/TPC index (values from 0 to 10 includes decimal)	float64
PGMG	Programming index (rated on a 5-point scale)	int64
LGS	Logic and problem-solving index (rated on a 5-point scale)	int64
DS	Design skills in using web design tools (rated on a 5-point scale)	int64

Table 2 Prediction case results with 30% testing data

Regression algorithm	R^2	MSE	RMSE
Lasso regression	0.96825	0.0029	0.0538
Ridge regression	0.93333	0.0038	0.0616
Linear regression	0.86392	0.0049	0.07
Decision tree	0.92145	0.0039	0.0717
Random forest	0.95673	0.0032	0.0596

Table 3 Prediction case results with 20% testing data

Regression algorithm	R^2	MSE	RMSE
Lasso regression	0.8432	0.0053	0.0728
Ridge regression	0.8148	0.0088	0.0938
Linear regression	0.7269	0.019	0.1378
Decision tree	0.8054	0.0050	0.1045
Random Forest	0.8327	0.0044	0.0821

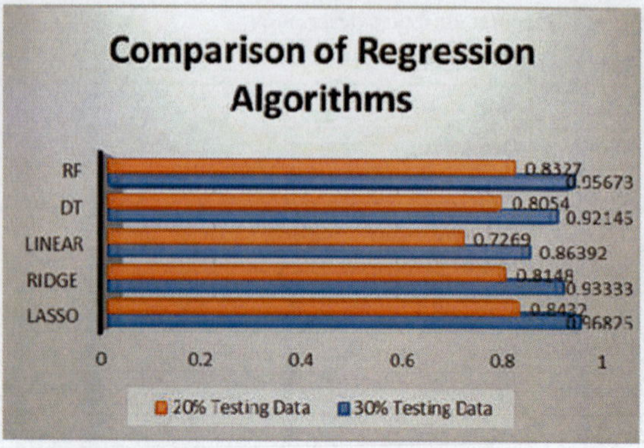

Fig. 2 Comparison of regression algorithms in 30 and 20% testing data

5 Conclusion

The motive behind this study is to predict the career prospects of non-computer engineering graduates in software firms. This research is done with five prediction algorithms and tested the results in 20 and 30% testing data. To minimize the error possibilities, in prediction, R^2, Mean Square Error and Root Mean Square Error scores were studied with two different test data ratios. Out of the five algorithms examined, Lasso Regression produced the best prediction results. The findings from these studies clearly point out that the students with programming index have more chance for employability in IT industry. On comparison with the actual placement data of three consecutive years, the attributes chosen for the data sets can be made optimum. The study can be extended with more attributes which may be relevant for the prediction process for getting more accuracy. A new method of prediction can also be tried with the data set.

References

1. Web developers and digital designers : occupational outlook handbook: U.S. bureau of labor statistics. https://www.bls.gov/ooh/computer-and-information-technology/web-developers.htm. Accessed 29 Dec 2021
2. Chilana PK, Singh R, Guo PJ (2016) Understanding conversational programmers: a perspective from the software industry. In: Proceedings of the 2016 CHI conference on human factors in computing systems, pp 1462–1472
3. Bydžovská H (2015) Are collaborative filtering methods suitable for student performance prediction? In: Portuguese conference on artificial intelligence, pp 425–430
4. Agrawal H, Mavani H (2015) Student performance prediction using machine learning. Int J Eng Res Technol 4(03):111–113
5. Kabakchieva D (2012) Student performance prediction by using data mining classification algorithms. Int J Comput Sci Manag Res 1(4):686–690
6. Kim B-H, Vizitei E, Ganapathi V (2018) GritNet: student performance prediction with deep learning. ArXiv Prepr. ArXiv18040740

7. Sweeney M, Rangwala H, Lester J, Johri A (2016) Next-term student performance prediction: a recommender systems approach. ArXiv Prepr. ArXiv160401840
8. Salal YK, Abdullaev SM, Kumar M (2019) Educational data mining: student performance prediction in academic. IJ Eng Adv Tech 8(4C):54–59
9. Bergin S, Reilly R (2005) Programming: factors that influence success. In: Proceedings of the 36th SIGCSE technical symposium on computer science education, pp 411–415
10. Ramaswami M, Rathinasabapathy R (2012) Student performance prediction. Int J Comput Intell Inform 1(4):231–235
11. Sahebi S, Brusilovsky P (2018) Student performance prediction by discovering inter-activity relations. Int Educ Data Min Soc
12. Wang X, Yu X, Guo L, Liu F, Xu L (2020) Student performance prediction with short-term sequential campus behaviors. Information 11(4):201
13. Hashim AS, Awadh WA, Hamoud AK (2020) Student performance prediction model based on supervised machine learning algorithms. In: IOP conference series: materials science and engineering, vol 928, no 3, p 032019
14. Tripathi A, Yadav S, Rajan R (2019) Naive Bayes classification model for the student performance prediction. In: 2019 2nd International conference on intelligent computing, instrumentation and control technologies (ICICICT), vol 1, pp 1548–1553
15. Thakar P, Mehta A (2017) A unified model of clustering and classification to improve students' employability prediction. Int J Intell Syst Appl 9(9):10
16. Bai A, Hira S (2021) An intelligent hybrid deep belief network model for predicting students employability. Soft Comput 25(14):9241–9254
17. Patro C, Pan I (2021) Decision tree-based classification model to predict student employability. In: Proceedings of research and applications in artificial intelligence. Springer, pp 327–333
18. Saini B, Mahajan G, Sharma H (2021) An analytical approach to predict employability status of students. In: IOP conference series: materials science and engineering, vol 1099, no 1, p 012007
19. Melin R, Fugl-Meyer AR (2003) On prediction of vocational rehabilitation outcome at a Swedish employability institute. J Rehabil Med 35(6):284–289
20. Baker RS, Yacef K (2009) The state of educational data mining in 2009: a review and future visions. J Educ Data Min 1(1):3–17
21. Wakelam E, Jefferies A, Davey N, Sun Y (2020) The potential for student performance prediction in small cohorts with minimal available attributes. Br J Educ Technol 51(2):347–370
22. Lim T-W, Khor K-C, Ng K-H (2019) Dimensionality reduction for predicting student performance in unbalanced data sets. Int J Adv Soft Compu Appl 11(2)

Sentiment Analysis and Its Applications in Recommender Systems

Bui Thanh Hung, Prasun Chakrabarti, and Prasenjit Chatterjee

Abstract E-commerce and social network make sentiment analysis be more important to businessmen. This is a process to detect opinion in text which is positive or negative. This research focuses on doing sentiment analysis and applies it to recommender system. Doing sentiment analysis, we apply both feature extraction with TF-IDF and word embedding to train the model by deep learning and machine learning. Based on sentiment analysis results, we predict the ratings number of the comments by Collaborative Filtering techniques to make recommendations to users.

Keywords Sentiment analysis · Machine learning · Deep learning · Collaborative filtering · Recommender system

1 Introduction

One of the most popular applications today is recommender system. This application is applied in some form in almost every major tech company. YouTube uses it to decide which video to play next on auto play, Amazon uses it to suggest products to customers, and Facebook uses it to recommend pages to like and people to follow. Because customers express their thoughts and feelings usually by their reviews, so to monitor and understand these reviews, sentiment analysis is becoming an essential tool with which Businesses can make faster and more accurate decisions by automatically sorting the sentiment behind reviews. From sentiment analysis, we can integrate into recommender system to predict the preference item of the user.

B. T. Hung (✉)
Data Science Department, Faculty of Information Technology, Industrial University of Ho Chi Minh city, Ho Chi Minh City, Vietnam
e-mail: buithanhhung@iuh.edu.vn

P. Chakrabarti
ITM SLS Baroda University, Vadodara, Gujarat 391510, India

P. Chatterjee
Department of Mechanical Engineering, MCKV Institute of Engineering, Howrah, West Bengal, India

P. Chatterjee et al. (eds.), *Computational Intelligence for Engineering and Management Applications*, Lecture Notes in Electrical Engineering 984,
https://doi.org/10.1007/978-981-19-8493-8_61

Machine learning is an advanced method to solve many problems of speech processing, image processing, and natural language processing, etc. Previous methods of solving sentiment analysis such as: combination method of rule-base and corpus-base (based on linguistics), Naïve Bayes, Support Vector Machine, Deep Neural Networks, … have been successfully developed and applied to many different problems. This research focuses on doing sentiment analysis and applies it to recommender system. The rest of this paper is structured as follows: Sect. 2 describes related works. The proposed model is described in details in Sect. 3. Section 4 shows the experiments. Finally, Sect. 5 summarizes our work and gives future directions.

2 Related Works

There are many ways to build a recommender system: Content Based, Collaborative Filtering, and Hybrid approach. Table 1 shows short summary about these methods [1–5].

Assume that *I* is the set of suggested item, *U* is the set of users, *u* is a specific user in the set *U*, and *i* is a specific item in *I* that we want to predict for *u* (based on *u*'s preference).

There are many ways to build recommender system. The new approach focuses on integrating sentiment analysis into it. In [6], they proposed sentiment analysis in recommender system. They used extracted features of twitter and Naïve Bayes, SVM to build the sentiment analysis model. In [7], from online social networks, they used faith and sentiment implication to improve recommendation. Preethi et al. proposed deep learning RNN to sentiment analysis and built Recommend Me application [8]. Xin Dong et al. proposed deep structure of a hybrid Collaborative Filtering model for recommender systems [9]. Bui et al. integrated sentiment analysis for recommender

Table 1 Summary of recommender system methods

Method	Database	Data Output	Process
Content based	The characteristics of the item in *I*	The ratings of *u* for the items in *I*	Describes the user *u* preferences, and rates *i* for the preference level of *u I* by the created model
Collaborative filtering	Scores of users in *U* for items in *I*	The *u*-rated scores for the items in *I*	Identify the user in *U* that is similar to *u* and then rate score *i* to *u*'s
Hybrid approach	The characteristics of the item in *I*. The knowledge about the suitability between the objects and the user's needs	A description of user needs and preferences *u*	Infer the fit between *I* and *u*'s needs

systems. They used matrix factorization for recommender systems and hybrid deep learning approach CNN-LSTM for sentiment analysis [10].

Our research focuses on sentiment analysis and integrates it in the recommender system. We use deep learning and machine learning model for sentiment analysis. We apply both feature extraction with TF-IDF and word embedding to train the model. We predict the ratings number of the comments by Collaborative Filtering techniques to make recommendations to users basing on sentiment analysis results. By integrating sentiment analysis, quality of recommender systems is improved.

3 The Proposed Model

Figure 1 presents architecture of our proposed model. We will describe each layer in detail as follows. In this model, there are two main functions: Sentiment Analysis and using Sentiment Analysis for Recommender System. From raw data, we will preprocess, after that, we extract TF-IDF features for Machine learning model and use pre-trained word embedding for deep learning model. The result of this step will be fed to calculate sentiment analysis scores. We normalize this score and use it for Recommender System by Collaborative Filtering method.

3.1 Sentiment Analysis

There are many approaches for sentiment analysis. And deep learning with word embedding is a good method [11–14]. Based on the previous researches, we apply TF-IDF feature to machine learning and word embedding to deep learning model. The approach consists of the following steps:

- Preprocessing each word of a review to its frequency value.
- We get one-hot encoding of a word by the previous process.
- We use pre-trained word2vector [15] for deep learning and extract TF-IDF features for machine learning model.
- Two machine learning models (SVM, Bayes) and two deep learning models (LSTM, Bi-LSTM) [16, 17] are used for training with input of the previous steps.
- The final layer will do activation function to make the result of a real number (from 0 to 1).

3.2 Recommender System

We normalize the probability of sentiment analysis result into a rating from 1 to 5 specifically as follows:

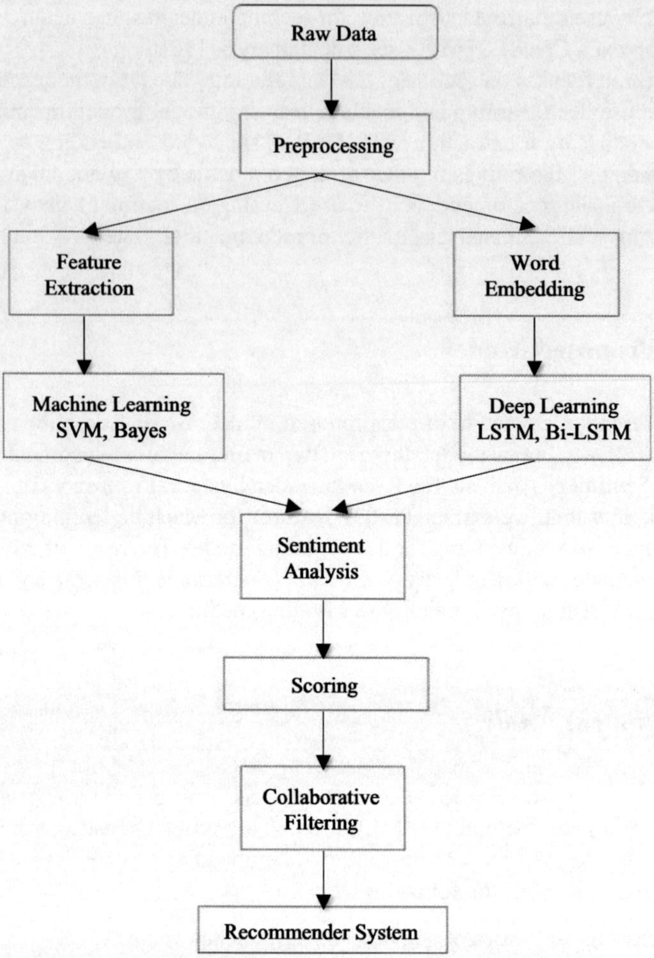

Fig. 1 The proposed model

1: 0.000–0.199
2: 0.200–0.399
3: 0.400–0.599
4: 0.600–0.799
5: 0.800–1.000.

After that, we put it into the content-based Filtering training model. Fig. 2 presents the process.

To produce a prediction value for the target user, we used a weighted average to combine the neighbor's item ratings. In Collaborative Filtering technique, an aggregation of some similar users' rating of the item is calculated as follows:

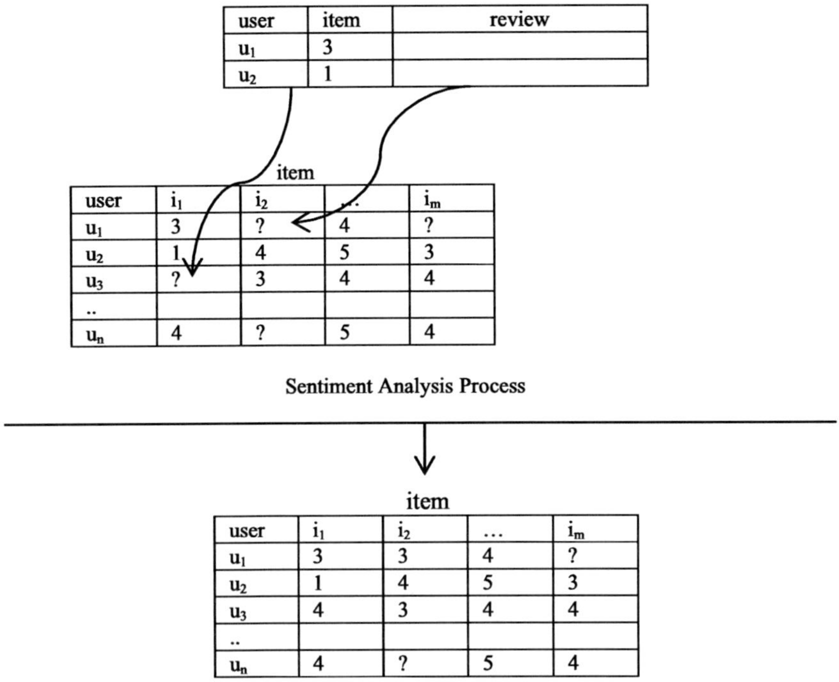

Fig. 2 Collaborative filtering recommender system with sentiment rating

$$r_{i,j} = \text{aggr}_{u' \in U} r_{u',i} \tag{1}$$

where the set of top *N* users which are most similar to user *u* rated item *i* is denoted by *U*.

With the set of items rated by both user *x* and user *y is* I_{xy}, the Pearson correlation similarity of two users *x*, *y* is calculated as follows:

$$\text{simil}(x, y) = \frac{\sum_{i \in I_{xy}} \left(r_{x,i} - \overline{r_x}\right)\left(r_{y,i} - \overline{r_y}\right)}{\sqrt{\sum_{i \in I_{xy}} \left(r_{x,i} - \overline{r_x}\right)^2} \sqrt{\sum_{i \in I_{xy}} \left(r_{y,i} - \overline{r_y}\right)^2}} \tag{2}$$

4 Experiments

Amazon food reviews are used in our experiments [18]. This data set is collected from October 1999 to October 2012 with 568,454 reviews. Table 2 described more detail about this data set.

Accuracy score is used to evaluate the result of sentiment analysis as follows:

$$\text{Accuracy} = \frac{\text{TN} + \text{TP}}{\text{TN} + \text{FN} + \text{TP} + \text{FP}} \quad (3)$$

where,

TN true negative
FN false negative
TP true positive
FP false positive.

The data is divided with ratio 8:2 of the training and testing data set. For preprocessing, we used Beautiful Soup and NLTK. The pre-trained word2vector is used in our model. We used Keras [19] and Tensorflow [20] for deep learning and Sklearn library for machine learning. We did experiments on 4 models: SVM + TF-IDF, Naive Bayes + TF-IDF, LSTM + Word Embedding, and Bi-LSTM + Word Embedding. Table 3 and Fig. 3 show the results.

From the results of Table 3 and Fig. 4, we saw that LSTM got the best score. We evaluated more detail this model by Epoch with accuracy and loss. Figs. 4, 5 and 6 show the results.

Table 2 Amazon food reviews data set

Users	256,059
Reviews	568,454
Products	74,258
Number of users has more than 50 reviews	260
Median of number of words per review	56
Time	Oct 1999 to Oct 2012

Table 3 The results in sentiment analysis

Method	Accuracy
SVM + TF-IDF	0.7956
Naive Bayes + TF-IDF	0.7532
LSTM + Word Embedding	0.8258
Bi-LSTM + Word Embedding	0.8123

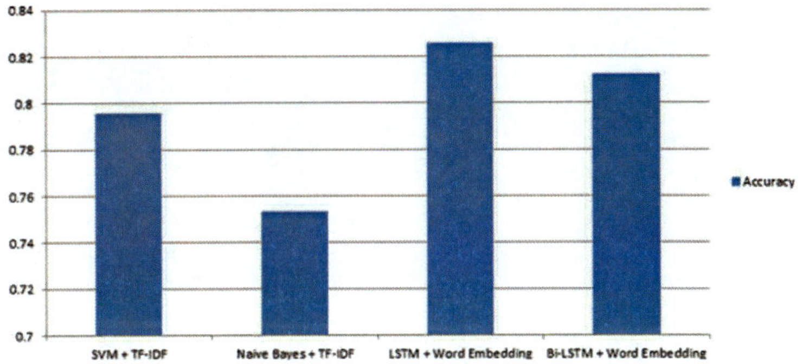

Fig. 3 Sentiment Analysis result

Fig. 4 Accuracy of LSTM model by Epoch

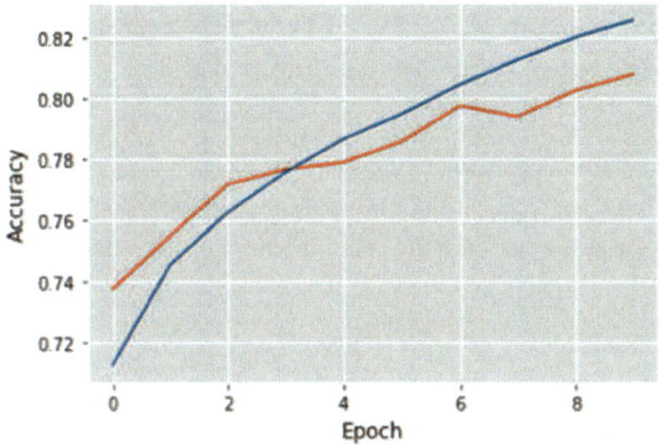

Fig. 5 Loss of LSTM model by Epoch

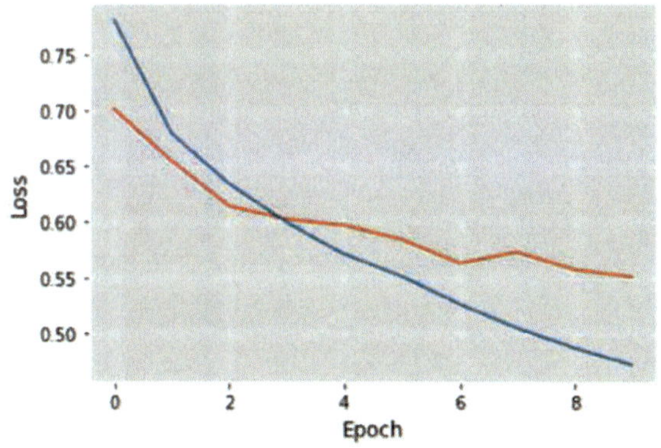

```
Epoch 9/10
900/900 [==============================] - 468s 520ms/step -
loss: 0.4863 - accuracy: 0.8201 - val_loss: 0.5574 - val_accu
racy: 0.8028
Epoch 10/10
900/900 [==============================] - 468s 520ms/step -
loss: 0.4715 - accuracy: 0.8258 - val_loss: 0.5510 - val_accu
racy: 0.8081
```

Fig. 6 The results of LSTM model

Table 4 Different recommender system performances

Model	RMSE
Recommender system-based sentiment analysis	**1.1245**
Popular (baseline)	1.6572

As we presented in Session 3, we integrated the result of the best model of sentiment analysis into recommender systems. RMSE metric—Root Mean Square Error is used to evaluate the recommender system as,

$$\text{RMSE} = \sqrt{\frac{1}{n}\sum\nolimits_{u,i}(p_{u,i} - r_{u,i})^2} \quad (6)$$

where,

u user

i item

n is the total number of ratings

$r_{u,i}$ is the actual rating

$p_{u,i}$ is the predicted rating.

We compare baseline with our proposed model, and the results are shown in Table 4.

5 Conclusion

We proposed integrating sentiment analysis in recommender system in this paper. We did experiments of sentiment analysis in four models based on deep learning and machine learning approach, in which deep learning model—LSTM + Word Embedding got the best score. We built recommender system based on sentiment analysis result with Collaborative Filtering technique. The result shows that integrating sentiment analysis is helpful to the recommender system. We would like to explore sentiment analysis deeply and integrate it into a recommender system with some additional functions to get the better results in the future.

References

1. Lops P, De Gemmis M, Semeraro G (2011) Content-based recommender systems: state of the art and trends. In: Recommender systems handbook. Springer US, pp 73–105
2. Bobadilla J, Ortega F, Hernando A, Gutierrez A (2013) Recommender systems survey. Knowl-Based Syst 46:109–32
3. Isinkaye FO, Folajimi YO, Ojokoh BA (2015) Recommendation systems: principles, methods and evaluation. Egypt Inform J 16(3):261–273
4. Schafer JB et al (2007) Collaborative filtering recommender systems. The adaptive web. Springer Berlin Heidelberg, pp 291–324
5. Sheetal G, Mukhopadhyay D et al (2015) Role of matrix factorization model in collaborative filtering algorithm: a survey
6. Yang X, Guo Y, Liu Y (2013) Bayesian-inference-based recommendation in online social networks. IEEE Trans Parallel Distrib Syst 24(4):642–651
7. Anto MP, Antony M, Muhsina KM, Johny N, James V, Wilson A (2016) Product rating using sentiment analysis. In: International conference on electrical, electronics, and optimization techniques (ICEEOT), 2016, pp 3458–3462
8. Preethi G, Krishna PV, Obaidat Mohammad S, Saritha V, Sumanth Y (2017) Application of deep learning to sentiment analysis for recommender system on cloud. In: International conference on computer, information and telecommunication systems (CITS), 2017, pp 93–97
9. Dong X, Yu L, Wu Z, Sun Y, Yuan L, Zhang F (2017) A hybrid collaborative filtering model with deep structure for recommender systems. In: AAAI, pp 1309–1315
10. Hung BT (2020) Integrating sentiment analysis in recommender systems. In: Reliability and statistical computing. Springer Series in Reliability Engineering, pp 252–264
11. Irsoy O, Cardie C (2014) Opinion mining with deep recurrent neural networks. In: Proceedings of the 2014 conference on empirical methods in natural language processing (EMNLP-14), pp 720–728
12. HungBT (2018) Domain-specific versus general-purpose word representations in sentiment analysis for deep learning models. In: The 7th international conference on frontiers of intelligent computing: theory and applications, FICTA
13. Hung BT, Semwal VB, Gaud N, Bijalwan V (2021) Hybrid deep learning approach for aspect detection on reviews. In: Proceedings of integrated intelligence enable networks and computing. Springer Series in Algorithms for Intelligent Systems
14. Li Y, Pan Q, Yang T, Wang S, Tang J, Cambria E (2017) Learning word representations for sentiment analysis". Cogn Comput 9(6):843–851
15. Mikolov T, Chen K, Corrado G, Dean J (2013) Efficient estimation of word representations in vector space. In: Proceedings of international conference on learning representations (ICLR-13): workshop track
16. Hochreiter S, Schmidhuber J (1997) Long short-term memory. Neural Comput 9(8):1735–1780
17. Hung BT (2021) Combining syntax features and word embeddings in bidirectional LSTM for vietnamese named entity recognition. Further Adv Internet of Things Biomed Cyber Phys Syst
18. McAuley J, Leskovec J (2013) From amateurs to connoisseurs: modeling the evolution of user expertise through online reviews. WWW
19. Chollet F, Keras (2015) https://github.com/fchollet/keras

20. Abadi M, Barham P, Chen J, Chen Z, Davis A, Dean J, Devin M, Ghemawat S, Irving G, Isard M, Kudlur M, Levenberg J, Monga R, Moore S, Murray DG, Steiner B, Tucker P, Vasudevan V, Warden P, Wicke M, Yu Y, Zheng X (2016) Tensorflow: a system for large-scale machine learning. In: Proceedings of the 12th USENIX conference on operating systems design and implementation, OSDI'16, pp 265–283

Two-Stage Model for Copy-Move Forgery Detection

Ritesh Kumari, Hitendra Garg, and Sunil Chawla

Abstract Digital images can be tempered or modified very easily with many tools available on the internet. Nowadays some software not only enhance and retouch but also copy-move objects of the image in such a way that is hard to differentiate. Image forgery detection deals with the recognition of tempering like splicing, copy-move, and other modifications done on the image. One of the most common ways of passive forgery detection is the block-based copy-move technique. In this paper, forgery detection is achieved in two parts. In the first part, Copy-Move forgery is classified by support vector machine (SVM) and features are extracted by local binary pattern (LBP) and discrete cosine transform (DCT). However, the region of forgery is calculated in the second part with the help of the Gray-level Co-occurrence Matrix (GLCM) through overlapping blocks. The whole experiment and evaluations are performed on the CoMoFoD image data set. The performance is measured in terms of accuracy and sensitivity. The promising result indicates the effectiveness of the model with enhanced accuracy of 97.82%.

Keywords Local binary pattern · Gray-level co-occurrence matrix · Copy-Move forgery detection

1 Introduction

Internet especially social media works on the phenomenon of good impression, in short, what you see is exactly what you believe and this is the main reason for

---

R. Kumari (✉) · H. Garg
Department of Computer Engineering and Applications, GLA University, Mathura, India
e-mail: myselfritesh@gmail.com

H. Garg
e-mail: hitendra.garg@gla.ac.in

S. Chawla
Department of Computer Science and Engineering, Chandigarh University, Mohali, India
e-mail: drskchawla1983@gmail.com

P. Chatterjee et al. (eds.), *Computational Intelligence for Engineering and Management Applications*, Lecture Notes in Electrical Engineering 984,
https://doi.org/10.1007/978-981-19-8493-8_62

spreading false information online like a fire. Software such as Pixir and Adobe Photoshop for image formatting and filtering are very popular nowadays which eventually increase the forged content. With the advancement of such software, it's hard to identify the real image. Many morphed images are circulated on the internet with false news and claims that ultimately spread rumors and misguide people.

Image forgery deals with the morphed, modified, and copied images to find authenticity. There are two basic types of forgery detection in the case of digital images. The first category is active and the second is passive or sometimes called the blind forgery detection method.

In an active detection approach, signature and watermarking are generally used. This kind of detection requires prior information for further processing. The drawback with watermarking is, the impression should be embedded at the time of image formation or before its circulation on the internet which is not feasible each time.

On the other hand, passive detection doesn't need any prerequisite. It is based on statistics of the image. Therefore, this technique is also known as blind image forgery detection. These passive methods can be further classified based on several characteristics. Detection of forged images can be done based on statistics like pixels, image formats, geometry, light, shadow, etc. In the proposed method, two research questions are focused mentioned below:

RQ1: Whether the given input image is real or manipulated?
RQ2: If the image suffers from any tempering, where is the region of forgery?

To resolve these questions, two different approaches are presented. In the first approach, classification of the image is done by applying LBP and DCT, followed by SVM. Whereas in the second approach, the forged area is identified by applying GLCM on that image. The implemented scheme of applying LBP and DCT on the image leads us to achieve the following benefits:

- It can detect multiple forgeries.
- Resolved the attacks such as scaling and compression.
- Once the model is trained, the computational time reduces.

The rest of the paper is structured in the manner given below; after the introduction, Sect. 2 consists of recent and literature studies of the topic. While in Sect. 3, the copy-move forgery technique which combines DCT and LBP in the first part and GLCM with the overlapping block is explained in the second part. The next section analyses the result and presents the comparison tables. Section 5 lightens on the conclusions and future work for CMFD.

2 Background

Manipulation of digital images is very common nowadays. This manipulation is sometimes done for personal use and sometimes it may be forged or misused with the wrong intention. Several methods have already been introduced till now which are listed below. Background study comprised of several review papers from [1–6].

Multiple data sets [1] are available to detect and study forgery cases. These data set contain images with one or multiple formats according to threats like splicing, blurring, rotation, noise, and scaling. Some of them are CASIA v 1.0 and v 2.0, CoMoFoD, MICC-F, GRIP, FAU, etc., where FAU and CoMoFoD contains JPEG images compression threats. Generally, most used data sets are selected for further studies as the same data set can compare multiple implemented approaches at the same time.

Forgery detection techniques depend upon the types of morphing or tempering. If the initial information about the input image is not needed during the examination, the type of forgery detection is called passive methods. Passive forgery detection uses characteristics like pixel, format, geometrical graphic, etc. Active forgery detection normally applied watermarking, codes, or patterns on a host image to save it from threats and attacks.

A detailed study on various classification approaches like SVM, KNN, and Naïve Bayes [2] are explained and compared for the passive approach. Deep learning using CNN [3] is another very effective method for copy-move forgery detection. As it learns during the training phase, its processing time is very less and results are accurate. Gajanan et al. [4] reviewed many passive forgery techniques with their merits and demerits. The complex mathematical models can be substituted with CNN, its Implementations and application are explained in [5]. In a similar study [6], attacks based on the pixel, format, physics, geometry, and camera are summarized.

Comparison is very important to study the performance of multiple classifiers in terms of sensitivity, accuracy, robustness, etc. Classification techniques like support vector machine (SVM), K-Nearest Neighbor (KNN), Naïve Bayes (NB), etc., are used to detect the manipulated images. A review on comparison of three classification techniques such as Naïve Bayes, SVM, and KNN is done in [7], which shows that KNN tops for splicing detection with the highest accuracy. In this study, the most common tempering method, that is, copy-move forgery detection has been analyzed.

Common techniques utilized for the detection of copy-move are block-oriented algorithms and key point-oriented methods. In key-point detection, the main features of an image are extracted and then matched with the original input image. Papers from [8–17] are included for deep study and recent advancements in forgery techniques for CMFD.

For the detection of copy-move tempering, there are multiple methods available like PCA, DWT, FMT, block matching via DCT, etc. A combination of two forgery detection, that is, copy-move and splicing have been implemented using block DCT coefficients [8] and SVM. It can also be detected by exploiting Shannon entropy [9].

While in block-oriented algorithms, the image is divided into blocks which are later compared with pixel value or coefficient values. First, the input image is transformed into 4×4 or 8×8 blocks. The RGB is converted into grayscale and lastly, block matching is compared based on pixel sensitivity, ROC, precision, f-measure, etc. The Superpixel segmentation [10] approach currently is very popular as it can change the fixed structure of pixels for any image. SIFT is used for the detection and matching of local features in the input image. Sometimes when it is hard to resolve geometrical image forging, Hilbert transforms [10] can be a good option as it is compatible with the rotational and scaled image and able to locate coordinates.

Another approach, using adaptive over-segmentation [11] can divide the image into non-overlapping blocks. In conventional methods, block sizes are non-overlapped and fixed, whereas in over-segmentation, the input image is categorized into non-overlapped and variable blocks so, the initial size of superpixel can be selected as per need.

For copy-move tempering, Cao et al. [12] presented a DCT approach with the selected number of features that have been sorted through lexicographers. DCT mainly prefers only passive forgery and doesn't account for digital watermarking or active forgery cases. Mahmood et al. [13] applied DCT and Gaussian RBF kernel PCA on overlapping square blocks to achieve a high detection rate. Mahdian and Saic [14] used the blur moment invariant with the help of principal component transform (PCT) for CMFD. Another technique, DWT-DCT (QCD) is presented in [15] to find cloning forgeries.

GLCM [16] is applied to each orientation of the image to extract features. It can classify characteristics like co-relation, variance, entropy, etc. Kaur et al. [17] applied LBP and DFrCT to detect forgery. This method increases the accuracy rate and detects multiple objects at the same time. In the next section, the proposed scheme is explained.

3 Proposed Methodologies

Copy-move forgery is the method where another object (s) from a different image is pasted on the original image. It may generate duplicity, hide real content, and spread false information. Copy-move forgery detection gives two different kinds of solution: block-based and key-based. Both approaches have some advantages and limitations. The experimental setup is divided into two parts. The proposed model is divided into two parts. Our first motive is to find, whether a given image is forged or not. For this purpose, the SVM classification model is used to identify, if the image is tempered. The second part is to locate the particular region in the image in case of SVM identify any forgeries. Figure 1 shows how training and testing are done using an SVM classifier. For the implementation of the proposed model CoMoFoD data set is used. This data set consists of many images with various threats.

To train the proposed model, as shown in Fig. 1, the features are first extracted from the CoMoFoD data set and then data is divided into two parts; training set (X_{tr}, Y_{tr}) and testing set (X_t, Y_t), where X is the input feature vector and Y is the Expected Output.

Data set used: Data set CoMoFoD is used for the implementation of the proposed technique. It contains 260 images (200 small size + 60 big size images) of 512 ×

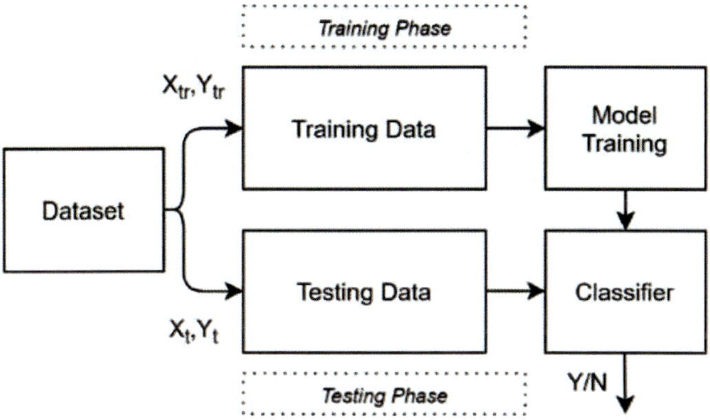

Fig. 1 The basic structure of the classification model

Table 1 Details of data set CoMoFoD

Data set	Image size	Image type	Authentic image	Forged Image (40 per image)	Total image
CoMoFoD	512 × 512	PNG	200	8000	10,400
	3000 × 2000		60	2400	

512 and 3000 × 2000 pixels. It is comprised of transformations and attacks like jpeg compression, noise, masking, distortion, and rotation (Table 1).

3.1 *Part-1: Copy-Move Classification*

This section belongs to the classification of the image that decides whether the image is forged or not. Figure 2 explains the basic process, where LBP and DCT are exploited for feature retrieval and SVM is used as a classifier.

Following are the steps used in copy-move classification.

Pre-processing: In this step, the image is split into Y, C_b, C_r channels, where Y is luminance and C_b, C_r represent the chrominance component. Y channel preserved quality content compared to C_b, C_r. Chrominance part helps us to identify tampering artifacts of image. Equation 1 shows how to calculate Y, C_b, C_r value from the R, G, B channel. For processing, the image is converted into Y, C_b, C_r channels. After pre-processing step, features are extracted from the image. After that feature vectors are passed into the classifier and then the classifier predicts that the given input image is forged or not (Fig. 3).

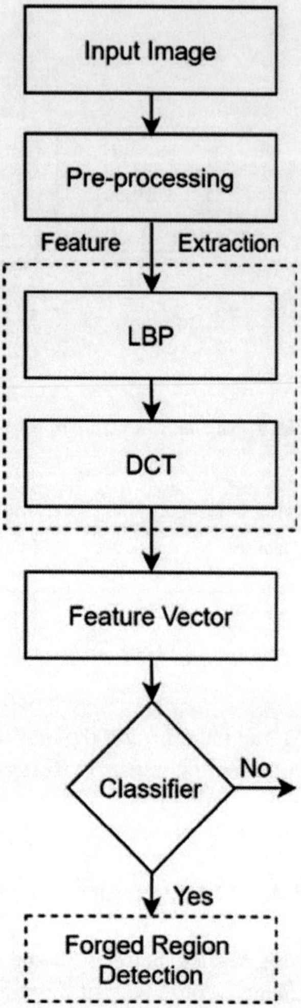

Fig. 2 Proposed model Part-1, used for classification

$$\begin{bmatrix} Y \\ C_b \\ C_r \end{bmatrix} = \begin{bmatrix} 16 \\ 128 \\ 128 \end{bmatrix} + \frac{1}{256}\begin{bmatrix} 65.738 & 129.057 & 25.064 \\ -37.495 & -74.494 & 112.439 \\ 112.439 & -94.154 & -18.285 \end{bmatrix}\begin{bmatrix} R \\ G \\ B \end{bmatrix} \tag{1}$$

Local Binary Pattern: It is widely used for measuring the change in illumination. The biggest benefit of using LBP [17] is the computational simplicity and prone to higher accuracy. It is used as a descriptor that correlates the minute change in block edgings at the time of forgery. The edging of an original image block is usually smooth, while copy-move breaks the continuity of edging. This statistical change is captured by LBP. In this process, it captures the local information between nearby pixels. The function for LBP is given below.

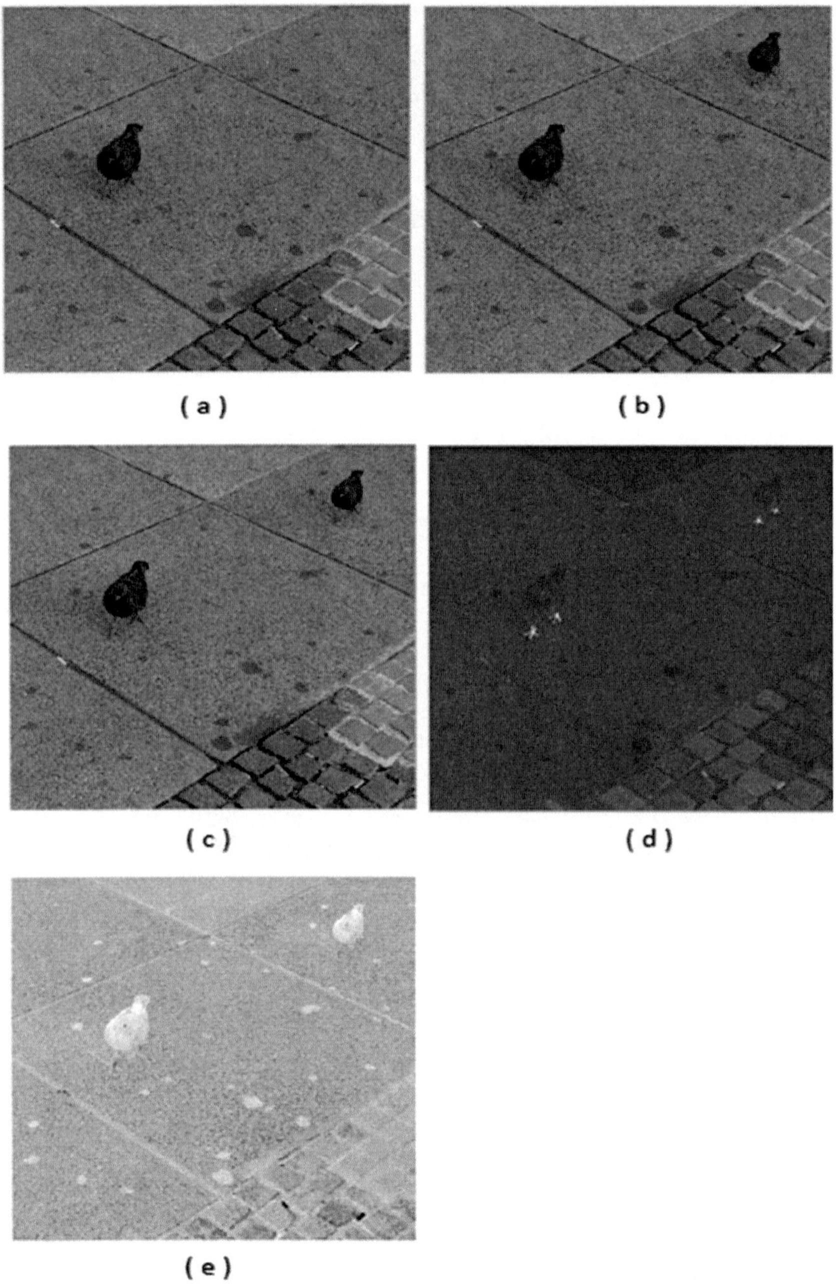

Fig. 3 **a** Original image, **b** Tempered image, **c**, **d**, and **e** Y, C_b, and C_r channel of the original image

Fig. 4 A simple example of LBP

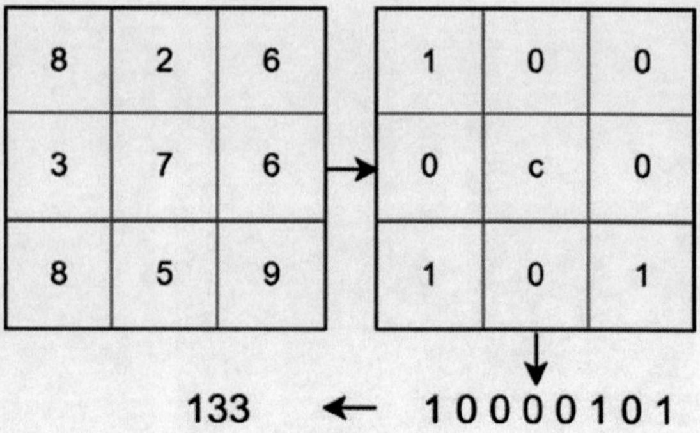

$$\text{LBP}_{M,N} = \sum_{i=0}^{M-1} T(m_i - m_c)2^i \tag{2}$$

where, M is the number of pixels, m_i is the neighboring pixel, m_c is the center pixel used as a threshold, N is the radius of the neighboring pixel, and $T(m_i - m_c)$ is the threshold function which can be expressed as,

$$T(m_i - m_c) = \begin{cases} 1, m_i - m_c \geq 0 \\ 0, m_i - m_c < 0 \end{cases} \tag{3}$$

LBP is then connected to DCT to convert the local statistical change into the frequency domain (Fig. 4).

Discrete Cosine Transformation: A fast transform that can locate the region where intensity abruptly changes and gives signs for tempering in terms of value for the frequency domain. DCT [8, 15] transforms the same size blocks and normally holds the same values. Here, each block holds relative information in terms of AC coefficients and the remaining elements or DC coefficients are mapped to zero. Quantization is performed to select only featured values.

The general equation for a 2D (N by M image) DCT is defined by the following equation:

$$F(u, v) = \left(\frac{2}{N}\right)^{\frac{1}{2}} \left(\frac{2}{M}\right)^{\frac{1}{2}} \sum_{i=0}^{N-1} \sum_{j=0}^{M-1} A(i) \,.\, A(j) \,.$$
$$\cos\left[\frac{\pi.u}{2.N}(2i+1)\right] \cos\left[\frac{\pi.v}{2.M}(2j+1)\right] . f(i, j) \tag{4}$$

For $N \times M$ input image, $f(i)$ is the intensity of pixel (i, j) and $F(u)$ is DCT coefficient.

Support Vector Machine: It is a well-known classifier for binary classification. It is derived to find the dimension of the hyperplane between two classes. Those training samples which lie on decision boundaries are known as support vectors. To implement the proposed method, SVM [8, 17] is used with radial basis function (RBF) kernel. RBS generally used in non-linear binary classification problems. Here, five-fold cross-validation is performed to find optimal accuracy.

3.2 Part-2: Forged Region Detection

The second part of the proposed model is to locate the forged region. The block diagram for this is shown in Fig. 5. After identifying that the given image is forged, we need to locate the tempered area on the image which is copied and pasted.

To find the forged area, the following steps are performed.

Step 1: First convert forged image into the grayscale format.

Step 2: Grayscale image is segmented into 6 × 6 overlapping blocks.

Step 3: Features are extracted from each block using GLCM feature extraction techniques.

Step 4: Each block's features are arranged into row format. This row arrangement of the block makes a feature matrix M of each block.

Step 5: These feature matrices are arranged in order (sorted) using the lexicographical sorting algorithm.

Step 6: In this step, the similarity between each block is calculated by Euclidean distance between M_i and M_{i+1} feature block matrix. Since distance is already sorted, so we calculate distance between adjacent matrixes.

Blocks with small distances represent two identical blocks. To show the matching blocks on the image, initially a blank image of the same size (equal to the original image) is created. After those, white patches (size of overlapping mask) are generated on the blank image, where two blocks are matched.

Gray-level Co-occurrence Matrix (GLCM): It studies the textural gap between spatial relations of pixels and is based on second-order statistics. It calculates the GLCM function by the occurrence of some specific value pixels and then transforms them into a matrix. It can calculate the homogeneity, correlation, contrast, and energy or uniformity.

$$\text{Contrast} = \sum_{i,j=0}^{N-1} \left(P_y(i-j)^2\right) \tag{5}$$

$$\text{Entropy} = \sum_{i,j}^{N-1} \left(\frac{P_y}{1+(i-j)^2}\right) \tag{6}$$

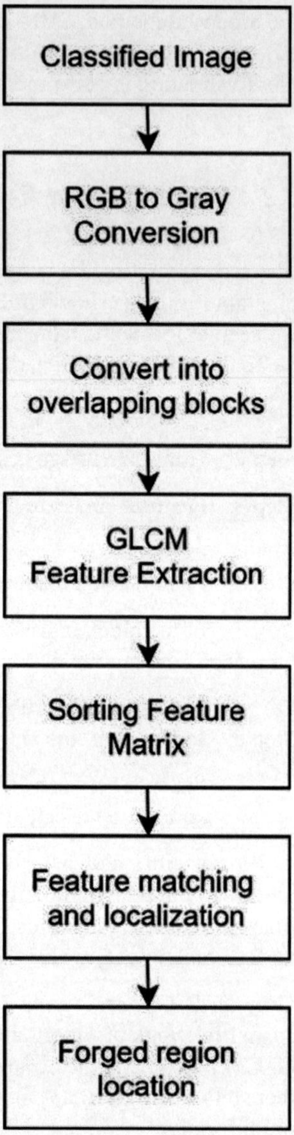

Fig. 5 Proposed model Part-2, forged region extraction

$$\text{Energy} = \sum_{i,j=0}^{N-1} \left(P_y\right)^2 \tag{7}$$

$$\text{Correlation} = \frac{\sum_{i,j=0}^{N-1}\left(P_y(i-\mu)(j-\mu)\right)}{\sigma^2} \tag{8}$$

It is used for feature extraction for finding the forged area as shown in Fig. 5. Once the features are extracted, matching and localization take place to identify the forged area.

4 Result Analysis

To perform the proposed model, MATLAB.2014a is used on Windows 10 (8 GB RAM).

As described in the previous section, the CoMoFoD data set is used. For one genuine image, only two forged images are taken for classification purposes. So, for 260 genuine images, we have 520 forged images. The whole data set is divided into three parts: 80% for training, 10% for testing, and the remaining 10% for validation. Fivefold cross-validation is used to find overall optimal performance. Table 2 shows the average value of confusion matrix after fivefold cross-validation.

The performance of the model is measured in Accuracy, Sensitivity, and Specificity as shown in Table 3.

$$\text{Accuracy} = \frac{\text{TP} + \text{TN}}{\text{TP} + \text{TN} + \text{FP} + \text{FN}} \times 100\% \tag{9}$$

$$\text{Sensitivity} = \frac{\text{TP}}{\text{TP} + \text{FN}} \times 100\% \tag{10}$$

$$\text{Specificity} = \frac{\text{TN}}{\text{TN} + \text{FP}} \times 100\% \tag{11}$$

$$\text{TPR} = \frac{\text{TP}}{\text{TP} + \text{FN}} \tag{12}$$

$$\text{FNR} = \frac{\text{FN}}{\text{FN} + \text{TP}} \tag{13}$$

$$\text{TNR} = \frac{\text{TN}}{\text{TN} + \text{FP}} \tag{14}$$

$$\text{FPR} = \frac{\text{FP}}{\text{FP} + \text{TN}} \tag{15}$$

Table 2 Confusion matrix

		Actual class	
		Genuine	Forged
Predicted class	Genuine	254	6
	Forged	11	509

Table 3 Performance measurement using accuracy, sensitivity, specificity

Accuracy	Sensitivity	Specificity
97.82%	97.69%	97.88%

Table 4 Performance measurement using TPR, TNR, FPR, FNR

TPR	TNR	FPR	FNR
0.976	0.978	0.021	0.023

The results are also analyzed using True Positive Rate, False Negative Rate, True Negative Rate, and False Positive Rate as shown below in Tablc 4. The performance output of the second part of the model is shown in Fig. 6. It shows five sample images: the first column indicates original images, the second column shows forged images, while the third column shows the mapping of copy-move forged area in white color. The mapping part (third column) images are enhanced by using region filling or dilation to show a clear visual of the forged area.

5 Conclusion

In this paper, we proposed a new multistage technique to identify the forged image. The whole model is divided into two parts. The first part is a classification model. For this, CoMoFoD data set is used. The second part of the model detects the area of the image where forgery happened. Though the main limitation still is, it doesn't deal with rotation. So, a study on combining two or more techniques and the use of deep learning would be promising for future research.

CNN is very effective and simple as used for both feature extraction and classification. The use of DCT for copy and move forgery is very promising as it can handle attacks like Gaussian blurring, noise, and multiple tempering. It also reduces computational complexity. While working with a data set, it is very important to extract only selected features rather than all or many. It not only increases accuracy, precision but also reduces computational complexity. Therefore, a big data set that contains all compression and attacks is a big need.

For future work, instead of designing a model into two parts, we propose to design a generalized model which should be capable to classify images based on types of forgery. For example, SVM can be trained to classify copy-move and spliced forgery detection.

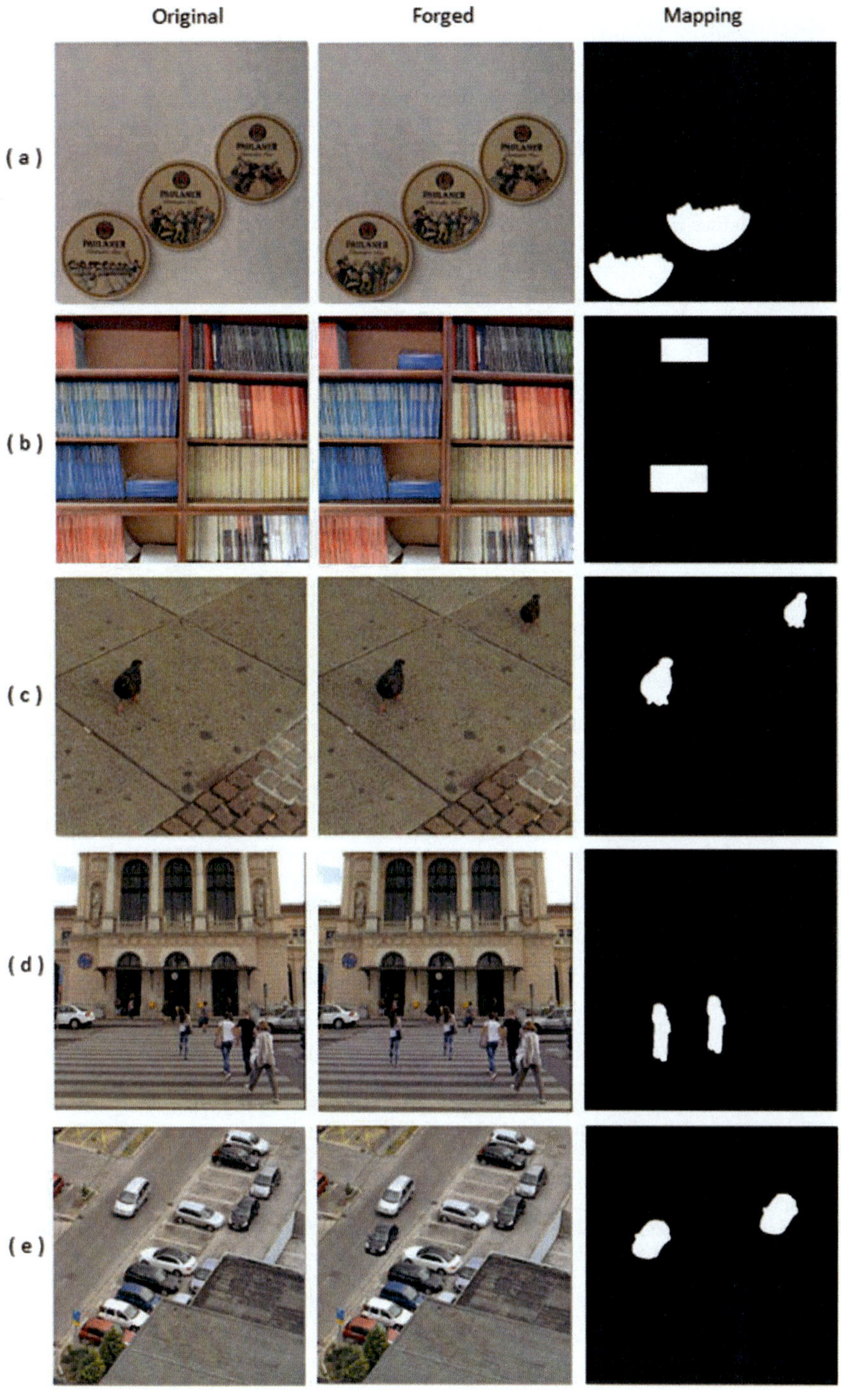

Fig. 6 Original, forged, and mapping of the forged area of five different images

References

1. Bansal A, Atri A (2020) Study of copy-move forgery detection techniques in images. In: 2020 8th International conference on reliability, infocom technologies and optimization (trends and future directions) (ICRITO) (2020), pp 630–635. https://doi.org/10.1109/ICRITO48877.2020.9197801
2. Prayla Shyry S, Meka S, Moganti M (2019) Digital image forgery detection. Int J Recent Technol Eng (IJRTE) 8(2S3). ISSN: 2277-3878. https://doi.org/10.35940/ijrte.B1121.0782S319
3. Abidin ABZ, Majid HBA, Samah ABA, Hashim HB (2019) Copy-Move image forgery detection using deep learning methods: a review. In: 2019 6th International conference on research and innovation in information systems (ICRIIS) (2019), pp 1–6. https://doi.org/10.1109/ICRIIS48246.2019.9073569
4. Birajdar GK, Mankar VH (2013) Digital image forgery detection using passive techniques: a survey. Digital Invest 10(3):226–245. ISSN 1742-2876. https://doi.org/10.1016/j.diin.2013.04.007
5. Shwetha B, Sathyanarayana SV (2017) Digital image forgery detection techniques: a survey. Accents Trans Inf Secur 2(5). ISSN (Online): 2455–7196. http://dx.doi.org/https://doi.org/10.19101/TIS.2017.25003
6. Farid H (2009) Image forgery detection: a survey. IEEE Signal Process Mag. Digital Object Identifier. https://doi.org/10.1109/MSP.2008.931079
7. Almawas L, Alotaibi A, Kurdi H (2020) Comparative performance study of classification models for image-splicing detection. Procedia Comput Sci 175:278–285. ISSN 1877-0509. https://doi.org/10.1016/j.procs.2020.07.041
8. Dua S, Singh J, Parthasarathy H (2020) Image forgery detection based on statistical features of block DCT coefficients. Procedia Comput Sci 171:369–378. ISSN 1877-0509. https://doi.org/10.1016/j.procs.2020.04.038
9. Savakar DG, Hiremath R (2020) Copy-Move image forgery detection using Shannon entropy. Appl Comput Vision Image Process Adv Intell Syst Comput 1155:76–90. https://doi.org/10.1007/978-981-15-4029-5_8
10. Huang HY, Ciou AJ (2019) Copy-move forgery detection for image forensics using the superpixel segmentation and the Helmert transformation. EURASIP J Image Video Process 2019:68. https://doi.org/10.1186/s13640-019-0469-9
11. Pun C, Yuan X, Bi X (2015) Image forgery detection using adaptive over-segmentation and feature point matching. IEEE Trans Inf Forensics Secur 10(8):1705–1716. https://doi.org/10.1109/TIFS.2015.2423261
12. Cao Y, Gao T, Fan L, Yang Q (2012) A robust detection algorithm for copy-move forgery in digital images. Forensic Sci Int 214(1–3):33–43. ISSN 0379-0738. https://doi.org/10.1016/j.forsciint.2011.07.015
13. Mahmood T, Nawaz T, Irtaza A, Ashraf R, Shah M, Mahmood MT (2016) Copy-Move forgery detection technique for forensic analysis in digital images. https://doi.org/10.1155/2016/8713202
14. Mahdian B, Saic S (2007) Detection of copy-move forgery using a method based on blur moment invariants. Forensic Sci Int 171(2–3):180–189
15. Ghorbani M, Firouzmand M, Faraahi A (2011) DWT-DCT (QCD) based copy-move image forgery detection. In: 2011 18th International conference on systems, signals and image processing (2011), pp 1–4
16. Babu SBGT, Rao CS (2021) An optimized technique for copy move forgery localization using statistical features. ICT Express. https://doi.org/10.1016/j.icte.2021.08.016
17. Kaur N, Jindal N, Singh K (2021) Efficient hybrid passive method for the detection and localization of copy-move. Turk J Elec Eng Comp Sci 29:561–582. TÜBİTAK.https://doi.org/10.3906/elk-2001-138

Mathematical Model for Broccoli Growth Prediction Based on Artificial Networks

Jessica N. Castillo and Jose R. Muñoz

Abstract Broccoli is considered a nutritious vegetable that contains Vitamin C and A, Calcium, Iron, and Sulfur. It also has properties that help to control diabetes and antimicrobial, sulforaphane that kills cancer cells and controls blood pressure. The data were collected from the Freire farm located in the canton of Latacunga. Due to the importance of broccoli, it was necessary to develop a mathematical model that predicts its growth, using the Multilayer Perceptron Neural Network, which is efficient when classifying and finding patterns. This model simulates the growth in two scenarios, the data were collected in the field, thus building the network with 10 input neurons, 7 in the hidden layer, and two output neurons. The result was the height of the stem and the diameter of the foil. When the mathematical model was validated, it became evident that it is efficient. The parameters of the inputs can be modified and the behavior of the broccoli can be evidenced without the need to carry out physical experiments, this saves time and money.

Keywords Mathematical model · Broccoli · Floret · Perceptron neural network

1 Introduction

In recent years, broccoli production has become a non-traditional product of major export, which benefits the country's economic system. Therefore, its production must be subject to a rigorous process and especially consider several exogenous and endogenous factors that may affect production. This vegetable must also be provided with sufficient amounts of phosphorus and potassium that make it appetizing in national and international markets, knowing the process of planting and cultivation is important because in this process it is known that the adversities that may occur,

J. N. Castillo (✉) · J. R. Muñoz
Escuela Superior Politécnica de Chimborazo, Riobamba, Ecuador
e-mail: jecastillo@espoch.edu.ec

J. R. Muñoz
e-mail: jose.munoz@espoch.edu.ec

P. Chatterjee et al. (eds.), *Computational Intelligence for Engineering and Management Applications*, Lecture Notes in Electrical Engineering 984,
https://doi.org/10.1007/978-981-19-8493-8_63

mainly have strategies or alternatives to mitigate potential problems that affect the production of broccoli, the use of chemical fertilizers that must be used properly in order not to cause toxic residues in the plant, as they are composed of nutrients that become integral elements of the chemical structure of the plant [1].

Farmers currently continue to use precarious sowing techniques, as they focus on obtaining a greater volume of production, leaving aside the quality and care of the soils that affect the production of this vegetable, for this method of planting farmers apply fertilizers in excess which represent 20–30% of the production costs of the crop, deteriorating the quality and fertility of the soils that have a suitable climate for agricultural production [2].

The farmers of the country should also consider the incidence of different factors that can help in the production process of broccoli, where the proper distribution of fertilizers will allow improvements that will safeguard the planting process, since unnecessary additions of fertilizers, phosphorus, nitrogen, and potassium applied to the soil generate disadvantages in production. Lighting is an essential source for plant photosynthesis, and it plays a crucial role in their growth and development. Plants can respond to the intensity and color of light through their photoreceptors, such as phototropins, cryptochromes, and phytochromes, which are precise in regulating their development and growth under different environmental conditions. Therefore, lighting systems for production in controlled environments are of utmost importance, and technological advances in this area will facilitate better generation and forecasting [3].

The research is based on the design of a mathematical model for the prediction of broccoli growth at Finca Freire in greenhouse #1 located in Latacunga using neural networks, for which parameters, propositions of real facts and the existing relationship between the operational variables involved in the broccoli production process will be considered, this model will allow consolidating a quality production.

Currently, Greenhouse 1 does not have a computer tool that predicts broccoli growth by modifying chemical parameters, and what would be the behavior of its development, so the main objective is to perform a physical analysis of growth, develop a mathematical model with neural networks in a Matlab interface. Considering that the Freire Farm relies on historical or empirical facts to make its production decisions, this generates inadequate alternatives to achieve the effective development of each stage of the broccoli legacy crop.

The design of a model that provides tools to support each stage of broccoli production, allows to maintain an adequate analysis of the real situation of the crop, providing alternatives that favor decision-making and mitigate possible problems that may arise.

2 Broccoli Production in Ecuador

Broccoli is currently considered a source of calcium and phosphorus, mainly consumed in Eastern countries due to its nutritional and health properties [4]. According to the Food and Agriculture Organization of the United Nations, broccoli

Table 1 Broccoli planting, harvesting, production, and sales in Ecuador 2020

Year	Sowings ha	Harvest ha	Production TM	Sales TM
2014	6.86800	6.86800	113.02900	112.70300
2015	7.81695	7.60605	107.38614	106.14937
2016	5.52026	5.51902	74.19008	73.11059
2017	7.21375	7.19250	114.27186	110.65646
2018	11.46184	11.43127	188.09455	186.75596
2019	9.92343	9.91901	169.72475	164.04817
Average	8.13404	9.08931	127.78273	125.57059
Maximum	11.46184	11.43127	188.09455	186.75596
Minimum	5.52026	5.52902	74.19008	73.11059

consumption reached a production of 38,241,388 tons worldwide in 2019, of which 81% is concentrated in three countries, China, mainland China, and India. Mexico is in fifth place with 2% and Ecuador in twenty-third place with 0.29%. In terms of hectares planted, India, China, and the United States occupy the top positions, while Ecuador has increased its production to 13th place in the list of 98 countries [5]. Ecuador is a country with a great biodiversity, spatially enjoys a climate that favors the production of this vegetable, but today this potential is not being exploited. This means that Ecuador needs more hectares of land to produce broccoli, compared to other countries that produce more in smaller tracts of land. Broccoli crops are transitory in Ecuador, between 2017 and 2020, an average of 9000 hectares of this vegetable were planted, achieving a harvest equivalent to 99.8% of the sowing. The average time between planting and harvesting takes 80–90 days [6]. By 2020, production in metric tons of broccoli decreased 10%, when in previous years, it had grown more than 55%. It is observed in the historical data that, in 2017, for production to grow 54%, the number of hectares planted had to increase by 31%. The same happened in 2018, with a ratio of 59% more in hectares planted for 65% more tons produced. In terms of sales, there was an average growth of 8% during the period analyzed.

2020 sold the fewest metric tons, 5676 fewer than in 2018 and only 96.7% of total production, when the average was 98.6% [7]. In Ecuador, the average price per kilogram of broccoli sold by producers varies between $0.25 and $0.29 cents in domestic markets (Table 1).

2.1 Broccoli

Broccoli is part of the Cruciferae family and its botanical name is Brassica oleracea var Italica, variety botrytis subvar. cymosa Lam. Broccoli provides vitamins and essential elements to the human diet, due to its antioxidant compounds that improve health, strengthens the immune system, keeps bones strong, helps to prevent cancer and heart disease. It contains a large amount of vitamins, minerals, and antioxidants. It is considered as the vegetable with the highest nutritional value per unit weight; it is characterized by its easy processing [8].

2.2 Mathematical Models

A model is an object, concept, or set of relationships, which is used to represent and study in a simple and understandable way a portion of empirical reality [9].

One of the most interesting tools we have today to predict and analyze the behavior of phenomena is based on the construction and simulation of the well-known mathematical models, which also serve as support for decision-making, allowing an approach to reality [10].

2.3 Neural Networks

In the case of artificial neurons, the sum of the inputs multiplied by their associated weights determines the "nerve impulse" received by the neuron (Fig. 1).

This terminology appears approximately since the 40 s, with the emergence of computer science, the neural model was developed over time to become a modern theory that focuses on learning and neural processing, the author Frank Rosenblatt was the first developer of neural networks in the 50 s who proposed a device called perceptron [11]. Nowadays and thanks to technological advances, neural networks are understood as a technology that includes models inspired by the understanding and behavior of the human brain, its continuous research has allowed neural networks

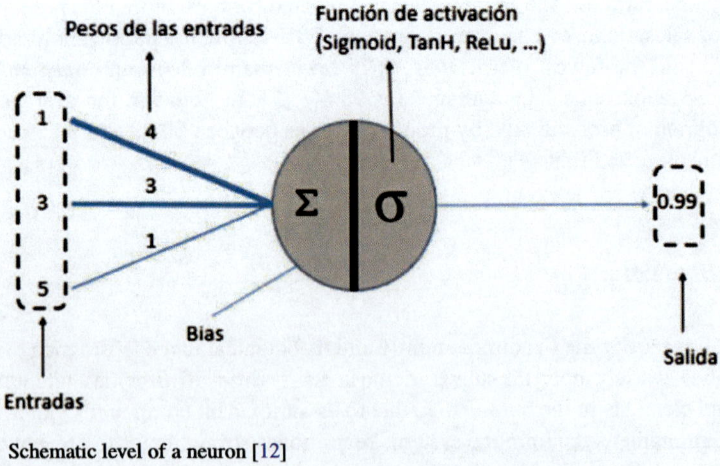

Fig. 1 Schematic level of a neuron [12]

Table 2 Human brain versus computer [12, 13]

Characteristics	Computer	Human Brain
Speed of the process	Entre $10^{-8}y10^{-9}$seg	$1/10^3$; $1/10^2$seg
Style of procedure	Sequential (serial)	Parallel
Number of processors	Few	Entre $1/10^{-11}y1/10^{-14}$
Links	Few	$1/10^{-4}$
Information storage	In fixed directions	Distributed
Failure threshold	Precise positions	Broad
Control domain	Centralized (dictatorial)	Own organization (democrática)
Energy required to perform a task	10^{-6} Julios	10^{-16} Julios

to participate in the resolution of various problems that arise in society today through the use of conventional algorithmic techniques, which have allowed the diagnosis of diseases, the approximation of functions, or the recognition of images, projection of phenomena, etc. [12] (Table 2).

2.4 Learning Algorithm

The learning algorithm is an important component within the neural network, it allows to modify and adapt the value of the weights, by running data through several layers of neural network algorithms, which pass a simplified representation of the data to the next layer. These algorithms progressively learn about the image as it passes through each neural network layer, where the first layers learn to detect low-level features such as edges, and later layers combine the features from the previous layers into a holistic representation [13].

In order to carry out the programming of a neural network, two alternatives can be considered [14].

2.5 Activation Functions

The activation functions are focused on limiting the output range of the neuron, which can be linear or nonlinear; they are a function of the investigated problem and are generated based on the researcher's criteria where the trial and error tests will be based, on the speed and accuracy required and on the selected learning algorithm. The ranges of the activation function used frame whether it is necessary to scale or transform the input data to fit the ranges required by the research [15].

The logistic activation function is the most commonly used due to its derivative characteristics, which allow solving prediction problems. The logistic activation

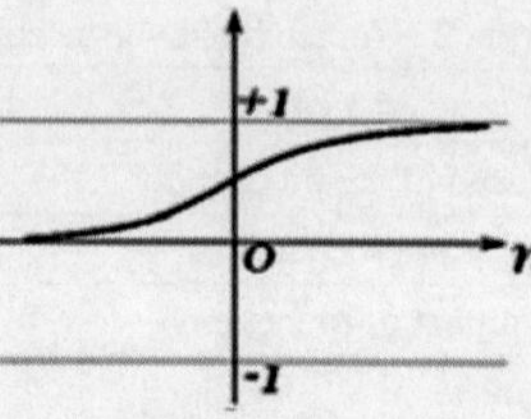

Fig. 2 Function Logsig (n)

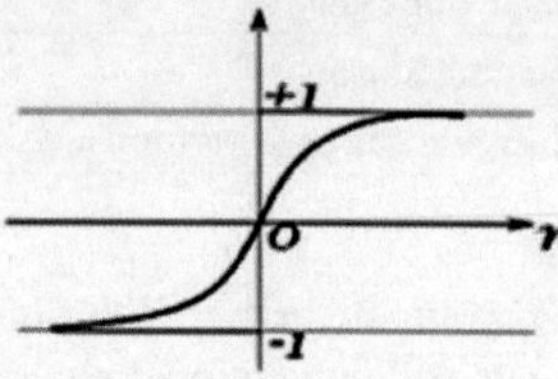

Fig. 3 Función Tansig (n)

function is composed of a large linear part and thus achieves maximum training speed and timely approximation to the desired results, in fact the logistic function is commonly used because its derivative is one of the easiest to solve [16] (Figs. 2 and 3).

$$\text{Function logsig } n = \frac{1}{1 + e^{-n}} \tag{1}$$

$$\text{Function tansig } n = \frac{2}{1 + e^{-2n}} - 1 \tag{2}$$

3 Development

The neural network for broccoli growth prediction is composed of 3 layers, an input layer, a hidden layer, and an output layer [17]. The input layer has 10 neurons and the output layer has 2 neurons. Figure 4 shows normal planting parameters in Scenario 1.

Figure 5 shows the data for Scenario 2 when the soil pH parameter is changed.

3.1 Mathematical Model of Broccoli Growth

The mathematical model for the prediction of broccoli growth is shown in Eq. 3, which was generated with the Multilayer Perceptron Neural Network. As can be seen, the variable a represents the output of the network, W represents the intensity

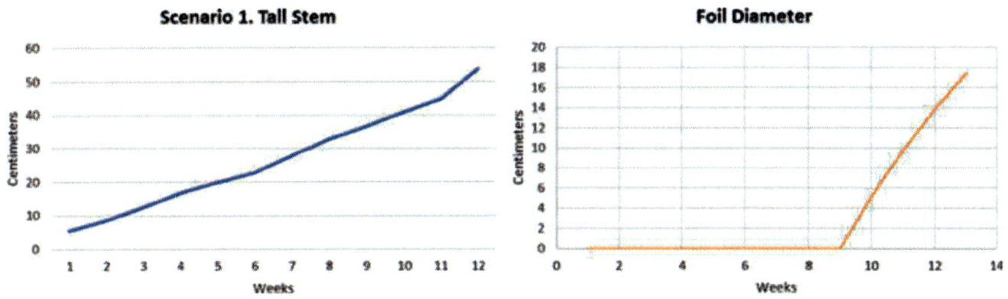

Fig. 4 Data obtained scenario 1

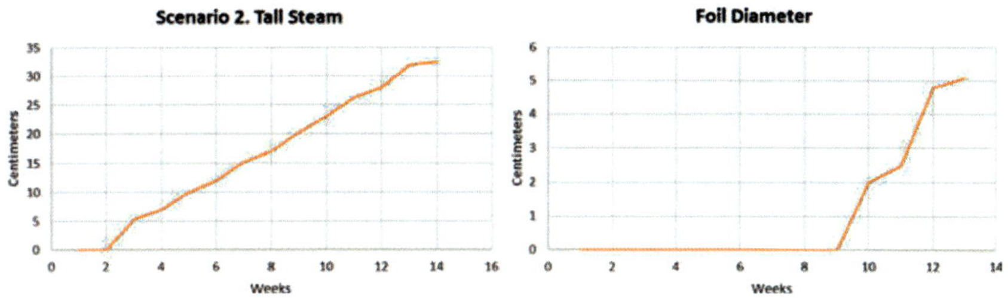

Fig. 5 Data obtained scenario 2

of each of the inputs, and b is the bias [18] (Fig. 6).

$$a = f(Wp + b) \tag{3}$$

Fig. 6 Perceptron neural network architecture

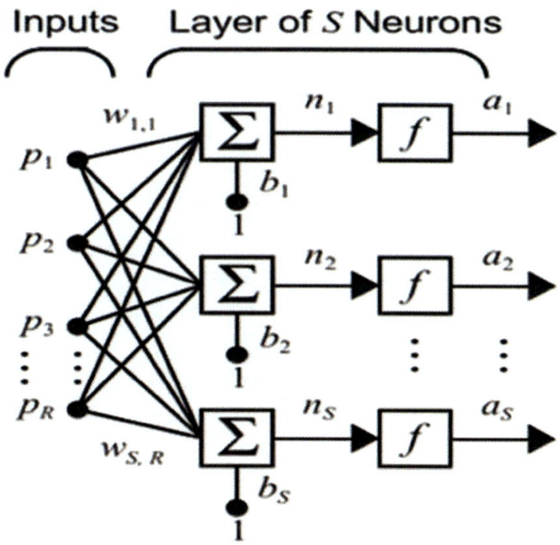

3.2 Neural Network Input Parameters

The input parameters are the main elements of the neural network, since thanks to them the network can identify the behavior of the phenomenon under study [19, 20]. For the prediction of broccoli growth, 10 input parameters were used which are described (Table 3).

With all these parameters described, the Broccoli Growth Equation, where a is the output, W (weights), f nonlinear functions (activation-functions), is:

$$a = \log\text{sig}(W_2(\tan\text{sig}(W_1 p + b_1) + b_2)) \tag{4}$$

where

$P =$ input values

$a =$ network output

$T =$ target values

$F\ (e) = T - a.$

The optimizer used for the nonlinear functions, is the derivative, thus the gradients were found.

The equation for the broccoli growth prediction was given from the following mathematical process described below:

$$\text{Logsig}(n) = \left(1 - a^2\right)\left(a^2\right) \tag{5}$$

$$F(x) = E\left[e^T e\right] = \frac{1}{Q}\sum_{q=1}^{Q}(t - a)^T(t - a) \tag{6}$$

Table 3 Neural network entries

Inputs	Units
N (Nitrogen)	Parts per million (ppm)
Fo (match)	
P (Potassium)	
Ma (magnesium)	
Ca (calcium)	
Az (sulfur)	
Te (Temperature)	Degrees Celsius
PH	PH index
Hu (humidity)	% de Humedad
Días	Number of growing days

Weight calculations

$$W^m(k+1) = W^m(K) - \frac{\alpha}{Q}\sum_{q=1}^{Q} S_q^m (a_q^{m-1})^T \tag{7}$$

Bias calculations

$$b^m(k+1) = b^m(K) - \frac{\alpha}{Q}\sum_{q=1}^{Q} S_q^m \tag{8}$$

1000 epochs were used to converge the error to the located tolerance parameter.

$$f(e) = (T-a)^2 \tag{9}$$

$$f(e) = (T-a)^T(T-a) \tag{10}$$

Sensitivity calculations are performed on the basis of the previous layer.

$$S^M = -2F^M(n^M)(t-a) \tag{11}$$

Previously, the equations for the learning of the neural network are shown in Fig. 5, which shows the learning of the network with a number of epochs = 1000 (Fig. 7).

On the ordinate axis, you can see the normalized values of stem growth and flower diameter; this helps to make the analysis process faster. Bearing in mind that the largest value of the stem and the diameter of the floret would be 1 and the largest almost zero. The formula used was,

$$x_{\text{norm}} = \frac{x - x_{\min}}{x_{\max} - x_{\min}} \tag{12}$$

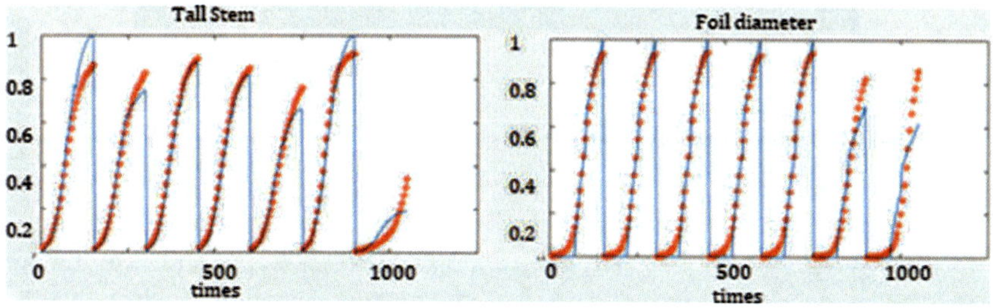

Fig. 7 Neural Network Learning

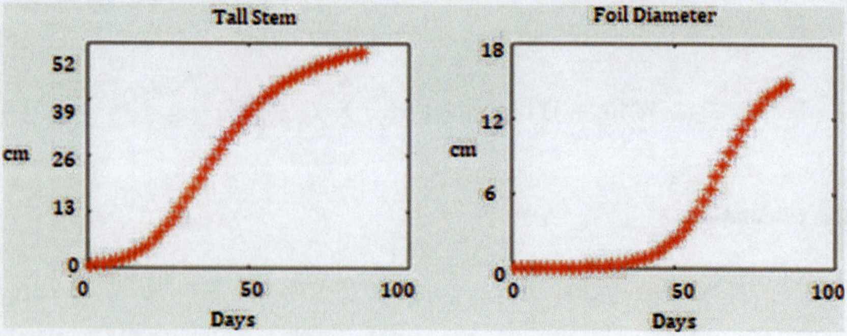

Fig. 8 Broccoli growth with optimal soil chemical levels

3.3 *Model Behavior*

The learning results of the neural network in the different scenarios are shown below.

3.3.1 Scenario 1

In order to know how the Perceptron Neural Network works, an exploratory analysis was carried out to evaluate the efficiency. In this way, we could compare with the prediction of the Network. The following shows the broccoli cr3ecimeinto in scenario 1 with normal values (Fig. 8).

The figure shows that the prediction of the neural network is very similar to the physically collected data of scenario 1.

3.3.2 Scenario 2

In scenario 2, the soil Ph parameter was changed to a lower parameter (Fig. 9).

As evidenced in the figure, broccoli is prone to show boron deficiencies when the soil reaction is close to neutral pH. While very acid soils may show symptoms of magnesium deficiencies so much so that it fails to grow in stem and floret.

3.4 *Validation of the Model*

The following shows the error produced by the Perceptron Neural Network at the time of learning; this error converges in approximately 144 epochs, i.e., it did not take many epochs for the NR to learn. It is important, as mentioned above, that the Network cannot be over-trained, and neither can too little data be entered.

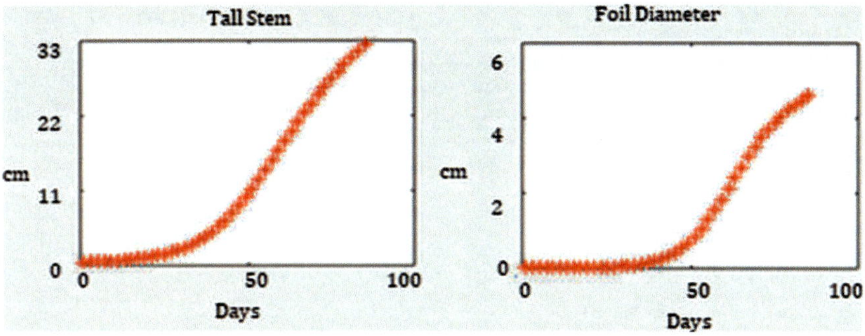

Fig. 9 Broccoli growth in acid soil

Fig. 10 Error

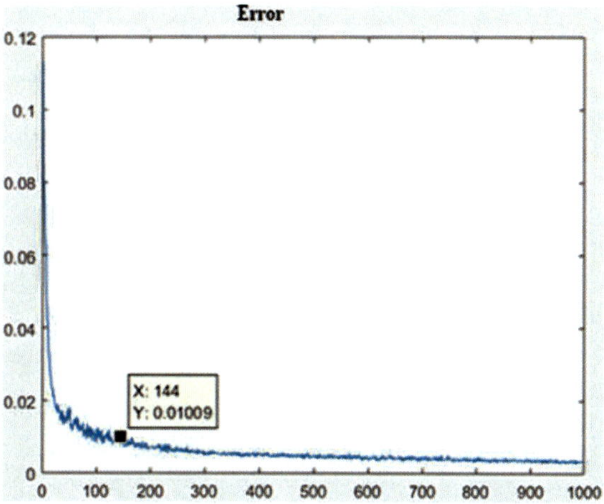

Figure 10 shows that the error converges approximately at 144, taking into account that in previous epochs, it reaches an error of 0.01 but does not stabilize at the required error. The more epochs the network needs, the more is the response time, but it must be taken into account that the more epochs the network has to approach an error of 0, which is the optimum, since each epoch that the network performs is a complete process of growth generating more knowledge and less error.

3.5 Mean Absolute Percentage Error—MAPE

It is the average of the absolute percentage errors expressed in percentage terms.

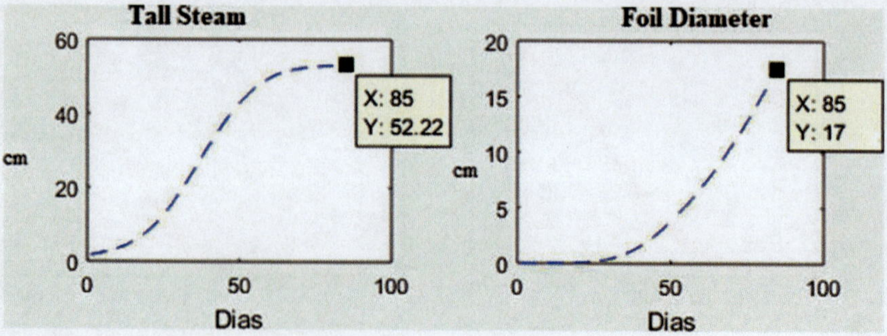

Fig. 11 Escenario 1 con valores obtenidos en RN

$$\text{MAPE} = \sum_{i=1}^{N} \left| \frac{y_i - \hat{y}}{y_i} \right| \times 100\% \tag{13}$$

y_i = actual historical value of the independent variable

$\hat{y}$ = estimated value of the independent variable.

Figure 11 shows the behavior of the legacy broccoli with its growth values, these data obtained help to find the Mean Absolute Percentage Error, thus validating the mathematical model of the Multilayer Preceptron Neural Network, implemented in the Matlab software.

To check the Mean Absolute Percentage Error, the formula 13 mentioned above is used.

$$\text{MAPE} = \left| \frac{53.8 - 52.22}{53.8} \right| \times 100\%$$

$$\text{MAPE} = \left| \frac{1.58}{53.8} \right| \times 100\%$$

$$\text{MAPE} = |0.029368| \times 100\%$$

$$\text{MAPE} = 2.936\%$$

As can be seen, the result shows that the Mean Absolute Percentage Error is approximately 3%, thus giving as a result that the learning of the Neural Network is efficient.

3.6 Scenario 3

In scenario 3, the iron parameter was decreased to 0.1, the behavior of legacy broccoli is shown in Fig. 12.

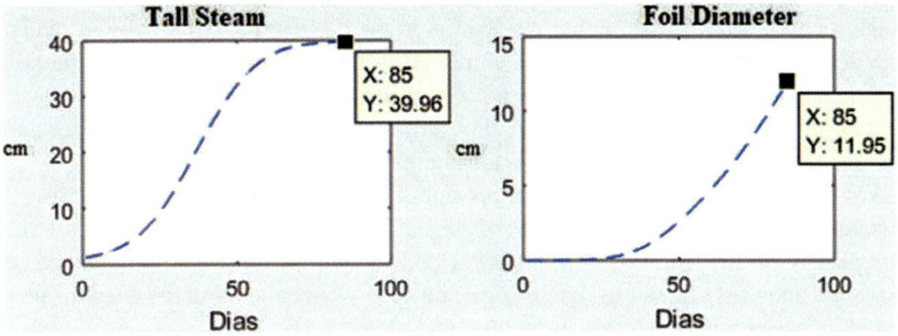

Fig. 12 Prediction of NR

When broccoli (Brassica) lacks iron in quantities outside acceptable ranges, it causes both structural and functional losses, basically limiting its metabolic characteristics, minimum productive yields, deficient planting densities, and its composition will affect the transport of macro and micronutrients, which will affect the final quality of the product.

Likewise, the absence of iron will generate phenotypic malformation in the stems because they will become hollow, also triggering an inappropriate circulation of raw and processed sap, which turns this crop into a rather weak vegetable, with small leaves and delays in their germinative processes. This also affects the growth of the stem and the phloem, affecting it permanently, affecting its pH and including malformation in its roots and leaves, generating innumerable losses in the quality of this primordial vegetable whose nutritional advantages are transcendental for the final consumers at a global level.

4 Conclusions

The development of neural networks in various scenarios has enabled to determine the effectiveness of chemicals in achieving optimal growth for broccoli crops. Additionally, the network provided data on the stem height (52.22 cm) and floret diameter (17 cm), which helped to identify the ideal timing for harvesting broccoli. According to the model, the optimal combination of chemicals, temperature, and humidity is determined in the first scenario.

Broccoli in Latacunga is grown throughout the year, but in the summer season, its harvest is greater than the winter so the climate is in optimal conditions (about 8–25 °C), in the winter time it harms the plantation because its climate is presented at low temperatures that can lead to 7 to – 1 °C.

Its way of sowing is with pre-developed plants in seedbeds of 4–5.5 cm in size which are transplanted in lots already prepared (plowed and disinfected soil where it will be grown).

Broccoli is grown with higher production in the highlands, specifically in sandy, clay soils (maintain humidity of 60% with a pH 7–7.5 and water at 90%); to maintain moisture, it is irrigated every 2 days for 4 h on hot days.

In broccoli, development is being checked every month after cultivation for any disease during the 3 months and also during that time, it is sprayed 12 times.

One of the most used networks for prediction is the Multilayer Perceptron Neural Network because it is efficient in classifying and finding patterns. Depending on the complexity of the problem, neurons are used. The parameters of the inputs can be modified and the behavior of the broccoli can be evidenced without the need to carry out physical experiments, this saves time and money.

References

1. Sánchez A, Vayas T, Mayorga F, Freire C (2020) Producción de brócoli en ecuador. (En línea). Available: https://blogs.cedia.org.ec/obest/wp-content/uploads/sites/7/2020/12/Brocoli-en-Ecuador.pdf
2. Toledo J (2003) Cultivo del Brócol. (En línea). Available: http://repositorio.inia.gob.pe/bitstream/20.500.12955/895/1/Toledo-Cultivo_brocoli.pdf
3. Navas J (2018) Modelos matemáticos. (En línea). Available: https://matema.ujaen.es/jnavas/web_modelos/pdf_mmb08_09/introduccion.pdf
4. Xeridia (2019) Redes Neuronales Artificiuales. (En línea). Available: https://www.xeridia.com/blog/redes-neuronales-artificiales-que-son-y-como-se-entrenan-parte-i
5. Universidad de Salamanca (2000) Redes neuronales. (En línea). Available: http://avellano.fis.usal.es/~lalonso/RNA/index.htm
6. Raquel F, Fernández J (2012) Las Redes neuronales artificiales. Netbiblo, España
7. Gónzalez M (2001) Aplicación de redes neuronales en el cálculo de sobretensiones y tasa de contorneamientos. (En línea). Available: https://www.tdx.cat/bitstream/handle/10803/6281/Capitulo_6a.PDF?sequence=9&isAllowed=y
8. Villacres PS (2015) Producción y exportación de brócoli en el Ecuador y su impacto en la generación de empleo y
9. Castillo J (2019) Modelo matemático del crecimiento de la Rosa freedom mediante Redes Neuronales. En línea
10. The MathWorks Inc. (2005) Neural Network Toolbox. (En línea). Available: http://matlab.izmiran.ru/help/toolbox/nnet/logsig.html
11. Ramos D, Terry E (2015) Generalidades de los abonos orgánicos: importancia del Bocashi como alternativa nutricional para suelos y plantas. Scielo 35(4)
12. Grammont H (2009) «La evolución de la producción agropecuaria en el campo mexicano: concentración productiva, pobreza y pluriactividad. Scielo 7(13)
13. Álvares F (2015) Implemenatación de nuevas tecnologías. UFG, San Salvador
14. Dongyu Q (2019) El Estado mundial de la agricultura y la alimentación. Heba Khamis, Roma
15. Sánchez A, Vayas T (2020) Producción de brócoli en el Ecuador. (En línea). Available: https://blogs.cedia.org.ec/obest/wp-content/uploads/sites/7/2020/12/Brocoli-en-Ecuador.pdf
16. Harper J, Kime L (2015) Producción de Brócoli. (En línea). Available: https://extension.psu.edu/produccion-de-brocoli
17. Basogain (2010) Redes Neuronales Artificiales y sus aplicaciones. EHU, Bilbao
18. Zarranz (2020), Inteligencia artificial. SERV, España
19. Ocaris F, Horacio F (2017) Las Redes Neuronales y la Evalaución del Riesgo de Crédito. Scielo 6(10)

20. Mercado D, Pedraza L, Martínez E (2010) Comparación de Redes Neuronales aplicadas a la predicción de seriess de tiempo. Scielo 13(2):88–95

Computational Optimization for Decision Making Applications

Overview of the Method Defining Interrelationships Between Ranked Criteria II and Its Application in Multi-criteria Decision-Making

Darko Božanić and Dragan Pamucar

Abstract The paper provided a detailed description of a new method for defining weight coefficients of criteria which is called Defining Interrelationships Between Ranked Criteria II (DIBR II). The method is based on defining significance of adjacent ranked criteria by decision-makers, respectively, experts. Based on the defined significance values, the weight coefficients of criteria are calculated using a simple mathematical apparatus. This approach eliminates certain shortcomings of previous methods which is used to calculate the weight coefficients of criteria. The application of the DIBR II method is presented using two illustrative examples.

Keywords Weight coefficients of criteria · Defining interrelationships between Ranked Criteria II (DIBR II) · Multi-criteria decision-making (MCDM)

1 Introduction

Within the last two decades, it has been developed a large number of multi-criteria decision-making methods (MCDM). Despite this fact, the best method has not yet been founded, but the decision-making processes are usually related to specific features of a problem [1–9], or the methods are used in order to select or compare the existing techniques, respectively, the MCDM methods [10].

A part of the MCDM methods can be used to define weights of criteria, a part to rank alternatives, and some of them are universal, so these can be used for both segments of decision-making. In this paper, the authors focused on defining the weight coefficients of criteria. The motivation for developing a new method, presented in this paper, is found in the fact that the people involved in a decision-making process, experts, usually do not understand the principles of the MCDM

D. Božanić (✉)
Military Academy, University of Defence, 11000 Belgrade, Serbia
e-mail: dbozanic@yahoo.com

D. Pamucar
Faculty of Organizational Sciences, University of Belgrade, 11000 Belgrade, Serbia

P. Chatterjee et al. (eds.), *Computational Intelligence for Engineering and Management Applications*, Lecture Notes in Electrical Engineering 984,
https://doi.org/10.1007/978-981-19-8493-8_64

methods, and it is often found to be very difficult to explain what is required of them: what to compare, how to compare two criteria, which scale to use when comparing and the like [11]. Due to the above, the authors of this paper tried to reformulate the thoughts of experts who do not understand the MCDM into a method, which made it significantly easier for experts to understand what they should do during the evaluation process of criteria.

The most common way to obtain data used for calculating weight coefficients is to compare the criteria. Typical and very often applied methods are the method of analytical hierarchical process (AHP) [12] and the best–worst method (BWM) [13]. Although these methods have so far confirmed their quality, in practice there are often problems related to the consistency of the comparisons made, primarily due to the need to define the relation between distant criteria [14]. The situation is similar with the Full Consistency Method (FUCOM) [15] and the Level-Based Weight Assessment (LBWA) [16]. However, the method Defining Interrelationships Between Ranked Criteria (DIBR) [17] eliminates the problem of comparing distant criteria and brings the way of comparison closer to the perception of decision-makers or experts. In the DIBR method, experts compare adjacent criteria so that the total significance interval of 100% is divided into two criteria. Starting from the idea of comparing adjacent criteria, a new method is developed based on the statement of experts on how many times is one criterion more important than another. Having in mind the initial idea on which is based and the principle of functioning, this method is called Defining Interrelationships Between Ranked Criteria II (DIBR II). For the needs of this method, a special mathematical apparatus is developed, and its application is presented in detail through two examples.

2 Overview of the Algorithm of the DIBR II Method

The DIBR II method is applied through seven steps which are presented further in this paper.

Step 1. *Defining criteria.* For the purpose of solving certain decision-making issues, it is defined a set C including n criteria, $C = \{C_1, C_2, ..., C_n\}$, based on which are later selected, respectively, ranked the existing alternatives.

Step 2. *Ranking criteria by their significance, respectively, weight.* The criteria defined from the set C are ranked from the most significant to the least significant. In order to present the method in a more simple way, the set of criteria is defined so that the criterion C_1 is the most significant, and the criterion C_n is the least significant, respectively, it is defined as follows: $C_1 > C_2 > C_3 > ... > C_n$.

Step 3. *Defining the relation of significance between criteria.*

Step 3.1. *Defining the relation of significance between the adjacent criteria.* For each, two adjacent criteria is defined their relation of significance ($\eta_{j,j+1}$ where $\eta_{j,j+1} \in \left\{\eta_{1,2}, \eta_{2,3}, \eta_{3,4}, ..., \eta_{n-1,n}\right\}$), according to the following: in order to compare the criteria C_1 and C_2, it is defined the significance relation $\eta_{1,2}$, respectively; in order to compare the criteria C_{n-1} and C_n, it is defined the significance relation $\eta_{n-1,n}$. For this, significance relation applies that $\eta_{j,j+1} \geq 1$.

The value $\eta_{j,j+1}$ shows how many times is the criterion C_j more significant than the criterion C_{j+1}. For example, the value $\eta_{1,2} = 2$ shows that the criterion C_1 is twice as significant as the criterion C_2, while the value $\eta_{2,3} = 1.3$ shows that the criterion C_2 is 1.3 times as more significant as the criterion C_3. Accordingly, in this step, the following relations are defined:

$$w_1 : w_2 = \eta_{1,2} : 1 \Rightarrow \frac{w_1}{w_2} = \eta_{1,2} \tag{1}$$

$$w_2 : w_3 = \eta_{2,3} : 1 \Rightarrow \frac{w_2}{w_3} = \eta_{2,3} \tag{2}$$

$$\ldots$$

$$w_{n-1} : w_n = \eta_{n-1,n} : 1 \Rightarrow \frac{w_{n-1}}{w_n} = \eta_{n-1,n} \tag{3}$$

Step 3.2. *Defining significance relations between the most and the least significant criteria.* In this step, the relation is defined as:

$$w_1 : w_n = \eta_{1,n} : 1 \Rightarrow \frac{w_1}{w_n} = \eta_{1,n} \tag{4}$$

This relation has the role of the control factor in evaluating the quality of the defined significance of the adjacent criteria.

Step 4. *Calculation of the relation of the most significant criterion with other criteria.* Based on the expressions (1)–(3), the value of the second and the other lower in range criteria is presented through the most significant criterion, as follows:

- Based on the expression (1), it has obtained the value of the weight coefficient w_2:

$$w_2 = \frac{w_1}{\eta_{1,2}} \tag{5}$$

- Based on the expressions (2) and (5), it has obtained the value of the weight coefficient w_2:

$$w_3 = \frac{w_2}{\eta_{2,3}} = \frac{w_1}{\eta_{1,2} \cdot \eta_{2,3}}$$
$$\ldots \tag{6}$$

- Based on the expression (3), it has obtained the value of the weight coefficient w_n:

$$w_n = \frac{w_1}{\eta_{1,2} \cdot \eta_{2,3} \cdot \ldots \cdot \eta_{n-1,n}} \tag{7}$$

Step 5. *Calculation of the value of the most significant criterion.* If it is as follows:

$$\sum_{j=1}^{n} w_j = 1 \tag{8}$$

Then, based on the expressions (5)–(8), follows that:

$$w_1 + \frac{w_1}{\eta_{1,2}} + \frac{w_1}{\eta_{1,2} \cdot \eta_{2,3}} + \cdots + \frac{w_1}{\eta_{1,2} \cdot \eta_{2,3} \cdot \cdots \cdot \eta_{n-1,n}} = 1 \tag{9}$$

Based on what is calculated the value of the most significant criterion:

$$w_1 = \frac{1}{1 + \frac{1}{\eta_{1,2}} + \frac{1}{\eta_{1,2} \cdot \eta_{2,3}} + \cdots + \frac{1}{\eta_{1,2} \cdot \eta_{2,3} \cdot \cdots \cdot \eta_{n-1,n}}} \tag{10}$$

Step 6. *Calculation of weight coefficients of other criteria.* Applying the expressions (5)–(7), other weight coefficients of criteria $w_2, w_3, \ldots, w_n$ are obtained.

Step 7. *Evaluation of quality of defined significance of the adjacent criteria.* The relations of significance of the adjacent criteria need to be checked, so as to avoid as much as possible subjectivity of the decision-makers.

Step 7.1. *Evaluation of quality of defined significance values.* The evaluation of quality of defined significance values is made based on the relation of the significance of the most and the least significant criteria ($\eta_{1,n}$). The value of the least significant criterion can be obtained from the relation (4):

$$w_n^k = \frac{w_1}{\eta_{1,n}} \tag{11}$$

where w_n^k presents the control weight coefficient of the criterion C_n.

The values w_n and w_n^k should be approximately equal. If their deviation amounts to 10% approximately, it can be concluded that the relations of significance of the adjacent criteria are defined with quality, and vice versa. Checking of the deviation is done by applying the expression:

$$d_n = \left| 1 - \frac{w_n}{w_n^k} \right| \tag{12}$$

where d_n presents the value of the deviation of the weight coefficients of the criterion C_n.

If the condition is met where $0 \leq d_n \leq 0.1$, then the evaluation of the significance relations of the adjacent criteria is defined with quality, respectively, and these meet the requirements. If $d_n > 0.1$, it is necessary to define new relations between the criteria. However, since the research is usually an extensive process engaging significant resources, an additional step can be applied in order to find an error. In such cases, it is applied the *Step 7.2.*

Step 7.2. *Additional evaluation of quality of the defined significance of the adjacent criteria.* Before returning to defining relations of significance of the adjacent criteria again, it is possible to make additional quality control. For this purpose, it repeated the step from the Step 7.1, which are defined the relations between the criteria C_{n-1} and C_{n-2}.

It is necessary for the decision-maker to define the relations:

$$w_1 : w_{n-1} = \eta_{1,n-1} : 1 \Rightarrow \frac{w_1}{w_{n-1}} = \eta_{1,n-1} \tag{13}$$

$$w_1 : w_{n-2} = \eta_{1,n-2} : 1 \Rightarrow \frac{w_1}{w_{n-2}} = \eta_{1,n-2} \tag{14}$$

Next, it is calculated as:

$$w_{n-1}^k = \frac{w_1}{\eta_{1,n-1}} \tag{15}$$

$$w_{n-2}^k = \frac{w_1}{\eta_{1,n-2}} \tag{16}$$

where w_{n-1}^k and w_{n-2}^k present the control weight coefficients of the criterion C_{n-1}, respectively, C_{n-2}.

Finally, it has obtained the value:

$$d_{n-1} = \left| 1 - \frac{w_{n-1}}{w_{n-1}^k} \right| \tag{17}$$

$$d_{n-2} = \left| 1 - \frac{w_{n-2}}{w_{n-2}^k} \right| \tag{18}$$

where d_{n-1} and d_{n-2} present the values of deviation of the weight coefficients of the criteria C_{n-1}, respectively, C_{n-1}.

If the conditions are met where $0 \leq d_{n-1} \leq 0.1$ and $0 \leq d_{n-2} \leq 0.1$, it can be concluded that there has been an error in defining the relations of significance between the most and the least significant criteria ($\eta_{1,n}$). In that case, the existing results can be accepted or the values can be defined again ($\eta_{1,n}$) and quality is checked again as well. Shortly, if $d_{n-1} > 0.1$ or $d_{n-2} > 0.1$, the complete procedure of defining the relations of significance and calculation of the weight coefficients must be repeated.

3 Application of the DIBR II Method

The application of the DIBR II method is provided with two examples of defining the weights of criteria.

Example 1

Step 1. In the first example, it has presented the calculation of the weight coefficients of the criteria for the selection of cars supply. For this purpose, it is defined the set of four criteria including: the criterion 1 (C1)—Price, the criterion 2 (C2)—Quality, the criterion 3 (C3)—Fuel consumption, the criterion 4 (C4)—Warranty.

Step 2. It has performed the ranking by the significance of the criteria according to $C_1 > C_2 > C_3 > C_4$.

Step 3.1. The relations of the adjacent criteria are defined as follows

$$w_1 : w_2 = 3 : 1 \Rightarrow \frac{w_1}{w_2} = 3.0;$$
$$w_2 : w_3 = 1.7 : 1 \Rightarrow \frac{w_2}{w_3} = 1.7;$$
$$w_3 : w_4 = 1.1 : 1 \Rightarrow \frac{w_3}{w_4} = 1.1.$$

Step 3.2. The relation of significance of the most and the least significant criteria is defined as follows:

$$w_1 : w_4 = 6 : 1 \Rightarrow \frac{w_1}{w_4} = 6.0$$

Step 4. It has made the calculation of the relations of the most significant criterion and the other criteria, as follows:

$$w_2 = \frac{w_1}{3}$$
$$w_3 = \frac{w_2}{1.7} = \frac{\frac{w_1}{3}}{1.7} = \frac{w_1}{3 \cdot 1.7}$$
$$w_4 = \frac{w_3}{1.1} = \frac{\frac{w_2}{1.7}}{1.1} = \frac{w_2}{1.7 \cdot 1.1} = \frac{\frac{w_1}{3}}{1.7 \cdot 1.1} = \frac{w_1}{3 \cdot 1.7 \cdot 1.1}$$

Step 5. It has made the calculation of the weight coefficient of the most important criterion as follows:

$$w_1 + \frac{w_1}{3} + \frac{w_1}{3 \cdot 1.7} + \frac{w_1}{3 \cdot 1.7 \cdot 1.1} = 1 \Rightarrow$$
$$w_1 \cdot (1 + \frac{1}{3} + \frac{1}{3 \cdot 1.7} + \frac{1}{3 \cdot 1.7 \cdot 1.1}) = 1 \Rightarrow$$

$$w_1 = \frac{1}{(1 + \frac{1}{3} + \frac{1}{3 \cdot 1.7} + \frac{1}{3 \cdot 1.7 \cdot 1.1})} \Rightarrow$$
$$w_1 = 0.586$$

Step 6. The calculation of the weight coefficients of the other criteria is made as follows:

$$w_2 = \frac{0.586}{3} = 0.195$$
$$w_3 = \frac{0.586}{3 \cdot 1.7} = 0.115$$
$$w_4 = \frac{0.586}{3 \cdot 1.7 \cdot 1.1} = 0.104$$

Summing the obtained values of the weight coefficients of the criteria $w_1 + w_2 + w_3 + w_4 = 0.586 + 0.195 + 0.115 + 0.104 = 1$, it has met the condition, where $\sum_{j=1}^{n} w_j = 1$.

Step 7.1. Calculation of the control weight coefficient of the criterion C_4 is made through the evaluation of significance values defined in the Step 3.2 and the obtained weight coefficient, as follows:

$$w_4^k = \frac{w_1}{\eta_{1,4}} = \frac{0.586}{6} = 0.0976$$

The value of the deviation of the weight coefficients of the criterion C_4 is calculated according to the expression (12) and amounts to:

$$d_4 = \left| 1 - \frac{0.104}{0.0976} \right| = |1 - 1.0695| = 0.0695$$

Considering that the condition is met, where $0 \leq d_4 \leq 0.1$, it can be stated that the evaluations of significance of the adjacent criteria are made with quality.

Example 2

Step 1. In this example, the calculation of the weight coefficients for ranking and evaluation of the effectiveness of social media is presented, taken over from [18]. It is to define the set of six criteria including: the criterion 1 (C1)—Presence, the criterion 2 (C2)—Purpose, the criterion 3 (C3)—Functionality, the criterion 4 (C4)—Target audience, the criterion 5 (C5)—Engagement, and the criterion 6 (C6)—Content richness.

Step 2. It has made the ranking by significance of the criteria as follows: $C_3 > C_4 > C_1 > C_5 = C_6 > C_2$.

Step 3.1. The relations of significance of the adjacent criteria are defined as follows

$$\frac{w_3}{w_4} = 1.3; \frac{w_4}{w_1} = 1.0\frac{w_1}{w_5} = 2.1; \frac{w_5}{w_6} = 1.4; \frac{w_6}{w_2} = 1.8.$$

Step 3.2. The relation of significance of the most and the least significant criteria is defined as follows:

$$\frac{w_3}{w_2} = 8.0$$

Step 4. It has made the calculation of the relation of the most significant criterion with the other criteria, as follows:

$$w_4 = \frac{w_3}{1.3}$$
$$w_1 = \frac{w_3}{1.3 \cdot 1} = \frac{w_3}{1.3}$$
$$w_5 = \frac{w_3}{1.3 \cdot 1 \cdot 2.1} = \frac{w_3}{2.73}$$
$$w_6 = \frac{w_3}{1.3 \cdot 1 \cdot 2.1 \cdot 1.4} = \frac{w_3}{3.822}$$
$$w_2 = \frac{w_3}{1.3 \cdot 1 \cdot 2.1 \cdot 1.4 \cdot 1.8} = \frac{w_3}{6.88}$$

Step 5. It has made the calculation of the weight coefficient of the most significant criterion as follows:

$$w_3 + \frac{w_3}{1.3} + \frac{w_3}{1.3} + \frac{w_3}{2.73} + \frac{w_3}{3.822} + \frac{w_3}{6.88} = 1 \Rightarrow$$
$$w_3 = \frac{1}{(1 + \frac{1}{1.3} + \frac{1}{1.3} + \frac{1}{2.73} + \frac{1}{3.822} + \frac{1}{6.88})} \Rightarrow$$
$$w_3 = 0.302$$

Step 6. The calculation of the weight coefficients of other criteria is made as follows:

$$w_4 = \frac{0.302}{1.3} = 0.232$$
$$w_1 = \frac{0.302}{1.3} = 0.232$$
$$w_5 = \frac{0.302}{2.73} = 0.111$$
$$w_6 = \frac{0.302}{3.822} = 0.079$$

$$w_2 = \frac{0.302}{6.88} = 0.044$$

The sum of calculated weight coefficients amounts to $w_1 + w_2 + w_3 + w_4 + w_5 + w_6 = 0.232 + 0.044 + 0.302 + 0.232 + 0.111 + 0.079 = 1$, confirming again that the condition is met where $\sum_{j=1}^{n} w_j = 1$.

Step 7.1. The calculation of the control weight coefficient of the criterion C_2 is made through the evaluation of significance defined in the Step 3.2 and the obtained weight coefficient of the most significant criterion, as follows:

$$w_2^k = \frac{w_3}{\eta_{3,2}} = \frac{0.302}{8} = 0.038$$

The value of the deviation of the weight coefficients of the criterion C_2 is calculated according to the expression (12) and it amounts:

$$d_2 = \left|1 - \frac{0.044}{0.038}\right| = |1 - 1.158| = 0.158$$

Considering that the condition is not met where $0 \leq d_2 \leq 0.1$, it can be concluded that there are errors in defining the relation of significance values. In order to define whether the procedure should be restarted, it is applied the Step 7.2.

Step 7.2. Additional evaluation of quality of the defined significance of the adjacent criteria is made through the criteria C_6 and C_5. The decision-maker defines first the relations:

$$\frac{w_3}{w_6} = 4$$
$$\frac{w_3}{w_5} = 3$$

The calculation of the control weight coefficient of the criteria C_6 and C_5 is made as follows:

$$w_6^k = \frac{0.302}{4} = 0.075$$
$$w_5^k = \frac{0.302}{3} = 0.101$$

The calculation of the values of deviations of the weight coefficients of the criteria C_6 and C_5 is made as follows:

$$d_6 = \left|1 - \frac{0.079}{0.075}\right| = |1 - 1.053| = 0.053$$

$$d_5 = \left|1 - \frac{0.111}{0.101}\right| = |1 - 1.099| = 0.099$$

Both necessary conditions $0 \leq d_6 \leq 0.1$ and $0 \leq d_5 \leq 0.1$ are met, so it can be concluded that there has been an error in defining the relation of significance between the most and the least significant criteria ($\eta_{3,2}$). In the repeated process of defining such a relation, the decision-maker corrected the opinion and defined that $\eta_{3,2} = 7.3$. Based on this, it follows that the new value is $d_2 = 0.062$, which is within the prescribed limits. Accordingly, it can be concluded that the relations by which are defined the criteria significance values are met.

4 Conclusion

The paper successfully presents a new approach in defining the weight coefficients of the criteria. This approach is based on a comparison of adjacent ranked criteria and it is called DIBR II. Through the application of a new approach to defining the weights of criteria, certain advantages over some other methods are noticed: a very simple mathematical apparatus for calculating the weights of the criteria is developed regardless of the number of criteria being compared; the obligations of experts or decision-makers when comparing the criteria in this method are very easy to explain, regardless of how much they are familiar or not with the MCDM methods; only adjacent ranked criteria are compared, thus avoiding the comparison of distant criteria; the number of mutual comparisons of criteria is reduced to a minimum; the scale for comparing criteria is not limited; a mathematical apparatus is developed to evaluate the quality of the comparison of adjacent criteria.

The new method should be further tested in solving specific problems, in order to confirm its quality.

References

1. Božanić D, Pamučar D, Đorović B (2013) Modification of analytic hierarchy process (AHP) method and its application in the defense decision-making. Tehnika 68:327–334
2. Alosta A, Elmansuri O, Badi I (2021) Resolving a location selection problem by means of an integrated AHP-RAFSI approach. Rep Mech Eng 2:135–142
3. Božanić D, Tešić D, Marinković D, Milić A (2021) Modeling of neuro-fuzzy system as a support in decision-making processes. Rep Mech Eng 2:222–234
4. Lukovac V, Zeljić-Drakulić S, Tomić L, Liu F (2021) Multicriteria approach to the selection of the training model of dangerous goods transport advisors in the ministry of defense and the Serbian Army. Vojnotehnički glasnik/Military Tech Courier 69:28–851
5. Pamučar D, Dimitrijević S (2021) Multiple-criteria model for optimal Anti Tank Ground missile weapon system procurement. Vojnotehnički glasnik/Military Tech Courier 69:792–827
6. Radovanovic M, Ranđelović A, Jokić Ž (2020) Application of hybrid model fuzzy AHP—VIKOR in selection of the most efficient procedure for rectification of the optical sight of the long-range rifle. Decis Making Appl Manage Eng 3:131–148

7. Dragan Pamučar, Darko Božanić, Dejan Kurtov, Fuzzification of the Saaty's scale and a presentation of the hybrid fuzzy AHP-TOPSIS model: an example of the selection of a brigade artillery group firing position in a defensive operation. Vojnotehnički glasnik/Military Tech Courier 64:966–986
8. Muhammad LJ, Badi I, Haruna AA, Mohammed IA (2021) Selecting the best municipal solid waste management techniques in Nigeria using multi criteria decision making techniques. Rep Mech Eng 2:180–189
9. Jokić Ž, Božanić D, Pamučar D (2021) Selection of fire position of mortar units using LBWA and Fuzzy MABAC model. Oper Res Eng Sci Theory Appl 4:115–135
10. Bandyopadhyay S (2021) Comparison among multi-criteria decision analysis techniques: a novel method. Progress Artif Intell 10:195–216
11. Božanić D (2017) Model of decision support in overcoming water obstacles in Army combat operations, (Only in Serbian: Model podrške odlučivanju pri savlađivanju vodenih prepreka u napadnoj operaciji Kopnene vojske). Doctoral dissertation, Belgrade, Serbia: University of Defence in Belgrade, Military Academy
12. Saaty TL (1980) The analytic hierarchy process. McGraw Hill, New York, USA
13. Rezaei J (2015) Best-worst multi-criteria decision-making method. Omega 53:49–57
14. Asadabadi MR, Chang E, Saberi M (2019) Are MCDM methods useful? A critical review of analytic hierarchy process (AHP) and analytic network process (ANP). Cogent Eng 6:1–11
15. Pamučar D, Stević Ž, Sremac S (2018) A new model for determining weight coefficients of criteria in MCDM models: full consistency method (FUCOM). Symmetry 10:393
16. Žižović M, Pamučar D (2019) New model for determining criteria weights: level based weight assessment (LBWA) model. Decis Making Appl Manage Eng 2:126–137
17. Pamucar D, Deveci M, Gokasar I, Işık M, Zizivic M (2021) Concepts in urban mobility alternatives using integrated DIBR method and fuzzy Dombi CoCoSo model. J Clean Prod 323:129096
18. Bobar Z, Božanić D, Đurić-Atanasievski K, Pamučar D (2020) Ranking and assessment of the efficiency of social media using the Fuzzy AHP-Z number model—fuzzy MABAC. Acta Polytechnica Hungarica 17:43–70

Rooftop Photovoltaic Panels' Evaluation for Houses in Prospective MADM Outline

Sarfaraz Hashemkhani Zolfani and Ramin Bazrafshan

Abstract Global warming is a critical issue that countries have a plan for reducing. So, one of the solutions for decreasing this is using electric vehicles (EVs). Therefore, the supply of fuels for EVs is a significant issue that societies will face in the future. Generating electricity by solar radiation with a photovoltaic panel is a good solution for supplying the demands. One of the issues that should consider is electricity because people intend to use EVs with inexpensive fuels. This issue put governments that encourage people to produce electricity on their rooftops by photoelectric panels. By this strategy, governments supply the demand for electricity at less price and less cost. This study surveys the best criteria for selecting the proper rooftop PV panels by maximum efficiency. The eight criteria are proposed by researchers as efficiency, temperature coefficient, material warranty, cost per watt of the panel, solar inverter and solar inverter cost per watt-peak, automatic protector, and smart panel. The numerical example for selection of best panels is proposed by these criteria and four alternatives. So, the researchers used the multiple criteria decision-making (MCDM) method for evaluating these options. It should be noted that one of the major issues considered in this study is the time vision. The alternatives evaluating are performed in two-time vision current and future. For this, two methods are combined: distance from the average solution (EDAS) and prospective multi-attribute decision-making (PMADM). The results show that alternatives one and four are the best in the future and current vision, respectively.

Keywords Prospective multi-attribute decision-making (PMADM) method · Combining of evaluation based on the distance from the average solution (EDAS) · Electrical vehicle (EV) · Rooftop photovoltaic panels

---

S. H. Zolfani (✉)
School of Engineering, Catholic University of the North, Larrondo 1281, Coquimbo, Chile
e-mail: sarfaraz.hashemkhani@ucn.cl

R. Bazrafshan
Department of Industrial Engineering and Management Systems, Amirkabir University of Technology (Tehran Polytechnic), Tehran, Iran

P. Chatterjee et al. (eds.), *Computational Intelligence for Engineering and Management Applications*, Lecture Notes in Electrical Engineering 984,
https://doi.org/10.1007/978-981-19-8493-8_65

1 Introduction

Due to air pollution and global warming, developed countries are preparing to use electric vehicles (EVs). Since fuel burns in the internal combustion engines, some harmful gases are released into the environment, causing pollution. These gases combine with air and produce what we call greenhouse gases. Greenhouse gases are one of the significant sources that cause climate change and global warming. This phenomenon causes many developed countries to have planned on using EVs. For example, Toronto of Canada targets 80% registered electric vehicles by 2040 [1].

Although the use of EVs decreases pollution, people less intended to use them. As a result of electricity's cost and high electricity consumption, therefore, price is an essential part of the large-scale adoption of new vehicle technologies [2]. EV ownerships tend to spend low budget for using electricity proportion to gasoline. The logical and literal solution for this issue is encouraging people to produce electricity in their houses. The governments should perform this policy and help or patronage people to use the PV panels. The photovoltaic converts light to electricity by utilizing solar cells' technology. These cells are placed on a panel and closely connected. By this decision policy, people generate electricity and sell it to governments or use it by own self and use the EVs at more inexpensive fuel costs. By this description, now, the main problem of people is in which panel is the better than other for generating electricity by high efficiency. This study considers a smart home; it has a rooftop PV, generates electricity for EVs charging, and introduces the proper criteria for selecting a good panel. Selecting a good rooftop PV is a challenging problem which people are faced with it. This research list some criteria for choosing the best rooftop PV and rank them by the MCDM. One of the issues that researchers should be considered in this study is time vision. Using rooftop PV panels is not prevalent in communities, but shortly, it is usual. To achieve this, the alternatives need to be ranked based on their suitability for future. Therefore, this study incorporates future considerations into decision-making using the PMADM method. The PMADM process provides an effective framework for weighing criteria in a future-oriented perspective.

2 Literature Review

The most important reason people have less intention to use an electric vehicle is the cost of electricity. Due to the use of EVs shortly, researchers are searching for how to reduce the cost of electricity. The use of PV is the best solution for this because the sun's radiation is free. PV energy production benefits the environment and does not compromise future generations [3]. By setting up rooftop PV can use solar energy in all houses and generate electricity. Many researchers emphasize the importance of PV and the proposed schedule and optimization model for using it. In continuing, some articles surveying this issue and emphasis on it are reviewed.

Ren et al. [4] used AHP and PROMETHEE to determine the optimal energy system for a two-floor residential building in Kitakyushu, Japan. They defined ten alternatives or energy system equipment as PV, gas engine, fuel cell, wind turbine, hybrid system, battery, air conditioner, gas boiler, electric heater, and the utility grid. They considered an economic and environmental factor for determining the criteria and defined four criteria, namely, investment cost (IC), running cost (RC), CO2 emissions (CE), and primary energy consumption (EC). The result shows that if considering cost, fossil fuels are better than renewable energy systems. The PV system is suitable for reaching a low carbon society economically. Yadav and Bajpai [5] survey and evaluate the performance of rooftop solar PV in northern India. They analyzed the performance of the five kWp rooftop photovoltaic and the effect of temperature on it. The results show that the annual energy yield is 7175.4 kWh, and when the temperature is high, the energy loss is maximum. In reviewing some articles that used MCDM in their research, [6] used TOPSIS to select the best renewable heating technologies in Denmark. They consider three alternatives such as solar heating, heat pumps, and wood pellet boilers—the criteria assessing by economic, technological, and environmental aspects. The result shows that solar heating is better than others. Wang et al. [7] presented the MCDM model by fuzzy AHP, DEA, and TOPSIS. They used them to find the best location for building a solar power plant in Vietnam. Zhang et al. [8] said that using the MCDM framework for sustainable planning of the energy systems should incorporate two types of information. For that, they consider expert assessments for the public impacts and the willingness to pay measures for the private ones in their analysis. The MCDM methods which they used were TOPSIS, EDAS, and WASPAS. Monte Carlo analysis is used to check the results, and the results show that biomass boiler or solar thermal installations are the best technology for Lithuania. Corcelli et al. [9] used rooftop panels under Mediterranean climatic conditions and said that these panels could reduce greenhouse gas emissions. They compared two rooftop systems, rooftop greenhouses (RTGs) and building-applied solar photovoltaic (BAVP), and declared that both systems decreased environmental impacts. Saleem et al. [10] surveys among six renewable energy (solar, wind, biomass, geothermal, ocean, and hydel energies) to find the less expensive technology for generating electricity. They used AHP, reported that solar energy technology is better than others, and said that solar energy is a good source for domestic sector electricity demand of Pakistan. Seddiki and Bennadji [11] gather information about electricity generation in a residential building. They used three methods in this research: Delphi to select a preliminary set of renewable energy, Fuzzy AHP to obtain the criteria's weights, and Fuzzy PROMETHEE to get a complete ranking of the alternatives. These methods are used for selecting the best renewable energy for electricity generation in a residential building. Li et al. [12] developed a new framework for sustainable renewable energy evaluation. They used ANP to evaluate the importance of each criterion and WSM, TOPSIS, PROMETHEE, ELECTRE, and VIKOR to scoring alternatives. This new method was used in China and showed that East and Northwest China are good places for setting photovoltaic panels. Kamari et al. [13] extracted the criteria for renewable energy systems by

MCDM. They consider economic, environmental, risk, social acceptance, and technical features for opting criteria. Rigo et al. [14] identify the most common MCDM methods in the renewable energy area and found that five categories of problems are solved by MCDM methods: source selection, location, sustainability, project performance, and technological performance. The AHP, TOPSIS, and ELECTRE are the MCDM methods that are used more than others.

2.1 Research Gap

By reviewing the literature find that one problem that exists is time vision. This means that researchers who work in this field do not consider the future in their evaluations. In this article, researchers are trying to consider future times in decision-making. The most MCDM method considers current time, but for viewing future time should use other methods. The PMADM is a framework that considers the future. This article combines the PMADM approach to the EDAS method for ranking alternatives.

3 Statement Problem

As referred to in the introduction section, the use of electric vehicles will be inclusive in the next few years for reducing air pollution. As the number of electric vehicles increases, the demand for electricity increases too. Supply the required electricity is a critical problem in the future. One of the reasonable solutions for supply these demands is generating electricity by solar energy. Since the photovoltaic system is a good instrument for generating electricity from solar radiation, we consider a house. So, many rooftop panels exist in the market; which one is better for using?

So, selecting the best rooftop photovoltaic panel for a house is our problem. Some criteria in the numerical example section are defined, which significantly impact the panel's performance. Figure 1 shows a house that sets up a rooftop panel on it. The sun radiation shines on a panel; the panel sends DC to the inverter, and then, the inverter transmutes it to AC and sends it to meter. The meter shows the amount of electricity generated by the PV and the amount of electricity consumption in the house. Finally, the electricity sends to the grid or in a home for use.

4 Methodology

There are many MCDM methods in the literature that evaluates criteria and ranking alternatives. Some methods need to determine the weight for criteria like ARAS [16], MARCOS [17], CoCoSo [18], MABAC [19], and some method is independent of weight like SECA [20]. All of these methods consider the current time, but we need

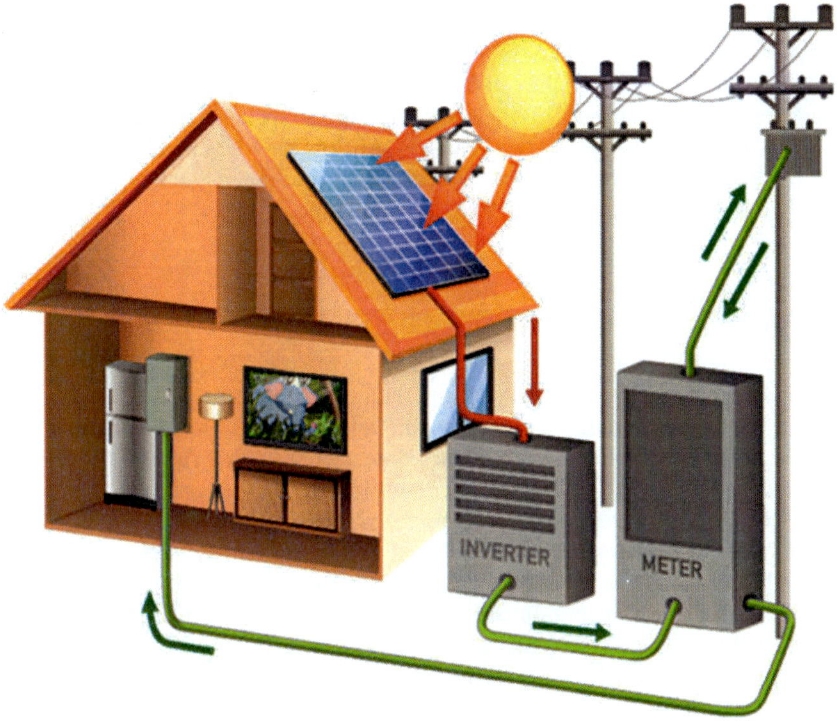

Fig. 1 Rooftop photovoltaic panel on a house [15]

a technique that considers the future. Sometimes, priorities in the future are different from the current time; if one does not consider this issue, it may get an improper decision and result badly.

Hashemkhani Zolfani et al. [21] developed the MADM method as prospective MADM or PMADM. This method considers the future in decision-making and helps decision-makers make good decisions for the future. This method is a good solution for viewing time vision in decision-making and help get a better decision for the future. After proposing this method, some researchers use the PMADM method in various fields, which are explained in the following.

Hashemkhani Zolfani and Masaeli [22] used this method in the health device industry of Iran and increased the medical device market share ten times more than before. Hashemkhani Zolfani and Derakhti [23] used PMADM for weighting system for machine tool selection. They applied text mining for the process of criteria selection. Hashemkhani Zolfani et al. [24] studied company goals and proposed a proper method basis on PMADM for managers to make a good decision and consider the future in their decision-making. They used the EDAS method on the PMADM framework and named it a vision-based weighting system (VIWES) in PMADM.

So, this study is about finding the best rooftop PV panel by future vision. The weight of the criteria is determined by householders who want to use this PV system. They allocate high weight to the cost and efficiency of the panel. EDAS is selected for ranking of the alternatives. This method needs less computation than other MCDM methods; this is a good reason for selecting EDAS. The PMADM framework was also used in this research. The researchers used the EDAS method for ranking alternatives and the PMADM method for weighting in future time.

5 Numerical Example

In this section, define some criteria which are essential in selecting a rooftop panel. In this research, eight criteria were introduced. The six obtained from the photovoltaic panels' brochure: panel efficiency, the power temperature coefficient, solar inverter, material panel warranty, cost per watt, and solar inverter cost per watt-peak. However, this research surveys the future, but these criteria may not be essential for costumer in the future. So, the researchers introduced two new criteria, which are crucial in foresight studies. These criteria are achieved by a focus on predicting the future and investigating the worldviews. In continuing, explained about each of these.

Panel efficiency is a criterion that has great importance in choosing a panel. Panel efficiency is a measure that distincts the efficiency of the panel in converting sunlight to electricity. As much as this efficiency be more considerable is better. This criterion is calculating by the below formulation.

$$\text{Panel Efficiency} = \frac{\text{max panel power}}{(\text{Area} * 1000\ \text{W/m}^2)} * 100$$

The power temperature coefficient is one of the essential criteria. Power temperature coefficient does with laws of thermodynamics. This coefficient is distinct in that the surface temperature of the panel affects the power output of the panel. If solar cells have a temperature coefficient of – 0.5%/°C, this means that panel will lose half of one percent of its power for every degree the temperature rises [25]. The solar panel is rated at 25 °C, but the roof temperature is not the same. Consider a 250 W panel installed on a roof whose temperature is 55 °C. The power loss calculation is assumed as follows.

$$\begin{aligned}
&55\text{–}25\ °\text{C} = 30\ °\text{C} \\
&30\ °\text{C} * (-0.5\%) = 15\% \\
&\text{power losses} = 15\% * 250 = 37.5\ \text{W} \\
&\text{panelpower in } 55\ °\text{C is} = 250 - 37.5 = 212.5
\end{aligned}$$

Since these coefficients are negative, we find that a lower coefficient is more efficient.

Material panel warranty is another criterion that has a significant impact on the selection of panels. Manufacturers guarantee their panels that work in how many years and protect customers against equipment failures. It needs to be mentioned that the most critical criteria that have a significant impact on customer opinion are the cost of the panel.

The solar inverter is one of the crucial instruments in selecting a rooftop panel. Solar inverters convert direct current (DC) electricity generated by a panel to alternating current (AC) for the home. This inverter has a vital role in the performance of solar power systems. There are three types of solar inverters: string inverters, power optimizer systems, and micro-inverters. String inverters are the most cost-effective than the two others. If a part of the panel is shaded, string inverters do not work genuinely. They optimize power output at the string level on the panel level and vice versa or for two other inverters [26]. The power optimizer system and micro-inverter monitor panel-level performance. Among the above inverters, the micro-inverter is increased safety because of lower DC voltage and increased design flexibility and yield when panels are in overshadowing.

The above criteria are routine in all rooftop PV panels, but when to use the PMADM method, we should consider criteria that belong to the future. According to searching by the researcher, dirtiness over time is one of the problems of this panel. This problem decreases the panels' efficiency. The main reason for this dirtiness is dust sitting on the surface of the panel. To maintain the rooftop from dust, researchers of this article propose an automatic protector which covers the panel in shadowing and at night. This protector increases the panels' efficiency and saves the time using and protecting them from tornadoes and hurricanes. This criterion increases the life of panels and protects them from wrecking. For quantified, this criterion is used years because the researcher considers that if the panel protects from dust and hurricane, the life of the panel is still further.

The smart panel is one of the criteria which is very important soon. The purpose of the smart panel here is the internet of things (IoT), meaning that the panel is equipped with Wi-Fi and transmits the information to the panel factory. With this information, operators check the efficiency and safety of panels and control them remotely and do maintenance operations. In Table 1, this criterion is quantified by the percent of operation of the factory. By these eight criteria, compare the alternative which is listed in Table 1. The alternatives are the manufacturers who produce the rooftop panels.

Since the solar inverter column is not numerical, we assign five numbers to it. Micro-inverter gets number 5, and string inverter gets number 1 in this research.

The www.mcdm.app is used to apply EDAS method and the following results are obtained.

Table 2 shows that the ranking of alternatives is different in the current and future visions. This shows us that ranking of alternatives depends on on-time vision.

Table 1 Amount of criteria

	Efficiency	Temperature coefficient	Material warranty (years)	Cost per watt	Solar inverter	Solar inverter cost per watt-peak	Automatic protector	Smart panel
Weights	Max	Min	Max	Min	Max	Min	Max	Max
Current vision	0.15	0.15	0.15	0.25	0.05	0.15	0.05	0.05
Future vision	0.1	0.1	0.1	0.2	0.05	0.05	0.25	0.15
A1	17%	– 0.39/°C	20	2.53$	string	0.05	8 years	0.65
A2	20%	– 0.35/°C	15	2.46$	Power optimizer	0.08	2 years	0.65
A3	20%	– 0.28/°C	20	2.96$	string	0.06	1 years	0.3
A4	15%	– 0.25/°C	10	2.74$	micro	0.29	5 years	0.4

Table 2 Rank of alternatives for current vision and future vision

Alternatives	EDAS for current vision	Ranking of current vision	EDAS for future vision	Ranking of future vision
A1	0.62	2	0.597	1
A2	0.352	4	0.371	4
A3	0.526	3	0.596	2
A4	0.709	1	0.419	3

6 Conclusion

Due to the plan of countries to use EVs in the future, nations' big problem is the supply of electricity. The main reason for using EVs is to reduce air pollution and global warming, but electricity is generated by burning fossil fuels. This is causing air pollution itself. So, generating electricity with renewable energy is an important issue. Sun radiation is a good source for generating electricity. Due to the high electricity demand, houses act as an instrument for producing electricity by sun radiation. Setting rooftop photovoltaic is a necessary instrument for generating electricity in homes. This rooftop PV has some properties which differ from others. So, selecting the best rooftop photovoltaic panel is an essential issue that this study pays attention to. The efficiency, temperature coefficient, material warranty, cost per watt of the panel, solar inverter, and solar inverter cost per watt-peak, automatic protector, and smart panel are considered in this research.

Researchers of this article consider two visions for evaluating these criteria. For this, they are used PMADM with the EDAS method and ranked the alternatives. The ranking of alternatives in the two-time vision is different, and this achievement helps the decision-makers to get better decisions for selecting rooftop PV. Evaluations show that A_1 gets rank one in future time and A4 gets rank one in current time. By surveying, criteria find that the intelligent panel that transfers information to the factory is a critical criterion in the future. One of the big problems that householders face is the fluctuation of panel efficiency and other problems that may impact panel efficiency. By smart panel factories, control panels online and they repair them in critical situations. The automatic protector is one of the criteria which increases the life of panel using. This criterion is the defined basis of years. It can plus it to the guaranty criterion because dust floating in the air has a negative impact on panel efficiency and life.

References

1. Tu R, Gai Y, Farooq B, Posen D, Hatzopoulou M (2020) Electric vehicle charging optimization to minimize marginal greenhouse gas emissions from power generation. Appl Energy 277
2. Borlaug B, Salisbury S, Gerdes M, Muratori M (2020) Levelized cost of charging electric vehicles in the United States. Joule 4:1–16
3. Mulcué-Nieto L, Echeverry-Cardona L, Restrepo-Franco A, García-Gutiérrez G, Jiménez-García F, Mora-López L (2020) Energy performance assessment of monocrystalline and polycrystalline photovoltaic modules in the tropical mountain climate: the case for Manizales-Colombia. Energy Rep 6:2828–2835
4. Ren H, Gao W, Zhou W, Nakagami K (2009) Multi-criteria evaluation for the optimal adoption of distributed residential energy systems in Japan. Energy Policy 37:5484–5493
5. Yadav S, Bajpai U (2018) Performance evaluation of a rooftop solar photovoltaic power plant in Northern India. Energy Sustain Dev 43:130–138
6. Yang Y, Ren J, Solgaard H, Xu D, Nguyen T (2018) Using multi-criteria analysis to prioritize renewable energy home heating technologies. Sustain Energy Technol Assess 29:36–43
7. Wang C, Nguyen V, Thai H, Duong D (2018) Multi-criteria decision making (MCDM) approaches for solar power plant location selection in Viet Nam. Energies 11(6)
8. Zhang C, Wang Q, Zeng S, Balezentis T, Streimikien D, Alisauskaite-Seskiene L, Chen X (2019) Probabilistic multi-criteria assessment of renewable micro-generation technologies in households. J Clean Prod 212:582–592
9. Corcelli F, Fiorentino G, Petit-Boix A, Rieradevall J, Gabarrell X (2019) Transforming rooftops into productive urban spaces in the Mediterranean. An LCA comparison of agri-urban production and photovoltaic energy generation. Resour Conserv Recycl 144:321–336
10. Saleem L, Ulfat I (2019) A multi criteria approach to rank renewable energy technologies for domestic sector electricity demand of Pakistan. Mehran Univ Res J Eng Technol 38(2):443–452
11. Seddiki M, Bennadji A (2019) Multi-criteria evaluation of renewable energy alternatives for electricity generation in a residential building. Renew Sustain Energy Rev 110:101–117
12. Li T, Li A, Guo X (2020) The sustainable development-oriented development and utilization of renewable energy industry—a comprehensive analysis of MCDM methods. Energy 212
13. Kamari M, Isvand H, Nazari M (2020) Applications of multi-criteria decision-making (MCDM) methods in renewable energy development: a review. RERA 1(1):47–54
14. Rigo P, Rediske G, Rosa C, Gastaldo N, Michels L, Neuenfeldt Júnior A, Siluk J (2020) Renewable energy problems: exploring the methods to support the decision-making process. Sustainability 12:95–101. https://doi.org/10.3390/su122310195
15. Howlader A, Sadoyama S, Roose L, Chen Y (2020) Active power control to mitigate voltage and frequency deviations for the smart grid using smart PV inverters. Appl Energy 285

16. Zavadskas E, Turskis Z (2010) A new additive ratio assessment (ARAS) method in multicriteria decision-making. Technol Econ Dev Econ 16(2):159–172
17. Stević Z, Pamučar D, Puška A, Chatterjeed P (2020) Sustainable supplier selection in healthcare industries using a new MCDM method: measurement of alternatives and ranking according to compromise solution (MARCOS). Comput Ind Eng 140
18. Yazdani M, Zaraté P, Zavadskas E, Turskis Z (2018) A Combined compromise solution (CoCoSo) method for multi-criteria decision-making problems. Manag Decis 57(3)
19. Pamucar D, Ćirović G (2015) The selection of transport and handling resources in logistics centers using multi-attributive border approximation area comparison (MABAC). Expert Syst Appl 42(6):3016–3028
20. Keshavarz-Ghorabaee M, Amiri M, Zavadskas E, Antucheviciene J (2018) Simultaneous evaluation of criteria and alternatives (SECA) for multi-criteria decision-making. Informatica 29(2)
21. Ecer F (2018) Third-party logistics (3PLs) provider selection via Fuzzy AHP and EDAS integrated model. Technol Econ Dev Econ 24(2):615–634
22. Hashemkhani Zolfani S, Maknoon R, Zavadskas E (2016) An introduction to prospective multiple attribute decision making (PMADM). Technol Econ Dev Econ 22(2):309–326
23. Hashemkhani Zolfani S, Masaeli R (2020) From past to present and into the sustainable future. In: PMADM approach in shaping regulatory policies of the medical device industry in the new sanction period. Sustainability Modeling in Engineering, pp 73–95
24. Keshavarz Ghorabaee M, Zavadskas E, Turskis Z, Olfat L (2017) Multi-criteria inventory classification using a new method of evaluation based on distance from average solution (EDAS). Informatica 26(3):435–451
25. Tindosolar (2021) Tindosolar. (Online) Available: https://www.tindosolar.com.au/learn-more/temperature-coefficient/
26. Thoubboron K (2021) Energysage. (Online) Available: https://news.energysage.com/string-inverters-power-optimizers-microinverters-compared/. Accessed 7 May 2021

A Proposed q-Rung Orthopair Fuzzy-Based Decision Support System for Comparing Marketing Automation Modules for Higher Education Admission

Sanjib Biswas, Dragan Pamucar, Akanksha Raj, and Samarjit Kar

Abstract With rapid developments in Information and Communication Technology (ICT) sector prevailing over the present age of Industry 4.0, the nature of traditional marketing has undergone a paradigm shift. The recent pandemic has reflected the need of more emphasis on technology-enabled marketing. It is said the battle is now on digital space. Digital Marketing (DM) has become a key strategic advantage for the organizations. The age of Industry 4.0 has enabled generation of a massive volume of data at lightning speed with extremely high variety and capturing with improved hardware technology. However, it is important to derive actionable insight out of the data for formulating appropriate decision on time. Here comes the importance of marketing automation tools that help in proper organization of captured data from various sources including social media on a real-time basis for aiding the strategic decision-making process and action planning. Likewise, all sectors, education industry, particularly higher educational institutes (HEI) have witnessed a transformational change in recent years in the admission process. With more reliance on digital media, the HEIs have felt the need to acquire, nurture, engage and optimize the potential inquiries or "leads" for enhancing the enrollment. In this context, the present paper aims to compare a set of recently developed marketing automation tools such as chat BOT, Whatsapp Business API, Google Tracking, etc. from the perspective of effective lead management and optimization for improving the return on investment (ROI). For this purpose, in the present paper we propose a novel extension of a very recently developed multi-criteria decision-making (MCDM) algorithm such as Preference Ranking on the Basis of Ideal-Average Distance (PROBID) with

S. Biswas (✉) · A. Raj
Decision Science and Operations Management Area, Calcutta Business School, South 24, Parganas, West Bengal 743503, India
e-mail: sanjibb@acm.org

D. Pamucar
Department of Logistics, Military Academy, University of Defence, 11000 Belgrade, Serbia

S. Kar
National Institute of Technology, Durgapur, West Bengal 713209, India

P. Chatterjee et al. (eds.), *Computational Intelligence for Engineering and Management Applications*, Lecture Notes in Electrical Engineering 984,
https://doi.org/10.1007/978-981-19-8493-8_66

q-Rung Orthopair Fuzzy (qROF) information wherein the criteria weights are determined by using qROF-based entropy method. We carry out the sensitivity analysis and validation test that reflect the stability and validity of the result.

Keywords Digital marketing · Automated lead management · Higher education admission · q-Rung Orthopair Fuzzy Sets (qROFS) · Preference Ranking on the Basis of Ideal-Average Distance (PROBID) · Entropy method

1 Introduction

The higher education institutions (HEI) have witnessed a transformational change in the last decade. The changes are evident not only in evolutions of smart and innovative teaching and learning with increasing use of high-end technology and contemporary curriculum but also in the behavior of the stakeholders like students and their parents, employers and community. To stay relevant in the marketplace, HEIs need to recognize the requirements of the stakeholders and follow an engaging and professional approach [1]. Co-creation of the value of the educational programs through seamless collaboration of HEIs, students, corporates, community and governing bodies is a mandate today. In this regard, digital technology has emerged as a key enabler and strategic advantage over the last few years for the HEIs [2].

Given the changes in the preferences of the students and their parents in the present knowledgeable, informed and technology-driven society, the concept of marketing has undergone a metamorphosis from traditional "push" mentality to "pull" approach with innovative content and seamless open communication and real-time engagement. Digital technology has become a critical enabler for formulating marketing strategy across the industries. The Higher Education Sector (HES) is not an exception in embracing the power of application digital technologies in promotional and branding campaigns. Marketing is no longer an outlandish area as against the customs and beliefs of HEIs [3]. There has been a change in focus as HEIs now look forward to self-sustainability through availing new opportunities, offering portfolio of products (aka programs) and revenue generations [4]. Students are given utmost importance and the education is centered on building an ecosystem that fosters effective collaboration among institutes, society, industries and government to develop resource-based operational capabilities and unique competencies. Formulation of effective marketing strategies and communications is therefore a critical success factor for the HEIs now. Marketing using digital media aka digital marketing (DM) has emerged as a strategic decision area for HEIs. HEIs are increasingly leveraging DM for reaching to a larger student community over a wide geographic coverage area, getting connected with the faculty members with diverse expertise from various institutions and universities and research organizations and, finally, networking with several corporate organizations.

According to a recent figure shared by [5], the internet penetration rate in India has reached the level of around 45% in 2021 (with an equivalent figure of nearly half of the total population in India) as compared with 4% in 2007. The other report [6]

confirmed a surge of 12% in smartphone sale in 2021 in India. The aforesaid two figures are indicative of the growing importance of DM in recent times. The recent pandemic has unveiled the importance of DM in a more prominent way. Post COVID-19, the HEIs have no choice other than utilizing DM to reach the students who are "socially distant but digitally connected" and attract them for possible enrollment [7]. In fact, digital technology has helped the HEIs continue providing education to the students amidst the gloomy environment shaken by the sound of ambulance and despondent news of increase in the number of infected cases and death toll. The authors [8] noted that in the post-pandemic era, online learning over internet in a virtual classroom becomes an integral part of HEIs.

DM entails the use of digital technologies in the marketing activities to gain insight about customers' needs and promote the offered products and services to match their requirements in an interactive, engaging, focused and measurable way [9–11]. The concept of DM has started garnering increasing attention since late 2000s and its importance got manifold with advent of the age of industry 4.0 from 2011. Various strategies of DM, such as search engine optimization (SEO), search engine marketing (SEM), content marketing through ads, emails and SMS blast, influencer marketing, social media marketing and optimization, display advertisements and animated branding, help the organizations to reach and engage and ultimately win over the potential customers in the most efficient way [12]. DM has helped the HEIs to expand their brand awareness and get engaged with a large number of students from various locations and backgrounds [13].

DM plays a crucial role in the admission process of the HEIs. The admission of quality students is the hallmark of a good HEI that helps in delivering quality outcome [14]. To enroll the students of better quality, HEIs promote their offerings and special features to attract and engage the talents. As a matter of fact, admission stands as a survival factor for many HEIs that aim to achieve self-sustainability. The admission process begins with targeting and connecting with the potential students through branding and continuous communication and engagement with them. Any potential candidate aka "lead" turns out be an applicant through effective marketing communication and branding. Through customized engagement and selection process, HEIs enroll the applicants. The process from attraction to conversion requires an arduous effort wherein DM acts as one of the cornerstones of success. Considering the fact that young minds prefer digital media, especially social media for information and subsequently for formation of the opinions, HEIs have been actively leveraging the social media and web for campaigning to attract the talents, engaging with the potential students throughout the admission process and communication [15–18]. DM supports the HEIs in positioning of brands, segmentation and targeting of markets at macro and micro level, outreach to potential students and engagement, discovery of latent expectations, communication and information sharing, faster operations, and Customer Relationship Management (CRM) in a cost effective way [3]. CRM acts as one of major support to HEIs for strengthening the relationship with students, promoting best practices and thereby creating an experience for the students and other stakeholders [19–21]. For HEIs, DM and CRM are mutually dependent and aligned with the information system to support the entire value chain. Hence, both

DM and CRM have been considered as important strategic decisions [22, 23]. In this regard, the Enterprise Resource Solution (ERP) plays a role of backend support for achieving integration of all activities across the value chain of the HEIs including admission of students and improving operational efficiency [24].

With the advancement in the information and communication technology, advanced systems for data capturing and analytics (powered by AI and machine learning algorithms) have been utilized by the HEIs for drawing actionable insights that provide impetus to strategic decision-making. It is evident from the actual practices followed by HEIs that there has been an increasing use of AI and machine learning-based algorithms, augmented reality (AR), virtual reality (VR) and seamless integration of cyber and physical space to support the DM and CRM as well as the admission process [25, 26]. In fact, the CRM process has been automated substantially. User-centric customized DM campaigns and marketing communication and branding followed by the automated lead capturing and engagement have become a key antecedent to success of HEIs in terms of number of enrollment. In this regard, a number of modules have been developed for automated lead capturing and management and subsequently supporting DM.

The present paper aims to compare some of the marketing automation modules using a Qrof-based multi-criteria group decision-making (MAGDM). We are inquisitive about the extent to which the modules differ from each other. For comparative ranking of the marketing automation modules, we propose a new integrated qROF-based intelligent framework of entropy and PROBID methods based on a number of attributes (that features the objectives of DM and CRM) and experts' opinions. The preliminary concepts and definitions of qROFS and the algorithms of the entropy and PROBID methods are described in the succeeding sections. Some of the benefits of our proposed qROF-Entropy-PROBID (qEP) model are:

- As compared with Intuitionistic Fuzzy Sets (IFS) or Pythagorean Fuzzy Sets (PyFS), qROFS provides wide options in terms of values of membership and non-membership functions to the decision-makers in a group decision-making scenarios and extends the opportunity to carry out more granular analysis with reasonably accurate and stable results [27, 28].
- The entropy method utilizes the decision matrix itself for deriving the criteria weights. Hence, there is no need to take subjective opinions for criteria rating which reduces the bias. Further, the entropy method determines criteria weights using objective information while combining various criteria having different units and distribution. The entropy method is a non-linear approach that works on probability concepts and withstand the presence of noise and inference in data and is able to generate reasonably accurate results with imprecise and uncertain data [29–35].
- The PROBID method has been introduced very recently by [36] that combines the favorable features of two other well-known MCDM algorithms such as the Technique for Order of Preference by Similarity to Ideal Solution (TOPSIS) and Evaluation Based on Distance from Average Solution (EDAS). PROBID not only considers all possible ideal solutions on or within the two extreme reference points

such as most positive ideal solution (PIS) and most negative PIS, i.e., negative ideal solution (NIS) but also consider the average solution as a benchmark. Thus, PROBID provides wide options to all kind of decision-makers with varying risk tolerance levels and withstands subjective variations. Further, it is free from rank reversal phenomenon though it utilizes distance measures like TOPSIS [36, 37].

In the extant literature, it is evident that there is a scarcity of work on comparison of marketing automation modules. As a matter of fact, there is a very limited work on marketing automation for HEIs. Further, PROBID is found to be used in a handful number of papers [36, 37] with no extensions using qROFS. To add, there is no evidence of utilizing qROF scores in calculating entropy values in the process of determining criteria weights. Our paper fills the gaps in the literature in the above-mentioned ways.

The reminder of the paper is presented as follows. Section 2 provides some preliminary concepts about qROFS, while in Sect. 3 we describe the case study and the research methodology including our proposed framework. Section 4 provides the findings of the data analysis and some discussion agenda based on the findings. Section 5 exhibits the sensitivity analysis and validation test. Finally, Sect. 6 provides some concluding remarks including the implications of the research and propose some of the future scope of research.

2 Preliminaries of qROFS

In many situations, the decision-makers rate the favorable element (that he/she wants to be included as a member to the selection set) and/or the non-favorable element (that he/she does not want to be included as a member to the selection set) in such a way that the sum of membership and non-membership values exceeds the unity [28]. To tackle such situations and provide the decision-makers wide options in expressing their views, qROFS was proposed by [27] as an advanced extension of the IFS family. qROFS has been increasingly applied in various real-life problems, for example, some of the applications are summarized in Table 1.

Table 1 Some applications of qROFS

Application area	References
Selection of collaborators for setting up food processing plant in India	[38]
Investment decision making	[39, 40]
Laptop selection	[41]
Sports management	[42]
Comparison of agri-farming choices	[43]
Enterprise risk planning and management	[44]

In this section, we present some of the primary definitions related to concepts and operators of qROFS.

Definition 1 The Pythagorean Fuzzy Sets (PyFS) [45]

$$\tilde{A}^{P,} = \{\langle x, \mu_{\tilde{A}^P}(x), \vartheta_{\tilde{A}^P}(x)\rangle : x \in U\}$$

U is the universe of discourse. $\mu_{\tilde{A}^P}(x)$ and $\vartheta_{\tilde{A}^P}(x)$ are the degrees of membership and degree of non-membership, respectively.

$$\mu_{\tilde{A}^P}(x) : U \to [0, 1]$$
$$\vartheta_{\tilde{A}^P}(x) : U \to [0, 1]$$
$$0 \leq (\mu_{\tilde{A}^P}(x)^2 + \vartheta_{\tilde{A}^P}(x)^2) \leq 1; \quad \forall x \in U$$

The degree of indeterminacy is derived as

$$\pi_{\tilde{A}^P}(x) = \sqrt{1 - \mu_{\tilde{A}^P}(x)^2 - \vartheta_{\tilde{A}^P}(x)^2}; \ \forall x \in U \tag{1}$$

Definition 2 A qROFS is defined as [27]

$$\tilde{A}^Q = \{\langle x, \mu_{\tilde{A}^Q}(x), \vartheta_{\tilde{A}^Q}(x)\rangle : x \in U\}$$

U is the universe of discourse. $\mu_{\tilde{A}^Q}(x)$ and $\vartheta_{\tilde{A}^Q}(x)$ are the degrees of membership and degree of non-membership, respectively.

$$\mu_{\tilde{A}^Q}(x) : U \to [0, 1]$$
$$\vartheta_{\tilde{A}^Q}(x) : U \to [0, 1]$$
$$0 \leq ((\mu_{\tilde{A}^Q}(x))^q + (\vartheta_{\tilde{A}^Q}(x))^q) \leq 1; \forall x \in U$$

The degree of indeterminacy is derived as

$$\pi_{\tilde{A}^Q}(x) = \sqrt[q]{1 - (\mu_{\tilde{A}^Q}(x))^q - (\vartheta_{\tilde{A}^Q}(x))^q}; \forall x \in U \tag{2}$$

When $q = 1$, $\tilde{A}^Q$ becomes an Atanassov's Intuitionistic Fuzzy Sets (IFS), and for $q = 2$, it gets converted into PyFS. In this paper, we shall use the qROFS in terms of a q-Rung Orthopair Fuzzy Number (qROFN) represented as $\mathbb{Q} = (\mu, \vartheta)$.

Definition 3 Basic operations on qROFN.

Let, $\mathbb{Q} = (\mu, \vartheta)$, $\mathbb{Q}_1 = (\mu_1, \vartheta_1)$, $\mathbb{Q}_2 = (\mu_2, \vartheta_2)$ are the three qROFNs. Some of the basic operators are defined as follows [27]

$$\mathbb{Q}^c = (\vartheta, \mu) \tag{3}$$

$$\mathbb{Q}_1 \oplus \mathbb{Q}_2 = (\sqrt[q]{\mu_1^q + \mu_2^q - \mu_1^q \mu_2^q}, \vartheta_1 \vartheta_2) \tag{4}$$

$$\mathbb{Q}_1 \otimes \mathbb{Q}_2 = (\mu_1 \mu_2, \sqrt[q]{\vartheta_1^q + \vartheta_2^q - \vartheta_1^q \vartheta_2^q}) \tag{5}$$

$$\alpha \mathbb{Q} = (\sqrt[q]{1 - (1 - \mu^q)^\alpha}, \vartheta^\alpha); \quad \alpha \text{ is a constant} \tag{6}$$

$$\mathbb{Q}^\alpha = (\mu^q, \sqrt[q]{1 - (1 - \vartheta^q)^\alpha}) \tag{7}$$

Definition 4 Score and Accuracy Function.

There has been a number of definitions available for the score and accuracy functions.

The Score Function (SF) is defined as [39]

$$S = \mu^q - \vartheta^q; S \in [-1, 1] \tag{8}$$

[46–48] have defined SF as

$$S' = \frac{(1 + \mu^q - \vartheta^q)}{2} \tag{9}$$

The definition by [49]

$$S'' = \mu^q - \vartheta^q + \left(\frac{e^{\mu^q - \vartheta^q}}{e^{\mu^q - \vartheta^q} + 1} - \frac{1}{2}\right)\pi^q \tag{10}$$

The authors [50] have provided an alternative definition of SF assuming that if $\mu = \vartheta$. In this paper, we use the definition given by [50] for calculating SF

$$S^* = \frac{\mu^q - 2\vartheta^q - 1}{3} + \frac{\lambda}{3}(\mu^q + \vartheta^q + 2), \lambda \in [0, 1] \tag{11}$$

The Accuracy Function (AF) is defined as [39]

$$H = \mu^q + \vartheta^q, \quad H \in [0, 1] \tag{12}$$

Rule for comparison:

$$S_1 > S_2 \Rightarrow \mathbb{Q}_1 > \mathbb{Q}_2$$
$$S_1 < S_2 \Rightarrow \mathbb{Q}_1 < \mathbb{Q}_2$$
$$S_1 = S_2$$
$$H_1 < H_2 \Rightarrow \mathbb{Q}_1 < \mathbb{Q}_2$$
$$H_1 > H_2 \Rightarrow \mathbb{Q}_1 > \mathbb{Q}_2$$

Definition 5 q-Rung Orthopair Fuzzy Weighted Averaging Operator (qROFWA) [39]

$$\text{q-ROFWA}(\mathbb{Q}_1, \mathbb{Q}_2, \mathbb{Q}_3, \ldots, \mathbb{Q}_r) = \left\langle (1 - \prod_{k=1}^{r} (1 - \mu_k^q)^{\alpha_k})^{1/q}, \prod_{k=1}^{r} \vartheta_k^{q^{\alpha_k}} \right\rangle \tag{13}$$

Here, α_k is the corresponding weight.

Definition 6 Distance Measure [51–53]

The normalized hamming distance is given by

$$d_h = \frac{1}{2}(|\mu_1^q - \mu_2^q| + |\vartheta_1^q - \vartheta_2^q| + |\pi_1^q - \pi_2^q|) \tag{14}$$

3 Materials and Methods

In this paper, we aim to compare 9 (nine) marketing automation modules used in the admission process of HEIs. These modules are supporting various stages of admission by six (6) ways such as lead acquisition, lead nurturing, brand equity, marketing optimization, ROI tracking and shortening lead to enrollment journey. Hence, in the present study, we have 9 alternatives $M_1, M_2, M_3, \ldots, M_9$ and 6 attributes $C_1, C_2, C_3, \ldots, C_6$. Tables 2 and 3 provide the details of the alternatives and attributes. Essentially, these attributes decide the purpose of using the modules. We use a multi-criteria group decision-making (MAGDM) setup wherein three experts having substantial experiences of 16 years, 22 years and 25+ years, respectively, in counseling, admission and digital marketing have assessed the utilities of the modules with respect to the attributes on a seven-point linguistic scale (1: Extremely Low and 7: Extremely High). We propose a new integrated Entropy-PROBID framework wherein the ratings are expressed in terms of qROFNs (see Table 4) following the work of [54].

The steps of the research methodology are shown in the flow diagram (see Fig. 1).

3.1 Conventional Entropy Method

Entropy measures the disorder. The entropy method [55] has been a widely used algorithm for determining criteria weights using objective information as evidenced in various applications in the extant literature, for example, carbon emission management in power sector [56], road safety planning [57], assessment of impact of online marketing [58], customer segmentation [59], analysis of demand in tourism [60],

Table 2 List of marketing automation modules

S/L	Module
M_1	Student Admission Portal 2.0
M_2	Document Verification, Offer Letter
M_3	FB Lead Ad
M_4	Google Tracking Parameter
M_5	Adv. Tracking Parameters (Benchmarking, Trend Analysis, Student Quality Index)
M_6	Advance Landing Page Builder
M_7	Chat BoT (Whatsapp and Web)
M_8	WhatsApp Business API
M_9	ERP Integration

Table 3 List of attributes

S/L	Attributes	Effect
C_1	Lead acquisition	(+)
C_2	Lead nurturing	(+)
C_3	Brand equity	(+)
C_4	Marketing optimization	(+)
C_5	ROI tracking	(+)
C_6	Lead to enrollment journey	(−)

Table 4 qROFN linguistic scale

Linguistic term	Code value	qROFN	
		μ	ϑ
Extremely high	7	0.85	0.25
Very high	6	0.75	0.35
High	5	0.65	0.45
Moderate	4	0.55	0.55
Low	3	0.45	0.65
Very low	2	0.35	0.75
Extremely low	1	0.25	0.85

crowd sourcing resource selection [61], portfolio selection [62, 63] and evaluation of bank performance [64] to mention a few. According to entropy method, more weight is assigned to the criterion or attribute that has higher entropy value. The algorithm is described below.

Suppose, $X = \left[x_{ij}\right]_{m \times n}$ is the decision matrix (DMTR), where m is the number of alternatives and n is the number of attributes.

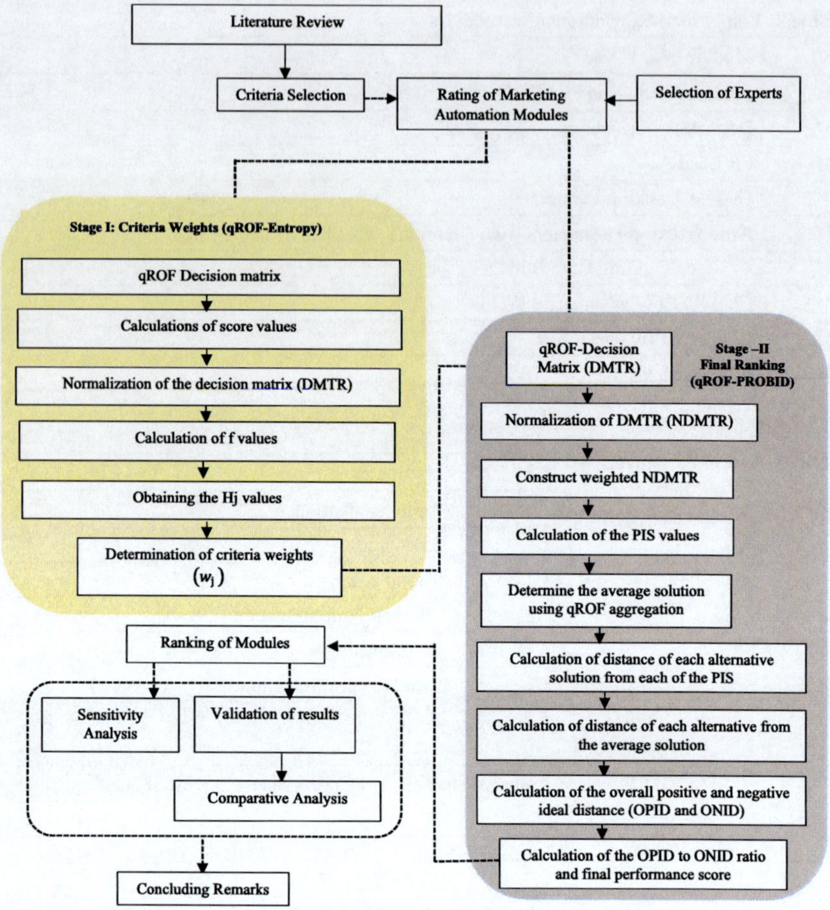

Fig. 1 Proposed research framework (qROF-Entropy-PROBID)

Step 1: Normalization of DMTR

The normalized decision matrix (NDMTR) is represented as $R = [r_{ij}]_{m\times n}$ where the elements r_{ij} are given by:

$$r_{ij} = \begin{cases} \dfrac{(x_{ij} - x_{j(\min)})}{(x_{i(\max)} - x_{j(\min)})};\ j \in j^+ \\ \dfrac{(x_{j(\max)} - x_{ij})}{(x_{i(\max)} - x_{j(\min)})};\ j \in j^- \end{cases} \quad (15)$$

Step 2: Calculation of entropy values

The entropy value for *i*th alternative for *j*th criterion is given by:

$$H_j = -k \sum_{i=1}^{m} f_{ij} \ln(f_{ij}) \quad (16)$$

where *k* is a constant and is defined by:

$$k = \frac{1}{\ln(m)} \quad (17)$$

and

$$f_{ij} = \frac{r_{ij}}{\sum_{i=1}^{m} r_{ij}} \quad (18)$$

If $f_{ij} = 0$
then,

$$f_{ij} \ln(f_{ij}) = 0 \quad (19)$$

Step 3: Calculation of criteria weight

The weight for each criterion is given by:

$$w_j = \frac{1 - H_j}{n - \sum_{j=1}^{n} H_j} \quad (20)$$

Here, higher is the value of w_j, more is the information contained in the *j*th criterion.

3.2 *Conventional PROBID Method*

PROBID method combines the features of two other popular distance-based MCDM methods such as Technique for Order Preference by Similarity to Ideal Solution (TOPSIS) and Evaluation based on Distance from Average Solution (EDAS). PROBID has been introduced very recently, and its applications are yet to be widely explored. Our limited search reveals two applications of PROBID method such as chemical processing [36] and comparison of videoconferencing platforms [37]. The computational steps for this method are given below.

Step 1. Normalization of DMTR

The DMTR $X = [x_{ij}]_{m\times n}$ can be converted into a normalized DMTR (NDMTR) $R = [r_{ij}]_{m\times n}$, where the elements are given by using the expression (15).

Step 2. Derive the weighted NDMTR

The weighted NDMTR (WNDMTR) is represented by $V = [v_{ij}]_{m\times n}$ where the elements are derived as

$$v_{ij} = r_{ij} w_j \tag{21}$$

$$i \in \{1, 2, \ldots, m\};\ j \in \{1, 2, \ldots, n\}$$

w_j is the weight of *j*th criterion.

Step 3. Calculation of the PIS values

Let, $P_{k(j)}; k = 1, 2, \ldots, m$ are the PIS values where $P_{1(j)}$ is the most favorable PIS value. Therefore, $P_{m(j)}$ is the most non-favorable PIS i.e., most NIS value. The *k*th PIS value is given by

$$P_{k(j)} = \left\{(\max(v_j, k);\ j \in j^+), (\min(v_j, k) j \in j^-)\right\} \tag{22}$$

j^+ is the set of maximizing criteria and j^- is the set of minimizing criteria from the criteria set for $j \in \{1, 2, 3, \ldots, n\}$.

Step 4. Determine the average solution

The average solution is given by

$$v_{\text{avg}(j)} = \frac{\sum_{k=1}^{m} P_{k(j)}}{m};\quad j \in \{1, 2, \ldots, n\} \tag{23}$$

Step 5. Calculation of distance of each alternative solution from each of the *m* number of PIS. Applying the standard formula of Euclidean distance of *i*th alternative from each of the PIS is derived as

$$d_{i(k)} = \sqrt{\sum_{j=1}^{n} (v_{ij} - P_{k(j)})^2};\quad i \in \{1, 2, \ldots, m\};\quad k \in \{1, 2, \ldots, m\} \tag{24}$$

Step 6. Calculation of distance of each alternative from the average solution

In the same way, the Euclidean distances are calculated as

$$d_{i(\text{avg})} = \sqrt{\sum_{j=1}^{n} (v_{ij} - v_{\text{avg}(j)})^2};\quad i \in \{1, 2, \ldots, m\} \tag{25}$$

Step 7. Calculation of the overall positive ideal distance (OPID)

The OPID represents the weighted sum distance of an alternative from the first half of the PIS values and is expressed as

$$d_{i\,(\text{pos-ideal})} = \begin{cases} \sum_{k=1}^{(m+1)/2} \frac{d_{i\,(k)}}{k}; & i \in \{1, 2, \ldots, m, \text{odd}\} \\ \sum_{k=1}^{m/2} \frac{d_{i\,(k)}}{k}; & i \in \{1, 2, \ldots, m, \text{even}\} \end{cases} \tag{26}$$

Step 8. Calculation of the overall negative ideal distance (ONID)

The ONID is the weighted sum distance of an alternative from the second half of the PIS values and is expressed as

$$d_{i\,(\text{neg-ideal})} = \begin{cases} \sum_{k=(m+1)/2}^{m} \frac{d_{i\,(k)}}{m-k+1}; & i \in \{1, 2, \ldots, m, \text{odd}\} \\ \sum_{k=\frac{m}{2}+1}^{m} \frac{d_{i\,(k)}}{m-k+1}; & i \in \{1, 2, \ldots, m, \text{even}\} \end{cases} \tag{27}$$

Unlike OPID, in case of ONID the weight increases as k approaches to m.

Step 9. Calculation to OPID to ONID ratio

The ratio OPID to ONID is given by

$$R_i = \frac{d_{i\,(\text{pos-ideal})}}{d_{i\,(\text{pos-ideal})}} \tag{28}$$

Step 10. Determine the final performance score

The final performance score is given by

$$\text{PS}_i = \frac{1}{1 + R_i^2} + d_{i\,(\text{avg})}; \quad i \in \{1, 2, \ldots, m\} \tag{29}$$

It may be noted that if $R_i \to 0$ PS_i increases which means the respective alternative solution becomes closer to most PIS. Hence, higher is the value of PS_i, more preferable is the corresponding alternative.

3.3 Steps of the Proposed QROF-Entropy-PROBID (qEP) Framework

The procedural steps of qEP framework are explained as follows.

General Steps

Step 1. Selection of the alternatives and the attributes (see Tables 2 and 3).

Step 2. Selection of the experts.

Step 3. Rating of the alternatives with respect to the attributes using qROFN linguistic rating scale (see Table 4) by each expert.

Step 4. Aggregation of the responses of the individual experts using expression (13) to form the Qrofn-based DMTR (qDMTR).

Let, the qDMTR is denoted as

$$\widetilde{X} = \left[\tilde{x}_{ij}\right]_{m \times n}. \tag{30}$$

Phase I: Attribute Weights

Step 5. Find out the scores of the elements of the qDMTR using expression (11).

Let, the score value based DMTR is given as $X = \left[x_{ij}\right]_{m \times n}$. It may be noted that, under normal circumstance, for the simplicity in calculation, q value is considered as 1 and λ value is generally taken as 0.8.

Step 6. Normalization of DMTR using expression (15).

Step 7. Calculation of entropy values of the attributes using expressions (16)–(19).

Step 8. Find out the weights of the attributes using expression (20).

Phase II: Ranking of the Alternatives

Step 9. Normalization of qDMTR using the expression (3).

Let, the normalized qDMTR is denoted as

$$\widetilde{R} = \left[\tilde{r}_{ij}\right]_{m \times n}. \tag{31}$$

Step 10. Find out weighted normalized DMTR using the expression (6).

Step 11. Find out the PIS matrix.

In this case, most PIS is given as $(\mu_{\max}, \vartheta_{\min})$.

Step 12. Derive the average solution based on weighted normalized DMTR using expression (13).

Step 13. Calculate the distances of the alternatives from the average solution and all PIS.

Here, we use the normalized Hamming distance using the expression (14).

Steps 14–17. Ranking of alternatives using the expressions (26)–(29).

4 Results

In this section, we present the brief summary of the results. First, we obtained the individual responses and aggregated the same. The qDMTR is given in Table 5, obtained by aggregating the individual responses of the experts using the expression (13) where we consider $q = 1$.

Now, we use expression (11) and derive the scores of the elements of the qDMTR (whose elements are qROFNs). The score values of the matrix shown in Table 5 is given in Table 6 (considering $q = 1$ and $\lambda = 0.8$). We proceed to follow the conventional steps to calculate the entropy values and finally attribute weights by using the expressions (15)–(20). Here, we have m (no of alternatives) = 9 and n (no of attributes) = 6. Therefore, $K = \ln(m) = 2.19722$. Table 7 provides the entropy values (H_j) and weights (w_j) for all attributes.

It is seen that $C_1 > C_6 > C_5 > C_4 > C_3 > C_2$ which implies that the experts believe that acquisition of leads i.e., identifying potential candidates who might take admission and enroll them into student role at the earliest are the two most important attributes of a successful admission process. We then move to compare the automation modules with respect to the attributes using qROF-PROBID approach. First, we normalize the qDMTR using the expression (3) as given in Table 8. The normalized qDMTR is then multiplied with the corresponding weights of the attributes (see Table 7), and the PIS values are calculated wherein most PIS is given as $(\mu_{\max}, \vartheta_{\min})$. Table 9 shows the PIS matrix. Next, the average solution is derived by using the expression (13) as exhibited in Table 10. It may be noted that the PIS 1 is the most PIS and PIS 9 is the last one, i.e., most NIS.

We now find out the distances of the alternatives with respect to the PIS points and average solution point. It may be noted that in this paper, we use normalized hamming distance measures (see expression (14)). Table 11 shows the distances. Finally, we obtain the OPID, ONID and final appraisal score values (PS_i) for the alternatives as given in Table 12.

It is observed that informative and engaging landing page, chat BOT options and social media advertisements hold higher significance as we find $M_6 > M_7 > M_3 > M_4 > M_8 > M_5 > M_1 > M_2 > M_9$.

5 Sensitivity Analysis and Validation

For any MCDM- or MAGDM-based analysis, it is important to examine the impact of any changing condition on the stability of the result [65–67]. In other words,

Table 5 Decision matrix with qROFN ($q = 1$)

Module	Attributes											
	C_1		C_2		C_3		C_4		C_5		C_6	
	μ	ϑ	μ	ϑ	μ	ϑ	μ	ϑ	μ	ϑ	μ	ϑ
M_1	0.2849	0.8153	0.6598	0.4425	0.6363	0.4678	0.5189	0.5815	0.2500	0.8500	0.6598	0.4425
M_2	0.2500	0.8500	0.7500	0.3500	0.6871	0.4138	0.5862	0.5144	0.4500	0.6500	0.8222	0.2797
M_3	0.8500	0.2500	0.5862	0.5144	0.6959	0.4069	0.7203	0.3806	0.8222	0.2797	0.7361	0.3699
M_4	0.5189	0.5815	0.6871	0.4138	0.3182	0.7820	0.8500	0.2500	0.8222	0.2797	0.3500	0.7500
M_5	0.3182	0.7820	0.7500	0.3500	0.4856	0.6148	0.8222	0.2797	0.8500	0.2500	0.6500	0.4500
M_6	0.8500	0.2500	0.6500	0.4500	0.8500	0.2500	0.7203	0.3806	0.7891	0.3129	0.7203	0.3806
M_7	0.7891	0.3129	0.8500	0.2500	0.7891	0.3129	0.7500	0.3500	0.8500	0.2500	0.7837	0.3251
M_8	0.3500	0.7500	0.7891	0.3129	0.8500	0.2500	0.8500	0.2500	0.7203	0.3806	0.7500	0.3500
M_9	0.2500	0.8500	0.2500	0.8500	0.4500	0.6500	0.4500	0.6500	0.2849	0.8153	0.8222	0.2797

Table 6 Score values of the elements of qDMTR

Module	Attributes					
	C_1	C_2	C_3	C_4	C_5	C_6
	B	B	B	B	B	NB
M_1	0.0449	0.4189	0.3946	0.2787	0.0100	0.4189
M_2	0.0100	0.5100	0.4467	0.3459	0.2100	0.5814
M_3	0.6100	0.3459	0.4548	0.4800	0.5814	0.4937
M_4	0.2787	0.4467	0.0782	0.6100	0.5814	0.1100
M_5	0.0782	0.5100	0.2454	0.5814	0.6100	0.4100
M_6	0.6100	0.4100	0.6100	0.4800	0.5483	0.4800
M_7	0.5483	0.6100	0.5483	0.5100	0.6100	0.5401
M_8	0.1100	0.5483	0.6100	0.6100	0.4800	0.5100
M_9	0.0100	0.0100	0.2100	0.2100	0.0449	0.5814

B = max type attributes and NB = min type attribute

Table 7 Weights of the attributes (entropy method)

Attributes	(+)	(+)	(+)	(+)	(+)	(−)
	C_1	C_2	C_3	C_4	C_5	C_6
H_j	0.7377	0.9397	0.9071	0.9011	0.8832	0.7525
w_j	0.2985	0.0687	0.1058	0.1125	0.1329	0.2817

sensitivity analysis is performed to test the stability in the results [68]. In this paper, we vary the values of q and λ to get different score values and utilize to change the qDMTR. We then use qEP framework to rank the modules under different scenarios. Table 13 exhibits the scheme for sensitivity analysis, and Table 14 shows the ranking results under various experimental cases.

The stability in the result gets reflected in considerable consistency in the ranking orders (see Table 14) and the pictorial representation of the outcome of the sensitivity analysis (see Fig. 2). To check the statistical significance, we carry out Spearman's Rank Correlation test (see Table 15) on the ranking results (see Table 14) and Friedman's test (see Table 17) using the appraisal scores of the modules under various experimental cases (see Table 16) with varying attribute weights of the sensitivity analysis. The test outcomes indicate that our model can produce stable and statistically consistent results.

It is also quite imperative to validate the result of the proposed model. We therefore follow the approaches available in the literature [69–71], i.e., comparison of the result of the proposed model with that of other existing methods. In the present paper, we compare qROFN score-based Proximity Index Value (PIV) method [72] and Simple Additive Weighting (SAW) model. Table 18 shows the comparison of the ranking results, and Table 19 reflects statistically that the results are consistent.

Table 8 Normalized qDMTR

Module	Attributes											
	C_1		C_2		C_3		C_4		C_5		C_6	
	μ	ϑ	μ	ϑ	μ	ϑ	μ	ϑ	μ	ϑ	μ	ϑ
M_1	0.2849	0.8153	0.6598	0.4425	0.6363	0.4678	0.5189	0.5815	0.2500	0.8500	0.4425	0.6598
M_2	0.2500	0.8500	0.7500	0.3500	0.6871	0.4138	0.5862	0.5144	0.4500	0.6500	0.2797	0.8222
M_3	0.8500	0.2500	0.5862	0.5144	0.6959	0.4069	0.7203	0.3806	0.8222	0.2797	0.3699	0.7361
M_4	0.5189	0.5815	0.6871	0.4138	0.3182	0.7820	0.8500	0.2500	0.8222	0.2797	0.7500	0.3500
M_5	0.3182	0.7820	0.7500	0.3500	0.4856	0.6148	0.8222	0.2797	0.8500	0.2500	0.4500	0.6500
M_6	0.8500	0.2500	0.6500	0.4500	0.8500	0.2500	0.7203	0.3806	0.7891	0.3129	0.3806	0.7203
M_7	0.7891	0.3129	0.8500	0.2500	0.7891	0.3129	0.7500	0.3500	0.8500	0.2500	0.3251	0.7837
M_8	0.3500	0.7500	0.7891	0.3129	0.8500	0.2500	0.8500	0.2500	0.7203	0.3806	0.3500	0.7500
M_9	0.2500	0.8500	0.2500	0.8500	0.4500	0.6500	0.4500	0.6500	0.2849	0.8153	0.2797	0.8222

Table 9 PIS matrix

PIS #	Attributes											
	C_1		C_2		C_3		C_4		C_5		C_6	
	μ	ϑ	μ	ϑ	μ	ϑ	μ	ϑ	μ	ϑ	μ	ϑ
1	0.4323	0.6612	0.1221	0.9092	0.1818	0.8636	0.1922	0.8556	0.2228	0.8317	0.3233	0.7440
2	0.4323	0.6612	0.1014	0.9233	0.1818	0.8636	0.1922	0.8556	0.2228	0.8317	0.1550	0.8857
3	0.3716	0.7069	0.0908	0.9305	0.1518	0.8843	0.1766	0.8664	0.2051	0.8442	0.1517	0.8895
4	0.1962	0.8506	0.0908	0.9305	0.1183	0.9093	0.1444	0.8886	0.2051	0.8442	0.1262	0.9117
5	0.1206	0.9177	0.0767	0.9412	0.1157	0.9109	0.1336	0.8970	0.1869	0.8569	0.1220	0.9173
6	0.1080	0.9292	0.0714	0.9455	0.1014	0.9228	0.1336	0.8970	0.1558	0.8795	0.1143	0.9222
7	0.0953	0.9409	0.0695	0.9466	0.0679	0.9498	0.0945	0.9279	0.0764	0.9444	0.1049	0.9336
8	0.0823	0.9527	0.0588	0.9554	0.0613	0.9555	0.0790	0.9408	0.0436	0.9732	0.0883	0.9463
9	0.0823	0.9527	0.0196	0.9889	0.0397	0.9743	0.0651	0.9527	0.0375	0.9786	0.0883	0.9463

Table 10 Average solution

Avg solution	Attributes											
	C_1		C_2		C_3		C_4		C_5		C_6	
	μ	ϑ	μ	ϑ	μ	ϑ	μ	ϑ	μ	ϑ	μ	ϑ
$\overline{M}$	0.6742	0.8322	0.3000	0.9410	0.3775	0.9141	0.4644	0.8973	0.4931	0.8854	0.5715	0.8975

Table 11 Distance values

Module	Distance from PIS #									Distance from Avg
	1	2	3	4	5	6	7	8	9	
M_1	0.0782	0.0625	0.0509	0.0330	0.0232	0.0186	0.0114	0.0113	0.0170	0.1960
M_2	0.0772	0.0615	0.0500	0.0278	0.0196	0.0170	0.0082	0.0112	0.0180	0.1956
M_3	0.0337	0.0180	0.0166	0.0237	0.0292	0.0342	0.0486	0.0547	0.0615	0.1690
M_4	0.0368	0.0491	0.0407	0.0281	0.0358	0.0393	0.0484	0.0553	0.0584	0.1737
M_5	0.0545	0.0387	0.0307	0.0181	0.0155	0.0170	0.0261	0.0340	0.0407	0.1805
M_6	0.0287	0.0129	0.0166	0.0292	0.0325	0.0375	0.0519	0.0598	0.0665	0.1667
M_7	0.0297	0.0175	0.0107	0.0233	0.0330	0.0378	0.0508	0.0587	0.0654	0.1677
M_8	0.0507	0.0350	0.0328	0.0216	0.0157	0.0151	0.0300	0.0377	0.0445	0.1775
M_9	0.0929	0.0771	0.0656	0.0434	0.0329	0.0270	0.0124	0.0044	0.0023	0.2101

Table 12 Final appraisal scores and ranking of the alternatives

Module	Si-Pos ideal	Si-neg ideal	Ri	Pi	Rank
M_1	0.1393	0.0357	3.899937	0.257734	7
M_2	0.1355	0.0345	3.929223	0.256388	8
M_3	0.0600	0.1194	0.502305	0.967528	3
M_4	0.0891	0.1192	0.747798	0.815026	4
M_5	0.0917	0.0738	1.242927	0.573486	6
M_6	0.0545	0.1295	0.420567	1.016435	1
M_7	0.0545	0.1278	0.426296	1.013907	2
M_8	0.0877	0.0803	1.092208	0.633553	5
M_9	0.1708	0.0220	7.765136	0.226376	9

Table 13 Sensitivity analysis scheme

Criteria	Original	Exp 1	Exp 2	Exp 3	Exp 4	Exp 5
	$\lambda = 0.8$	$\lambda = 0.9$	$\lambda = 0.10$	$\lambda = 0.11$	$\lambda = 0.12$	$q = 1$
	$q = 1$	$q = 2$	$q = 3$	$q = 5$	$q = 9$	$\lambda = 0.9$
C_1	0.2985	0.3033	0.3026	0.2881	0.2398	0.2984
C_2	0.0687	0.0690	0.0710	0.0767	0.0812	0.0687
C_3	0.1058	0.1084	0.1130	0.1237	0.1436	0.1057
C_4	0.1125	0.1135	0.1205	0.1467	0.2184	0.1125
C_5	0.1329	0.1342	0.1350	0.1330	0.1185	0.1329
C_6	0.2817	0.2717	0.2578	0.2318	0.1984	0.2819
Sum	1.0000	1.0000	1.0000	1.0000	1.0000	1.0000

Table 14 Ranking of modules using varying weights of the attributes

Module	Ranking					
	$\lambda = 0.8$					$q = 1$
	$q = 1$	$q = 2$	$q = 3$	$q = 5$	$q = 9$	$\lambda = 0.9$
	Original	Exp 1	Exp 2	Exp 3	Exp 4	Exp 5
M_1	7	7	8	8	8	7
M_2	8	8	7	7	7	8
M_3	3	3	3	3	3	3
M_4	4	4	4	4	4	4
M_5	6	6	6	6	6	6
M_6	1	1	2	2	2	1
M_7	2	2	1	1	1	2
M_8	5	5	5	5	5	5
M_9	9	9	9	9	9	9

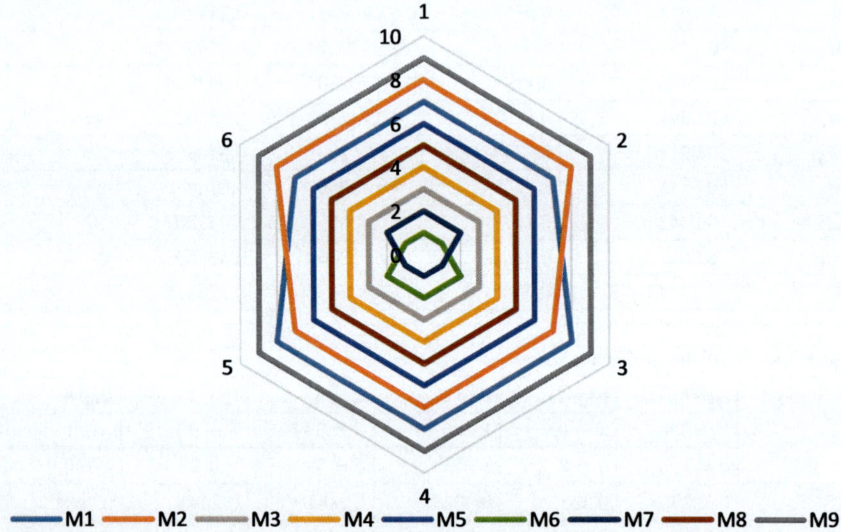

Fig. 2 Pictorial representation of the result of sensitivity analysis

6 Conclusion and Future Scope

In this paper, we address a contemporary issue of importance of marketing automation tools in the admission process of the HEIs and present a new qROFN-based Entropy-PROBID framework for MAGDM. We compare the automation modules namely Student Admission Portal 2.0, Document Verification and Offer Letter, FB

Table 15 Spearman's rank correlation test

		Original	Exp 1	Exp 2	Exp 3	Exp 4	Exp 5
Spearman's rho	Original	1.000					
	Exp 1	1.000**	1.000				
	Exp 2	0.967**	0.967**	1.000			
	Exp 3	0.967**	0.967**	1.000**	1.000		
	Exp 4	0.967**	0.967**	1.000**	1.000**	1.000	
	Exp 5	1.000**	1.000**	0.967**	0.967**	0.967**	1.000

** Correlation is significant at the 0.01 level (2-tailed)

Table 16 Appraisal scores of the modules under different attribute weights

Module	Pi score					
	$\lambda = 0.8$					$q = 1$
	$q = 1$	$q = 2$	$q = 3$	$q = 5$	$q = 9$	$\lambda = 0.9$
M_1	0.2577	0.2134	0.1875	0.2043	0.2670	0.2577
M_2	0.2564	0.2111	0.1911	0.2202	0.3051	0.2564
M_3	0.9675	0.9389	0.9149	0.8917	0.9020	0.9675
M_4	0.8150	0.8149	0.8159	0.8382	0.8481	0.8151
M_5	0.5735	0.5669	0.5841	0.6482	0.7253	0.5735
M_6	1.0164	0.9932	0.9766	0.9640	0.9712	1.0164
M_7	1.0139	0.9898	0.9824	0.9861	0.9978	1.0138
M_8	0.6336	0.6368	0.6736	0.7556	0.8291	0.6335
M_9	0.2264	0.1630	0.1268	0.1561	0.2071	0.2264

Table 17 Friedman's test

Chi-square	11.634
df	5
Asymp. Sig.	0.040*

The null hypothesis H_0: there is no significant difference among the various results

* Significant at 0.01 level

Lead Ad, Google Tracking Parameter, Adv. Tracking Parameters (Benchmarking, Trend Analysis, Student Quality Index), Advance Landing Page Builder, Chat BoT (Whatsapp and Web), WhatsApp Business API and ERP Integration based on the utility attributes such as Lead Acquisition, Lead Nurturing, Brand Equity, Marketing Optimization, ROI Tracking and Lead to Enrollment Journey. We observe that more than the promotional campaign (i.e., ADs) it is important to give focus on lead acquisition and quickly convert the lead into possible enrollment through the admission

Table 18 Comparison of result of qEP framework with others

Module	Ranking		
	$q = 1$ and $\lambda = 0.8$		
	qROF-PROBID	qROF-Score-PIV	qROF-Score_SAW
M_1	7	8	8
M_2	8	7	7
M_3	3	3	4
M_4	4	6	2
M_5	6	5	6
M_6	1	2	1
M_7	2	1	3
M_8	5	4	5
M_9	9	9	9

Table 19 Spearman's rank correlation among the result of qEP and others

Test statistic	Method	qROF_Score_PIV	qROF_Score_SAW
Spearman's rho	qROF_PROBID	0.917**	0.933**

** Correlation is significant at the 0.01 level (2-tailed)

process. The finding provides an important implication for the strategic decision-makers that promotional campaigns are required to attract the potential candidates and continuous engagement is essential for conversion. Further, the ranking order suggests that landing page and chat BOT are essential tools that help to answer the queries of the potential students throughout the day and initiate the admission process. In effect, HEIs can capture the lead and influence them to get engaged in the admission process and eventually quickly convert them. The present paper reveals the need to embrace the digital marketing and automation tools. However, the present study is an early and small-scale study which may be further converted into a large-scale empirical causal model to investigate the impact of individual attributes and automation tools on the actual conversion possibility. Further, the efficacy of our model may be tested in various other complex problems. Nevertheless, the present study is a one of its kind attempt so far that provides methodological as well as theoretical impetus to the researchers and practitioners.

References

1. Bhattacharjee M, Bandyopadhyay G, Guha B, Biswas S (2020) Determination and validation of the contributing factors towards the selection of a B-School—an Indian perspective. Decis Mak: Appl Manag Eng 3(1):79–91
2. Seres L, Pavlicevic V, Tumbas P (2018) Digital transformation of higher education: competing

on analytics. In: Proceedings of INTED2018 conference, 5–7 March 2018, pp 9491–9497
3. Biswas S (2020) Exploring the implications of digital marketing for higher education using intuitionistic fuzzy group decision making approach. BIMTECH Bus Perspect 2(1):33–51
4. Camilleri M (2020) Higher education marketing communications in the digital era. In: Strategic marketing of higher education in Africa. Routledge, pp 77–95
5. Statista Report. Internet penetration rate in India from 2007 to 2021. https://www.statista.com/statistics/792074/india-internet-penetration-rate/. 18 March 2022
6. The Indian Express Report (2022) India's smartphone market grew 12 per cent in 2021. Realme was big winner: Canalys. https://indianexpress.com/article/technology/tech-news-technology/india-smartphone-market-grew-12-per-cent-in-2021-realme-was-big-winner-canalys-7739045/. Accessed 12 Feb 2022
7. Ramadhan A, Gunarto M (2021) Analysis of digital marketing strategies in the era of the COVID-19 pandemic in private higher education. In: Proceedings of the 11th annual international conference on industrial engineering and operations management, Singapore, March 2021, pp 5674–5684
8. Rashid S, Yadav Ss (2020) Impact of Covid-19 pandemic on higher education and research. Indian J Hum Dev 14(2):340–343
9. Lamberton C, Stephen AT (2016) A thematic exploration of digital, social media, and mobile marketing: research evolution from 2000 to 2015 and an agenda for future inquiry. J Mark 80(6):146–172
10. Ghotbifar F, Marjani MR, Ramazani A (2017) Identifying and assessing the factors affecting skill gap in digital marketing in communication industry companies. Indep J Manag Prod 8(1):001–014
11. Bala M, Verma D (2018) A critical review of digital marketing. Int J Manag IT Eng 8(10):321–339
12. Kannan PK (2017) Digital marketing: a framework, review and research agenda. Int J Res Mark 34(1):22–45
13. Fierro I, Cardona Arbelaez DA, Gavilanez J (2017) Digital marketing: a new tool for international education. Pensamiento Gestión 42:241–260
14. McClea M, Yen DC (2005) A framework for the utilization of information technology in higher education admission department. Int J Educ Manag 19(2):87–101. https://doi.org/10.1108/09513540510582390
15. Salem O (2020) Social media marketing in higher education institutions. SEA–Pract Appl Sci 8(23):191–196
16. del Rocío Bonilla M, Perea E, del Olmo JL, Corrons A (2020) Insights into user engagement on social media. Case study of a higher education institution. J Mark High Educ 30(1):145–160
17. Kusumawati A (2019) Impact of digital marketing on student decision-making process of higher education institution: a case of Indonesia. J E-Learn High Educ 1(1):1–11
18. Dhote T, Jog Y, Gavade N, Shrivastava G (2015) Effectiveness of digital marketing in education: an insight into consumer perceptions. Indian J Sci Technol 8(S4):200–205
19. Hilbert A, Schönbrunn K, Schmode S (2007) Student relationship management in Germany—foundations and opportunities. Manag Rev 18(2):204–219
20. Trocchia PJ, Finney RZ, Finney TG (2013) Effectiveness of relationship marketing tactics in a university setting. J Coll Teach Learn 10(1):29–38
21. Rigo GE, Pedron CD, Caldeira M, Araújo CCSD (2016) CRM adoption in a higher education institution. JISTEM—J Inf Syst Technol Manag 13:45–60
22. Daradoumis T, Rodriguez-Ardura I, Faulin J, Juan AA, Xhafa F, Martinez-Lopez F (2010) Customer Relationship Management applied to higher education: developing an e-monitoring system to improve relationships in electronic learning environments. Int J Serv Technol Manag14(1):103–125
23. Prieto GPA, Piedra N (2021) Exploring success factors of CRM strategies in higher education institutions: a case study of CRMUTPL for conversion of prospects in university students. In: 2021 XVI Latin American conference on learning technologies (LACLO). IEEE, pp 474–477

24. Soliman M, Karia N (2016) Enterprise Resource Planning (ERP) systems in the Egyptian higher education institutions: benefits, challenges and issues. In: International conference on industrial engineering and operations management, Kuala Lumpur, Malaysia, pp 1935–1943
25. Marcinkowski F, Kieslich K, Starke C, Lünich M (2020) Implications of AI (un-)fairness in higher education admissions: the effects of perceived AI (un-)fairness on exit, voice and organizational reputation. In: Proceedings of the 2020 conference on fairness, accountability, and transparency, pp 122–130
26. Nair KR, Kumar S (2021) Impact of Industry 4.0 on digital marketing for higher education. NVEO Nat Volatiles Essent Oils J NVEO 8(5):5178–5208
27. Yager RR (2016) Generalized orthopair fuzzy sets. IEEE Trans Fuzzy Syst 25(5):1222–1230. https://doi.org/10.1109/Tfuzz.2016.2604005
28. Shaheen T, Ali MI, Toor H (2021) Why do we need q-rung orthopair fuzzy sets? Some evidence established via mass assignment. Int J Intell Syst 36(10):5493–5505
29. Zou ZH, Yi Y, Sun JN (2006) Entropy method for determination of weight of evaluating indicators in fuzzy synthetic evaluation for water quality assessment. J Environ Sci 18(5):1020–1023
30. Liu P, Zhang X (2011) Research on the supplier selection of a supply chain based on entropy weight and improved ELECTRE-III method. Int J Prod Res 49(3):637–646
31. Song Y, Zhang J (2016) Discriminating preictal and interictal brain states in intracranial EEG by sample entropy and extreme learning machine. J Neurosci Methods 257:45–54
32. Karmakar P, Dutta P, Biswas S (2018) Assessment of mutual fund performance using distance based multi-criteria decision making techniques—an Indian perspective. Res Bull 44(1):17–38
33. Zhang J, Li L, Zhang J, Chen L, Chen G (2021) Private-label sustainable supplier selection using a fuzzy entropy-VIKOR-based approach. Complex Intell Syst 1–18. https://doi.org/10.1007/s40747-021-00317-w
34. Zhang Q, Ding J, Kong W, Liu Y, Wang Q, Jiang T (2021) Epilepsy prediction through optimized multidimensional sample entropy and Bi-LSTM. Biomed Signal Process Control 64:102–293
35. Biswas S, Majumder S, Dawn SK (2021) Comparing the socioeconomic development of G7 and BRICS countries and resilience to COVID-19: an entropy–MARCOS framework. Bus Perspect Res. 22785337211015406
36. Wang Z, Rangaiah GP, Wang X (2021) Preference ranking on the basis of ideal-average distance method for multi-criteria decision-making. Ind Eng Chem Res 60(30):11216–11230
37. Biswas S, Pamucar D, Chowdhury P, Kar S (2021) A new decision support framework with picture fuzzy information: comparison of video conferencing platforms for higher education in India. Discrete Dyn Nat Soc 2021. https://doi.org/10.1155/2021/2046097
38. Garg H (2021) A new possibility degree measure for interval-valued q-rung orthopair fuzzy sets in decision-making. Int J Intell Syst 36(1):526–557
39. Liu P, Wang P (2018) Some q-rung orthopair fuzzy aggregation operators and their applications to multiple-attribute decision making. Int J Intell Syst 33(2):259–280. https://doi.org/10.1002/int.21927
40. Garg H, Ali Z, Mahmood T (2021) Algorithms for complex interval-valued q-rung orthopair fuzzy sets in decision making based on aggregation operators, AHP, and TOPSIS. Expert Syst 38(1):e12609
41. Khan MJ, Kumam P, Shutaywi M (2021) Knowledge measure for the q-rung orthopair fuzzy sets. Int J Intell Syst 36(2):628–655
42. Khan MJ, Alcantud JCR, Kumam P, Kumam W, Al-Kenani AN (2021) An axiomatically supported divergence measures for q-rung orthopair fuzzy sets. Int J Intell Syst 36(10):6133–6155
43. Riaz M, Hamid MT, Afzal D, Pamucar D, Chu YM (2021) Multi-criteria decision making in robotic agri-farming with q-rung orthopair m-polar fuzzy sets. PLoS ONE 16(2):e0246485
44. Cheng S, Jianfu S, Alrasheedi M, Saeidi P, Mishra AR, Rani P (2021) A new extended VIKOR approach using q-rung orthopair fuzzy sets for sustainable enterprise risk management assessment in manufacturing small and medium-sized enterprises. Int J Fuzzy Syst 23(5):1347–1369

45. Yager RR (2013) Pythagorean fuzzy subsets. In: 2013 joint IFSA world congress and NAFIPS annual meeting (IFSA/NAFIPS), June 2013. IEEE, pp 57–61
46. Wang R, Li Y (2018) A novel approach for green supplier selection under a q-rung orthopair fuzzy environment. Symmetry 10(12):687. https://doi.org/10.3390/sym10120687
47. Wei G, Gao H, Wei Y (2018) Some q-rung orthopair fuzzy Heronian mean operators in multiple attribute decision making. Int J Intell Syst 33(7):1426–1458. https://doi.org/10.1002/int.21985
48. Wang H, Ju Y, Liu P (2019) Multi-attribute group decision-making methods based on q-rung orthopair fuzzy linguistic sets. Int J Intell Syst 34(6):1129–1157. https://doi.org/10.1002/int.22089
49. Peng X, Dai J, Garg H (2018) Exponential operation and aggregation operator for q-rung orthopair fuzzy set and their decision-making method with a new score function. Int J Intell Syst 33(11):2255–2282
50. Peng X, Dai J (2019) Research on the assessment of classroom teaching quality with q-rung orthopair fuzzy information based on multiparametric similarity measure and combinative distance-based assessment. Int J Intell Syst 34(7):1588–1630
51. Liu P, Liu P, Wang P, Zhu B (2019) An extended multiple attribute group decision making method based on q-rung orthopair fuzzy numbers. IEEE Access 7:162050–162061
52. Rani P, Mishra AR (2020) Multi-criteria weighted aggregated sum product assessment framework for fuel technology selection using q-rung orthopair fuzzy sets. Sustain Prod Consum 24:90–104
53. Kumar K, Chen SM (2022) Group decision making based on q-rung orthopair fuzzy weighted averaging aggregation operator of q-rung orthopair fuzzy numbers. Inf Sci. https://doi.org/10.1016/j.ins.2022.03.032
54. Pinar A, Boran FE (2020) A q-rung orthopair fuzzy multi-criteria group decision making method for supplier selection based on a novel distance measure. Int J Mach Learn Cybern 11(8):1749–1780
55. Shannon CE (1948) A mathematical theory of communication. Bell Syst Tech J 27(3):379–423
56. Cui X, Zhao T, Wang J (2021) Allocation of carbon emission quotas in China's provincial power sector based on entropy method and ZSG-DEA. J Clean Prod 284:124683
57. Petrov AI (2022) Entropy method of road safety management: case study of the Russian Federation. Entropy 24(2):177
58. Wu ZH, Chen HJ (2021) The influence of e-marketing on performance of real estate enterprises: based on super-efficiency DEA and grey entropy methods. Math Probl Eng 2021. https://doi.org/10.1155/2021/7502676
59. Teimouri HB, Gharib J, Hossein Zadeh A, Pouya A (2021) An integrated entropy/VIKOR model for customer clustering in targeted marketing model design (case study: IoT technology services companies). Adv Math Finance Appl6(4):1–22
60. Ruiz Reina MÁ (2021) Entropy method for decision-making: uncertainty cycles in tourism demand. Entropy 23(11):1370
61. Pramanik PKD, Biswas S, Pal S, Marinković D, Choudhury P (2021) A comparative analysis of multi-criteria decision-making methods for resource selection in mobile crowd computing. Symmetry13(9):1713
62. Biswas S, Bandyopadhyay G, Guha B, Bhattacharjee M (2019) An ensemble approach for portfolio selection in a multi-criteria decision making framework. Decis Mak: Appl Manag Eng 2(2):138–158
63. Gupta S, Bandyopadhyay G, Bhattacharjee M, Biswas S (2019) Portfolio selection using DEA-COPRAS at risk–return interface based on NSE (India). Int J Innov Technol Expl Eng (IJITEE) 8(10):4078–4086
64. Laha S, Biswas S (2019) A hybrid unsupervised learning and multi-criteria decision making approach for performance evaluation of Indian banks. Accounting 5(4):169–184
65. Pamucar D, Torkayesh AE, Biswas S (2022) Supplier selection in healthcare supply chain management during the COVID-19 pandemic: a novel fuzzy rough decision-making approach. Ann Oper Res 1–43. https://doi.org/10.1007/s10479-022-04529-2

66. Biswas S, Pamucar D, Kar S, Sana SS (2021) A new integrated FUCOM–CODAS framework with Fermatean fuzzy information for multi-criteria group decision-making. Symmetry 13(12):2430
67. Biswas S, Pamučar DS (2021) Combinative distance based assessment (CODAS) framework using logarithmic normalization for multi-criteria decision making. Serb J Manag 16(2):321–340
68. Biswas S, Anand OP (2020) Logistics competitiveness index-based comparison of BRICS and G7 countries: an integrated PSI-PIV approach. IUP J Supply Chain Manag 17(2):32–57
69. Biswas S, Pamucar D, Kar SK (2022) A preference-based comparison of select over-the-top video streaming platforms with picture fuzzy information. Int J Commun Netw Distrib Syst (forthcoming). https://doi.org/10.1504/IJCNDS.2022.10043309
70. Pamucar D, Žižović M, Biswas S, Božanić D (2021) A new logarithm methodology of additive weights (LMAW) for multi-criteria decision-making: application in logistics. Facta Univ Ser: Mech Eng 19(3):361–380
71. Biswas S (2020) Measuring performance of healthcare supply chains in India: a comparative analysis of multi-criteria decision making methods. Decis Mak: Appl Manag Eng 3(2):162–189
72. Mufazzal S, Muzakkir SM (2018) A new multi-criterion decision making (MCDM) method based on proximity indexed value for minimizing rank reversals. Comput Ind Eng 119:427–438

Normalization of Target-Nominal Criteria for Multi-criteria Decision-Making Problems

Irik Z. Mukhametzyanov

Abstract A generalization of methods for normalizing of target-nominal criteria is proposed, which ensures consistency with the main methods of normalizing criteria for benefits and costs for multi-criteria decision-making problems.

Keywords Multi-criteria decision-making · Target-nominal criteria · Normalization

1 Introduction

In this study, a generalization of normalization methods of target-nominal criteria is proposed, which ensures agreement with the main methods of normalization of benefit and costs attributes within the framework of a standard decision-making problem—choosing the best solution from a set of alternatives (homogeneous objects) defined by some a set of attributes [1, 2].

A class of models is considered in which the ranking of alternatives is performed based on the performance indicators of alternatives obtained by aggregating the normalized values of attributes. Aggregation of normalized attribute values transforms the original multi-criteria decision-making problem with different-sized and differently directed criteria to a one-dimensional problem of ranking alternatives in descending or ascending order of the performance indicator of alternatives:

$$Q = F(A, C, D, \omega, \text{'norm'}, \text{'dm'}, \text{'par'}), \quad (1)$$

where A_i are alternatives (objects) ($i = 1, \ldots, m$), C_j are criteria or objects properties ($j = 1, \ldots, n$), a_{ij} are elements of decision matrix D, w_j are weight or importance of criteria ($j = 1, \ldots, n$), 'norm' is the normalization method of decision matrix, 'dm' is the distance metric in n-dimensional space of criteria, 'par' are other parameters of the aggregation method F, Q_i are performance indicator of alternatives.

I. Z. Mukhametzyanov (✉)
Ufa State Petroleum Technological University, Ufa, Russia
e-mail: izmukhametzyanov@gmail.com

P. Chatterjee et al. (eds.), *Computational Intelligence for Engineering and Management Applications*, Lecture Notes in Electrical Engineering 984,
https://doi.org/10.1007/978-981-19-8493-8_67

The target value of the attribute for each of the criteria can be of three types:

(1) smaller-is-better,
(2) larger-is-better,
(3) nominal-is-best.

Accordingly, the criteria are referred to as cost criteria, benefit criteria and target-nominal criteria or ***tn***-criteria. Nominal the best: a characteristic with a specific target value. The nominal value of an attribute is some intermediate value between the highest and the lowest. Typically, this value represents optimal value or optimal performance or can be specified by the customer.

In contrast to the cost/benefit criteria, for which many normalization methods have been developed [3], there is a limited list of normalization methods for target-nominal attributes. There are only a few options for target normalization, which are focused on the simultaneous processing of all types of criteria. These are analogs of **Max** and **Max–Min** normalization: nominal-is-best method was proposed by Wu [4], linear method-ideal by Zhou et al. [5] and target-based normalization method by Jahan et al. [6], and variants using analogs of the Harrington desirability function [7] for a two-sided specification based on an exponential normalization function [8, 9].

These two variations of normalization for tn-criteria cannot be used in conjunction with other normalization methods (except for **Max** and **Max–Min),** since the range of normalized values will differ significantly, which entails the priority of individual attributes during aggregation.

The apparent universality of applying target normalization to criteria of all types is relative. The above procedures narrow the domains of normalized values and have partial or weak agreement with the domains of the benefit or cost criteria. Target normalization and normalization for target-nominal criteria are not identical concepts, since the latter refers to the concept of an optimal value between the largest and the smallest values.

As shown in our study, despite the fact that target-normalization formulas are similar to some of the linear normalization methods (**Max** and **Max–Min**), the degree of data compression is different, and these are methods of different classes. Therefore, it is undesirable to use target-based normalization methods in conjunction with methods of normalization of benefit and cost criteria. In addition, existing target-normalization methods cannot be used in conjunction with methods such as **Sum**, **Vec**, **dSum** or **Z-score**.

This study fills this gap. This study proposes a generalization of normalization methods for target- nominal criteria, ensuring consistency with the main linear methods of normalization of benefit and cost attributes.

2 Normalization of Target-Nominal Criteria

2.1 Feature of Normalization of Multidimensional Data

When processing multidimensional data, linear transformation of values is most often used:

$$r_{ij} = \frac{a_{ij} - a_j^*}{k_j}, \tag{2}$$

where: a_{ij}, r_{ij} are the natural and normalized values of the jth attribute of the ith alternative, respectively, and a_j^* and k_j are some pre-assigned numbers that determine the displacement of the normalized values and the degree of their compression, respectively. All quantities on the right-hand side of Eq. (2) have the same measurement, which provides a conversion to dimensionless values. Additionally, as k_j, a value (optionally) not less than the numerator is used, which ensures the mapping of the natural values of the attributes of the alternatives into some fixed sub-domain of the values of the set of the segment [0; 1]. Selection of the range of values [0; 1] is due to a universal set intuitively understandable for a person with categories from 0 (bad) to 1 (excellent).

The most commonly used normalizations: **Max, Sum, Vec, Max–Min, dSum, Z-score** and domains of normalized values for the decision matrix of dimension [7 × 5] are shown in Fig. 1.

The characteristic features of multidimensional normalization (even within the framework of one normalization method) are obvious [10]:

(1) different compression of data for different attributes,
(2) displacement of domains of normalized values for various attributes (except **Max–Min**).

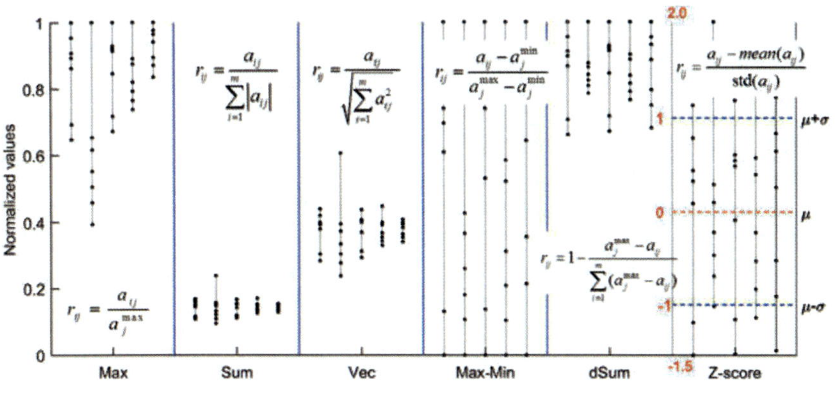

Fig. 1 Domain's configuration for various linear multivariate normalization methods

These are the main factors influencing the subsequent ranking. It is these effects that need to be adjusted (matched) when choosing a normalization method, choosing an inversion method and performing additional data transformation. As a consequence, you cannot apply different normalization methods to different attributes. For example, if we applied **Sum**-normalization to benefit attributes, then we should apply the same method to ***tn***-attributes and perform cost criteria inversion without displacement of domains. Unfortunately, a significant number of studies have been and are being carried out without due attention to these problems.

2.2 Goal Inversion

Each of the criteria is associated with a common goal in the context of a task. For selection problems, the target value of the attribute is either the maximum of the set of available alternatives: $a_j^{\max} = \max_i a_{ij}$ (benefit attributes), or the minimum $a_j^{\min} = \min_i a_{ij}$ (cost attributes), or some intermediate a_j^t value: $a_j^{\min} < a_j^t < a_j^{\max}$, which is defined as the target-nominal attributes. In the case, when the target-nominal value of the attribute is the best, normalization is performed so that the normalized nominal value is the largest for the direction of maximization (3) or the smallest for the direction of minimization (4):

$$\max\ Q_i\colon r_j^t \geq r_{ij},\ \forall i, \tag{3}$$

$$\min\ Q_i\colon r_j^t \leq r_{ij}\ ,\ \forall i. \tag{4}$$

Therefore, the values of the attributes for the target-nominal criteria are normalized taking into account the choice made in the direction of the goal either to a maximum or a minimum. But this, as shown below, is not necessary if an adequate procedure for inverting the values of the criterion is used in accordance with the general goal.

When aggregating normalized attributes, it is necessary that the directivity of the criteria is the same. Either for all attributes, the best value should be the largest (direction of maximization) or the smallest (minimization). Matching the goals of individual criteria and the best performance of the alternatives is achieved by inverting the goal from minimum to maximum or vice versa at the normalization stage by inverting the values.

The choice of direction to maximize or minimize the performance indicator does not affect the ranking result. As a rule, such a choice is determined by the ratio of the number of criteria for which "smaller-is-better" or "larger-is-better," following the principle of reducing the number of algebraic data transformations. For example, in the study [11], 7 out of 8 criteria are cost criteria, and one is a benefit criterion. It is obvious that it is rational to invert the criterion of benefit, and the best alternative will correspond to the minimum value of the performance indicator of alternatives.

The currently used concept of matching permissible pairs to normalize the attributes of benefit and cost is not very productive and in some cases even harmful. As shown in [12], the inversion of the optimization direction in these pairs is based on the transformation $1 - r$ and $1/r$, the first of which produces a significant displacement of the normalized values (anti-phase), and the second transformation is nonlinear and violates the disposition of natural values. In the same study, a universal and efficient inversion algorithm was proposed for both benefit attributes and costs and ***tn***-attributes, both for linear and nonlinear transformations—the ReS-algorithm:

(1) for natural values of attributes

$$\tilde{a}_{ij^*} = -a_{ij^*} + a_{j^*}^{\max} + a_{j^*}^{\min}, \ \forall j^* \in C_j^{\text{inv}}, \tag{5}$$

(2) for the normalized values of attributes

$$r_{ij} = \text{Norm}(a_{ij}), \ \ \forall j = 1, \ldots, n, \tag{6}$$

$$\tilde{r}_{ij^*} = -r_{ij^*} + r_{j*}^{\max} + r_{j^*}^{\min}, \ \forall j^* \in C_j^{\text{inv}}, \tag{7}$$

where Norm() is any of the linear normalization methods, and C_j^{inv} is the group of attributes for which the inversion needs to be performed.

Inversion of the goal from minimum to maximum or vice versa is advisable to perform after normalization by inverting the values.

2.3 *Overview and Critique of the Application of Target-Based Normalization Techniques*

This section presents various options for target-based normalization, focused on the simultaneous processing of all types of criteria.

For selection problems, it is advisable to take the attribute closest to the target-nominal value equal to a_j^t. In such a case, the maximum normalized value is the attribute value of one of the alternatives. This will not contradict the formulation of the problem in which one of the alternatives for each attribute is preferable.

One of the target-based normalization options in the case of maximizing the performance indicator of alternatives, called the ideal-linear method, has the form [5]:

$$r_{ij} = \frac{\min\limits_i(a_{ij}, a_j^t)}{\max\limits_i(a_{ij}, a_j^t)} = \begin{cases} \dfrac{a_{ij}}{a_j^t}, & a_{ij} < a_j^t \\ \dfrac{a_j^t}{a_{ij}}, & a_{ij} \geq a_j^t \end{cases}. \tag{8}$$

The normalization formula represents the analog of the **Max** normalization on the interval $[a_j^{\min}; a_j^t)$ and inverse ***i*-Max** (inverse Max) normalization on the interval $[a_j^t; a_j^{\min}]$. Therefore, some of the values are normalized according to the linear algorithm, and the values exceeding $a_j{}^t$ are normalized using a nonlinear (hyperbolic) transformation that changes the dispositions of natural values. The compression ratio (slope of the straight line) for this method does not match any of the linear normalization methods. This means that the method is not consistent with the benefit and cost attribute normalization methods.

The next variant of the target-based normalization in the case of maximizing the performance indicator of alternatives has the form [6]:

$$r_{ij} = 1 - \frac{\left|a_{ij} - a_j^t\right|}{\max(a_{ij}, a_j^t) - \min(a_{ij}, a_j^t)}. \tag{9}$$

Normalization formula (9) represents an analog of the **Max–Min** normalization method on the interval $[a_j^{\min}; a_j^t)$ and inverse normalization ***i*-Max–Min** on the interval $[a_j^t; a_j^{\min}]$. However, the stretch–contraction ratio of the natural values at these two intervals is generally different and is equal to the length of the gap. Therefore, the proportions of normalized data of the same attribute to the left and to the right of the target value a_j^t will differ, and this may affect the final rating. The compression ratio for this method also does not match any of the linear normalization methods. This means that the method is not consistent with normalization of benefit and cost attributes.

The third variant of target-based normalization (nominal-is-best method) has the form [4]:

$$r_{ij} = 1 - \frac{\left|a_{ij} - a_j^t\right|}{a_j^{\max} - a_j^{\min}}. \tag{10}$$

The normalization formula (10) also represents an analog of the **Max–Min** normalization method on the interval $[a_j^{\min}; a_j^t)$ and inverse normalization ***i*-Max–Min** on the interval $[a_j^t; a_j^{\min}]$. However, unlike the variant of formula (9), the data stretch–compression ratio is fixed. Therefore, the proportions of normalized data of the same attribute do not change in absolute value, and formula (10) is preferable.

Figure 2 shows a graphical illustration of target-based normalization by Eqs. (8)–(10) for the case of maximizing the performance indicator of alternatives (solid line) and for the case of minimize the performance indicator of alternatives (dashed line). Initial data: $a = (267, 164, 220, 48, 210, 78, 215)$ correspond to the third attribute according to the illustration in Fig. 1.

The target-nominal score is 215. The latter was obtained by inverting the values using the ReS-algorithm by Eq. (7).

The fourth of the presented graphs in Fig. 2 represents target-based normalization using a nonlinear transformation (9) in the form:

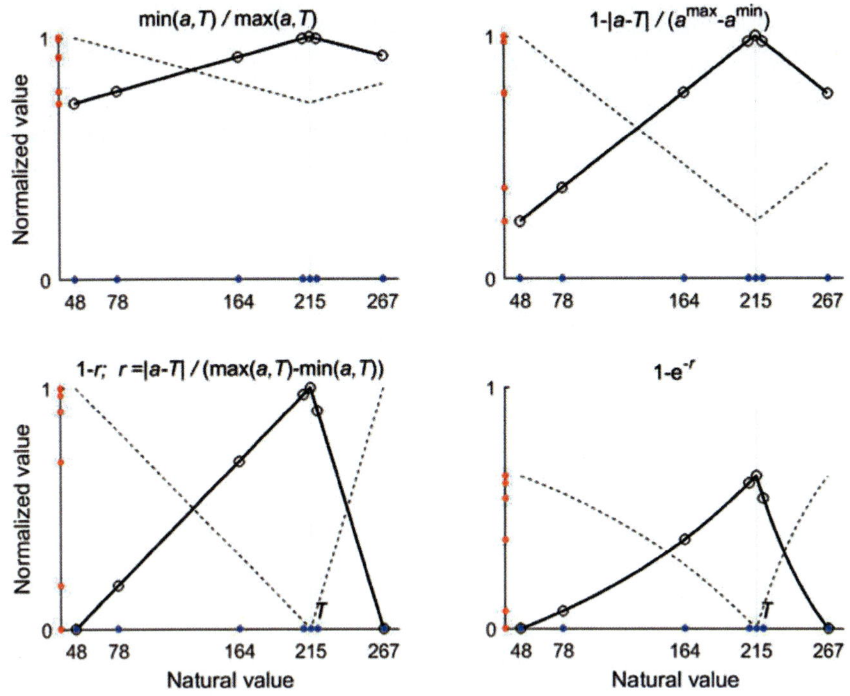

Fig. 2 Illustration of different variants of target-based normalization methods

$$r_{ij} = 1 - \exp\left(-\frac{\left|a_{ij} - a_j^t\right|}{\max(a_{ij}, a_j^t) - \min(a_{ij}, a_j^t)}\right). \quad (11)$$

For all the cases presented in the target-based normalization formulas, there is a module-function, which provides a break in the normalization line at the target point a_j^t and determines the applicability of the formulas for all types of criteria.

Only formula (10) produces normalized values, the compression of which corresponds to the compression of the benefit criteria in the **Max–Min** method. For this reason, we do not recommend applying transformation (8) in conjunction with the **Max**-normalization for benefit criteria, and we do not recommend applying transformations (9) and (10) in conjunction with the **Max–Min** normalization for benefit criteria. Despite the fact that the target-based normalization formulas are similar to the **Max** and **Max–Min** normalization methods, the degree of data compression is different, and these are methods of different classes. In addition, these methods cannot be used in conjunction with **methods** such as **Sum**, **Vec**, **dSum** or **Z-score**.

2.4 Generalization of Are Normalization Methods of Target-Nominal Criteria

The author proposes a generalization of the formulas for the normalization of ***tn***-criteria. Generalization is achieved by using the difference between the target and the current value in the modulus formula while maintaining the compression stretch coefficients and the displacement a_j^* (for normalization methods with displacement):

$$r_{ij} = -\frac{\left|a_j^t - a_{ij}\right|}{k_j} + \frac{a_j^t - a_j^*}{k_j}, \tag{12}$$

where k_j and a_j^* are defined by Eq. (2) according to the following detail for the normalization methods:

$$\textbf{Max}: a_j^* = 0, \quad k_j = a_j^{\max} \tag{12.1}$$

$$\textbf{Sum}: a_j^* = 0, \quad k_j = \sum_{i=1}^{m} |a_{ij}| \tag{12.2}$$

$$\textbf{Vec}: a_j^* = 0, \quad k_j = \left(\sum_{i=1}^{m} a_{ij}^2\right)^{0.5} \tag{12.3}$$

$$\textbf{Max–Min}: a_j^* = a_j^{\min}, \quad k_j = a_j^{\max} - a_j^{\min} \tag{12.4}$$

$$\textbf{dSum}: a_j^* = a_j^{\max} - k_j, \quad k_j = \sum_{i=1}^{m} (a_j^{\max} - a_{ij}) \tag{12.5}$$

$$\textbf{Z} - \textbf{score}: a_j^* = a_j^{\text{mean}}, \quad k_j = \text{std}_i(a_{ij}). \tag{12.6}$$

Indeed, if $a_{ij} < a_j^t$, then we use linear normalization formula as:

$$r_{ij} = \frac{a_{ij} - a_j^*}{k_j}. \tag{13}$$

If $a_{ij} \geq a_j^t$, then the formula is:

$$r_{ij} = \frac{2a_j^t - a_{ij} - a_j^*}{k_j}. \tag{14}$$

The advantage of formula (12) over target-based normalization by Eqs. (8)–(10) is that the compression and displacement ratio is consistent with the methods of linear normalization of the attributes of benefits and costs in accordance with the parameters of normalization (12.1)–(12.6).

Figure 3 is a graphical illustration of are normalization methods of ***tn***-criteria for six different variants of linear normalization for the goal maximization case.

Initial data: $\boldsymbol{a} = (267, 164, 220, 48, 210, 78, 215)$ correspond to the third attribute according to the illustration in Figs. 1 and 2. The target-nominal score is 215.

The abbreviation of the main normalization method of target-nominal criteria has been added the prefix "*t*-" meaning "target," for example, *t*-**Max**, *t*-**Vec**, etc.

The disadvantage of formula (12) when used together with the corresponding linear normalization method is that the range of values according to (12) is smaller, and the largest value is shifted to zero. It is necessary to equalize the upper values when maximizing or lower values when minimizing. This is important since the contribution of ***tn***-criteria to the integral indicator should not be lower. In some cases, formula (12) allows negative values. The result is determined by the position

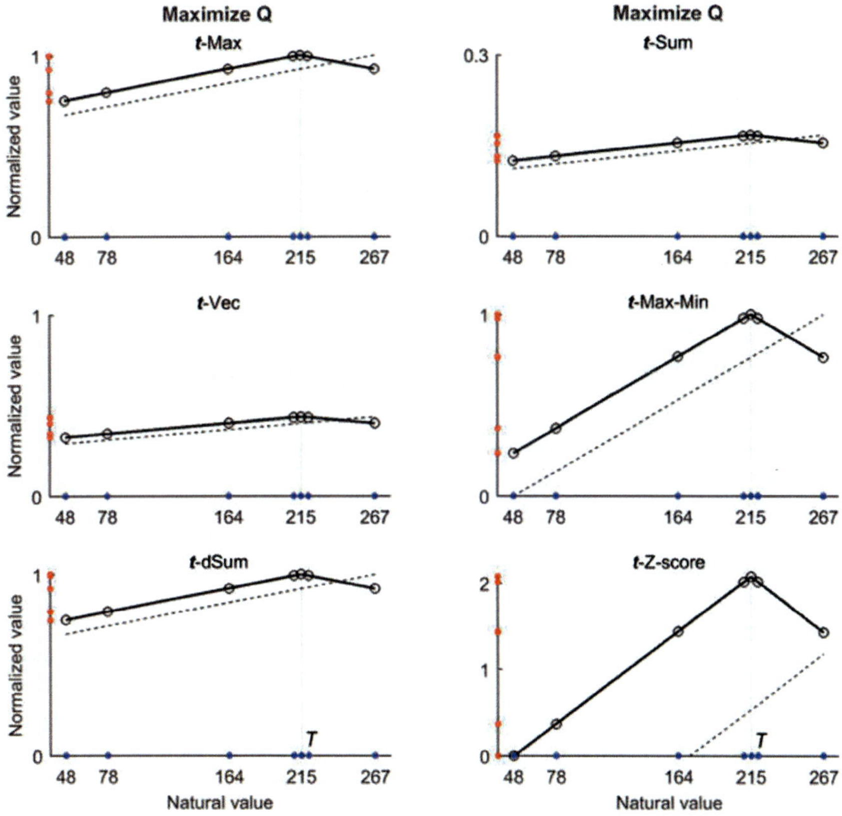

Fig. 3 Illustration of the agreement of the generalized method of normalization of target-nominal criteria with the methods of normalization **Max**, **Sum**, **Vec**, **dSum**, **Max–Min**, **Z-score** (maximization)

of the target value a_j^t in the interval $[a_j^{\min}; a_j^{\max}]$. You can fix the problem with the usual data shift.

Displacement of normalized values within the area [0; 1] is performed by parallel shifting values along the ordinate axis. An illustration of the shift is shown in Fig. 3 for all methods of normalization with respect to the base dashed line. The displacement is determined by the value $\Delta r_j = r_j^{\max} - r_j^t$. Then the formula (12) is transformed to the form:

$$r_{ij} = r_{ij} + r_j^{\max} - r_j^t, \tag{15}$$

which should be used as a formula in programming; i.e., three mathematical operations are performed sequentially:

(1) calculate r_{ij} and r_j^t by Eq. (12),
(2) calculate $r_j^{\max}$ corresponding to the $a_j^{\max}$ value by Eq. (13),
(3) calculate the new value r_{ij} by Eq. (15).

Parallel transfer obviously preserves the range (swing) of the normalized values, which is important for the final result when aggregating the normalized values.

For the case of minimizing the performance indicator of alternatives, the target-nominal normalization is obtained by inverting the values using the ReS-algorithm in accordance Eq. (7). Figure 4 shows a graphical illustration of the normalization of target-nominal criteria for the case of minimizing the performance indicator of the alternatives.

2.5 *Comparative Analysis of Normalization of Target-Nominal Criteria Using Different Methods*

Figure 5 shows a graphical illustration of the relative position of the domains of normalized values for the target-nominal criteria of the third attribute for various variants of linear normalization defined by formulas (8)–(10) and the proposed generalization of the target-nominal normalization in accordance with formulas (12)–(15) presented in Sect. 2.3.

Initial data: $\boldsymbol{a} = (267, 164, 220, 48, 210, 78, 215)$ correspond to the third attribute according to the illustration in Figs. 1, 2 and 3. The target-nominal score is 215. Normalization is performed in the case of maximizing of the performance indicator of alternatives.

Normalization according to formula (8) corresponds to the normalization of ***t*Max**, but the degree of data compression is somewhat lower. Normalization by formula (9) is identical to ***t*Max-Min** normalization. Normalization by formula (10) corresponds to ***t*Max-Min** normalization, but the degree of compression is lower. For the **Sum**, **Vec**, **dSum**, **Z-score** normalization methods, target-normalization formulas (8)–(10) cannot be used since the degree of data compression and domain shifting are very

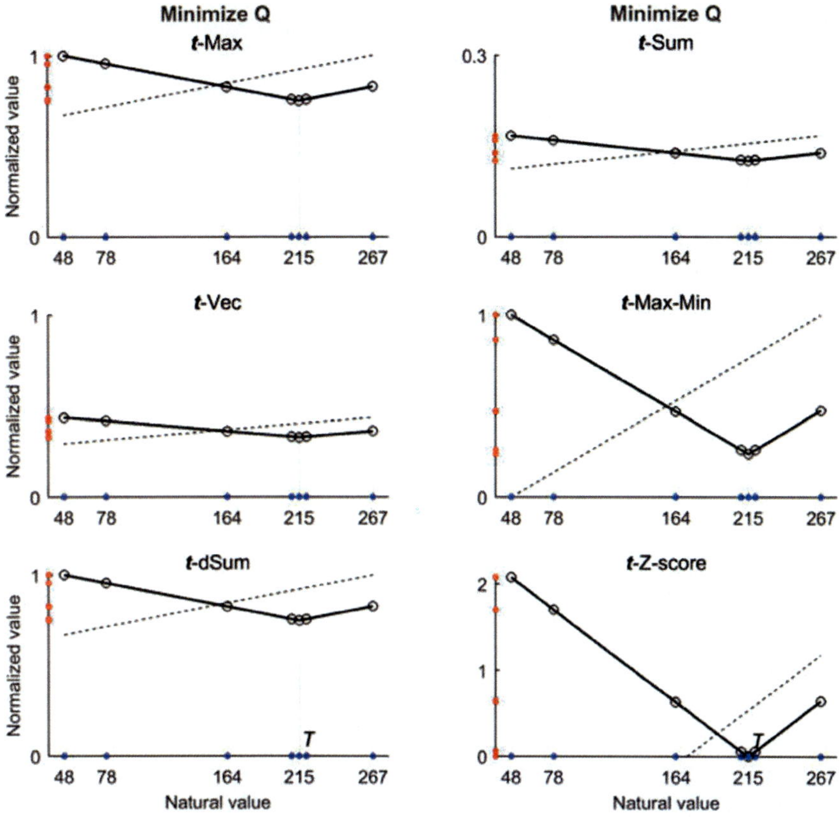

Fig. 4 Illustration of the inversion of normalization of target-nominal criteria (minimization)

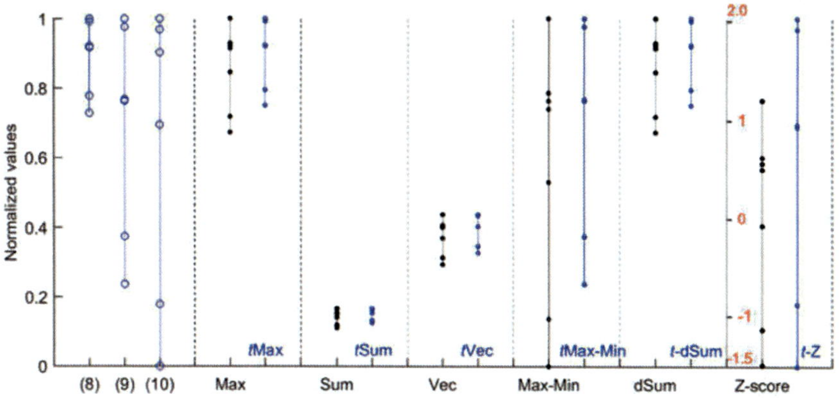

Fig. 5 Relative position of the domain of normalized values of the target-nominal criterion (the third attribute in Fig. 1) with target normalization by Eqs. (8)–(10) and for various linear and target-nominal normalization (maximization)

different. This entails the priority of the contribution of individual criteria in the indicator of the effectiveness of alternatives and the distortion of the ranking. Therefore, in the case of using linear methods **Sum**, **Vec**, **dSum** or **Z-score** when normalizing profit attributes, the generalization proposed by the author in the form of formulas (12)–(15) is an adequate method for normalizing target-nominal criteria.

3 Conclusion

In the tasks of multi-criteria decision-making, it is relevant to jointly agreed conversion of values for three types of criteria—costs, benefits and nominal values. The nominal value of an attribute is some intermediate value between the largest and the smallest represents the optimal characteristic or performance or can be declared by the customer.

The existing methods for transforming attributes of the nominal type based on the target-normalization approach produce areas of normalized values that can differ significantly from the range of values of the benefit and cost criteria. For a number of target-normalization formulas similar to the **Max** and **Max–Min** normalization methods used for the benefit and cost criteria, the degree of data compression is different, and these are methods of different classes. The class of target-normalization transformations is limited: target-normalization methods, for example, cannot be used in conjunction with methods such as **Sum**, **Vec**, **dSum** or **Z-score**. Target normalization and normalization for target-nominal criteria are not identical concepts.

This study proposes a generalization of the normalization methods for target-nominal criteria that is consistent with the basic linear normalization methods for benefit attributes and cost attributes.

The author finds research on the harmonization of normalized scales in multivariate normalization very relevant. The critical issue is to find the balance between the compression of the data of different measurements and the bias of the normalized values.

References

1. Hwang CL, Yoon K (1981) Multiple attributes decision making: methods and applications. A state-of-the-art survey. Springer, Berlin, Heidelberg, p XI
2. Tzeng G-H, Huang J-J (2011) Multiple attribute decision making: methods and application. Chapman and Hall/CRC
3. Jahan A, Edwards KL (2015) A state-of-the-art survey on the influence of normalization techniques in ranking: improving the materials selection process in engineering design. Mater Des 65:335–342. https://doi.org/10.1016/j.matdes.2014.09.022
4. Wu H-H (2002) A comparative study of using grey relational analysis in multiple attribute decision making problems. Qual Eng 15:209–217. https://doi.org/10.1081/QEN-120015853

5. Zhou P, Ang BW, Poh KL (2006) Comparing aggregating methods for constructing the composite environmental index: an objective measure. Ecol Econ 59(3):305–311
6. Jahan A, Bahraminasab M, Edwards KL (2012) A target-based normalization technique for materials selection. Mater Des 35:647–654. https://doi.org/10.1016/j.matdes.2011.09.005
7. Harrington J (1965) The desirability function. Ind Qual Control 21(10):494–498
8. Shih H-S, Shyur H-J, Stanley Lee E (2007) An extension of TOPSIS for group decision making. Math Comput Modell 45(7–8):801–813. https://doi.org/10.1016/j.mcm.2006.03.023
9. Jahan A, Mustapha F, Ismail MY, Sapuan SM, Bahraminasab M (2011) A comprehensive VIKOR method for material selection. Mater Des 32(3):1215–1221. https://doi.org/10.1016/j.matdes.2010.10.015
10. Mukhametzyanov IZ (2022) Elimination of the domains' displacement of the normalized values in MCDM tasks: the IZ-method. Int J Inf Technol Decis Mak (in press). https://doi.org/10.1142/S0219622023500037
11. Rezk H, Mukhametzyanov IZ, Al-Dhaifallah M, Ziedan A (2021) Optimal selection of hybrid renewable energy system using multi-criteria decision-making algorithms. Comput Mater Contin 68(2):2001–2027. https://www.techscience.com/cmc/v68n2/42160
12. Mukhametzyanov IZ (2020) ReS-algorithm for converting normalized values of cost criteria into benefit criteria in MCDM tasks. Int J Inf Technol Decis Mak 19(5):1389–1423. https://doi.org/10.1142/S0219622020500327

Assessment of Factors Influencing Employee Retention Using AHP Technique

Mohini Agarwal, Neha Gupta, and Saloni Pahuja

Abstract In the current competitive, challenging and dynamic scenario, there is a shift from financial to technology to human resources. There is a need for understanding the importance of HR as assets and for realizing the strategic role of Human Resource Management, thus contributing to adding business value. Human resources meet all the criteria of being a source of sustained competitive advantage. As a general saying that Competitor can copy any model but cannot copy the human resource and its knowledge keeping this in mind, all are in a race to hire and retain star performers. Thus, the organisation should have investment orientation towards its employees and be ready to provide avenues and opportunities to retain its employees further taking them as strategic partners. Keeping into account the importance and sensitivity of the issue of retention to any organisation, this study will help to determine the strategies that will help the organisation to retain its potential employees for which Analytical Hierarchical Process have been used to determine the importance attached with various factors.

Keywords Business value · Competitive advantage · Employee retention · Human resources · Investment orientation · Strategic partner

1 Introduction

In this competitive scenario, every organisation wants to retain star performers and that is the biggest challenge HR managers now a days are facing. All are in race to have committed and competent workforce. They add value to the firm further helping

M. Agarwal · N. Gupta · S. Pahuja (✉)
Amity School of Business, Amity University, Noida, Uttar Pradesh, India
e-mail: Spahuja1@amity.edu

M. Agarwal
e-mail: magarwal@amity.edu

N. Gupta
e-mail: ngupta10@amity.edu

P. Chatterjee et al. (eds.), *Computational Intelligence for Engineering and Management Applications*, Lecture Notes in Electrical Engineering 984,
https://doi.org/10.1007/978-981-19-8493-8_68

the firm to gain competitive advantage. It is the human resources that are utilising all other resources and meet all the criteria's of being source of sustained competitive advantage. So, in case management can organise them well and able to leverage best out of them, they can reflect extraordinary performance leading to high-performance work systems.

HR plays a key role in framing the retention strategies along with the management. The organisation's most important asset is its human resource. Human resource differs from one organisation to another. As it is said, any idea or product can be copied by the competitors, but one thing that cannot be copied is the human resource. Each human is unique in its nature, style of accomplishing a task, characteristic and behaviour. This is because of the thought process and interpretation of the problem. Thus, it becomes a vital responsibility of the organisation to keep its potential employees.

Employees want to work in organisation where they work is recognised and appreciated. Thus, for ensuring long-term stay and commitment of employee's organisation, effective retention strategies should be framed. There are various factors that should be given importance to reduce the attrition or to enhance the employee retention. Also, management should have an investment orientation towards its employees. Investing in your employees, enhancing their skills and making them competent enhance their ability to contribute well leading to create High Performance Work Systems, and hence right organisational culture will be created where there is knowledge sharing and development.

From the organisation point of view, it costs a lot when an able employee leaves the organisation. This is because the hiring, training and development costs are associated with him. The organisation faces both financial loss and loss of talent too. Getting the right replacement is bit more challenging, time taking and expensive exercise. Existing employees are aware of the organisations' expectations and work accordingly. But a new employee takes time to understand and meet the organisation's expectations. Thus, the settled working culture gets disturbed with the attrition of a single employee.

Employee retention to the organisation can be predicted (see Fig. 1) and is important:

- As existing potential employees know the process and importance of maintaining the quality of service/production, losing them can lead to gaps in quality of work.
- Over a time with settled work, culture employees feel committed to the organisation. There is transformation in level of needs which pushes them to work better for the organisation.
- The big amount is invested on employees in form of training and development so as to enhance their capabilities and make them learn and understand the organisations' expectations.
- It helps the organisation to preserve the talents and ideas that can give a competitive edge.
- The environment has less of storming stage and more of performing stage which creates a friendly working culture.

Fig. 1 Importance of employee retention

Representative maintenance issues are rising as the most basic workforce administration difficulties of the quick future. Scopes have demonstrated that in the future, triumphant associations will be those which adjust their authoritative conduct to the substances of the current workplace where life span and achievement rely on development, innovativeness and adaptability. Maintenance is a perplexing idea, and there's no single equation for keeping representatives with an association.

There are factors that need to be studied which impacts employee retention. These factors act like an external and internal envelope, and if these are higher in any workplace, then the retention will also increase. Thus, focus should be made on these. In any start-up, since the business itself is at very risky stage, the organisation should follow some of these factors so that it can gain trust of its capable employees and retain them. As employees are a great asset, retaining the best employees is very important as committed and competent employees are inimitable.

Therefore, this paper is designed to answer the following research questions:

RQ1: What are the factors that influence the employee's retention?
RQ2: How to prioritise the employee retention factor?

The rest of the paper is organised as follows: After the brief introduction in Sect. 1, an extensive literature review to identify the factors that influence employee's retention in Sect. 2. The research methodology and factors in Sect. 3. Section 4 contains the case study to illustrate the present ideology with discussion with conclusion in Sect. 5. Lastly, the list of references has been supplemented.

2 Literature Review

Research in the field of employee retention has always been an important aspect for organisations and has been carried out from an ancient time. Yet many challenges exist for organisations and researchers keep on exploring this area. Fitz-enz [1] identified that employee retention does not depend on a single factor. There are lot of factors which retain the employees in the organisation. The factors that help in retaining employees are compensation, rewards, job security, working environment, training, and development and the like. Organisations should encourage these activities to create health cognitive culture and sense of belongingness which will further enhance the employee retention [2]. Effective leadership also contributes to achieving higher performance at all levels and increase the employee motivation [3]. Osteraker [4] identified that organisation success mainly depends on employee satisfaction and retention. He divided the factors leading to employee retention into three categories that is social, mental and physical.

Dissatisfied employees with low morale make more mistakes leading to increase in the cost and turnover rate [5]. Clarke [6] and Wright et al. [7] highlighted the importance of HR practices and stressed on realising the value of HR as an asset. Effective adoption and implementation of HR practices leads to increase in retention.

Walker [8] identified several factors that increase employee retention such as remuneration on performance-based work, safety for difficult work, giving training and development, friendly environment and employee-employer relations that should be good, whereas Kehr [3] developed three factors for employee retention that is power, accomplishment and connection.

Focus should be on selective hiring, development and maintenance of the resources too to retain them. Retention of employees enhances the organisational efficiency, and employee retention relates to the success of the firm [9]. Right and transparent communication and seniors' support always help the employees to grow and perform well [10].

Employee turnover effects the customer service and creates interruption in the service [11]. Das and Baruah [12] mentioned various techniques like rewards and recognition, work life balance, participation in decision-making and employee retention used by organisations to promote satisfaction and reduce employee turnover.

Sharma [13] discussed the employee's preferences about the organisations they want to work in and listed that career development opportunities are preferred followed by compensation, type/kind of work, company culture and family circumstances. Also suggested that employees should be trained and developed as per the requirements on continuous basis as skill development further increases the satisfaction and motivates employees to contribute more as it enhances their capabilities. Skill development and learning culture also reduce the attrition rate and help the organisations to enhance the employee performance leading to competitive advantage.

Pahuja and Dalal [14] revealed that if the firm adopts effective retention strategies and focus on retaining its employees, it will have more satisfied and committed workforce which will add value to the firm. Study described how effective HRM practices will lead to retention of star performers. Kundu and Lata [15] also discussed the impact of supportive work environment on employee retention. Study revealed that supportive work environment practices impact the retention of the employee. If employees feel they are working in supportive environment, they tend to stay longer and thus firm can retain employees by creating supportive environment.

Raj and Brindha [9] identified that if organisation focus on creating long-term career options, then they can make talented people stay for long term. Study also highlighted that retention can be achieved through respecting employee opinion and ideas, recognising them, giving them performance-based rewards, listening to their ideas, concerns and thus assisting them with various career advancement programmes. Alkhyeli and Ewijk [16] present four fundamental elements with demonstrated effect on educators' activity fulfilment in a thorough model for the UAE: inspiration, school initiative style, work qualities and social knowledge. Acknowledgement, pay and independence scored as the instructors' most noteworthy concerns. In this manner, concentrating on these perspectives ought to be a need for private school management to decrease teacher turnover.

Singh and Dhillon [17] explained the factors that lead to retention of employees in Indian IT sector. Relevance of retaining the employees was highlighted and firm should frame effective retention strategies to retain its employees. Study also suggested various factors that companies should take care of for retaining the start performers like rewards and recognition, good compensation structure, support learning, training and development opportunities, skill recognition, pleasant working environment, providing flexi-timing facilities, annual performance appraisal and mentoring and coaching sessions.

Sitati et al. [18] identified that employee recognition is one of the parameters which positively impact the employee retention. Findings revealed that employee recognition along with some opportunities to grow in their career and increase in employee involvement leads to better satisfaction and retention of employees in hotel industry. Both recognition and career development are important for them to stay for longer period. And both the variables are positively related with employee retention. Also, the way reward management is handled and implemented in the organisation impacts the employee motivation and thus retention.

Ndede [19] evaluated the linkage between the motivation and the retention. Study revealed significant positive impact of work climate and reward system on employee retention. It analysed the impact of leadership style on retention in banking industry. Also, it was found that there is a significant positive link between the motivation and the employee retention in banking industry.

Pahuja et al. [20] highlighted the importance of employee relations in retention and for gaining competitive edge. Maintaining good relations is as equally important to retain the good employees as other factors like fair compensation, training and development, employee participation, career development, recognition and performance-based appraisals.

Mohan and Rani [21] discussed the retention issues at Admaark especially in concern with female employees. Study highlighted the relevance of retaining employees by investing in them. It also suggested various measures to handle and reduce the attrition rate like flexi timing, creating health work culture, career development plans, constant mentoring and flexible recruitment policies. Also, it focussed on how transparency in appraisal system and linking the same with performance can lead to higher motivation and satisfaction of employees (Table 1).

3 Research Methodology

In this study, exploratory as well as analytical research design has been adopted with the objective of identifying the factors from the literature which play a significant role in employee retention. Analytic Hierarchy Process (AHP) has been used for generating weights for each evaluation criterion according to the decision-maker's pairwise comparisons of the criteria. The higher the weight, the more important the corresponding criterion becomes. These identified factors based on their importance has been used for decision-making and developing strategies.

3.1 Analytical Hierarchy Process (AHP)

AHP is a technique of decision-making in complex environments in which many variables or criteria are considered in the prioritisation and selection of alternatives or projects [25]. AHP has been extensively studied and is currently used in decision-making for complex scenarios, where people work together to make decisions when human perceptions, judgments and consequences have long-term repercussions [26]. AHP is an effective tool for measuring the discrete and continuous paired comparisons in multi-level hierarchy structures. It has a special feature of considering the consistency in the judgments. In this process, the decision-maker carries out simple pairwise comparison of attributes and the resulting judgments are used to determine the overall weights of the attributes considered. After structuring the hierarchy, AHP evaluation utilised pairwise comparison to get decision-makers' judgments and it applied relative importance matrix to be the matrix. In the domain of HR, not much of research has been applied AHP to date. Some of the researchers are as follows: Divkolaii [27] presents an observational examination to decide various elements impacting efficiency of HR of Islamic Republic of Iran Broadcasting (IRIB) in area of Mazandaran, Iran. He utilises AHP to rank 17 significant factors and confirms that individual qualities were the most significant components followed by the board-related elements and natural elements. Hussain and Jalal [28] took a mind-boggling errand of site choice for small and medium ventures. They proposed micro factors that impact site determination choice with AHP analysis to rank which factor is generally critical.

Table 1 Factors of employee retention

Factor	Description	Authors
Skill recognition (F1)	The skills of the employees need to be identified and used efficiently so that the employees are doing the works where they have done a specialisation	Das and Baruah [12], Raj and Brinda [9], Singh and Dhillon [17], Sitati et al. [18], Pahuja et al. [20]
Learning and working climate (F2)	The environment of the organisation has a great impact on the employee retention since it's like a second home for the employees because they spend their most of the time of a day in the workplace. Thus, if the workplace is happening and interactive, then they will be comfortable and energetic towards work. This is a learning place and also thus the employees should be given proper training and help in personality development	Fit-enz [1], Sheridan [22], Osteraker [4], Walker [8], Sharma [13], Kundu and Lata [15], Singh and Dhillon [17], Ndede [19], and Mohan and Rani [21]
Job flexibility (F3)	Most of the employees in the exit interview have always suggested about flexibility in the workplace. Theory Y also says that the employees should be given some flexibility assuming that they will be comfortable and work more efficiently. Since the employees are most important resource for the organisation, they should have the flexibility of time, workload, etc. so that the organisation can make them feel that they mean to the organisation	Singh and Dhillon [17], Alkhyeli and Ewijk [16], and Mohan and Rani [21]
Superior-subordinate relationship (F4)	When an employee can work great under a supervisor that means their bond is strong, it is always good to have a positive bond within superior and subordinate. This relation will make a great team which can work together and generate great ideas. This will in fact help the organisation to grow faster because internal relations are stronger and will be united towards a single objective	Osteraker [4], Walker [8], Kehr [3], Cole et al. [10], Pahuja and Dalal [14], Ndede [19], and Pahuja et al. [20]

(continued)

Table 1 (continued)

Factor	Description	Authors
Compensation (F5)	The organisation needs the employees to grow. The organisation grows with the employees' growth. Thus, it is the responsibility of the organisation to give salary, incentives, fringe benefits and non-monetary benefits which in other words called compensation to employees as a motivation technique. The employees will be always motivated and will retain in the organisation	Fit-enz [1], Wright et al. [7], Walker [8], Sharma [13], Raj and Brinda [9], Pahuja and Dalal [14], Singh and Dhillon [17], Alkhyeli and Ewijk [16], Pahuja et al. [20], and Mohan and Rani [21]
Organisational commitment (F6)	The vision and mission should be clear to the employees so that they can understand and deal objectively. They should work as it is their home where compensation is just a thing of value, but responsibility is everything. For this, the organisation should make the employees a part of it. Then, the commitment will be visible from the employees through employee retention	Kehr [3] and Pahuja and Dalal [14]
Communication (F7)	The direction of flow of communication in an organisation impacts the employee retention in an organisation. If the flow of communication is not clear among the employees, then the employees will be uncomfortable and may not retain in the workplace. Their performance may be hampered as the information is not clear enough	Kehr [3], Cole et al. [10], Pahuja and Dalal [14], Raj and Brindha [9], Alkhyeli and Ewijk [16], and Pahuja et al. [20]
Employee motivation (F8)	This is an important factor for employee retention. Employees may be monotonous in their work, thus job rotation should be there. The employees should be given proper incentives so that they are more committed towards their workplace. The workplace should be attractive and interactive so that they are more energetic	Jassim [23], Dess and Shaw [5], Raikes and Vernier [24], Singh and Dhillon [17], Sitati et al. [18], Ndede [19], and Mohan and Rani [21]

(continued)

Table 1 (continued)

Factor	Description	Authors
Participation in decision-making (F9)	This enhances the chances of employee retention as the employee feel committed towards the work. The people feel not as a worker but as an associate of the organisation. This will improve the relationship strength between the employees and employers. That will in fact decrease the attrition rate	Das and Baruah [12], Sitati et al. [18], Pahuja et al. [20], and Mohan and Rani [21]
Training and development (F10)	Training and development will keep the human resource more updated and competitive every time. It is process of enlarging current skills and overall development of the employees. It is a win–win situation where the employees can walk step with changing time, and the organisation is also able to retain its best employees. The employees are free of risk of being laid off and remain committed towards organisation	Raikes and Vernier [24], Sharma [13], Singh and Dhillon [17], Sitati et al. [18], Pahuja et al. [20], and Mohan and Rani [21]

With the ideology to explore the factors that are important to employee retention and their weightage, authors used ten factors and assess them based on the responses of 250 employees using AHP technique. Following are the steps for performing the algorithm:

Step 1: Prepare a decision hierarchy model and collect the input from the respondents by a pairwise comparison method using a preference scale as described in Table 2.

Table 2 Preference scale

Importance intensity	Definition
1	Equally prefer both
3	Moderately prefer one over the other
5	Strongly prefer one over the other
7	Very strongly prefer one over the other
9	Extremely prefer one over the other
2, 4, 6, 8	Intermediate preference points

Table 3 Random consistency index value (RI)

N	1	2	3	4	5	6	7	8	9	10
RI	0	0	0.525	0.882	1.115	1.252	1.341	1.404	1.452	1.484

Step 2: Determine the hierarchy consistency utilizing the pairwise comparison matrix A and derive the preference of factors using the below given equation:

$$Aw = \lambda_{\max} w \quad (1)$$

where,

$$\sum_{i=1}^{n} a_i w_i = 1 \text{ and } \lambda_{\max} = \frac{\sum_{j=1}^{n} a_j w_j}{n}$$

where, $n \rightarrow$ order of the pairwise comparison matrix,
$A \rightarrow$ definite pairwise comparison matrix of order n,
$\lambda_{\max} \rightarrow$ the principal eigenvalue of A and
$w \rightarrow$ priority vector.

The consistency ratio (CR), a quantity of judgments in the pairwise comparison matrix, is obtained as follows:

$$\mathrm{CR} = \frac{\mathrm{CI}}{\mathrm{RI}} \quad (2)$$

where, $\mathrm{CI} = \frac{(\lambda_{\max} - n)}{(n-1)}$ is the consistency index, and RI is the average random consistency index for different matrix order, as shown in Table 3.

Consistency ratio less than 0.1 indicates the matrix is consistent and if greater than 0.1, the comparison matrix is considered inconsistent, and it needs to be revised.

Step 3: Finally, the weights of each factor are obtained by the normalised principal eigenvectors.

4 Case Study

The application of AHP begins with a problem being decomposed into a hierarchy of criteria to be more easily analysed and compared in an independent manner. After this logical hierarchy is constructed, the decision-makers can systematically assess the alternatives by making pair-wise comparisons for each of the chosen criteria. This comparison may use concrete data from the alternatives or human judgments to input information [29]. To illustrate the application of AHP in human resource sector, five private and public organisations are selected using purposive sampling technique wherein 10 human resource executives (experts) from each of the five organisations with an experience of 10–15 years have been selected and their responses are collected for study. The objective and purpose of the study has been communicated to the expert, and then based on their preference of the identified factors, a pairwise comparison decision matrix has been prepared as given in Table 4.

The above pairwise comparison matrix has been normalised to compute the criteria weights.

$$\lambda_{max} = \frac{10.717 + 10.698 + 10.982 + 10.648 + 10.848 + 11.509 + 11.982 + 11.960 + 11.628 + 11.586}{10} = 11.256$$

$$\text{Consistency Index (C.I.)} = \frac{\lambda_{max} - n}{n - 1} = \frac{11.256 - 10}{9} = 0.140$$

$$\text{Consistency Ratio (C.R.)} = \frac{\text{Consistency Index (C.I.)}}{\text{Random Index (R.I.)}} = \frac{0.140}{1.49} = 0.094$$

From the above calculation and Tables 4, 5 and 6, it can be observed that the data is consistent as consistency ratio i.e. 0.094 < 0.1 [22] (Table 7).

Figure 2 describes the graphical representation of each identified factor of employee retention.

Table 4 Pairwise comparison matrix

Factors	F1	F2	F3	F4	F5	F6	F7	F8	F9	F10
F1	1	1	1	0.333	0.2	0.333	0.2	0.143	0.111	0.2
F2	1	1	0.333	0.25	0.333	0.143	0.2	0.333	0.111	0.2
F3	1	3	1	1	0.5	0.333	0.333	0.2	0.125	0.143
F4	3	4	1	1	0.333	0.25	0.333	0.333	0.167	0.111
F5	5	3	2	3	1	0.2	0.25	0.333	0.2	0.143
F6	3	7	3	4	5	1	0.25	0.25	0.333	0.333
F7	5	5	3	3	4	4	1	0.333	0.25	0.5
F8	7	3	5	3	3	4	3	1	0.333	0.5
F9	9	9	8	6	5	3	4	3	1	0.5
F10	5	5	7	9	7	3	2	2	2	1
SUM	40	41	31.333	30.583	26.367	16.26	11.567	7.926	4.631	3.63

Table 5 Normalised pair-wise comparison matrix

Factors	F1	F2	F3	F4	F5	F6	F7	F8	F9	F10	Criteria weight
F1	0.025	0.024	0.032	0.011	0.008	0.021	0.017	0.018	0.024	0.055	0.023
F2	0.025	0.024	0.011	0.008	0.013	0.009	0.017	0.042	0.024	0.055	0.023
F3	0.025	0.073	0.032	0.033	0.019	0.021	0.029	0.025	0.027	0.039	0.032
F4	0.075	0.098	0.032	0.033	0.013	0.015	0.029	0.042	0.036	0.031	0.04
F5	0.125	0.073	0.064	0.098	0.038	0.012	0.022	0.042	0.043	0.039	0.056
F6	0.075	0.171	0.096	0.131	0.19	0.062	0.022	0.032	0.072	0.092	0.094
F7	0.125	0.122	0.096	0.098	0.152	0.246	0.086	0.042	0.054	0.138	0.116
F8	0.175	0.073	0.16	0.098	0.114	0.246	0.259	0.126	0.072	0.138	0.146
F9	0.225	0.22	0.255	0.196	0.19	0.185	0.346	0.378	0.216	0.138	0.235
F10	0.125	0.122	0.223	0.294	0.265	0.185	0.173	0.252	0.432	0.275	0.235

5 Discussion and Conclusion

On analysing the collected responses, the highest weight is given to employee motivation (23.50%) which means the company must bring some changes in polices to motivate the employees. Employee motivation is directly proportional to employee retention. It is observed that those organisations who are empowering their personnel policies to motivate their employees can retain their best employees. Other factors like training and development (23.50%), participation in decision-making (14.60%), communication (11.60%) and organisational commitment (9.40%) are also highly weighted which indicates employees need to be updated with dynamic environment and time. Literature as well the factors of employee motivation and training and development are considered as the influential factors for employee retention [18, 21].

The feeling of committing employee to organisation is necessary as it can increase their commitment and loyalty towards organisation. The organisation must make policies which will keep the employees motivated and in turn can increase the rate of employee retention. The retention of the best of employees can enhance the organisation performance and efficiency. It can also give a competitive edge as everything can be copied, but human resource of an organisation and its talent and knowledge cannot be copied. The presented work is limited to the application of employee retention aspect only, whereas the domain of human resource development can focus on other aspects. Furthermore, some other multi-criteria decision-making techniques such as TOPSIS, DEMATEL, EDAS, etc. can be applied to evaluate the various factors of employee retention.

Table 6 Calculation for consistency

Factors	F1	F2	F3	F4	F5	F6	F7	F8	F9	F10	Weighted sum value	Ratio
F1	0.023	0.023	0.032	0.013	0.011	0.031	0.023	0.021	0.026	0.047	0.252	10.717
F2	0.023	0.023	0.011	0.01	0.019	0.013	0.023	0.049	0.026	0.047	0.244	10.698
F3	0.023	0.068	0.032	0.04	0.028	0.031	0.039	0.029	0.029	0.034	0.354	10.982
F4	0.07	0.091	0.032	0.04	0.019	0.024	0.039	0.049	0.039	0.026	0.429	10.648
F5	0.117	0.068	0.065	0.121	0.056	0.019	0.029	0.049	0.047	0.034	0.604	10.848
F6	0.07	0.16	0.097	0.161	0.278	0.094	0.029	0.037	0.078	0.078	1.082	11.509
F7	0.117	0.114	0.097	0.121	0.223	0.376	0.116	0.049	0.059	0.117	1.388	11.982
F8	0.164	0.068	0.161	0.121	0.167	0.376	0.348	0.146	0.078	0.117	1.747	11.96
F9	0.211	0.205	0.258	0.242	0.278	0.282	0.463	0.438	0.235	0.117	2.731	11.628
F10	0.117	0.114	0.226	0.362	0.39	0.282	0.232	0.292	0.47	0.235	2.72	11.586

Table 7 Criteria weightage

Factors	Attribute criteria	Criteria weights	Percentage of weight (%)
F1	Skill recognition (SR)	0.023	2.30
F2	Learning and working climate (LWC)	0.023	2.30
F3	Job flexibility (JF)	0.032	3.20
F4	Superior-subordinate relationship (S-SR)	0.04	4.00
F5	Compensation (CS)	0.056	5.60
F6	Organisation commitment (OC)	0.094	9.40
F7	Communication (CM)	0.116	11.60
F8	Participation in decision-making (PDM)	0.146	14.60
F9	Training and development (T&D)	0.235	23.50
F10	Employee motivation (EM)	0.235	23.50

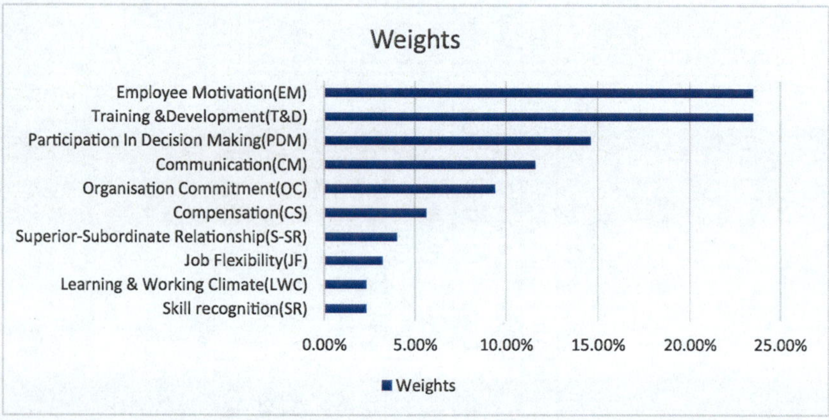

Fig. 2 Graphical representation of weights

References

1. Fitz-enz J (1990) Getting and keeping good employees. In Personnel 67:25–29
2. Stein N (2000) Winning the war to keep top talent. Fortune 141(11):132–136
3. Kehr HM (2004) Integrating implicit motives, explicit motives, and perceived abilities. The compensatory model of work motivation and volition. Acad Manag Rev
4. Osteraker MC (1999) Measuring motivation in a learning organization. J Work Learn 11(2):73–77
5. Dess G, Shaw J (2001) Voluntary turnover, social capital, and organizational performance. Acad Manag Rev 26(3):446–457
6. Clarke KF (2001) What businesses are doing to attract and retain employee—becoming an employer of choice. Employee Benefits J 34–37

7. Wright PM, Dunford BB, Snell SA (2001) Human resources and the resource-based view of the firm. J Manag 27(6):701–721
8. Walker JW (2001) Perspectives. Hum Resour Plan 24:6–10
9. Raj SR, Brindha G (2017) A study on employee retention strategies with special reference to Chennai IT industry. Int J Civ Eng Technol 8(6):38–43
10. Cole MS, Bruch H, Vogel B (2006) Emotion as mediators of the relations between perceived supervisor support and psychological hardiness on employee criticism. J Organ Behav 27(4):463–484. https://doi.org/10.1002/job.381
11. Hsiao WH, Chang TS, Huang MS, Chen YC (2011) Selection criteria of recruitment for information systems employees: using the analytic hierarchy process (AHP) method. Afr J Bus Manag 5(15):6200
12. Das BL, Baruah M (2013) Employee retention: a review of literature. IOSR J Bus Manag 14(2):8–14. https://doi.org/10.9790/487X-1420816
13. Sharma V (2016) Exploring employee retention in IT industry in India: study based on multi response analysis. IJCEM 2(12):86–97
14. Pahuja S, Dalal R (2016) Study of human resource system, competitive advantage status and their relation in Indian Public Sector Banks. IIMS J Manag Sci 7(3):292–300
15. Kundu SC, Lata K (2017) Effects of supportive work environment on employee retention: mediating role of organizational engagement. Int J Organ Anal 25(4):703–722. https://doi.org/10.1108/IJOA-12-2016-1100
16. Alkhyeli HE, Ewijk AV (2018) Prioritisation of factors influencing teachers' job satisfaction in the UAE. Int J Manag Educ 12(1):1–24
17. Singh RS, Dhillon M (2018) Factors affecting employee retention in Indian IT sector. Res Rev Int J Multidiscip 03(11):583–587. https://doi.org/10.5281/zenodo.1499913
18. Sitati N, Were S, Waititu GA, Miringu A (2019) Effect of employee recognition on employee retention in hotels in Kenya. Direct Res J Soc Sci Educ Stud 6(8):108–117. https://doi.org/10.5281/zenodo.3458598
19. Ndede CO (2020) Motivation on employee retention in commercial banks in Homabay County, Kenya. https://doi.org/10.5281/zenodo.3963839
20. Pahuja S, Kadiyala MC, Mittal S (2020) Employee relations as a tool of strategic competitive environment. IGI Global
21. Mohan MK, Rani RS (2021) Employee retention at Admaark: a case study. https://doi.org/10.5281/zenodo.4505608
22. Sheridan JE (1992) Organizational culture and employee retention. Acad Manag J 35(5):1036–1049
23. Jassim RK (1998) Competitive advantage through the employee. Unpublished doctoral dissertation, Reg. Eng. College of Technology at Jeddah
24. Lucille R, Jean-François V (2004) Rewarding and retaining key talent: are you ready for the recovery? www.towersperrin.com
25. Saaty TL (2008) Decision making with the analytic hierarchy process. Int J Serv Sci 1(1):83–98
26. Bhushan N, Rai K (2007) Strategic decision making: applying the analytic hierarchy process. Springer
27. Divkolaii M (2014) An investigation on factors influencing on human resources productivity. Manag Sci Lett 4(5):883–886
28. Hussain AS, Jalal ABM (2015) Micro-factors influencing site selection for small and medium enterprises (SMEs) in Saudi Arabia: Al-Hassa area using Analytical Hierarchy Process (AHP) analysis. Eur Sci J 11:115
29. Saaty TL (1984) The analytic hierarchy process: decision making in complex environments. In: Quantitative assessment in arms control. Springer, Boston, MA, pp 285–308